# Chemistry

## CHRIS CONOLEY
## PHIL HILLS

4

Collins
ADVANCED
SCIENCE

**Published by HarperCollins*Publishers* Limited**

77-85 Fulham Palace Road
Hammersmith
London
W6 8JB

**www.CollinsEducation.com**
**Online support for schools and colleges**

First published 2002
10 9 8 7 6 5 4 3 2 1

ISBN 000 713597 1

British Library Cataloguing in Publication Data:
A catalogue record for this book is available from the
British Library.

Publishing manager: Martin Davies
Publishing coordinator: Pat Winter
Project manager: John Ormiston
Proofreader: Peter Millward
Indexer: Jacqueline Kaye
Picture researcher: Caroline Thompson
Series designer: Caroline Grimshaw
Book designer: Ken Vail Graphic Design
Artists: Illustrated Arts, Peter Harper

Printed and bound by Scotprint

**You might also like to visit**
**www.fireandwater.com**
**The book lover's website**

# Contents

# Preface to second edition

The first edition of the **Collins Advanced Science** series represented a breakthrough in comprehensive support and appeal to A-level pupils and teachers that received wide critical acclaim. With the advent of AS and A2 level specifications, the second edition has been fully revised and updated to build on the strengths of the first edition.

The text has been augmented to increase the breadth of examinations board coverage. Each of the major examination boards' specifications are now represented, including the International Baccalaureate. With that in mind, examination questions from across the boards have been included for student practice at the ends of chapters.

## For the teacher

Each chapter in the second edition of **Collins Advanced Science: Chemistry** includes an opening page that identifies the AS and A2 nature of each section in the chapter, relating to the majority of examination specifications. The section headings can be checked for the detail against individual specifications.

To support the textbook, a comprehensive **Teacher Support CD-ROM** is available. This CD-ROM includes answers to the examination questions contained within the text, illustrations for use with OHPs and white boards, further questions for in-class and homework use, together with suggestions for practical exercises.

**Collins Advanced Science: Chemistry Teacher Support CD-ROM**
ISBN 000 713960 8

## For the student

Maintaining the book's strong support for self study, the design of the second edition updates its visual appeal. The content is written in an accessible style and there are plenty of interesting real-life applications of the subject, frequent self-test questions and 'Remember This' boxes to aid understanding. The Assignments at the end of each chapter give practice and portfolio evidence for Key Skills requirements.

## Dedication

To my wife Anne for her unfailing support, and to my children Simon, Elizabeth and Victoria, who have been so understanding when the writing has intruded into family life. Also to my parents Pat and Vera Conoley with much love. *Chris Conoley*

To my mum, David and Peter for their patience, understanding and support throughout the duration of this project. Thanks also to the students and staff at King Edward VI School for their ideas and support. *Phil Hills*

### Examination questions

The publisher thanks examination boards for their permission to reproduce examination questions. Questions are acknowledged as follows:

| | |
|---|---|
| AQA | The Assessment and Qualifications Alliance |
| Edexcel | Edexcel Foundation |
| IBO | International Baccalaureate Organization |
| OCR | Oxford Cambridge and RSA Examinations |
| WJEC | Welsh Joint Education Committee |

# 01

# Reactions, equations, energy and equilibria

# Reactions, equations, energy and equilibria

The Humber Bridge is the largest single span suspension bridge in the world. It contains 27 500 tonnes of steel, all supplied by Corus (previously British Steel): 16 500 tonnes make up the deck section and 11 000 tonnes are in the cables that support it.

Each part of such a large construction needs a different type of steel with different properties to meet particular needs, which may include rigidity, flexibility, strength in tension and ease of being welded. The properties of steel depend on its composition. Steels contain a little carbon in the iron, and can include very small amounts of manganese, niobium, vanadium and titanium. Phosphorus and sulphur are impurities from iron ore that can make steel brittle. The number of times the steel sheet is rolled will also affect its final properties. It is the job of the metallurgist to obtain the balance of composition and treatment of the steel that suits a specific need.

For a gas or oil pipeline, a low carbon content and some niobium and vanadium in the steel make a clean, strong weld that will not crack or leak, while for reinforcing structures in buildings, more carbon gives steel greater rigidity and strength.

**The Humber bridge** – the largest of its kind in the world

**In the manufacture of steel**, enormous amounts of water are used to cool the rolling mills. The water is recycled many times and can pick up contaminants, including cyanide, nitrite and chloride, from the steel. So chemists regularly check that the water eventually discharged meets environmental requirements

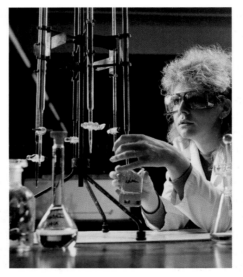

Laser welding is a recent development in working with sheet steel. In conventional 'oxyacetylene' welding, the parts to be joined are heated extensively to melt and join the surfaces. Then, when the steel cools and shrinks, there is inevitably distortion at the join where the steel does not cool evenly. With laser welding, the beam is focused to a small spot that moves along the joint. Only the surfaces of the sheets are heated, very intensely and for a very short time, so the process avoids distortion and saves a great deal of money.

## INTRODUCTION

Nearly all the materials you touch, the polymers and dyes in clothes, the food on your plate, or the paracetamol in the bathroom cabinet, have at some stage involved industrial chemists. This chapter will give you an insight into the world of the chemical industry. You will discover that organizing the economics of a large manufacturing plant depends on understanding the chemistry of industrial processes.

Also, throughout this chapter you will find out more about chemical equations, the chemical reactions that they represent and how energy is transferred in reactions. You will also meet reversible reactions that can form dynamic equilibria.

Fig 1.1 The artificial aorta that is fitted to a man's heart is made of a polyester fibre called Dacron

Fig 1.2 Oral rehydration powders contain sodium chloride, glucose, trisodium citrate and potassium chloride. As the treatment for victims of dehydration, this mixture dissolved in water has saved the lives of millions of people, especially children

# 1 CHEMICAL FORMULAS AND CHEMICAL EQUATIONS

Iron and steel have transformed the world around us in a way no other chemicals have. Think of their use in various means of travel – bridges, railways, cars, the Channel Tunnel and ships – in buildings such as skyscrapers, factories and warehouses, as well as in leisure products such as golf clubs and garden furniture.

Anyone fortunate enough to tour an iron- and steel-making plant comes away impressed by the large scale of it all. Yet the process is fairly simple and uses only four major raw materials: iron ore, coal (made into coke), limestone and air. The principles of iron making have not changed much since Roman times, so where does the chemist fit in today? Chemists have the best understanding of the manufacturing process, so they can, for example, calculate the exact amounts of raw materials required and advise on the best reaction conditions. Later in this chapter, you can read about the role of chemists in optimizing the efficiency of processes, controlling the properties of steels, understanding reactions in equilibrium and avoiding industrial pollution.

In the blast furnace, coke, which is carbon, reacts with air to produce carbon monoxide. This can be represented by a word equation:

carbon + oxygen → carbon monoxide

However, it is more convenient and much more informative to use formulas in an equation. A **molecular formula** tells you how many atoms of each element there are in a molecule of the compound. Oxygen has two atoms per molecule: $O_2$. But not all substances are made up of molecules. For example, sodium chloride is a giant lattice made up of sodium and chloride ions. For these it is more correct just to use the term **formula** or

Fig 1.3 The Queen Victoria blast furnace in Corus's Scunthorpe iron-making works

**Fig 1.4** The base of the blast furnace: hot air at about 750 ºC is blown into it

2 atoms of carbon + 1 molecule of oxygen → 2 molecules of carbon monoxide

**Fig 1.5** Combining 2 carbon atoms with 1 oxygen molecule

---

**✔ REMEMBER THIS**

Two other state symbols are:
(l) which means the liquid state; and
(aq) which means the substance is dissolved in water to form an aqueous solution.

---

**❓ TEST YOURSELF**

**A** Write balanced equations for the following reactions that are all associated with iron and steel making:
**a)** Calcium carbonate ($CaCO_3$) decomposing to calcium oxide (CaO) and carbon dioxide ($CO_2$).
**b)** Manganese (Mn) reacting with oxygen ($O_2$) to form manganese(II) oxide (MnO).
**c)** Phosphorus (P) reacting with oxygen to form phosphorus(V) oxide ($P_4O_{10}$).

**SEE QUESTION 1.**

---

**formula unit**. So the equation can be written:

$$C + O_2 \rightarrow CO$$

This equation is unbalanced. There are more oxygen atoms on the left side of the equation than there are on the right, where one atom of oxygen has 'disappeared'. Atoms are not made or destroyed in chemical reactions, and so in any chemical equation the number of atoms of each element must be the same on both sides. This makes for a *balanced* equation.

It would be simple to balance the equation above by making CO into $CO_2$. But this changes the chemical carbon monoxide into carbon dioxide, a different gas with completely different properties. Instead, as shown in Fig 1.5 the balanced equation for the reaction is:

$$2C + O_2 \rightarrow 2CO$$

Chemical equations often include **state** symbols. These show the physical state of the reactants, usually at room temperature. So this equation may be written:

$$2C(s) + O_2(g) \rightarrow 2CO(g)$$

where (s) means the solid state and (g) means the gaseous state.

---

**⬤ EXAMPLE**

**Q** One type of iron ore is haematite. The chemical name for haematite is iron(III) oxide and its formula is $Fe_2O_3$. When haematite reacts with carbon monoxide in the blast furnace, carbon dioxide and iron are produced.

**Work out the balanced equation for this reaction.**
**A Step 1.** Write the word equation:

iron(III) oxide + carbon monoxide → iron + carbon dioxide

**Step 2.** Change the words into formulas:
$$Fe_2O_3 + CO \rightarrow Fe + CO_2$$
**Step 3.** Balance the equation by making the number of atoms of each element the same on both sides and add state symbols. (Remember that the formulas are fixed.)
$$Fe_2O_3(s) + 3CO(g) \rightarrow 2Fe(s) + 3CO_2(g)$$
As you get used to balancing equations, you will probably do these steps in your head.

---

## 2 RELATIVE ATOMIC MASS

The masses of different atoms can be compared using the **relative atomic mass scale**, also called the **$A_r$ scale** (r stands for 'relative'). On this scale, the isotope carbon-12 is given an $A_r$ of exactly 12. The isotope carbon-12 is the standard and the $A_r$ values of all other atoms are measured relative to this standard. Most elements exist as two or more isotopes and the $A_r$ is an average of the masses of all these different isotopes taking into account the percentages in which they occur naturally. This leads to the definition:

**Relative atomic mass of an element is the average mass of the atom (taking into account all of its isotopes and their abundance) compared to 1/12 the mass of one atom of carbon-12.**

Historically, the standard for the $A_r$ scale was hydrogen. As the lightest element, it was given an $A_r$ of 1. Carbon-12 is now used because it is easier to handle than gaseous hydrogen and is an abundant isotope. Table 1.1 shows some approximate relative atomic masses.

Notice in Table 1.1 that the relative atomic masses have no units. The relative atomic mass just shows *how many times heavier* one atom is compared to another. So one atom of silicon is 28 times heavier than one atom of hydrogen, and calcium atoms are twice as heavy as neon atoms.

**Table 1.1 The approximate relative atomic masses of some elements**

| Element | Symbol | Relative atomic mass, $A_r$ |
|---------|--------|------------------------------|
| Hydrogen | H | 1 |
| Carbon | C | 12 |
| Oxygen | O | 16 |
| Neon | Ne | 20 |
| Silicon | Si | 28 |
| Sulphur | S | 32 |
| Calcium | Ca | 40 |
| Iron | Fe | 56 |
| Copper | Cu | 64 |

**TEST YOURSELF**

**B** Work out how many times heavier the following are:
**a)** Fe atoms than H atoms;
**b)** S atoms than O atoms;
**c)** Fe atoms than O atoms.

YOU CAN FIND OUT MORE ABOUT ISOTOPES IN CHAPTER 2.

A FULL LIST OF RELATIVE ATOMIC MASSES IS GIVEN IN APPENDIX 2.

# 3 AMOUNTS AND THE MOLE

When chemists use the word *amount*, they are talking about the number of particles in a substance. The particles can be atoms, molecules, ions, electrons, etc. Amount may also mean other things, for instance the mass of a substance, which has units of grams or kilograms. To describe the amount of something such as eggs, you would probably use the unit dozens. When chemists talk about the amount of substance they use the unit **moles**.

The $A_r$ of copper is 64, and so each copper atom is twice as heavy as a sulphur atom. Imagine that 64 grams of copper and 64 grams of sulphur ($A_r$ of 32) are weighed out. The mass of each substance is the same, but to a chemist the amount of each substance is very different: 64 grams contains twice as many sulphur *atoms* as copper atoms: there are double the *amount* of sulphur atoms.

Using relative atomic masses, you can work out amounts of atoms or molecules in different masses of substances. Weigh out 32 grams of sulphur, and you have the same number of atoms as there are in 64 grams of copper. Also, weigh out the relative atomic mass in grams of different elements, and you are weighing the same *amount* of atoms each time. This amount is called the mole:

**The relative atomic mass in grams of any element contains one mole of atoms.**

The mole is a unit in the same way that the gram is a unit. The shortened form of the mole unit is **mol**. (Care: This is not short for molecule!) This is its definition:

**One mole is the amount of substance that contains as many particles as there are atoms in exactly 12 g of carbon-12.**

These are some examples:
The amount of atoms in 32 g sulphur is 1 mol.
In 32 g of copper there is 0.5 mol of copper atoms.
32 g of oxygen atoms is 2 mol of oxygen atoms.

To work out how many moles of atoms are in a particular mass of substance, use this equation:

$$\text{amount in moles} = \frac{\text{mass in grams}}{\text{mass of one mole (in grams)}}$$

**REMEMBER THIS**

The mass of one mole of a substance is the **molar mass**.

---

## ⬤ EXAMPLE

**Q** How many moles are there in 56 g silicon?

**A** Amount in moles = $\dfrac{\text{mass in grams}}{\text{mass of one mole (in grams)}}$

moles of silicon $= \dfrac{56\ \text{g}}{28\ \text{g}}$

$= 2$ moles

**Q** What is the mass of 0.25 mol of iron?

**A** Mass in grams $=$ amount in moles $\times$ mass of 1 mole
mass of iron $= 0.25$ mol $\times 56$ g
$= 14$ g

---

### ❓ TEST YOURSELF

**C**
**a)** Work out how many moles of atoms there are in:
56 g of iron;    20 g of calcium;
128 g of copper;  4 g of sulphur.
**b)** Give the mass of:
3.0 mol of calcium atoms;
0.3 mol of neon atoms.

**SEE QUESTIONS 2 AND 3.**

**Fig 1.6** Working out the relative molecular mass of carbon dioxide

### ❓ TEST YOURSELF

**D** Give the relative formula mass ($M_r$) of:
**a)** $P_4O_{10}$;  **b)** $O_2$;  **c)** $CaSiO_3$.

### ❓ TEST YOURSELF

**E** Give the $M_r$ of: **a)** $Fe(OH)_2$;
**b)** $Al_2(SO_4)_3$; **c)** $(CH_3CO)_2O$.

One mole of substance is $6.02 \times 10^{23}$ particles. This is a very large amount of atoms or molecules to try and imagine. It is so large that even all the sand grains around the whole coastline of the British Isles do not make a mole. The number of atoms or molecules in one mole is called the **Avogadro constant**.

## 4 RELATIVE MOLECULAR MASS

We saw on page 4 that carbon dioxide is made up of molecules each with the molecular formula $CO_2$. $Fe_2O_3$ is the formula unit of iron(III) oxide. It is not a molecular formula because this compound is not made up of separate molecules. Most equations that chemists deal with involve either molecules or formula units.

The **relative molecular mass, $M_r$**, is calculated using the relative atomic mass scale. Again, we use the standard carbon-12 as a comparison for the masses of molecules or formula units. Suppose we want to find the $M_r$ of carbon dioxide (see Fig 1.6).

$$\begin{aligned}\text{The } M_r \text{ of } CO_2 &= A_r \text{ of C} + ((A_r \text{ of O}) \times 2)\\ &= 12 + (16 \times 2) = 44\end{aligned}$$

$$\begin{aligned}\text{The } M_r \text{ of } Fe_2O_3 &= ((A_r \text{ of Fe}) \times 2) + ((A_r \text{ of O}) \times 3)\\ &= (56 \times 2) + (16 \times 3) = 160\end{aligned}$$

In this case, the $M_r$ refers to the formula $Fe_2O_3$ and is called the **relative formula mass**.

During your Advanced course, you will come across many formulas with brackets, for example, $Ca(OH)_2$. Fig 1.7 shows what this means.

Notice that the particles in $Ca(OH)_2$ are ions and have positive and negative charges. The charges on ions do not affect their $A_r$ values.

**Fig 1.7** Diagram to explain the parts of the formula $Ca(OH)_2$

So:

$$\begin{aligned}M_r \text{ of } Ca(OH)_2 &= A_r \text{ of Ca} + 2 \times (A_r \text{ of O} + A_r \text{ of H})\\ &= 40 + 2 \times (16 + 1)\\ &= 40 + 2 \times 17 = 74\end{aligned}$$

# 5 USING EQUATIONS IN INDUSTRY

We have seen that the chemical formula CO means that 1 molecule contains 1 atom of carbon combined with 1 atom of oxygen. It also means that 1 mole of carbon atoms is combined with 1 mole of oxygen atoms. This is sometimes called the **stoichiometric ratio**. In the equation for the reaction of coke in the blast furnace:

$$2C \quad + \quad O_2 \quad \rightarrow \quad 2CO$$

This means:

$$2 \text{ atoms C} \quad + \quad 1 \text{ molecule } O_2 \quad \rightarrow \quad 2 \text{ molecules CO}$$

If instead of 2 atoms of carbon we had 2 million atoms, then:

2 million C atoms + 1 million $O_2$ molecules → 2 million CO molecules

We now use moles:

$$\begin{array}{ccccc} 2 \text{ mol carbon} & + & 1 \text{ mol oxygen} & \rightarrow & 2 \text{ mol carbon} \\ \text{atoms} & & \text{molecules} & & \text{monoxide molecules} \end{array}$$

The carbon monoxide formed in the blast furnace **reduces** (removes oxygen from) the iron(III) oxide to iron:

$$Fe_2O_3(s) \quad + \quad 3CO(g) \quad \rightarrow \quad 2Fe(s) \quad + \quad 3CO_2(g)$$
So: $\quad$ 1 mol $Fe_2O_3$ + 3 mol CO → 2 mol Fe + 3 mol $CO_2$

Now that we know the amounts of the chemical substances in this equation, we can work out the masses of the substances involved using the $A_r$ scale.

The mass of 1 mol of $Fe_2O_3$ is $(56 \times 2)$ g + $(16 \times 3)$ g $\quad$ = 160 g
The mass of 1 mol of CO $\quad$ is 12 g + 16 g $\qquad\qquad$ = 28 g
The mass of 1 mol of Fe $\quad$ is 56 g $\qquad\qquad\qquad$ = 56 g
The mass of 1 mol of $CO_2$ $\quad$ is 12 g + $(16 \times 2)$ g $\qquad$ = 44 g

To work out the masses of substances involved in converting 160 g of iron(III) oxide to iron, we can use the following steps:

**Step 1.** Write the balanced equation:

$$Fe_2O_3(s) \quad + \quad 3CO(g) \quad \rightarrow \quad 2Fe(s) \quad + \quad 3CO_2(g)$$

**Step 2.** Convert the equation to amounts:

$$1 \text{ mol } Fe_2O_3 \quad + \quad 3 \text{ mol CO} \quad \rightarrow \quad 2 \text{ mol Fe} \quad + \quad 3 \text{ mol } CO_2$$

**Step 3.** Work out the amount being used:

$$\text{Moles of } Fe_2O_3 \text{ in 160 g} \quad = \frac{\text{mass in grams}}{\text{mass of one mole (in grams)}} = \frac{160}{160} = 1 \text{ mol}$$

**Step 4.** Scale the amounts in the equation:
In this case, only 1 mol $Fe_2O_3$ is used, so the amounts do not need to be scaled.

$$1 \text{ mol } Fe_2O_3 \quad + \quad 3 \text{ mol CO} \quad \rightarrow \quad 2 \text{ mol Fe} \quad + \quad 3 \text{ mol } CO_2$$

**Step 5.** Convert amounts (moles) to masses:

160 g $Fe_2O_3$ + 3 × 28 g (= 84 g) CO →
$$2 \times 56 \text{ g (= 112 g) Fe} + 3 \times 44 \text{ g (= 132 g) } CO_2$$

So, from 160 g of $Fe_2O_3$, in theory 112 g of iron could be produced and 132 g of carbon dioxide given off. In the iron and steel industry, amounts this size are ridiculously small. A typical blast furnace can produce up to 10 000 tonnes a day of molten iron.

## ✔ REMEMBER THIS

Extraction of a metal from its ore always includes a reduction reaction. Many metal ores contain oxygen, but for those that don't, a suitable definition of reduction is to be found in Chapter 5.

### ? TEST YOURSELF

**F** Coke also reacts directly with iron ore in the hotter part of the blast furnace to give iron and carbon monoxide.
**a)** Work out the balanced equation for this reaction.
**b)** How many tonnes of carbon monoxide are produced in this part of the furnace for every tonne of iron produced?

**FOR INFORMATION ABOUT SIGNIFICANT FIGURES, REFER TO APPENDIX 1.**

Suppose we now want to find out how much iron can be produced from 10 tonnes of iron(III) oxide. We can still use the same method. The first step is identical, so we can start at Step 2.

**Step 2.** Convert the equation to amounts:

$$1 \text{ mol } Fe_2O_3 \text{ gives } 2 \text{ mol } Fe$$

Note: In this example we do not need to know the moles of CO or $CO_2$.

**Step 3.** Work out the amount being used (1 tonne = 1 000 000 g):

$$\text{Moles of } Fe_2O_3 \text{ in 10 tonnes} = \frac{\text{mass in grams}}{\text{mass of one mole (in grams)}}$$

$$= \frac{10\,000\,000 \text{ g}}{160 \text{ g}}$$

$$= 62\,500 \text{ mol}$$

**Step 4.** Scale the amounts in the equation:

$$1 \text{ mol } Fe_2O_3 \text{ produces} \qquad 2 \text{ mol } Fe$$
$$62\,500 \text{ mol } Fe_2O_3 \text{ produces } 125\,000 \text{ mol } Fe$$

**Step 5.** Convert amount (moles) of Fe to a mass:

$$125\,000 \text{ mol} \times 56 \text{ g} = 7\,000\,000 \text{ g}$$
$$= 7.0 \text{ tonnes of Fe (to two significant figs)}$$

## 6 YIELD AND PERCENTAGE YIELD

You should now be able to use balanced equations to calculate the masses of substances involved in reactions. We have found that 10 tonnes of iron(III) oxide can give 7.0 tonnes of iron. The mass of iron produced is known as the **yield**. But Corus could never achieve this yield in a real blast furnace. Therefore we say that the calculation from the balanced equation gives the **theoretical yield**, which is the maximum amount if all the reactants are converted into products. There are many reasons why the theoretical yield is not achieved: reactions may not be finished in the time available, or some of the product may be lost during the purification procedure.

The **actual yield** from a reaction can be found only by doing the reaction. From 10.0 tonnes of iron(III) oxide, it is going to be less than 7.00 tonnes of iron. Percentage yield is a convenient way of expressing how close the actual yield is to the theoretical yield.

### ? TEST YOURSELF

**G** What is the percentage yield of iron produced if 10.0 tonnes of iron(III) oxide yields 6.7 tonnes of iron?

**H** Another ore of iron has the formula $Fe_3O_4$. In a blast furnace, 10.0 tonnes of this ore makes 6.50 tonnes of iron.
**a)** Write down the balanced equation for the reaction of $Fe_3O_4$ with carbon monoxide to make iron and carbon dioxide.
**b)** How many tonnes of iron can be made from 10.0 tonnes of $Fe_3O_4$?
Note: This will be the theoretical yield.
**c)** Calculate the percentage yield.

**SEE QUESTION 4.**

$$\textbf{Percentage yield} = \frac{\text{actual yield}}{\text{theoretical yield}} \times 100\%$$

Industrial processes aim for 100 per cent yield. The closer they get, the less is the waste of raw material.

## 7 RATE OF REACTION

**YOU CAN FIND OUT MORE ABOUT RATES OF REACTION IN CHAPTERS 8 AND 26**

The aim of chemical industry is to get as close as possible to a 100 per cent yield. However, in producing any chemical economically, the yield from a reaction is just one of the factors involved. Another very important factor is the *rate* of reaction. It is no use having a high yield of iron in a process that is extremely slow.

Fig 1.8 An explosion is just a reaction with a very rapid rate

Fig 1.9 The owner of this fishing trawler knows that it will be a long time before his boat starts leaking because of rust

As a simple definition:

**The rate of a reaction is the amount of substance formed per unit of time.**

The rate could be in moles per second. For industry, it is more convenient to give the rate as the mass of substance, such as kilograms or tonnes, formed per unit of time. Reactions can be very fast and uncontrolled, such as a gas explosion (Fig 1.8). Reactions can also be very slow, as in the rusting of iron (Fig 1.9).

An industrial chemist needs to control the rate of a reaction. This could either mean speeding it up or slowing it down. In industry, the rate often needs to be increased so that the amount of product formed in, say, a day is enough to make a profit.

## FACTORS THAT AFFECT RATE OF REACTION

Chemists control the rate of a reaction by changing the conditions of the reaction. These are some of the conditions.

- **Temperature:** an increase in temperature makes a reaction go faster (except for those involving enzymes).
- **Pressure:** increasing pressure increases the rate of reactions that involve gases.
- **Concentration:** increasing the concentration normally increases the rate of reaction.
- **Surface area of reactant:** an increase in surface area increases the rate of reaction.
- **Catalyst:** the addition of a catalyst usually increases the rate of a reaction.

YOU CAN FIND OUT MORE ABOUT CATALYSTS ON PAGES 179–180 OF CHAPTER 8 AND PAGES 223–227 OF CHAPTER 10

Reactions involve the rearrangement of atoms when bonds are broken and others are made. This rearrangement seldom takes place spontaneously; the particles normally need to be colliding with each other. All the conditions listed above will change the collision rate between particles. And if the collision rate is increased then the rate of reaction will increase.

Changing temperature, pressure, concentration and surface area can control the rate of reaction. However, these factors can also reduce the yield of an industrial process, so there is often a compromise between rates of reaction and yield. Remember, the first concern is the economics of the process, that is, making product quickly enough to give the maximum profit. We look at this compromise in more detail in Section 14.

# 8 CHEMICAL REACTIONS AND ENERGY CHANGES

All chemical reactions involve a change of energy, meaning *the transfer of energy to or from chemicals*. This is crucial not only to industry, but also to the reactions of life itself. In many chemical reactions, *energy is given out by the reactants as they form products*, causing the temperature of the surroundings to rise. Such reactions are known as **exothermic reactions**.

In the blast furnace (see Fig 1.13), blasts of hot air at 750 °C are blown in at the base and start a reaction between coke and oxygen:

$$C + O_2 \rightarrow CO_2$$

This reaction is highly exothermic, raising the temperature at the base of the furnace to about 2000 °C. This is because the stored energy in carbon and oxygen is greater than the stored energy in carbon dioxide. Stored energy is known as **enthalpy**, symbol **$H$**. It is not possible to measure enthalpy, but **enthalpy changes** can easily be found by measuring temperature changes during reactions at constant pressure. Enthalpy changes are given the symbols $\Delta H$; $\Delta$ is a Greek letter pronounced 'delta' and is used by chemists to mean 'change of'.

In the coke and oxygen reaction, the energy released is 394 kilojoules (kJ) for every mole of carbon that reacts. So we write:

$$C(s) + O_2(g) \rightarrow CO_2(g) \quad \Delta H = -394 \text{ kJ mol}^{-1}$$

Notice that $\Delta H$ has a negative sign. This is because as the reaction proceeds, *energy is lost from the reactants*. This energy heats up the surroundings, in this case the contents of the blast furnace. We show the reaction and its enthalpy changes in an **energy level diagram**, see Fig 1.10.

Not all reactions are exothermic. **Endothermic reactions** are the opposite of exothermic reactions. In endothermic reactions *energy is taken in by the reactants to form products*. The energy comes from the surroundings, which lose energy and cool down, so there is a drop in temperature.

The carbon dioxide produced at the base of the blast furnace reacts with more coke to produce carbon monoxide. This is an endothermic reaction.

$$CO_2 \text{ (g)} + C \text{ (s)} \rightarrow 2CO \text{ (g)} \quad \Delta H = +173 \text{ kJ mol}^{-1}$$

The plus sign shows that carbon monoxide *takes in energy when it is formed* (Fig 1.11). This is one of the reasons why the blast furnace becomes cooler towards the top (see also Fig 1.13).

**Fig 1.10** Energy level diagram for the exothermic reaction C (s) + O$_2$ (g) → CO$_2$ (g)

**Fig 1.11** Energy level diagram for the endothermic reaction CO$_2$ (g) + C (s) → 2CO (g)

---

**?** **TEST YOURSELF**

**1** The following reactions are all associated with iron and steel manufacture. For each reaction, draw its energy level diagram and say whether the reaction is exothermic or endothermic.
**a)** $S(s) + O_2(g) \rightarrow SO_2(g)$
$\Delta H = -297$ kJ mol$^{-1}$;
**b)** $CaCO_3(s) \rightarrow CaO(s) + CO_2(g)$
$\Delta H = +178$ kJ mol$^{-1}$;
**c)** $Fe_2O_3(s) + 3CO(g) \rightarrow 2Fe(s) + 3CO_2(g)$
$\Delta H = -27$ kJ mol$^{-1}$.

**SEE QUESTION 5.**

# 9 THE BLAST FURNACE

The blast furnace smelts the iron ore. Smelting means that the ore is melted, mixed with the other reactants and reduced to the metal. The reaction of iron ore with coke, limestone and hot air, produces the iron (Fig 1.12).

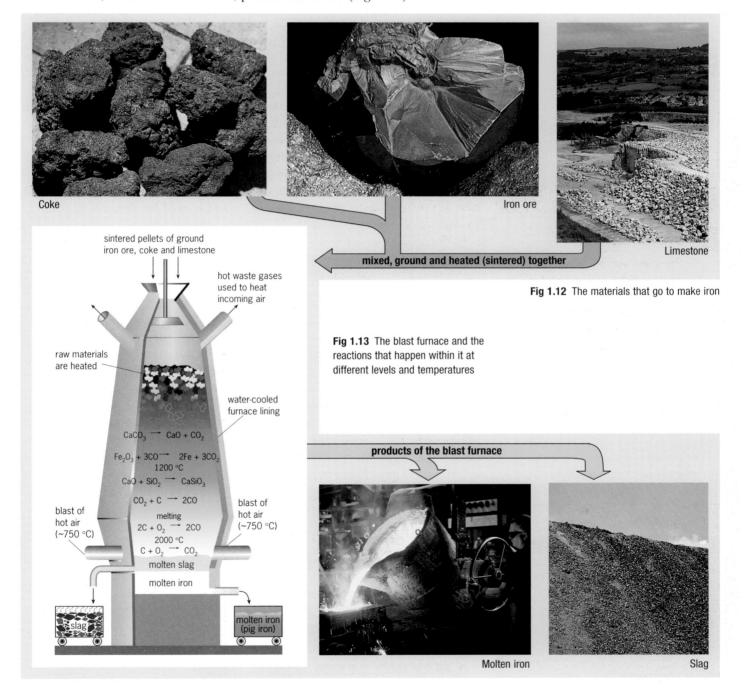

**Fig 1.12** The materials that go to make iron

**Fig 1.13** The blast furnace and the reactions that happen within it at different levels and temperatures

The iron-making process needs high-quality iron ore that contains at least 60 per cent iron. Most iron ores contain impurities such as sand ($SiO_2$), sulphur compounds and phosphorus compounds so, if not pure enough, the ore has to be pre-refined to increase the percentage of iron.

The blast furnace lining has to last for many years to save the cost of replacing it and to minimize shutdown time for replacement work, since the loss of production may cost millions of pounds.

**? TEST YOURSELF**

**J** The temperature of the furnace may reach 2000 °C, especially near the base. Most materials melt at this temperature. What properties does the lining material of the blast furnace need to have?

**K** **a)** Coke, iron ore and limestone are mixed and ground together before they enter the blast furnace. Suggest why this is done.
**b)** The powder produced is heated (sintered) to make loosely packed pellets. What is the advantage of having pellets rather than powder in a blast furnace?

## CONDITIONS IN THE BLAST FURNACE

Some of the blast furnace reactions require high temperatures, so a great input of energy is needed. Fortunately, the reaction of carbon with oxygen to give carbon dioxide provides the energy, since it is an extremely exothermic reaction.

The blasts of hot air, which give the blast furnace its name, are enriched with oxygen. The concentration of oxygen, higher than in air, speeds up the reaction. The faster the rate of reaction, the more the energy produced by the exothermic reactions. In this way, the high temperatures are easier to maintain.

The carbon dioxide formed near the base travels upwards through the melt and solids, and is converted into carbon monoxide (see Fig 1.13). It is the carbon monoxide rather than the solid carbon that reduces most of the iron ore, since gases react faster than solids. The use of pelletized reactants ensures that carbon monoxide comes into very close contact with the iron(III) oxide. This helps to speed up the reaction because the surface area of the ore has been increased.

Finally, there are the reactions involving limestone. Limestone (calcium carbonate) decomposes into calcium oxide (lime) and carbon dioxide in the heat of the furnace:

$$CaCO_3 \rightarrow CaO + CO_2$$

The calcium oxide removes impurities such as sand (silicon(IV) oxide) as liquid slag. The reaction for sand is:

$$CaO + SiO_2 \rightarrow CaSiO_3$$

Slag and iron are both liquids at the high temperatures of the furnace. Conveniently, the less dense slag floats on top of the more dense iron, so they are tapped off separately at different levels.

The production of iron is a **continuous process**. In continuous processes, raw materials are constantly added and the products are continually removed. A continuous process is a very efficient and cost-saving way to produce materials in large quantities. By comparison, in **batch processes**, the reactor vessel must be closed down and reset to make another batch. This is expensive because there is 'dead time' when no product is being produced. However, a range of products can be made in the same vessel, although care is exercised to ensure no contamination occurs. For small quantities, the batch process is more cost effective.

**? TEST YOURSELF**

**L** Explain why companies prefer to use continuous processes instead of batch processes whenever possible.

**SEE QUESTION 6.**

## DISPOSAL OF WASTE

The waste gases and slag could cause pollution problems, but ways have been found to minimize them. Much of the slag goes to build roads and make cement. Some of it is even used to insulate houses: Rockwool, used for non-flammable loft insulation, is produced by blowing air into the molten slag to make it light and fluffy.

## 10 MAKING STEEL FROM IRON

Iron from a blast furnace is impure, brittle and not very strong, so most iron produced is changed immediately into steel. Steel is the name given to countless **alloys** that contain iron, carbon and, usually, small amounts of other elements.

**An alloy is a mixture of two or more elements, at least one being a metal, which are mixed when they are molten and allowed to cool down to form a uniform solid.**

**Fig 1.14** Different types of steel have properties that iron lacks, giving steel a very wide range of uses
A  fractured bones are held together by permanent steel pins
B  the Thames Barrier is mainly steel and concrete
C  well over half the components of a typical car are made of steel, including the body shell, clutch plates and engine
D  the Fruit Drop at Blackpool is made of steel to meet high safety standards

Alloys have different properties from the elements that form them. It is the addition of different elements to iron that changes the properties of the alloy, and so makes steels such versatile materials.

## CARBON STEELS

As the name suggests, carbon steels are alloys of mainly iron with some carbon. The sheet steel used in car bodies contains just 0.2 per cent of carbon, an amount that makes it easy to bend and shape.

As the carbon content of steel is increased, the steel becomes stronger and more rigid. Most of the steel used to construct a bridge, which needs some flexibility, contains between 0.3 and 0.6 per cent carbon. The steel used in drill bits has to be very hard and contains up to 1.5 per cent carbon. It is tempting to think that by increasing the carbon content the steel would carry on strengthening. Unfortunately, this is not the case, and at just 4 per cent carbon, steel becomes very brittle.

## ALLOY STEELS

Alloy steels are steels that contain one or more other metals. These metals include manganese, tungsten, chromium and vanadium. Stainless steels are probably the best-known alloy steels, containing at least 12 per cent chromium. The chromium increases steel's rust resistance. A common stainless steel is called 18–8 and contains 18 per cent chromium and 8 per cent nickel, a composition you will find in some cutlery. Steels that contain tungsten are

### ? TEST YOURSELF

**M** Desulphurization is a process that can be carried out on the impure molten iron from the blast furnace. Powdered magnesium is added to the molten iron and it reacts with the sulphur to form magnesium sulphide (MgS).

**a)** Write a balanced equation for this reaction.

**b)** How much magnesium would need to be added to remove 0.32 kg of sulphur?

**c)** In practice, much more magnesium would be required than the theoretical value you have calculated. Explain why.

**d)** Suggest why the magnesium is added in powdered form.

**e)** Liquid magnesium sulphide is less dense than the molten iron. Suggest how these two liquids may be separated.

very hard wearing. Adding molybdenum (with certain other elements) enables drill bits to retain their cutting edge, even when hot. Spacecraft use titanium steel because it can withstand the high temperatures of re-entry.

### STEEL MAKING

The main process for converting iron into steel is the **basic oxygen process**. Impure iron from a blast furnace is known as pig iron and contains about 4 per cent carbon, together with other elements such as silicon, manganese and phosphorus. To convert the iron into steel, the carbon content is lowered and other elements are removed by reacting them with oxygen. An oxygen lance is lowered into the basic oxygen furnace (Fig 1.16) and oxygen is blown into the molten iron at twice the speed of sound.

The impurity elements form oxides. These are some of the reactions:

$$2C + O_2 \rightarrow 2CO$$
$$2Mn + O_2 \rightarrow 2MnO$$
$$4P + 5O_2 \rightarrow P_4O_{10}$$
$$Si + O_2 \rightarrow SiO_2$$

Being a gas, carbon monoxide bubbles out of the liquid mixture. Both $SiO_2$ and $P_4O_{10}$ are acidic oxides and are removed by adding lime (calcium oxide), which is a basic oxide – hence the name 'basic' in the basic oxygen process. They react to form a slag that floats on top of the molten steel.

These chemical reactions with oxygen are highly exothermic and generate great heat, so the reactions inside the furnace keep the contents molten. In fact, the temperature has to be stopped from becoming too high. Scrap iron and steel are added to prevent the mixture from overheating, because as the mixture melts it takes in energy – melting is an endothermic process.

If the temperature rose unchecked, then the rate of the exothermic reactions would increase, causing yet more energy to be released and making the reaction faster still. Eventually, the process would be uncontrollable and the furnace lining would be damaged.

It is the job of industrial chemists, known as metallurgists, to ensure that the reaction conditions of the steel making process are optimized. They also establish and monitor the best mix of elements for a particular type of steel to match customer demand.

oxygen
water-cooled hood to collect fumes

lined steel shell

tap hole: the furnace is tipped to remove liquid steel

molten pig iron, scrap steel, and calcium oxide (lime)

**Fig 1.15** The arrangement for the basic oxygen process

**Fig 1.16** The basic oxygen furnace for making steel being charged with molten iron

## STEEL – MATCHING SUPPLY AND DEMAND

Steel manufacture is a business, and so the production of steels must be tied closely with long-term planning to sell it. An industrial plant the size of an iron and steel works cannot run economically if it has to start and stop manufacture to match short-term demands. A blast furnace works efficiently only if it is running all the time. Also, if furnace production is stopped, the furnace has to cool down and then be cleaned out. Starting it up again is a costly and difficult process that requires a lot of energy. But if the production of steel outstrips demand, the only economic solution may be to close down the whole steel works.

So to keep steel works operating, the staff who are in charge of sales constantly search for new markets.

## SAVING ON ENERGY

We have seen that steel making has very expensive energy requirements. About 15 per cent of the cost of making steel is spent on the energy that the process consumes, so companies are continually on the look-out for ways to make savings.

Building the steel works next to the blast furnace allows molten iron to be used as soon as it is made. Since the iron doesn't cool much between processes, its energy is not wasted. Also, the energy from hot waste gases and slag can be transferred to other parts of the operation. The waste gases can even be burnt to release energy.

# 11 REACTIONS IN EQUILIBRIUM

The reactions we have so far discussed appear **irreversible** and for practical purposes they are. Let's take burning sulphur in oxygen as an example:

$$S(s) + O_2(g) \rightarrow SO_2(g)$$

The single arrow tells us that this reaction is irreversible and **goes to completion**.

But many reactions can be reversed. A **reversible reaction** is one that can take place in either direction. Such a reaction may form an **equilibrium** if the circumstances allow. Equilibrium means a state of balance. Understanding chemical **equilibria** is fundamental, not only to theoretical chemistry, but to life itself.

The oxygen we breathe in forms an equilibrium with haemoglobin in red blood cells. In the lungs oxygen reacts with haemoglobin to produce oxyhaemoglobin, which is transported to the body's tissues. Here, a different set of conditions reverses the reaction and releases the oxygen. We can represent the equilibrium as follows:

$$\text{haemoglobin} + \text{oxygen} \rightleftharpoons \text{oxyhaemoglobin}$$

In the lungs, which have high concentrations of oxygen, the position of the equilibrium is shifted to the side of the product, oxyhaemoglobin. The concentration of oxygen is much lower in the body tissue, so when the red blood cell reaches it the position of the equilibrium shifts in favour of the reactants and so oxygen is released. Understanding the changes in conditions that cause an equilibrium to shift is particularly important in industry, as we shall see in Section 12.

 **REMEMBER THIS**

The reaction of sulphur and oxygen to produce sulphur dioxide occurs in the blast furnace and although the sulphur is present in only small amounts, the sulphur dioxide produced enters the atmosphere and contributes to acid rain.

## UNDERSTANDING DYNAMIC EQUILIBRIA

One of the first reversible reactions to be investigated was that between hydrogen and iodine more than a hundred years ago.

$$H_2(g) + I_2(g) \rightleftharpoons 2HI(g)$$

**Fig 1.18** The changes in concentrations with time for the reaction:
$$H_2(g) + I_2(g) \rightleftharpoons 2HI(g)$$
when the same concentrations of $H_2(g)$ and $I_2(g)$ are mixed

**Fig 1.17** Experimental data produced in 1894 for the reaction:
$$H_2(g) + I_2(g) \rightleftharpoons 2HI(g)$$
carried out at 448 °C

### ? TEST YOURSELF

**N** **a)** In the experiment that Bodenstein performed, why does rapid cooling apparently stop the reaction?

**b)** If the starting mixture contained 1 mol $H_2$ and 1 mol $I_2$, how many moles of HI would be present if the reaction went to completion?

**c)** Predict the amount of HI that would actually be present in the mixture from **b)** when it reached equilibrium? Assume the same conditions as those shown in Fig 1.17.

A German chemist called Max Ernst Bodenstein (1871–1942) sealed mixtures of hydrogen and iodine in glass bulbs and kept them, at the same temperature, in a thermostatically controlled bath. He repeated this procedure with glass bulbs of hydrogen iodide. At certain time intervals he would stop the reactions in the bulbs by cooling them rapidly and analyse their contents. The results of his experiments are shown in Fig 1.17.

The vertical axis represents the amount of HI present in the reaction mixtures from the start of the reaction. The lower curve (drawn in blue) shows reaction mixtures of 0.5 moles of $H_2$ and 0.5 moles of $I_2$. If the reaction went to completion there would be 1 mole of HI in the flask and zero moles of $H_2$ and $I_2$. But this did not happen. Only 0.78 moles of HI were present. The red, upper curve shows that by starting with 1 mole of HI in the flask, at the same temperature, the same amount was left after 84 minutes (i.e. 0.78 mol). It also did not matter how much time elapsed after 84 minutes: the amounts of all three chemicals were identical. After 84 minutes the reaction had reached its **equilibrium position**.

The concentrations of the gases in the flask do not change once equilibrium has been established (Fig 1.18). But does this mean that reactions stop when equilibrium is reached? The answer is no. This is where the term **dynamic** comes in! Although the concentrations of reactants and products remain constant, both the forward and reverse reactions are still proceeding. This means that, at a molecular level, $H_2$ molecules are colliding with $I_2$ molecules to form HI molecules:

$$H_2(g) + I_2(g) \rightarrow 2HI(g) \text{ (forward reaction)}$$

and HI molecules collide to give the reverse reaction:

$$2HI(g) \rightarrow H_2(g) + I_2(g) \text{ (reverse reaction)}.$$

As the concentrations of reactants and products remain the same, the forward and reverse reactions must occur at the same rate.

> **A dynamic equilibrium is reached when the forward and reverse reactions occur at the same rate.**

The forward and reverse reactions still continue so we call the equilibrium *dynamic*.

In Fig 1.19 you can see that when the reaction between $H_2$ and $I_2$ starts the rate of reaction is at its fastest. As the reaction continues, the rate of the forward reaction begins to slow. At the same time, the concentration of HI increases, which means that the rate of the reverse reaction increases. When the two rates are equal, a dynamic equilibrium is established.

**? TEST YOURSELF**

**O** Predict what would happen if you removed some HI from the equilibrium mixture:
**a)** to the rate of the reverse reaction;
**b)** to the position of the equilibrium.

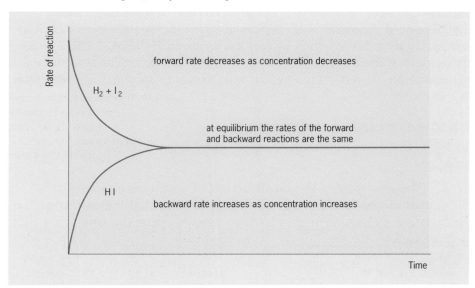

**Fig 1.19** Graphs of rate of reaction with time for the forward and reverse reactions of:
$H_2(g) + I_2(g) \rightleftharpoons 2HI(g)$

## THE IMPORTANCE OF A CLOSED SYSTEM

The bulbs that Bodenstein used to do his experiments were sealed so that none of the gases could escape. This is an example of a **closed system**, in which no material is exchanged with the surroundings. Any loss of reactant or product affects their equilibrium concentrations, and from Section 7 (page 9) you may remember that rate of reaction depends on concentration. A lower concentration means the slowing down of a reaction so the equilibrium is destroyed.

The equilibrium mixture which Bodenstein investigated is an example of a **homogeneous equilibrium**, because the components are all in the same phase – the gaseous phase. If we consider a **heterogeneous equilibrium**, where the components are in different phases, the importance of a closed system may be easier to understand.

If we heat calcium carbonate in a closed vessel the following equilibrium is established:

$$CaCO_3(s) \rightleftharpoons CaO(s) + CO_2(g)$$

If we now let some of the carbon dioxide escape, the rate of the reverse reaction becomes slower, so the position of the equilibrium shifts towards the

**YOU CAN FIND OUT MORE ABOUT HOW TO CALCULATE THE POSITION OF AN EQUILIBRIUM IN CHAPTER 18 AND CHAPTER 28.**

products. If we take the lid off completely, all the carbon dioxide escapes and in this **open system** an equilibrium cannot be established, so all of the calcium carbonate reacts to form products. The blast furnace is just such an open system, which is why all of the calcium carbonate reacts irreversibly.

## 12 THE EFFECTS OF CHANGING CONDITIONS ON EQUILIBRIUM SYSTEMS

The key to achieving the best yields from industrial processes that involve reversible reactions is to understand the effects of changing the conditions on the position of equilibrium. The position of equilibrium is the proportion of the concentration of products to the concentration of reactants in the equilibrium. If the products are present in high concentrations and there are hardly any reactants left, then we say that the position of the equilibrium lies to the right. If there is a high concentration of reactants still present and the concentration of products is very low, then we say the position of equilibrium lies to the left.

### LE CHATELIER'S PRINCIPLE

In 1884, Henri Le Chatelier, a French chemist, formulated his famous principle:

**The position of the equilibrium of a system changes to minimize the effect of any imposed change in conditions.**

| Change | Type or part of system | Effect on equilibrium |
|---|---|---|
| Pressure increasing | If there are more moles of Gaseous products than gaseous reactants | Shift to left |
| | If there are more moles of gaseous reactants than gaseous products | Shift to right |
| Temperature increasing | If $\Delta H^{\ominus}_{\text{forward reaction}}$ is positive, i.e. the forward reaction is endothermic | Shift to right |
| | If $\Delta H^{\ominus}_{\text{forward reaction}}$ is negative, i.e. the forward reaction is exothermic | Shift to left |
| Concentration increasing | Of reactants | Shift to right |
| | of products | Shift to left |
| Catalyst added | | No effect |

**Fig 1.20** A summary of Le Chatelier's principle. $\ominus$ in $\Delta H^{\ominus}$ means that the conditions are standard conditions – usually 100 kPa, 1 mol $dm^{-3}$ and 298 K

The conditions referred to in Le Chatelier's principle are temperature, pressure and concentration of a substance in an equilibrium mixture, and when a change in one or more of these is imposed this affects the position of equilibrium. Le Chatelier's principle applies to any reaction that is in dynamic equilibrium and allows chemists to manage and manipulate equilibrium reactions.

Fritz Haber discovered one of the most important industrial reactions in history in 1909. He succeeded in synthesizing 100 g ammonia ($NH_3$) from nitrogen and hydrogen. The reaction was important because it allowed the manufacture of a nitrogen compound (ammonia) from nitrogen in the air. Before the **Haber process** was found, nitrogen was thought to be too unreactive to be used to manufacture a compound economically. The reaction on which the Haber process is based is:

$$N_2(g) + 3H_2(g) \rightleftharpoons 2NH_3(g)$$

and an understanding of Le Chatelier's principle explains how Haber succeeded in manufacturing a nitrogen compound when so many chemists had tried and failed previously.

## ! SCIENCE FEATURE Solving the nitrate crisis

**Fig 1.21** Fritz Haber (1868–1934) in 1905

In 1909, Germany was preparing for war. Realizing that British control of the seas would make it very difficult to import Chilean nitrate, German scientists were set the problem of producing nitrogen compounds from nitrogen in the air.

Nitrogen gas is so plentiful it might seem surprising that this route had not been used before to manufacture its compounds. But nitrogen is very unreactive and many chemists had tried, but none were successful in discovering a cost-effective process. In 1909, Fritz Haber, a German chemist, had succeeded in producing 100 g of ammonia ($NH_3$) from nitrogen and hydrogen. Between 1910 and 1913, Haber carried out more than 6000 experiments to find the most effective catalyst. In 1913 Carl Bosch, a chemical engineer employed by BASF, scaled up the process to produce the first industrial plant.

Today the rising population of the world relies on fertilizers produced using the Haber process. It was Haber's understanding of equilibria that made this process possible.

In the twenty-first century we will probably face an oil crisis. At the beginning of the twentieth century scientists were predicting that there would be a nitrate crisis. Already, valuable deposits of nitrates from bird droppings in Peru were exhausted. Although there were still vast supplies of sodium nitrate in the deserts of Chile, these supplies could not keep up with demand. So what was so important about nitrates and why were they in such high demand?

Nitrogen compounds are naturally present in the soil. If they were not, most plants would not be able to grow. Farmers have, for thousands of years, been replacing the nitrogen compounds taken out of the soil through harvesting crops by adding fertilizer – usually animal manure. However, a huge population increase in the nineteenth century in Europe and North America made these sources insufficient to support the more intensive agriculture required, so nitrate compounds were imported from wherever they could be found.

Nor was this the only use of nitrates. Explosives such as trinitrotoluene (TNT) and nitroglycerine required nitric acid, and this too needed nitrates as the raw material.

**Fig 1.22** Fertilizers manufactured from ammonia support the lives of millions of people through the intensive growing of crops

## TEMPERATURE CHANGES AND THE POSITION OF EQUILIBRIUM

When Haber started his research, several methods of making ammonia and nitrates had been invented. However, they were all costly in terms of energy and raw materials. Haber was interested in the most efficient way to make nitrogen compounds from nitrogen in the air. As you can read in the Feature

**? TEST YOURSELF**

**Q** Le Chatelier's principle suggests that lowering the temperature should give a much higher yield. Why did Haber consider using high temperatures? Hint: consider rates of reaction.

**? TEST YOURSELF**

**R** Predict the effect of increasing the temperature on the following equilibria. Explain your reasoning.
**a)** $2SO_2(g) + O_2(g) \rightleftharpoons 2SO_3(g)$
$\Delta H^{\ominus} = -98$ kJ mol$^{-1}$
**b)** $4NH_3(g) + 5O_2(g) \rightleftharpoons 4NO(g) + 6H_2O(l)$
$\Delta H^{\ominus} = -908$ kJ mol$^{-1}$
**c)** $CH_4 + H_2O(g) \rightleftharpoons CO(g) + 3H_2(g)$
$\Delta H^{\ominus} = +210$ kJ mol$^{-1}$

**? TEST YOURSELF**

**S** Use Le Chatelier's principle to predict the effect of increasing the pressure on each of these equilibria:
**a)** $CH_4(g) + H_2O(g) \rightleftharpoons CO(g) + 3H_2(g)$;
**b)** $CO(g) + H_2O(g) \rightleftharpoons CO_2(g) + H_2(g)$;
**c)** $2NO_2(g) \rightleftharpoons N_2O_4(g)$.

Box, it was crucial to find a process that would do this, because nitrogen compounds are not only essential fertilizers, but also the necessary starting point for manufacturing explosives.

At 1000 °C, the position of equilibrium lay well towards the reactants. In fact, the reaction mixture contained only a tiny percentage of ammonia (0.000 5–0.001 2%).

Using Le Chatelier's principle, we can predict the effect of an increase in temperature on the position of equilibrium:

$$N_2(g) + 3H_2(g) \rightleftharpoons 2NH_3(g) \quad \Delta H^{\ominus} = -92 \text{ kJ mol}^{-1}$$

According to Le Chatelier's principle, raising the temperature will cause the equilibrium to shift to absorb energy. This is because the change imposed on the system by increasing the temperature increases the energy of the system. This equilibrium is **exothermic** in the forward direction – the sign of $\Delta H$ tells you this. So raising the temperature shifts the equilibrium in the endothermic direction (i.e. to the left) so that energy is absorbed.

Another reaction that involves nitrogen occurs in the high temperatures of the car engine. As we have seen, under normal conditions nitrogen is very unreactive, but because the petrol engine can reach temperatures of 1000 °C nitrogen monoxide is one of the compounds produced. It is an atmospheric pollutant, which you can read more about on page 225, Chapter 10.

$$N_2(g) + O_2(g) \rightleftharpoons 2NO(g) \quad \Delta H^{\ominus} = +90 \text{ kJ mol}^{-1}$$

This reaction is, endothermic so the forward reaction absorbs energy; thus, raising the temperature will cause the equilibrium to shift to the right. Even at the high temperatures in an engine the yield is less than 1 per cent, but this is still sufficient to cause pollution problems.

## PRESSURE CHANGES AND THE POSITION OF EQUILIBRIUM

Haber expected that increasing the pressure would shift the position of the equilibrium to the right. As he wrote:

> **To begin with, it was clear that a change to the use of maximum pressure would be advantageous. It would improve the point of equilibrium and probably the rate of reaction as well. The compressor which we then possessed allowed gas to be compressed to 200 atmospheres.**

He predicted this using Le Chatelier's principle. This time the *imposed change in condition* is an increase in pressure, so the equilibrium will shift to minimize this. In the equilibrium:

$$\begin{array}{ccc} N_2(g) & + & 3H_2(g) & \rightleftharpoons & 2NH_3(g) \\ 1 \text{ mole} & + & 3 \text{ moles} \;(= 4 \text{ moles}) & \rightleftharpoons & 2 \text{ moles} \end{array}$$

there are fewer molecules of gas on the right hand side – 2 moles instead of 4 moles. In a gas, pressure depends on the number of molecules, and therefore increasing the pressure causes the equilibrium to shift to the right.

By contrast, in this equilibrium:

$$\begin{array}{ccc} H_2(g) & + & I_2(g) & \rightleftharpoons & 2HI(g) \\ 1 \text{ mole} & + & 1 \text{ mole} \;(= 2 \text{ moles}) & \rightleftharpoons & 2 \text{ moles} \end{array}$$

the number of molecules of gas is the same on both sides, so increasing the pressure has no effect on the equilibrium position. (Also, note that a change in pressure has no effect on equilibria that do not involve gases.)

## CONCENTRATION CHANGES AND THE POSITION OF EQUILIBRIUM

Equilibria in the liquid phase are not affected by changes in pressure, because there is no significant change in volume, but they are affected by changes in concentration.

Esters are sweet-smelling organic compounds. They are mainly used as solvents, but some of them can also be used to flavour food to give, for example, the taste of raspberry or pear. In addition, they are added to plastics such as polyvinylchloride (PVC) to make them more flexible. One industrial reaction to produce an ester is

$$CH_3COOH(l) + C_2H_5OH(l) \rightleftharpoons CH_3COOC_2H_5(l) + H_2O(l)$$
$$\text{ethanoic acid} + \text{ethanol} \rightleftharpoons \text{ethyl ethanoate ester} + \text{water}$$

According to Le Chatelier's principle, increasing the concentration of either of the reactants ethanol or ethanoic acid will shift the equilibrium to the right. This is because the imposed change in condition is an increase in concentration, so the equilibrium will shift to the product side to minimize this effect. In industry, because an increase in either reactant can be used to shift the equilibrium to form more ester, the cheaper reactant is used in excess.

Another way to manage the reversible reaction is to *remove* one of the products the moment it has formed. This prevents the reaction from achieving equilibrium, but it nevertheless will 'keep trying' to, as long as the product continues to be removed from the reaction mixture. In esterification, water is the easier product to remove.

> **THIS REACTION IS DISCUSSED IN MORE DETAIL IN CHAPTER 18.**

## THE EFFECT OF ADDING A CATALYST ON THE POSITION OF EQUILIBRIUM

A catalyst does not affect the position of equilibrium, but it does increase the rate of both the forward and reverse reactions, decreasing the time taken to reach equilibrium. Haber carried out more than 6000 experiments just to find the most effective catalyst, because he needed to speed up the reaction between nitrogen and hydrogen to obtain ammonia more quickly.

## 13 OBTAINING AN ECONOMIC YIELD IN THE HABER PROCESS

Le Chatelier's principle suggests that low-temperature and high-pressure conditions would maximize ammonia production (see Fig 1.23):

$$N_2(g) + 3H_2(g) \rightleftharpoons 2NH_3(g) \quad \Delta H^\ominus = -92 \text{ kJ mol}^{-1}$$

However, the chemical industry does not use these conditions for economic reasons. In choosing the right temperature, manufacturers must consider:
● the yield, that is, the percentage of ammonia in the equilibrium mixture;
● the rate of reaction.

The rate of reaction is related to temperature. If the temperature is too low the reaction proceeds too slowly. Therefore, manufacturers use a combination of a moderate temperature and a catalyst to ensure that the rate of ammonia production is high.

High pressure gives a fast rate of reaction for a gas-phase reaction, but using higher pressures drastically increases the cost of building and running the plant because the pipes that carry the gaseous mixtures need to withstand high pressures, and much more energy is used to maintain these pressures.

> **SEE QUESTIONS 7 AND 8.**

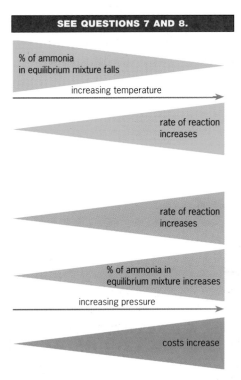

**Fig 1.23** The effects of changing conditions on the Haber process

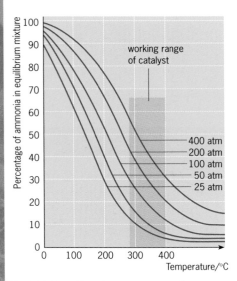

Fig 1.24 The effect of temperature on the percentage of ammonia in the equilibrium mixture

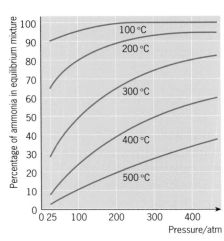

Fig 1.25 The percentage of ammonia in the equilibrium mixture as pressure changes

Fig 1.26 Flow diagram of the ammonia manufacturing process

## ❓ TEST YOURSELF

**T** In the steam-reforming stage shown in Fig 1.26, hydrogen is made by reacting methane and steam at 850 °C and 30 atmospheres pressure.

$CH_4(g) + H_2O(g) \rightleftharpoons CO(g) + 3H_2(g)$
$\Delta H^\ominus = +210$ kJ mol$^{-1}$

**a)** Use Le Chatelier's principle to explain why such a high temperature is used.

**b)** What is the effect on the position of equilibrium of using an increased pressure?

**c)** Why is a higher pressure of 30 atmospheres used? (Hint: Think about the rate of the reaction.)

**U** In the synthesis reaction vessel in Fig 1.26 ammonia is formed.

**a)** Explain why the iron catalyst used is finely divided, rather than in large lumps?

**b)** What is the effect of reducing the concentration of ammonia in the equilibrium mixture by removing it in the ammonia condenser?

Most manufacturers use:
● a temperature between 400 and 450 °C. This is a compromise temperature at which the rate of reaction is fast enough to give a reasonable yield in a short time.
● a pressure of about 200 atmospheres. Higher pressures are very costly in terms of running a plant and also more expensive to construct, as thicker-walled pipes are required.
● a finely divided iron catalyst, which speeds up both the forward and reverse reactions so that the equilibrium is attained faster.

A flow diagram for the complete process of ammonia manufacture is shown in Fig 1.26. Using the conditions outlined, the percentage yield of ammonia is about 15% in a single run. The unreacted nitrogen and hydrogen are recycled, which eventually gives yields of up to 98%.

## 14 ECONOMICS AND THE UK CHEMICAL INDUSTRY

With exports worth in excess of £12 billion per year, the chemical industry as a whole is the largest export earner in the UK. So it is not surprising that some of Britain's largest companies are in the chemical industry. They contribute towards our quality of life by supplying the products we buy and the raw materials used to make them, and by providing jobs and earning income for the nation.

The chemical principles of energy changes, yield and rate of reactions have been described for making iron and steel and for the manufacture of ammonia. For any industrially produced chemical, the same factors influence manufacturing costs. Other economic decisions that have to be made include which raw materials to start from, and which chemical reactions to use. These are some of the economic principles applied:
● **Raw materials.** These must be as cheap as possible.
● **Location of the plant.** This should be as near as possible to its raw materials. It should have good rail and road links. Iron and steel plants

are often close to ports for easy export and import. It should also be near an appropriately skilled workforce.
- **Type of process.** Research chemists are constantly seeking processes that minimize costs.
- **Energy costs.** The enormous amount of energy that iron and steel making requires accounts for a significant part of the total cost of a finished product. In all chemical production processes, energy transfer is carefully controlled to optimize energy use.
- **Safety.** Before any new plant is built or any existing plant modified, the safety of the workers on the site and of the people who live in the area is given top priority. All potential risks are assessed and minimized in the design, from the possibility of a pipe leaking to the threat of an explosion.
- **Pollution control.** A growing public awareness of the problems of the emission of gases and other pollutants means that industry must take steps to reduce pollution. However, this increases process costs.

The costs of a process include both fixed and variable costs. The fixed costs are incurred whether one tonne or one million tonnes are manufactured. These include the capital costs of setting up the plant, the rent, council taxes and payment of any loans, and depreciation in value of the plant. Variable costs change with output. If there is no production, then these will not be incurred. Examples of variable costs include the costs of raw materials, fuel, labour and transport.

## DO THE BENEFITS OF A CHEMICAL PLANT OUTWEIGH DISADVANTAGES?

This is a question to which there is no right answer. We are ready to enjoy the tremendous benefits that manufactured chemicals have brought us – the polymers in our clothing, the paint on our buildings, the materials that make these buildings, the fertilizers that help us to grow more food for more people, the medicinal drugs that combat so many potentially fatal diseases – the list goes on.

The chemical industry is part of our society and we have come to need many of its products. However, for every chemical process there are costs as well as benefits. But any *cost/benefit analysis* depends on the viewpoint of the person doing the analysis.

We can easily recognize the costs of damage to the environment through chemical pollution. There is no doubt that nowadays we are more aware than we used to be of pollution and its consequences and that chemists have improved the detection methods. We have seen that pressure to take action has led to more effective laws to control emissions of toxic and other waste products.

Now that we understand much more about the consequences of pollution, a main concern in designing new chemical plants is the possible effect on the environment. Also, as a society, we need to decide which chemicals are essential to maintain a healthy and comfortable lifestyle, and which are not.

## MINIMIZING ENVIRONMENTAL DAMAGE

Returning to iron and steel making, we said at the outset of this chapter that no other process had so transformed the world around us. Yet mining the iron ore can devastate large areas of land with its quarries and large spoil heaps. Likewise, slag from the blast furnace and steel making used to be piled high. Now it is used to make cement; so not only is the Humber bridge made of British steel, the concrete in it is also made of British steel slag!

The chemist is at the forefront of pollution control, both in its detection and in the development of control procedures. The waste gases from the blast furnace are harmful, so it is chemists who have developed ways to reduce these emissions. Carbon monoxide is a waste gas from iron and steel manufacture and is toxic to all vertebrates. Any trace of sulphur, particularly in iron sulphide ores, leads to the production of sulphur dioxide, which causes acid rain (see more about atmospheric pollutants on pages 163–171).

The chemical industry still suffers from a negative public image, despite the essential nature of many of its products. But a world without manufactured chemicals would now be unimaginable.

## SUMMARY

After studying this chapter, you should know and understand the following.

- A **molecular formula** tells you how many atoms of each element there are in a molecule of the compound.
- The term **formula** applies to ionic compounds as well as to molecules. It is the number of atoms of the different elements that make up the smallest complete unit of a substance.
- In the **relative atomic mass ($A_r$)**, scale, the average mass of the atom of an element (taking into account all of its isotopes and their abundances) is compared to 1/12 the mass of one atom of carbon-12.
- The **relative molecular mass ($M_r$)** is calculated using the relative atomic mass scale using the same standard, carbon-12. The $A_r$ values of each atom in the molecule are added together to give the $M_r$. When dealing with ionic compounds we use the term **relative formula mass**.
- The **amount** is the number of particles (atoms, molecules, ions, electrons, etc) in a substance. The unit of amount is the **mole** (mol).
- The **mole** is defined as the amount of substance that contains as many particles as there are atoms in exactly 12 g of carbon-12.
- The relative atomic mass in grams of any element contains one mole of atoms.

$$\text{amount in moles} = \frac{\text{mass in grams}}{\text{mass of one mole (in grams)}}$$

- The mass of one mole is also known as the **molar mass**.
- The **Avogadro constant** is the number of atoms or molecules in one mole which is $6.02 \times 10^{23}$.
- The **theoretical yield** is the maximum amount obtainable if all the reactants are converted into products. This is calculated from the balanced equation:

$$\text{percentage yield} = \frac{\text{actual yield}}{\text{theoretical yield}} \times 100\%$$

- The **rate of reaction** is affected by temperature, pressure, concentration, surface area of reactants, and catalysts.
- The **enthalpy change**, $\Delta H$, is a measure of the transfer of energy into or out of a reacting system at constant pressure.
- **Exothermic reactions** give out energy and cause the temperature of the surroundings to rise. $\Delta H$ is negative.
- **Endothermic reactions** take in energy and cause the temperature of the surroundings to fall. $\Delta H$ is positive.
- Iron ore, coke, limestone and air are the raw materials used in a blast furnace. The iron ore is reduced to iron by carbon monoxide. A high carbon content makes the iron very brittle.
- Steels are alloys of iron with carbon and other elements such as manganese and vanadium. The percentage of the elements alloyed with iron gives a particular steel its unique properties.
- A **reversible reaction** is one that can take place in either direction.
- A **dynamic equilibrium** is reached when the forward and reverse reactions occur at the same rate. This will only occur in a **closed system**.
- **Le Chatelier's principle** states that **the position of the equilibrium of a system changes to minimize the effect of any imposed change in conditions**.
- The **Haber process** is used to manufacture ammonia. Applying Le Chatelier's principle gives industrial chemists an understanding of the effect of changing temperature and pressure on the equilibrium position.

## ? EXAM QUESTIONS

**1**

**a)** The hydrocarbon in Fig 1.Q1 is constituent of the fuel used in road vehicles. Give its molecular formula.

$$H_3C-\overset{\displaystyle \underset{|}{CH_3}}{\underset{|}{C}}-CH_2-\overset{\displaystyle \underset{|}{CH}}{\underset{|}{CH}}-CH_3$$

**Fig 1.Q1**

**b)** Using the molecular formula, write an equation for the complete combustion of the hydrocarbon in Fig 1.Q1 to form carbon dioxide and water.

**2** The table below gives the accurate masses of two atoms.

| | $^1H$ | $^{12}C$ |
|---|---|---|
| Mass/g | $1.6734 \times 10^{-24}$ | $1.9925 \times 10^{-23}$ |

**a)** Calculate accurate values for the mass of one mole of each atom. (The Avogadro constant $L = 6.0225 \times 10^{23}$.)

**b)** Why is $^{12}C$ referred to when defining the relative atomic mass of an element?

**3** In the human body, oxygen, carbon and hydrogen are the three most abundant elements by mass. If a person weighs 50.0 kg calculate the moles of atoms of each element and state which one is the most abundant in terms of its atoms.

| Element | Per cent by mass |
|---|---|
| Oxygen | 65.0 |
| Carbon | 18.0 |
| Hydrogen | 10.0 |

**4** In 1979 the Nobel prize for Chemistry was shared by Herbert Brown (American) and George Wittig (German) for their work on the development of boron and phosphorus compounds. Brown used the gas diborane, $B_2H_6$, to prepare new organic compounds. One of the methods of diborane preparation involves this reaction:

$$3NaBH_4 + 4BF_3 \rightarrow 3NaBF_4 + 2B_2H_6$$

If 56.7 g of $NaBH_4$ reacts with an excess of $BF_3$ to produce 20.7 g of $B_2H_6$, what is the percentage yield of diborane?

**5** One of the ways to provide hot food in war-torn or starvation areas is by the reaction of magnesium and water.

$$Mg(s) + 2H_2O(l) \rightarrow Mg(OH)_2(s) + H_2(g)$$
$$\Delta H = -352.9 \text{ kJ mol}^{-1}$$

The food to be heated is stored in a pouch made of aluminium and plastic. When required it is placed inside a bag that contains magnesium powder, and a small amount of water is added. The reaction releases enough energy to make the food hot.

**a)** Explain why, at a given temperature, magnesium powder reacts faster than strips of magnesium ribbon.

**b)** Sketch an energy level diagram for this reaction.

**6**

**a)** Iron is extracted from the oxide $Fe_2O_3$ by reduction with carbon and carbon monoxide in a blast furnace.
**(i)** In the blast furnace, carbon monoxide is produced from carbon in a two-stage process. Write equations to show these two stages.
**(ii)** Write an equation for the reduction of $Fe_2O_3$ by carbon and an equation for the reduction of $Fe_2O_3$ by carbon monoxide.
**(iii)** Limestone is added to the blast furnace to remove impurities present in the iron ore. Identify the main impurity removed by limestone. Write an equation or equations to show how limestone acts to remove this impurity.

**b)** Stage the main impurity in iron obtained from the blast furnace. Explain how this impurity is removed in the conversion of impure iron into steel.

**c)** Identify a gas that is released from the blast furnace which leads to environmental problems. State the environmental problem.

**d)** Identify another gas that leads to environmental problems when sulphide ores are used in the extraction of metals other than iron. State the environmental problem.

AQA Unit 2 Foundation Physical and Inorganic Chemistry Jan 2001 Q4

**7** Ammonia, $NH_3$, is made industrially from its elements by the Haber process. This is an exothermic equilibrium reaction.

$$N_2(g) + 3H_2(g) \rightleftharpoons 2NH_3(g) \quad \Delta H = -92 \text{ kJ mol}^{-1}$$

**a)** State **three** reaction conditions that are used in the Haber process.

**b)** Describe and explain the effect of increasing the pressure on the **rate** of this reaction.

**c)** Describe and explain how the **equilibrium position** of this reaction is affected by:
**(i)** increasing the temperature;
**(ii)** increasing the pressure.

**d)** Why is the temperature used described as a compromise?

**e)** Some of the ammonia from the Haber process reacts with carbon dioxide to make the fertilizer urea.

$$\_NH_3(g) + \_CO_2(g) \rightarrow \_NH_2CONH_2(s) + \_H_2O(l)$$
urea

**(i)** Copy, then balance the above equation.
**(ii)** Calculate the maximum mass of urea that could be obtained from 1.00 kg of ammonia.

OCR 2813/1 AS How Far, How Fast? Jan 2001 Q7

## ? EXAM QUESTIONS

**8** Sulphur trioxide, $SO_3$, is made industrially by the Contact process. This is an example of dynamic equilibrium:

$$2SO_2(g) + O_2(g) \rightleftharpoons 2SO_3(g) \quad \Delta H = -197 \text{ kJ mol}^{-1}$$

**a)** State **two** features of a reaction with a *dynamic equilibrium*.

**b)** Use Le Chatelier's principle to explain what happens to the **equilibrium** position of this reaction as:
**(i)** the temperature is raised;
**(ii)** the pressure is increased.

**c)** Use your answer to **b)** to deduce the theoretical conditions for this equilibrium to provide a high yield.

**d)** Explain what happens to the **rate** of this reaction as:
**(i)** the temperature is raised;
**(ii)** the pressure is increased.

**e)** The conditions often used in the Contact process are 400 °C and normal atmospheric pressure. Using your answers to **b)**, **c)** and **d)**, comment on this choice of:
**(i)** temperature; **(ii)** pressure.
[The Key Skills Assignment below may help you.]

OCR 2813/01 AS How Far, How Fast? Specimen Paper Q4

# KEY SKILLS **ASSIGNMENT**

## MANUFACTURING SULPHURIC ACID

**In the nineteenth century it was said that the consumption of sulphuric acid is a barometer of a nation's commercial prosperity. Sulphuric acid is still a key industrial chemical to this day being used to manufacture paints, pigments, dyes, soaps and detergents, plastic and fibres, and some fertilizers.**

The raw materials for the manufacture of sulphuric acid are sulphur, air and water. The process used to make sulphuric acid involves three stages:

1. Molten sulphur is burnt in air to give sulphur dioxide:

$$S(s) + O_2(g) \rightarrow SO_2(g) \quad \Delta H^\oplus = -297 \text{ kJ mol}^{-1}$$

2. Sulphur dioxide is reacted with oxygen to give sulphur trioxide:

$$SO_2(g) + \frac{1}{2}O_2(g) \rightleftharpoons SO_3(g) \quad \Delta H^\oplus = -98 \text{ kJ mol}^{-1}$$

3. The sulphur trioxide is dissolved in concentrated sulphuric acid:

$$SO_3(g) + H_2SO_4(l) \rightleftharpoons H_2S_2O_7(l) \quad \Delta H^\oplus -230 \text{ kJ mol}^{-1}$$
$$H_2S_2O_7(l) + H_2O(l) \rightarrow H_2SO_4(l)$$

**1 a)** Explain the meaning of $\Delta$ and $\rightleftharpoons$.
**b)** Draw an energy level diagram for the equation in Stage 1 and state whether the reaction is exothermic or endothermic.
**c)** If 64 tonnes of sulphur is burnt in air what is the theoretical yield of sulphur dioxide?

Stage 2 is known as the **contact process**. The oxidation of sulphur dioxide into sulphur trioxide is a slow one and manufacturers use vanadium(V) oxide ($V_2O_5$) as a catalyst. $\Delta H$ is negative which means the reaction is exothermic in the forward direction. By Le Chatelier's principle the equilibrium position is shifted to the right by using a low temperature.

**2 a)** State Le Chatelier's principle and use it to explain why a low temperature shifts the equilibrium to the right in the Stage 2.
**b)** In practice a moderate temperature of 450 °C is used. Why is this done?

The Stage 2 equilibrium equation tells us that there are more molecules of gas on the left than on the right (1.5 moles > 1 mole). So a high pressure will shift the equilibrium position to the right, because this is the direction that minimizes the increase in pressure. However, despite Le Chatelier's prediction, a high pressure is not used. This is because the equilibrium position already lies well to the right, so the extra cost of building the plant and maintaining these higher pressures is not necessary. A slightly higher pressure than atmospheric is used simply to push the gases around the plant.

**3 a)** In the contact process (Stage 2) the concentration of reactants is kept high. Explain why.
**b)** The sulphur trioxide formed is continually removed. Explain why this is done.
**c)** The conversion of sulphur dioxide into sulphur trioxide in modern chemical plants is 99.5%. What are the economic advantages and environmental benefits of this yield?

# 02

# The nucleus and radioactivity

The eight most abundant elements of the Earth's crust are oxygen, silicon, aluminium, iron, calcium, sodium, potassium and magnesium. Hydrogen is the ninth, at less than 1 per cent. Yet 92 per cent of all atoms in the Universe are hydrogen, 7 per cent are helium, and all the other elements comprise only 1 per cent. So the Earth is very rich in its range of elements, much richer than some other parts of the Universe, and we may ask: Why are the elements so unevenly distributed?

The reason lies in the way the Universe, and later the Earth, first began. It is generally thought that the Universe started with tremendous fury in the Big Bang, at least 15 thousand million years ago, when a super-hot, super-dense 'soup' of sub-atomic particles exploded with unimaginable force. It was then that hydrogen atoms were first created. Some of these fused to give helium, and so a second element was born.

After about a thousand million years, stars began to form. Gaseous hydrogen with a little helium came together in local areas, contracting and becoming denser. The temperature rose to levels at which other light elements could form, and even more heat was produced. Stars continue to be born in this way from the dispersed material of the Universe.

The heavier elements appeared only in the relatively few very massive stars. Those with a much larger mass than the Sun became unstable after a few million years and exploded, propelling atoms out into the cosmos, to become the raw material of a second generation of stars with the heavier elements already formed.

It is thought that the Solar System was formed from such recycled material about 4500 million years ago. The Sun itself is a gigantic fusion reactor, its heat and light helping to sustain life on Earth. The Earth itself contains some 90 light and heavy elements from the remains of earlier stars. In its intensely hot core, it retains some of the heat from when it was formed. More heat is produced by radioactive elements, which decay and release energy. In fact, as well as the heat of the core and from the Sun, this decay process is significant in maintaining the Earth's present-day heat balance.

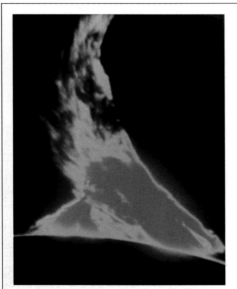

**Jet of glowing gas.** Evidence of nuclear fusion at the Sun's core is given by jets of glowing gas on the surface of the Sun

### A HISTORY OF IDEAS ABOUT THE ELEMENTS

In 580 BC, the Ancient Greek philosopher Thales suggested that water was the fundamental 'element' from which all matter in the Universe was composed. Aristotle, 200 years later, added earth, air and fire, proposing that these four elements made up the world, with a fifth, 'aether', making up the heavens. This idea persisted for almost 2000 years until Robert Boyle published a book called *The Sceptical Chymist* in 1661.

Boyle's book was a turning point in chemistry. It put forward the first

Fig 2.1 Robert Boyle (1627–1691)

Fig 2.2 The title page of Boyle's book, *The Sceptical Chymist*. In it he first introduced the idea of the 'chemist'. He also proposed the idea that elements cannot be changed into anything simpler and that they could be discovered by experimentation

Fig 2.3 John Dalton (1766–1844) and the symbols he used to represent the elements that were known at the time

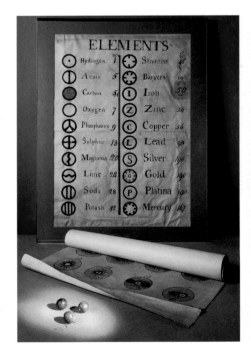

modern concept of an **element** as something that cannot be changed into anything simpler. The book also marked the beginning of chemistry as an experimental science, since Boyle was urging chemists to carry out practical investigations rather than merely observe, think and make deductions, as the Greeks had done.

His advice was heeded. In 1803, the English chemist, John Dalton, summed up a 150 years of progress through experiment with his **atomic theory**. He used the Greek word 'atomos', which means 'cannot be cut', to describe the particles that make up elements. The idea that all matter was made up of tiny, indivisible particles had been proposed in 450 BC, again by an Ancient Greek, Democritus. But now, Dalton could support it with experimental results. These are the main points of his theory:

- Atoms are indivisible and indestructible.
- All atoms of the same element have identical mass and identical chemical properties.
- When atoms react, they join together to form 'compound atoms' (now called molecules).

Today, we still hold Dalton's view of an element as a substance composed of only one type of atom. So the element iron is made up of only iron atoms, which are different from copper atoms, which make up the element copper.

We now know that an atom can be subdivided into smaller parts called **sub-atomic particles**, discovered after Dalton's time. But the atom is still the smallest particle with all the *chemical* properties of an element, and Dalton's atomic theory provided the foundation on which other scientists have built.

# 1 BUILDING MODELS OF THE ATOM

A **model** is the scientist's attempt to explain observations. If the model put forward explains all known observations, then the scientific community accepts it. Dalton's model of the atom suggested that atoms could not be sub-divided into smaller particles. This model was accepted until the end of the nineteenth century, when new discoveries and observations meant the model had to be changed.

## THE DISCOVERY OF THE ELECTRON

In 1897, J. J. Thomson was working at the Cavendish Laboratories in Cambridge. He was experimenting with a ray that was emitted from the **cathode** (the conducting material connected to the negative pole of a battery). This ray, called the **cathode ray**, had been discovered in the 1870s when an electric current was passed through a gas at low pressure (Fig. 2.4).

glass tube      gas at low pressure

cathode      stream of electrons: the cathode ray moves from cathode to anode      anode

**Fig 2.4** A simplified diagram of a cathode ray tube. The high voltage between the cathode and anode strips electrons from the atoms of the gas

many electrons with negative charge

spherical cloud of positive charge

**Fig 2.5** Thomson's plum pudding model of the atom. Negative and positive charges cancel out

**? TEST YOURSELF**

**A** In Thomson's model of the atom, how many electrons would be needed to account for the mass of the hydrogen atom?

Thomson found evidence that cathode rays were made up of particles, which he called **electrons**. He calculated the mass of an electron to be 2000 times less than the mass of hydrogen atom and, as they were attracted to the positive **anode**, he knew they must be negatively charged. (In 1906, this work gained him a Nobel Prize.) Since these particles must be being produced from the atoms inside the apparatus, he realized he had discovered a **sub-atomic particle**. This meant that a new model of the atom was required to explain the observation.

As no other sub-atomic particles had been discovered, Thomson's model of the atom was made up of negatively charged electrons. It was known that atoms had no net charge, that they were neutral, so the electrons were thought by Thomson to be embedded in a sphere of positive charge, which had no mass. To account for the mass of atoms, Thomson assumed that they must all have thousands of electrons. His model was known as the 'plum pudding' model because the electrons were pictured like raisins in a pudding.

## RUTHERFORD'S MODEL OF THE ATOM

Ernest Rutherford, a New Zealander working under Thomson at Cambridge, began investigating a newly discovered phenomenon, **radioactivity**. Over several years of work, he found that the radioactivity, which a French scientist called Henri Becquerel had first observed (see the Extension box), was composed of three main types of radiation. Rutherford named the first two **alpha** (α) and **beta** (β) radiation. When the third was eventually discovered, it was called **gamma** (γ), see Fig 2.6.

**Fig 2.6** The different penetrating properties of the three radiations. A lead block has to be about 10 cm thick to halt gamma rays

α particles    β particles    γ rays

lead block    radioactive source    sheet of paper    thin aluminium sheet    thick block of lead

## EXTENSION The discovery of radioactivity

While Thomson was experimenting on cathode rays, another momentous discovery was being made. The Frenchman Henri Becquerel was investigating crystals that phosphoresce – that is, chemicals that glow in the dark after being exposed to sunlight. Becquerel assumed that sunlight was causing the crystals to give out X-rays, which had just been discovered (see page 35). X-rays were not understood then, which is why they were called 'X'. But they were known to make a dark, blurred image, called 'fogging', on photographic film, even when it was sealed in opaque (light-proof) black paper.

Becquerel decided to test his assumption. He put crystals of a uranium salt on top of a photographic film, which was sealed in black paper that kept out all light. He let the Sun shine on the crystals and, as he expected, when the film was developed, it had become fogged.

On 26 February 1896, clouds shut out the Sun, so Becquerel just put the crystals, unexposed to sunlight, in a drawer on top of some sealed photographic film. Three days later, he decided to develop the film anyway, just to check that there was no image. To his amazement, he found a clear area of darkening on the film below the crystals (Fig 2.7). He concluded that some unknown kind of radiation that is not dependent on the Sun and that could pass through the black paper must have come from the crystals. It took the work of others to discover more about this radiation.

**Fig 2.7** Becquerel's photographic film on which he had put crystals unexposed to sunlight: the first recorded evidence of radioactivity

With her husband Pierre, Marie Curie began to study this phenomenon. She investigated other uranium compounds and found that they, too, gave off this new radiation. She called it **radioactivity**. (Radioactivity is the spontaneous break-up of atomic nuclei, giving off rays called radiation.)

The Curies observed that pitchblende, an ore known to contain uranium, was much more radioactive than they expected from the uranium in the ore alone. The Curies later discovered two new radioactive elements, radium and polonium. It was radium that gave pitchblende its high radioactivity. The radiation of radium had so much energy that it burned the skin and, while uranium fogged film after a few hours, radium did so instantly. Marie Curie also discovered another radioactive element, thorium.

**Fig 2.8** Marie Curie (1867–1934) was the first person to receive two Nobel Prizes. The first was for physics (1903), which she shared with her husband and Becquerel for work on radioactivity. The second was for chemistry (1911) for her discovery of radium and polonium (named after her homeland, Poland). At the time of her discoveries it was not realized that the radiation to which she was exposed would cause the leukaemia that was to end her life

**Fig 2.9** Deflection of the three types of radiations by charged plates

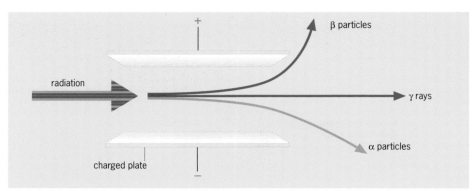

### TEST YOURSELF

**B** Fig 2.9 shows how the three different types of radiation are affected by an electric field.
**a)** What do you deduce about the charge on the different types of radiation?
**b)** Why do you think the beta particles are deflected so much more than the alpha particles?

Rutherford showed that alpha radiation was composed of particles that were four times heavier than hydrogen. In fact, they are the same as positively charged helium atoms. (We refer to them as helium nuclei.) The beta particles were soon found to be very high-energy electrons, which of course are negatively charged. Gamma rays were identified later as electromagnetic rays similar to X-rays (see page 42), but of much higher energy. Being electromagnetic rays, gamma rays were found to have no particle properties, and therefore no mass. You can find out more about radioactivity in the Extension box on page 31.

## RUTHERFORD'S SCATTERING EXPERIMENT

Rutherford moved to Manchester University to set up his own research group, which included Hans Geiger and Ernest Marsden. In their investigations into the structure of the atom, they fired alpha particles at gold foil, using the equipment shown in Fig 2.10(a). From Thomson's model of the atom, it was expected that a few alpha particles would be deflected slightly from a straight route when they passed through foils. Even the thinnest foils were about 2000 atoms thick, and some of the alpha particles were sure to collide with electrons. Marsden was asked to see if any alpha particles were scattered through large angles.

Rutherford said, 'I may tell you that I did not believe they would be, since we knew that the alpha particle was a very fast, massive particle, with a great deal of energy.'

But then, a few days later, Geiger announced that some alpha particles had been recorded as bouncing back from the gold foil. Rutherford recalled, 'It was quite the most incredible event that has happened to me in my life. It was almost as incredible as if you fired a fifteen inch [artillery] shell at a piece of tissue paper, and it came back and hit you.'

It was now clear to Rutherford that Thomson's model of the atom could not be correct. If it were, then the alpha particles should have shot through the foil with hardly any deflection. But one in 8000 had apparently been halted in its tracks and deflected back. Most particles passed through without any deflection, and Rutherford concluded that they were travelling through empty space. The positively charged alpha particles deflected through larger angles must have encountered (and been repelled by) another positive charge of considerable mass, which he called the **nucleus**.

From the angles of deflection and other data, Rutherford calculated that the radius of the gold nucleus was $10^{-14}$ metre. (We now know that the radius is closer to $10^{-15}$ metre.) He calculated, too, that the radius of the atom was approximately $10^{-10}$ metre. This means that the nucleus was 100 000 times smaller in diameter than the atom – an astonishing difference. If a full stop on this page represented the nucleus, then the outer limit of the atom would be about 25 metres away. No wonder that only a very few alpha particles were seen to come back.

### TEST YOURSELF

**C** A tennis ball has a radius of 3.1 cm. Assume it represents a nucleus. What is the radius of the atom on this scale to the nearest kilometre?

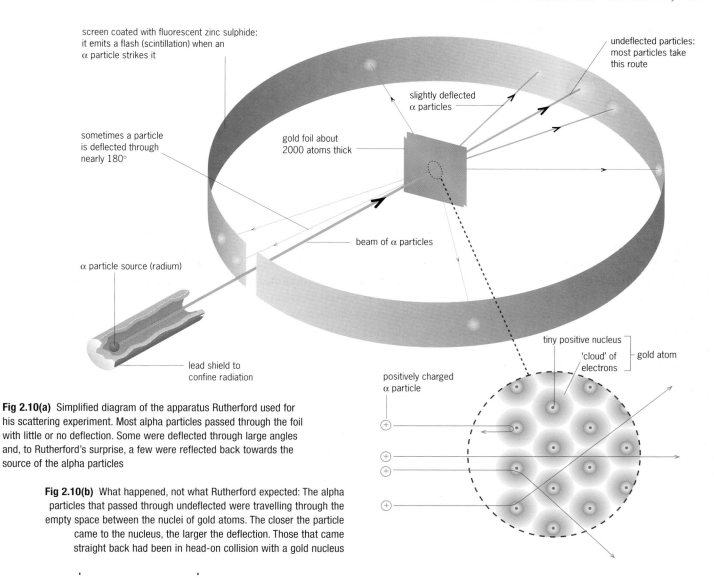

screen coated with fluorescent zinc sulphide: it emits a flash (scintillation) when an α particle strikes it

undeflected particles: most particles take this route

slightly deflected α particles

sometimes a particle is deflected through nearly 180°

gold foil about 2000 atoms thick

beam of α particles

α particle source (radium)

lead shield to confine radiation

**Fig 2.10(a)** Simplified diagram of the apparatus Rutherford used for his scattering experiment. Most alpha particles passed through the foil with little or no deflection. Some were deflected through large angles and, to Rutherford's surprise, a few were reflected back towards the source of the alpha particles

**Fig 2.10(b)** What happened, not what Rutherford expected: The alpha particles that passed through undeflected were travelling through the empty space between the nuclei of gold atoms. The closer the particle came to the nucleus, the larger the deflection. Those that came straight back had been in head-on collision with a gold nucleus

tiny positive nucleus

'cloud' of electrons

gold atom

positively charged α particle

## RUTHERFORD PROPOSES HIS MODEL

Rutherford now proposed his model of the atom – a minute, central nucleus containing all the positive charge of the atom and almost all of the mass, surrounded by empty space in which electrons orbited the nucleus, rather like planets round the Sun.

**Fig 2.11(a)** Ernest Rutherford (1871–1937) in his laboratory

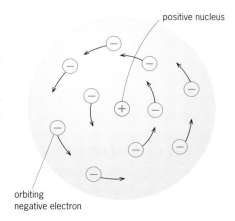

positive nucleus

orbiting negative electron

**Fig 2.11(b)** Rutherford's model of the atom proposed after the scattering experiment shown in **Fig 2.10(a)**

This model of the atom raised some difficult questions. In particular, since opposite charges attract, why didn't the electrons just fall into the nucleus? The Danish scientist, Niels Bohr, came up with the answer to this question, and you can read more about this in Chapter 3. The idea of the minute, central, positive nucleus is fundamental to all models of the atom accepted today.

## PROTONS AND NEUTRONS

The proton was the next sub-atomic particle to be discovered, and again the alpha particle was the tool used in research. When alpha particles were fired through hydrogen gas, positive particles emerged which were about the same mass as the hydrogen atom. In 1919, Rutherford said that these same particles could be knocked out of other atoms, suggesting that they must be present in the other atoms. He gave the name **protons** to these positive particles. The model of a nucleus made up of protons was now taking shape.

A year later, Rutherford suggested that protons were not the only type of particle in the nucleus, and that there must be particles of equal mass but neutral charge in the nucleus as well. In suggesting this, Rutherford was almost a lone voice in the scientific community. Then, in 1932, James Chadwick, another of his research team, was able to show that this neutral particle did exist.

Chadwick fired alpha particles at beryllium atoms. Instruments that detected charged particles were unable to detect any radiation, as in Fig 2.12(a). Yet when paraffin wax was put between the beryllium and the detector, a shower of protons was detected, as in Fig 2.12(b). Rutherford guessed that the protons were coming from the paraffin because some sort of radiation was hitting it. He said this radiation was like the effect of an invisible man who cannot be seen directly, but who is known to be there because he collides with other people in a crowd.

The invisible particle in the radiation of Chadwick's experiment – with no charge and with its mass equal to that of the proton – was named the **neutron**. In this way, protons and neutrons became known to scientists.

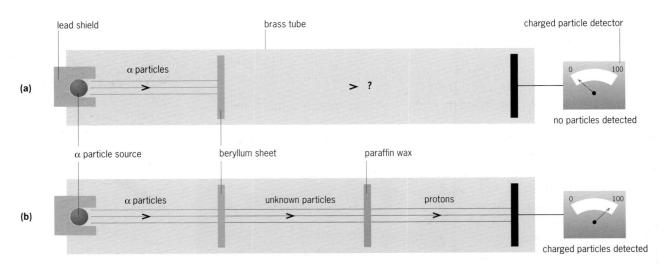

**Fig 2.12** A simplified diagram of Chadwick's apparatus. The unknown particles were neutrons, which knocked protons from the paraffin wax

# 2 DESCRIBING THE ATOM

## ATOMIC NUMBER (PROTON NUMBER)

Five years before Rutherford announced the discovery of the proton, Henry Moseley, another of his research team, had been working with X-rays. These were discovered in 1895 by Wilhelm Röntgen in Germany. Röntgen had been trying to make substances fluoresce by bombarding them with cathode rays (streams of high-energy electrons). Fluorescing substances absorb energy of one wavelength, usually outside the visible spectrum, and give out energy at another wavelength that is visible. (For more information about the spectrum, refer to Chapter 3.) Some substances emitted what Röntgen called an X-ray. The X-ray could penetrate some matter, and blackened photographic film, as Röntgen was able to show by making an image of his wife's hand.

Moseley bombarded different metal targets with cathode rays (Fig 2.14), and measured the frequency of the X-rays that emerged from the anode. Frequency increased with the increase in mass of the metal atom.

**Fig 2.13** The first X-ray image of a human, taken by Röntgen of his wife's hand, showing her wedding ring, a coin and a compass

**Fig 2.14** Moseley's experiment, in which he bombarded different metal targets with a beam of high-energy electrons – cathode rays – and measured the frequency of X-rays produced

From Moseley's results, Rutherford calculated the positive charge on the nucleus, which he called the **atomic number**:

> **The atomic number is the number of protons in the atom.**

It has the symbol **Z**, and is also known as the **proton number**.

The atomic number tells us the following:

- What the element is: only atoms of the same element have the same atomic number.
- The element's numbered position in the Periodic Table: the Periodic Table is arranged in atomic number order (see Chapter 5).
- The number of electrons in a neutral atom: the charge on an electron is equal, but opposite, to the charge on a proton

Information we now know about the three fundamental particles in an atom is summarized in Table 2.1.

**? TEST YOURSELF**

**D** Sodium is the eleventh element in the Periodic Table. Work out its atomic number, the number of protons in its nucleus and the number of electrons that orbit outside the nucleus.

**Table 2.1 Some properties of the three main sub-atomic particles**

| Particle | Symbol | Relative mass | Relative charge | Location |
|----------|--------|---------------|-----------------|----------|
| Proton | $_1^1 p$ | 1 | +1 | In nucleus |
| Neutron | $_0^1 n$ | 1 | 0 | In nucleus |
| Electron | $_{-1}^0 e$ | 0.000 55[a] | −1 | Orbiting nucleus |

[a]The mass of the electron is so small relative to the proton and neutron that chemists often take it as zero

# MASS NUMBERS (NUCLEON NUMBERS) AND ISOTOPES

The mass of the atom is almost entirely the mass of the protons and neutrons in its nucleus. Even the 92 electrons in an atom of uranium have hardly any effect on its mass. Since protons and neutrons contribute to an atom's mass, we call the total of both in an atom the **mass number**.

> **The mass number is the total number of protons and neutrons in the atom.**

It has the symbol **A**.

As protons and neutrons are constituents of the nucleus, they are referred to as **nucleons**. So $A$ is also known as the **nucleon number**. (Note that the mass number is not the *mass* of a nucleus.)

$$\text{Mass number} = \text{number of protons} + \text{number of neutrons}$$
$$A \quad = \quad Z \quad + \quad N$$

Chemists summarize information about an atom in this way:

mass number —— $A$
atomic number —— $Z$ $X$ —— element symbol

The atomic number $Z$ is always the same for a particular element, and it indicates the position of the element in the Periodic Table. It is often left out of symbol notation, since the element symbol itself identifies the atom.

## ● EXAMPLE

**Q** An atom of phosphorus has 16 neutrons. Write the full symbol notation for this atom.

**A** First you need to find out the atomic number of phosphorus. The atomic number, $Z$, for phosphorus is 15, and its symbol is P.

$A = Z + N$, so $A = 15 + 16 = 31$

The full symbol notation is $^{31}_{15}\text{P}$.

It specifies a particular atomic number and mass number, and hence represents an atom with a particular number of protons and a particular number of neutrons. It is called a **nuclide**.

**Q** How many protons, neutrons and electrons does the following atom have?

Fluorine, $^{19}_{9}\text{F}$

**A** $Z = 9$, so the number of protons = 9. The atom is neutral, so the number of electrons = 9.

$A = 19$, so $Z + N = 19$.

Since $Z = 9$, $N = 10$.

The number of neutrons ($N$) = 10.

Now try question E in the margin.

## ? TEST YOURSELF

**E** Work out the name, mass number, atomic number, and number of protons, electrons and neutrons in these nuclides:
a) $^{56}_{26}\text{Fe}$;
b) $^{200}_{80}\text{Hg}$.

**SEE QUESTION 1.**

The **mass spectrometer** made it possible to determine the masses of atoms very accurately (see Chapter 9), and it became clear that not all atoms of the same element had the same mass. This meant that there had to be a change to Dalton's atomic theory, which said that all atoms of an element had an identical mass.

The accurate masses gave rise to the idea of the **isotope**. In Greek, isotope means 'same place', but it is more helpful to think of it as meaning 'alternative'. So we cannot say that an element has only one isotope. An element has to have more than one nuclear arrangement (more than one **nuclide**) before its atoms can be described as isotopes. A good example is chlorine, which has two isotopes.

Table 2.2 shows that isotopes are atoms of the same element and so have the same atomic number. The only difference between the isotopes is their mass number, which indicates that the nucleus of one isotope has more neutrons than that of another. Thus:

**Isotopes are atoms with the same atomic number, but different mass numbers.**

## RELATIVE ISOTOPIC MASS

Chapter 1 explains that the relative atomic mass, $A_r$, compares the masses of atoms to the mass of one atom of carbon-12, an isotope of carbon whose relative atomic mass is taken as exactly 12. Its symbol is $^{12}_6C$, or $^{12}C$, because, as we have seen, once the element is known, the atomic number is fixed. In nature, most elements are composed of isotopes.

**The relative mass of an isotope (its relative isotopic mass) is the mass of one atom of that isotope compared to $\frac{1}{12}$ the mass of one atom of carbon-12.**

Some elements have only one nuclide and, as most of the mass of an atom is made up of protons and neutrons, the mass number is very nearly the same as the relative atomic mass. For example, naturally occurring fluorine has just one nuclide, $^{19}F$, and the $A_r$ of fluorine is 18.998 40. But for all except the most accurate work, 19 is used.

## RELATIVE ATOMIC MASS

The two isotopes of chlorine are $^{35}Cl$ and $^{37}Cl$. If these two isotopes had the same abundance, then the $A_r$ of chlorine would be 36. But $^{35}Cl$ has a 75 per cent abundance and $^{37}Cl$ has a 25 per cent abundance.

This means that for every 100 atoms, 75 have a relative isotopic mass of 35, and 25 have a relative isotopic mass of 37.

The $A_r$ is calculated as follows.

$$A_r \text{ chlorine} = \frac{(75 \times 35)}{100} + \frac{(25 \times 37)}{100}$$

$$= \frac{3550}{100}$$

$$= 35.5$$

This leads to the statement:

**The relative atomic mass of an element is the average mass of the atom (taking into account all of its isotopes and their abundance) compared to $\frac{1}{12}$ the mass of one atom of carbon-12.**

| Table 2.2 The two naturally occurring isotopes of chlorine | | |
|---|---|---|
| Isotope | $^{35}_{17}Cl$ | $^{37}_{17}Cl$ |
| Mass number, A | 35 | 37 |
| Atomic number, Z | 17 | 17 |
| Number of protons | 17 | 17 |
| Number of neutrons, N | 18 | 20 |
| Number of electrons | 17 | 17 |

### ? TEST YOURSELF

**F** Most of the carbon in nature is the isotope $^{12}_6C$. But two other isotopes exist, $^{13}_6C$ and $^{14}_6C$. $^{14}_6C$ is radioactive and is used to estimate the age of very old objects. Draw out and complete a table similar to Table 2.2 for the isotopes of carbon.

### ? TEST YOURSELF

**G** Study Table 2.3 and work out the $A_r$ of bromine and magnesium.

**SEE QUESTIONS 2 AND 3.**

| Table 2.3 Isotopes of bromine and magnesium | | |
|---|---|---|
| Element | Isotope | Abundance |
| Bromine | $^{79}Br$ | 50% |
| | $^{81}Br$ | 50% |
| Magnesium | $^{24}Mg$ | 78.6% |
| | $^{25}Mg$ | 10.1% |
| | $^{26}Mg$ | 11.3% |

## ▶ EXTENSION Nuclear fusion

**Fig 2.15** The Sun can be thought of as an immense fusion reactor that provides energy which reaches all parts of the Solar System

This chapter began by describing the immense energy that powers the Sun and builds up heavier elements, starting from hydrogen. This energy comes from the nucleus, or rather countless billions of nuclei. When atoms could be weighed very accurately, their mass was found to be less than the sum of the masses of their sub-atomic particles. Some of the mass appeared to be 'missing', and the missing amount is called the **mass defect**. Einstein explained this in perhaps the most famous equation of all time,

$$E = mc^2$$

$E$ is the energy in joules, $m$ is the missing mass in kilograms and $c$ is the speed of light in metres per second ($3 \times 10^8 \, \text{m s}^{-1}$). The missing mass is released as energy when nucleons (the protons and neutrons) fuse to form a nucleus. This energy is called the **binding energy**, and the resulting nucleus is in a lower, more stable energy state than the separate protons and neutrons. The speed of

light is an extremely large number squared, so it can be seen that just a small mass produces a very large amount of energy.

Let's consider deuterium, which is an isotope of hydrogen ($^2_1\text{H}$). It is made up of one proton and one neutron.

The 'missing' mass of its nucleus is $3.965 \times 10^{-30}$ kg.

Energy released when the proton and neutron fuse = $mc^2$,

$$= 0.003\,965 \times 10^{-27} \times (3.00 \times 10^8)^2$$
$$= 3.57 \times 10^{-13} \, \text{J (to three significant figures)}$$

This result is a very small amount of energy, but it is only for one nucleus. Consider a mole ($6.02 \times 10^{23}$) of protons and neutrons which fuse; the energy released is $2.15 \times 10^{11}$ J (215 million kJ). Compare this to burning one mole of carbon in oxygen, which gives a mere 394 kJ. The energy released in chemical reactions is clearly tiny compared to the awesome amount released from nuclear reactions. In the intense heat and pressure at the core of the Sun, nuclei fuse and release the tremendous amounts of energy we see and feel on Earth.

### NUCLEAR FUSION AS AN ENERGY SOURCE?

It is the dream of scientists to provide a cheap, unlimited source of power that has harmless by-products, by fusing nuclei and harnessing the vast amounts of energy released. Scientists see deuterium as the most likely candidate since, for every 20 000 hydrogen atoms in water molecules, three are deuterium atoms, so the supply from the oceans of the world is limitless. The reaction proposed is:

$$^2_1\text{H} + ^2_1\text{H} \rightarrow ^3_2\text{He} + ^1_0\text{n} + \text{energy}$$

The energy released in this reaction is 290 million kJ per mole.

In the process, deuterium nuclei stripped of their electrons would have to be brought together. A temperature of 40 million K is needed to overcome the repulsion forces of their positive charges, and to sustain such a high temperature on Earth is the challenge. To try and achieve this, scientists are using very powerful lasers and heating with an electric current, but making the dream a reality is likely to be some years away.

**Fig 2.16** The Tokamak fusion test reactor and deuterium plasma inside it. At the temperatures that exist inside stars, the electrons and nuclei of atoms separate to give a new state of matter called a plasma, and nuclei can fuse. For a fusion reactor to work on Earth, it needs to keep a plasma at a high enough temperature and density for long enough for the nuclei to fuse

# 3 DESCRIBING NUCLEAR REACTIONS

## REPRESENTING SUB-ATOMIC PARTICLES

We have used the standard notation for mass number, symbol $A$, and atomic number, symbol $Z$, to represent atoms of elements and their isotopes. Soon we will be looking at nuclear reactions using this notation. These reactions involve atoms, and some also include sub-atomic particles. So that we can balance nuclear equations, we use a system of notation for the sub-atomic particles that matches the one we use for atoms, shown in Table 2.4.

The diagram in the margin explains what the numbers mean for the sub-atomic particles in Table 2.4.

We know that protons and neutrons determine the mass numbers and atomic numbers of atoms. The information above shows us that the numbers given to the sub-atomic particles are connected with these mass numbers and atomic numbers, and we will find this is useful in writing nuclear equations.

| contributes to the mass number of an element | contributes to the mass number of an element | the negligible mass of an electron: nearly zero |
|---|---|---|
| $_1^1 \text{p}$ | $_0^1 \text{n}$ | $_{-1}^0 \text{e}$ |
| contributes to the atomic number of an element | does not contribute to an element's atomic number | the charge on an electron |

## BALANCING NUCLEAR EQUATIONS

The following is an example of a **nuclear equation**:

$$\underset{\text{hydrogen}}{_1^1 \text{H}} + \underset{\text{deuterium}}{_1^2 \text{H}} \rightarrow \underset{\text{helium isotope}}{_2^3 \text{He}} + \text{energy}$$

Notice that the mass numbers and the atomic numbers balance. In any nuclear equation, atomic numbers and mass numbers must balance in this way. The major fusion reaction of the Sun is thought to be between nuclei of deuterium and nuclei of another hydrogen isotope, tritium, to give helium and a neutron.

$$_1^2 \text{H} + _1^3 \text{H} \rightarrow _2^4 \text{He} + _0^1 \text{n} + \text{energy}$$

**Table 2.4 Properties of sub-atomic particles**

| Particle | Symbol | Relative mass | Relative charge | Location |
|---|---|---|---|---|
| Proton | $_1^1 \text{p}$ | 1 | +1 | In nucleus |
| Neutron | $_0^1 \text{n}$ | 1 | 0 | In nucleus |
| Electron | $_{-1}^0 \text{e}$ | 0.000 55 | −1 | Orbiting nucleus |

### ? TEST YOURSELF

**H** Our Sun is a relatively small star. In stars that are about 30 times larger than the Sun, much higher temperatures allow helium to fuse to form other elements. Complete this nuclear equation and work out the identity of element X.
$$_?^4 \text{He} + _?^4 \text{He} \rightarrow _?^? \text{X} + \text{energy}$$

## EXTENSION Elements formed in stars

The nuclear fusion reaction in question H is just one of many reactions of large stars in which the nuclei of light elements fuse to make heavier elements. All these reactions give out energy. But when iron nuclei fuse, they *take in* energy rather than releasing it, and this fusion marks the end for a star. Iron builds up right at its centre and, when fusing, iron nuclei take up too much of the star's energy, the core collapses with incredible speed and force, and its density increases hundreds of millions of times.

Then, the collapsing matter rebounds. In a stupendous explosion, the outer layers of the dying star are blown off into the cosmos and, for a short time, we see the star as a supernova, shining as brightly as an entire galaxy. Look at Fig 2.17 to see Supernova 1987A as its matter was blasted into space, to become part of a new generation of stars.

Our own Sun is a second-generation star containing recycled remnants from past supernovae, and the many elements found on Earth come from giant stars that exploded as supernovae billions of years ago.

**Fig 2.17** On 23 February 1987, a brilliant star-like object about 170 000 light years away was first seen from the Earth. A supergiant star had collapsed in on itself and then exploded. It became known as Supernova 1987A. The photo shows the Tarantula nebula, an enormous cloud of ionized gas, and below it, the exploding star

# 4 THREE RADIATIONS

Radioactivity is the spontaneous breakdown (disintegration) of a nucleus and the emission of radiation. 'Spontaneous' means something that happens without anything seeming to cause it.

Some naturally occurring isotopes are unstable and will break down. As a result, they release at least one of the three principal types of radiation mentioned earlier in the chapter – alpha ($\alpha$), beta ($\beta$), and gamma ($\gamma$) radiation. All three can knock electrons off atoms they collide with, and so they are called **ionizing radiations**.

| Table 2.5 Some properties of the three main types of radiation (See also Fig 2.6, page XXX) | | | | | | |
|---|---|---|---|---|---|---|
| Type | Nature | Speed | Charge | Relative mass | Distance travelled in air | Stopped by |
| $\alpha$ | Helium nucleus, $He^{2+}$ or $^4_2He$, i.e. 2 protons + 2 neutrons | 10% light speed | +2 | 4 | A few centimetres | Paper, skin or clothing |
| $\beta$ | High-speed electron | 90% light speed | −1 | 0.000 55 | A few metres | Thin aluminium sheet |
| $\gamma$ | Electromagnetic wave of high energy | Light speed | 0 | 0 | A few kilometres | 10 cm sheet of lead or several metres of concrete |

## WHAT MAKES SOME ISOTOPES UNSTABLE?

Some nuclei are stable (non-radioactive), while others decay until they achieve stability. For example, a helium nucleus is very stable, and many other proton–neutron combinations are also stable. If we plot the number of protons against the number of neutrons for stable isotopes, as in Fig 2.18, a zone of stability emerges. Radioactive isotopes are usually outside this zone, and move nearer to it as they decay.

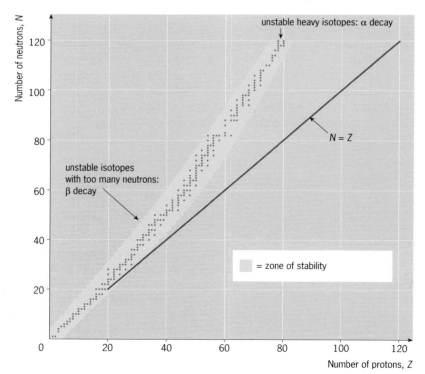

**Fig 2.18** For stable nuclei, the number of neutrons plotted against the number of protons. The stable nuclei lie in the zone of stability. Most radioactive nuclei fall outside this zone and, when they decay, the resulting nuclei move nearer to it.

# NUCLEAR EQUATIONS AND RADIOACTIVE DECAY

## ALPHA DECAY

Any atom with an atomic number greater than 82 will be radioactive, and most of these elements decay by emitting alpha radiation. At the upper limit of the stability zone is lead, $^{206}_{82}$Pb. The next element, bismuth, $^{209}_{83}$Bi, is unstable.

Americium, with an atomic number of 95, is the radioactive isotope used in smoke detectors. The nuclear equation for its **alpha decay** is:

$$^{241}_{95}\text{Am} \quad \rightarrow \quad ^{237}_{93}\text{Np} \quad + \quad ^{4}_{2}\text{He}$$

americium      neptunium      alpha particle

Study this equation carefully. We have already seen that mass numbers and atomic numbers need to balance on each side of the equation. There are two other points to note:

1   In alpha decay equations, we do not write He$^{2+}$. This is because the alpha particle comes from the nucleus of a larger atom – it is not a helium atom which has lost electrons. However, alpha particles do eventually gain electrons and become helium atoms.
2   In nuclear equations, we always write the atomic number, even though we know the name of the element.

Now look closely at what happens to americium. It becomes a new element because its atomic number decreases by 2. The event of one element changing into another in this way is called a **transmutation**.

**? TEST YOURSELF**

**I** Ernest Rutherford and Frederick Soddy were the first people ever to observe a transmutation. They discovered that radium ($^{226}$Ra) decayed by releasing an alpha particle to give an element never before known. Use the Periodic Table in Appendix 2 to help you decide what this element is called, and write a nuclear equation to describe this decay.

## ! SCIENCE FEATURE  Smoke detectors use radioactivity

Smoke can be the first sign that a house is on fire, and detecting smoke before the fire has a chance to take hold can be a real life saver.

One type of smoke detector makes use of the radioactivity of americium-241 to detect smoke. A small sample of americium-241 in the chamber of the detector gives off alpha radiation, which keeps the air in the chamber permanently ionized. This allows a small current to keep flowing between two electrodes, as shown in Fig 2.19.

Smoke entering the sensing chamber interferes with the ionized particles and allows them to recombine with electrons. This reduces the current, and when the current falls, the smoke alarm is set off.

A smoke detector, well sited inside a house, will often respond to a level of smoke concentration that the people in it are not aware of. Smoke detectors save hundreds of lives every year, and also people's homes.

**Fig 2.19** The smoke-sensing chamber of a smoke detector

**? TEST YOURSELF**

**J** Read the Feature box on smoke detectors.
   **a)** Why is alpha radiation used in smoke detectors rather than beta or gamma radiation? Hint: Look at Table 2.5 on the opposite page.
   **b)** The useful life of a smoke detector is about 10 years. Why should it not be thrown away with household waste?

**?** **TEST YOURSELF**

**K** Strontium-90 used to be produced in nuclear explosions. It decays by beta radiation. Write the nuclear equation for this decay, and state what new element has formed.

**L** Study Fig 2.18 on page 40 and decide where the following three isotopes are placed on the graph. Then predict which type of radioactive decay is most likely to occur in each, and write the balanced nuclear equation. Hint: work out the number of protons and neutrons to find the isotope's position in relation to the zone of stability.
**a)** Uranium-235: $^{235}_{92}U$
**b)** Iodine-131: $^{131}_{53}I$
**c)** Silicon-32: $^{32}_{14}Si$

**M** 15 million potassium-40 atoms decay inside each person's body every hour. How many grams is this? Hint: How many atoms in a mole?

**SEE QUESTIONS 4–6.**

## BETA DECAY

Lighter elements with isotopes that have too many neutrons lie to the left of the zone of stability (Fig 2.18, page 40). Such an isotope decays by emitting a beta particle (a high-speed electron). When a nucleus releases a beta particle, a neutron changes to a proton, and this brings the atom of the new element nearer to the stability zone.

The next nuclear equation is the equation for **beta decay**. Note that in this equation, the electron is given values of zero and –1. Look back to page 39 to remind you of this numbering.

$$^{1}_{0}n \rightarrow ^{1}_{1}p + ^{0}_{-1}e \quad (^{0}_{-1}e \text{ is a beta particle})$$

The nuclear reaction of carbon-14 is a good example of beta decay, and beta particles are detected in radiocarbon dating:

$$^{14}_{6}C \rightarrow ^{14}_{7}N + ^{0}_{-1}e$$

Since a neutron changes into a proton, the atomic number of the new element is increased by one, but the mass number stays the same.

## GAMMA RAYS

Gamma rays arise when the nucleus has excess energy – when it is in an excited state. The nucleus is usually excited after a nuclear decay, for instance after releasing alpha or beta particles. It loses its excess energy by emitting gamma rays, which are very high-energy electromagnetic radiation. Gamma radiation on its own does not result in the formation of a new element.

---

**!** **SCIENCE FEATURE** Radioactivity made useful

Radioactive materials are being used in an increasing number of ways.

**Treating cancer**
Cancerous cells are abnormal cells that divide at a rapid rate, producing tumours that invade surrounding tissues. Some types of radiation can cause a cancer. At the same time, cancerous cells are easily destroyed by carefully controlled doses of radiation, which do not affect other surrounding cells. Over a period of weeks, a tumour in a patient can be made inactive by exposure to doses of gamma rays. The radioisotope generally used for this treatment is cobalt-60.

**Tracers for diagnosis and treatment**
A radioisotope has the same chemical properties as any other atom of the same element, and molecules of a

**Fig 2.20** The head of a patient with bone cancer is being scanned. A short-lived radiotracer that concentrates more strongly in cancerous bone than in normal bone is used to diagnose whether or not the cancer is spreading

substance 'labelled' with a radioisotope can therefore be traced because their radiation can be detected. For example, iron-59 can be introduced into haemoglobin to follow the production of red blood cells in bone marrow. Iodine-131 as a label in sodium iodide ($Na^{131}I$) is used to investigate the activity of the thyroid gland and to diagnose and treat diseases of this gland. Labelled iodine also helps to detect liver and brain tumours. Sodium-24 as sodium chloride ($^{24}NaCl$) solution is put into the bloodstream to follow blood flow and locate clots and obstructions in vessels.

**Tracers in industry and agriculture**
Labelled sodium iodide in boreholes indicates the route taken by water underground. Barium sulphate, which is insoluble in water, can be labelled with barium-140, and the labelled compound $^{140}BaSO_4$ is used to monitor how silt is deposited in rivers. Phosphorus-32 in phosphate compounds helps to trace the uptake of fertilizers by plants.

# SUMMARY

Having studied this chapter, you should be able to know and understand the following.

- An atom has a minute central nucleus of protons and neutrons which is surrounded by electrons.
- A proton has a relative mass 1, and a relative charge of +1.
- A neutron has a relative mass of 1, and no relative charge.
- An electron has a relative mass of $5.5 \times 10^{-4}$, and a relative charge of −1.
- Atomic (proton) number = number of protons in the atom.
- Mass (nucleon) number = number of protons and neutrons in the atom.

- Isotopes have the same atomic number, but different mass numbers.
- The relative atomic mass of an element is the average mass of the atom (taking into account all the naturally occurring isotopes and their abundance) compared to $\frac{1}{12}$ the mass of one atom of carbon-12.
- Radioactivity is the spontaneous disintegration of the nucleus of an atom. Three kinds of radiation emitted from the nucleus are: alpha particles (helium nuclei), beta particles (electrons) and gamma rays (electromagnetic radiation).
- The stability of a nucleus depends on the number of protons and neutrons in it. A plot of number of protons against number of neutrons shows a zone of stability.

# ? EXAM QUESTIONS

**1**

**a)** Give the relative mass and relative charge of a neutron.

**b)** In terms of the number of their fundamental particles, what do two isotopes of an element have in common and how do they differ?

**c)** Give the complete atomic symbol, including mass number and atomic number, for an atom of the isotope with 22 neutrons and 19 electrons.

AQA CHM1 Atomic Structure, Bonding and Periodicity Jan 2001 Q1a–c

**2** The relative atomic mass of a sample of iron may be found by using a mass spectrometer to determine the isotopic composition.
In such a determination the following isotopic composition was found.

| Isotope | Percentage composition |
|---------|------------------------|
| $^{54}$Fe | 5.8 |
| $^{56}$Fe | 91.6 |
| $^{57}$Fe | 2.2 |
| $^{58}$Fe | 0.33 |

Calculate the relative atomic mass of this sample of iron, giving your answer to two decimal places.

Edexcel A/AS Unit Test 1 Jan 2001 Q6b(ii)

**3** [**Note:** In **question 3**, 'average molar mass' is the relative atomic mass in grams.]
Naturally occurring gallium contains 60.2% of $^{69}$Ga and 39.8% of $^{71}$Ga by mass. Calculate the average molar mass of gallium. Your answer should be given to three significant figures and include appropriate units.

Edexcel Chemistry Nuffield Curriculum 2000 Specimen Unit Test 1 Q2

**4**

**a)** **(i)** State the term which indicates the number of protons in the nucleus of an element.
**(ii)** Define the term *relative atomic mass* of an element.

**b)** The isotope $^{24}_{11}$Na is a β-emitter. Write an equation for a single-step radioactive decay of this isotope.

WJEC Chemistry Modular Paper CH1 Jan 2001 Q12a,b

**5**

**a)** State what is meant by the terms *mass number* and *isotopes*.

**b)** The atom $^{14}$C is a β-emitter and $^{238}$U is an α-emitter. Write equations for a single-step radioactivity decay of each of these atoms.

WJEC Specimen Papers 2001/02 Paper CH1 Q11a,b

**6** Fig 2.Q6 shows how radiation from a radioactive source is affected by an electric field.

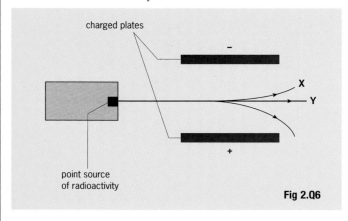

charged plates

–

X

Y

+

point source
of radioactivity

**Fig 2.Q6**

Identify **each** of the radiations labelled **X** and **Y** as α, β or γ radiation by ticking the correct column:

| Radiation | α-radiation | β-radiation | γ-radiation |
|---|---|---|---|
| X | | | |
| Y | | | |

WJEC Chemistry Modular CH1 Jan 2001 Q1

# 03

# A close look at electrons

# A close look at electrons

The giant gas clouds in interstellar space – the space between stars – contain vast amounts of very thinly dispersed atoms, ions and molecules. We know what this material consists of because each type of atom, ion or molecule absorbs and emits energy of particular wavelengths across the electromagnetic spectrum, depending on the arrangement of its electrons. From Earth, we can detect and record the patterns of these absorbed and emitted wavelengths: each type of particle has its own pattern, which reveals its presence, rather like a fingerprint. Hydrogen, for example, emits radio waves at a wavelength of 21 cm.

Chemists are very interested in the molecules of interstellar space. So far, they have identified over 100. Some are also common on Earth, such as hydrogen chloride, carbon monoxide, water and ethanol. Others are unusual, and several were discovered in space before they were even identified in the laboratory. An example is a three-member carbon ring, $C_3H_2$, an interstellar molecule widespread in our Galaxy and common in others.

It is thought that many small molecules and ions are formed in space when atoms are bombarded by cosmic radiation. These include the diatomic molecule carbon monoxide, CO, the ion $HCS^+$ and the three-carbon ring of $C_3H_2$. Also, there seems to be a size limit to these molecules, probably because the intense radiation disrupts larger molecules. However, the very existence of the smaller molecules in interstellar space may point the way to understanding the chemistry that led to life on Earth, and may perhaps reveal it elsewhere in the Universe.

**Interstellar clouds.** These contain some very interesting molecules, which are identified by their spectra. Some may represent the forerunners of the molecules of life itself

## INTRODUCTION

The glorious colours of fireworks, the discovery of helium in the Sun before it was discovered on Earth (page 54), our knowledge of the atoms and molecules in distant stars – all these are connected with electrons. To a chemist, the most interesting part of an atom is its electrons, because it is the electrons, not the protons or neutrons, that account for every chemical reaction.

Scientists in the nineteenth century investigated the light that materials absorb and emit, and went on to discover new elements. In the twentieth century, our understanding of the nature of light led scientists to release the energy of lasers (see page 55).

**Fig 3.1** The spectacular colours of fireworks are caused by the electrons in the atom releasing energy at different wavelengths in the visible spectrum

# 1 ALL THE COLOURS OF THE RAINBOW

In 1666, Sir Isaac Newton first recorded the fact that, when visible light was passed through a prism, it was split to produce a **continuous spectrum** of rainbow colours that contained all the wavelengths of light.

white light

slit in screen          prism

**Fig 3.2(a)** Light from a white light source passes through a slit and then through a prism. It is split into its different wavelengths, producing a continuous spectrum – a rainbow effect

**Fig 3.2(b)** Isaac Newton, who used a prism to split visible light and produce a continuous spectrum

Robert Bunsen is best known for inventing the Bunsen burner. But, with Wilhelm Kirchhoff, he invented a far more important instrument for the progress of chemistry – the spectroscope. When put in a flame, compounds that contain sodium were known to colour the flame yellow; potassium compounds coloured it lilac. The spectroscope took these observations and Newton's visible spectrum a step further.

In the spectroscope, light passes through a narrow slit and a prism, and forms a spectrum. When Bunsen and Kirchhoff viewed the sodium flame with this light, they saw bright lines (images of the slit) in the yellow part of the spectrum and some less bright lines in other parts. They viewed light from compounds of other metals in the same way, and soon realized that each element had its own characteristic *fingerprint* of lines in different parts of the spectrum.

With this technique, they discovered the elements rubidium and caesium while examining the spectrum of a lithium ore. They found strong red and blue lines (Fig 3.4) which could not be accounted for by the elements known to be in the ore. Rubidium is taken from the Latin for 'red' and caesium is named after the Latin for 'blue'.

**Fig 3.3** Robert Bunsen (centre) and Wilhelm Kirchhoff (left)

Sodium

400          500          600          700     nm

Rubidium

400          500          600          700     nm

Caesium

**Fig 3.4** The line emission spectra of sodium, and of rubidium and caesium, both discovered by Bunsen and Kirchhoff

As elements were being discovered and identified by their spectral fingerprints, astronomers were soon pointing spectroscopes at the stars to look for characteristic lines of elements and finding that the cosmos was made up of the same elements as the Earth.

## 2 THE ELECTROMAGNETIC SPECTRUM

The 'light' our eyes detect is just a small part of a very much wider **electromagnetic spectrum**. This spectrum is made up of all the **electromagnetic radiations**, and includes X-rays used by dentists, the microwave radiation used to heat food in microwave ovens and radio waves that bring us radio and television signals. Fig 3.5 is the **continuous spectrum** of electromagnetic radiation, called continuous because all the wavelengths are represented. We shall now look more closely at electromagnetic radiation to help us understand electrons and how they are arranged in atoms.

As its name suggests, electromagnetic radiation is made up of two components: an electrical and a magnetic one. Electromagnetic radiation is one of the ways energy is transmitted through space. It is why we feel the warmth of the Sun on Earth, and why we suffer sunburn if we have too much ultraviolet radiation.

**Fig 3.5** The electromagnetic spectrum. Notice that visible light – the radiation our eyes can detect – is only a very small part of the spectrum

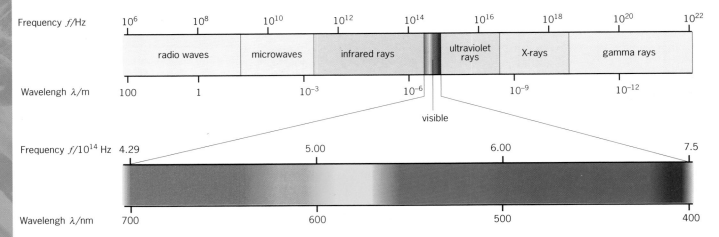

## 3 LIGHT AS WAVES

Electromagnetic radiation is usually thought of as **waves** and, like any wave, it can be described by its **wavelength, frequency** and **speed**. Wavelength is measured in metres (m), and its symbol is the Greek letter $\lambda$ (lambda). Frequency, $f$, is measured in hertz (Hz), which means cycles per second, and which can also be written as $s^{-1}$. The speed of electromagnetic waves is the speed of light, with the symbol $c$, which is in metres per second ($m\,s^{-1}$).

Fig 3.6 shows what these terms mean, and that when the frequency is higher, the wavelength is smaller. The terms are related in a simple equation:

$$c = \lambda \times f$$
$$(m\,s^{-1}) \quad (m) \quad (s^{-1})$$

**Fig 3.6** Two electromagnetic waves, showing their wavelength $\lambda$ and frequency $f$. The wavelength is the distance between two identical points on the waves. Here, it is between the crests. The speed of these waves is identical because all electromagnetic radiation travels at the speed of light, $c$, so the shorter the wavelength, the higher the frequency

**(a)** In one second, three waves have passed a particular point. So the frequency is three cycles per second, or 3 hertz (this is far lower than the frequency of actual electromagnetic radiation)

**(b)** In one second, six cycles have passed the same fixed point. So the frequency is 6 cycles per second, or 6 hertz (again, far lower than frequencies found in the electromagnetic spectrum)

## 4 LIGHT AS PARTICLES

Up to the beginning of the twentieth century, scientists agreed on a model of electromagnetic energy in which it was radiated and absorbed by matter in the form of waves. The model did not explain all the observations, notably the fact that heated objects emit radiation that has particular spectra.

In 1900, Max Planck proposed that particles such as atoms and molecules absorb and emit energy in discrete (separate) amounts or packets called '**quanta**'. Planck's equation is:

$$E = hf$$

where $E$ is the energy in joules (J), $h$ is Planck's constant, with a value of $6.626 \times 10^{-34}$ J s, and $f$ is the frequency in hertz (Hz or s$^{-1}$).

A **quantum of energy** is a precise packet of energy. These packets can have different energy values, depending on their source, but you cannot have parts of packets, only whole ones.

### THE PHOTOELECTRIC EFFECT

In 1905, Albert Einstein used Max Planck's equation to explain another phenomenon that baffled scientists, the **photoelectric effect**. This is the release of electrons by some metals when light is shone on them. It was found that the light had to be above a minimum frequency (and so a minimum energy) before electrons were emitted. It did not matter how intense (bright) the electromagnetic radiation was below this *threshold frequency*; only with a high-enough frequency were electrons released. Then, the more intense the light, the greater the number of electrons released. According to the old *wave-only* model, the electrons would absorb even low-energy radiation and eventually release it. Einstein developed Planck's idea of quanta, saying that some of the properties of electromagnetic radiation could be explained only if light were thought of as consisting of *particles*, which we now call **photons**. He also said that each photon is associated with a particular quantum (amount) of energy, linked to its frequency by Planck's equation: $E = hf$.

**Fig 3.7** Albert Einstein. By the age of 26, while an assistant in a Swiss patent office, he had written three papers in his spare time that were to change twentieth-century science

The photoelectric effect could now be explained. Negative electrons are held in the atom by electrostatic forces of attraction to the positive protons in the nucleus. When a photon collides with an electron, the photon gives up its energy to that electron. If the energy is high enough, the electron breaks free of the atom. The more intense the radiation, the greater the number of photons, so more electrons are released from materials that the radiation reaches.

**Fig 3.8** One of the uses of the photoelectric effect is in automatic cameras, in which a photocell releases electrons. The brighter the light, the more electrons released

## THE WAVE–PARTICLE NATURE OF LIGHT

In 1905, Einstein wrote three papers. One was on his famous theory of special relativity, another an explanation of Brownian motion. His third, an explanation of the photoelectric effect, gained him a Nobel Prize in 1921, three years after his great friend Max Planck received one. All three papers had a profound effect on twentieth-century science and Einstein was seen by many as one of the two greatest scientists that ever lived, the other being Isaac Newton.

We have noted some observations suggesting that light is made up of particles and others suggesting it is made up of waves. So we now think of light as having a *dual* nature, of both particles and waves. This **wave–particle duality** is not just confined to light; matter, too, can sometimes behave as if it were made up of waves, as you will see on page 55.

## 5 THE ATOMIC EMISSION SPECTRUM OF HYDROGEN

When hydrogen is placed in a discharge tube at low pressure and with a high voltage between the two plates, some of the bonds in the hydrogen molecules ($H_2$) are broken to give separate hydrogen atoms, see Fig 3.9(a). When the radiation emitted from the discharge tube is passed through a spectroscope, a series of characteristic lines shows up in the visible spectrum.

**Fig 3.9(a)** (Below) A hydrogen discharge tube showing the colour emitted by hydrogen atoms
**Fig 3.9(b)** (Below right) The line emission spectrum of excited hydrogen atoms. When the radiation passes through the slit, it is separated by the prism and detected by the photographic film as separate line images of the slit

## THE BALMER SERIES

You read earlier about Bunsen and Kirchhoff's work in identifying elements by their characteristic spectral lines. These spectra are called **line emission spectra** for the following reason. When an element's atoms are given energy, they emit radiation as discrete (separate) lines, at specific frequencies (and hence energies) for that element. For example, hydrogen, which has just one electron, has several prominent lines in the visible part of its emission spectrum (Fig 3.10, page 52). These are known as the **Balmer series**, after a Swiss music teacher who worked out a mathematical relationship between the lines.

## OTHER SERIES

Other series of lines for hydrogen are found in different parts of the electromagnetic spectrum. The **Lyman series** is found in the ultraviolet section and the **Paschen series** in the infrared. Both these series are named after their discoverers.

Like the photoelectric effect, the different series of lines (characteristic for an element) was another nineteenth-century mystery that could not be explained by the theories of that time. An explanation had to wait for the ideas of Planck and Einstein to be developed.

## 6 NIELS BOHR'S MODEL OF THE HYDROGEN ATOM

In Chapter 2 we were left with the Rutherford model of the atom. However, there are problems with this model. Rutherford proposed that electrons orbited the nucleus, rather like planets round the Sun. Planetary motion was well understood by this time – the Sun's gravity tends to pull planets towards it, while their acceleration, caused by being in a circular orbit, creates a balancing force. (For an object to travel at constant speed and travel in a circular motion, constantly changing direction, it has to accelerate constantly.)

Negative electrons in circular motion are attracted to the positive protons in the nucleus by electrostatic forces. If they orbited the nucleus like planets, their acceleration would keep them from falling into the nucleus. But electrons are charged particles, and *accelerating* charged particles were known to emit electromagnetic radiation and lose energy. If Rutherford's model was correct, instead of a few separate lines, a continuous spectrum should have been observed, with the atom emitting light all the time and the electron losing its energy and falling into the nucleus, causing the hydrogen atom to collapse.

Clearly, this does not happen!

Another model of the atom was needed to explain the observations. As we saw in Chapter 2, a model is intended to explain all the observations known at the time. If it does so, then the scientific community accepts it. Niels Bohr used the quantum ideas developed by Planck and Einstein to propose his model for the hydrogen atom.

Bohr's model still had the hydrogen electron orbiting the nucleus. But the orbits, or **energy levels** the hydrogen electron could occupy, were *quantized*, that is, they had fixed energy values. With this model, hydrogen's line emission spectrum could be explained.

**Fig 3.10** The Balmer series for the hydrogen atom, and its line emission spectrum. The lines in the visible spectrum are produced when the excited electron falls back to the $n = 2$ energy level. The higher the energy level from which it falls, the higher the frequency of the photon emitted. Note that the energy levels eventually merge at $n = \infty$ (infinity). This is when the electron has escaped from the attraction of the nucleus so that the atom has become ionized

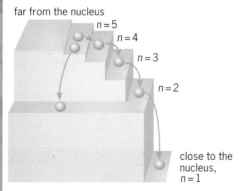

**Fig 3.11(a)** Niels Bohr (1885–1962). For his work on the structure of atoms, Bohr received a Nobel Prize in 1922

**Fig 3.11(b)** A staircase model for the energy levels of a hydrogen atom. Note the ball can fall down one level at a time, or fall more than one level

**? TEST YOURSELF**

**D** Look back to Fig 3.9(b). It shows the lines moving closer together towards one end of the spectrum for hydrogen. Are they closer together at the higher energy end of the spectrum or the lower energy end?

In Bohr's model, the electron normally occupies the lowest possible energy level, called the **ground state**. This is the energy level of the electron when it is not excited (see Fig 3.10). Raising the electron to an excited state (giving it energy), say by an electric discharge, causes it to move up to a higher energy level by absorbing a quantum of energy (Fig 3.10). When it returns to the lower, ground state energy level, it releases this quantum of energy as a photon of light, of a specific frequency, so giving a line in the emission spectrum.

Imagine that the hydrogen electron is like a ball on a staircase, as in Fig 3.11(b). The ball can rest on any step, but it cannot stop in between. This is the case with the electron. The ball needs energy to go up the step, and when it falls back down it releases this energy. The lines in an emission spectrum are closer together at one end because the higher energy levels that the electron can occupy are also closer together. This happens as the electron moves away from the nucleus, as shown in Fig 3.11(b).

When the electron is closest to the nucleus, it is at its lowest energy level. Moving the electron away from the nucleus requires energy, and the further away it is moved, the more energy it requires. So the more energy an electron receives, the higher it can rise through the energy levels, and the more energy it will release as it falls back down again.

Each energy level is given a number, called the **principal quantum number**, $n$. The term 'principal quantum number' is still used to describe the main energy levels of electrons in an atom. When the hydrogen's electron is in the $n = 1$ level, it is not excited, so this is the ground state.

The Balmer series of lines is for the energy transitions when the excited hydrogen electron falls back from higher energy levels to $n = 2$, see Fig 3.10. We see the lines because they are in the visible spectrum. The higher the energy level the electron falls from, the higher the frequency of the emitted photon. Note that the energy levels eventually merge at $n = \infty$ (infinity). This is when the atom has become ionized and lost its electron.

Neon is used everywhere in advertising signs. An electric current is passed through the gas at low pressure. The fast-moving electrons of the electric current excite electrons in the neon atoms into higher energy levels and, when they return to lower energy levels, orange–red light is emitted. The colour can be varied by adding other atoms such as argon or mercury, or by colouring the glass tube the neon is in.

Street lights usually contain sodium or mercury and they work on the same principle as neon lights. When excited mercury atoms return to their ground state, the radiation they emit has frequencies in the ultraviolet, yellow, green and blue parts of the spectrum.

Sodium street-lights are now being used to replace mercury ones because the radiation emitted by excited electrons in sodium atoms when they return to their lower levels is centred upon yellow. This has longer wavelengths than the light from mercury and is not as readily scattered by fog, so it can illuminate further. Also, sodium atoms require less energy to excite their electrons, and sodium is not as toxic as mercury.

Fluorescent lights in the home or office also contain low-pressure mercury vapour. The inside of the lighting tube is coated with zinc sulphide, which absorbs the energy of ultraviolet light when its electrons are excited and, on returning to the ground state, produces many frequencies of light in the visible range. These combine to give white light.

The advantage of fluorescent lights over filament lamps is that they use less energy, since nearly all the energy is radiated in the visible spectrum. They feel cool to touch, while ordinary light bulbs with tungsten filaments waste energy as they become very hot.

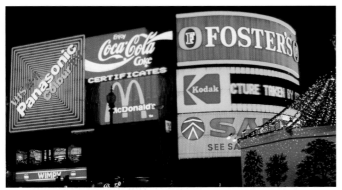

**Fig 3.12** Excited electrons in atoms of neon produce the colours for these advertising signs at Piccadilly Circus

The Lyman series of lines is found in the ultraviolet part of the spectrum. It is caused by the excited hydrogen electron returning from higher levels to the $n = 1$ ground-state energy level. As this is the lowest level nearest to the nucleus, far more energy is released when an electron excited to a particular energy level returns to $n = 1$ than when it returns to $n = 2$, and so the lines show up in the more energetic ultraviolet part of the spectrum.

Each line of hydrogen's emission spectrum represents one electron transition, a movement from a higher to a lower level. As a large numbers of hydrogen atoms are involved, all the possible transitions are represented, which gives the full spectrum of lines.

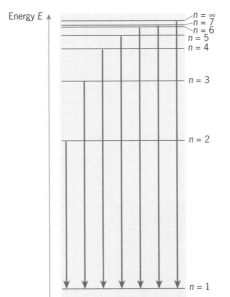

**Fig 3.13** The Lyman series for a hydrogen atom. In this series, the transitions of the electron are to $n = 1$

## IONIZATION ENERGY

From the Lyman series we can work out the energy needed to remove an electron completely from a hydrogen atom. This is the ionization energy for hydrogen. It is the energy required to take the electron from the ground state, at $n = 1$, to where the energy levels converge, at $n = \infty$, when the electron is free of the attraction of the nucleus. We shall look more closely at ionization energy on page 56–57.

✓ **REMEMBER THIS**

A substance fluoresces when it takes in light of one wavelength, usually outside the visible spectrum, and gives out light of another, often in the visible spectrum. This is described on page 35.

**SEE QUESTION 1.**

? **TEST YOURSELF**

**E** The Paschen series is one of the series of spectral lines for hydrogen, this time in the infrared part of the spectrum and for movement of the electron to the $n = 3$ level.
**a)** Draw a diagram similar to the Lyman series diagram in Fig 3.13 to show the first three electron transitions of the Paschen series.
**b)** Which transition in this series will result in the lowest energy line in the infrared spectrum?

## SUMMARY OF THE BOHR MODEL OF THE ATOM

● Electrons exist only in certain permitted energy levels and in these levels they do not emit energy.
● Electrons move to higher energy levels by absorbing quanta of energy. They return to lower energy levels by emitting these quanta as photons of light, which show up as lines in different parts of the electromagnetic spectrum.

The Bohr model of the atom successfully explained the lines in the emission spectrum of the hydrogen atom. It worked for hydrogen, the simplest atom with just one electron, but it did not predict accurately the spectral lines of atoms with several electrons.

The Bohr model is important because it used the idea of *quantized* energy levels to explain atomic structure and provided a foundation on which others could build. The Nobel Prize went to Bohr in 1922, one year after Einstein had received the prize for his explanation of the photoelectric effect.

**? TEST YOURSELF**

**F** Read the box on the absorption spectrum of the Sun below. In the spectrum, what is the colour of the helium line that Janssen observed?

▶ **EXTENSION Absorption spectrum of the sun reveals its elements**

### DISCOVERING SODIUM

Almost 50 years before Kirchhoff and Bunsen used emission spectra to identify elements (see page 47), the German scientist Josef von Fraunhofer was experimenting with light. When he passed sunlight through a very fine slit and a high-quality prism, some wavelengths were missing from the sunlight's absorption spectrum. They showed up on his spectrum as black lines in the otherwise continuous spectrum. These lines represented energies that were being absorbed by some unknown material.

Fraunhofer recorded the wavelengths of several hundred black lines, which are now named after him. Two were close together in the yellow part of the spectrum at almost 600 nm. At exactly the same wavelengths, Bunsen and Kirchhoff found two characteristic yellow lines in the emission spectrum of sodium (look back to Fig 3.4, on page 47). They concluded that sodium must be in the Sun's atmosphere. Using this hypothesis, they soon discovered other elements in the Sun.

**Fig 3.14** Fraunhofer found over 600 dark lines in the visible spectrum of the Sun. These represent wavelengths missing from a continuous spectrum of light from the Sun. Some of the lines have been matched with the elements responsible for them

To explain the Fraunhofer lines more fully, we need to look at the Bohr model of the atom again. When excited, electrons fall from higher energy levels to lower levels: they *emit* energy to give a line *emission* spectrum. In reverse, they reach the higher (excited) energy levels by *absorbing* quanta of energy (photons).

So, in an *absorption* spectrum, the quanta of energy the electrons absorb in going to these higher energy levels will be represented as black lines for the wavelengths missing from the spectrum.

For Fraunhofer's spectrum for sunlight, the light has passed from the interior of the Sun through the Sun's atmosphere. Assuming that there is sodium vapour in the Sun's atmosphere and that its electrons are being excited to higher energy levels, then the corresponding points in the spectrum for the light reaching Earth will be missing – hence the dark lines for sodium.

### DISCOVERING HELIUM

In 1868, the French astronomer Pierre Jules Janssen took his spectroscope to India to view a total eclipse of the Sun. In the spectrum of light from the Sun's corona, he saw a bright line very close to the two sodium lines. This line could not be accounted for by the line spectra of any known element. In the same year the English scientist, Norman Lockyer, also observed this line and suggested that it was caused by a new element. It was given the name helium, after the Greek word *helios* for the Sun.

At the time, Lockyer was ridiculed for his suggestion that he had identified a new element. Yet this discovery later helped helium to be discovered on Earth. In 1895, a quarter of a century after helium was found on the Sun, it was isolated by the Scottish chemist William Ramsey. He found the gas trapped in a uranium ore, where it had been formed as a product of radioactive decay.

## ! SCIENCE FEATURE Lasers — amplifying the energy from electron transitions

The letters in the word laser stand for **l**ight **a**mplification by **s**timulated **e**mission of **r**adiation. Lasers, invented in 1960, have revolutionized medicine, technology and science. The ruby laser was the first and works by using a burst of ultraviolet light to excite electrons in atoms of a ruby rod. In the same instant, a few electrons fall back to lower energy levels and produce photons. These photons are bounced back and forth by mirrors at the ends of the rod and stimulate other excited electrons to fall back to lower levels, emitting a simultaneous burst of photons of the same frequency. This laser pulse is very intense and, provided one of the mirrors is partially transparent, it can pass through. The ruby laser produces a wavelength in the red part of the spectrum. Other lasers use other materials and produce wavelengths from infrared to X-ray.

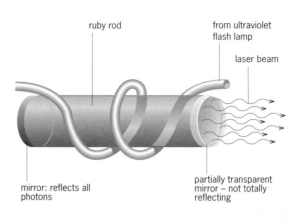

ruby rod
from ultraviolet flash lamp
laser beam
mirror: reflects all photons
partially transparent mirror – not totally reflecting

**Fig 3.15** A ruby laser emitting laser light

## ▶ EXTENSION Electrons as waves producing images of smaller and smaller objects

In 1923, Louis de Broglie proposed that, just as light could behave as a particle or a wave, so could matter. He suggested that electrons, regarded as particles, also had wave-like properties. Four years later, a beam of electrons was diffracted (bent and scattered) by a metal crystal to form a pattern. This diffraction could only be explained by assuming that the electrons were behaving as waves.

The electron microscope uses the wave-like behaviour of electrons. It allows us to see images of very small objects in much greater detail than we can see with ordinary microscopes, which rely on visible light. Light is visible to us between wavelengths of about 400 to 700 nm. To form an image of an object, the object cannot be smaller than half the shortest wavelength, that is, 200 nm (or a twenty thousandth of a millimetre).

Electrons travelling at high speed have very short wavelengths, and they allow images of objects measuring $2 \times 10^{-3}$ nm to be made and magnified. The electron microscope, then, is a very helpful tool in biological and chemical research.

In 1986, the scanning tunnelling microscope (STM) took electron microscopy one stage further. With the aid of a computer to process the data and enhance the image, scientists used the charge on the electrons round atoms to make images of separate atoms and molecules. This technique, and the newer technique using the scanning probe microscope (SPM), hold out the prospect of producing new materials and structures by controlling and rearranging atoms and molecules.

**Fig 3.16(a)** (Left) An electron micrograph of AIDS viruses budding off from the surface of a human T-lymphocyte, a white blood cell. Such images of organisms that cause disease help scientists to seek a way to combat infection

**Fig 3.16(b)** (Above) A scanning tunnelling microscope was used to produce this piece of 'atomic art'. To make this molecular person, 28 carbon monoxide (CO) molecules were positioned on a piece of platinum by using a charged probe wire

# 7 A MODERN MODEL OF HOW ELECTRONS ARE ARRANGED

Once it was realized that electrons could behave like waves, a new model of the atom was possible. This model is based on some very complicated mathematics to describe the wave properties of electrons. Erwin Schrödinger takes the credit for devising an equation that describes the energy levels for electrons in hydrogen and other atoms. (The mathematics is beyond the scope of this book.) His model is known as the **quantum mechanical model**.

## ELECTRON SHELLS

Electron shells correspond to the energy levels that Bohr first identified in his model of the atom. The first shell has a **principal quantum number** $n = 1$, the second shell's principal quantum number is $n = 2$, and so on, as in Table 3.1. The first shell is closest to the nucleus and has the lowest energy. As the principal quantum number (or shell number) increases, so does the energy associated with it.

Each shell can hold a maximum number of electrons. You first met shells in your GCSE course, and you may remember that the arrangement of electrons is known as its **electron configuration**, also referred to as **electronic structure**.

---

### ✔ REMEMBER THIS

The energy level of an electron is given a principal quantum number, and this number is also given to the shell it is in.

---

**Table 3.1 Principal quantum numbers for electrons, and the maximum number of electrons each shell can contain**

| Principal quantum number, $n$ | Shell | Max. no. of electrons in shell |
|---|---|---|
| 1 | first | 2 |
| 2 | second | 8 |
| 3 | third | 18 |
| 4 | fourth | 36 |

---

### ? TEST YOURSELF

**G** If you think like a mathematician, you may be able to work out a simple formula containing the principal quantum number, $n$, which can be used to predict the maximum number of electrons in the $n$th shell.
Hence, calculate the number of electrons in the fifth shell.
Hint: the same formula can be used to predict the maximum number in the first four shells.

---

### ? TEST YOURSELF

**H** Make drawings like that in Fig 3.17 of the following atoms:
a) aluminium (Z = 13),
b) carbon (Z = 6),
c) chlorine (Z = 17).
Hint: refer also to Table 3.1.

---

### ⬤ EXAMPLE

**Q** Magnesium has an atomic (proton) number (Z) of 12. Work out the arrangement of electrons in the shells of a magnesium atom.

**A** The atomic number tells you there are 12 electrons. (The atomic number is the number of protons in an atom, and as the atom has no net charge, it has the same number of electrons. If you need to remind yourself about this, look back to page 35.)

The shells of lowest energy are filled first, so:
● shell 1 will take 2 electrons, which is all it can hold;
● shell 2 will take the next 8 electrons, the maximum it can hold;
● shell 3 will take the remaining two.

So the electron configuration of magnesium is 2,8,2. You can use a *dot-and-cross* diagram to represent this, as shown in Fig 3.17.

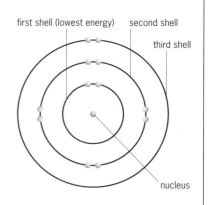

**Fig 3.17** The arrangement of the electrons in the shells of a magnesium atom. This is known as a dot-and-cross diagram because the electrons are represented by dots (as here), or crosses, or both. The first shell is at the lowest energy. The second and third shells are at increasingly higher energies. It requires little additional energy to remove electrons from the third shell

---

## EVIDENCE FOR SHELLS FROM SUCCESSIVE IONIZATION ENERGIES

Earlier in this chapter, we met the term ionization energy, which is the energy that an electron must be given to remove it from an atom (or ion). Now let's look at this term more closely. In order to measure ionization energy, the atoms have to be separated, and this means they must be in the gaseous state.

For convenience, the value for ionization energies is given for one mole of atoms (or ions):

> **The energy required to remove one mole of electrons from one mole of gaseous atoms to form one mole of ions with a single positive charge is called the first ionization energy.**

So the first ionization energy is the energy required to do this:

$$X(g) \rightarrow X^+(g) + e^- \qquad \Delta H = \text{first ionization energy in kJ mol}^{-1}$$

where X is any element, (g) tells you that it is gaseous, and $e^-$ is the symbol for an electron.

The **second ionization energy** is the energy required to remove a second mole of electrons:

$$X^+(g) \rightarrow X^{2+}(g) + e^-$$

Let's look again at magnesium with its electron configuration of 2,8,2.

First ionization energy:   $Mg(g) \rightarrow Mg^+(g) + e^-$

$$\Delta H = +738 \text{ kJ mol}^{-1}$$

Second ionization energy: $Mg^+(g) \rightarrow Mg^{2+}(g) + e^-$

$$\Delta H = +1451 \text{ kJ mol}^{-1}$$

Third ionization energy:  $Mg^{2+}(g) \rightarrow Mg^{3+}(g) + e^-$

$$\Delta H = +7733 \text{ kJ mol}^{-1}$$

The ionization energy increases as each successive electron is removed. This is the reason: the positive nuclear charge stays the same, because there are still twelve protons in the nucleus, and each time an electron is removed, the remaining ones are attracted more strongly by the nucleus. Notice that there is a large jump in the energy required to remove the third electron. This is because we are breaking into the second shell, which is closer to the nucleus. This shell is also less **shielded** by inner electron shells from the positive charge of the nucleus, because there is only one full shell between it and the nucleus.

From these values, you can see that there is a large range in the ionization energies of magnesium. It would be difficult to choose a scale on which to plot them. But if we convert the ionization energies into logarithms to the base 10 ($\log_{10}$), we condense the scale and make it more manageable, as in Fig 3.18. Don't worry if you are not familiar with logarithms. Using them here is just a way of making the pattern of the graph easier to see.

**? TEST YOURSELF**

**I** **a)** Write an equation for the eleventh ionization energy of magnesium (omitting the energy value). **b)** The energy required to remove the eleventh electron is almost 170 000 kJ mol$^{-1}$, but the tenth ionization energy is 'only' 35 500 kJ mol$^{-1}$. Explain this huge jump in energy to remove the eleventh electron.

**? TEST YOURSELF**

**J** Sketch a graph like Fig 3.18 for $\log_{10}$ of the successive ionization energies of sodium, and explain its shape.
Hint: sodium has an electron configuration of 2,8,1.

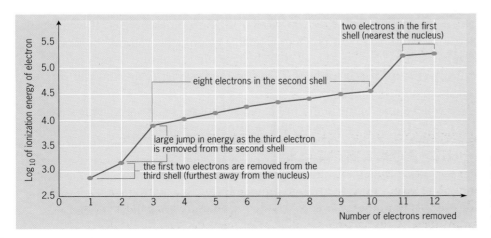

**Fig 3.18** Graph of the successive ionization energies (as logarithms to base 10) of the magnesium atom against the number of electrons removed. The values for electrons closest to the nucleus are on the right

## ARRANGING ELECTRONS IN SUBSHELLS

Let us now return to the atomic emission spectra of elements. When looked at in finer detail, the lines on these spectra are seen to be divided into more lines. Each finer line represents the energy level of a **subshell**. So electrons arranged in shells are also subdivided into subshells, or **sub-levels**. The fine lines result from electron transitions (movement of electrons) between the subshells. The subshells are known by letters:

- **s** subshell contains 2 electrons
- **p** subshell contains 6 electrons
- **d** subshell contains 10 electrons
- **f** subshell contains 14 electrons

The letters were given in the early twentieth century and refer to spectral lines. Some of the lines were **s**harp, hence **s**, some were more spread out or **d**iffuse, and some of the lines were brighter and called **p**rincipal lines. (Note: In this book we do not go into detail about the f subshell.)

We can now see in Table 3.2 how the shells are split into subshells.

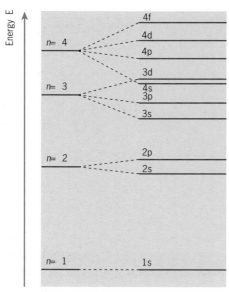

**Fig 3.19** The energies of the various subshells in an atom with many electrons

**Table 3.2 The system for arranging electrons in subshells**

| Principal quantum number, $n$ | Shell number | Subshells | Maximum number of electrons | |
|---|---|---|---|---|
| 1 | 1 | 1s | 2 | Total = 2 |
| 2 | 2 | 2s | 2 | Total = 8 |
| | | 2p | 6 | |
| 3 | 3 | 3s | 2 | Total = 18 |
| | | 3p | 6 | |
| | | 3d | 10 | |
| 4 | 4 | 4s | 2 | Total = 32 |
| | | 4p | 6 | |
| | | 4d | 10 | |
| | | 4f | 14 | |

Notice that the subshell (sub-level) takes the number of the principal quantum number or shell.

The energies of the subshells are shown in Fig 3.19. Notice that the 3d subshell has a higher energy than the 4s subshell. This has important consequences for the chemistry of the transition elements, which you can read more about in Chapter 24.

## EVIDENCE FOR SUBSHELLS FROM FIRST IONIZATION ENERGIES

The evidence for subshells also comes from ionization energies. If we plot the first ionization energy of different elements against their atomic numbers, a regular pattern emerges (Fig 3.20). The repeating pattern of a property is called **periodicity**. It forms the basis of the Periodic Table, and you can find out more about it in Chapter 5.

Several features of the graph in Fig 3.20 point to the existence of subshells. Let's look at the eight elements lithium to neon, which have electrons in shell 2. We would expect an increase in the first ionization energy as the atomic number increases. This is because the number of protons in the nucleus increases, which increases the nuclear charge. If the electrons were all in the same energy level, we would expect the graph to be a straight line, showing a steady increase in the first ionization energy as the number of protons in the nucleus increases.

### ? TEST YOURSELF

**K** **a)** What is meant by the first ionization energy of an element?
**b)** Explain why the first ionization energy of neon is higher than that of lithium.

**SEE QUESTION 2.**

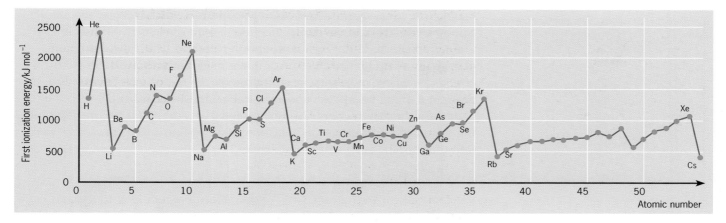

**Fig 3.20** Graph showing the periodicity of first ionization energies as they vary with atomic number (proton number)

The 2,3,3 pattern of first ionization energies tells us that the electrons removed from this second electron shell are not arranged such that they are all in the same energy level.

From reading this section on subshells, you may be able to explain the 2,3,3 pattern. Refer back to Fig 3.20 as you read on.

The first part of the 2,3,3 pattern is caused by electrons being taken from the 2s subshell. There is an increase in first ionization energy from Li to Be because Be has an extra proton attracting the outer electrons. Then, instead of a further increase from Be to B, there is a decrease in first ionization energy as the electron in B is taken from the 2p subshell: this electron is at a higher energy level than the 2s subshell and further from the nucleus.

In moving from B to N, there is the expected increase in first ionization energy, but another dip occurs at O. After N, the electrons in the 2p subshell start to pair up. (This pairing of electrons in subshells is covered in the next section.) There is more electrostatic repulsion between paired electrons, which means that the fourth electron in the 2p subshell is easier to remove, explaining the dip at O.

You can see the same pattern repeated for elements starting with Na when the next shell is being filled. After Mg, there is a drop at Al, and another drop at S. The 3p subshell is complete at Ar.

After this, there is an interruption to the 2,3,3 pattern. There is the expected increase from K to Ca as electrons are taken from the same 4s subshell. Then we have electrons removed from the slightly higher energy 3d subshell (refer back to Fig 3.19) before the 3,3 pattern returns, beginning with Ga, as the 4p subshell electrons are removed.

# 8 ATOMIC ORBITALS

Another great scientist of the twentieth century was Werner Heisenberg, whose work has clarified our current model of the way electrons are arranged in atoms. In 1927, he said that you could determine either the speed of an electron or its position, but not both at the same time. This he summed up in the **Heisenberg uncertainty principle**. While it applies to any particle, it becomes important only when the particle is very tiny. Heisenberg's mathematics has enabled electron arrangements to be worked out in even more detail.

We have seen that lines appear in the emission spectra of excited atoms, representing the wavelengths (energies) of photons emitted by excited electrons returning to lower energy levels. When this light is passed through a

**? TEST YOURSELF**

**L** Read about the 2,3,3 pattern for first ionization energies of the elements lithium to neon and look at Fig 3.20. Then answer the following questions:
**a)** Why is there an increase in first ionization energy from sodium to magnesium?
**b)** Explain why there is a drop at (i) aluminium and (ii) sulphur.

**? TEST YOURSELF**

**M** The first ionization energy of beryllium (Be) is 900 kJ mol$^{-1}$, while that of boron (B) is 801 kJ mol$^{-1}$. Explain this, using information about subshells in the text.

**SEE QUESTION 3.**

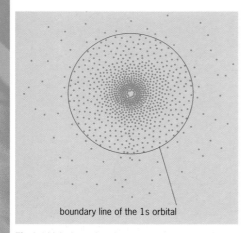

Fig 3.21(a) boundary line of the 1s orbital

**Fig 3.21(a)** A section through the 1s orbital. If we could map the position of an electron as a dot at regular intervals, then over a period of time 90 per cent of the dots could be enclosed by drawing a boundary that is spherical. This drawing represents a slice through the sphere. Notice that the electron charge is concentrated close to the nucleus and tails off further out

magnetic field, even more lines show up, and these are evidence of what we can now call **atomic orbitals**.

An atomic orbital is a region around the atom in which there is a high probability of finding an electron at any moment in time.

The existence of electrons in orbitals helps the chemistry of atoms and molecules to fall into place, and so is tremendously useful.

In the s subshell there is only one orbital, the s orbital. If we draw around the region of the atom where the electron spends most of its time, then we find the s orbital is spherical, see Fig 3.21(a).

1s          2s          3s

**Fig 3.21(b)** The boundary surface of 1s, 2s and 3s orbitals. In each case, the imaginary 'surface' is the limit of the space in which the 1s, 2s and 3s electrons are likely to be found. Also in each case, the orbital is centred on the nucleus, and the boundary represents a 90 per cent probability of finding an electron within it

There are three p orbitals in the p subshell. They are dumb-bell shaped and arranged at right angles to each other, with the nucleus in the centre of each dumb-bell.

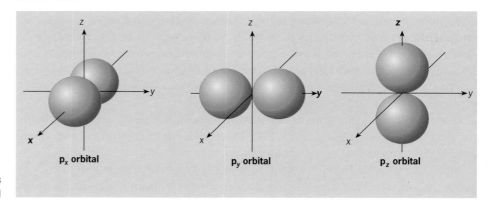

$p_x$ orbital          $p_y$ orbital          $p_z$ orbital

**Fig 3.22** The three p orbitals found in the p subshell

The d subshell contains five d orbitals. We shall look at the shapes of d orbitals in Chapter 24.

## SPINNING ELECTRONS

Each atomic orbital can hold a maximum of two electrons. As the electrons move in their orbitals they are also spinning on their own axes. If there are two electrons, one spins clockwise, while the other spins anticlockwise. This keeps the repulsion of the electrons to a minimum. Electron spin is often represented by box diagrams, in which each box is an atomic orbital and arrows represent the electrons spinning in opposite directions, as in Fig 3.23.

## ADDRESSING ELECTRONS

So we are now in a position to give *addresses* to electrons. In the same way that a letter can be addressed with the country, the town, the street and the house

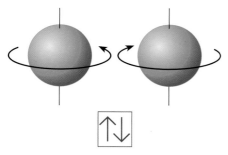

**Fig 3.23** Two electrons in the same atomic orbital spin in opposite directions, which is represented by a box diagram with arrows showing the opposite spin of each electron. The box is the atomic orbital

number so that it arrives at only one unique destination, so we can give a
unique 'address' to represent the state of an electron in an atom. The idea of a
unique state for every electron is known as the **Pauli exclusion principle**.
According to this principle, an electron is precisely defined by:

● its shell – the main energy level;
● its subshell – the sub-energy level division of the shell;
● its atomic orbital – each subshell possesses at least one atomic orbital,
  and all orbitals in a subshell possess the same energy;
● and its direction of spin – a maximum of two electrons of opposite spin
  can occupy one atomic orbital: orbitals in the same subshell are singly
  filled first, with electrons of parallel spin.

Fig 3.24 shows the electron configuration of the magnesium atom in its
ground state. Remember, this is the lowest energy state of the atom, so the
electrons occupy orbitals in subshells of the lowest possible energy.

  Notice that:

● the s subshells contain one s atomic orbital;
● the p subshells contain three p orbitals;
● each orbital contains a maximum of two electrons with opposite spin.

**Fig 3.24** The arrangement of electrons in the
atomic orbitals of a magnesium atom. The atom is
in its ground state, so no electrons have been
excited into higher energy orbitals

## 9  THE ELECTRON CONFIGURATIONS OF THE FIRST 36 ELEMENTS

Let us now look at how electrons fill orbitals in an atom. The orbitals of
lowest energy are occupied first, and this is known as the **aufbau
principle**. Aufbau is a German word for building up. So when you are
building up the electron configuration of an element, remember that each
electron goes into the lowest energy orbital available.

### Hydrogen (Z = 1)

The lowest energy orbital is in the first shell that has a principal quantum
number $n = 1$ and is an s orbital, so it is written 1s.

1s

$\boxed{\uparrow}$  electron configuration $1s^1$

### Helium (Z = 2)

1s

$\boxed{\uparrow\downarrow}$  electron configuration $1s^2$

Note: When you say $1s^2$, it is 'one s two', not 'one s squared'!

  The next atom has three electrons, and the aufbau principle tells us that
this electron will occupy the next orbital of lowest energy, which is in the $n =
2$ shell, so it will start with 2 and be an s orbital, and therefore it is 2s.

### Lithium (Z = 3)

1s   2s

$\boxed{\uparrow\downarrow}$ $\boxed{\uparrow}$   electron configuration $1s^2\,2s^1$

### Beryllium (Z = 4)

1s   2s

$\boxed{\uparrow\downarrow}$ $\boxed{\uparrow\downarrow}$   electron configuration $1s^2\,2s^2$

Next, we come to filling the p subshell. Notice that each orbital is occupied singly at first. This is because two electrons in the same orbital exert a repulsion that raises the energy of a doubly filled orbital above that of a singly filled one. Notice that the spins are all in the same direction. The repulsion that results helps to keep the electrons apart.

**Boron (Z = 5)**

electron configuration $1s^2 2s^2 2p^1$

**Carbon (Z = 6)**

1s   2s        2p

electron configuration $1s^2 2s^2 2p^2$

**Nitrogen (Z = 7)**

1s   2s        2p

electron configuration $1s^2 2s^2 2p^3$

**Oxygen (Z = 8)**

1s   2s        2p

electron configuration $1s^2 2s^2 2p^4$

**Fluorine (Z = 9)**

1s   2s        2p

electron configuration $1s^2 2s^2 2p^5$

**Neon (Z = 10)**

1s   2s        2p

electron configuration $1s^2 2s^2 2p^6$

The next eight elements follow the same pattern, filling up the s and p orbitals of the third shell.

Potassium and calcium come next. The 3d subshell is available, but remember (Fig 3.19, page 58) that the 4s subshell has a lower energy, so it is filled first.

**Potassium (Z = 19)**

electron configuration $1s^2 2s^2 2p^6 3s^2 3p^6 4s^1$

Notice that we do not write $3d^0$ in the electron configuration. We leave it out. Similarly, we can draw the box diagram for calcium like this:

**Calcium (Z = 20)**

electron configuration $1s^2 2s^2 2p^6 3s^2 3p^6 4s^2$

**EXAMPLE**

**Q** What is the electron configuration of sulphur (Z = 16)?

**A** The atomic number is 16, so there are sixteen electrons. We already know that the arrangement in the first two shells is:

$$1s^2 2s^2 2p^6$$

This takes ten electrons, leaving six electrons to place. Remember to start with the lowest energy orbital available, the 3s. This orbital takes two electrons:

$$1s^2 2s^2 2p^6 3s^2$$

leaving another four to make the electron configuration for sulphur:

$$1s^2 2s^2 2p^6 3s^2 3p^4$$

**SEE QUESTION 4.**

**? TEST YOUSELF**

**N** Write down the electron configuration of sodium (Z = 11), chlorine (Z = 17) and aluminium (Z = 13).

# FILLING THE d ORBITALS (Z = 21–30)

**Table 3.3  Electron configurations of titanium (Z = 21) to zinc (Z = 30)**

| Z | element | electron configuration | 1s | 2s | 2p | | | 3s | 3p | | | 3d | | | | | 4s |
|---|---|---|---|---|---|---|---|---|---|---|---|---|---|---|---|---|---|
| 21 | Sc | $1s^2 2s^2 2p^6 3s^2 3p^6 3d^1 4s^2$ | ↑↓ | ↑↓ | ↑↓ | ↑↓ | ↑↓ | ↑↓ | ↑↓ | ↑↓ | ↑↓ | ↑ | | | | | ↑↓ |
| 22 | Ti | $1s^2 2s^2 2p^6 3s^2 3p^2 3d^2 4s^2$ | ↑↓ | ↑↓ | ↑↓ | ↑↓ | ↑↓ | ↑↓ | ↑↓ | ↑↓ | ↑↓ | ↑ | ↑ | | | | ↑↓ |
| 23 | V | $1s^2 2s^2 2p^6 3s^2 3p^2 3d^3 4s^2$ | ↑↓ | ↑↓ | ↑↓ | ↑↓ | ↑↓ | ↑↓ | ↑↓ | ↑↓ | ↑↓ | ↑ | ↑ | ↑ | | | ↑↓ |
| 24 | Cr | $1s^2 2s^2 2p^6 3s^2 3p^6 3d^5 4s^1$ | ↑↓ | ↑↓ | ↑↓ | ↑↓ | ↑↓ | ↑↓ | ↑↓ | ↑↓ | ↑↓ | ↑ | ↑ | ↑ | ↑ | ↑ | ↑ |
| 25 | Mn | $1s^2 2s^2 2p^6 3s^2 3p^6 3d^5 4s^2$ | ↑↓ | ↑↓ | ↑↓ | ↑↓ | ↑↓ | ↑↓ | ↑↓ | ↑↓ | ↑↓ | ↑ | ↑ | ↑ | ↑ | ↑ | ↑↓ |
| 26 | Fe | $1s^2 2s^2 2p^6 3s^2 3p^6 3d^6 4s^2$ | ↑↓ | ↑↓ | ↑↓ | ↑↓ | ↑↓ | ↑↓ | ↑↓ | ↑↓ | ↑↓ | ↑↓ | ↑ | ↑ | ↑ | ↑ | ↑↓ |
| 27 | Co | $1s^2 2s^2 2p^6 3s^2 3p^6 3d^7 4s^2$ | ↑↓ | ↑↓ | ↑↓ | ↑↓ | ↑↓ | ↑↓ | ↑↓ | ↑↓ | ↑↓ | ↑↓ | ↑↓ | ↑ | ↑ | ↑ | ↑↓ |
| 28 | Ni | $1s^2 2s^2 2p^6 3s^2 3p^6 3d^8 4s^2$ | ↑↓ | ↑↓ | ↑↓ | ↑↓ | ↑↓ | ↑↓ | ↑↓ | ↑↓ | ↑↓ | ↑↓ | ↑↓ | ↑↓ | ↑ | ↑ | ↑↓ |
| 29 | Cu | $1s^2 2s^2 2p^6 3s^2 3p^6 3d^{10} 4s^1$ | ↑↓ | ↑↓ | ↑↓ | ↑↓ | ↑↓ | ↑↓ | ↑↓ | ↑↓ | ↑↓ | ↑↓ | ↑↓ | ↑↓ | ↑↓ | ↑↓ | ↑ |
| 30 | Zn | $1s^2 2s^2 2p^6 3s^2 3p^6 3d^{10} 4s^2$ | ↑↓ | ↑↓ | ↑↓ | ↑↓ | ↑↓ | ↑↓ | ↑↓ | ↑↓ | ↑↓ | ↑↓ | ↑↓ | ↑↓ | ↑↓ | ↑↓ | ↑↓ |

Table 3.3 shows the electron configurations of the next ten elements. The 4s subshell has already been filled at calcium, so now the 3d subshell, which is at a slightly higher energy level (see Fig 3.24), can begin to fill. Electrons spinning in the same direction first occupy the 3d orbitals singly, since this helps to minimize electron repulsion.

Notice that the chromium atom and the copper atom have only one electron in the 4s atomic orbital. The reason for this is explained in Chapter 24.

## GALLIUM (Z = 31) TO KRYPTON (Z = 36)

From gallium to krypton the 4p orbitals are filled because they have the next highest energy level (see Fig. 3.19).

### Gallium (Z = 31)

| 1s | 2s | 2p | | | 3s | 3p | | | 3d | | | | | 4s | 4p | | |
|---|---|---|---|---|---|---|---|---|---|---|---|---|---|---|---|---|---|
| ↑↓ | ↑↓ | ↑↓ | ↑↓ | ↑↓ | ↑↓ | ↑↓ | ↑↓ | ↑↓ | ↑↓ | ↑↓ | ↑↓ | ↑↓ | ↑↓ | ↑↓ | ↑ | | |

electron configuration $1s^2\, 2s^2\, 2p^6\, 3s^2\, 3p^6\, 3d^{10}\, 4s^2\, 4p^1$

### Krypton (Z = 36)

| 1s | 2s | 2p | | | 3s | 3p | | | 3d | | | | | 4s | 4p | | |
|---|---|---|---|---|---|---|---|---|---|---|---|---|---|---|---|---|---|
| ↑↓ | ↑↓ | ↑↓ | ↑↓ | ↑↓ | ↑↓ | ↑↓ | ↑↓ | ↑↓ | ↑↓ | ↑↓ | ↑↓ | ↑↓ | ↑↓ | ↑↓ | ↑↓ | ↑↓ | ↑↓ |

electron configuration $1s^2\, 2s^2\, 2p^6\, 3s^2\, 3p^6\, 3d^{10}\, 4s^2\, 4p^6$

**? TEST YOURSELF**

Write down the electron configurations of arsenic (Z = 33), selenium (Z = 34) and bromine (Z = 35).

**YOU CAN FIND OUT ABOUT THE ELECTRONIC CONFIGURATIONS OF IONS IN CHAPTER 4.**

# 10 ELECTRON CONFIGURATIONS AND THE PERIODIC TABLE

When we know the electron configurations of the elements, we can organize and explain the chemical properties of the elements, because it is the electrons that determine these properties. Long before electrons were even known to exist, Mendeleev grouped together elements with similar properties in his Periodic Table. You can read much more about this in Chapter 5. However, from your GCSE course you will know about the Periodic Table, so let's see how electron configurations fit in.

**Fig 3.25**

Fig 3.25 shows the electron configuration for the outermost subshell of the first 36 elements. In Period 4, containing the transition elements, electron configurations show the two outermost subshells, because it is the d subshell that is being filled. Notice that the elements in each of the eight main groups end with the same number of electrons in their outer subshell. So Group 7 always has a $p^5$ subshell and Group 2 always has a filled $s^2$ orbital. The Periodic Table is often divided into blocks, as in Fig 3.26, according to the subshell being filled.

**Fig 3.26** The Periodic Table, showing the different blocks labelled according to which outer orbital is being filled

So when we know how electrons are arranged in atoms, we have a better understanding of how the Periodic Table is arranged. It groups elements together with similar properties, and we can now see that it is the electron configuration that determines this similarity in properties.

## SUMMARY

When you have studied this chapter, you should be able to understand the following ideas.

■ Evidence for the arrangement of electrons in energy levels comes from the line emission spectra of atoms.

■ Energy levels are quantized, which means that each level has a fixed amount of energy.

■ An electron can move from a lower to a higher energy level by absorbing a quantum (packet) of energy. It releases the same quantum of energy when it falls back, as a photon of light of a particular frequency (which gives a line in the emission spectrum).

■ The arrangement of electrons in an atom is called its electron configuration.

■ Energy levels are also known as shells and subshells.

■ Shells are numbered 1, 2, 3 and so on. The shell number is also known as the principal quantum number.

■ The higher the shell number, the higher is its energy and the further the shell is from the nucleus.

■ Successive ionization energies of atoms provide evidence of shells.

■ Within each shell there are subshells s, p, d and f, of different energies. Shell 1 only contains an s subshell. Shell 2 contains two subshells, s and p, while shell 3 contains three subshells s, p and d.

■ A plot of first ionization energy against atomic number provides evidence of subshells. In each subshell the electrons are arranged in atomic orbitals.

■ Atomic orbitals are regions in the atom where there is a high probability of finding the electron.

■ There is one s atomic orbital in the s subshell, three p orbitals in the p subshell and five d orbitals in the d subshell.

■ Each atomic orbital can accommodate two electrons of opposite spins.

■ The Periodic Table is arranged in blocks, s, p, d and f, according to the subshell that is being filled.

## ? EXAM QUESTIONS

**1** When an electrical discharge passes through gaseous hydrogen at low pressure, electromagnetic radiation is emitted.

**a)** Explain what processes within a hydrogen atom cause radiation to be emitted.

**b)** If the radiation in **a)** is passed through a spectrometer, several series of converging lines are observed.
**i)** Explain why there are several series of lines.
**ii)** Why does each series of lines converge?

**c)** The convergence limit of the Lyman series of lines occurs at a wavelength of $1.00 \times 10^{-7}$ m.
**i)** What does the limit represent?
**ii)** Calculate the energy, in kJ mol$^{-1}$, of the convergence limit.
($c = 3.00 \times 10^8$ m s$^{-1}$; $L = 6.02 \times 10^{23}$ mol$^{-1}$; $h = 6.34 \times 10^{-34}$ J s)

**d)** State one use of *flame* emission spectroscopy.
OCR 2815/04 Methods of Analysis and Detection Specimen Paper Q4

**2** Sketch a graph of the first ionization energies of the elements lithium to neon. Explain as fully as you can the reasons for the shape of this graph.

**3** Fig 3.Q3 shows the variation of the first ionization energies of the elements **neon** to **potassium**.

**Fig 3.Q3**

**a)** Define the term **first ionization energy** with reference to neon.

**b)** Explain the general trend in the first ionization energies of the elements **sodium** to **argon**.

**c)** Explain why the first ionization energy of **neon** is greater than the first ionization energy of **argon**.
Edexcel 6081/P.01 Module Test 1 Jan 2001 Q1d

**4**

**a)** What is meant by the phrase 'the ground state of an atom'?

**b)** Identify the following elements from their electron configurations. One of them is in an excited state. State which one and give its ground state electron configuration.
A $1s^2\,2s^2\,2p^2$        C $1s^2\,2s^2\,2p^6\,3s^1$
B $1s^2\,2s^2\,2p^4\,3s^1$        D $1s^2\,2s^2\,2p^6\,3s^2\,3p^2$

# KEY SKILLS **ASSIGNMENT**

## THE COLOURS OF FIREWORKS

**We think fireworks were invented by the Chinese over a thousand years ago. They used black powder (we call it gunpowder) to produce loud bangs. But the brilliant colours we see today only found their way into fireworks in the nineteenth century. Before this time a firework would have produced one colour – gold – due to the burning of charcoal in gunpowder.**

Fig 3.A1 The spectacular colours of fireworks are due to excited electrons

It is the movement between energy levels of electrons in compounds that causes the colours we see. A yellow colour is obtained with sodium compounds. The electrons in a sodium ion are excited to higher energy levels by absorbing energy. This energy is supplied when the fuel used in a firework is oxidized in an exothermic reaction. When the electrons return to lower energy levels they release some of this energy at a wavelength of 589 nm, which is yellow in the visible spectrum.

**1** **a)** What is the frequency of the yellow light emitted by the sodium ion?
**b)** What is the energy of (i) one photon of this yellow light, (ii) one mole of these photons?
**c)** Explain what is meant by the following terms used in this Assignment: (i) excited electron, (ii) exothermic, (iii) energy level.

**2** Before the firework is let off, the electrons in the sodium ion are in their ground state.
**a)** What is meant by ground state?
**b)** What is the ground state electron configuration using s, p, d notation of (i) the sodium atom, (ii) the sodium ion?

Different colours are produced using other metal salts. Strontium in strontium carbonate produces colours at 606 nm and 636 nm, while barium nitrate produces a green colour (lines in its emission spectrum between 505 and 535 nm are green). Blue is a difficult colour to produce well, but it is achieved using copper salts.

The colour of a firework is intensified using compounds that donate chlorine. While the firework is alight they cause the very unstable compounds BaCl, SrCl and CuCl to be produced in the flame. One of the chlorine donors is the polymer PVC.

**3** **a)** What colour is produced by strontium? Hint: Look back at Fig 3.5.
**b)** Strontium and barium are both in Group 2 of the Periodic Table. What are their outer shell configurations in s, p, d notation?
**c)** Copper is a transition metal. To which block of the Periodic Table does it belong?

**4** **a)** You may have carried out a flame test. Research and write down the procedure you would follow to carry out a flame test on a potassium compound.
**b)** Explain the result you would expect to observe in terms of energy levels and electrons.

**5** **a)** Mercury chloride was once used to intensify the colours of fireworks. Why is it no longer used?
**b)** Should we be concerned about any of the metal salts that colour fireworks and so are dispersing into the atmosphere? Explain your answer.

# 04

# Chemical bonding and the shapes of molecules and ions

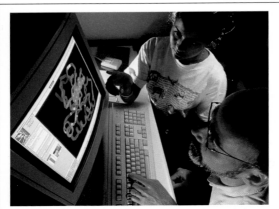

**Computers aid drug design.** The geometry of biological molecules is studied on screen, and this helps to design drugs that are likely to fit receptor sites and alter biological activity

Most of the drugs used to treat illnesses today have been discovered after a process of search and trial. Many thousands of compounds are screened for any biological activity before a promising compound is found. Now, this time-consuming and enormously expensive process is being dramatically speeded up by computer-aided drug design, based on knowing the three-dimensional shape of chemicals and on how a molecule's shape affects its biological activity.

Drugs work on complex chemical receptor sites around the body, rather like keys fitting into locks. Researchers first use X-ray crystallography to work out the shape and structure of the receptor site they are interested in. Once this is known, they can start to design a drug that fits into the site and so may turn out to be useful. It is here that computers help, since molecular shapes are easily visualized using computer graphics.

During on-screen design, different groups of atoms are made to probe the surface of the site, and gradually the shape of the active part of a drug, the pharmacophore, can be built up. Investigators then see if the designed compound is known or if it can be synthesized. Databases of known compounds are searched for compounds that have the right pharmacophore.

While computers will not replace laboratory chemistry, they greatly reduce the cost and time of finding suitable compounds to investigate further.

## 1 NOBLE GASES AND STABILITY

Until 1962, chemists thought that the elements in Group 0 were inert (totally unreactive and so unable to form compounds). The group was even called 'the inert gases'. However, in 1962, the British chemist Neil Bartlett synthesized the first Group 0 compound to contain xenon, xenon hexafluoroplatinate. This caused a sensation, and chemists rushed to make more. Three years later there were textbooks devoted to noble gas chemistry. Only helium and neon deserve the title 'inert' as they still seem not to form any compounds, so Group 0 now bears the name of 'noble'.

The noble gases were discovered more than 50 years before chemists made compounds of them. Why did it take so long? It is largely because chemists learnt about elements through their reactions, so unreactive elements remained unknown for longer.

### OUTER SHELL STABILITY OF NOBLE GASES
In 1916, American chemists Gilbert Lewis and Irvin Langmuir separately realized that the outer shells of all the noble gases (except helium) contained eight electrons, and they suggested this electron

configuration made the noble gases unreactive. They put forward the idea that when atoms of other elements formed compounds they gained, lost or shared electrons to make a noble gas outer shell of eight – or two for the lighter elements close to helium.

The ionization energy of the noble gases also confirms this stability (see Fig 3.20, page 59); as you go across a period, the first ionization energy rises to a maximum at Group 0. As the configuration is so stable a great deal of energy is needed to remove an electron.

## 2 IONIC BONDING

A **chemical bond** is an electrostatic force of attraction between two atoms or ions. **Ions** are formed when atoms lose or gain electrons and become charged.

> **An ionic bond is the electrostatic attraction that forms between oppositely charged ions.**

Sometimes ionic bonds are called **electrovalent bonds**.

A common ionic compound is sodium chloride, an industrially important raw material. The dot-and-cross diagram of Fig 4.1 shows how the atoms of sodium and chlorine achieve noble gas configurations. Usually, only the outer electrons are shown, as the inner shells stay the same.

The sodium atom loses one electron, leaving a noble gas core of 2,8. However, because it still has 11 protons it now has a +1 positive charge. It has formed a positive ion: positive ions are called **cations**.

The chlorine atom gains one electron to make the noble gas configuration of 2,8,8: it has one more electron than protons in the nucleus, giving it a charge of –1. Negative ions are called **anions**. Opposite charges attract by electrostatic attraction and this holds the sodium ions and chloride ions together.

Sodium ions and chloride ions are roughly spherical, and the negative charge is distributed evenly all over the spheres. Each sodium ion attracts several chloride ions and vice versa, so the ionic bonding is not just between one sodium ion and one chloride ion. The ions form an ordered three-dimensional structure known as a **lattice**, shown in Fig 4.2(a). The formula unit of this lattice is NaCl (see Chapter 1, page 3). As the ionic bonds in the lattice are strong, it takes a lot of energy to separate the ions, and so sodium chloride has a high melting point.

**? TEST YOURSELF**

**A** What is the outer shell electron configuration (using s and p orbital notation) of the noble gases? Hint: look back at Chapter 3, page 60. (Electron configuration means the number of electrons and their arrangement.)

**Fig 4.1** Dot-and-cross diagram for the outer electrons of sodium chloride. The transfer of an outer shell electron from sodium to chlorine gives two oppositely charged ions. Both ions have a noble gas configuration

**? TEST YOURSELF**

**B** When sodium and chlorine form ions, which noble gas configuration does each have?

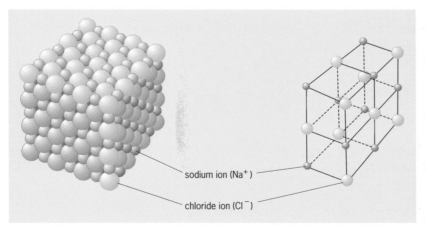

**Fig 4.2(a)** Part of a sodium chloride lattice. Each ion is surrounded by six oppositely charged ions (find out more about ionic lattices and their properties in Chapters 21 and 22)

**Fig 4.2(b)** Crystals of sodium chloride

? **TEST YOURSELF**

**C** Work out dot-and-cross diagrams for:
a) potassium fluoride;
b) magnesium oxide;
c) calcium chloride;
d) sodium oxide.
State the formula of each compound.

? **TEST YOURSELF**

**D** a) Fluorine, chlorine, bromine and iodine are all in Group 7. How many electrons are there in the outer shell of each?
b) How many electrons are in the outer shell of:
i) Group 1 elements;
ii) Group 2 elements?

? **TEST YOURSELF**

**E** Sodium ions and chloride ions are isoelectronic with which noble gases?

○ **EXAMPLE**

**Q** Work out the dot-and-cross diagrams for calcium fluoride and aluminium oxide.

**A**

**Calcium fluoride**

**Fig 4.3**

Calcium must lose two electrons to achieve a noble gas configuration, and fluorine needs only one. So there are two fluoride ions for each calcium ion, which gives a formula of $CaF_2$.

**Aluminium oxide**

**Fig 4.4**

From Fig 4.4, the formula is $Al_2O_3$.

## IONIC BONDING AND ATOMIC ORBITALS

So far we have seen the formation of ions in terms of dot-and-cross diagrams. This is a good model to explain how ions are formed, but how does it fit in with the atomic orbital model of the atom on pages 59–61? Let's look at some examples.

In sodium chloride:

$$Na \quad + \quad Cl \quad \rightarrow \quad Na^+ \quad + \quad Cl^-$$
$$1s^22s^22p^63s^1 \quad 1s^22s^22p^63s^23p^5 \rightarrow \quad 1s^22s^22p^6 \quad 1s^22s^22p^63s^23p^6$$
$$\downarrow 1 \text{ electron moves} \uparrow$$

We can write electron configurations using a simplified notation; we represent the inner shells (sometimes known as the noble gas core) by the appropriate noble gas:

$$[Ne]3s^1 + [Ne]3s^23p^5 \rightarrow [He]2s^22p^6 + [Ne]3s^23p^6$$

Both ions have attained noble gas configurations on the right: they are **isoelectronic** with the noble gases, meaning that they have the same electron configuration.

The 3s electron is removed from the sodium atom because the 3s orbital is the highest occupied energy orbital in the outer shell, and so the electron needs the least amount of energy to transfer it. Electrons in the inner shells are not removed during a normal chemical reaction: their orbitals are at a much lower energy because they are part of a stable full shell closer to the nucleus, so their removal would require very high energies.

For calcium fluoride:

$$Ca \quad + \quad F \quad \rightarrow \quad Ca^{2+} \quad + \quad F^{-}$$
$$1s^22s^22p^63s^23p^64s^2 \quad 1s^22s^22p^5 \quad 1s^22s^22p^63s^23p^6 \quad 1s^22s^22p^6$$
$$[Ar]4s^2 \qquad [He]2s^22p^5 \qquad [Ne]3s^23p^6 \qquad [He]2s^22p^6$$

$$F \qquad\qquad F^{-}$$
$$1s^22s^22p^5 \qquad\qquad 1s^22s^22p^6$$
$$[He]2s^22p^5 \qquad\qquad [He]2s^22p^6$$

## THE ENERGY CONSIDERATIONS OF IONIC BONDING

Notice that ionic bonds are formed between metal atoms, which lose electrons, and non-metal atoms, which gain electrons. Look back at Fig 3.20, page 59, which shows that metals have low first ionization energies. So the energy required to remove electrons from metals to attain a noble gas structure is relatively low. Non-metals, particularly oxygen and those in Group 7, have a strong affinity (attraction) for electrons. **Electron affinity** can be measured and is the energy change when *gaseous atoms* attract electrons. Fluorine has a very large electron affinity:

$$F(g) + e^- \rightarrow F^-(g) \quad \Delta H = -328 \text{ kJ mol}^{-1}$$

To be strictly correct, this is the **first electron affinity** of fluorine; the definition of first electron affinity is:

> **The first electron affinity is the energy change when one mole of gaseous atoms accepts one mole of electrons to form one mole of singly charged anions.**

Look at the first ionization energy of sodium:

$$Na(g) \rightarrow Na^+(g) + e^- \quad \Delta H = +496 \text{ kJ mol}^{-1}$$

We can see that the energy released by fluorine gaining an electron is not enough to remove an electron from sodium. So why do sodium and fluoride ions form the compound sodium fluoride? It is because of the very large amount of energy released when solid sodium fluoride forms from these gaseous ions:

$$Na^+(g) + F^-(g) \rightarrow Na^+F^-(s) \quad \Delta H = -918 \text{ kJ mol}^{-1}$$

> **The energy released when one mole of solid sodium fluoride is formed from its gaseous ions is called the lattice energy.**

It is this highly exothermic **lattice energy**, caused by the strong electrostatic attraction of oppositely charged ions, which makes sodium fluoride much more stable than the elements sodium and fluorine. This shows that, while ions being formed need to achieve full noble gas shells, we have to look at all the energy changes to see the full picture.

**? TEST YOURSELF**

**F** Write out the full and simplified electron configurations (using s, p, d notation) to show how ionic bonding arises in:
**a)** potassium fluoride;
**b)** magnesium chloride;
**c)** aluminium oxide.

**SEE QUESTION 1.**

**? TEST YOURSELF**

**G** What does the minus sign tell you about the first electron affinity of fluorine?

> ✔ **REMEMBER THIS**
>
> Energy is released when gaseous atoms each attract an electron. Since the atoms lose energy, the value for the first electron affinity is negative. However, the value of the **second electron affinity** of oxygen is endothermic. This is because energy is required to overcome the repulsion of the negative charge on the $O^-$ ion as it attracts a second electron.
> $$O^-(g) + e^- \rightarrow O^{2-}(g)$$
> $$\Delta H = +798 \text{ kJ mol}^{-1}$$

**YOU CAN READ MORE ABOUT LATTICE ENERGY, ELECTRON AFFINITY AND FIRST IONIZATION ENERGY IN THE FORMATION OF IONIC COMPOUNDS IN CHAPTER 22.**

# 3 COVALENT BONDING AND THE FORMATION OF MOLECULES

Many of the compounds we come across in everyday life are covalent compounds. The wood or plastic of a table top, the clothes we wear, nearly all the food we eat and most of the body's chemicals have atoms that are joined together by **covalent bonds**.

So what is a covalent bond? The answer again lies in the stability of noble gas electron configurations: with the exception of helium, there are eight electrons in the outer shell of a noble gas. For elements in Groups 4 to 7 in the Periodic Table, the loss of four or more electrons to achieve a noble gas configuration requires a great deal of energy, which could not be paid back by the formation of a stable lattice. So ions with a charge of 4+ or more do not occur under the usual conditions of a chemical reaction. Instead, the atoms of these elements can achieve noble gas configurations by *sharing* electrons.

Let us look at the example of methane, $CH_4$ (Fig 4.5), found in natural gas. Its molecule contains one carbon atom and four hydrogen atoms.

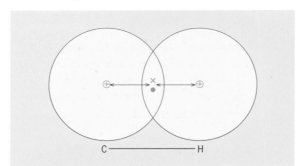

**Fig 4.5** Dot-and-cross diagram for methane

By sharing outer shell electrons, hydrogen has the noble gas configuration of helium (i.e. 2) and carbon has the neon configuration (2,8).

Now we can answer the question: What is a covalent bond?

A covalent bond is a pair of electrons shared between two atoms.

The force that holds the carbon and hydrogen atoms together in this bond is the electrostatic attraction between their positive nuclei and the shared pair of negative electrons.

We represent the shared pair of electrons as a line, as shown for the C–H bond in Fig 4.6 and the methane molecule in Fig 4.7.

**Fig 4.6** The C–H bond: electrostatic forces between the shared pair of electrons and the nuclei hold the carbon and hydrogen atoms together in the covalent bond

**Fig 4.7** The methane molecule

The bonding in a fluorine molecule (Fig 4.8) is also covalent. Fluorine is in Group 7, so has 7 electrons in its outer shell.

**Fig 4.8** The fluorine molecule

Each fluorine atom has achieved an **octet** (8 electrons) through sharing a pair of electrons. In this molecule there is one **bonding pair** of electrons and three **non-bonding pairs** or **lone pairs** on each atom. While lone pairs play no part in the bond, they are important in determining the shape of a molecule. We look at this more closely on page 77.

---

### ● EXAMPLE

**Q** Use dot-and-cross diagrams to show the covalent bonding in water, $H_2O$.

**A** The oxygen atom has 6 electrons (it is in Group 6). Draw out the separate atoms first, and then combine them in the molecule.

2H• + ×O× → H×O×H

**Fig 4.9**

---

### ? TEST YOURSELF

**H** Draw electron dot-and-cross diagrams for the following molecules:
**a)** ammonia, $NH_3$;
**b)** hydrogen, $H_2$;
**c)** chlorine, $Cl_2$.

---

## MULTIPLE BONDS

Sometimes sharing one pair of electrons is not enough to make up a noble gas configuration. Carbon dioxide is a very common molecule held together by **double bonds** (Fig 4.10).

**Fig 4.10** The carbon dioxide molecule

**Fig 4.11** Bonds in carbon dioxide

Carbon shares two pairs of electrons with each oxygen, so all three atoms achieve the noble gas octet. Count the electrons round each atom to be convinced of this. As in Fig 4.11, double bonds are shown by two lines.

Nitrogen, $N_2$, forms 79 per cent of the Earth's atmosphere. It is a very unreactive molecule because its very strong **triple bond** needs to be broken before it can react.

×N× + •N• → ×N×N× N≡N

**Fig 4.12** The nitrogen molecule

**Q** Methanal, HCHO, reacts with phenol to make the glue that holds plywood together. Draw out the dot-and-cross diagram for methanal and hence show the covalent bonds.

**A Step 1.** Draw dot-and-cross diagrams for the separate atoms of the molecule, Fig 4.13.

**Fig 4.13**

**Step 2.** Arrange the atoms in the order they are in the molecule (if you know this) and put in shared pairs of electrons for those atoms that can only form a single bond – in this case, the hydrogen atoms, Fig 4.14.

**Fig 4.14**

**Step 3.** Now make up noble gas configurations for the other atoms using one, two or three shared pairs, Fig 4.15.

**Fig 4.15**

Now you can draw the bonds for each of the shared pairs, Fig 4.16.

**Fig 4.16**

---

**I** Draw dot-and-cross diagrams for the following molecules and draw separate diagrams to show the covalent bonds.
**a)** oxygen;
**b)** ethene, $C_2H_4$;
**c)** hydrogen cyanide, HCN.

---

## 4 COORDINATE (DATIVE COVALENT) BONDS

When atoms bond together, the electron pair is usually made up of one electron from each atom. But often, both the electrons in a covalent bond come from just one of the atoms. This bond is called a **dative covalent bond**, or a **coordinate bond**.

The poisonous gas carbon monoxide is a good example; see Fig 4.17(a). As in Fig 4.17(b), an arrow is sometimes used to show which atom has donated the pair of electrons. But since the bond, once formed, is exactly the same as any other covalent bond, we usually draw it as in Fig 4.17(c).

(a)

(b) C⇌O    (c) C≡O

**Fig 4.17(a–c)** The carbon monoxide molecule

Another example is the ammonium ion, formed when an ammonia molecule bonds with a proton ($H^+$). Note that when a hydrogen atom loses its electron, a nucleus that consists of one proton is left.

**Fig 4.18** Formation of the ammonium ion

---

**J** Draw the dot-and-cross diagram for the oxonium ion, $H_3O^+$. Hint: This is made up of water and a proton.

---

The whole species has a charge of +1, because there is still one more proton than there are electrons.

# 5 EXCEPTIONS TO THE OCTET RULE

We work out how a molecule is bonded using noble gas configurations. Since the atoms of the noble gases (apart from helium) contain eight outer shell electrons, this is sometimes called the **octet rule**. However, some atoms can expand their octets.

## MORE ELECTRONS THAN AN OCTET

At the start of this chapter we saw that the first noble gas compound was made in 1962, more than 50 years after noble gases were discovered. Chemists had not tried earlier since they were convinced that an atom with eight outer shell electrons would be inert (unreactive). One of the first compounds to be synthesized was xenon tetrafluoride. Fig 4.19(b) shows the dot-and-cross diagram for this molecule.

Before chemists found that noble gases could expand their octets, many other compounds were known that broke the octet rule, for example phosphorus(V) chloride, $PCl_5$ (Fig 4.20), and sulphur dioxide (Fig 4.21).

**Fig 4.20** The phosphorus(V) chloride molecule    **Fig 4.21** The sulphur dioxide molecule

## FEWER ELECTRONS THAN AN OCTET

Some compounds have less than an octet of electrons. Aluminium chloride as a gas exists as $AlCl_3$, as in Fig 4.22(a).

(Aluminium chloride has another surprise: it has covalent bonds between metal and non-metal atoms. We shall return to this on page 84.)

Though there are only six electrons round the aluminium, we can produce an imaginary dot-and-cross diagram with a stable octet, Fig 4.22(b). But this does not agree with the experimental evidence obtained using X-rays, which shows three single covalent bonds round the aluminium.

When aluminium chloride gas is cooled, the formula of the molecule is found to be $Al_2Cl_6$, as in Fig 4.22(c). There are two dative covalent bonds, shown with arrows: you can see that every atom now has its noble gas octet.

Many compounds of boron are called electron deficient because boron has just 6 electrons round its atom, as in boron trifluoride, $BF_3$, shown in Fig 4.23.

**(a)**

**(b)**    or

**(c)**

**Fig 4.19(a)** Xenon tetrafluoride crystals. One of the first noble gas compounds to be made

**Fig 4.19(b)** The xenon tetrafluoride molecule

## ❓ TEST YOURSELF

**K** **a)** Draw a dot-and-cross diagram for sulphur(VI) fluoride, $SF_6$.
**b)** Draw, using lines, the bonds in the sulphur dioxide molecule. You should also show the lone pair on the sulphur atom.

**L** $BF_3$ reacts with $NH_3$ to form the molecule $BF_3NH_3$. In this molecule, boron has eight electrons in its outer shell. Draw the dot-and-cross diagram for this molecule and a diagram of the molecule showing the bonds as lines.

**SEE QUESTION 2.**

**Fig 4.22(a–c)** The aluminium chloride gas molecule (a) as a gas with six electrons round the aluminium atom, (b) with an octet round Al – *this form does not exist*, (c) in the cooled gas

**Fig 4.23** The boron trifluoride molecule

## EXTENSION Discovering crystal structures with X-ray crystallography

X-Ray crystallography is one of the most effective techniques used to establish the structure of compounds. It has its origins back in 1912, when the German scientist Max von Laue first suggested that atoms in crystals might diffract (bend and scatter) X-rays. Crystals diffract X-rays because the wavelength of X-rays is about the same as the distance between the nuclei in a crystal. As the X-rays scatter, some of the waves constructively interfere – they follow parallel wave paths, see Fig 4.24(a) – and this leads to spots on a photographic film, as shown in the diffraction pattern of Fig 4.24(b).

**Fig 4.24(a)** X-rays that follow parallel wave paths (are in phase) constructively interfere to reinforce each other

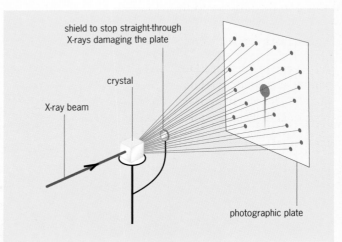

**Fig 4.24(b)** Arrangement for obtaining the X-ray diffraction pattern of a crystal

The pattern of spots must now be matched to the positions of atoms or ions. The English scientists William and Lawrence Bragg, a father-and-son team, worked out the mathematical calculations to match patterns to three-dimensional positions, and received the Nobel Prize for Physics in 1915.

Nowadays, the X-ray diffraction pattern is usually detected electronically, and computers carry out the calculations.

X-ray crystallography is a very important tool for the chemist. It has been used recently to confirm the structure of the new form of carbon, buckminsterfullerene (see page 163), and has determined the structure of well over 1000 proteins.

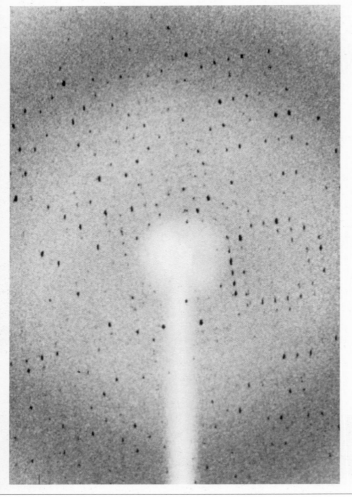

**Fig 4.24(c)** X-ray diffraction pattern for lysozyme. This is an enzyme, found in body fluids (such as tears and saliva), that destroys bacteria. The X-ray image helped researchers to work out the structure of lysozyme (see Fig 19.33) and the part of the molecule that acts against bacteria

## 6 THE SHAPES OF MOLECULES AND IONS

We now have a way of working out the *bonds* in a molecule; and the model we have used will also tell us about a molecule's *shape*, which is often important in determining the properties of the molecule. For example, the taste of substances is thought to depend on the shape of its molecules and on how they fit into taste receptor molecules on the tongue. In the same way, the shape of drug molecules can determine just how effective they are in treating illness (see the Opener to this chapter).

### ▶ EXTENSION Dorothy Hodgkin, pioneer in X-ray crystallography

At the age of 11 years, Dorothy Hodgkin started at secondary school in Beccles, Suffolk. The school allowed pupils to follow physics and chemistry for one period each week. However, because she was female she could not do physics as it clashed with domestic science. Fortunately, this was the only time that Hodgkin's gender affected her scientific career.

At Oxford University she joined a newly formed X-ray crystallography group, and by 1945 she had used the technique to work out the structure of penicillin. She pioneered the use of the computer to help interpret the X-ray images, and by 1956,

Fig.4.25 Dorothy Mary Crowfoot Hodgkin (1910–1994)

after eight years of work, she had unravelled the complex structure of vitamin B12, a molecule of more than 90 atoms. Once its structure was known, chemists could synthesize the vitamin and use it to combat pernicious anaemia. It was for this that she became the only British female scientist to receive the Nobel Prize, which in 1964 was awarded to her for Chemistry.

By 1969 she had achieved a lifetime ambition: 25 years earlier she had been given a tiny sample of insulin, and she was able to work out the structure of this complex molecule, which contains more than 800 atoms.

### ELECTRON-PAIR REPULSION AND SHAPES OF MOLECULES AND IONS

Dot-and-cross diagrams are used to represent two types of electron pairs. A pair of electrons involved in bonding is called a **bonding pair**, while a pair of electrons not involved in bonding is known as a **non-bonding pair** or **lone pair**. We now look at a model of molecular shape. In this model, pairs of electrons repel each other and move as far apart from each other as possible.

Beryllium chloride (Fig 4.26(a)) has two bonding pairs and no lone pairs round its central atom – the only pairs we are interested in here since the three lone pairs round each chlorine atom do not affect the molecule's shape. As Fig 4.26(b) shows, the position of minimum repulsion has a bond angle of 180°. We describe the shape as **linear** because the atoms lie in a straight line.

In this model of molecular shape, we group together the bonding pairs of electrons in double and triple bonds, so that one group is the equivalent of one electron pair. Carbon dioxide, with two double bonds (Fig 4.27(b)), then has a linear shape because it has two groups of electrons.

Fig 4.26(a, b) Beryllium chloride is a linear molecule

Fig 4.26(c) Ball and stick model of beryllium chloride

**Fig 4.27(a, b)** Carbon dioxide is a linear molecule   **Fig 4.27(c)** Ball-and-stick model of carbon dioxide

### ? TEST YOURSELF

**M** Explain why hydrogen cyanide, HCN, is a linear molecule. You may have already drawn the dot-and-cross diagram in answer to question **I(c)**.

**Fig 4.28(a)** Ball-and-stick model of boron trichloride

**Fig 4.28(b–c)** Boron trichloride is a trigonal planar molecule

Boron trichloride has three bonding pairs and no lone pairs round the central boron (Fig 4.28).

**Fig 4.29**
The electrons shown in red are the two extra electrons, which give the carbonate ion its 2– charge. Explain what shape you would expect this ion to have.

SOMETIMES THE MODEL OF ELECTRON PAIRS REPELLING EACH OTHER IS CALLED THE VALENCE SHELL ELECTRON PAIR REPULSION (VSEPR) THEORY, THE VALENCE SHELL BEING THE OUTERMOST SHELL.

The furthest apart these bonding pairs can get is 120°, which gives a **trigonal planar** shape. Trigonal tells us that the atoms are where the points of a triangle would be, and planar means that the molecule would lie flat on a plane.

Ethene, $C_2H_4$, is used in industry to make a wide range of important chemicals including polythene and antifreeze. The double bond is one group of electrons and counts as one bonding pair (Fig 4.30), so there are the equivalent of three bonding pairs of electrons around each carbon atom. This gives a trigonal planar shape round each carbon atom.

**Fig 4.30(a, b)** Ethene is a trigonal planar molecule

**Fig 4.30(c)** Ball-and-stick model of ethene

Methane (Fig 4.31) has four bonding pairs round the carbon. This molecule does not lie in one plane, but has a three-dimensional structure in which all the bond angles are 109.5°. The shape is tetrahedral because the hydrogen atoms are at the points of a tetrahedron. All carbon atoms with four bonding pairs have a tetrahedral structure.

**Fig 4.31(a)** Three ways to represent the structure of methane, a tetrahedral molecule

**Fig 4.31(b)** Ways to represent the directions of bonds

**Fig 4.31(c)** Ball-and-stick model of methane

**Fig 4.32(a)** The ethane molecule: the hydrogen atoms are arranged tetrahedrally around the carbon atoms. The carbon atom can rotate about its single bond, so the hydrogen atoms at one end could be in any position relative to those at the other end. But they tend to remain in the position shown in (b), in which the repulsion of the bonding pairs of electrons is the least, and which is the most stable arrangement. However, the energy difference between this and other positions is very small, which results in free rotation about the carbon–carbon single bond at room temperature

**Fig 4.32(b)** Ball-and-stick model of ethane

Phosphorus(V) fluoride, $PF_5$ (Fig 4.33), has five bonding pairs and no lone pairs, so the shape that puts the bonding pairs furthest apart is **trigonal bipyramidal** (two triangular pyramids base to base).

Sulphur hexafluoride, $SF_6$ (Fig 4.34), has an **octahedral shape** in which all bond angles are 90° because it has six bonding pairs and no lone pairs.

Dot-and-cross diagram of $PF_5$    The molecule showing the bond angles    Shape diagram: the molecule is inside a trigonal bipyramid with the fluorine atoms at the points

**Fig 4.33(a)** Three ways to represent the phosphorus(V) fluoride molecule

**Fig 4.33(b)** Ball-and-stick model

Dot-and-cross diagram of $SF_6$    Shape diagram: all bond angles are 90°    The molecule inside an octahedron

**Fig 4.34(a)** Three ways to represent the sulphur hexafluoride molecule

**Fig 4.34(b)** Ball-and-stick model

## THE SHAPES OF COMPLEX IONS

Many metal ions form **complex ions**, often more simply known as **complexes**. The central metal ion is surrounded by species – atoms, ions or molecules. These species donate lone pairs of electrons to form dative covalent (coordinate) bonds and are called **ligands**. It is common for six ligands to be bonded to a central metal cation. Since there are six bonding pairs of electrons these also have an octahedral shape.

One example is the hydrated magnesium ion, $[Mg(H_2O)_6]^{2+}$, Fig 4.35. You can read more about complexes and their shapes in Chapter 24.

**Fig. 4.35** The $Mg^{2+}$ ion accepts lone pairs of electrons from the water molecules to form six dative covalent bonds. The six bonding pairs repel each other to form an octahedral shape.

> **! SCIENCE FEATURE** Sulphur hexafluoride gas replaces polychlorinated biphenyls
>
> Sulphur hexafluoride (also known as sulphur(VI) fluoride) is an interesting gas for which new uses are being found, as it is non-toxic, odourless, colourless and very unreactive. As a result of its insulating properties, it is now used in high voltage transformers and circuit breakers, replacing polychlorinated biphenyls (PCBs).
>
> PCBs are manufactured chemicals that are very widespread pollutants in the environment today and break down extremely slowly. They are thought to cause cancer, accumulating in the fatty tissues of humans and other animals at the top of the food chain. They even appear in breast milk. Although most production ended by the late 1970s, about 30 per cent of the 1.2 million tonnes made is still around today. Disposing of PCBs in old transformers will remain a problem for many years to come.
>
> Sulphur hexafluoride is also used to trace air flow in the ventilation systems of buildings, and is used to dilute oxygen breathed in during lung radiography. It is a fully fluorinated compound (FFC) and it is increasingly used to replace ozone-depleting chlorofluorocarbons (CFCs).
>
> However, its very stability could cause future problems in the atmosphere. Molecule for molecule, it is the most 'warming' greenhouse gas ever evaluated. Its concentration in the atmosphere is very small, yet increases by 8 per cent each year, and it is believed to persist in the atmosphere for thousands of years. At the moment there is no restriction on its use, but perhaps there should be.
>
>
>
> **Fig. 4.36** Sulphur hexafluoride can be used to dilute the oxygen breathed in by a person having a lung radiograph

**? TEST YOURSELF**

**Q** Using the electron-pair repulsion theory, predict the shape of $[Be(H_2O)_4]^{2+}$ and draw it.

**SEE QUESTIONS 3–5.**

## REPULSION BY LONE PAIRS

So far, we have looked at compounds with no lone (non-bonding) pairs of electrons around the central atom. When we name the shape of a molecule, we are looking at the position of the atoms only. We do not include the lone pairs, yet they are critical in determining that shape.

### AMMONIA

Ammonia has three bonding pairs of electrons and one lone pair (Fig 4.37) around the central atom. According to the electron-pair repulsion theory, we could expect the shape to be based on a tetrahedron. However, if we look at the arrangement of the atoms, the actual shape of the ammonia molecule is best described as **pyramidal**.

Although ammonia's shape is based on a tetrahedron, the bond angles are not the expected 109.5°. The three bonds have been squeezed closer together because **the lone electron pair exerts a greater repulsion than the bonding pairs**.

**Fig 4.37(a)** Three ways to represent the ammonia molecule

**Fig 4.37(b)** Ball-and-stick model of ammonia

**Fig 4.37(c)** The ammonia molecule showing the bonding pairs and the lone pair as electron charge clouds

Think of the pairs of electrons as electron clouds and the reason for the greater repulsion becomes clear. The bonding-pair clouds are not as spread out as the lone-pair cloud, because the bonding electrons are held between the nuclei of two atoms and attracted to both, so the bonding-pair clouds are relatively thin. Only one nucleus holds the lone-pair cloud, so its electrons are pulled closer to the nucleus, it spreads out more and is therefore fatter than the bonding-pair clouds, as in Fig 4.37(c). This is why lone pairs have greater repulsion.

## WATER

In the ammonia molecule there are two types of repulsion:
● bonding pair–bonding pair
● lone pair–bonding pair
In the water molecule (Fig 4.38) there is an extra repulsion:
● lone pair–lone pair
We can put these three types of repulsion in order of increasing repulsion:

● bonding pair–bonding pair repulsion
● lone pair–bonding pair repulsion          increasing repulsion
● lone pair–lone pair repulsion

You can see that because of this, the by the two lone pairs in water exert the greatest repulsion and narrow its bond angle with respect to that in ammonia.

**Fig 4.38(a)** Three ways to represent the water molecule

**Fig 4.38(b)** Ball-and-stick model of water

## 7 COVALENT BONDING AND THE OVERLAP OF ATOMIC ORBITALS

The simple dot-and-cross model agrees with much of the experimental evidence. But how does our atomic orbital model of atoms in Chapter 3 treat covalent bonding? A covalent bond is formed by the overlap of two atomic orbitals (one from each atom). Each orbital must be occupied by a single electron, so that when they overlap the bond formed contains two electrons.

### σ AND π BONDS

In hydrogen chloride, a covalent bond forms when an s orbital overlaps with a p orbital, as in Fig 4.39. Note that the s orbital is spherical and the p orbital has two lobes. This kind of covalent bond is known as a **σ bond (sigma bond)**. A σ bond is formed when two atomic orbitals overlap at one point: there is always a σ bond between two atoms if they are covalently bonded. A σ bond gives the greatest possible electron density between the two nuclei.

In double and triple bonds there is a sideways overlap between two p orbitals, and a **π bond (pi bond)** forms (Fig 4.40). Ethene has a carbon–carbon double bond made up of a σ bond and a π bond. The π bond

**SEE QUESTIONS 6–9.**

**Fig 4.39** Hydrogen chloride: a σ bond forms by the overlap of two singly filled atomic orbitals

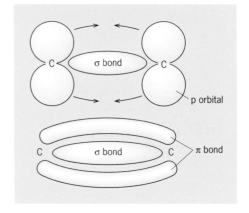

**Fig 4.40** A carbon–carbon double bond: a π bond forms by the sideways overlap of two p orbitals. This arrangement is found in molecules such as ethene, $C_2H_4$, but for clarity the following have been omitted: the two σ bonds on each carbon atom, and the other atoms that would be joined to double-bonded carbon atoms

is what makes ethene such a versatile industrial chemical. Being above and below the plane of the nuclei, the π bond is more loosely held by the carbon nuclei than the σ bond and therefore allows ethene to be very reactive. To find out more about ethene, see Chapter 14.

---

**▶ EXTENSION** A closer look at the orbitals in carbon

Let us look at just how carbon forms its covalent bonds. It has the electron configuration $[He]2s^2 2p^2$, with two singly occupied p orbitals. So on the atomic orbital overlap model it should form only two bonds. But we know that carbon forms four bonds. To do this it promotes a 2s electron to a 2p orbital and, together, these become four $sp^3$ orbitals, as shown below. These mixed orbitals all have the same energy and are sometimes called **hybrid** orbitals.

Outer shell orbitals in ground state

Carbon is excited and promotes an s electron to a p orbital

In singly bonded carbon compounds such as methane, for example, carbon's one 2s orbital and three 2p orbitals together form four $sp^3$ orbitals.

$sp^3$

Then, as in Fig 4.41, the σ bonds in methane form when the carbon $sp^3$ orbitals overlap with hydrogen's 1s orbital. The $sp^3$ orbitals repel each other equally, to give a tetrahedral shape to the methane molecule.

**Fig 4.41** In methane, each of the four σ bonds is formed when a hydrogen 1s orbital overlaps with a carbon $sp^3$ orbital

In ethene, there are three σ bonds and a π bond around each carbon, as shown in Fig 4.42. In this case, after the 2s electron is excited to the 2p orbital, three $sp^2$ orbitals form, leaving a single 2p orbital.

**Fig 4.42** The bonding in ethene

$sp^2$        2p

The $sp^2$ orbitals repel each other to give the trigonal planar shape shown in Fig 4.29. The p orbitals overlap to give a π bond, as in Fig 4.43.

**Fig 4.43** The $sp^2$ orbitals are all in one plane, while the p orbitals overlap to form the p bond

**? TEST YOURSELF**

**S** Explain why $sp^2$ orbitals are given that name.

---

# 8 POLAR COVALENT BONDS AND ELECTRONEGATIVITY

**Fig 4.44** In a hydrogen molecule the electrons in a pair are equally shared

We know that a covalent bond is a shared pair of electrons, but is the pair of electrons shared equally? If we look at the hydrogen molecule, $H_2$, the electrons are shared equally between the two atoms.

The force of attraction between the two nuclei and the electrons in the covalent bond holds the hydrogen molecule together: being the same, both atoms exert the same force, so the electrons are shared equally.

However, in hydrogen chloride the electrons are not shared equally. The chlorine atom attracts the shared pair more than the hydrogen atom. This means that chlorine has a slight excess of negative charge, which is not balanced out by the protons in the nucleus. We say that the chlorine has a partial negative charge, which is shown by the symbol δ–. Hydrogen has a δ+ charge, because the electrons in the covalent bond are nearer the chlorine atom, which gives hydrogen a partial positive charge. This bond is called a **polar covalent bond**, or just a **polar bond**.

> **TO REMIND YOURSELF ABOUT THE ELECTROSTATIC FORCES OF ATTRACTION THAT ARE PRESENT IN A COVALENT BOND, SEE FIG. 4.6, PAGE 72.**
>
> **SEE QUESTIONS 10 AND 11.**

$$\overset{\delta+}{H} \overset{\times}{\underset{\bullet}{-}} \overset{\delta-}{Cl}$$

**Fig 4.45** Electrons in this covalent bond spend more time nearer the chlorine atom, giving it a slightly negative charge and the hydrogen a slightly positive charge

In any covalent bond between two different atoms the sharing of electrons is likely to be unequal, because different atoms have different powers to attract bonding pairs of electrons. The power of an atom in a molecule to attract electrons to itself is called its **electronegativity**. The most electronegative element is fluorine. It is given a value of 4.0 on a scale devised by one of the twentieth century's most famous scientists, Linus Pauling. Next comes oxygen at 3.5 and the third most electronegative elements are chlorine and nitrogen, both having a value of 3.0.

---

### ▶ EXTENSION Pauling and his work on chemical bonds

In a life that spanned much of the twentieth century, the American Linus Pauling was one of the most influential chemists. He worked briefly with Niels Bohr (see Chapter 3) and in 1954 he received the Nobel Prize for Chemistry for his work on the nature of the chemical bond. His book *The Nature of the Chemical Bond* has deeply influenced scientists. It explains how atoms combine and helped predict the way compounds react. He put forward the ideas of hybrid orbitals and the partial ionic character of covalent bonds.

Shortly after the Second World War, helped by his wife, he spearheaded a campaign for nuclear disarmament. This led in 1963 to a limited test ban treaty. For this he received the Nobel Peace Prize in 1963 and became the first person (and to date the only person) to be the sole recipient of two Nobel prizes.

**Fig 4.46** Linus Carl Pauling, 1901–1995

---

In the Periodic Table (Fig 4.47), we see that electronegativity increases up a group and across a period.

> ### ? TEST YOURSELF
>
> **T** Use Fig 4.47 to decide whether the following bonds are polar. Where appropriate show the negative and positive poles using δ– and δ+. Rank the polarity of these bonds putting the most polar first: C–Cl, P–H, H–F, O–H, F–F, N–H, C–I.

**Fig 4.47** The Periodic Table showing electronegativity values

**Fig 4.48** The difference in electronegativities causes hydrogen chloride to be polar

**Fig 4.50** Inner shell electrons and protons in carbon–chlorine and carbon–fluorine

Fig 4.48 shows the difference in electronegativity between hydrogen and chlorine in a hydrogen chloride molecule. Chlorine is the more electronegative element, so it attracts electrons more and has a partial negative charge, $\delta-$. In this molecule the difference between the electronegativities is 1.9. The larger the difference between the electronegativities, the more polar is the covalent bond, and a value of 1.9 means that the hydrogen chloride molecule is very polar.

## CORE CHARGE

Why do different atoms have different powers to attract electrons in a covalent bond? To answer this question we must look at the **core charge** of atoms and at their size.

**The core charge is the negative charge of the inner electron shells plus the positive charge due to the protons in the nucleus.**

On page 70 we saw that the inner shells of an atom can be called the **noble gas core**. As seen in Fig 4.50, chlorine has a noble gas core of 10 electrons in its two inner shells. There are 17 positive protons in the nucleus, so the outer bonding electrons experience a core charge of 7+. Fluorine also has a core charge of 7+, because it has two electrons in its inner shell and nine protons in the nucleus. It is more electronegative than chlorine because fluorine is a smaller atom, so the outer bonding electrons are closer to its nucleus. With a core charge of 7+ and being a small atom, fluorine is the strongest of all the atoms in attracting electrons to itself.

A carbon–chlorine bond is polar because the shared pair is attracted more to the larger core charge. However, a bond with even greater polarity is the carbon–fluorine bond, because the fluorine atom is so small (see Fig 4.50).

To summarize, the two factors that help to determine electronegativity are the core charge of an atom and its size. Small atoms with high core charges are the most electronegative atoms.

## THE IONIC CHARACTER OF MANY COVALENT BONDS

The more polar a covalent bond, the greater the electron shift towards the more electronegative element. In some compounds this shift is so great that, instead of covalent bonds, ions are formed.

In fact, very few covalent bonds share electrons exactly equally. Most have some ionic character. The greater the difference in electronegativity between the two atoms in a bond, the more ionic is its character. When the electronegativity difference is more than about 2.0, the bond is more ionic than covalent, and we call it ionic. Thus there are pure covalent bonds and pure ionic bonds, but most compounds lie between these two extremes.

## THE COVALENT CHARACTER OF IONIC BONDS

In the same way that covalent bonds have some ionic character, so all ionic bonds have some degree of covalent character. We think of ions as spheres of negative charge that surround the nucleus, but negative ions (anions) can be **polarized** by positive ions (cations). Negative ions have more electrons than protons, so the electron cloud is held more loosely than in an atom. A positive ion may cause some distortion of this electron cloud and this is called **polarizing** the negative ion.

To work out how much covalent character there is likely to be in an ionic bond, first we need to know the sort of cation that has the most polarizing power over an anion, and the sort of anion that is most easily polarized.

**Fig 4.51(a)** There is no polarization of the anion by the cation, so there is no covalent character

**Fig 4.51(b)** The cation is polarizing the anion, so there is some covalent character in the ionic bond

**A cation has high polarizing power if:**
● it has a **high positive charge** (to attract the electrons in the anion)
● it is a **small cation** (the nucleus has more attraction for the electrons in the anion)
These cations have a **high charge density.**

**An anion is easily polarized if:**
● it has a **high negative charge** (outer electrons are more loosely held by the nucleus)
● it is a **large anion** (the nucleus is further away from the outer electrons, so has less of an attraction for them)
These ideas were first proposed by the Polish chemist Kasimir Fajans in 1923 and are often known as Fajans' rules. You can read more about the ionic and covalent character of compounds in Chapter 21.

> **REMEMBER THIS**
>
> Ionic radius and the charge on an ion can be treated as a single property of the ion, called **charge density**. You can read more about this in Chapter 22.

> **? TEST YOURSELF**
>
> **V** Which of the following compounds show the most covalent character?
> Sodium chloride, magnesium chloride or aluminium chloride.

**SEE QUESTIONS 13 AND 14.**

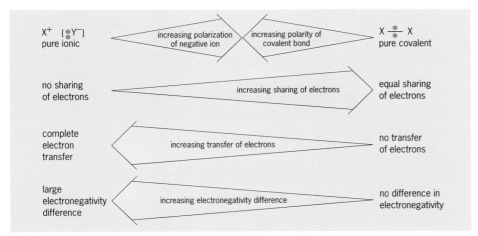

**Fig 4.52** Ionic and covalent are two extremes of bonding while in-between a gradual transition takes place

**EXAMPLE**

**Q** Explain why the covalent character of the chlorides in Group 1 of the Periodic Table increases on going up the group.

**A** All Group 1 ions have a 1+ charge. On going down the group, an extra electron shell is added, so the cations are larger, which means they become less polarizing.

The anion is the chloride ion in each compound. It is going to be most readily polarized by the smallest positive ion, which in this case is lithium, and least distorted by the largest ion, which is the francium ion. Thus, covalent character increases on going up Group 1.

## SUMMARY

When you have finished studying this chapter, you should understand and be able to use the following ideas.

- Usually when atoms bond together in a compound they lose, gain or share electrons to make stable noble gas configurations of outer shell electrons.
- Ionic bonding involves the complete transfer of electrons between atoms to create oppositely charged ions, which are held together by strong electrostatic forces of attraction. Lattice energy is the energy released when one mole of lattice is formed from its gaseous ions.
- A covalent bond is a shared pair of electrons between two atoms. It is an electrostatic attraction between the pair of negative electrons and the positively charged nuclei. Double and triple bonds can also form between atoms. When the pair of electrons in a covalent bond comes from one atom it is called a dative covalent (coordinate) bond. Sometimes atoms in a covalent compound have more or less than the usual noble gas octet.
- The shape of molecules and ions can be determined by counting up the number of electron pairs around

the central atom. Bonding pairs exert the least repulsion on each other; lone pair–bonding pair repulsion is greater; while two lone pairs have the most repulsion.
- A single covalent bond is formed by the overlap of two atomic orbitals to form a σ bond. π bonds occur in double or triple bonds (together with σ bonds). π bonds are formed by the sideways overlap of p orbitals.
- Ionic and covalent bonds are two extreme forms of bonding. Most bonds are neither purely ionic nor purely covalent. Pure covalent bonds are formed between atoms with the same electronegativities. Polar covalent bonds arise when there is a difference in electronegativities between the atoms in the bond and the pair of electrons is not equally shared. Electronegativity is the power of an atom within a molecule to attract bonding pairs of electrons to itself.
- All ionic compounds have some degree of covalent character. Small highly charged cations are best at distorting the electron cloud around anions. Large highly charged anions are the most easily distorted.

## ? EXAM QUESTIONS

**1** Write the full electron configurations, in terms of s and p electrons, of the following ions:
$K^+$, $Ca^{2+}$, $Cl^-$, $S^{2-}$.
With which noble gas are these ions isoelectronic?

**2** In the gaseous state, phosphorus(V) chloride exists as a molecule with the formula $PCl_5$. However, when it is a solid it is ionic with the formula $PCl_4^+PCl_6^-$. Draw dot-and-cross diagrams for the $PCl_5$ molecule and the ions $PCl_4^+$ and $PCl_6^-$. Then work out, draw and label the shapes of these three species.

**3** Describe and explain the shape of the following species:
$NH_3$, $NH_4^+$, $NH_2^-$.
Your answer should include reference to the bond angles involved.

**4** Ethanoic acid ($CH_3COOH$) is found in vinegar.
**a)** Draw a dot-and-cross diagram for this molecule.
**b)** Draw another diagram to show the covalent bonds and bond angles.

**5** Sulphur dioxide is an important atmospheric pollutant.
**a)** Draw the dot-and-cross diagram for sulphur dioxide. Hint: sulphur is bonded to each oxygen by a double bond.

**b)** Draw the shape of sulphur dioxide and estimate the bond angle.

**6** State which **one** of the following has a trigonal planar shape.
**A.** $H_2O$  **B.** $NH_3$  **C.** $BCl_3$  **D.** $SF_6$
WJEC Chemistry (Modular) Paper CH1 Jan 2001 Q3

**7** Consider the following shapes.

A        B        C        D

Indicate by a letter (**A**, **B**, **C** or **D**) the shape of the following ions or molecules: **(i)** $H_2O$; **(ii)** $NH_3$; **(iii)** $CO_2$; **(iv)** $CO_3^{2-}$.
Edexcel 6241/P.01 Unit Test 1 Jan 2001 Q1a

**8** [In question **8** the term **valence shell** refers to the outermost shell of the electrons.]
**a)** Use the Valence Shell Electron Pair Repulsion (VSEPR) theory to deduce the shapes of the following three molecules, which all have carbon as the central atom: **i)** HCN; **ii)** $H_2CO$; **iii)** $H_2CCl_2$. (Draw shapes.)

# ? EXAM QUESTIONS

**b)** Silicon is below carbon in Group 4 of the Periodic Table. It can form a complex ion with chlorine with the formula $[SiCl_6]^{2-}$.
**(i)** How many pairs of outer shell electrons are around the central silicon atom in the complex ion?
**(ii)** Deduce the shape of the complex ion.
*IBO Chemistry Standard Level Paper 3 Aug 2000 QA2*

**9**
**a)** Explain the meaning of the following terms.
**(i)** An ionic bond. **(iii)** A $\pi$ (pi) bond.
**(ii)** A covalent bond. **(iv)** A dative bond.
**b)** Outline the basic principle for predicting the shape of a simple molecule.
*Edexcel AS/A Chemistry 2000 Specimen Unit Test 1 Q4a,b*

**10** Many of the symptoms of hay fever are caused by the production of histamine in the body when pollen grains enter the nose.
**a)** Estimate the five bond angles shown in this molecule.
**b)** How many $\sigma$ (sigma) bonds and how many $\pi$ (pi) bonds are there in a molecule of histamine?

**11** Explain how the electron-pair repulsion theory can be used to predict the shapes of the molecules $H_2O$ and $PF_5$. Illustrate your answer with diagrams of the molecules on which the bond angles are shown.
*AQA CHM1 Unit 1: Atomic Structure, Bonding and Periodicity Jan 2001 Q6b*

**12**
**a)** Draw electron dot structures for methane, ammonia and water. Use these structures to explain why the molecular bond angles decrease in the order:

methane > ammonia > water

**b)** Draw electron dot structures for $N_2$ and $F_2$ and explain why $F_2$ is much more reactive than $N_2$.
**c)** Compare the polarity of the bonds N–Cl and C–Cl. Are the molecules $NCl_3$ and $CCl_4$ polar or non-polar? In all your answers give your reasons.
*IBO Chemistry Standard Level Paper 2 May 1999 Q5a,b,d*

**13** The compounds $NH_3$, $BF_3$ and HI all have covalent bonding and simple molecular structures. The Pauling electronegativity values shown in Table 4.Q13 can be used to predict polarity in these compounds.

**Table 4.Q13**

| | | | H 2.1 | | | |
|---|---|---|---|---|---|---|
| Li 1.0 | Be 1.5 | B 2.0 | C 2.5 | N 3.0 | O 3.5 | F 4.0 |
| Na 0.9 | | | | | | Cl 3.0 |
| K 0.8 | | | | | | Br 2.8 |
| | | | | | | I 2.5 |

**a)** Explain the term *electronegativity*.
**b)** The electronegativity values in the table can be used to predict the polarity of a bond. Copy the structures below and show the polarity of each bond by adding $\delta+$ and $\delta-$ to each bond. The first structure has been completed for you.

$$\delta-O–H\,\delta+ \qquad H–N \qquad F–B \qquad H–I$$

**c)** Using outer electron shells only, draw 'dot-and-cross' diagrams for molecules of $NH_3$ and $BF_3$.
**d)** The diagrams below show the shapes of molecules of $NH_3$ and $BF_3$. For each diagram, state the bond angle in each molecule and state the name of each shape.

**e)** Explain why $NH_3$ has polar molecules whereas molecules of $BF_3$ are non-polar.
**f)** Polar molecules of $NH_3$ form hydrogen bonds. Draw a diagram to show this hydrogen bonding. [Note: you may need to refer to page 289.]
**g)** $NH_3$ reacts with HI to form the ionic compound $NH_4I$, made up of $NH_4^+$ and $I^-$ ions.

$$NH_3 + HI \rightarrow NH_4I$$

**(i)** Explain why the H–N–H bond angle in $NH_3$ is less than that in $NH_4^+$.
**(ii)** Describe a simple test to confirm the presence of $I^-$ ions in an acidified solution of $NH_4I$.
[Note: you may need to refer to Chapter 23.]
*OCR 2811 Foundation Chemistry Jan 2002 Q4*

**14** $CCl_2F_2$ is no longer used in refrigerators as it can damage the ozone layer.
**a)** Draw the Lewis structure for $CCl_2F_2$. (This can be a dot-and-cross diagram.)
**b)** Predict the shape of a molecule of $CCl_2F_2$. (Answer in words.)
**c)** State, giving your reasons, whether a molecule of $CCl_2F_2$ is polar or non-polar.
*IBO Chemistry Standard Level Paper 2 Nov 1999 Q2*

# KEY SKILLS **ASSIGNMENT**

## INSECT PHEROMONES

A pheromone is a chemical secreted by an animal. It affects the behaviour of other individuals of the same species – we can say that the pheromone communicates an instruction. The pheromones may be secreted by:

- ants laying a trail to tell other ants the way to a good source of food;
- female moths sending out a sex attractant to attract mates; or
- the queen bee of a hive producing a chemical to tell the workers that she is in the hive and they are not to start producing new queens.

Our understanding of insect pheromones has increased rapidly since the first pheromone – from a female silkworm moth – was isolated in 1959.

**Fig 4.A1** The structure of the female silkworm moth sex pheromone

A pheromone has its effect because of the shape of its molecules. They fit into receptor sites in receiving creatures and trigger complex chemical reactions that produce the desired response.

**1** **a)** The atoms in the silkworm sex attractant are joined together by chemical bonds. Name these bonds and explain how they hold the atoms together.
**b)** Work out the likely bond angles labelled 1, 2 and 3. Give reasons for your answers.
**c)** Explain why the C–C bonds in this compound are non-polar, while the O–H bond is polar.
**d)** There are two $\pi$ bonds in this pheromone, together with 46 $\sigma$ bonds. Describe the difference between these two types of bond.

The sex pheromones of moths are among the simplest known. They have linear carbon chains of $C_{10}$ to $C_{20}$ atoms. They differ in their terminal groups of atoms, which are called **functional groups**. Two of the functional groups found at the ends of moth sex pheromones are:

—O—H    alcohol functional group: look for this on the end of the silkworm pheromone, see Fig 4.A1

—C (with H and double-bond O)    aldehyde functional group

**2** **a)** Copy the alcohol and aldehyde functional groups and mark on the polarity of any polar bonds using $\delta+$ and $\delta-$.
**b)** Predict, giving your reasons, the bond angles in the aldehyde functional group.

The other way moth sex pheromones can differ is in the position of the double bonds and the shape they give to the molecule. Notice that the double bonds in the silkworm pheromone are slightly different: one double bond has the hydrogen atoms on the same side, while the other has the hydrogen atoms on opposite sides. Double bonds do not allow the carbon atoms to rotate, so the hydrogen atoms are fixed in position, which affects the shape of the pheromone.

**Fig 4.A2** The female silkworm moth. Female moths use pheromones to attract males

We can draw a **skeletal formula** that shows the carbon skeleton shape clearly. The single lines in Fig 4.A3 represent single bonds and the carbon atoms are at the corners of the zigzags. You can see that the two different arrangements of the hydrogen atoms around the double bond do affect its shape. (You can read more about skeletal formulas on page 197.)

—OH

**Fig 4.A3** The skeletal formula of the silkworm moth sex pheromone

**3** Using information in Fig 4.A4, suggest why the carbon skeleton is reckoned to be a zigzag shape.

**Fig 4.A4** There is free rotation about the C–C bond at room temperature, but this is the position of least repulsion of bonding pairs

In moths, a mixture of two, three or four different chemicals, similar to those ones described above, make a species-specific pheromone cocktail to attract the males of the same species.

**4** Research into insect pheromones has already proved very important in agriculture. Why do you think this is? Look through scientific journals, such as *New Scientist*, to find one important use for insect pheromones in agriculture and describe it briefly in your own words.

# 05

# The Periodic Table, redox and oxidation number

**Seeking elements.** Research is carried out at CERN, the European centre for Particle Physics, into the nature of the elements and their subatomic particles

Ever since Mendeleev proposed his theory of the periodic classification of elements, scientists have set themselves the task of discovering yet more elements. During the past 30 years or so, the search has been on to discover elements that are not normally found on Earth. Scientists have bombarded atoms with high-energy particles in an effort to achieve the alchemist's dream of transmutation, the changing of one element into another.

Once an element has been made and conclusively identified, the next step is to find it a name. This seemingly simple task has led to disputes between different groups of research scientists. Each group suggests a name and a symbol. Suddenly an element has not one name and one symbol, but possibly three names and three symbols, and it can take more than ten years before there is any agreement between scientists.

In 1994 element number 111 was discovered. Could it be another ten years before it is given a name? Until then, this element will be known as the uninspiring 'unununium', symbol Uuu. Let's hope the scientists make their minds up soon.

**Fig 5.1** Dmitri Mendeleev – father of the Periodic Table. By bringing together all the facts known then about the elements, Mendeleev made a system of classification that led to many predictions and the discovery of new elements

## INTRODUCTION

For thousands of years, people have been curious about the nature of materials in the world around them, and have had theories on the simplest forms of matter, or 'elements', that make up these materials. However, only in the past 200 years, since modern experimental work began, have chemists been sure that they have identified **elements** – substances that could not chemically be changed to anything simpler. Only then could the chemists start to devise useful theories on how elements are related to each other, theories that would help scientists progress in their knowledge and understanding of materials.

The greatest breakthrough in this recent endeavour was presented to the world of chemistry on 17 February 1869, when the Russian chemist, Dmitri Mendeleev, published his work on the properties of elements. Assembling the observations and discoveries of earlier workers, he had assembled the symbols of the chemical elements in the order of their atomic masses. Mendeleev's particular arrangement became the first modern Periodic Table, and was to prove invaluable to chemists. It greatly accelerated the discovery of new elements and understanding of their properties, and enabled scientists to predict the full range of elements that were eventually found or artificially made.

This chapter describes how the Periodic Table developed. We explore its importance, both in explaining chemical trends in the elements, long before the structure of the atom was understood, and in predicting the discovery of unknown elements.

# 1 EARLY IDEAS ABOUT THE ELEMENTS

More than 2000 years ago the Ancient Greeks pictured all the material in the world as being composed of four 'elements' – earth, water, air and fire. The idea may seem far-fetched now, but the Ancient Greeks were thinkers and not experimenters. It is interesting, however, that their model reflected the three states of matter – solid, liquid and gas – and fire, which has the capacity to transform materials.

**Fig 5.2** A caricature of about 1800 showing alchemists trying to discover the Philosopher's Stone, a process they believed would transform metals into gold

Civilizations, some of them older than the Ancient Greeks, were able to produce metallic elements such as copper and mercury by mixing their ores with charcoal and heating them in a simple furnace. Other elements, including gold and silver, were found naturally in their elemental state. The drive to produce metals and their alloys arose not because people were interested in chemistry for its own sake, but because of the need to make tools and weapons that were hard-wearing and shatterproof.

The early Arab civilization was noted for its practical interest in chemicals, and has given us words currently used in chemistry such as alchemy and alkali. In the Middle Ages, particularly, alchemists in Europe tried hard to find a way to make gold from lead and other metals, not knowing that this was chemically impossible.

The pace of discovery of the elements is shown in Fig 5.4. Notice that where the graph is nearly vertical, several elements were discovered at the same time. Typically, this happened when a new theory or idea was proposed, or new apparatus was developed.

**Fig 5.3** Fire, charcoal and copper ore were used in the process of smelting copper shown in this engraving from 1574

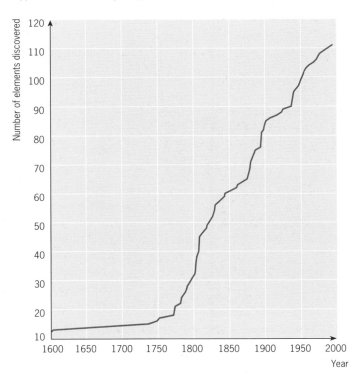

**Fig 5.4** Graph to show the number of elements known from 1600 to the present. Notice how the graph steepens as the rate of new discoveries increases, for example at the time of Mendeleev's work

**? TEST YOURSELF**

**B** What development happened in 1898 which allowed five elements to be discovered?
Hint: look back to page 31.

**FOR A DEFINITION OF RELATIVE ATOMIC MASS, SEE CHAPTER 2, P.37.**

For example, towards the end of the eighteenth century, glassware designed for gas preparation led to the discovery of oxygen, chlorine and hydrogen, and air was shown to be a mixture of gases, not a single substance. Later, the new technique of electrolysis enabled Humphry Davy to discover another group of elements – sodium, potassium, magnesium, barium and strontium.

## DALTON'S ATOMIC THEORY

In Dalton's atomic theory (see Chapter 2), he proposed that an atom was the smallest part of an element and could not be split, and that atoms of a particular element had a characteristic mass. We now call this mass the element's **relative atomic mass**, but in Dalton's time it was known as the atomic weight.

Through the early 1800s, chemists continued their search for new elements, gradually improving their techniques, but working in an unsystematic way, as they lacked a basis for predicting the existence of unknown elements.

## DÖBEREINER AND HIS TRIADS

Johann Döbereiner was among the scientists who were looking for ways to classify the elements. In 1829, he reported on patterns he had found among the elements known at the time. He noticed that elements with similar properties could be grouped into threes, or triads, and that there was a mathematical pattern in the values for their relative atomic masses. One of his triads was lithium, sodium and potassium, elements that Davy had discovered using electrolysis.

## NEWLANDS' OCTAVES

Döbereiner had begun to bring order to the elements, showing patterns that existed in their relative atomic masses. Then, in 1864, John Newlands arranged the elements in the order of their relative atomic mass. He, too, found a pattern, noticing that elements with similar chemical properties were eight positions away from each other, rather like the notes in a musical octave. (At this time, the noble gases, which would have added another position, had not been discovered.)

Look at Table 5.1. If you start with lithium as '1' and count eight elements, you come to sodium. The two elements are very similar physically and chemically. Count another eight elements and you reach potassium, also very similar to lithium and sodium.

**Table 5.1 Newlands' octaves**

| Hydrogen | Lithium | Beryllium | Boron | Carbon | Nitrogen | Oxygen |
|---|---|---|---|---|---|---|
| Fluorine | Sodium | Magnesium | Aluminium | Silicon | Phosphorus | Sulphur |
| Chlorine | Potassium | Calcium | Chromium | Titanium | Manganese | Iron |

Though the rule of octaves works for some elements, there are exceptions. Look at sulphur and iron, which are, again, eight elements apart and yet have very different properties. It needed another system of grouping to make sense of these differences, but Newlands was too preoccupied with the idea of an octave to devise an explanation.

**? TEST YOURSELF**

**C** Look at the following triads. The relative atomic mass is written underneath each element.

| Lithium | Sodium | Potassium |
|---|---|---|
| 7 | 23 | 39 |
| Sulphur | Selenium | Tellurium |
| 32 | 79 | 128 |
| Chlorine | Bromine | Iodine |
| 35.5 | 80 | 127 |
| Calcium | Strontium | Barium |
| 40 | 88 | 137 |

What is the mathematical connection between the atomic masses in these triads?
Suggest an explanation for this connection.

# 2 MENDELEEV AND THE DEVELOPMENT OF THE PERIODIC TABLE

The Russian chemist Mendeleev was fascinated by the elements. He wrote down each known element and its properties on a separate card, and began to put them in a logical order (see Table 5.2). He ignored hydrogen, for it did not seem to fit anywhere, and started with lithium. Guided by relative atomic masses, he put beryllium, boron, carbon, nitrogen, oxygen and fluorine in a column. The next element was sodium. He put it next to lithium as both were reactive soft metals. Then the elements magnesium, aluminium, silicon, phosphorus, sulphur and chlorine duly took their place.

Mendeleev continued his list with potassium and calcium. The obvious place for the next card, for titanium, seemed to be below calcium. Then Mendeleev made a brilliant decision. He recognized that a gap had to be left between calcium and titanium for an element that was yet to be discovered. The chemical properties of titanium were more like those of carbon and silicon than of boron and aluminium, and so Mendeleev placed titanium next to silicon.

Mendeleev concluded, 'The elements, if arranged according to their atomic weights [now known as relative atomic masses], exhibit an evident periodicity of properties.' Mendeleev's original Periodic Table is shown in Fig 5.5.

Notice that Mendeleev first placed the elements in columns. Soon afterwards, he decided on the arrangement of the Periodic Table that is familiar to us. Each element had its own number and fixed position in the table, and Mendeleev left several gaps for element that were then unknown, but which he believed existed.

Mendeleev's table made a considerable impact among chemists of his time, because it was so useful to them. It revealed clear patterns and trends in the properties of elements, both known and unknown. For the first time, chemists could guess at the total number of elements that might exist and, in noting the gaps, they were spurred on to make discoveries of new elements.

Mendeleev used his table to predict the properties of scandium and gallium, and also germanium, which he first named 'eka-silicon'. Within 20 years, all three elements had been discovered. Table 5.3 shows the properties he expected germanium to have, and how close his predictions proved to be. Altogether, Mendeleev predicted properties for 10 unknown elements, and was later proved correct in eight cases.

| Table 5.2 Mendeleev's early lists of elements | | |
| --- | --- | --- |
| Lithium | Sodium | Potassium |
| Beryllium | Magnesium | Calcium |
| Boron | Aluminium | ? |
| Carbon | Silicon | Titanium |
| Nitrogen | Phosphorus | Vanadium |
| Oxygen | Sulphur | |
| Fluorine | Chlorine | |

$$
\begin{array}{lllll}
 & & & \text{Ti}=50 & \text{Zr}=90 & ?=180. \\
 & & & \text{V}=51 & \text{Nb}=94 & \text{Ta}=182. \\
 & & & \text{Cr}=52 & \text{Mo}=96 & \text{W}=186. \\
 & & & \text{Mn}=55 & \text{Rh}=104,4 & \text{Pt}=197,4 \\
 & & & \text{Fe}=56 & \text{Ru}=104,4 & \text{Ir}=198. \\
 & & & \text{Ni}=\text{Co}=59 & \text{Pl}=106_6, & \text{Os}=199. \\
\text{H}=1 & & & \text{Cu}=63,4 & \text{Ag}=108 & \text{Hg}=200. \\
 & \text{Be}=9,4 & \text{Mg}=24 & \text{Zn}=65,2 & \text{Cd}=112 & \\
 & \text{B}=11 & \text{Al}=27,4 & ?=68 & \text{Cr}=116 & \text{Au}=197? \\
 & \text{C}=12 & \text{Si}=28 & ?=70 & \text{Sn}=118 & \\
 & \text{N}=14 & \text{P}=31 & \text{As}=75 & \text{Sb}=122 & \text{Bi}=210 \\
 & \text{O}=16 & \text{S}=32 & \text{Se}=79,4 & \text{Te}=128? & \\
 & \text{F}=19 & \text{Cl}=35,5 & \text{Br}=80 & \text{I}=127 & \\
\text{Li}=7 & \text{Na}=23 & \text{K}=39 & \text{Rb}=85,4 & \text{Cs}=133 & \text{Tl}=204 \\
 & & \text{Ca}=40 & \text{Sr}=87,6 & \text{Ba}=137 & \text{Pb}=207. \\
 & & ?=45 & \text{Ce}=92 & & \\
 & & ?\text{Er}=56 & \text{La}=94 & & \\
 & & ?\text{Yt}=60 & \text{Di}=95 & & \\
 & & ?\text{In}=75,6 & \text{Th}=118? & &
\end{array}
$$

**Fig 5.5** Mendeleev's Periodic Table, published in his book *Principles of Chemistry* in 1869. In it he first set out his periodic law, which earned him international fame. The question marks on their own before the equal signs are for unknown elements. Mendeleev published his version of the Periodic Table with horizontal periods and vertical groups in 1870

| Table 5.3 Properties of 'eka-silicon', germanium and tin | | | | |
| --- | --- | --- | --- | --- |
| | Silicon | Predicted properties for eka-silicon | Actual values for germanium (eka-silicon) | Tin |
| Atomic mass | 28 | 72 | 72.59 | 118 |
| Density/g cm$^{-3}$ | 2.3 | 5.5 | 5.3 | 7.3 |
| Appearance | grey non-metal | grey metal | grey metal | white metal |
| Formula of oxide | $SiO_2$ | $EkO_2$ | $GeO_2$ | $SnO_2$ |
| Formula of chloride | $SiCl_4$ | $EkCl_4$ | $GeCl_4$ | $SnCl_4$ |
| Reaction with acid | none | very slow | slow with concentrated acid | slow |

| Group Series | 0 | I | II | III | IV | V | VI | VII | VIII | | |
|---|---|---|---|---|---|---|---|---|---|---|---|
| 1 | | Hydrogen H 1.008 | — | — | — | — | — | — | | | |
| 2 | Helium He 4.0 | Lithium Li 7.03 | Beryllium Be 9.1 | Boron B 11.0 | Carbon C 12.0 | Nitrogen N 14.04 | Oxygen O 16.00 | Fluorine F 19.0 | | | |
| 3 | Neon Ne 19.9 | Sodium Na 23.05 | Magnesium Mg 24.3 | Aluminium Al 27.0 | Silicon Si 28.4 | Phosphorus P 31.0 | Sulphur S 32.06 | Chlorine Cl 35.45 | | | |
| 4 | Argon Ar 38 | Potassium K 39.1 | Calcium Ca 40.1 | Scandium Sc 44.1 | Titanium Ti 48.1 | Vanadium V 51.4 | Chromium Cr 52.1 | Manganese Mn 55.0 | Iron Fe 55.9 | Cobalt Co 59 | Nickel N (Cu) 59 |
| 5 | | Copper Cu 63.6 | Zinc Zn 65.4 | Gallium Ga 70.0 | Germanium Ge 72.3 | Arsenic As 75 | Selenium Se 79 | Bromine Br 79.95 | | | |
| 6 | Krypton Kr 81.8 | Rubidium Rb 85.4 | Strontium Sr 87.6 | Yttrium Y 89.0 | Zirconium Zr 90.6 | Niobium Nb 94.0 | Molybdenum Mo 96.0 | — | Ruthenium Ru 101.7 | Rhodium Rh 103.0 | Palladium Pd (Ag) 106.5 |
| 7 | | Silver Ag 107.9 | Cadmium Cd 112.4 | Indium In 114.0 | Tin Sn 119.0 | Antimony Sb 120.0 | Tellurium Te 127 | Iodine I 127 | | | |
| 8 | Xenon Xe 128 | Caesium Cs 132.9 | Barium Ba 137.4 | Lanthanum La 139 | Cerium Ce 140 | — | — | — | — | — | — |
| 9 | — | — | — | — | | | | — | | | |
| 10 | — | — | — | Ytterbium Yb 173 | — | Tantalum Ta 183 | Tungsten W 184 | — | Osmium Os 191 | Indium Ir 193 | Platinum Pt (Au) 194.9 |
| 11 | — | Gold Au 197.2 | Mercury Hg 200.0 | Thallium Tl 204.1 | Lead Pb 206.9 | Bismuth Bi 208 | — | — | | | |
| 12 | — | — | Radium Rd 224 | — | Thorium Th 232 | — | Uranium U 239 | | | | |

**Fig 5.6** By 1905, Mendeleev's Periodic Table had a few new elements, but still looked similar to his earlier version

### ? TEST YOURSELF

**D** Look at a modern Periodic Table (opposite page).
**a)** Cobalt and nickel are not in the order of their atomic masses. Explain why this is so.
**b)** Write down the names of two more pairs of elements that are not arranged in order of relative atomic mass in the Periodic Table.

### ✔ REMEMBER THIS

In older versions of the Periodic Table, the group numbers are roman numerals – Group 7 is Group VII, for example.

### ✔ REMEMBER THIS

In some modern versions of the Periodic Table, Groups 1, 2, 3, 4, 5, 6 and 7 in Fig 5.7 are referred to as 1A, 2A, 3A, 4A, 5A, 6A and 7A.

Mendeleev positioned elements on the basis of their properties. Where elements did not fit in order of their relative atomic masses, he assumed that these had been determined incorrectly. So, for example, he placed cobalt before nickel, though cobalt has a greater mass.

### ATOMIC (PROTON) NUMBER

We have seen that, on the basis of their properties, a few elements in the Periodic Table had to be placed out of sequence in terms of their relative atomic masses. We now know that relative atomic mass depends on the overall structure of the nucleus, and that it is the order of *atomic number* of elements that determines the trends in their properties.

**The atomic number of an element is the number of protons found in one atom of that element.**

### PERIODS AND GROUPS IN THE PERIODIC TABLE

The table Mendeleev devised in 1905 (Fig 5.6) has much in common with a modern version of the Periodic Table (see Fig 5.7). *Horizontal rows* of elements represent **periods**, numbered 1 to 7. Note how the seven modern periods compare to Mendeleev's series, and how they differ.

The *vertical columns* 1 to 7 and 0 in Fig 5.7 represent the **groups** of elements. Elements in a group have very similar chemical properties. Notice in Fig 5.6 that each group has two columns. These correspond to a group in the s and p block and another group taken from the d-block elements (Fig 5.8, see later). We just refer to Group 1 as being all the elements in the vertical column headed by lithium, and similarly, Group 7 is all the elements headed by fluorine.

**Fig 5.7** A modern version of the Periodic Table, with 109 elements (relative atomic mass in brackets = mass number of most stable isotope)

# ELECTRON CONFIGURATION

As you may have read in Chapter 3, the electron configuration of an atom describes the arrangement of its electrons. Electrons occupy a series of shells round the nucleus, numbered 1, 2, 3, 4, and so on. Each shell has particular subshells labelled s, p, d, f, etc. Each subshell has its own energy level. Within a shell the energy levels of the subshells increase in the order s, p, d, f, etc. The chart sums up this information:

| Shell | Subshells | | | |
|-------|-----------|------|------|------|
| 1 | 1s | | | |
| 2 | 2s, | 2p | | |
| 3 | 3s, | 3p, | 3d | |
| 4 | 4s, | 4p, | 4d, | 4f |

increasing energy level of subshell →

A maximum number of electrons can occupy each subshell:

| Subshell | s | p | d | f |
|----------|---|---|----|----|
| Maximum number of electrons | 2 | 6 | 10 | 14 |

In each shell, electrons fill the subshells in the same order. Usually, a subshell has to be fully occupied before further electrons start to fill the next level. We indicate the number of electrons in a subshell by a raised number and describe the electrons in an atom as:

$1s^2\ 2s^2\ 2p^6\ 3s^2\ 3p^6\ 3d^{10}\ 4s^2\ 4p^6\ 4d^{10}$, etc.

The chemical properties of an element depend on its electron configuration. Since elements are arranged in the Periodic Table in order of their atomic

## ? TEST YOURSELF

**G** Write down the electron configurations for each of the following elements: nitrogen, chlorine, calcium and titanium.

**H** For each of these electron configurations, state the group to which the element belongs:
**a)** $1s^2 2s^2 2p^6 3s^2 3p^6 3d^{10} 4s^2 4p^5$;
**b)** $1s^2 2s^2 2p^6 3s^2 3p^6 3d^{10} 4s^2 4p^1$;
**c)** $1s^2 2s^2 2p^6 3s^2 3p^6 3d^{10} 4s^2 4p^6 4d^{10} 5s^2 5p^2$;

**GO BACK TO PAGE 61 IF YOU WISH TO FIND OUT MORE ABOUT ELECTRON CONFIGURATIONS.**

**SEE QUESTION 1.**

## ? TEST YOURSELF

**I** In which block of the Periodic Table would you place the elements with the following electron configurations?
**a)** $1s^2 2s^2 2p^6 3s^2 3p^3$;
**b)** $1s^2 2s^2 2p^6 3s^2 3p^6 3d^{10} 4s^2 4p^5$;
**c)** $1s^2 2s^2 2p^6 3s^2 3p^6 3d^{10} 4s^2 4p^6 4d^6 5s^2$;
**d)** $1s^2 2s^2 2p^6 3s^2 3p^6 3d^{10} 4s^2 4p^6 4d^{10} 5s^2$.

number and chemical properties, it is no surprise to see that elements are also ordered according to their electron configurations. Another strength of Mendeleev's Periodic Table is that it reflected electron configurations before electrons were even discovered.

We can see that electron configurations of the outer shells of elements in the Periodic Table show regular trends or **periodicity**. (We look again at periodicity in section 3.) The elements of any particular group have the same number of electrons in the same type of outer shell. For example, the Group 6 elements oxygen, sulphur, selenium, tellurium and polonium all have six electrons in their outer shell, namely $ns^2 np^4$, where $n$ is the period number.

If an element is in Period 4, then its outer electrons occupy either 4s or 4p orbitals; if an element is in Period 6, then its outer electrons occupy either 6s or 6p orbitals.

## 'BLOCKS' IN THE PERIODIC TABLE

Refer to Fig 5.8 as you read this section. The **s block** contains the elements with an electron configuration that ends with electrons in an s subshell, meaning that the outermost electrons are in s subshells. A typical s-block element is strontium, with an electron configuration of $1s^2 2s^2 2p^6 3s^2 3p^6 3d^{10} 4s^2 4p^6 5s^2$, so its outer electrons are in an s subshell.

The **p block** contains elements that have an electron configuration that ends with electrons in a p subshell, so the outermost electrons are in a p subshell. A typical p-block element is chlorine, with an electron configuration of $1s^2 2s^2 2p^6 3s^2 3p^5$, so the outer electrons are in a p subshell.

The **d block** contains elements in Periods 4, 5 and 6. They are often called the transition elements or transition metals. Most of the d-block elements have an incomplete set of d electrons in their second-to-outermost subshell, and either one or two electrons in the outermost s subshell. A typical element of the d block is iron, with the electron configuration $1s^2 2s^2 2p^6 3s^2 3p^6 3d^6 4s^2$, where $3d^6$ is an incomplete subshell of shell 3.

Ten electrons can fill a d subshell, and the d block is also 10 elements wide. In Period 4, for example, scandium on the far left has one electron in the d subshell, and on the right zinc has 10. Electrons are added one at a time to the 3d subshell as you go across the d-block part of the Period.

The f block contains elements with an incomplete set of electrons in an f subshell. These elements are often called the lanthanides and actinides, named after the elements lanthanum and actinium just before them in Periods 6 and 7.

| s block | 1 | 2 |
|---|---|---|
| 1s | H | |
| 2s | Li | Be |
| 3s | Na | Mg |
| 4s | K | Ca |
| 5s | Rb | Sr |
| 6s | Cs | Ba |
| 7s | Fr | Ra |

**d block**

| | | | | | | | | | | |
|---|---|---|---|---|---|---|---|---|---|---|
| 3d | Sc | Ti | V | Cr | Mn | Fe | Co | Ni | Cu | Zn |
| 4d | Y | Zr | Nb | Mo | Tc | Ru | Rh | Pd | Ag | Cd |
| 5d | La | Hf | Ta | W | Re | Os | Ir | Pt | Au | Hg |
| 6d | Ac | (4f, see below) | | | | | | | | |
| | | (5f, see below) | | | | | | | | |

| p block | 3 | 4 | 5 | 6 | 7 | 0 |
|---|---|---|---|---|---|---|
| 1s | | | | | | He |
| 2p | B | C | N | O | F | Ne |
| 3p | Al | Si | P | S | Cl | Ar |
| 4p | Ga | Ge | As | Se | Br | Kr |
| 5p | In | Sn | Sb | Te | I | Xe |
| 6p | Ti | Pb | Bi | Po | At | Rn |

**f block**

| | | | | | | | | | | | | | | |
|---|---|---|---|---|---|---|---|---|---|---|---|---|---|---|
| Lanthanides 4f | Ce | Pr | Nd | Pm | Sm | Eu | Gd | Tb | Dy | Ho | Er | Tm | Yb | lu |
| Actinides 5f | Th | Pa | U | Np | Pu | Am | Cm | Bk | Cf | Es | Fm | Md | No | Lr |

**Fig 5.8** The blocks of the Periodic Table

# 3 PATTERNS IN THE PERIODIC TABLE

Examining the Periodic Table, we have seen that the properties of elements are often repeated at regular intervals. This characteristic is called **periodicity**. Now we have seen that the fundamental structure of atoms, represented by their electron configurations, also shows similar patterns in periodicity

## ATOMIC RADII

Look at Fig 5.9, which shows the atomic radii of the elements.

| 1 H 30 | | | | | | | | | | | | | | | | | 2 He |
|---|---|---|---|---|---|---|---|---|---|---|---|---|---|---|---|---|---|
| 3 Li 152 | 4 Be 111 | | | | | | | | | | | 5 B 88 | 6 C 77 | 7 N 70 | 8 O 66 | 9 F 64 | 10 Ne |
| 11 Na 186 | 12 Mg 160 | | | | | | | | | | | 13 Al 143 | 14 Si 117 | 15 P 110 | 16 S 104 | 17 Cl 99 | 18 Ar |
| 19 K 231 | 20 Ca 197 | 21 Sc 160 | 22 Ti 146 | 23 V 131 | 24 Cr 125 | 25 Mn 129 | 26 Fe 126 | 27 Co 126 | 28 Ni 124 | 29 Cu 128 | 30 Zn 133 | 31 Ga 122 | 32 Ge 122 | 33 As 121 | 34 Se 117 | 35 Br 114 | 36 Kr |
| 37 Rb 244 | 38 Sr 215 | 39 Y 180 | 40 Zr 157 | 41 Nb 143 | 42 Mo 136 | 43 Tc 136 | 44 Ru 133 | 45 Rh 134 | 46 Pd 138 | 47 Ag 144 | 48 Cd 149 | 49 In 168 | 50 Sn 140 | 51 Sb 141 | 52 Te 137 | 53 I 133 | 54 Xe |
| 55 Cs 262 | 56 Ba 217 | 57 La 188 | 72 Hf 157 | 73 Ta 143 | 74 W 137 | 75 Re 137 | 76 Os 134 | 77 Ir 135 | 78 Pt 138 | 79 Au 144 | 80 Hg 155 | 81 Tl 171 | 82 Pb 175 | 83 Bi 146 | 84 Po 140 | 85 At 140 | 86 Rn |
| (87) Fr 270 | 88 Ra 220 | 89 Ac 200 | | | | | | | | | | | | | | | |

**Fig 5.9** Atomic (covalent) radii of the elements measured in picometres $(10^{-12}$ m)

The atomic radius of an isolated atom cannot be measured as it is not possible to know where exactly its outer boundary is. There are several definitions for the atomic radius of an atom, but we shall use the following one.

> **The atomic radius of an element X is half of the distance between the nuclei of two atoms joined by a single covalent bond X–X.**

This is also referred to as the **covalent radius** of an element. The force between the negative electrons and the positive nucleus is the force that holds the electrons in the atom. The larger the force, the stronger is the attraction between the electrons and the nucleus and so the smaller the covalent radius.

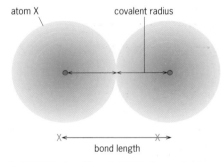

**Fig 5.10** The bond length X–X in $X_2$ is twice the covalent radius of X

**SEE QUESTION 2.**

### CHANGE IN ATOMIC RADIUS WITHIN A GROUP

Look at any group in the Periodic Table. Going down the group, the number of (positively charged) protons per atom increases, and hence the nuclear charge increases. At the same time, the number of electrons per atom increases. You might expect that, with this increase in nuclear charge, the electrons would be held more tightly, to make the atomic radius smaller. However, as the atomic number increases, and there are more occupied electron shells, the inner electron shells shield the outermost electrons from the full positive charge, and so the outermost electrons move further away from the nucleus. Sometimes we refer to an 'effective' nuclear charge, which takes this shielding into consideration.

**K** You can estimate the bond length of a covalent bond by adding together the two atomic radii of the atoms concerned. For example, the bond length of H–Cl can be estimated by adding the atomic radius of chlorine to that of hydrogen.
**a)** Use the information in the Periodic Table of Fig 5.9 to calculate the following bond lengths: (i) the O–H bond in a water molecule, (ii) the bond length of the N–H bond in ammonia, $NH_3$.
**b)** Explain why you cannot use the data in the Periodic Table to estimate the bond length of the carbon–carbon double bond in ethene, $C_2H_4$.
**c)** The Periodic Table does not have a value for the atomic radius of helium. Why not?

**L** Study the information below which is for elements in Group 5.
Nitrogen:  a.r. 70;  n.c. +7;
  e.c. $1s^2\ 2s^2\ 2p^3$
Phosphorus: a.r. 110;  n.c. +15;
  e.c. $1s^2\ 2s^2\ 2p^6\ 3s^2\ 3p^3$
Arsenic:  a.r. 121;  n.c. +33;
  e.c. $1s^2\ 2s^2\ 2p^6\ 3s^2\ 3p^6\ 3d^{10}\ 4s^2\ 4p^3$
Antimony:  a.r.141;  n.c. +51
  e.c. $1s^2\ 2s^2\ 2p^6\ 3s^2\ 3p^6\ 3d^{10}\ 4s^2$
  $4p^6\ 4d^{10}\ 5s^2\ 5p^3$
a.r. = atomic radius in picometres
$(1 \times 10^{-12}$ m);
n.c. = nuclear charge from protons;
e.c. = electron configuration.
Use this information to explain the change in atomic radius in Group 5.

**YOU CAN READ MORE ABOUT X-RAY CRYSTALLOGRAPHY ON PAGE 76.**

**SEE QUESTION 3.**

To summarize: as the atomic number of an element in a group increases,
● the shielding effect increases;
● the positive charge of the nucleus increases;
● so the effective nuclear charge is reduced;
● and the attraction of the nucleus for the outermost electrons reduces;
● so the atomic radius increases.

## CHANGE IN ATOMIC RADIUS ACROSS A PERIOD

Look at one of the short periods and notice that, as you go across the Periodic Table, the atomic radius tends to become smaller as the atomic number increases. The nuclear charge is effectively shielded only by the electrons in the inner shells, and not by the electrons in the outer shell. Going across the period, the nuclear charge increases, but the same outer electron shell is filling up, so there is no increase in shielding by inner electron shells. This means that as the atomic number increases the electrons in the outer shell are more strongly attracted towards the nucleus, and so the atomic radius decreases.

**Fig 5.11** A silicon atom has a larger atomic radius than a carbon atom, because although it has more protons it also has an extra shell of shielding inner electrons

To summarize across a period:
● the shielding effect of inner shells remains constant;
● the nuclear charge increases;
● so there is an increasing force of attraction between the outer electrons and the nucleus;
● so the atomic radius decreases.

**Fig 5.12** A beryllium atom has a smaller atomic radius than lithium because it has an increased nuclear charge, but no extra shielding inner electron shells

## IONIC RADII

The ionic radii of ions have been determined experimentally from X-ray crystallography studies on ionic crystals.
A sodium ion, $Na^+$, is formed when a sodium atom loses one electron. Its ionic radius is 95 picometres, which is smaller than the atomic radius

sodium atom          sodium ion

**Fig 5.13** A sodium ion has a smaller radius than a sodium atom as the ion has one more proton than electrons and so the excess positive charge pulls the outer electrons closer to the nucleus. In addition, a sodium ion has one less shielding shell of electons of sodium. This is because there is one less occupied shell and the ion has one proton more than its number of electrons, so the outer electrons are held with a greater attraction and are drawn closer towards the nucleus.

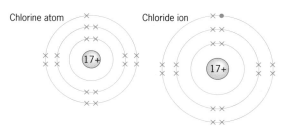

**Fig 5.14** A chloride ion has a larger radius than a chlorine atom as the ion has one more electron than protons and so the outer electrons experience less electrostatic attraction from the nucleus

A chloride ion, Cl⁻, is formed when a chlorine atom gains one electron. A chloride ion has a larger radius than a chlorine atom since it has one electron more than its number of protons, so the outer electrons are held with less attraction.

## FIRST IONIZATION ENERGY

The Periodic Table features trends in the energy required to remove electrons from an atom. In Chapter 3 we defined the first ionization energy as the energy required for one mole of gaseous atoms to lose one mole of electrons, to form one mole of gaseous ions with a single positive charge.

The process can be represented by the equation below, in which the electron is designated by the symbol e⁻. It is important to include the state symbols for this process because ionization energy only refers to gaseous atoms becoming gaseous ions.

$$X(g) \rightarrow X^+(g) + e^- \qquad \Delta H_{IE} = \text{positive (first ionization energy measured in kJ mol}^{-1})$$

The first ionization energy is necessarily endothermic, since electrons are firmly attracted electrostatically to the positive nucleus. The greater the attraction the larger the first ionization energy.

Look at Fig 5.15, which shows the first ionization energies for the first 54 elements of the Periodic Table. Many of these first ionization energies were determined using atomic emission spectroscopy.

**? TEST YOURSELF**

**M**
**a)** A sodium cation and a magnesium cation have the same number of electrons. Explain why the ionic radius of Mg²⁺ is much smaller than that of Na⁺.

**b)** Look at the table below, which shows the radius of some ions and atoms.

| Species | | | | | | |
|---|---|---|---|---|---|---|
| P³⁻ | S²⁻ | Cl⁻ | Ar | K⁺ | Ca²⁺ | Sc³⁺ |
| **Radius/pm** | | | | | | |
| 212 | 184 | 181 | 154 | 133 | 99 | 81 |

(i) Explain why a sulphide ion has a larger radius than a sulphur atom.
(ii) The species in the table are isoelectronic, meaning that they all have the same electron configuration. Explain the trend shown by the radii of these species.

**YOU CAN READ MORE ABOUT ATOMIC EMISSION SPECTROSCOPY IN CHAPTER 27.**

**Fig 5.15** Graph of first ionization energy against atomic number

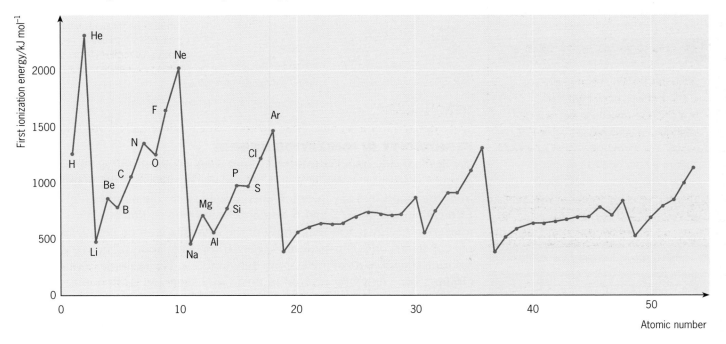

**YOU CAN READ MUCH MORE ABOUT THE REASONS FOR THE VARIATION OF THE FIRST IONIZATION ENERGIES OF ELEMENTS ON PAGE 56 IN CHAPTER 3 AND IN CHAPTER 22.**

The line joining the values for the first ionization energies of the elements shows a regular up-and-down pattern with peaks and troughs. It is difficult to remove an electron from the gaseous atom of the elements at the peaks, because there is a strong attraction between the outer electrons and the positive nucleus. The electron configurations of these elements are often referred to as being stable. Much less energy is needed to remove electrons from elements that appear in a trough, because there is a much weaker attraction between the outer electrons and the positive nucleus.

## CHANGE IN FIRST IONIZATION ENERGY IN A GROUP

The first ionization energies decrease down a group. This is because as the atomic number increases:
● the positive charge of the nucleus increases;
● but there are more shielding inner electron shells;
● so the effective nuclear charge decreases;
● and the electrostatic attraction between the outer electrons and the positive nucleus becomes weaker;
● so less energy is needed to remove an outer electron from the atom.

## CHANGE IN FIRST IONIZATION ENERGY ACROSS A PERIOD

The general trend of the first ionization energy is that it increases across a period so that all the elements of Group 0, the noble gases, are at peaks. The reason for this trend is that as the atomic number increases:
● the nuclear charge increases;
● but the number of inner shielding electron shells does not change;
● so the effective nuclear charge increases;
● and the electrostatic attraction between the outer electrons and the positive nucleus increases;
● so more energy is needed to remove an outer electron from the atom.
As a result, the first ionization energy of the noble gases (group 0) in a period will always be the highest. The electron configurations of the noble gases are very stable. They are shown in Table 5.4.

**? TEST YOURSELF**

**N** Explain why the first ionization energy of sodium is much lower than the first ionization energy of chlorine.

**? TEST YOURSELF**

**O** Why does helium have a higher first ionization energy than any other element?

**? TEST YOURSELF**

**P** a) Explain in terms of electron configuration why an element in Group 3 has a lower first ionization energy than the elements in Groups 2 or 4 of the same period.
b) Explain in terms of electron configuration why an element in Group 6 has a lower first ionization energy than the elements in Groups 5 or 7 of the same period.

**? TEST YOURSELF**

**Q** Look at the first ionization energies for the elements scandium to zinc in the first series of the transition block in Fig 5.15. The element with the highest first ionization energy is zinc. Account for this in terms of its electron configuration.

| Table 5.4 Electron configuration of the noble gases | | |
|---|---|---|
| **Atom** | **Atomic number** | **Electron configuration** |
| He | 2 | $1s^2$ |
| Ne | 10 | $1s^2\, 2s^2\, 2p^6$ |
| Ar | 18 | $1s^2\, 2s^2\, 2p^6\, 3s^2\, 3p^6$ |
| Kr | 36 | $1s^2\, 2s^2\, 2p^6\, 3s^2\, 3p^6\, 3d^{10}\, 4s^2\, 4p^6$ |
| Xe | 54 | $1s^2\, 2s^2\, 2p^6\, 3s^2\, 3p^6\, 3d^{10}\, 4s^2\, 4p^6\, 4d^{10}\, 5s^2\, 5p^6$ |

## PERIODICITY IN IONIZATION ENERGY

We have already noted that the first ionization energy of the noble gases is always the highest within a period. This is not the only pattern that emerges from the graph of first ionization energy against atomic number (Fig 5.15). Elements in groups 1, 3 and 6 also appear in troughs; that is, their first ionization energy is less than those of the elements with an atomic number plus or minus one. This pattern is repeated in every period. These troughs are associated with particular electron configurations. For example, elements in Group 1 have only one electron in their outer shell and so have the maximum number of inner shielding electrons compared to outer shell electrons for the period.

**SEE QUESTIONS 4 AND 5.**

**YOU CAN READ MORE ABOUT FIRST IONIZATION ENERGY AND ELECTRON CONFIGURATION ON PAGES 56 AND 61 IN CHAPTER 3.**

# 4 METALS, NON-METALS AND METALLOIDS

The 110 or more known elements can be divided into three main groups: the metals, the non-metals and the metalloids or semi-metals. Metals and non-metals often have opposite chemical and physical properties.

## METALS

Metals make up over 80 per cent of all the elements and occupy the centre and left-hand side of the Periodic Table. They have many physical properties in common. A metal is generally a good conductor of heat and electricity and it often has a lustre – a shine. It is likely to be malleable, that is, it can be shaped by hammering, and ductile, meaning that it can be drawn into a wire. Most metals have relatively high melting and boiling points.

**Fig 5.16** Metals. Clockwise from top left: zinc, silver foil, lead shot, copper crystals

**Fig 5.17** Non-metals. Clockwise from top left: sulphur, bromine, phosphorus, carbon, iodine

## CHEMICAL PROPERTIES OF METALS

The most striking property of a metal atom is its ability to lose electrons. Commonly, metal atoms lose one, two or three electrons to form cations. They will *lose* electrons so as to obtain a stable electron configuration like that found in one of the noble gases. (You will have met this idea in Chapter 3, page 61.)

## NON-METALS

Non-metals occupy the right-hand side of the Periodic Table. Non-metal solids are normally poor conductors of heat and electricity, rarely have a lustre and are normally brittle and non-ductile. Many non-metals are gases at room temperature and pressure, although other non-metals have very high melting and boiling points.

## CHEMICAL PROPERTIES OF NON-METALS

Non-metals, unlike metals, tend to *gain* electrons to become anions or to make covalent bonds. It is clear, then, that metals will tend to react with non-metals, and a metal atom 'gives away' electrons to a non-metal atom. So an *ionic* substance is formed when a non-metal reacts with a metal. When a non-metal reacts with another non-metal a covalent substance is formed, since it is impossible for one non-metal atom to gain electrons and another to lose them to make ions.

## METALLOIDS

If an element cannot be classified as a metal or a non-metal, it is called a metalloid or semi-metal and it will possess some of the properties of metals

> **THE RELATIONSHIP BETWEEN THE STRUCTURE AND BONDING OF ELEMENTS AND THEIR PHYSICAL PROPERTIES IS EXPLORED IN MORE DETAIL IN CHAPTER 7**

**Table 5.5 The main properties of metals and non-metals**

**Metals**

Readily form cations by loss of electrons

Normally have 1, 2 or 3 electrons in their outer shell of electrons

Normally form basic or amphoteric oxides

React with non-metals to form ionic compounds

Do not react with other metals

Are normally reducing agents

**Non-metals**

Readily form anions by gain of electrons

Normally have 4, 5, 6, 7 or 8 electrons in their outer shell

Normally form acidic or neutral oxides

React with other non-metals to form molecular compounds

React with metals to form ionic compounds

Are normally oxidizing agents

and some of the properties of non-metals. A good example of a metalloid is silicon, which is shiny, but brittle and a poor conductor of heat and electricity.

Since a metalloid element has some of the properties of a metal and some of a non-metal, it is quite likely to be able to gain electrons or to lose electrons.

# 5 REDOX REACTIONS

We have described how metal atoms lose electrons when they react and that non-metal atoms gain electrons. So when a metal reacts with a non-metal, electrons are transferred from metal atoms to non-metal atoms. A reaction that involves a transfer of electrons is called a **redox reaction**.

## REACTION BETWEEN SODIUM AND CHLORINE

Consider the reaction of sodium with chlorine. Sodium is a typical metal, and chlorine is a typical non-metal. When they react, one electron is transferred from a sodium atom to a chlorine atom to form a sodium ion and a chloride ion. This transfer of electrons results in both the sodium ion and the chloride ion obtaining a stable electron configuration like that of a noble gas.

The reaction may be written as follows:

$$Na \rightarrow Na^+ + e^-$$
$$\tfrac{1}{2}Cl_2 + e^- \rightarrow Cl^-$$

Taken together, these equations become:

$$Na + \tfrac{1}{2}Cl_2 + e^- \rightarrow Na^+ + Cl^- + e^-$$

You can see that the electron can be cancelled out on both sides of the equation, so that this reaction can be written as:

$$Na + \tfrac{1}{2}Cl_2 \rightarrow Na^+ + Cl^-$$

The equation becomes more familiar if the product is given its normal formula which ignores the charges on the ions:

$$Na + \tfrac{1}{2}Cl_2 \rightarrow NaCl$$

The final equation hides the fact that during the reaction an electron is transferred from the sodium to the chlorine and that this is a redox reaction.

**In a redox reaction, one particle loses electrons and another particle gains them.**

## OXIDATION AND REDUCTION

Redox is a shorthand word for reduction–oxidation.

In the process of **oxidation**, a particle loses electrons – it is an electron donor. In **reduction**, a particle gains electrons – it is an electron acceptor. A particle here can be an atom or a molecule or an ion.

Looking at the reaction between sodium and chlorine, you can see that sodium is oxidized and chlorine is reduced. Redox is best remembered by using OIL RIG:

**Oxidation Is Loss and Reduction Is Gain of electrons.**

It is impossible to have a redox reaction in which only reduction or only oxidation takes place. Where there are two reactants, when an atom, molecule or ion loses electrons, it follows that the other reactant must gain electrons.

**Fig 5.18** The redox reaction between aluminium powder and iodine powder is highly exothermic. The purple vapour in the figure is iodine vaporized during the reaction

**Fig 5.19** Sodium burning in chlorine

**SEE QUESTION 6.**

Since the main property of metal atoms is to lose electrons, and since 'oxidation is loss', **metals are always oxidized** by other substances when they react. The most reactive of the metals have atoms that lose electrons easily and they are oxidized easily. Conversely, the noble metals, such as gold, have atoms that lose electrons with difficulty and are not oxidized easily.

Since non-metal atoms gain electrons they are normally reduced when they react with other substances. A reactive non-metal, such as fluorine, is reduced very easily, but a non-reactive non-metal is reduced with difficulty.

**? TEST YOURSELF**

**R** The most reactive metal in Group 1 is francium, Fr. Suggest why this might be so, in terms of the ease of loss of electrons from a francium atom.

Fig 5.20 In an oxidation reaction, magnesium gives fireworks and flares a brilliant white colour

Fig 5.21 In the thermite process, a very vigorous redox reaction, aluminium reduces iron oxide to iron

**? TEST YOURSELF**

**S** For each of the following processes decide whether it is oxidation, reduction or neither oxidation nor reduction:

**a)** $Na \rightarrow Na^+ + e^-$
**b)** $Ca \rightarrow Ca^{2+} + 2e^-$
**c)** $Cl_2 + 2e^- \rightarrow 2Cl^-$
**d)** $H_2O(l) \rightarrow H_2O(g)$
**e)** $Fe^{2+} \rightarrow Fe^{3+} + e^-$
**f)** $2I^- \rightarrow I_2 + 2e^-$
**g)** $O_2 + 4e^- \rightarrow 2O^{2-}$

## ▶ EXTENSION Oxidation and reduction in organic chemistry

The term oxidation was originally used exclusively to describe the reaction of a substance with oxygen to form an oxide. This definition has been extended to take account of recent knowledge about the redistribution of electrons during a reaction. Nevertheless, in organic chemistry, it is still useful to consider oxidation as the addition of oxygen to an atom or molecule or the removal of hydrogen from an atom or molecule.

In an equation or reaction scheme, oxidation is often shown by the symbol [O]. For example, the oxidation of ethanol, $CH_3CH_2OH$, to make ethanal, $CH_3CHO$, can be represented as:

$$C_2H_5OH + [O] \rightarrow CH_3CHO + H_2O$$

In organic chemistry, it is often easier to consider reduction as the addition of hydrogen to an atom or molecule, or the removal of oxygen. In an equation, reduction is often shown by the symbol [H]. For example, the reduction of nitrobenzene, $C_6H_5NO_2$, to form phenylamine, $C_6H_5NH_2$, can be represented as:

$$C_6H_5NO_2 + 6[H] \rightarrow C_6H_5NH_2 + 2H_2O$$

## OXIDIZING AGENT OR OXIDANT

The **oxidizing agent** or **oxidant** is the species (atom, molecule or ion) that gains electrons. That is, it is the substance that does the oxidizing. To be an oxidizing agent, a species must be able to accept electrons easily. Many non-metals have electron configurations that are only one, two or three electrons short of the electron configurations of a noble gas. This means that the atoms of these non-metals can gain electrons to make a stable octet of electrons. Consequently, such non-metals are oxidizing agents.

**Fig 5.22** Magnesium burning in chlorine

SEE QUESTION 7.

## REDUCING AGENT OR REDUCTANT

The **reducing agent** or **reductant** is the species (atom, molecule or ion) that loses electrons. That is, it is the substance that does the reducing (and is itself oxidized). To be a good reducing agent, a species must be able to lose electrons easily. Many metals have electron configurations with an outer shell that contains only one, two or three electrons. This means that the atoms of these metals can easily lose electrons to make a stable octet of electrons. Consequently, such metals are reducing agents.

## HALF EQUATIONS

It is often convenient to write the equation for a redox reaction as two half equations, one for the oxidation part and the other for the reduction part. A half equation is a balanced equation for an oxidation or a reduction that shows the atom, ion or molecule gaining or losing electrons. A half equation always includes electrons, e⁻.

Consider the reaction of magnesium with chlorine to form magnesium chloride. This is the reaction of a metal with a non-metal, so it is a redox reaction. Magnesium is the reducing agent and chlorine the oxidizing agent. First, think of the oxidation process. It involves the loss of electrons. Magnesium atoms can lose electrons to form magnesium ions. Since each atom has two electrons in its outer shell, it needs to lose two electrons to achieve a stable octet. The half equation for this process is:

$$Mg \rightarrow Mg^{2+} + 2e^-$$

**Fig 5.23** Electron configurations to show the half equation Mg → Mg²⁺ + 2e⁻

Fig 5.23 represents this half equation with the full electron configurations of a magnesium atom and a magnesium ion.

Now consider the reduction process. This must involve chlorine because, being a non-metal, it can gain electrons. A chlorine atom has seven electrons in its outer shell, so it needs to gain one electron to achieve a stable octet and form a chloride ion. Cl is not written in the half equation, since chlorine exists as molecules, Cl₂, at room temperature. The half equation therefore is:

$$Cl_2 + 2e^- \rightarrow 2Cl^-$$

**Fig 5.24** Electron configurations to show the half equation Cl₂ + 2e⁻ → 2Cl⁻

Notice that this is a balanced equation. The atoms balance and the charges balance (each side of the equation has two negative charges).

Do not try to balance the symbol e⁻, just its charge. Remember, e⁻ is an electron, not an atomic symbol. Only atoms and charges are balanced in an equation.

There is a second point to note: the number of electrons in this half equation is the same as in the magnesium equation. That is, it states exactly how the redox reaction is taking place. Two electrons are being transferred from a magnesium atom to a chlorine molecule. When the two half equations are added together, the electrons cancel out.

The two half equations added together give:

$$Mg + Cl_2 + 2e^- \rightarrow Mg^{2+} + 2Cl^- + 2e^-$$

So, the full equation is:

$$Mg + Cl_2 \rightarrow Mg^{2+} + 2Cl^-$$

Even the full equation is often simplified still further to show the two ions formed as the formula unit $MgCl_2$:

$$Mg + Cl_2 \rightarrow MgCl_2$$

## BALANCING REDOX REACTIONS USING HALF EQUATIONS

It is fundamental to a redox reaction that the number of electrons lost by one species always equals the number of electrons gained by another. Knowing this can help you to construct balanced equations for redox reactions.

You will have noticed that in two of the half equations in question **T**, the number of electrons lost is not equal to those gained. This apparent imbalance can be very useful in constructing a balanced equation. For example, the half equations associated with the reaction of magnesium with nitrogen to form magnesium nitride are:

Oxidation $\quad\quad Mg \rightarrow Mg^{2+} + 2e^-$
Reduction $\quad\quad N_2 + 6e^- \rightarrow 2N^{3-}$

The oxidation half equation involves two electrons, but the reduction half equation involves six. It is important to realize that the half equations do not each need to involve the same number of electrons. They need only show the oxidation and the reduction parts, each as a self-contained reaction. Since the electrons gained by the nitrogen molecule have to be lost from the magnesium, this would suggest that three magnesium atoms are involved, which together would lose the six electrons required by nitrogen. So, the half equations could be written as:

Oxidation $\quad\quad 3Mg \rightarrow 3Mg^{2+} + 6e^-$
Reduction $\quad\quad N_2 + 6e^- \rightarrow 2N^{3-}$

Now each half equation has the same number of electrons. Therefore, when the two half equations are added together the electrons are the same on both sides (and can be cancelled out):

$$3Mg + N_2 + 6e^- \rightarrow 3Mg^{2+} + 2N^{3-} + 6e^-$$

So the full equation is written as:

$$3Mg + N_2 \rightarrow Mg_3N_2$$

with the formula unit being used for magnesium nitride instead of the separate ions.

The equations for these types of redox reaction are quite easy to work out without this analysis of the half equations. But for more complicated redox reactions, this method of balancing equations is invaluable.

# 6 OXIDATION NUMBER

So far, we have looked only at the reactions of metals and non-metals to form ionic compounds. In these, it is easy to see the transfer of electrons in terms of electrons being gained and lost (which is precisely what happens during the formation of an ionic bond). However, non-metals can react with one another, and these too are redox reactions. One of the non-metals is oxidized and the other is reduced. The compound formed is covalent, involving the sharing of electrons. So, how can we tell which element is oxidized, since there is no actual loss of electrons to produce ions? We use a concept called **oxidation number**.

## OXIDATION NUMBER IN IONIC SUBSTANCES

For bonding to take place, there must be a redistribution of electrons to create attraction between the atoms involved. The driving force for this redistribution is the formation of a stable set of outer electrons, often an octet. Each atom in a substance is assigned an oxidation number. This is the number of outer electrons of the atom that are involved in forming a stable set of electrons. The oxidation number is given a positive or a negative sign to indicate whether electrons have been added or removed from the atom to achieve the stable set.

Consider sodium chloride with the formula NaCl. It is composed of the sodium ion and the chloride ion. To form the stable octet of the sodium ion, a sodium atom **loses one electron**. So, in sodium chloride, sodium has an oxidation number of +1.

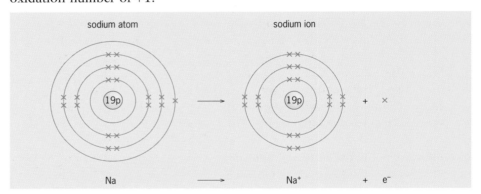

**Fig 5.25** The formation of the sodium ion

**Fig 5.26** The formation of the chloride ion

To form the stable octet of the chloride ion, a chlorine atom has to **gain one electron**. So, in sodium chloride, chlorine has an oxidation number of –1. Note that the oxidation number corresponds to the charge on the ion. This is not a coincidence and makes it easy to calculate the oxidation number in simple ions:

**oxidation number = (no. of outer electrons in the atom before bonding) – (no. of outer electrons in the atom or ion after bonding)**

It is assumed for this equation that a positive ion, such as $Mg^{2+}$, has 0 electrons in its outer shell.

**The oxidation number is the charge on the ion.**

### ● EXAMPLE

**Q** What is the oxidation number of magnesium and oxygen in magnesium oxide?

**A** The formula of magnesium oxide is MgO. It is an ionic compound composed of a magnesium ion $Mg^{2+}$ and an oxide ion $O^{2-}$.

So, the oxidation number of magnesium is +2, and the oxidation number of oxygen is –2.

A magnesium atom has two outer electrons and a magnesium ion has no outer electrons. Therefore, the equation for oxidation number also gives +2 for magnesium.

## OXIDATION NUMBER IN COVALENT COMPOUNDS

The simple equation for oxidation number does not work with compounds bonded covalently, because the electrons are shared rather than transferred. So the equation must be modified.

Given the connection between electronegativity and polar covalent bonds, first the more electronegative atom in the bond is identified. Once this is done, all the electrons being shared are assumed to be in the outer shell of this atom. Now it is possible to use the equation to calculate the oxidation number of the element. Another way is to pretend that the polar covalent bond is actually ionic and then work out the charge on the ions formed.

Take, for example, covalent hydrogen fluoride (Fig 5.27). Since fluorine is the more electronegative atom, it will form a polar covalent bond in which the fluorine end of the molecule is negative. To work out the oxidation number of the fluorine atom, assume that all the electrons being shared in the covalent bond are in the outer shell of the fluorine atom, and proceed as follows:

| | |
|---|---|
| Number of electrons in outer shell of fluorine | = 7 |
| Number of electrons in outer shell of fluorine after bonding (fluorine is the more electronegative atom) | = 8 |
| Therefore: Oxidation number of fluorine | = 7 – 8 = –1 |

Now consider the hydrogen atom. In this case, the hydrogen atom starts with one electron in its outer shell, but after bonding it has none, since it has a lower electronegativity than fluorine. Therefore, the hydrogen atom has an oxidation number of $1 - 0 = +1$.

Remember that this is a theoretical exercise in that the bond between the fluorine atom and the hydrogen atom is always *covalent*.

The more electronegative atom in a bond has a negative oxidation number that corresponds to the number of electrons it is contributing to the covalent bonds it forms. The less electronegative atom in a bond has a positive oxidation number that corresponds to the number of electrons it has contributed to the covalent bonds it forms.

**? TEST YOURSELF**

**V** What is the oxidation number of each of the atoms in the following ionic compounds?
**a)** Aluminium oxide, $Al_2O_3$, containing the ions $Al^{3+}$ and $O^{2-}$;
**b)** Calcium chloride, $CaCl_2$, containing the ions $Ca^{2+}$ and $Cl^-$;
**c)** Magnesium nitride, $Mg_3N_2$; containing the ions $Mg^{2+}$ and $N^{3-}$;
**d)** Copper(II) chloride, $CuCl_2$;
**e)** Potassium sulphide, $K_2S$;
**f)** Barium fluoride.

**YOU CAN READ ABOUT ELECTRONEGATIVITY ON PAGE 82 AND ABOUT POLAR COVALENT BONDS ON PAGE 72.**

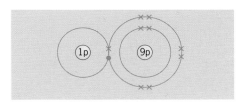

**Fig 5.27** Dot-and-cross diagram for hydrogen fluoride

**Fig 5.28** Distortion of the electron cloud towards fluorine in hydrogen fluoride

**? TEST YOURSELF**

**W** **a)** Draw the dot-and-cross diagram for tetrachloromethane (carbon tetrachloride).
**b)** Which atom is more electronegative, carbon or chlorine?
**c)** Work out the oxidation number for each atom.

## EXAMPLE

**Q** What are the oxidation numbers of each atom in sulphuric acid?

$$H-O-\overset{\overset{\displaystyle O}{\|}}{\underset{\underset{\displaystyle O}{\|}}{S}}-O-H$$

**Fig 5.29** The displayed formula of sulphuric acid

**A** The most electronegative atom is oxygen. Each oxygen atom has two covalent bonds round it, so it has an oxidation number of –2.

The least electronegative atom is hydrogen. Each hydrogen atom has just one covalent bond connecting it to another atom, so it has an oxidation number of +1.

This leaves the oxidation number of sulphur to be found, as follows:

Sum of the known oxidation numbers

$$= 4 \times (-2) + 2 \times (+1) = -6$$

(Note that the oxidation number of every atom is used.)

Given the sum of all the oxidation numbers of the atoms in a compound is zero, it follows that:

Oxidation number of sulphur

$$= 0 - (-6) = +6$$

The displayed formula confirms that the sulphur atom is indeed surrounded by six covalent bonds.

## OXIDATION NUMBERS FROM DISPLAYED FORMULAS

Another way to work out the oxidation number of an atom in a compound is to draw the displayed formula. First, decide which is the most electronegative atom. This atom will have a negative oxidation number. Then, count the number of covalent bonds around the atom to find the oxidation number. Do the same for the least electronegative atom. This atom will have a positive oxidation number. To work out the oxidation number of any other atom, an additional rule is needed, namely:

**The sum of the oxidation numbers in a compound is zero.**

## OXIDATION NUMBERS AND ELEMENTS

So far, the discussion of oxidation numbers has concentrated on atoms in compounds. But what about the oxidation number of an atom in an element? Consider oxygen, which exists as a diatomic molecule, $O_2$. The bonding is covalent, but since both atoms have the same electronegativity, the molecule is non-polar. This means it is impossible to apply any of the rules used so far. In fact, the oxidation number of each atom in this case is taken as zero.

## OXIDATION NUMBERS IN COMPOUNDS

Table 5.6 shows some simple rules about oxidation numbers that should help you to calculate them in compounds.

**Table 5.6**

| | |
|---|---|
| 1  The oxidation number of an atom in an element is always 0. | 8  The oxidation number of hydrogen in a compound is +1, unless the hydrogen atom is bonded to a metal ion, in which case it has an oxidation number of –1. |
| 2  The oxidation number of a simple ion is taken as its charge. | |
| 3  The oxidation number of fluorine in a compound is always –1. | 9  The oxidation number of a metal in a compound is always positive. |
| 4  The oxidation number of a Group 1 element in a compound is always +1. | 10  The sum of the oxidation numbers of all the atoms and ions in a compound is always zero. |
| 5  The oxidation number of a Group 2 element in a compound is always +2. | |
| 6  The oxidation number of oxygen in a compound is nearly always –2. | 11  The sum of the oxidation numbers of all the atoms in an ion is always the charge on the ion. |
| 7  The oxidation number of a Group 7 element in a compound is often –1. | |

## EXAMPLE

**Q** What are the oxidation numbers of potassium, chlorine and oxygen in potassium chlorate(V), $KClO_3$?

**A** Being in Group 1, potassium must have an oxidation number of +1. Since oxygen almost always has an oxidation number of –2, it is assumed to be –2 in this case. This leaves the chlorine oxidation number to be found. To work this out, first add up all the known oxidation numbers and then see what the chlorine oxidation number must be to end up with zero as the total:

Sum of the known oxidation numbers

$$= (+1) + 3 \times (-2) = -5$$

Sum of all the oxidation numbers

$$= -5 + \text{oxidation number of chlorine} = 0$$

Therefore, the oxidation number of the chlorine must be +5.

This explains why $KClO_3$ is called potassium chlorate(V). The (V) refers to the oxidation state of the chlorine in the compound.

## ? TEST YOURSELF

**a)** What is the oxidation number of each atom in nitric acid (Fig 5.30)?

**Fig 5.30** The displayed formula of nitric acid

**b)** (i) What is the oxidation number of each atom in chloric(V) acid (Fig 5.30)?

**Fig 5.31** The displayed formula of chloric(V) acid

(ii) What is the significance of the (V) in chloric(V) acid?

## ? TEST YOURSELF

**a)** What is the oxidation number of each of the elements in potassium sulphate, $K_2SO_4$?

**b)** What is the oxidation number of each of the elements in potassium sulphite, $K_2SO_3$?

**c)** What is the oxidation number of oxygen in $F_2O$?

**d)** What is the oxidation number of oxygen in $H_2O_2$?

**e)** What is the oxidation number of sulphur in $Na_2S_2O_3$?

**f)** What is the oxidation number of sulphur in $Na_2S_4O_6$?

## ✔ REMEMBER THIS

When an atom can have several oxidation states in different compounds, the oxidation state is normally included in the names of the compounds, in $KClO_3$, for instance. Chlorine normally has an oxidation number of –1. So it is important to state that chlorine's oxidation number is +5, hence the name potassium chlorate(V).

Oxidation numbers should be whole numbers, so why does the sulphur in question **Y(f)** have an oxidation number of +2.5? The reason is that the compound has at least two different sulphur atoms with different oxidation numbers. So, what is worked out is an *average* oxidation number over all four sulphur atoms.

## OXIDATION NUMBERS OF ATOMS IN IONS

Many ions have both ionic and covalent bonds present, but this poses little difficulty in determining the oxidation number of each element. What has to be remembered is that:

**The sum of the oxidation numbers of all the atoms is the same as the charge on the ion.**

## ● EXAMPLE

**Q** What is the oxidation number of chromium in the dichromate ion $Cr_2O_7{}^{2-}$?

**A** The oxidation number of oxygen is almost always –2.

The sum of all the known oxidation numbers (due to oxygen) is

$$7 \times (-2) = -14$$

The sum overall of the oxidation numbers must be the charge on the ion, –2.

So, the sum of the oxidation numbers of the two chromium atoms

$$= -2 - (-14) = +12$$

This means that each chromium atom has an oxidation number of +6, which explains why the dichromate ion is more accurately referred to as the dichromate(VI) ion.

The displayed formula method can also be used to work out the oxidation numbers of atoms in ions. The charge on an atom also contributes to the oxidation number.

## ? TEST YOURSELF

**a)** What is the oxidation number of iron in the ferrate ion $FeO_4{}^{2-}$?

**b)** Suggest the systematic name for this ion.

**c)** What is the oxidation state of manganese in $MnO_4{}^-$ and $MnO_4{}^{2-}$?

**d)** What is the oxidation number of the following?

(i) copper in $CuCl_4{}^{2-}$.

(ii) nitrogen in $NH_4{}^+$.

(iii) sulphur in $S_2O_3{}^{2-}$.

## THE THIOSULPHATE ION

The thiosulphate ion has the formula $S_2O_3^{2-}$. The oxidation number of each oxygen atom is $-2$, which makes $+2$ the oxidation number of each sulphur atom.

The displayed formula of the thiosulphate ion (Fig 5.32) shows that the two sulphur atoms are not identical. One is bonded only to a sulphur atom, while the other is bonded to three oxygen atoms as well as a sulphur atom. Using the displayed formula to determine the oxidation numbers shows this difference between the sulphur atoms. One sulphur atom has an oxidation state of $+4$. This is the atom attached to the oxygen atoms. The other sulphur atom has an oxidation number of 0, because covalent bonds between atoms of the same element are not counted.

**Fig 5.32** The displayed formula of the thiosulphate ion

---

### EXTENSION Naming inorganic compounds

As the number of inorganic compounds discovered and synthesized increased all the time, it became clear that the naming of them had to be standardized. Traditional names, such as ammonia, are of no use if they do not convey information about the structures, and help chemists to remember them: 'ammonia' offers no clue to the fact that the compound contains nitrogen and hydrogen.

Chemists recognized that the oxidation number should be stated in the name of a compound if there could be any doubt as to its value. For example, the oxidation number of fluorine in any compound is always $-1$, so there is no reason to specify it. However, in different compounds chromium can have oxidation numbers that range from $+2$ to $+6$. So, it is helpful to specify the oxidation numbers in their names. Hence, $CrO$, $Cr_2O_3$ and $CrO_3$ are distinguished as follows: chromium(II) oxide, chromium(III) oxide and chromium(VI) oxide.

Under the rules introduced for naming compounds, the name of a simple inorganic chemical starts with the name of the least electronegative element (often a metal), followed by the name of the most electronegative element or the name of the anion involved. So, $K_2CrO_4$ is named potassium chromate(VI). The (VI) refers to the oxidation number of chromium in the anion.

---

### ? TEST YOURSELF

**A**  **a)** Which substance is the oxidizing agent in the combustion of carbon?
**b)** Which substance is the reducing agent in the combustion of carbon?

SEE QUESTIONS 8 AND 9.

### ? TEST YOURSELF

**B**  Sodium reacts with hydrogen to give a substance called sodium hydride, NaH. Explain in terms of electron transfer why this is a redox reaction.

## REDOX REACTIONS AND OXIDATION NUMBER

The oxidation number is a most useful way of explaining whether a reaction is an example of a redox reaction. During oxidation, the oxidation number of an element increases; during reduction, it decreases. Do not confuse this with what happens to the oxidizing agent and to the reducing agent. The oxidizing agent can be identified because it contains an element for which the oxidation number decreases during the reaction. The reducing agent can be identified because it contains an element for which the oxidation number increases during the reaction.

During the combustion of coal at a power station, carbon reacts to form carbon dioxide. This is a redox reaction between two non-metals:

$$C(s) + O_2(g) \rightarrow CO_2(g)$$

During the reaction, the oxidation number of carbon changes from 0 in the element to $+4$ in carbon dioxide. This means that carbon is oxidized. At the same time, the oxidation number of oxygen has changed from 0 to $-2$, so oxygen has been reduced.

Consider next the reaction of two non-metals that does not involve oxygen, in which it is not so easy to spot the oxidizing agent. The hydrogen chloride is made by reacting hydrogen and chlorine together:

$$H_2(g) + Cl_2(g) \rightarrow 2HCl(g)$$

During the reaction, the oxidation number of the hydrogen atom changes from 0 to $+1$ (oxidation), while the oxidation number of the chlorine atom changes from 0 to –1 (reduction). That is, chlorine is the oxidizing agent. The non-metal, which has the higher electronegativity, will be the oxidizing agent in a reaction between two non-metals. (Electronegativity values are given on page 83.)

## DISPROPORTIONATION

In question **C′(b, iv)** you should find something unusual when you look at the change of oxidation number of chlorine. The numbers should go up and down. The oxidation number of chlorine in the chlorine molecule is 0, while in sodium chloride it is –1 and in sodium chlorate(I), NaOCl, it is +1. This means that chlorine is both the oxidizing and the reducing agent at the same time. A reaction in which a substance can be both oxidized and reduced at the same time is called a disproportionation reaction.

# 7 PREDICTIONS USING THE PERIODIC TABLE

One of the uses of the Periodic Table is that it enables you to make informed guesses about the chemical properties of an element. The Periodic Table allows you to:
- predict whether an element is a metal or non-metal;
- predict the charge of any ion that may be formed;
- predict the relative reactivity of the element compared to other elements in its group
- predict whether the element is a reducing or oxidizing agent.

Take as an example the element radium. Radium is highly radioactive, so it would be impossible to study its chemistry in a school laboratory. However, with the aid of the Periodic Table we can make some sensible predictions about this element. Firstly, it is in Group 2. Its atomic number is 88 and its electron configuration is:

$$1s^2\ 2s^2\ 2p^6\ 3s^2\ 3p^6\ 3d^{10}\ 4s^2\ 4p^6\ 4d^{10}\ 4f^{14}\ 5s^2\ 5p^6\ 5d^{10}\ 6s^2\ 6p^6\ 7s^2.$$

(It would have been easier to state that all its inner shells were full and that the outer electrons were $7s^2$!) This means that, to obtain a stable electron configuration, a radium atom must lose two electrons. In a reaction, the radium atom will always form a radium ion:

$$Ra \rightarrow Ra^{2+} + 2e^-$$

This property of forming an ion establishes that radium is a metal, which reacts with non-metals to form ionic compounds. Now let us make some predictions about radium.

It would be reasonable to suggest that radium could react with chlorine to give $RaCl_2$, and that its oxide would have the formula RaO. The effective nuclear charge for radium is quite small compared to that of other elements in Group 2, because the large number of electrons in its six inner shells shield the outer electrons from the nuclear charge. So we can reasonably guess that radium will be the most reactive metal in Group 2. As a metal, radium will probably be a good conductor of heat and electricity and is likely to be ductile and malleable.

**? TEST YOURSELF**

**C′** **a)** Phosphorus(III) chloride reacts with excess chlorine to form phosphorus(V) chloride:

$$PCl_3(l) + Cl_2(g) \rightarrow PCl_5(s)$$

(i) What are the changes in the oxidation number for each element?
(ii) Which substance is the oxidizing agent?
**b)** For each of the following reactions, use the change in oxidation numbers to deduce the oxidizing and reducing agents.
(i) $2Cu^{2+} + 4I^- \rightarrow 2CuI + I_2$
(ii) $MnO_2 + 4HCl \rightarrow MnCl_2 + Cl_2 + 2H_2O$
(iii) $2Cu + 4HCl + O_2 \rightarrow CuCl_2 + 2H_2O$
(iv) $2NaOH + Cl_2 \rightarrow NaOCl + NaCl$

**SEE QUESTION 10.**

**? TEST YOURSELF**

**D′** Work out the oxidation numbers of every chlorine atom or ion in the following equation. Use your answers to explain why the following reaction is an example of disproportionation.

$$6NaOH + 3Cl_2 \rightarrow 5NaCl + NaClO_3 + 3H_2O$$

**SEE QUESTIONS 11–13.**

**? TEST YOURSELF**

**E′** Use the Periodic Table to make some predictions about selenium, Se.

## SUMMARY

As a result of studying this chapter you should understand that:

- In the Periodic Table, elements are arranged in order of increasing atomic number.
- Columns of elements in the Periodic Table are in the same group.
- Rows of elements in the Periodic Table are called periods.
- Elements in a group have atoms with the same number of electrons in their outer shell.
- Elements in the same period have atoms with the same number of electron shells.
- The first ionization energy decreases and the covalent radius increases down a group, because of the increased shielding effect by the inner electron shells.
- The first ionization energy tends to increase and the covalent radius decreases across a period, because of the increasing nuclear charge.
- A positive ion is smaller than the atom from which it is formed and a negative ion is larger than the atom from which it is formed.

- Properties of elements, such as ionization energy and atomic (covalent) radius, show a periodic variation (periodicity) with increasing atomic number.
- The Periodic Table can be used to predict the properties of elements.
- Reduction is the gain of electrons by a species. Reduction can be recognized by a decrease in the oxidation number of an atom.
- Oxidation is the loss of electrons from a species. Oxidation can be recognized by an increase in the oxidation number of an atom.
- Metal atoms lose electrons and are oxidized during a reaction. Metals are reducing agents.
- Non-metal atoms gain electrons and are reduced during a reaction between a metal and a non-metal. Non-metals are often oxidizing agents.
- During a redox reaction there is a transfer of electrons.

## ? EXAM QUESTIONS

**1** An element has the electron configuration $1s^2\ 2s^2\ 2p^6\ 3s^2\ 3p^6\ 3d^{10}\ 4s^2\ 4p^1$.
**a)** To which period does the element belong?
**b)** To which block does the element belong?
**c)** To which group does the element belong?
**d)** How many electrons does it have in its outer shell?
**e)** The element forms a chloride and an oxide. Write down the formula of both compounds and state the type of bonding present in these compounds.

**2** Covalent radii can be used to estimate bond lengths.
**a)** What is the meaning of the term covalent radii?
**b)** Estimate the bond length of each of the bonds in the following molecules:
**(i)** $CCl_4$; **(ii)** $HI$; **(iii)** $C_2H_6$; **(iv)** $H_3BO_3$.
[See Fig 5.9, page 97.]

**3** The atomic radii of some of the elements in Groups 1–7 of the Periodic Table are shown in Table 5.Q3. Some radii have been omitted.

**Table 5.Q3**

| | Group | | | | | | |
|---|---|---|---|---|---|---|---|
| | **1** | **2** | **3** | **4** | **5** | **6** | **7** |
| **Period 2** | | | | | | | |
| Element | Li | Be | B | C | N | O | F |
| Atomic radius/nm | 0.134 | 0.125 | 0.090 | 0.077 | 0.075 | 0.073 | 0.071 |
| **Period 3** | | | | | | | |
| Element | Na | Mg | Al | Si | P | S | Cl |
| Atomic radius/nm | 0.154 | 0.145 | 0.130 | 0.118 | 0.110 | | 0.009 |
| **Period 4** | | | | | | | |
| Element | K | Ca | Ga | Ge | As | Se | Br |
| Atomic radius/nm | 0.196 | 0.174 | | 0.122 | 0.122 | 0.117 | 0.114 |

**a) (i)** State the trend shown in atomic radius across a period.
**(ii)** Explain this trend.
**b) (i)** State the trend shown in atomic radius down a group.
**(ii)** Explain this trend.

# ? EXAM QUESTIONS

**c)** Mendeleev studied periodic data to make predictions for the properties of elements that had yet to be discovered. Using the data above, suggest values for the atomic radius of:
**(i)** S; **(ii)** Ga.

OCR Foundation Chemistry 2000 Specimen Paper Q4a–c

**4** Astatine, At, follows iodine in the group for which the properties are given in Table 2.Q4. Astatine remained undiscovered until 1940. Use the Periodic Table in Fig 5.7 to predict the listed properties for this element.
**a)** How many electrons does an atom of astatine have in its outer shell?

**Table 2.Q4**

|  | Fluorine | Chlorine | Bromine | Iodine |
|---|---|---|---|---|
| Relative atomic mass | 19 | 35.5 | 79.9 | 127 |
| Density/g cm$^{-3}$ | 1.11 at its boiling point | 1.56 at its boiling point | 2.93 at its boiling point | 4.93 at its boiling point |
| Melting point/°C | −220 | −101 | −7.2 | 113.5 |
| Boiling point/°C | −188 | −35 | 59 | 184 |
| Appearance at room temperature | colourless gas | green gas | orange liquid | purple solid |
| Formula of hydrogen halide | HF | HCl | HBr | HI |
| Formula of sodium halide | NaF | NaCl | NaBr | NaI |
| Electrode potential/V | 2.87 | 1.36 | 1.09 | 0.54 |
| First ionization energy/kJ mol$^{-1}$ | 1680 | 1260 | 1140 | 1010 |

**b)** Use your knowledge of trends within a group to predict a numerical value for astatine for each of the following quantities:
**(i)** density at its boiling point;
**(ii)** melting point;
**(iii)** boiling point;
**(iv)** electrode potential;
**(v)** first ionization energy.
**c)** Suggest the appearance of astatine at room temperature and pressure.
**d)** What is the formula of sodium astatide and hydrogen astatide?
**e)** What type of bonding will be found in sodium astatide?
**f)** Explain why the covalent radius of the halogens increases with increasing atomic number.
**g)** Explain why the radius of an astatide ion is larger than the radius of an astatine atom.

**5** Nitrogen and phosphorus are the first two members of Group 5.
**a)** Explain why both of these elements have a higher first ionization energy than the elements with either one less atomic number or one more atomic number.
**b)** Which of these elements has the higher first ionization energy? Explain your answer.
**c)** The atomic radius of phosphorus is larger than the atomic radius of nitrogen. Explain why.

**6** Sodium street lamps contain neon gas at a very low pressure. When an electric current is passed through the street lamp, neon gas is ionized. Later, sodium ions are ionized.
**a)** Write down an equation, including state symbols, to show the process associated with the first ionization energy of **(i)** neon and **(ii)** sodium.
**b)** Which of the two elements neon or sodium has the higher ionization energy?
Explain your answer.
**c)** Why is the ionization of sodium referred to as oxidation?

**7** Rubidium, Rb, atomic number 37, is a very reactive metal.
**a)** How many electrons are there in the outer shell of rubidium?
**b)** What is the formula of a rubidium ion? Hence write down the electron configuration of a rubidium ion.
**c)** Is rubidium a reducing agent or an oxidizing agent? Explain your answer.

**8**
**a)** Define oxidation and reduction in terms of:
**(i)** electron transfer;
**(ii)** change in oxidation number.
**b)** Sulphur can be oxidized by concentrated nitric acid as shown in the equation:

$$S(s) + 4HNO_3(l) \rightarrow SO_2(g) + 4NO_2(g) + 2H_2O(l)$$

**(i)** What is the oxidation state of nitrogen in nitric acid?
**(ii)** What is the oxidation number of nitrogen in nitrogen dioxide?
**(iii)** What is the change in oxidation number of the sulphur atom during the reaction?

## ? EXAM QUESTIONS

**c)** The two half equations for a redox reaction are:

$$MnO_4^-(aq) + 8H^+(aq) + 5e^- \rightarrow Mn^{2+}(aq) + 4H_2O$$
$$2Cl^-(aq) \rightarrow Cl_2(g) + 2e^-$$

Write down the balanced equation for the redox reaction between $MnO_4^-$ and $Cl^-$.

**9** Household bleach contains the ion $ClO^-$. The concentration of $ClO^-$ ions can be determined by using their ability, under acid conditions, to oxidize iodide ions to give iodine. The liberated iodine can then be estimated by titration with a standard solution of sodium thiosulphate, $Na_2S_2O_3$.

**a) (i)** Define oxidation in terms of electrons.
   **(ii)** Give the oxidation state of chlorine in ClO.

**b)** Write an equation for the reaction between thiosulphate ions and iodine.

*AQA Unit 2 Foundation Physical and Inorganic Chemistry, Jan 2001 Q 2a,b*

**10**
**a)** Define the term **oxidation number**.
**b)** Copy the equation below, which shows the disproportionation of chlorine:

$$Cl_2(g) + H_2O(l) \rightarrow HClO(aq) + HCl(aq)$$

   **(i)** Underneath the chlorine-containing species write the oxidation number of chlorine in each case.
   **(ii)** Use these oxidation numbers to explain the term **disproportionation**.

*Edexcel A/AS Unit test 1 Jan 2001 Q5a,b*

**11** The Periodic Table can be used to make predictions. Consider the undiscovered element atomic number 119. Suppose it has a symbol of Y.
**a)** In which group would it be placed?
**b)** How many electrons would it have in its outer electron shell?

**c)** Which element would it most likely resemble in properties?
**d)** The element forms a chloride.
   **(i)** What is the most likely formula for this chloride?
   **(ii)** What type of bonding is most likely in the chloride?
**e)** The element forms an oxide.
   **(i)** What is the most likely formula for this oxide?
   **(ii)** What type of bonding is most likely in the oxide?

**12** By using its position in the Periodic Table predict some of the physical and chemical properties of francium, Fr.

**13** The atomic radii of the elements Li to F and Na to Cl are shown in Table 2.Q13 below.

**Table 2.Q13**

| Element | Li | Be | B | C | N | O | F |
|---------|-----|-----|-----|-----|-----|-----|-----|
| Atomic radius/nm | 0.134 | 0.125 | 0.090 | 0.077 | 0.075 | 0.073 | 0.071 |

| Element | Na | Mg | Al | Si | P | S | Cl |
|---------|-----|-----|-----|-----|-----|-----|-----|
| Atomic radius/nm | 0.154 | 0.145 | 0.130 | 0.118 | 0.110 | 0.102 | 0.099 |

**a)** Using **only** the elements in this table, select:
   **(i)** an element with **both** metallic and non-metallic properties;
   **(ii)** the element with the largest first ionization energy;
   **(iii)** an element with a giant molecular structure.
**b)** Explain what causes the general **decrease** in atomic radii across each period.
**c)** Predict and explain whether a sodium **ion** is *larger*, *smaller* or the *same size* as a sodium **atom**.

*OCR Foundation Chemistry Jan 2001 Q6*

# KEY SKILLS **ASSIGNMENT**

## FINDING OUT ABOUT NEW ELEMENTS

**The Periodic Table in the 1930s (below) looked different from the version we use today. To start with, it contained only 88 elements. It finished at element 92 and had four gaps – at atomic numbers 43, 61, 85 and 87.**

**1** **a)** Briefly explain how knowledge of the Periodic Table helped scientists to discover new elements.

**b)** The four elements for which chemists in the 1930s left gaps are (i) technetium, (ii) promethium, (iii) astatine and (iv) francium. For each, note its position in a modern Periodic Table, and predict two physical and two chemical properties. Explain the reason for each prediction.

At that time, the elements beyond uranium (atomic number 92) were unknown. But scientists were convinced that they did exist and that, sooner or later, they would be discovered. The elements of atomic number 93 and beyond are now known as the **transuranium elements**, meaning the elements across the period from uranium. From the positions of these elements in the Periodic Table, scientists could quite confidently predict their properties.

**2** **a)** Note down the block of a modern Periodic Table in which uranium is present.

**b)** Explain why scientists supposed that elements beyond 92 were radioactive.

To start with, the scientists made the mistake of looking for another set of d-block elements. This was because they had placed uranium in the Periodic Table under tungsten (symbol W, atomic number 74).

They assumed that the next element, number 93, would resemble rhenium, which is next to tungsten. They did realize, though, that the element would be highly radioactive and expected it to pose experimental and safety problems.

The investigators believed that it would be possible to produce element 93 by carrying out a **transmutation**. This is a process of changing one element into another, achieved by bombarding target atoms with a beam of subatomic particles. A transmutation to produce element 93 had to wait until 1940, when Edwin McMillan and Philip Abelson succeeded in making and identifying it.

$$^{238}_{92}U + ^{1}_{0}n \rightarrow ^{239}_{92}U$$
uranium-238 + neutron → uranium-239

This decays with a half-life (the time taken for half the element to decay) of 23 minutes to give:

$$^{239}_{92}U \rightarrow ^{239}_{93}Np + ^{0}_{-1}e$$

The new element was named neptunium (Np) and it was found to have properties just like those of uranium. However, it was a surprise to find that its properties differed markedly from those of rhenium. The half-life of neptunium–239 is 2.3 days.

Soon afterwards, element number 94, plutonium (Pu), was discovered by Glenn Seaborg, and it too had properties similar to uranium, but not like those of osmium (number 76), as had been predicted.

$$^{239}_{93}Np \rightarrow ^{239}_{94}Pu + ^{0}_{-1}e$$

Researchers at that time realized that their Periodic Table was incorrect and needed refining. They concluded that the elements after actinium were actually f-block elements rather than d-block elements, and formed a series (atomic numbers 89–104 inclusive) called the **actinides**, because they have similar properties to actinium.

In 1944, elements 95 and 96 were discovered. These were found to resemble the corresponding f-block elements europium and gadolinium, and provided further proof that the transuranium elements were f-block elements.

| 1 H | | | | | | | | | | | | | | | | | 2 He |
|---|---|---|---|---|---|---|---|---|---|---|---|---|---|---|---|---|---|
| 3 Li | 4 Be | | | | | | | | | | | 5 B | 6 C | 7 N | 8 O | 9 F | 10 Ne |
| 11 Na | 12 Mg | | | | | | | | | | | 13 Al | 14 Si | 15 P | 16 S | 17 Cl | 18 Ar |
| 19 K | 20 Ca | 21 Sc | 22 Ti | 23 V | 24 Cr | 25 Mn | 26 Fe | 27 Co | 28 Ni | 29 Cu | 30 Zn | 31 Ga | 32 Ge | 33 As | 34 Se | 35 Br | 36 Kr |
| 37 Rb | 38 Sr | 39 Y | 40 Zr | 41 Nb | 42 Mo | (43) | 44 Ru | 45 Rh | 46 Pd | 47 Ag | 48 Cd | 49 In | 50 Sn | 51 Sb | 52 Te | 53 I | 54 Xe |
| 55 Cs | 56 Ba | 57 La | 72 Hf | 73 Ta | 74 W | 75 Re | 76 Os | 77 Ir | 78 Pt | 79 Au | 80 Hg | 81 Tl | 82 Pb | 83 Bi | 84 Po | (85) | 86 Rn |
| (87) | 88 Ra | 89 Ac | 90 Th | 91 Pa | 92 U | 93 | 94 | 95 | | | | | | | | | |

| | 58 Ce | 59 Pr | 60 Nd | (61) | 62 Sm | 63 Eu | 64 Gd | 65 Tb | 66 Dy | 67 Ho | 68 Er | 69 Tm | 70 Yb | 71 Lu |
|---|---|---|---|---|---|---|---|---|---|---|---|---|---|---|

**Fig 5.A1** The Periodic Table in the 1930s. Notice that four elements were undiscovered and that the actinides (atomic numbers 90 and 92) are in the wrong place

Extensive studies of the chemistry of plutonium show it to have chemical properties typical of a metal. It is a very reactive metal that will reduce most non-metals, such as oxygen, nitrogen, the halogens and hydrogen. It has several common oxidation states.

In terms of its physical properties, plutonium is rather a poor metal, having a low thermal and electrical conductivity, and being quite brittle. As it is highly radioactive, it is self-heating, which damages the metal.

Those f-block elements with only a small number of 5f electrons have high electrical conductivity, since these electrons are delocalized (see Chapter 7). The f-block elements, with a large number of 5f electrons, have a low electrical conductivity since their 5f electrons are localized.

**3** Why did investigators think that neptunium would resemble the element rhenium?

**4** The electron configuration for plutonium is $[Rn]5f^67s^2$.

**a)** Suggest why a lot of research work has been carried out on the chemistry of uranium and plutonium since the 1940s.
**b)** Suggest possible oxidation states for plutonium.
**c)** Predict the level of electrical conductivity (high, low or intermediate) of plutonium.
Since the 1940s, many other elements have been made, as shown in Table 5.A1.

**Table 5.A1 Some of the transuranium elements which have been artificially synthesized**

| Name | Symbol | Atomic number | Nuclear reaction |
|---|---|---|---|
| Neptunium | Np | 93 | $^{238}_{92}U + ^{1}_{0}n \rightarrow ^{239}_{93}Np + ^{0}_{-1}e$ |
| Plutonium | Pu | 94 | $^{238}_{92}U + ^{2}_{1}H \rightarrow ^{238}_{93}Np + 2^{1}_{0}n$ |
| | | | $^{238}_{93}Np \rightarrow ^{238}_{94}Pu + ^{0}_{-1}e$ |
| Americium | Am | 95 | $^{239}_{94}Pu + ^{1}_{0}n \rightarrow ^{240}_{95}Am + ^{0}_{-1}e$ |
| Berkelium | Bk | 97 | $^{241}_{95}Am + ^{4}_{2}He \rightarrow ^{243}_{97}Bk + 2^{1}_{0}n$ |
| Californium | Cf | 98 | $^{242}_{96}Cm + ^{4}_{2}He \rightarrow ^{245}_{98}Cf + ^{1}_{0}n$ |
| Einsteinium | Es | 99 | $^{238}_{92}U + 15^{1}_{0}n \rightarrow ^{253}_{99}Es + 7^{0}_{-1}e$ |
| Unniquadium | Unq | 104 | $^{249}_{98}Cf + ^{12}_{6}C \rightarrow ^{257}_{104}Unq + 4^{1}_{0}n$ |
| Unipentium | Unp | 105 | $^{249}_{98}Cf + ^{15}_{7}N \rightarrow ^{260}_{105}Unp + 4^{1}_{0}n$ |

Sometimes, only minute quantities of these new elements have been made. It seems extraordinary, but the scientists who first made element 109 produced only one atom of it, and within $5 \times 10^{-3}$ seconds it had decayed!

**5** To which block of the Periodic Table does element 109 belong?

As soon as a new element is made, its name and symbol have to be decided on. To start with, this caused no difficulty. Like uranium, the two elements 93 and 94 were named after planets, neptunium and plutonium. The next two were named in a similar way to the elements directly above them in the Periodic Table.

**6 a)** What are the names of elements 95 and 96 and those directly above them in the Periodic Table? Explain how their names relate to each other.
**b)** Describe, with reasons, the level of electrical conductivity (high, low or intermediate) you would expect for element 101.

**7** Scientists need to calculate the amount of radioactive isotope that remains after a period of time. It is proposed to leave a sample of 1.00g of neptunium-239 in a storage tank for 23 days. How much of the neptunium will remain?

**8** What is the atomic number of the second element in the **fourth** transition series?

**9 a)** Suggest some of the problems involved in studying the chemistry of the transuranium elements.
**b)** The transuranium elements have been synthesized by bombarding target atoms with particles. Study the information given for the synthesis of the transuranium elements. Using this information, copy and complete the following table.

| Bombarding particle | Element bombarded | New element | Other product |
|---|---|---|---|
| $^{1}_{0}n$ | ? | $^{240}_{?}Am$ | ? |
| $^{4}_{2}He$ | ? | $^{?}_{97}Bk$ | ? |
| ? | $^{238}_{92}U$ | ? | $^{0}_{-1}e$ |
| $^{12}_{6}C$ | $^{249}_{98}Cf$ | ? | ? |
| ? | ? | $^{260}_{105}Unp$ | $^{1}_{0}n$ |

Hint: Chapter 2, p.41, may help you.

**10** Find out about how fermium was synthesized, and write a short explanation.

**11** The early work on the transmutation of elements was connected with the Manhattan Project, which was concerned with the development of an atomic bomb during the Second World War.

Working in a small group, find out the reasons why the scientists thought this project was so important. Discuss whether you think it was right or wrong for scientists to work in this type of research.

# 06

# Gases, liquids, solids and solutions

Airbags are a familiar addition to car safety these days. But a crash happens very fast, so how does the airbag inflate in time to protect the driver from injury? One type of airbag uses a very fast gas-producing chemical reaction to blow up the bag.

First, the rapid deceleration of the car triggers an electric current, and the energy from this current explodes a detonator cap. The explosion makes the solid substance sodium azide ($NaN_3$) decompose and release nitrogen gas extremely rapidly. The nitrogen inflates the airbag, cushioning the driver and reducing the likelihood of injury.

This is the equation for the decomposition of sodium azide:

$$2NaN_3(s) \rightarrow 2Na(s) + 3N_2(g)$$

Sodium azide is a solid, so its particles are arranged very close together. The nitrogen it produces is a gas with widely spaced particles. For a relatively small amount of solid, the volume of gas that inflates the airbag is very large. At the same time, because the particles are now widely spaced, they can be compressed and so cushion the driver. The whole process takes about 40 milliseconds. Then, the specially designed bag deflates rapidly, so that the driver has vision and movement after the crash. This also applies where airbags are fitted for other passengers.

Airbags work, therefore, because of the different arrangements of particles in the different states of matter, an application of chemistry that regularly saves lives.

## 1 GASES

The Greeks thought that air was one of four elements that made up the Earth (see the start of Chapter 2). Later, alchemists produced what they called 'airs' or 'vapours' in their efforts to change common metals, such as iron and lead, into gold. It was a Dutch scientist, Jan van Helmont, who realized that not all these 'airs' were the same, though they would fill any container they were put into. The Greeks had a name for the substance that they thought the gods had changed into the four elements of Earth – chaos. In 1624, van Helmont used this word for 'airs'. When pronounced in Dutch the word became 'gas'. It is now recognized as a **state of matter**, together with the other two states, solid and liquid.

### VOLUMES OF GASES

Towards the end of the eighteenth century, chemists were experimenting with reactions that involved gases. The French chemist Joseph Gay-Lussac studied a large number of these reactions and realized that when gases reacted their volumes were in a simple whole-number ratio, provided he measured the volumes at the same temperature and pressure. If the product was a gas this volume too was in a simple ratio to the reactants.

When Gay-Lussac reacted 5 $dm^3$ of hydrogen gas with 5 $dm^3$ of chlorine gas, he produced 10 $dm^3$ of hydrogen chloride. The simple whole-number ratio for this reaction is:

| 1 volume of hydrogen | : | 1 volume of chlorine | : | 2 volumes of hydrogen chloride |

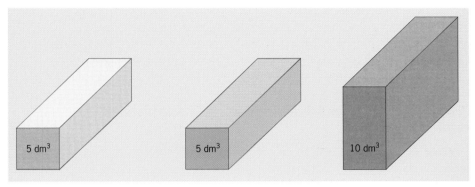

**Fig 6.1**

In 1808, Gay-Lussac summed up his experimental observations in his law of combining volumes:

> **When gases react, they do so in volumes that bear a simple whole-number ratio to each other and to the volumes of any gaseous products, provided the temperature and pressure are the same when the volumes are measured.**

## AVOGADRO'S LAW

In 1811, an Italian named Amedeo Avogadro put forward a theory to explain the experimental findings of Gay-Lussac:

> **Equal volumes of different gases contain equal numbers of molecules at the same temperature and pressure.**

It took about 50 years of confusion before Avogadro's theory was accepted. One of the stumbling blocks was John Dalton, who in 1803 had produced his atomic theory (see page 29). He thought that all gaseous elements were made up of separate atoms. He did not understand, for instance, that hydrogen and chlorine were made up of $H_2$ molecules and $Cl_2$ molecules. If Dalton had been right, and hydrogen and chlorine had been made up of atoms, then Avogadro's theory would have suggested that these atoms were split.

 **REMEMBER THIS**

A decimetre is 10 cm. So a cubic decimetre ($dm^3$) is $10 \times 10 \times 10 = 1000$ $cm^3$. It is the same as a litre.

**Fig 6.2** Joseph Gay-Lussac (1778–1850) is most famous for his experiments with gases. He also made the highest balloon flight of his day. At over 7 kilometres, the record stood for a very long time

**?  TEST YOURSELF**

**A** What is the simple whole-number ratio of gas volumes for the reaction of 5 $dm^3$ of hydrogen with 2.5 $dm^3$ of oxygen producing 5 $dm^3$ of steam?

**Fig 6.3** Amedeo Avogadro (1776–1856), lawyer and later a professor of physics. It wasn't until four years after his death that another distinguished Italian, Stanislao Cannizzaro, rediscovered Avogadro's hypothesis and used it to explain Gay-Lussac's observations

Let's try to follow Dalton's argument to see if it is logical:

| 1 volume | + | 1 volume | → | 2 volumes of |
| of hydrogen | | of chlorine | | hydrogen chloride |

Since, according to Avogadro's theory, each volume must contain the same number of particles if they are measured at the same temperature and pressure, then:

| 1 million atoms | + | 1 million atoms | → | 2 million |
| of hydrogen | | of chlorine | | molecules of |
| | | | | hydrogen chloride |

So:  1 atom H  +  1 atom Cl  →  2 molecules HCl

**Fig 6.4**

***This is wrong***. If it were correct, there would have to be two hydrogen atoms created from one. This confusion held up progress in chemistry for half a century.

However, if we accept that hydrogen and chlorine are made up of diatomic molecules, then the observations of Gay-Lussac are explained easily by Avogadro:

| 1 million | + | 1 million | → | 2 million |
| molecules $H_2$ | | molecules $Cl_2$ | | molecules HCl |

So:  1 molecule $H_2$  +  1 molecule $Cl_2$  →  2 molecules HCl

**Fig 6.5**

Avogadro's theory is also called Avogadro's law, because gases at low pressures obey it.

## MOLAR GAS VOLUME

We first introduced the **mole** in Chapter 1. It is the unit chemists use for the **amount** of a substance and is $\mathbf{6.02 \times 10^{23}}$ **particles**. It follows from Avogadro's law that the volume of one mole of any gas – **the molar gas volume $V_m$** – must be the same under identical conditions of temperature and pressure, because we are dealing with the same number of particles.

At 273 K (0 °C) and a pressure of 101 kilopascals, kPa (1 atmosphere), 1 mole of gas occupies 22.4 dm³. These conditions of temperature and pressure are called **standard temperature and pressure (s.t.p.)**. So:

**The molar gas volume at s.t.p. is 22.4 dm³.**

**At 298 K and 101 kPa (sometimes called room conditions), the molar gas volume is 24 dm³.**

**? TEST YOURSELF**

**B** Dalton thought that the formula of water was HO. It was known that two volumes of hydrogen reacted with one volume of oxygen to give two volumes of steam. Explain how this observation suggests that the formula of water is $H_2O$.

**✔ REMEMBER THIS**

One mole is the amount of substance that contains as many particles as there are atoms in exactly 12 g of carbon-12. The number of particles in a mole is $6.02 \times 10^{23}$ and is called the **Avogadro constant**. You can find out more about the mole on page 5.

**? TEST YOURSELF**

**C** What is 298 K in Celsius (°C)?

## EXAMPLE

**Q** What is the volume at s.t.p. of:
**a)** 1 mol hydrogen, **b)** 2 mol hydrogen chloride, **c)** 0.5 mol oxygen?

Remember: mol is the shortened form of the mole unit.

**A  a)** Since one mole of any gas occupies 22.4 dm$^3$ at s.t.p., the volume occupied by hydrogen must be 22.4 dm$^3$.

**b)** The amount of gas is 2 moles, so the volume of hydrogen chloride is double the molar volume at s.t.p., which is 44.8 dm$^3$ (22.4 × 2).

**c)** Oxygen will occupy half the molar gas volume, since there is 0.5 mole present.

$$22.4 \times 0.5 = 11.2 \text{ dm}^3$$

## ? TEST YOURSELF

**D** The volume of an oxygen cylinder is 6 dm$^3$. How many moles of gas are left in the cylinder when it is empty under room conditions (298 K and 101 kPa)? Hint: When the cylinder is 'empty', the oxygen gas left in it is at the pressure and temperature of the surrounding room.

## CALCULATIONS FROM CHEMICAL EQUATIONS

An understanding of the laws of Gay-Lussac and Avogadro is very useful to chemists. It allows them to use equations to work out the volumes of gases required for a particular reaction, or to predict the volumes of gases that are produced by a given reaction.

**SEE QUESTION 1.**

## EXAMPLE

**Q** A coal-fired power station uses fuel that contains 0.5 per cent sulphur by mass. What volume of sulphur dioxide at 298 K and 101 kPa is released into the atmosphere by burning 100 tonnes of coal?

**A  Step 1.** Write the balanced equation for the reaction:

$$S(s) + O_2(g) \rightarrow SO_2(g)$$

(Remember: (s) = solid, (l) = liquid, (g) = gas.)

**Step 2.** Convert the equation to amounts:

$$1 \text{ mol S gives 1 mol SO}_2$$

Note: In this example we do not need to know the number of moles of $O_2$.

**Step 3.** Work out the amount being used (1 tonne is 1 000 000 g):
100 tonnes of coal contains 0.5 per cent S.
So the mass of S = 0.5 tonnes = 500 000 g.

$$\frac{\text{Moles of S}}{\text{in 0.5 tonnes}} = \frac{\text{mass (in grams)}}{\text{mass of one mole (in grams)}}$$

$$= \frac{500\,000 \text{ g}}{32 \text{ g}}$$

$$= 15\,625 \text{ mol (ignoring sig. figs)}$$

**Step 4.** Scale the amounts in the equation:

$$1 \text{ mol S produces 1 mol SO}_2$$
$$15\,625 \text{ mol S produces } 15\,625 \text{ mol SO}_2$$

**Step 5.** Calculate the volume of gas:

1 mole of gas occupies 24 dm$^3$ at 298 K and 101 kPa.

$$\text{Volume of SO}_2 = 15\,625 \times 24 = 375\,000 \text{ dm}^3$$
$$= 380\,000 \text{ dm}^3 \text{ (to two significant figures)}$$

**Q** What volume of oxygen is required to burn 600 cm$^3$ of natural gas under room conditions? Assume that natural gas is pure methane.

**A  Step 1.** Write the balanced equation for the reaction:

$$CH_4(g) + 2O_2(g) \rightarrow CO_2(g) + 2H_2O(l)$$

**Step 2.** Convert the equation to volumes:

Since:  1 mol $CH_4$ reacts with 2 mol $O_2$
1 volume $CH_4$ reacts with 2 volumes $O_2$

**Step 3.** Use the simple whole number ratio of volumes to work out the actual volume used:

$$600 \text{ cm}^3 CH_4 \text{ reacts with } 2 \times 600 \text{ cm}^3 O_2$$

So the volume of oxygen used is 1200 cm$^3$.

## ? TEST YOURSELF

**E** Petrol is mainly a mixture of hydrocarbons, of which octane, $C_8H_{18}$, is typical. Calculate the volume of carbon dioxide produced by burning 1.0 kg of octane at room conditions.

## ? TEST YOURSELF

**F** Explosives work by producing large numbers of gaseous molecules. Nitroglycerine detonates according to this equation:

$$4C_3H_5(NO_3)_3(l) \rightarrow 12CO_2(g) + 10H_2O(g) + 6N_2(g) + O_2(g)$$

A sample of nitroglycerine explodes to produce 1 dm$^3$ of oxygen. What will be the total volume of gas produced during the explosion, assuming that pressure and temperature remain constant?

# 2 THE GAS LAWS

Gas is in some ways the simplest of the three states of matter. Through doing countless experiments, scientists have investigated the physical properties of different gases. They have found that, with minor variations, all gases behave in a similar way under room conditions and that there are definite relationships between pressure, volume, temperature and amount. These relationships have been defined in what we call the **gas laws**.

## THE RELATIONSHIP BETWEEN PRESSURE AND VOLUME: BOYLE'S LAW

If you have read the beginning of Chapter 2 (page 29) you will know that Robert Boyle had a deep influence on modern chemistry in urging chemists to do experiments to find elements. He also conducted his own experiments on gases.

In 1662 he carried out a series of experiments using a glass tube 5 metres high, shaped like a letter J and closed at the short end. He trapped an amount of air in the short end of the J tube by pouring mercury into the other end. He found that the more mercury he added, the smaller the volume of air became. He had thus discovered that the volume of air depended on the pressure of the mercury upon it, a finding he summarized in what we now know as **Boyle's law**:

> **The volume of a fixed amount of gas is inversely proportional to its pressure at constant temperature.**

We can express this mathematically as:

$$p \propto \frac{1}{V} \quad \text{or} \quad pV = \text{a constant}$$

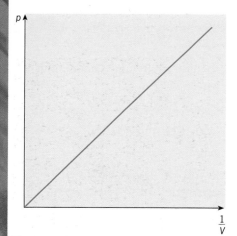

Fig 6.6 Graph showing the effect of pressure on the volume of a fixed amount of gas at constant temperature

(a) At atmospheric pressure, the trapped gas occupies a particular volume

(b) The pressure is now greater than atmospheric pressure because of the different heights of the mercury, and the volume of trapped air has decreased

(c) The pressure has increased even more and the volume of the gas is even smaller

Fig 6.7 Robert Boyle's J tube experiment. Boyle was the first to carry out quantitative experiments on gases. He measured the volume of gas and the difference in heights of the mercury columns to work out the pressure. Note that the temperature is kept constant throughout the experiment

We can show this graphically by plotting pressure against 1/volume as in Fig 6.6.

So if we double the pressure on a gas the volume will halve, and if the volume increases four times then the pressure on the gas must have decreased to a quarter, provided the temperature stays constant.

► **EXTENSION** Breathing and Boyle's law

When we are at rest we breathe in and out about twelve times a minute. As we breathe in, the volume of our chest cavity expands, increasing the volume of our lungs. This causes the pressure of air inside the lungs to decrease. Since the pressure of the air outside our bodies is now greater than that inside, air flows in. The reverse happens when we breathe out: the chest volume reduces, which increases the pressure of air in the lungs. The air is now at a higher pressure than that outside, so air flows out.

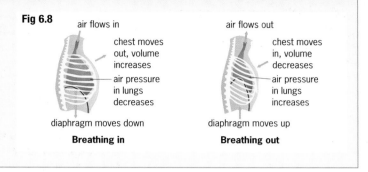

**Fig 6.8**

air flows in
chest moves out, volume increases
air pressure in lungs decreases
diaphragm moves down
**Breathing in**

air flows out
chest moves in, volume decreases
air pressure in lungs increases
diaphragm moves up
**Breathing out**

Pressure $p$ times volume $V$ is always equal to the same constant for a fixed amount of gas at the same temperature. So we can write:

$$p_1V_1 = k = p_2V_2 \text{ or } p_1V_1 = p_2V_2$$

where $p_1$ and $V_1$ are one set of conditions, and $p_2$ and $V_2$ are another set.

We can use this equation, which comes from Boyle's law, to work out the effect of changing the pressure on a particular volume of gas. It can also be used to find the effect on the pressure exerted by a gas of changing the volume of that gas.

● **EXAMPLE**

**Q** A 400 cm$^3$ mixture of petrol vapour and air is taken into the cylinder of a car engine at 200 °C and a pressure of 100 kPa. The piston compresses this gaseous mixture to 50 cm$^3$. What is the pressure of the compressed gas if the temperature does not change?

**A**

| Initial conditions | Final conditions |
| --- | --- |
| $p_1 = 100$ kPa | $p_2 = ?$ |
| $V_1 = 400$ cm$^3$ | $V_2 = 50$ cm$^3$ |

Since $\quad p_1V_1 = p_2V_2$:

$100 \times 400 = p_2 \times 50$

$p_2 = 800$ kPa

**? TEST YOURSELF**

▌ A 5 cm$^3$ bubble of gas rises up from ocean floor, where it is at a pressure of 2000 kPa. The pressure just below the surface is 100 kPa. What is the volume of the bubble when it reaches this point?

## THE RELATIONSHIP BETWEEN VOLUME AND TEMPERATURE: CHARLES' LAW

During the eighteenth century, several experimenters, including Boyle, noticed that temperature had an effect on the volume of a gas. In about 1800, two French scientists carried out experiments on the relationship between volume and temperature, as a spin-off from their ballooning activities. Jacques Charles was the first and Gay-Lussac, working independently, was the second.

Both workers discovered that, when a fixed amount of gas was kept at constant pressure, the volume varied in proportion to the temperature. Gay-Lussac also noticed that if the volume of gas at 0 °C was taken, then for every 1 °C drop in temperature the volume decreased by 1/273 under the same conditions of pressure and amount. This suggested that at −273 °C there would be a zero volume of gas.

It was Lord Kelvin, a British scientist, who realized the significance of this 50 years later. He called −273 °C the **absolute zero** temperature, below

**Fig 6.9** During 1783, hot-air balloon flights became possible thanks to the enterprising Montgolfier brothers. In the same year, Jacques Charles filled a balloon with hydrogen, as seen here, for the first ascent in a hydrogen balloon and the second-ever manned balloon flight

**TEST YOURSELF**

**J** What are the values of the following temperatures on the absolute temperature scale:
**a)** 25 °C;
**b)** 100 °C;
**c)** –50 °C?
Hint: Temperature in K = 273 + temperature in °C.

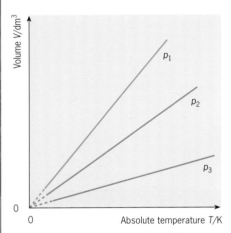

**Fig 6.10** Three graphs plotted at three different pressures ($p_1$, $p_2$, $p_3$) demonstrate Charles' law. All gases condense before absolute zero and the dashed portion of the graph shows the extrapolation, assuming that gases do not condense

**TEST YOURSELF**

**K** A sample of carbon dioxide in a syringe occupies 50.0 cm³ at 20.0 °C. What will the volume of $CO_2$ be at 100.0 °C if the pressure is constant?

which it was impossible to go. This gave rise to a new temperature scale – the absolute temperature scale. Absolute zero has a value of 0 K (zero kelvin), and a 1 K rise in temperature is the same as a 1 °C rise in temperature. This means that at 0 °C the temperature on the Kelvin scale is 273 K.

The relationship between volume and temperature is called **Charles' law**:

> **The volume of a fixed amount of gas is directly proportional to its absolute temperature at constant pressure.**

This can be expressed mathematically:

$$V \propto T \quad \text{or} \quad V = kT$$

where $k$ is a constant and T is absolute temperature. Note that $k$ is not the same as the constant in Boyle's law.

For a fixed amount of gas at the same pressure, the same constant will apply, so we can write:

$$\frac{V_1}{T_1} = k = \frac{V_2}{T_2} \quad \text{so} \quad \frac{V_1}{T_1} = \frac{V_2}{T_2}$$

**EXAMPLE**

**Q** A balloon is filled with 1250 cm³ of helium gas at 25.0 °C. Overnight the temperature cools to 10.0 °C. What is the new volume of the balloon, assuming the pressure is constant?

**A** The temperatures must first be converted into absolute temperatures. To do this, 273 is added to the temperature in Celsius so:

$$25\,°C = 273 + 25 = 298\ K, \text{ and } 10\,°C = 273 + 10 = 283\ K$$

**Initial conditions**  **Final conditions**
$V_1 = 1250$ cm³     $V_2 = ?$ cm³
$T_1 = 298$ K       $T_2 = 283$ K

Since:
$$\frac{V_1}{T_1} = \frac{V_2}{T_2}$$

$$\frac{1250}{298} = \frac{V_2}{283}$$

$$V_2 = \frac{1250 \times 283}{298}$$

$$V_2 = 1190 \text{ cm}^3 \text{ (to three significant figures)}$$

## THE IDEAL GAS EQUATION

Another way of expressing Avogadro's law on page 119 is:

> **The volume of a gas is directly proportional to its amount at constant temperature and pressure.**

This can be expressed mathematically as:

$$V \propto n$$

where $n$ = amount in moles.

So the volume of a gas is related to three other properties: amount, temperature and pressure.

Avogadro's law     $V \propto n$       (constant $p$ and $T$)
Boyle's law        $V \propto 1/p$     (constant $n$ and $T$)
Charles' law       $V \propto T$       (constant $n$ and $p$)

Combining these:

$$V \propto \frac{nT}{p} \quad \text{or} \quad V = \frac{RnT}{p}$$

where $R$ is a constant called the **gas constant**. Rearranging this equation gives the **ideal gas equation**:

$$pV = nRT$$

When $p$ is measured in pascals (Pa), $V$ in cubic metres ($m^3$), $n$ in moles (mol) and $T$ in kelvins (K), **$R$ is 8.31 J K$^{-1}$ mol$^{-1}$**. These units are internationally agreed and are called **SI units**, after the French words *Système Internationale*.

Gases that obey the ideal gas equation exactly are called **ideal gases**. In reality, no gas is an ideal gas, but the variations from ideal behaviour are quite small over wide ranges of temperature and pressure. This enables us to use the ideal gas equation as a very useful tool to relate the four properties of volume, temperature, pressure and amount.

---

### ● EXAMPLE

**Q** At 90.0 °C how many moles of nitrogen are present in a flask of volume 750.0 cm$^3$ at 100 kPa pressure?

**A** First convert the units into SI units.

The volume in cm$^3$ must be converted into m$^3$. There are $100^3$ cm$^3$ in 1 m$^3$ (i.e. $10^6$ cm$^3$), so:

$$V = 750.0 \times 10^{-6} = 7.50 \times 10^{-4} \text{ m}^3$$

$$p = 100 \times 10^3 \text{ Pa}$$

$$T = 273 + 90.0 = 363 \text{ K}$$

Then substitute into the ideal gas equation:

$$pV = nRT \text{ so } n = \frac{pV}{RT}$$

$$n = \frac{100 \times 10^3 \times 7.50 \times 10^{-4}}{8.31 \times 363}$$

$$n = 0.0249 \text{ mol}$$

---

### ? TEST YOURSELF

**a)** What is the volume in cm$^3$ of 2 mol fluorine gas at 27 °C and 100 kPa?

**b)** The balloon used by Jacques Charles for his historic flight contained 1300 moles of hydrogen. What was its volume at 17 °C and 100 kPa?

---

## USING THE IDEAL GAS EQUATION TO CALCULATE RELATIVE MOLECULAR MASS

Nowadays, mass spectrometers are used to measure relative molecular mass $M_r$ accurately. However, when this instrument is not available the $M_r$ of gases and volatile liquids can be calculated using the ideal gas equation:

$$pV = \frac{mRT}{M_r}$$

One way of doing this is to use a gas syringe (Fig 6.11). If the substance is a gas, it can be passed into a syringe of known mass and the syringe reweighed to give the mass of gas. The volume of gas is read from the syringe and the temperature and pressure are measured. Then these values are simply inserted into the ideal gas equation.

---

### ✔ REMEMBER THIS

Amount in moles:

$$n = \frac{\text{mass in grams}}{\text{mass of one mole (in grams)}}$$

and $M_r$ is numerically equal to the mass of one mole: see pages 6 and 7.

Also, see Chapter 11, page 237, for information about the mass spectrometer.

**Fig 6.11** Using a gas syringe to measure the $M_r$ of a gas

---

### EXAMPLE

**Q** The gas propane contains hydrogen and carbon only. When 100.0 cm³ of it is passed into a syringe, the mass is found to be 0.178 g. The pressure and temperature are 298 K and 100 kPa, respectively. Calculate the $M_r$.

**A** The measurements are converted into SI units, with the exception of the mass, which is left as grams. (The SI unit for mass is the kilogram.)

$$p = 100 \times 10^3 \text{ kPa}, V = 100 \times 10^{-6} \text{ m}^3, m = 0.178 \text{ g},$$

$$M_r = ?, R = 8.31 \text{ J K}^{-1}\text{mol}^{-1} \text{ and } T = 298 \text{ K}.$$

Insert these values into the ideal gas equation:

$$pV = \frac{mRT}{M_r}$$

$$100 \times 10^3 \times 100 \times 10^{-6} = \frac{0.178}{M_r} \times 8.31 \times 298$$

Rearranging: $\qquad M_r = \dfrac{0.178 \times 8.31 \times 298}{100 \times 10^3 \times 100 \times 10^{-6}}$

$$M_r = 44.1$$

---

### ? TEST YOURSELF

**M** A gaseous compound containing sulphur and fluorine is used to trace leaks in air conditioning systems. If 100.0 cm³ has a mass of 0.586 g at 27.0 °C and 100.0 kPa pressure, what is the $M_r$ of the compound?

### ? TEST YOURSELF

**N** Often a small volume of air is in the syringe. Why is it important to know the exact volume of this air?

The same method is used to determine the $M_r$ of a volatile liquid, only this time the end of the syringe is sealed with a self-sealing rubber cap through which a known mass of the liquid under investigation is injected. Since the liquid is volatile it has a low boiling point, and if the syringe is heated to above the boiling point the volume of the gas produced can be measured on the syringe. In practice, there is usually a small volume of air in the syringe. So long as this volume is known at the temperature of the experiment, it can be subtracted from the volume of the vaporized sample.

**Fig 6.12** The apparatus used to measure the $M_r$ of a volatile liquid

## EXAMPLE

**Q** The following results were obtained from a sample of volatile liquid injected into a syringe:

| | |
|---|---|
| Mass of liquid sample | 0.180 g |
| Volume of air | 10 cm$^3$ |
| Volume of vapour + air | 100 cm$^3$ |
| Temperature | 80 °C |
| Pressure | 100 600 Pa |

Calculate the $M_r$ of the sample.

**A** Volume of vapour = 100 cm$^3$ − 10 cm$^3$ = 90 cm$^3$.

Once again, the measurements are converted into SI units, with the exception of the mass which is left as grams.

$p = 100\,600$ Pa, $V = 90 \times 10^{-6}$ m$^3$, $m = 0.180$ g,

$M_r = ?$, $R = 8.31$ J K$^{-1}$ mol$^{-1}$ and $T = 273 + 80 = 353$ K.

Insert these values into the ideal gas equation:

$$pV = \frac{mRT}{M_r}$$

$$100\,600 \times 90 \times 10^{-6} = \frac{0.180}{M_r} \times 8.31 \times 353$$

Rearranging:
$$M_r = \frac{0.180 \times 8.31 \times 353}{100\,600 \times 90 \times 10^{-6}}$$

$$M_r = 58$$

**? TEST YOURSELF**

**O** In a gas syringe, 0.121 g of an alcohol is vaporized to produce 80 cm$^3$ of gas at 370 K at 101 kPa pressure. Calculate the $M_r$ of the alcohol.

**✔ REMEMBER THIS**

A scientist puts forward a model to explain observations and, so long as this model explains all the known observations, it is accepted by the scientific community. (See Chapter 2, page 30).
Ideal gases obey the ideal gas equation exactly under all conditions.

## 3 A KINETIC–MOLECULAR MODEL FOR HOW GASES BEHAVE

So far, we have considered experimental observations of the behaviour of gases that relate pressure, volume, temperature and amount through the ideal gas equation. The concise statements of these observations we have called **laws**, because they are true for all gases under a range of conditions.

However, scientists soon began to produce **models** to explain gas behaviour. The simple model that we accept today for the behaviour of ideal gases is called the **kinetic–molecular model** or the **kinetic theory**, after the Greek work *kinein*, to move. Several notable scientists developed the model in the nineteenth century, in particular Ludwig Boltzmann, an Austrian, and James Maxwell, a Scot.

In the kinetic–molecular model, ideal gases are assumed to be made up of particles (atoms, molecules or ions) that:
- are very widely separated;
- have negligible (zero) volume;
- exert no force of attraction on each other;
- move continuously and randomly;
- have perfectly **elastic collisions** with each other and the container walls (this means that there is no net gain or loss of energy in collisions);
- have an average kinetic energy that is directly proportional to the absolute temperature of the sample.

**✔ REMEMBER THIS**

**Kinetic energy** is the energy due to the motion of the particles. Not all molecules move at the same speed, or velocity, at any particular temperature above 0 K, so molecules possess a range of kinetic energies at a particular temperature.
For example, in any sample of gas, some molecules will be moving very fast, while others will be moving fairly slowly. This range of speeds leads to a range of energies. You can read more about this in Chapters 10 (page 186) and 26, in which we discuss the Maxwell–Boltzmann distribution of molecular energies.

Now that we have a model we can see how well it fits the experimental observations – the true test of any model.

**Gases can be compressed** – the kinetic–molecular model explains this because the particles are widely spaced.

**Boyle's law** – the collisions of particles with the container walls cause pressure. Two factors determine the pressure: the number of collisions in a certain time on a specific area, and the force of these collisions. If we keep the temperature and amount of gas particles constant and decrease the volume of the container that the gas is in, more particles will collide with a specific area of container wall, so the pressure will increase.

**Fig 6.13** Boyle's law explained by the kinetic theory. There is a fixed amount of gas at constant temperature. If the pressure is doubled, the volume is decreased by half

**Charles' law** – According to the kinetic–molecular model, the average kinetic energy of the particles is proportional to the absolute temperature. As the temperature of a gas increases, so does the average kinetic energy of its particles. This means that the particles will collide with the container walls more often and with greater force. So doubling the absolute temperature doubles the average kinetic energy of particles in a gas, which leads to a doubling of volume provided the pressure stays constant.

**Fig 6.14** Charles' law explained by the kinetic theory. The amount of gas and pressure remain the same, but as the temperature doubles the average kinetic energy of the molecules also doubles to cause a doubling of the volume

## REAL GASES

So far, we have looked at ideal gases that obey the ideal gas equation exactly under any conditions. In the calculations, we have assumed that gases behave ideally, and at room conditions this assumption is a good one. However, the kinetic–molecular model assumes that particles in a gas have negligible (zero) volume. This is reasonable at room temperature and pressure, when the particles are very widely spaced. But as the particles in a gas squeeze closer together, their volume becomes significant. The kinetic–molecular model also assumes that there are no attractive forces between molecules. This, again, is reasonable provided that the particles are far apart, but, as they draw closer together, forces of attraction are increasingly important. The higher the temperature, the faster the particles move, which also tends to make the forces of attraction between particles negligible.

**REMEMBER THIS**

If there were no attractive forces between particles, gases would never liquefy and liquids would never solidify.

> **Thus the conditions that allow a gas to approach ideal gas behaviour are high temperatures and low pressures.**

Conditions in which real gases do not show ideal behaviour are those that bring the particles close together.

**Fig 6.15(a)** At low pressures the particles are far apart, so their volume is negligible compared with the volume of the gas

**Fig 6.15(b)** At high pressures the gas particles take up a larger proportion of the volume, so the volume of the particles becomes significant. Also, because the particles are closer together, attractive forces can operate

At low temperatures the speed of the particles is slower, so forces of attraction between them can operate. At the boiling point of a substance the forces of attraction are great enough to allow the gas to condense to a liquid. So the closer the temperature of a gas is to its boiling point, the more it will deviate from ideal behaviour. This can be seen for nitrogen gas from the graphs in Fig 6.16. If nitrogen were an ideal gas it would obey the ideal gas equation, $pV = nRT$. So if we had one mole of nitrogen:

$$\frac{pV}{RT} = 1$$

We can plot $pV/RT$ against pressure, and at any pressure the value should be 1. However, at a low temperature of 203 K the graph shows wide deviation from this value, while at higher temperatures the deviation from ideal behaviour is much less.

**Fig 6.16** Graphs of $pV/RT$ against pressure for nitrogen gas at three different temperatures. Notice that as the temperature increases, the gas begins to approach ideal behaviour

## 4 HOW PARTICLES ARE ARRANGED IN LIQUIDS AND SOLIDS

The kinetic–molecular model can be extended to include the liquid and solid states of matter. The particles in a liquid are much closer together than they are in a gas, and because of this they exert considerable attractive forces on each other. The particles are also free to move, but are much slower than in the gas.

Fig 6.17 The particles in a liquid are close together and in constant motion

Fig 6.18 The particles in a solid are very close together and fixed into position, so they can vibrate but not move out of position

---

In the solid state, particles are even closer than liquids, so while it is possible to compress liquids slightly, solids are almost incompressible. The particles are also fixed into position, held by attractive forces, vibrating but not free to move out of position. This three-dimensional arrangement of particles gives solids a definite shape and is called a lattice. Most solids are crystalline, which means that the particles are arranged in an orderly way. You can read more about some of the different attractive forces in crystalline lattices in Chapters 7 and 21.

However, not all solids are crystalline. Glass may have a fixed shape and so we call it a solid, but the arrangement of its particles has the disorder associated with a liquid.

## THE KINETIC–MOLECULAR MODEL USED TO EXPLAIN CHANGES IN STATE

It takes energy to change a substance from a solid to a liquid and then to a gas, as shown in Fig 6.19. As energy is transferred to a solid, the particles vibrate more until they break out of their fixed lattice positions. This is when the solid melts. There is no change in temperature during this time, because all the energy transferred to the substance goes into overcoming the forces that hold the particles in the lattice. The energy needed to change one mole of substance from a solid to a liquid at the melting point is called the **enthalpy change of fusion ($\Delta H_m$)**. When all the solid has melted, the temperature begins to rise and the particles have more energy and move faster.

At the boiling point, the liquid begins to change into a gas and the temperature again stays constant, since all the energy taken in by the liquid is used to overcome the forces that hold the particles close together in the liquid state. The **enthalpy change of vaporization ($\Delta H_b$)** is the energy that must be supplied to change one mole of liquid into a gas at a particular temperature. When all the liquid has boiled to form a gas, the temperature rises again as the particles move faster. The process of changing state is a physical process because no chemical bonds are broken.

---

**? TEST YOURSELF**

**Q** Explain the following in terms of the kinetic–molecular model:
**a)** 1300 dm³ of steam condenses to only 1 dm³ of water.
**b)** Gases are easily compressed, but liquids are only slightly compressible.
**c)** The density of a liquid is much higher than that of a gas.
**d)** A liquid takes the shape of the container it is in.
**e)** Solids do not flow, whereas gases and liquids do.

---

**? TEST YOURSELF**

**R** In Fig 6.19, notice that the flat portion of the heating curve is much less at the melting point ($T_m$) than at the boiling point ($T_b$). What does this tell you about the relative magnitudes of $\Delta H_m$ and $\Delta H_b$?

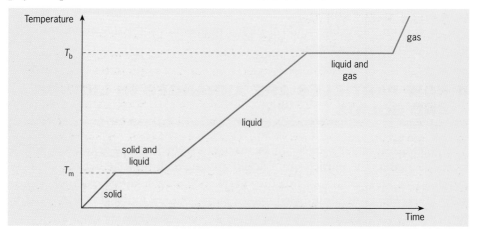

Fig 6.19 A plot of temperature against time as energy is supplied at a constant rate to a substance that starts in the solid state

## ! SCIENCE FEATURE  Liquid crystals: a new state of matter

You often come across liquid crystal displays (LCDs): the numbers displayed on a digital watch use liquid crystals, you may have seen liquid crystal thermometers that change colour depending on the temperature, and laptop computer screens also use liquid crystals. LCDs are commonplace now, but they have only been around for about 30 years.

**Fig 6.20(a)** A laptop computer with a liquid crystal display

**Fig 6.20(b)** A digital watch showing its LCD

**Fig 6.20(c)** The nematic arrangement of molecules in a liquid crystal state. Much of the order of the crystalline state has disappeared, but the molecules are still, on average, parallel to each other

When crystals melt, the ordered crystalline arrangement of their particles breaks down into the disordered liquid state. However, some crystals melt to give particles in an ordered state. This effect was first noticed over 100 years ago by an Austrian botanist, Friedrich Reinitzer. He was heating cholesteryl benzoate to try to determine its molecular structure, and noticed that at 145 °C a cloudy liquid was produced. The liquid went completely clear at 179 °C. This process was reversible on cooling. Reinitzer had discovered a new state of matter that was intermediate between the liquid and solid states.

The liquid crystal state was an interesting curiosity, but thought to be of no practical use since there were no stable compounds that showed the liquid crystal state at room temperature. In the late 1960s, the search was on to replace the heavy cathode-ray tubes in aircraft. (Your television screen is part of a cathode-ray tube.) There was renewed interest in trying to produce chemicals that showed liquid crystal properties at room temperature. In 1972 at Hull University, the world's first stable, room-temperature liquid crystal was made.

Liquid crystals all contain long, thin, rigid organic molecules. There are three different arrangements of molecules in the liquid crystal state. The one that is used in digital watches is called nematic: an electric field alters the orientation of some of the liquid crystal molecules in a thin film; this causes parts of the display to go dark and so gives the shapes of different numbers. Another arrangement of liquid crystals gives different colours at different temperatures and these are used in thermometers.

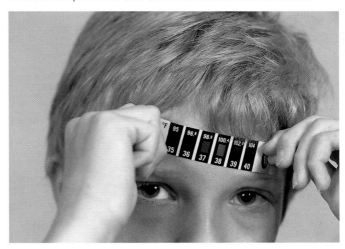

**Fig 6.20(d)** Using a liquid crystal thermometer

Liquid crystals are now very big business and a lot of research is underway to produce new molecules that show this interesting state of matter.

## VAPOUR PRESSURE

Although a liquid changes into a gas at its boiling point, we know that a liquid can change into a gas at temperatures below the boiling point. If a glass of water is left on a windowsill for a week, some of it changes into a gas and the volume of water decreases. The process is **evaporation**. At any temperature, the particles in a liquid have a range of kinetic energies. Some particles have enough energy to break free of the forces that hold them in the liquid state and they evaporate. Since it is the particles with the most kinetic energy that evaporate, the particles with lower kinetic energies are left in the liquid, so the average kinetic energy of the liquid drops and the liquid temperature falls. On a windy day in winter, weather forecasters talk about the wind chill factor. This is because air moving over the skin causes water to evaporate from the skin faster, so the apparent temperature we feel is much colder than the air temperature.

SEE QUESTION 2.

If the liquid is in a closed vessel, then the rate at which the particles escape from the liquid surface equals the rate at which the particles rejoin the liquid. When the rate of evaporation is the same as the rate of condensation, we say that the liquid is in **equilibrium** with its vapour (or gas). The pressure caused by the vapour above the liquid at equilibrium is called its **vapour pressure**. As the temperature increases, so does the average kinetic energy of the particles in the liquid. This means that more particles have sufficient energy to escape into the gas and so the vapour pressure increases. When the vapour pressure of a liquid is the same as the external pressure above its surface, then the liquid boils. For example, water boils when its vapour pressure equals atmospheric pressure. This is why water boils at 70 °C on the top of Mount Everest, where the atmospheric pressure is very much less than it is at sea level.

! **SCIENCE FEATURE** Sublimation and freeze-dried coffee

When a solid changes straight to a gas without passing through the liquid state, the change of state is known as

**sublimation.** The reason that some solids can do this is that they possess relatively high vapour pressures. Iodine and carbon dioxide are two examples of solids with high vapour pressures, so at atmospheric pressure their solid forms sublime.

**Fig 6.21** A flask containing solid iodine in equilibrium with its purple vapour

While we expect ice to melt rather than sublime, it can change straight to water vapour: you may have noticed that frozen puddles gradually shrink in freezing weather. This phenomenon is used in the production of instant coffee. The coffee is brewed and then frozen in a container from which air is removed by a vacuum pump. Lowering the pressure causes the ice in the frozen brewed coffee to sublime. When nearly all the ice has been removed, the coffee is said to be freeze-dried

and is ready for packaging. This method of removing water leaves the flavour molecules intact, and so freeze-dried instant coffee has a much better flavour than coffee that is dried by slow heating.

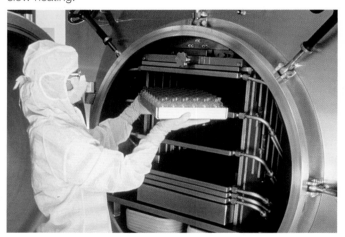

**Fig 6.22** In the freeze-dry process, the sample is put in a chamber and frozen very fast. Then the chamber is evacuated and the sample is slowly heated, when any ice turns straight to vapour

## 5 PHASE DIAGRAMS

Solid, liquid and gas are the three states of matter, but these states can be called by a different name – **phases**.

**A phase is defined as a homogeneous (uniform) portion of matter separated from other portions of matter by a boundary surface.**

A mixture of gases is a single phase, because there is no boundary surface between the different gases. However, oil and water have a boundary layer so there are two phases (Fig 6.23). A solid completely dissolved in a liquid is one phase, because the solution is homogeneous. A mixture of solids has as many phases as there are solids in the mixture, because each solid is separated from the other solids by a definite boundary surface.

**Fig 6.23** Oil floating on water gives two liquid phases, because there is a definite boundary surface

We have seen that both temperature and pressure affect which phase of matter a particular substance exists in. A **phase diagram** shows whether solid, liquid or gas exists at a particular temperature and pressure for a closed system. (A closed system means that no matter can escape or enter.)

## WATER

The three areas in the phase diagram of Fig 6.24 represent ice, water and water vapour. The lines that separate these areas indicate the temperatures and pressures at which two phases can exist in equilibrium.

The line between the water and water vapour phases gives the variation of the vapour pressure of water with temperature. At a pressure of $1.01 \times 10^5$ Pa (normal atmospheric pressure at sea level), water boils at $100\,°C$. The line between the liquid and solid phases shows how the melting point varies with pressure.

For most substances the melting point increases with an increase in pressure, but water is unusual because its melting point decreases as pressure increases.

The line between the solid and vapour phases gives the conditions needed for ice to sublime and explains why it sublimes in the freeze-drying process in instant coffee production described in the Science Feature box.

There is one point on the phase diagram where it is possible for all three phases to coexist in equilibrium. This is called the **triple point**. Notice also that the line between the liquid and vapour phases ends at the **critical point**. Beyond this particular temperature it is not possible to liquefy water vapour. We will return to the critical point later.

## THE UNUSUAL BEHAVIOUR OF WATER – HYDROGEN BONDING

Water is a remarkable liquid. Indeed, if you compare its boiling point to those of the hydrides of other Group 6 elements you could be forgiven for thinking that it should not really be a liquid at all at room temperature (see Fig 6.25).

**Fig 6.25** Boiling points of the Group 6 hydrides

Water's surprisingly high boiling point and melting point results from a special intermolecular bonding known as **hydrogen bonding**. Hydrogen bonds form when hydrogen is covalently bonded to the very electronegative elements oxygen, fluorine and nitrogen.

In water, the hydrogen–oxygen covalent bond is very polar. Hydrogen has only one electron, so when the oxygen atom in water pulls the bonded pair of electrons towards itself, it leaves the hydrogen atom's nucleus of a single proton more exposed. The resulting $\delta+$ on the hydrogen atom is attracted to the lone pair ($\delta-$) of the oxygen atom to form a very strong intermolecular attraction (Fig 6.26).

**? TEST YOURSELF**

**S** How many phases are there in:
**a)** a saturated solution if there is excess solute?
**b)** solder, which is a homogenous mixture of lead and tin?
Hint: Are there any boundary surfaces in the system you are considering?

**YOU CAN FIND MORE ABOUT WHAT HAPPENS WHEN CONDITIONS ARE ALTERED FOR A SYSTEM IN EQUILIBRIUM BY READING CHAPTER 1.
THE LATTICE STRUCTURE OF ICE IS THE RESULT OF HYDROGEN BONDING, WHICH YOU CAN READ MORE ABOUT IN CHAPTER 21.**

**Fig 6.24** Phase diagram for water

**SEE QUESTION 3.**

**? TEST YOURSELF**

**T** If water followed the trend of the other Group 6 hydrides, what would you predict its boiling point to be in:
**a)** Kelvin;
**b)** Celsius?

**✓ REMEMBER THIS**

Electronegativity is the power of an atom within a molecule to attract electrons to itself. You can read more about this and about polar bonds in Chapter 4, pages 82–84.

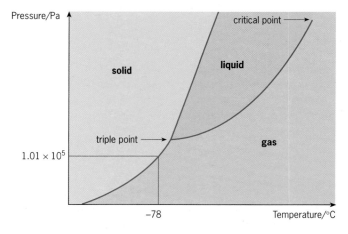

**Fig 6.26** Hydrogen bonding in water

Hydrogen bonds are very strong compared to other intermolecular forces. This means that for water to boil these strong hydrogen bonds have to be broken before the molecules can escape into the vapour, which leads to the relatively high boiling point of 100 °C.

Hydrogen bonding also explains why water expands on freezing. As water cools, the molecules come closer together and so the density increases until, at 4 °C, the density is at a maximum. Below this temperature, the water molecules start to move apart because of the increasing number of hydrogen bonds they form. When water freezes, the water molecules are held apart by hydrogen bonds in a three-dimensional lattice. Thus the ice is less dense than water, which explains why ice is one of the very few solids to float on its liquid.

## CARBON DIOXIDE

The phase diagram for carbon dioxide is similar to that for water, except that the solid–liquid line has a positive slope. This is because when liquid carbon dioxide freezes the volume decreases as the particles come closer together to form a solid lattice, as happens with most solids.

In contrast, as we have seen above, as water approaches 0 °C, the particles move slightly apart, so the water expands on freezing and becomes less dense. If pressure is increased, the equilibrium of ice and liquid shifts to produce more liquid, and hence the melting point decreases and a negative slope is obtained.

**? TEST YOURSELF**

**U** Hydrogen bonding is also present in ammonia and hydrogen fluoride. Draw a diagram similar Fig 6.26 to show this hydrogen bonding.

**FOR IN FORMATION ABOUT HYDROGEN BONDING IN ALCOHOLS, SEE CHAPTER 13, PAGE 289; AND FOR INFORMATION ABOUT HYDROGEN BONDING IN PROTEINS AND HOW THIS MAKES YOUR HAIR CURL, SEE CHAPTER 19, PAGE 423.**

**? TEST YOURSELF**

**V** Ice is less dense than liquid water, so ice floats. Most solids sink in their liquids. Imagine what would happen if ice sank in the oceans. For one thing, it would probably never melt, as it would be away from the warming of the Sun's rays. Describe the effects you think this would have on our weather systems and on life on Earth.

**Fig 6.27** Phase diagram for carbon dioxide

A point to note is that at atmospheric pressure ($1.01 \times 10^5$ Pa) solid carbon dioxide sublimes at –78 °C instead of melting. This is why it is called dry ice. Dry ice is used to keep objects cold and can be packed around them because it does not turn into a liquid when it warms up. Solid carbon dioxide also produces the mist you see at pop concerts: as it sublimes, the cold $CO_2$ gas makes water vapour in the air condense to water droplets, forming clouds of mist.

## THE CRITICAL POINT

We have already seen that the liquid–vapour line finishes at the critical point. The temperature at this point is known as the **critical temperature**. At this temperature, no pressure, however high, can force the particles in a gas to condense to a liquid: they have too much kinetic energy to liquefy.

During the early part of the nineteenth century, experimenters managed to liquefy many of the gases that had resisted liquefaction. However, nitrogen, oxygen and hydrogen appeared to be 'permanent gases' – it did not seem possible to liquefy them. This was because scientists could not reach temperatures below their critical temperatures. For hydrogen it is 33 K. Hydrogen was eventually liquefied in 1898 by the Scotsman James Dewar, who a year later managed to solidify it at 14 K.

**TEST YOURSELF**

**a)** What is the critical temperature of hydrogen on the Celsius scale?

**b)** A gaseous element discovered on Earth in 1895 was finally condensed in 1908. It has a critical temperature of 5 K. What is the name of this gas? See page 54 to check your answer.

## 6 MIXTURES OF TWO COMPLETELY MISCIBLE LIQUIDS

When we considered the phase diagram for water we were dealing with a **one-component system**, because in all the phases the chemical substance was water. The carbon dioxide phase system is also a one-component system.

However, two liquids that are completely **miscible** in each other form a **two-component system**. **Miscible liquids** dissolve completely in each other in all proportions to form a homogeneous mixture. As the mixture is homogeneous it is one phase, but it contains two components. A good example of a mixture of two miscible liquids is ethanol and water.

Chemists express the proportions of the two liquid components in a mixture in terms of **mole fractions**.

**The mole fraction is the ratio of the number of moles of a particular component to the *total* number of moles present in the system.**

So for a mixture of two miscible liquids A and B:

$$\text{Total number of moles} = n_A + n_B$$

The mole fraction of liquid A, $x_A = \dfrac{n_A}{n_A + n_B}$

The mole fraction of liquid B, $x_B = \dfrac{n_B}{n_A + n_B}$

Note that symbol $x$ is used for the mole fraction, and in a two-component system:

$$x_A + x_B = 1$$

**TEST YOURSELF**

**X** A mixture of ethanol and water contains 20 moles of ethanol and 5 moles of water. What are the mole fractions of ethanol and water in this mixture?

## RAOULT'S LAW

Suppose we start with pure liquid A that is volatile. It will have a vapour above its liquid surface, which will exert a vapour pressure. If we now start mixing in another volatile liquid, B, the vapour above the mixture will contain some molecules of B. When only A is present, the vapour pressure will be that of pure A: $p^0_A$. As it is diluted with liquid B, the mole fraction of A falls, and so also does the *number of molecules* of A in the vapour. This then reduces the vapour pressure of A to $p_A$. Meanwhile, as the mole fraction of B is increased, so does the number of molecules of B in the vapour, therefore increasing the contribution to the total vapour pressure made by B: $p_B$. You can see the effect of changing the proportions (mole fractions) of A and B in Fig 6.28.

**? TEST YOURSELF**

Y When the mole fraction of A in Fig 6.28 is 0.8, what is the mole fraction of B?

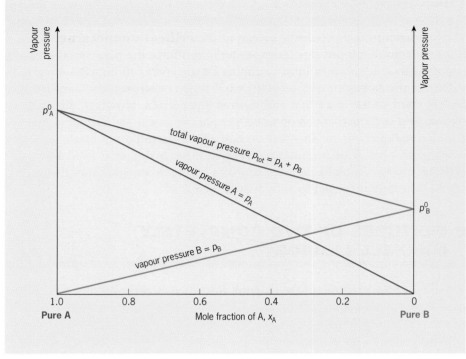

**Fig 6.28** Vapour pressure–composition diagram for two volatile miscible liquids at constant temperature

Notice that in Fig 6.28 the total vapour pressure above the mixture ($p_{tot}$) is the sum of the vapour pressures due to liquid A ($p_A$) and liquid B ($p_B$):

$$p_{tot} = p_A + p_B$$

Raoult's law states that:

**The vapour pressure due to liquid A ($p_A$) equals the vapour pressure of pure A ($p^0_A$) multiplied by the mole fraction of A ($x_A$) present in the mixture:**

$$p_A = p^0_A \times x_A$$

Raoult's law applies to **ideal mixtures** in which there is:
● no enthalpy change when the liquids are mixed;
● no change in volume on mixing;
● no change in intermolecular forces between molecules of the pure substances and between the different molecules in the mixture.
Any ideal mixture obeys Raoult's law.

## HOW BOILING POINT VARIES WITH COMPOSITION FOR AN IDEAL MIXTURE OF TWO MISCIBLE LIQUIDS

The vapour pressure of a liquid increases with an increase in temperature. At the temperature at which the vapour pressure equals atmospheric pressure, the liquid boils. At a particular temperature, if the vapour pressure of a liquid is already high, the boiling point will be at a temperature that is not much higher. For a liquid with a low vapour pressure, the temperature will need to be raised much more before it can boil. In Fig 6.28 the total vapour pressure above the liquid mixture varies according to its composition, so the boiling point also changes, which is shown in Fig 6.29.

The boiling point line is a curve, not a straight line. This is because vapour pressure increases at a faster rate than the rate of increase of temperature.

**Fig 6.29** The variation of boiling point with composition for an ideal mixture of liquids A and B at constant pressure. Notice that in Fig 6.28 liquid A has the higher vapour pressure when pure, so it boils at a lower temperature than liquid B

# FRACTIONAL DISTILLATION

Fig 6.30 is a phase diagram for mixtures of liquids A and B. In this diagram there are two boiling point–composition curves. The one for the liquid phase is identical to that in Fig 6.29, but as the liquid mixture boils it turns into a vapour phase, which is richer in the more volatile component – in this case, A. So the composition of the vapour is different from that of the boiling liquid, which is why fractional distillation works. The gap between the two curves does not represent any phase and should be regarded as a 'no-go area'.

**Fig 6.30** Phase diagram for liquid mixtures of A and B. The two curves are boiling point–composition curves – one for the liquid and one for the vapour. For any given temperature at which the liquid is boiling (lower curve), it produces a vapour that contains a greater proportion of A, because A is more volatile than B

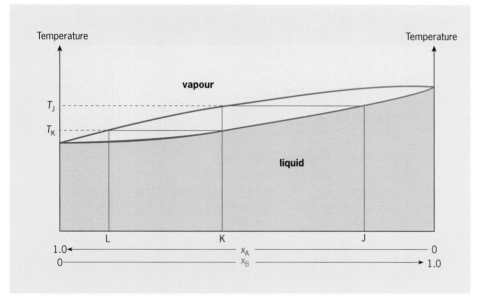

**Fig 6.31** Phase diagram for mixtures of liquids A and B, showing what happens when a mixture of composition J is boiled and its vapour condensed to give a liquid of a different composition K. The boiling and condensing process is repeated to give a liquid mixture of composition L. Each time, the condensed vapour is richer in the more volatile liquid A

Study Fig 6.31: imagine that you have a liquid mixture of A and B with a composition J. This mixture boils at a boiling point $T_J$. Compared with the mixture, the vapour formed on boiling is richer in A: the composition of the vapour at $T_J$ can be found by drawing a horizontal line to the vapour composition curve. If the vapour is now separated and condensed, the liquid has a composition K. This process is a simple distillation in which the liquids are boiled and the vapour condensed. When the liquid mixture is boiled again it will boil at temperature $T_K$. It will be even richer in component A and will have a composition of L. This process of separating and condensing the vapour can continue until pure A is obtained.

In practice, boiling the liquid mixture and condensing its vapour in a separate piece of apparatus would involve many separate distillations and so would be too time-consuming. A fractionating column does this in one piece of apparatus, vaporizing and condensing the mixture many times. Apparatus for fractional distillation in a laboratory is shown in Fig 6.32 (see page 138). As vapour enters the fractionating column it begins to cool and condense. The condensed liquid falls back, but is reheated by vapour coming from the flask, so it re-boils. As the vapour rises up the tower it becomes richer in the more volatile component A through a series of redistillations. The longer the column, the closer to pure A the composition of the vapour becomes. The glass beads are there to provide a large surface area on which the liquid and vapour can boil and condense. The liquid in the flask becomes richer in component B as A boils off.

**FOR FRACTIONAL DISTILLATION OF CRUDE OIL SEE PAGE 202.**

**? TEST YOURSELF**

**Z** As the composition of the liquid in the flask becomes richer in B, what happens to its boiling point?

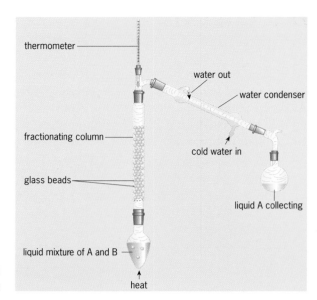

**Fig 6.32** Fractional distillation apparatus

Fractional distillation is used to separate liquid mixtures with components of different boiling points. Nitrogen, oxygen and the noble gases are obtained industrially from liquid air by fractional distillation. Crude oil is separated in oil refineries into fractions containing mixtures of hydrocarbons with closely similar boiling points.

## NON-IDEAL LIQUID MIXTURES

For a liquid mixture to be ideal it must obey Raoult's law. Very few mixtures do this exactly. Some have a positive deviation from the law, showing higher-than-predicted vapour pressures when mixed (see Fig 6.33), and some mixtures of liquids have a negative deviation, producing a lower-than-expected vapour pressure (see Fig 6.34).

**Fig 6.33** Vapour pressure–composition curve showing a slight positive deviation

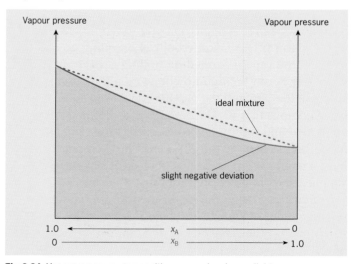

**Fig 6.34** Vapour pressure–composition curve showing a slight negative deviation

Figs 6.33 and 6.34 show relatively minor deviations from the ideal mixture and have no maximum or minimum vapour pressures. Consequently, they have no maximum or minimum boiling points. These mixtures can be separated by fractional distillation.

## AZEOTROPIC MIXTURES

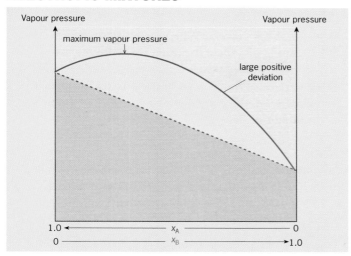

**Fig 6.35** Vapour pressure–composition curve showing a large positive deviation. Note that there is a maximum vapour pressure

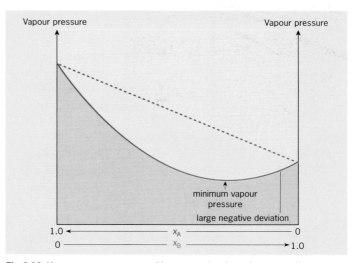

**Fig 6.36** Vapour pressure–composition curve showing a large negative deviation. Note that there is a minimum vapour pressure

When liquid mixtures have a large deviation from the ideal they also have maximum or minimum vapour pressures in their composition curves, as shown in Figs 6.35 and 6.36. Those with maximum vapour pressures have minimum boiling points and vice versa. This is shown in Figs 6.37 and 6.38. In each case, we can see what happens if we start with a mixture of composition $Z$ and fractionally distil it. Notice that only one pure component separates, to leave a constant boiling mixture called an azeotrope. Azeotropes boil to give a vapour of the same composition. Nitric acid and water form a maximum boiling point azeotrope, while mixtures of benzene and ethanol give a minimum boiling point azeotrope.

**SEE QUESTION 4.**

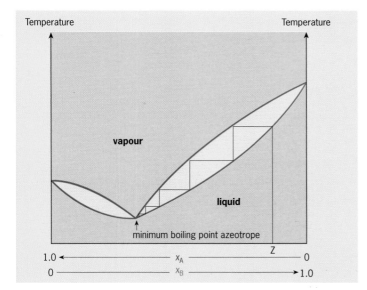

**Fig 6.37** Phase diagram for mixtures of liquids A and B with a large positive deviation, showing what happens when a mixture of composition $Z$ is fractionally distilled. Note that the azeotrope produced has a minimum boiling point, and this will boil off first, leaving, in this case, pure liquid B in the distillation flask

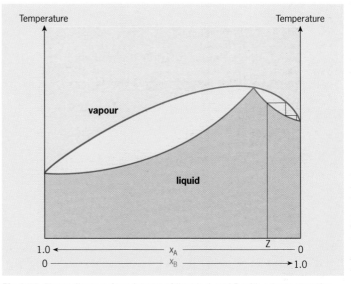

**Fig 6.38** Phase diagram for mixtures of liquids A and B with a large negative deviation, showing what happens when a mixture of composition $Z$ is fractionally distilled. Note that the azeotrope produced has a maximum boiling point, and this will be left in the distillation flask, while pure liquid B is distilled first from the top of the column

# 7 AQUEOUS SOLUTIONS AND CONCENTRATION

Aqueous solutions have water as the solvent. When we write equations and we want to show that a reactant is dissolved in water, we write the state symbol (aq). So a dilute solution of sulphuric acid is written: $H_2SO_4$(aq).

Water covers most of the surface of our planet, and many reactions take place in aqueous solution. Nearly all the chemical reactions of our bodies occur in the aqueous environment of our cells.

## CONCENTRATION

Concentrations can be expressed in terms of the mass of a substance in a certain volume of solution. The units often used are **grams per cubic decimetre** or **g dm$^{-3}$**. So if 40 grams of sodium hydroxide are present in one cubic decimetre the concentration is 40 g dm$^{-3}$. Chemists, however, are usually interested in the *amount* of a substance present. When we considered gases at the beginning of this chapter, it was not long before we came across the mole. It is the same with solutions. We need to know how many moles of solute are present in a particular volume of solution, because this tells us the number of particles that are there.

The concentration of a solute in a solution is measured in **moles per cubic decimetre – mol dm$^{-3}$.**

**✔ REMEMBER THIS**

One mole is the amount of substance that contains as many particles as there are atoms in exactly 12 g of carbon-12. The number of particles in a mole is $6.02 \times 10^{23}$ and is called the **Avogadro constant**. You can find out more about the mole on page 5 of Chapter 1. Now work through the Examples.

**SEE QUESTION 5.**

## EXAMPLE

**Q** Work out how many moles of solute are dissolved in these solutions:

**a)** 250 cm$^3$ of 1.0 mol dm$^{-3}$ NaOH(aq);

**b)** 20 cm$^3$ 0.50 mol dm$^{-3}$ HCl(aq).

**A**

**a)** 1 dm$^3$ = 1000 cm$^3$

Number of moles NaOH in 1000 cm$^3$ = 1 mol

So number of moles NaOH in 250 cm$^3$ = $1 \times \dfrac{250}{1000}$

$= 0.25$ mol

**b)** Number of moles of HCl in 1000 cm$^3$ = 0.50 mol

So number of moles HCl in 20 cm$^3$ = $0.50 \times \dfrac{20}{1000}$

$= 0.010$ mol

**Q** What is the concentration (in mol dm$^{-3}$) of the following aqueous solutions?

**a)** 0.25 mol HCl dissolved in 50 cm$^3$;

**b)** 5.85 g of NaCl dissolved in 250 cm$^3$.

**A**

**a)** There are 0.25 mol HCl in 50 cm$^3$

So in 1000 cm$^3$, the number of moles = $0.25 \times \dfrac{1000}{50}$

$= 5.0$ mol

Concentration of HCl in solution = 5.0 mol dm$^{-3}$

**b)** Moles of NaCl in 5.85 g

$\dfrac{\text{mass in grams}}{\text{mass of one mole (in grams)}} = \dfrac{5.85\text{g}}{58.5\text{g}} = 0.100$ mol

There are 0.100 mol NaCl in 250 cm$^3$

So in 1000 cm$^3$, the number of moles = $0.100 \times \dfrac{1000}{250}$

$= 0.400$ mol

Concentration of NaCl solution = 0.400 mol dm$^{-3}$

## ? TEST YOURSELF

**A** **a)** Calculate how many moles of solute are present in these solutions:
i) 100 cm$^3$ of 2 mol dm$^{-3}$ $H_2SO_4$(aq);
ii) 25.0 cm$^3$ of 5.00 mol dm$^{-3}$ $HNO_3$(aq);
iii) 5 dm$^3$ of 0.2 mol dm$^{-3}$ KCl(aq);
iv) 10 cm$^3$ of 0.5 mol dm$^{-3}$ $Na_2CO_3$(aq);
v) 50 cm$^3$ of 0.050 mol dm$^{-3}$ $NaNO_3$(aq).
**b)** Calculate the concentration of the following aqueous solutions:
i) 0.40 mol $CaCl_2$ dissolved in 250 cm$^3$;
ii) 16 g $CuSO_4$ dissolved in 100 cm$^3$ ($A_r$: Cu = 64, S = 32, O = 16).

# PREPARING SOLUTIONS OF KNOWN CONCENTRATION

Chemists often need to make up very accurately a solution of known concentration. To do this a volumetric flask is used. On its long neck is etched a line that indicates a precise volume (Fig 6.39(a)). The required amount of substance to be dissolved to make up the solution is weighed accurately on a balance. It is then dissolved in a small volume of distilled water in a beaker (Fig.6.39(b)) and the solution is poured into the volumetric flask (Fig.6.39(c)). Distilled water is then used to rinse any remaining solution into the flask. More distilled water is added until the meniscus in the neck of the flask just lies on the etched line (Fig 6.39(d)).

**Fig 6.39(a)** A 250 cm³ volumetric flask

**Fig 6.39(b)** A known mass of solute is poured into a beaker followed by a small volume of distilled water to dissolve it

**Fig 6.39(c)** The solution is poured in and the beaker rinsed with more distilled water. The rinsings are also poured into the flask to ensure all the dissolved solute has been transferred

**Fig 6.39(d)** The solution is topped up with distilled water until the bottom of the meniscus just sits on the etched line

---

## EXAMPLE

**Q** What mass of sodium hydroxide is required to make up 250.0 cm³ of a 0.50 mol dm⁻³ sodium hydroxide solution [$A_r$: Na = 23, O = 16, H = 1].

**A Step 1.** Work out the amount (moles) required:

$$1 \text{ dm}^3 = 1000 \text{ cm}^3$$

Number of moles NaOH required in 1000 cm³      = 0.50 mol

So number of moles NaOH required in 250 cm³   $= 0.50 \times \dfrac{250}{1000}$

$= 0.125 \text{ mol}$

**Step 2.** Convert amount (moles) of NaOH into a mass:

Mass in grams = amount in moles × mass of 1 mole

Mass of 1 mol of NaOH = 23 + 16 + 1 g = 40 g

So mass of sodium hydroxide = 0.125 mol × 40 g

$= 5.0 \text{ g}$

Therefore 5.0 g must be dissolved in 250 cm³ to make a 0.5 mol dm⁻³ solution.

---

## ✓ REMEMBER THIS

The concentration of a solution is expressed per dm³ of solution and not per dm³ of solvent. If one mole of solid is added to 1 dm³ of water the volume of the solution will probably be slightly greater than 1 dm³, so the concentration will not be exactly 1 mol dm⁻³.

## ? TEST YOURSELF

**B** What mass of solute is required to make up the following solutions?
**a)** 500 cm³ of 2.0 mol dm⁻³ CuSO₄(aq);
**b)** 2.0 dm³ of 0.125 mol dm⁻³ Na₂CO₃(aq);
**c)** 250 cm³ of 0.25 mol dm⁻³ KOH(aq).

# 8 TITRATIONS

**Fig 6.40(a)** A pipette is used to measure an accurate volume of base solution (alkali)

**Fig 6.40(b)** The base solution (alkali) has been coloured using a few drops of indicator. The acid solution is added from a burette until the alkali is completely neutralized. This is the end point and is shown by the indicator changing colour. At least two titrations need to be done to confirm the exact volume of acid solution required

A **titration** is an experimental technique that uses the known volumes of two solutions to determine the amounts of substances that are reacting very accurately. This technique is often referred to as **volumetric analysis**.

In a titration one of the solutions is of known concentration. If we know the balanced equation for a reaction, then we know how many moles of each reactant will react. Using this information we can calculate the concentration of the other solution.

Acids are **neutralized** by bases. If we wish to discover the concentration of a dilute hydrochloric acid solution we can perform an **acid–base titration.** Sodium hydroxide of known concentration is used as the base to neutralize exactly the hydrochloric acid. The reaction is followed using an **acid–base indicator**, which changes colour at the **neutralization point** or **end point** or **equivalence point**. Fig 6.40 shows how an acid–base titration may be performed.

⬤ **EXAMPLE**

**Q** What is the concentration of a dilute solution of hydrochloric acid if 26.1 cm$^3$ is exactly neutralized by 25.0 cm$^3$ of 0.100 mol dm$^{-3}$ sodium hydroxide?

**A** **Step 1.** Write the balanced equation:

$$NaOH(aq) + HCl(aq) \rightarrow NaCl(aq) + H_2O(aq)$$

**Step 2.** Convert the equation into amounts:

$$1 \text{ mol NaOH} + 1 \text{ mol HCl} \rightarrow 1 \text{ mol NaCl} + 1 \text{ mol H}_2O$$

**Step 3.** Work out the amount of sodium hydroxide being used:

Number of moles NaOH in 1000 cm$^3$ = 0.100 mol

So number of moles NaOH in 25.0 cm$^3$ = $0.100 \times \dfrac{25.0}{1000} = 0.002\,50$ mol

**Step 4.** Scale the amounts in the equation:

1 mol NaOH reacts with 1 mol HCl

0.002 50 mol NaOH reacts with 0.00250 mol HCl

So there are 0.00250 mol HCl in 26.1 cm$^3$ solution.

**Step 5.** Scale the amount in the known volume to 1 dm$^3$:

0.002 50 mol HCl in 26.1 cm$^3$ solution

In 1 dm$^3$ (1000 cm$^3$) moles HCl = $0.002\,50 \times \dfrac{1000}{26.1}$

Therefore concentration of HCl = 0.095 8 mol dm$^{-3}$.

**Q** Clover leaves contain ethanedioic acid, (COOH)$_2$. 1.01 g of clover leaves are crushed and the ethanedioic acid is extracted using a small volume of distilled water. This solution is exactly neutralized by 17.2 cm$^3$ of 0.200 mol NaOH. What is the mass of ethanedioic acid in this sample of leaves?

**A** **Step 1.** Write the balanced equation:

$$(COOH)_2(aq) + 2NaOH(aq) \rightarrow (COO^-Na^+)_2(aq) + 2H_2O(l)$$

## EXAMPLE (continued)

**Step 2.** Convert the equation into amounts:

1 mol $(COOH)_2$ + 2 mol NaOH → 1 mol $(COO^-Na^+)_2$(aq) + 2 mol $H_2O$(l)

**Step 3.** Work out the amount of sodium hydroxide being used:

Number of moles NaOH in 1000 $cm^3$ = 0.200 mol

So number of moles NaOH in 17.2 $cm^3$ = $0.200 \times \dfrac{17.2}{1000}$ = 0.003 44 mol

**Step 4.** Scale the amounts in the equation:

1 mol $(COOH)_2$ reacts with 2 mol NaOH

0.001 72 mol $(COOH)_2$ reacts with 0.003 44 mol NaOH

So there are 0.001 72 mol $(COOH)_2$ in the clover leaf solution.

**Step 5.** Convert the amount (moles) into a mass:

Mass in grams = amount in moles × mass of 1 mole

Mass of 1 mol of $(COOH)_2$ = (12 + (16 × 2) + 1 g) × 2 = 90 g

So the mass of ethanedioic acid in 1.01 g clover leaves = 0.001 72 mol × 90 g

= 0.155 g $(COOH)_2$

SEE QUESTIONS 6, 7 AND 8.

### ? TEST YOURSELF

**C'** Vinegar contains ethanoic acid, $CH_3COOH$. What is the concentration in mol $dm^{-3}$ and g $dm^{-3}$ if 25.0 $cm^3$ of 1.00 mol $dm^{-3}$ sodium hydroxide is neutralized by 26.3 $cm^3$ vinegar?
The balanced equation for this reaction is:

$CH_3COOH$(aq) + NaOH(aq) →
$CH_3COO^-Na^+$(aq) + $H_2O$(l)

## 9 ENTROPY

**Entropy** is a measure of the disorder of a system. Think of your bedroom – when it is tidy everything has a place and there is a high degree of order. We could say that the entropy of your room is low. However, it is all too easy for the entropy to rise as things are moved out of place and it becomes untidy.

When particles are in a solid they cannot move out of position, they can only vibrate. In the solid state there is usually a high degree of order and a limited number of ways to arrange the particles, so the entropy of a solid is usually low. The liquid state tends to have a higher entropy than the solid state, because the particles can move about, so many more arrangements are possible. A gas has widely spaced particles that move very rapidly in all directions. All gases have high entropies.

The entropies of different substances can be determined for a particular temperature. Entropy is given the symbol **S** and is measured in joules per kelvin per mole ($J K^{-1} mol^{-1}$). Table 6.1 gives the entropy values for some substances at 298 K and 101 kPa.

You can see from Table 6.1 that water vapour has a much higher entropy than water liquid. This is to be expected, since water molecules are more spread out in the vapour and more arrangements are possible. Diamond has a lower entropy than graphite, which means that diamond has a more ordered structure. Neon and helium are both gases and so have relatively high entropies. However, neon has heavier atoms. This tends to bring its energy levels closer together, which in fact provides more possibilities of arranging quanta of energy.

Entropy, then, is not just about the number of possible ways of arranging particles. It is also about how the energies of the particles are arranged in the energy levels, an aspect of entropy that we shall continue in Chapter 8.

**Table 6.1**

| Substance | Entropy (S)/$J K^{-1} mol^{-1}$ |
|-----------|---------------------------------|
| He(g) | 126.0 |
| Ne(g) | 146.2 |
| $H_2O$(l) | 69.9 |
| $H_2O$(g) | 188.7 |
| $C_{(graphite)}$ | 5.7 |
| $C_{(diamond)}$ | 2.4 |

### ? TEST YOURSELF

**D'** Predict whether each of the following leads to an increase or decrease in the entropy of the substances involved:
**a)** Ice melting.
**b)** Water vapour condensing.
**c)** Cooling oxygen from 60 °C to 20 °C.
**d)** Dissolving sugar in water.
**e)** Subliming solid iodine.
**f)** Freezing liquid bromine.

## SUMMARY

After studying this chapter, you should know and understand the following points and be able to use the equations.

- Equal volumes of gases contain equal numbers of molecules at the same temperature and pressure (Avogadro's law).
- 1 mole of any gas at standard temperature and pressure occupies $22.4$ dm$^3$. Under room conditions the molar gas volume is $24$ dm$^3$.
- Boyle's law relates the pressure and volume of a fixed amount of gas and is expressed mathematically as:
  $p \propto 1/V$, where $p$ = pressure and $V$ = volume.
- Charles' law relates volume and temperature of a gas and is expressed mathematically as:
  $V \propto T$, where $T$ = absolute temperature in kelvin.
- The ideal gas equation relates pressure, volume and temperature for a fixed amount of gas:
  $pV = nRT$.
- The kinetic–molecular model explains the behaviour of ideal gases by making a number of assumptions, such as zero intermolecular forces and that the volume of particles is negligible.
- Real gases approach ideal behaviour when at a high temperature or low pressure.
- The kinetic–molecular model can be extended to include solids and liquids.

- Vapour pressure is the pressure above a liquid when it is in equilibrium with its vapour.
- A phase is a homogeneous portion of matter separated from other portions by a definite boundary surface.
- Hydrogen bonding is a strong intermolecular force between hydrogen bonded to an F, N or O atom and the lone pair of another F, N or O atom. It accounts for the anomalous properties of water.
- An ideal liquid mixture obeys Raoult's law. Raoult's law states that:
  $p_A = p^0_A \times x_A$.
- Fractional distillation separates two liquids of different boiling points in a homogeneous mixture.
- Non-ideal mixtures show either positive or negative deviations from Raoult's law.
- An azeotrope is a constant boiling mixture. Azeotropes occur when there is a maximum or minimum vapour pressure.
- The concentration of a solute in a solution is measured in mol dm$^{-3}$ or g dm$^{-3}$.
- Titrations are a valuable way to analyse amounts in solutions.
- Entropy is a measure of the disorder of a system. Solids usually have low entropies, whereas gases have high entropies.

## ? EXAM QUESTIONS

**1** Silver ethanedioate, $Ag_2C_2O_4$, decomposes violently on heating:

$$Ag_2C_2O_4 \rightarrow 2Ag + 2CO_2$$

Calculate the maximum volume of carbon dioxide produced at room temperature and pressure when 0.010 mole of silver ethanedioate is completely decomposed by heat. [The volume of one mole of carbon dioxide at room temperature and pressure is 24 dm$^3$.]

WJEC CH1 June 2001 Q11f(ii)

**2**
**a)** Describe, using a kinetic-molecular model, the processes that take place in each of the following changes: **(i)** melting; **(ii)** a liquid evaporating.

**b)** Outline the basic assumption of the kinetic theory as applied to an ideal gas. Explain why a gas behaves most like an 'ideal gas' at low pressures and high temperatures.

OCR Specimen Paper 2815/05 Gases, liquids and solids 2000 Q5

**3**
**a)** **(i)** Sketch and label the phase diagram of water. Include the boiling point and freezing point of water.
**(ii)** Explain the atypical feature of this diagram.

**b)** **(i)** Describe and explain what phase or phases are present in clouds.
**(ii)** Suggest and explain how clouds can form out of clear blue skies. Indicate, on your phase diagram, the phase change that occurs when clouds form.

OCR Specimen Paper 2815/05 Gases, liquids and solids 2000 Q1

# ? EXAM QUESTIONS

**4**

**a)** **(i)** State Raoult's Law.
**(ii)** Explain, with a suitable example, what combinations of liquids obey Raoult's Law.

**b)** Fig 6.Q4 shows the vapour pressure of a mixture of propan-1-ol and methylbenzene.

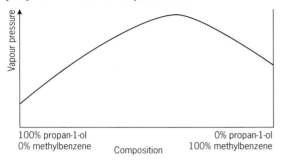

**Fig 6.Q4**

Explain the shape of this vapour pressure curve.
OCR Specimen Paper 2815/05 Gases, liquids and solids 2000 Q4

**5** Ammonium nitrate can be prepared by the reaction between ammonia and nitric acid:

$$NH_3 + HNO_3 \rightarrow NH_4NO_3$$

**a)** The concentration of a nitric acid solution is 2.00 mol $dm^{-3}$. Calculate the volume of this solution that would be required to react with exactly 20.0 g of ammonia.

**b)** A sample of ammonium nitrate decomposed on heating as shown in the equation below.

$$NH_4NO_3 \rightarrow 2H_2O + N_2 + \tfrac{1}{2}O_2$$

On cooling the resulting gases to 298 K, the volume of nitrogen and oxygen together was found to be 0.0500 $m^3$ at a pressure of 95.0 kPa.
**(i)** State the ideal gas equation and use it to calculate the total number of moles of nitrogen and oxygen formed. (The gas constant $R = 8.31$ J $mol^{-1}$ $K^{-1}$)
**(ii)** Using your answer to part **b)**(i), deduce the number of moles of ammonium nitrate in the sample.

AQA Unit 1 Atomic structure, bonding and periodicity Jan 2001 Q3

**6**

**a)** What is the name given to the number of molecules in one mole of carbon dioxide?

**b)** **(i)** State the ideal gas equation.
**(ii)** Calculate the volume of 1.00 mol of carbon dioxide gas at 298 K and 100 kPa. (The gas constant $R = 8.31$ J $mol^{-1}$ $K^{-1}$)
**(iii)** Calculate the mass of carbon dioxide gas at 273 K and 500 kPa contained in a cylinder of volume 0.00500 $m^3$.

**c)** Hydrogen can be made by the reaction of hydrochloric acid with magnesium according to the equation:

$$2HCl + Mg \rightarrow MgCl_2 + H_2$$

What mass of hydrogen is formed when 100 $cm^3$ of hydrochloric acid of concentration 5.0 mol $dm^{-3}$ reacts with an excess of magnesium?

AQA Unit 1: Atomic structure, bonding and periodicity June 2001 Q2a–c

**7** Ethanedioic acid is a solid at room temperature. It is soluble in cold water and is toxic.
In an exercise to determine the number of moles of water of crystallization, $x$, in one mole of ethanedioic acid, $(COOH)_2.xH_2O$, 1.48 g of ethanedioic acid was dissolved in water and made up to exactly 250 $cm^3$ of solution.
25.0 $cm^3$ of the ethanedioic acid solution was then titrated against sodium hydroxide of concentration 0.100 mol $dm^{-3}$. A titre of 23.5 $cm^3$ of sodium hydroxide was required.

$$(COOH)_2 + 2NaOH \rightarrow (COONa)_2 + 2H_2O$$

**a)** Describe how the ethanedioic acid would be dissolved and made up to exactly 250 $cm^3$ of solution. In your answer include the names of each piece of apparatus required and how each would be prepared for the procedure. Include a brief description of each step in the process.

**b)** Calculate the number of moles of sodium hydroxide used in the titration.

**c)** Calculate the number of moles of ethanedioic acid in 25 $cm^3$ of solution.

**d)** **(i)** From your answer to **c)(ii)** and the mass of ethanedioic acid used, calculate the relative molecular mass of ethanedioic acid.
**(ii)** Calculate a value for $x$.

**e)** Suggest one appropriate safety precaution that should be taken in this exercise, based on the fact that ethanedioic acid is toxic.

Edexcel A/AS Specimen Paper 6243/02 Unit 3B Q1(not a)

**8** Sodium hydrogensulphate, $NaHSO_4$, used in domestic cleaners, can be neutralized by sodium hydroxide solution.

$$NaHSO_4 + NaOH \rightarrow Na_2SO_4 + H_2O$$

0.005 mole of sodium hydrogensulphate is neutralized with sodium hydroxide solution of concentration 0.100 mol $dm^{-3}$. State which **one** of the following is the volume of sodium hydroxide required:

**A** 20.0 $cm^3$ **B** 50.0 $cm^3$ **C** 100.0 $cm^3$ **D** 500.0 $cm^3$

WJEC CH1 June 2001 Q6

# KEY SKILLS **ASSIGNMENT**

## SUPERCRITICAL FLUIDS

**Above a certain temperature it is not possible to condense a gas, no matter how much pressure is applied. This phenomenon was discovered in 1821 when Baron Charles Cagniard de la Tour sealed an amount of a liquid and its vapour in a cannon. He heated the cannon, tilted it backwards and forwards, and measured the temperature at which the sloshing sound of the liquid disappeared. This temperature is called the critical temperature. The pressure at which this happens is the critical pressure and both combine to define the critical point. At this critical point, the two phases of gas and liquid become indistinguishable from each other and have the properties of both.**

As the liquid was heated up inside the cannon it expanded, which reduced its density. Also, more molecules escaped into the vapour phase, so the density of the gas increased, until at the critical temperature the densities of the gas and liquid become the same and we say that the substance is a **fluid**.

**Fig 6.A1(left)** A fluorinated hydrocarbon close to the critical point. The phase boundary between the liquid and gas phases is clear

**Fig 6.A1(centre)** The critical temperature has almost been reached and the phase boundary is disappearing

**Fig 6.A1(right)** A supercritical fluid is present as the temperature rises above the critical point

**1 a)** What assumptions are made in the kinetic–molecular model about the particles in an ideal gas?
**b)** Explain in terms of intermolecular forces why cooling a gas below its critical temperature causes it to condense.
**c)** In the sealed cannon experiment, explain why the pressure increased as the liquid and vapour were heated.
**d)** Draw a phase diagram for carbon dioxide and mark the critical point on it.

**2** Some gases, such as nitrogen and hydrogen, were known as 'permanent gases' in the nineteenth century because they could not be liquefied even at very high pressures. Why was this?

Supercritical fluids (SCFs) will fill the volume of any container, just like a gas, but, like liquids, they have solvent properties. For more than a century the phenomenon of SCFs was regarded as a scientific curiosity of little practical value.

However, in the early 1960s, the German chemist Kurt Zosel used supercritical carbon dioxide to remove caffeine from wet coffee beans and opened a treasure chest of important industrial applications that we are still discovering to this day.

**3** Before supercritical $CO_2$ was used to decaffeinate coffee, liquid organic solvents were used and some of these contained chlorine.
**a)** Find out and describe the environmental problems of using chlorine-containing organic solvents.
**b)** As carbon dioxide is removed as a by-product during the fractional distillation of liquid air, it does not have any significant impact on the greenhouse effect. It has several advantages, one of which is that it is non-toxic. What are some of its other advantages?

**4** Supercritical $CO_2$ is used as a solvent to remove cholesterol from egg yolk. Why are food manufacturers very interested in this process?

Carbon dioxide becomes supercritical at just 31 °C and so can be used to dissolve complex molecules that would be destroyed at higher temperatures. This is opening up possibilities for extracting drugs from plants and micro-organisms. Already the chemicals responsible for some odours and flavours can be extracted from foodstuffs. The possibility of removing cooking oil from potato chips to produce healthier, oil-free chips is also under investigation.

Water, too, exhibits unusual properties when it becomes supercritical above 374 °C and $218 \times 10^5$ Pa. In particular, non-polar organic compounds that would not normally dissolve in polar water are able to dissolve. Supercritical water can be used to carry out reactions with molecular oxygen that convert hazardous organic compounds into small harmless species, such as $CO_2$, $H_2O$ and $Cl^-$. The notorious polychlorinated biphenyls (PCBs) can be destroyed using this technique, and much research is being carried out to destroy military stockpiles of chemical weapons safely using supercritical water. (You can read more about PCBs on page 80 of Chapter 4.)

**5 a)** Draw a dot-and-cross diagram for the chloride ion.
**b)** What properties of compounds that contain chloride ions make them safer alternatives to compounds made up of molecules that contain chlorine?

**6** Why are processes that involve supercritical water expensive?

**Fig 6.A2** A flame is produced spontaneously in a mixture of supercritical water and methane at $1000 \times 10^5$ Pa and 500 °C when oxygen is injected

# 07

# Structure and bonding of the elements

# Structure and bonding of the elements

**The geodesic dome.** Designed by Buckminster Fuller for Expo 67

**Fullerenes.** Sir Harry Kroto holding buckminsterfullerene, $C_{60}$, in his left hand, with other fullerenes he discovered

The United States pavilion at the 1967 World Fair Expo in Montreal was a giant geodesic dome, a revolutionary structure designed by the pioneering American engineer and architect Robert Buckminster Fuller.

Little did he know that 18 years later his design was to be repeated at atomic level in the Nobel prize-winning discovery of forms of carbon that made up the class of structures named after him – the buckminsterfullerenes. These are net-like spheres (and tubes), each of many atoms, possibly opening up new areas of organic chemistry.

Until Harry Kroto and his team at Sussex University discovered the chemical class of fullerenes in 1985, chemists believed that carbon had only two crystalline forms: graphite and diamond. Now there were three. Scientists were astonished, because – so they thought – carbon had been exhaustively researched over many decades. Equally astonishing was the discovery that fullerenes are present in soot.

**Fullerene model.** A football is a model of a fullerene

Like Buckminster Fuller's dome, the design of which is held in place by natural forces, the structure of a crystalline form such as a fullerene is constrained by the forces that hold its component particles in place. It is this definite internal arrangement that determines the physical properties of all elements in the crystalline state.

## 1 STRUCTURE AND THE PERIODIC TABLE

On pages 95–96, you can read how elements in the same group of the Periodic Table have similar chemical properties because they have a similar arrangement of their outer electrons. For example, all the elements in Group 7 have an outer electron configuration that ends in $p^5$. The periodicity of properties, such as ionization energy, is explained in terms of the regular recurring pattern of electron configurations from one period to another.

This chapter takes certain physical properties of elements and relates them to the structure and bonding of elements in the solid state. We look particularly at the electrical conductivity, melting points and boiling points of the elements in the second and third periods of the Periodic Table.

# 2 KINETIC–MOLECULAR MODEL FOR SOLIDS

In any discussion of solid-state structures we need to look at the kinetic–molecular model of matter as it applies to solids.

Fig 7.1 shows a typical arrangement of the particles in a solid. These are their common characteristics:

- The particles may be ions, atoms or molecules.
- They are closely packed in an ordered pattern, occupying relatively fixed positions.
- This ordered arrangement of particles is called a **lattice**.
- The particles in a solid are fixed in position, but are able to vibrate.
- The particles in a solid are attracted to one another.

We look at the nature of some of the attractive forces later in the chapter.

## TYPES OF SOLID

When a liquid substance is cooled sufficiently, it freezes and forms a solid. Normally, the particles of the substance take up ordered positions and form a crystalline solid. Even elements that are gases at room temperature and pressure solidify when the temperature is low enough. For helium, the temperature has to be lowered to within 3 degrees of absolute zero before it becomes a solid.

## CRYSTALLINE SOLIDS

In a **crystalline** solid, the particles are arranged in a definite repeating pattern throughout the solid. Such a solid is said to be homogeneous. This three-dimensional repeating pattern of particles in crystals is usually referred to as a **crystal lattice**.

Many elements form crystalline solids and have a crystal lattice that can be described geometrically. An example is diamond, one of the allotropes of carbon.

A crystalline solid has a definite melting point because all the attractive forces that hold the particles in the lattice have the same strength. So the same amount of energy is required to overcome the forces between every pair of neighbouring particles.

## AMORPHOUS SOLIDS

Some elements do not always form crystalline structures. They form **amorphous** solids. Carbon black or soot is an example. In an amorphous solid, the particles do not take up a definite regular pattern during freezing. This is usually the case when the freezing process is so fast that the particles have no time to become arranged in an orderly way.

FOR MORE INFORMATION ON THE KINETIC–MOLECULAR MODEL OF MATTER, SEE PAGE 127.

**Fig 7.1** The arrangement of particles in a solid

## ? TEST YOURSELF

**A** Describe the arrangement of particles in a liquid and in a gas.

**Fig 7.2** The arrangement of particles in an amorphous solid. It is very much like the arrangement in a liquid, but particles are free to move in a liquid and not in a solid

**Fig 7.3** Diamond has a highly ordered crystal lattice, which gives it a geometric shape. Diamond is seen here on graphite, another structured form of carbon. (Soot is an amorphous form of carbon)

Metallic glasses are amorphous solids, formed when liquid metal cools at a rate of a million degrees per second. The particles occupy completely random positions within the amorphous solid and so do not form a crystal lattice. Nevertheless, the positions of the particles are relatively fixed.

Usually, amorphous solids don't have a definite melting point, but soften gradually on heating. This is because the forces that hold their particles together have different strengths, and so different amounts of energy are needed to break them.

An element with an amorphous form always has at least one crystalline structure that is more stable than the amorphous form. Hence, an amorphous solid will slowly change into a crystalline solid. In this change of form, the particles are rearranged from a disordered state into an ordered pattern. However, if the forces between the particles in the amorphous form of a solid are strong, it could take a very long time to change from its amorphous form to its crystalline form.

# 3 PATTERNS ACROSS THE PERIODIC TABLE

**YOU CAN READ ABOUT THE CRYSTALS OF COMPOUNDS IN CHAPTER 21.**

We have already described the periodicity in properties, such as first ionization and atomic radii, that depend on the electron configuration of the elements (see Chapter 5). The periodicity shown by the physical properties of elements, such as boiling point or electrical conductivity, depends on the structure and bonding rather than the electronic configuration.

## MELTING AND BOILING POINTS

Fig 7.4 Melting points of elements in Periods 2 and 3

Fig 7.5 Boiling points of elements in Periods 2 and 3

Figs 7.4 and 7.5 show how the melting points and boiling points of the elements vary with atomic number (proton number). Although we can see a pattern – the melting point tends to rise to a maximum around Group 4 and then falls to the noble gases – it cannot really be described as a periodic function. In fact, if the melting point graph extended to elements with higher atomic numbers, we would see even less of a pattern.

Put simply, it is the strength of the attraction between the particles in an element's crystal lattice that determines the value of the melting point, rather than the electron configuration of the element. The greater the force between the particles, the higher the melting point. Much the same can be said of the boiling point, but in this case it is the magnitude of the attraction between the particles in the liquid phase.

Figs 7.4 and 7.5 show that, in terms of the melting and boiling points, there are broadly three types of element:
● metals that have reasonably high melting points;
● non-metals that have very high melting points (such as carbon and silicon);
● non-metals that have very low melting points.
The difference in the melting points of elements depends on their crystal structure and the strength of the bonding between particles (atoms, molecules or ions) of the elements.

## ELECTRICAL CONDUCTIVITY

Metals are good electrical conductors and non-metals are poor electrical conductors. For electrical conduction to occur, there must be charged particles that are free to move. In the case of an element, the charged particles are its electrons. Provided electrons in an element are free to move when subjected to a potential difference (i.e. when a voltage is applied), the element is a good electrical conductor. There is a complication: the electrical conductivity of elements can change under varying conditions of temperature and pressure, but this is dealt with on page 154.

## METALLIC AND NON-METALLIC ELEMENTS

We can understand the differences in the melting points, boiling points and electrical conductivities of the elements by looking at the types of particle which make up the crystal lattices.

A metal crystallizes into a giant structure of closely packed metal *ions*. No molecules are present. By contrast, a typical non-metal is very different: its structure normally contains *covalent bonds*, so it consists of individual molecules. Hence, the arrangement of particles in a non-metal is described as a **molecular lattice**.

The molecules may be **giant**, and we can consider the whole crystal as being one molecule, as in diamond and graphite. Alternatively, the lattice may be made up of repeating units of **simple** molecules, as in the case of solid iodine, which has a lattice that contains a repeating pattern of iodine molecules, $I_2$. The noble gases are monatomic, so the crystal lattice of a solid noble gas is based on a repeating pattern of atoms.

Across the whole range of crystalline elements, there is a wide variation in the strengths of the attractive forces between the particles that form their different crystal structures. Evidence for this is the wide variation in the melting and boiling points of these elements shown in Figs 7.4 and 7.5. Solids with small molecules, such as hydrogen, nitrogen and oxygen, have low melting points because the forces between them in the crystalline state are very weak. Graphite and diamond have very strong forces of attraction – covalent bonds – between their atoms, which gives them extremely high melting points.

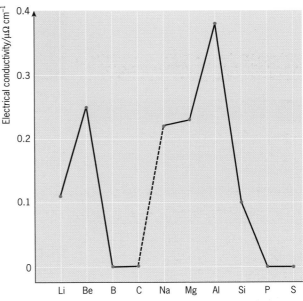

**Fig 7.6** The electrical conductivity of non-gaseous elements in Periods 2 and 3 at 298 °C

**TEST YOURSELF**

**B** Describe the changes in arrangement, motion and the strength of attraction between particles when a solid melts into a liquid

**SEE QUESTIONS 1 AND 2.**

# 4 METALLIC BONDING

Look at a lump of any pure metallic element. It is not obvious that it is crystalline: it does not have the regular faces that a crystal of diamond has. Nevertheless, a powerful microscope reveals its crystalline structure.

**Fig 7.7** The surface of stainless steel showing its crystalline nature

**Fig 7.8** Metal crystals growing on zinc in lead nitrate solution

We can see crystals forming in the displacement reaction between zinc and aqueous silver nitrate or lead(II) nitrate:

$$Zn(s) + 2Ag(NO_3)_2(aq) \rightarrow Zn(NO_3)_2(aq) + 2Ag(s)$$

$$Zn(s) + Pb(NO_3)_2(aq) \rightarrow Zn(NO_3)_2(aq) + Pb(s)$$

## CLOSE PACKING

The existence of crystals in a piece of metal means that the particles in the metal must have a regular arrangement. These particles are, in fact, positive ions and not metal atoms. All of these ions are the same size, so they can close pack in a manner similar to the most space-saving way of packing spheres in a layer: each sphere is surrounded by six others (Fig 7.9).

In some metals, close packing is achieved through a repeating pattern in which each positive ion is in contact with six others in the same layer. Such a

**Fig 7.9** Bubbles packed together in a close-packed array

---

**? TEST YOURSELF**

**C** Write down the ionic equation for the displacement reaction between:
**a)** zinc and aqueous silver ions;
**b)** zinc and aqueous lead(II) ions.

There are three different particle arrangements for metals. Each one is an example of close packing

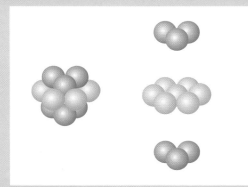

**Fig 7.10** Hexagonal close packing. The arrangement of each layer is like that of the bubbles in Fig 7.9

**Fig 7.11** Cubic close packing. The arrangement is also like that of the bubbles in Fig 7.9

**Fig 7.12** Body-centred cubic close packing

crystal will have not just one but many layers of positive ions. Each additional layer is arranged to maximize the number of ions that can be packed.

However, there is more than one way of close-packing positive ions in a three-dimensional metallic crystal lattice, as Figs 7.10 to 7.12 show.

The **coordination number** is the number of nearest neighbours that a particle has when it is in a crystal lattice. A **unit cell** is the simplest pattern of particles for which repetition in three dimensions produces a crystal lattice.

Since metal ions are closely packed, it is not surprising that most metals have high densities.

## POSITIVE IONS IN A SEA OF DELOCALIZED ELECTRONS

Since they all have the same charge, why don't the metal ions repel one another? Clearly, they don't, otherwise metals would not be strong and mostly hard. The answer lies in the electrons that have been removed to form the positive ions. These electrons are free to move throughout the metal crystal. The electrons are said to be **delocalized**. The delocalized electrons are often referred to as a 'sea of electrons', because they occupy the whole space between the closely packed metal ions (Fig 7.13). The positive metal ions are not repelled from one another because each is electrostatically attracted towards the sea of delocalized electrons.

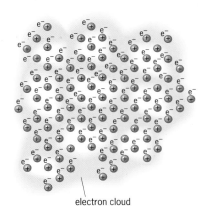

electron cloud

**Fig 7.13** The sea of electrons model

## ELECTRICAL CONDUCTIVITY OF METALS

All metals are good electrical conductors because of the presence of delocalized electrons throughout the metal lattice. It is only the electrons in the outer shell that can be delocalized. The inner electrons are still localized around the nucleus of the metal atom. This means that, for example, 1 mole of sodium has 1 mole of delocalized electrons, since it forms $Na^+$, whereas 1 mole of magnesium has 2 moles of delocalized electrons, since it forms $Mg^{2+}$.

When an electrical potential is placed across a strip of metal, electrons move from the negative to the positive area of the metal, and as a result a current flows. Normally, there is a resistance to the flow of electrons in a metal. This results in some of the electrical energy being lost as heat.

**SEE QUESTION 3.**

**?** **TEST YOURSELF**

**F** Which of the following metals is likely to have the greatest electrical conductivity: aluminium, magnesium or sodium? Explain your answer.

**✔ REMEMBER THIS**

In transition elements, d electrons can also be delocalized throughout the metal lattice. (See Chapter 24.)

ELECTRONS AND ENERGY LEVELS ARE COVERED IN CHAPTER 3.

SEE QUESTION 4.

**!** **SCIENCE FEATURE** Superconductors

Superconductors are materials that conduct electricity with little or no resistance. So, when a current flows through a superconductor, there should be little or no heating effect. The superconducting materials presently available work only at extremely low temperatures. Many superconductors have metal oxide structures. A few metals, such as aluminium and mercury, can be superconducting, but unfortunately only at temperatures just above absolute zero (1.2 K for aluminium and 4.2 K for mercury).

Superconductors offer the promise of loss-free power transmission, super-fast electronic circuits, powerful electromagnets and ultra-sensitive magnetic detectors. This will be fulfilled only if superconducting materials can be developed that work at around 20 °C.

**Fig 7.14** Levitating magnets, an example of superconductivity. This is a disc of yttrium–barium–copper oxide floating above a nitrogen-cooled cylinder of superconducting ceramic

## OTHER PHYSICAL PROPERTIES OF METALS

All the other typical physical properties of metals can be explained by reference to the model of bonding in which the positive ions are immersed in a sea of delocalized electrons.

● A metal is strong because when its structure is deformed by applied stress and the positive ions move, the delocalized electrons move as well, thus maintaining their attraction to the positive ions.

● Kinetic energy is transferred easily from one delocalized electron to another, which explains why metals are such good thermal conductors.

● The presence of delocalized electrons also explains why metals are shiny and have a lustre. These electrons are easily promoted to higher energy levels, from which they fall to lower levels, emitting the light that makes the metal shine.

● Metals have fairly high melting and boiling points because the attraction between the delocalized electrons and the positive ions is strong. The greater the number of delocalized electrons the stronger this attraction, so magnesium has a higher melting point than sodium.

## 5 GIANT MOLECULES

Some elements have extremely high melting points. This is because the particles – atoms – in the crystal lattice are strongly attracted to one another. In a giant molecule the atoms are held together by strong covalent bonds. The whole crystal of a giant molecule should be considered as one molecule. For such a structure to melt, each atom must be able to move freely. For this to happen, every covalent bond must be broken. Covalent bonds are very strong, so need a large amount of energy to break. This means the melting point, and consequently the boiling point, is very high. Carbon and silicon, which have the highest melting points in their respective periods, exist as giant molecules. Elements that have a giant molecular structure are also said to have a macromolecular or giant covalent structure.

## ALLOTROPY OF CARBON

Diamond and graphite are different crystalline forms of carbon that both have a giant molecular structure. These crystalline forms are called **allotropes**. An element is said to have allotropes (show allotropy) when it exists in the same state in more than one structural form. Oxygen, for example, can exist in two forms in the gaseous state: dioxygen ($O_2$) and ozone ($O_3$). The allotropes of an element have the same or similar chemical properties, but their physical properties are different, because each allotrope has its own crystal structure. In the case of diamond and graphite, the physical properties are markedly different. For example, graphite is a good conductor of electricity, but diamond is an extremely poor conductor. These differences in properties are associated with the different internal crystal structures.

## DIAMOND

Diamond is the hardest substance known and is the least compressible. It is a better conductor of heat at room temperature than any other material, and when completely pure it is transparent. These extreme properties make diamond technologically very useful. Its hardness makes it useful as an industrial abrasive and as a cutting tool in industry and surgery, and because of its excellent heat conductivity it is used as a heat sink to cool electronic components rapidly. Certain impurities implanted in diamond can make it a semiconductor (as for fullerenes, see the Assignment, page 164). These applications are a direct result of the internal structure of a diamond crystal (Fig 7.16), coupled with the fact that the carbon atoms in diamond are more closely packed than the atoms in any other material.

The atoms in a crystal of diamond are bonded into one giant molecule. Each carbon atom is covalently bonded to four other carbon atoms, and so each carbon atom has four nearest neighbours. That is, its coordination number is 4. When just four bonding pairs of electrons surround an atom, they are arranged tetrahedrally (see page 78). This is the case with diamond. All outer electrons of the carbon atoms are involved in the formation of covalent bonds. There is no possibility of delocalized or mobile electrons. As a consequence, diamond cannot conduct electricity.

? **TEST YOURSELF**

**G** Diamond and graphite both burn in oxygen to give a gaseous product. What is the name of this gaseous product?

SEE QUESTIONS 1 AND 7.

**Fig 7.15** A diamond-tipped oil drill part

**Fig 7.16** The internal structure of diamond

**Fig 7.17** A unit cell for diamond

✔ **REMEMBER THIS**

In diamond each carbon atom is $sp^3$ hybridized, whereas in graphite each carbon atom is $sp^2$ hybridized. (You can read about $sp^3$ and $sp^2$ hybridization on page 82.)

Top view of one
layer of graphite –
like a honeycomb

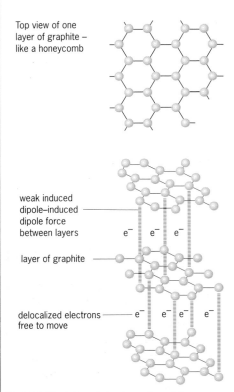

weak induced
dipole–induced
dipole force
between layers

layer of graphite

delocalized electrons
free to move

**Fig 7.18** The internal structure of graphite

**SEE QUESTIONS 5 AND 6.**

**Fig 7.19** A unit cell for graphite

# GRAPHITE

Figs 7.18 and 7.19 show that the arrangement of carbon atoms in graphite is considerably different from that in diamond. In graphite, each carbon atom is covalently bonded to three other carbon atoms. All the bond angles are 120°, which is what would be expected if only three bonding electron pairs surrounded the carbon atom.

However, a check on the number of electrons reveals that there is one electron per carbon atom left over. These electrons are delocalized and explain the electrical conductivity of graphite. Although graphite is composed of giant molecules, its atoms are arranged in layers that can slide past one another. Weak van der Waals forces hold the layers together (see page 158).

The internal structure of graphite accounts for its:

● extremely high melting point, since many very strong covalent bonds have to be broken to allow the carbon atoms to move freely;
● electrical conductivity, since the delocalized electrons are free to move when subjected to a potential difference;
● brittleness, since it is easy to split one layer from another as only the weak van der Waals forces need to be broken.

Graphite is used to make electrodes (particularly for high-temperature work, since they will not melt), electrical contacts and brushes for electrical motors. It is also used in lubricants and in pencil leads. Composite materials that contain graphite are used to make racquets, fishing rods and golf clubs.

**Fig 7.20** When graphite is ground and compressed, it forms the 'lead' in pencils

## EXTENSION Stability of graphite and diamond

Given its structure, it is surprising that graphite is the most stable form of carbon. The logical conclusion of this would be that diamond should change into graphite at room temperature. Fortunately, this does not happen, because to change the structure of the diamond lattice, all of its covalent bonds would have to be broken. This would need an enormous amount of energy, which is not available at room temperature. Chemists refer to this property as having a huge **activation energy**.

In the same way, it is difficult to convert graphite into diamond, because it has a very large activation energy. To convert graphite into diamond requires an extremely high temperature and pressure. In fact, a pressure of approximately 60 000 atmospheres and a temperature of 1500 °C are needed before diamond becomes more stable than graphite.

**Fig 7.21** Reaction profile for the conversion of diamond to graphite. The profile is not drawn to scale because of the magnitude of the activation energy. To convert diamond to graphite requires extremely high temperatures to provide sufficient energy to overcome the activation energy

The conversion of graphite to diamond is endothermic:

$$C_{(graphite)} \rightarrow C_{(diamond)} \quad \Delta H = +1.9 \text{ kJmol}^{-1}$$

Natural diamonds are produced by this combined action of heat and pressure deep under the Earth's surface. Artificially, graphite and carbon-containing compounds, such as coal, or even peanut butter, are processed at about 100 000 atmospheres and 2000 °C to make industrial diamonds. Despite the commercial importance of industrial diamonds, only about 100 tonnes per year are needed world-wide. Most of these industrial diamonds are used in abrasive coatings and in cutting tools. They are quite small and dull, unlike the clear, sparkling diamonds used in jewellery.

# FULLERENES

Since the mid-1980s, a new class of allotropes of carbon has been identified. They were given the name 'fullerenes' (see the Opener). The most widely studied fullerene is $C_{60}$, which is a self-contained molecule that exists in both crystalline and amorphous forms. An interesting aspect of the $C_{60}$ molecule, apart from its sheer size, is its shape, which resembles a football with hexagonal and pentagonal panels. So, originally called buckminsterfullerene, $C_{60}$ is now almost always known as 'buckyball'. Since the discovery of $C_{60}$, many other arrangements have been identified: for example, $C_{70}$, $C_{72}$ and $C_{84}$.

**Fig 7.22** Fullerene $C_{72}$

**YOU CAN READ MORE ON FULLERENES IN THIS CHAPTER'S ASSIGNMENT ON PAGE 163.**

# SILICON

Silicon is in Group 4 of the Periodic Table. From this we can reasonably predict that the silicon atom forms four covalent bonds. Silicon forms a giant molecular structure in which every silicon atom is surrounded by four covalent bonds, similar to the structure of diamond.

Since silicon and diamond have the same internal structures, their physical properties should be similar – and they are. Silicon has a very high melting point, it is very hard and in its pure state it is a very poor conductor of electricity.

**Fig 7.23** The internal structure of silicon. Note the similarity to diamond, the only difference being the longer covalent bond

## ! SCIENCE FEATURE Semiconductors

The biggest use for silicon is in semiconductor components for electronic circuits. You may wonder how it is that a non-metallic element with a diamond-like structure can be made to conduct electricity. **Semiconductors** are materials that, to put it simply, have been altered to allow electrons to pass through them, but not as easily as through metals. We know that metals conduct because they have a sea of delocalized electrons and that an insulator such as diamond does not conduct because all its electrons are locked into covalent bonds.

In silicon, it is possible to excite electrons into vacant higher energy orbitals. When this happens, these electrons become free to move. But at room temperature there is not sufficient energy to promote more than a small fraction of electrons and therefore the electrical conductivity stays exceedingly low.

A way to improve the electrical conductivity of silicon is to provide *extra electrons* that are easier to excite. This improvement is made by adding trace amounts of an element such as arsenic, a process called **doping**. The added element is called a **dopant**. As described in the Assignment (page 164), doping is a powerful way to increase the electrical conductivity of certain materials in the solid state.

Boron is also used to dope silicon, but the mechanism is not the same as for arsenic. With boron, the number of electrons is *reduced*, which causes the formation of electron vacancies, known as **holes**. This encourages electrons to move to fill the holes, and thereby generates new holes. These holes are filled by other electrons moving in, and so on. There is thus a flow of charge through the silicon and so it conducts electricity.

---

## ? TEST YOURSELF

**H** **a)** Which feature of the structure of graphite makes it useful as a lubricant?

**b)** Explain why graphite is a suitable material with which to make high-temperature crucibles.

**c)** Which allotrope of carbon would you expect to have the higher melting point? Explain your answer.

## ? TEST YOURSELF

**I** **a)** Explain why silicon has a very high melting point.

**b)** Silicon carbide, SiC, is used as an abrasive on grinding wheels. It has a high melting point and does not conduct electricity:

(i) What type of structure do you think silicon carbide has?

(ii) Draw the unit cell for silicon carbide.

## ? TEST YOURSELF

**J** The electrical conductivity of silicon increases as the temperature increases. Explain this property in terms of the ease with which electrons can be excited.

## ? TEST YOURSELF

**K** **a)** What is the number of electrons in the outer shell of a silicon atom and of an arsenic atom?

**b)** Explain how arsenic can provide extra electrons when added in trace amounts to silicon.

## ? TEST YOURSELF

**L** **a)** Write down the electron configuration for boron.

**b)** Explain how adding trace amounts of boron to silicon can produce electron holes.

## ⚠ SCIENCE FEATURE The solar cell

Most of the world's electrical energy requirement is met by burning fossil fuels, using the heat of nuclear reactors and harnessing water power, all of which have environmental consequences.

Fossil fuels produce lots of carbon dioxide when they are burnt, which is assumed to contribute to the greenhouse effect. In addition, many fossil fuels have a large sulphur content, which results in sulphur dioxide emissions that lead to acid rain. Sulphur dioxide emissions can be reduced, but carbon dioxide ones cannot. Furthermore, fossil fuels are a finite resource, which will eventually run out.

With nuclear fuels the main problems are how to safely dispose of nuclear waste and decommission nuclear power stations when they reach the end of their useful life.

Although hydroelectric power stations operate cleanly, both building and running them have ecological and hydrological consequences.

The Earth receives more energy in sunlight in two days than is stored in all the known energy resources. So, if just a fraction of this energy could be harnessed it would provide a major alternative supply of electricity. Many solar energy plants already exist around the world, but many more are needed if they are to become a real alternative to the other sources.

Solar cells are used to convert radiant energy into electrical energy. One form of solar cell consists of two joined layers of silicon. One layer is doped with arsenic or phosphorus and is called n-type silicon (arsenic and phosphorus each have five electrons in their outer shells). This layer has *extra* electrons available for conduction, generated by the small number of dopant atoms. The other layer is doped with boron and is called p-type silicon. This layer has a *shortage* of electrons (a surplus of holes) because boron has only three outer electrons. The cell is connected into an external circuit.

Electrons flow from the n-type to the p-type layer. Eventually, equilibrium is established with a potential difference between the two layers. When sunlight falls on the cell, the equilibrium is disturbed and electrons move from the p-type to the n-type layer. The electrons return to the p-type layer via the external circuit, and thereby generate an electric current.

**Fig 7.24** Aurora 101, which won the race in Australia for solar-powered cars

**Fig 7.25** Solar cells on the Hubble Telescope

**Fig 7.26** Solar cells on a roof

arsenic atom in place of silicon

extra electron due to presence of arsenic atom

**Fig 7.27** n-type silicon

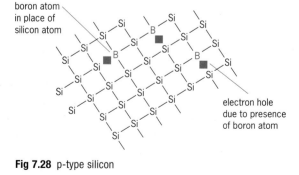

boron atom in place of silicon atom

electron hole due to presence of boron atom

**Fig 7.28** p-type silicon

sunlight

p-type silicon

n-type silicon

+ve external circuit

−ve

**Fig 7.29** A solar cell

# 6 SIMPLE-MOLECULE LATTICES

Some non-metals, such as solid nitrogen, oxygen, chlorine and iodine, form crystal lattices that consist of the simple molecules ($N_2$, $O_2$, $Cl_2$, $I_2$) held in place by attractive forces between the molecules. These forces are known as **intermolecular forces** because they are forces between molecules rather than within a molecule. The intermolecular forces are weak between simple molecules, and therefore the non-metals mentioned have relatively low melting points and boiling points. (The intermolecular forces are often referred to as **van der Waals** forces.) These weak forces explain why many non-metals are gases at room temperature and atmospheric pressure. It is easy to separate the molecules, but difficult to atomize them, because the covalent bonds within each molecule (the intramolecular bonds) are so much stronger.

Elements with a simple molecular structure do not conduct electricity, since they do not have electrons or ions that are free to move. All the electrons are localized.

## DIATOMIC MOLECULES

As mentioned before, many gases, including chlorine, oxygen, nitrogen and hydrogen, form diatomic molecules. They cannot have a permanent dipole as both atoms must have equal electronegativity. So it is difficult to imagine forces of attraction between these neutral molecules.

An induced dipole–induced dipole attraction is set up between molecules. A temporary dipole is set up in a *molecule* when the electrons in the covalent bond that links the atoms move more to one end of the molecule than to the other. It is important to realize that the dipole is temporary and the electrons may also move more to the other end of the molecule. As a result of the temporary dipole, one end of the molecule carries a very small positive charge and the other end a very small negative charge. This small charge separation is not permanent and so in another instant in time the charges can switch ends. Look at Fig 7.30, which shows the way an intermolecular force may be set up in a hydrogen molecule. Once a temporary dipole is set up within a molecule, it induces a similar dipole in neighbouring molecules and a very weak intermolecular force is set up.

The intermolecular force is known as an **induced dipole–induced dipole interaction**. When the temperature is sufficiently low, this force may be strong enough to hold the hydrogen molecules in a solid crystalline lattice.

The melting point of hydrogen is very low, since the two electrons in a hydrogen molecule are firmly attracted to the nuclei of their atoms. The larger the molecule, the greater the intermolecular force, because there are more electrons and the outer electrons are under a weaker attractive force from the nuclei. So it is easier to produce a temporary dipole and the induced dipole–induced dipole attraction becomes stronger. In Group 7 this is illustrated by the change in physical state at room temperature and pressure. Fluorine, $F_2$, a small molecule with few electrons, has weak induced dipole–induced dipole attractions, whereas bromine, $Br_2$, a liquid, has more electrons, a larger molecule and stronger induced dipole–induced dipole attractions. In the case of iodine, $I_2$, the strength of the intermolecular force is sufficient to make iodine a solid at room temperature.

It is important to understand that in these types of simple molecular crystal lattice it is attraction *between molecules* that holds the lattice together. The atoms within the molecule are covalently bonded to one another, but the strength of this covalent bonding does not determine the magnitude of the melting point.

**A TEMPORARY DIPOLE CAN ALSO BE KNOWN AS AN INSTANTANEOUS DIPOLE.**

attractive force between electrons and protons

Once an instantaneous dipole forms, attractive forces distort the electron cloud on a neighbouring molecule, producing a temporary induced dipole

The distortion (greatly exaggerated in the figure) results in only extremely weak forces of attraction

**Fig 7.30** The formation of induced dipoles in hydrogen

**SEE QUESTIONS 7, 8 AND 9.**

## ? TEST YOURSELF

**M** Which halogen is likely to have the highest melting point? Explain your answer.

**? TEST YOURSELF**

**N** Suggest why phosphorus and sulphur are solids at room temperature and atmospheric pressure, but nitrogen and oxygen are gases.

**SEE QUESTION 10.**

An instantaneous dipole forms on the left atom, causing an attractive force on the nucleus of the right atom, producing a temporary induced dipole

attractive force between electrons and protons

The distortion (greatly exaggerated in the figure) results in only extremely weak forces of attraction

**Fig 7.33** The formation of induced dipoles in helium

**SEE QUESTION 6.**

**? TEST YOURSELF**

**O** Which of the noble gases has the highest melting point?

# PHOSPHORUS AND SULPHUR

Phosphorus and sulphur both form simple molecular lattices, but a discussion of the structures is complicated since both elements exhibit allotropy. Suffice it to say that phosphorus forms a lattice in which $P_4$ molecules are arranged in a fixed pattern, while sulphur has a molecule with the formula $S_8$. The $P_4$ and $S_8$ molecules are held in position in their respective lattices by weak induced dipole–induced dipole interactions.

Phosphorus and sulphur are both solids at room temperature and atmospheric pressure because the induced dipole–induced dipole interactions are much stronger than those in the diatomic molecules of nitrogen and oxygen, since there are more electrons in molecules of phosphorus and sulphur. Sulpher has a higher melting point than phosphorus because it has larger molecules with more electrons, so the intermolecular force is stonger.

**Fig 7.31** The $P_4$ molecule is a regular tetrahedron

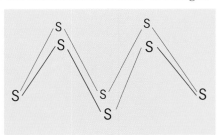

**Fig 7.32** The puckered ring of $S_8$. All S–S lengths are identical

# THE NOBLE GASES

The noble gases exist as monatomic molecules; that is, just as atoms. All the noble gases have low melting points. Helium has the lowest melting point of any element, and therefore it must have the weakest intermolecular forces between its atoms

The electrons in an atom are in constant motion and do not occupy set positions or set orbits (see page 56). Therefore, it is possible for both of the electrons in a helium atom to arrive simultaneously on the same side of the atom, as in Fig 7.33. This side of the helium atom thus becomes very slightly negatively charged, while the other side becomes very slightly positively charged.

The electron cloud around the helium nucleus becomes distorted (it is no longer spherical), and it causes a similar distortion in neighbouring helium atoms. In other words, a temporary or instantaneous dipole is induced in neighbouring atoms.

A very weak attraction then exists between the slightly positive side of one helium atom and the slightly negative side of a neighbouring helium atom. This is a very weak intermolecular force because it relies on an asymmetric distribution of electrons, which has a low probability since both electrons are close to the nucleus, so are firmly attracted to the nucleus. At the same time, they repel each other.

Once this force between helium atoms in the solid phase has been broken, the attraction between helium atoms in the liquid phase is negligible, and so helium boils only 4 degrees higher than the temperature at which it melts.

The larger the noble gas atom (hence the greater the number of electrons), the greater the likelihood of this asymmetric distribution of electrons. This is because the outer electrons are further away from the nucleus and are shielded from the nuclear charge. This means that the outer electrons can occupy a larger region and so undergo a smaller electron–electron repulsion, even when asymmetrically distributed. Therefore, the melting and boiling points of the noble gases increase with increasing atomic (proton) number.

# 7 A SUMMARY OF THE STRUCTURE AND BONDING OF ELEMENTS IN PERIOD 3

We have seen that the physical properties of elements in period 3 vary considerably. Table 7.1 summarizes the physical properties.

| Table 7.1 Summary of the physical properties of period 3 elements | | | | | | | | |
|---|---|---|---|---|---|---|---|---|
| | **Sodium** | **Magnesium** | **Aluminium** | **Silicon** | **Phosphorus** | **Sulphur** | **Chlorine** | **Argon** |
| Bonding | Metallic | Metallic | Metallic | Covalent | Covalent | Covalent | Covalent | |
| Structure | Giant | Giant | Giant | Giant | Simple | Simple | Simple | Simple |
| Melting point | Low | High | High | Very high | Low | Low | Very low | Very low |
| Electrical conductivity of solid | High | High | High | Low | Very low | Very low | Very low | Very low |

## SUMMARY

After studying this chapter, you should know the following.

- A crystalline solid has a regular arrangement of particles called a lattice that can be produced by the repetition in three dimensions of a unit cell. Crystalline solids have definite melting points.
- An amorphous solid has a closely packed structure that is disordered, and therefore does not have a unit cell. Amorphous solids melt over a range of temperatures, and many soften on heating.
- The structure and bonding within a crystal lattice determine the physical properties of an element.
- The melting point of a solid is determined by the strength of the force of attraction between particles in a lattice.
- A metal consists of closely packed positive ions in a sea of delocalized electrons. It has a giant structure.
- Metals conduct electricity because of the movement of the delocalized electrons in the presence of a potential difference.
- Metals are typically hard, strong, shiny and good thermal conductors. Many have high melting points. These properties can be explained by the structure and bonding.
- Non-metals form molecular lattices.

- Non-metals that form a molecular lattice have a regular arrangement of simple molecules held together by weak intermolecular forces, called van der Waals forces (induced dipole–induced dipole interactions). Such non-metals have low melting points and low boiling points and do not conduct electricity.
- The strength of the induced dipole–induced dipole interaction in a diatomic element depends on the molecular size (indicated by the number of electrons in the molecule).
- A non-metal that forms a giant molecular or macromolecular lattice has a regular arrangement of atoms held together by strong covalent bonds. The whole crystal is assumed to be the molecule. Such non-metals have high melting points.
- Non-metals are generally poor electrical conductors because they have no free electrons.
- Different forms of the same element in the same state are known as allotropes. Allotropes have different physical properties, but similar chemical properties.
- Diamond, graphite and fullerene ($C_{60}$) are three allotropes of carbon. Carbon and diamond have a giant molecular structure. Graphite has mobile delocalized electrons, so it conducts electricity. Diamond does not conduct electricity.

# ? EXAM QUESTIONS

**1** Explain why the melting point of the elements increase in the order:

Sodium < magnesium < aluminium < silicon

IBO Chemistry Standard Level Paper 2 May 1999 Q5c

**2**

**a)** State which of the elements neon, sodium, magnesium, aluminium and silicon has the lowest melting point and explain your answer in terms of the structure and bonding present in that element.

**b)** State which of the elements neon, sodium, magnesium, aluminium and silicon has the highest melting point and explain your answer in terms of the structure and bonding present in that element.

AQA Unit 1: Atomic Structure, Bonding and Periodicity Jan 2001 Q4

**3**

**a)** Describe the change in electrical conductivity of the solid elements across Period 3.

**b)** How do you account for these changes in terms of the structure and bonding of these solid elements?

**c)** Suggest why it is difficult to measure the electrical conductivity of an element such as argon or chlorine.

**4** Pure boron is a transparent crystalline solid. It has a melting point that is above 2000 °C. Boron crystals are made up of boron icosahedra linked together by forces of attraction. Look at the Fig 7.Q4, which shows one of the boron icosahedra.

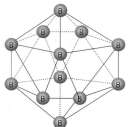

**Fig 7.Q4**

**a)** What is the formula of a boron icosahedron?

**b)** What is the coordination number of each boron atom in the icosahedron?

**c)** What deduction can you make from the fact that boron forms a crystalline solid rather than an amorphous one?

**d)** Would you expect boron to be a good electrical conductor? Explain your answer.

**e)** What deduction about the forces of attraction between boron icosahedra can you make from the magnitude of the melting point of boron?

**5** A number of compounds that contain boron–nitrogen bonds have been prepared. One of these is boron nitride, BN.

**a)** Boron nitride is prepared by the direct combination of nitrogen and boron at high temperature. Construct an equation for this reaction.

**b)** Boron nitride can exist as a giant molecule that has a layered structure.
Look at Fig 7.Q5, which shows a part of this layered structure.

**(i)** An allotrope of carbon has a similar structure: which allotrope?

**(ii)** Predict some of the physical properties you would expect boron nitride to possess.

**Fig 7.Q5**

**c)** When the layered structure of boron nitride is subjected to very high temperature and very high pressure it is changed into a very hard material.

**(i)** Suggest which allotrope of carbon this form of boron nitride resembles.

**(ii)** Suggest a structure for the unit cell of this very hard form of boron nitride.

**6** Many elements exist in several allotropic forms.

**a)** What does the term allotropy mean?

**b)** Give the name of an element that has two allotropes in the gas phase. What are the names of these two allotropes?

**c)** **(i)** Give the name of an element that shows allotropy in the solid state.

**(ii)** Give the names of two of these allotropes.

**(iii)** Would you expect these allotropes to give the same product when burnt in excess air? Explain your answer.

**7** Fig 7.Q7 shows the structure of crystalline iodine.

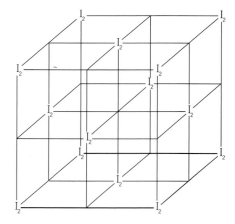

**Fig 7.Q7**

**a)** What type of lattice is this?

**b)** Describe the nature of the forces of attraction that exists between:

## ? EXAM QUESTIONS

**(i)** two iodine atoms in the iodine molecule,
**(ii)** the iodine molecules.
**c)** When heated, iodine sublimes.
  **(i)** What is sublimation?
  **(ii)** Describe the changes in the arrangement of iodine molecules and the forces between them when iodine is heated until it sublimes.

**8**
**a)** Name the type of force that holds the particles together in an ionic crystal.
**b)** Describe the bonding in a metal.
**c)** Name the type of force that exists between molecules of chlorine.

AQA Unit 1: Atomic Structure, Bonding and Periodicity June 2001 Q3 a, d, e (iii)

**9** How are the hydrogen molecules held together in liquid hydrogen?

AQA CH01 Feb 1997 Q4a (ii)

**10** Give an explanation for each of the following observations in terms of structure and bonding:
**a)** Diamond has a higher melting point than graphite.
**b)** Aluminium is a better electrical conductor than silicon.
**c)** Magnesium has a higher melting point than sodium.
**d)** Sulphur has a higher boiling point than oxygen.
**e)** The range of temperature over which helium is a liquid is much less than the range of temperature over which sodium is a liquid.
**f)** The melting points of the elements in Group 7 increase with increasing atomic (proton) number.

# KEY SKILLS **ASSIGNMENT**

## BUCKMINSTERFULLERENES

Fullerenes are a newly discovered class of carbon molecules that consist of spherical caged structures, and also tubes. The carbon atoms in a fullerene are held together by covalent bonds. $C_{60}$ was the first fullerene to be discovered, but now many more are known. The molecules are called fullerenes after the American architect Robert Buckminster Fuller, whose geodesic domes have the same structural features.

**1** Predict the type of intermolecular forces that exist between molecules in a crystal of fullerene $C_{60}$.

**Fig 7.A1** The mass spectrum of laser-vaporized graphite showing an *m/e* value of 720

Evidence for the $C_{60}$ molecule was first found in 1985 by Sir Harry Kroto. He vaporized graphite in a helium atmosphere and then analysed the resulting carbon species using mass spectroscopy, finding a positive ion that had a mass corresponding to 60 carbon atoms. In 1996, Sir Harry and the two American chemists Richard Smalley and Robert Curl were awarded the Nobel Prize for Chemistry for their pioneering work on fullerenes.

**2** Look at Fig 7.A1 and refer to mass spectrometry covered on page 237.
**a)** Explain how the value of *m/e* would suggest a carbon species with the formula $C_{60}$.
**b)** About 1 per cent of all carbon atoms contain the isotope carbon-13. Look carefully at the peaks in the mass spectrum of fullerene, which shows that the 720 peak is quite broad. Use the presence of the isotope carbon-13 to suggest why this may be so.
**c)** Suggest an identity for the peak at *m/e* = 840.
Sir Harry Kroto deduced that he had found a new form of carbon. In 1990 Wolfgang Kratschmer announced that he had prepared and analysed this new allotrope of carbon.

He vaporized graphite in an inert atmosphere by passing a very high current through it, and obtained a fluffy black soot.

X-Ray crystallography showed that it contained a spherical $C_{60}$ molecule. $C_{60}$ gives a red-wine coloured solution with benzene,

# KEY SKILLS **ASSIGNMENT**

whereas other forms of carbon are insoluble in this solvent. When benzene was added to the black soot, followed by filtering and evaporation of the solvent under reduced pressure, pure mustard-coloured crystals of the buckminsterfullerene were formed.

**3** Buckminsterfullerene $C_{60}$ is made by vaporizing graphite using an extremely high electric current.
**a)** Explain why a very high electric current is needed to vaporize graphite.
**b)** Suggest a reason why buckminsterfullerenes are not made by vaporizing diamond using an extremely high electric current.
**c)** Buckminsterfullerenes must be made in an absence of air. Suggest why.
**d)** In an experiment, a sample of 1.204 grams of graphite was subjected to vaporization in an inert atmosphere. The yield of purified buckminsterfullerene was only 0.0320 milligrams. Calculate the percentage yield of the buckminsterfullerene.

**4** Buckminsterfullerene $C_{60}$ is stable in the dark, but fragments in ultraviolet light. Suggest what type of carbon–carbon bond fission must take place.

Interest in these molecules has been intense and chemists have tried to introduce different atoms and functional groups into fullerenes in an effort to stabilize the spherical molecule and to give it different properties. Fullerene $C_{60}$ has been fully hydrogenated (reacted with hydrogen) to form $C_{60}H_{60}$.

**5** Under certain conditions it is possible to fully hydrogenate $C_{60}$ to form $C_{60}H_{60}$.
**a)** i) Explain how this suggests the presence of double bonds in fullerene.
ii) Suggest how many double bonds are present in fullerene.
**b)** Suggest another chemical test that could be used to determine whether $C_{60}$ contains a double bond.
**c)** Suggest a spectroscopic technique that could be used to establish the presence of a carbon–carbon double bond.
Hint: The hydrogenation of alkenes is covered in Chapter 14.

Fully fluorinated fullerene $C_{60}F_{60}$ has been prepared. By analogy with poly(tetrafluoroethene), Teflon, chemists argued that it should be highly water repellent and resistant to other chemicals. $C_{60}F_{60}$ is believed also to have excellent lubricating properties, so could we perhaps think of $C_{60}F_{60}$ as a future miniature ball bearing of diameter 0.7nm? Unfortunately, no, since $C_{60}F_{60}$ has been found to react with water to form hydrogen fluoride.

**6** What type of reaction takes place between fluorinated fullerene $C_{60}F_{60}$ and water to make hydrogen fluoride? Fullerene $C_{60}$ does not conduct electricity as it has no delocalized electrons. However, it can be made to conduct by introducing metal atoms into the spaces between the spherical molecules. This is a doping process and potassium atoms have been successfully introduced to form $K_3C_{60}$. When cooled below 18 K, $K_3C_{60}$ becomes a superconductor. The superconductivity is a result of potassium atoms donating electrons to the $C_{60}$ molecule.

Other dopant metals have been tried with $C_{60}$ in an effort to extend the temperature range of this type of superconductor. Success in this endeavour would revolutionize electrical engineering.

**7** **a)** Explain why electrical engineers are excited at the prospect of making superconducting fullerenes.
**b)** Why are research scientists trying to develop materials that are superconductive at room temperature?
**c)** Calculate the percentage by mass of potassium in crystals of the doped fullerene $K_3C_{60}$.

**8** List as many differences as you can in the structure and properties of the three allotropes of carbon (graphite, diamond and fullerene $C_{60}$).

**9** Write a report for a scientific magazine in which you explore the possible applications of fullerenes and their derivatives. You will need to do background research using the Internet and contemporary scientific literature.

# 08

# Fuels, energy changes and reaction rates

**Hydrogen fuel** This is dangerous to transport, as the *Hindenburg* disaster of 1937 showed. The *Hindenburg* was a hydrogen-filled airship that blew up, killing 35 people

**Hydrogen fuel** Hydrogen has become a fuel in space exploration: the US space shuttle piggy-backs a lift on giant tanks that contain 1.74 million litres of liquid hydrogen and 0.65 million litres of liquid oxygen

World supplies of oil are likely to run out before the end of the twenty-first century, but by then other energy sources should have been developed. A possibility is hydrogen, derived from water. Hydrogen has several advantages: when it burns it produces no harmful products, it gives out more energy per gram than natural gas or petrol, and existing car engines need only minor modifications to use it as a fuel.

The reason we don't start splitting water now to obtain hydrogen is that the process requires energy. In the UK, this would have to come from burning fossil fuels, which would cancel out any environmental advantages. Alternatively, more efficient solar cells might provide the energy to split water by electrolysis. The possibility of producing hydrogen from the photosynthetic reactions of plants has been investigated, but so far with little success.

Another problem is how to carry hydrogen around, since to stay liquid and occupy a convenient volume it would have to be stored below −253 °C. One promising area of research is the use of metal hydrides, solid compounds of a metal and hydrogen: the hydrogen could be produced from them as needed by heating.

**Hydrogen fuel** A car with an experimental engine that runs on hydrogen

**YOU CAN READ MORE ABOUT GREENHOUSE GASES AND SOME OF THE OTHER HARMFUL BY-PRODUCTS OF USING PETROL AS A FUEL ON PAGE 221.**

## ? TEST YOURSELF

**A** Magnesium produces energy on burning. Why don't we class magnesium as a fuel?

## ✔ REMEMBER THIS

kJ mol$^{-1}$ is the energy released per mole of equation. This means the mole quantities given in the equation release so many kilojoules.

$\Delta H$ (pronounced 'delta H') is the **enthalpy change** of the reaction. It is measured under conditions of constant pressure.

## WHAT IS A FUEL?

A fuel is a substance that releases energy that can do work. Most fuels release this energy during combustion reactions. So what properties make an ideal fuel? Petrol is a popular fuel, but is it ideal?

For its mass, petrol produces a lot of energy, which makes the expense of transporting and storing it worthwhile. It doesn't produce solid waste that needs to be dumped. But an ideal fuel produces no harmful by-products when it burns, and petrol does. Even if the by-products were just carbon dioxide and water, we now know that carbon dioxide is a major greenhouse gas and its continued output at today's levels is leading to global warming.

## 1 HOW A FUEL RELEASES ITS ENERGY: EXOTHERMIC REACTIONS AND ENTHALPY CHANGES

When natural gas burns it releases its energy to the surroundings, which could be the water in a central heating system. This is an **exothermic reaction** between methane (in natural gas) and oxygen.

$$CH_4(g) + 2O_2(g) \rightarrow CO_2(g) + 2H_2O(l) \ \Delta H^{\ominus}_{298} = -890 \text{ kJ mol}^{-1}$$

The stored energy of the reactants, methane and oxygen, is higher than the stored energy of the products of the reaction, carbon dioxide and water, and

Fig 8.1 Energy level diagram for methane burning

the difference in energy is released to the surroundings when methane and oxygen react. We can see this in Fig 8.1.

## STANDARD CONDITIONS

The symbol  tells us that the conditions for this enthalpy change are standard conditions at a particular temperature. Standard conditions are 100 kPa (1 atmosphere) pressure and, for solutions, a concentration of 1 mol dm$^{-3}$. The particular temperature for these standard conditions is usually 298 K (25 °C). A small 298 following $\Delta H^{\ominus}$ (pronounced 'delta H standard') gives the standard temperature. The substances in the chemical reaction (the reactants and products) are known as the **system**, and all substances in the system are in their **standard states**. The standard state is the most stable physical state at 1 atmosphere pressure and the particular temperature, usually 298 K (25 °C). So the standard state of methane is a gas and the standard state of water is a liquid.

> **Standard enthalpy change of a reaction is the enthalpy change under standard conditions at 298 K and 100 kPa, with all substances in their standard states.**

This refers to the mole quantities given in the balanced equation for the reaction.

The **surroundings** are everything that is not the system. It could be the reaction vessel, the air around the reaction vessel, the mercury in a thermometer and anything else that is not a reactant or product. But the surroundings must be associated thermally with the system, which means that heat can transfer between system and surroundings.

## 2 STANDARD ENTHALPY CHANGE OF COMBUSTION

Methane burning in oxygen is called a **combustion reaction**. The enthalpy change is an enthalpy change of reaction, but if we burn one mole of methane completely in oxygen it can also be referred to as an **enthalpy change of combustion**.

> **Standard enthalpy change of combustion $\Delta H^{\ominus}_{c,298}$ is the enthalpy change when one mole of a substance burns completely in oxygen under standard conditions (298 K and 100 kPa) with all reactants and products in their standard state.**

So the standard enthalpy change of combustion of methane is written:

$$\Delta H^{\ominus}_{c,298} (CH_4) = -890 \text{ kJ mol}^{-1}$$

and because the standard temperature is 298 K, we often write:

$$\Delta H_c^{\ominus} = -890 \text{ kJ mol}^{-1}$$

In the definition for standard enthalpy change of combustion we say *burns completely* to show clearly that only carbon dioxide and water are formed. With not enough oxygen, there may be incomplete combustion, which would produce carbon monoxide or even carbon (soot). The enthalpy changes for both these reactions will affect any measurement of enthalpy change of combustion.

In practice, when methane burns in oxygen, the temperature is much higher than 25 °C, so adjustments must be made to calculate the value under standard conditions.

## ? TEST YOURSELF

**B** Why do we need to include the state symbols in the equation in the Example below? Hint: The answer is on page 167.

## ○ EXAMPLE

**Q** Explain fully what is meant by:

$$\Delta H_{c,298}^{\ominus} (H_2) = -286 \text{ kJ mol}^{-1}$$

**A** This is the standard enthalpy change of combustion of hydrogen. It is the enthalpy change when one mole of hydrogen is completely burnt in oxygen under standard conditions of 100 kPa pressure and 298 K, with reactants and products in their standard states at this temperature.

$$H_2(g) + \tfrac{1}{2}O_2(g) \rightarrow H_2O(l) \quad \Delta H_c^{\ominus} = -286 \text{ kJ mol}^{-1}$$

Notice that this includes the state symbols. Also, $\tfrac{1}{2}O_2$ does not mean half a molecule of oxygen, but half a mole.

We could have written:

$$2H_2(g) + O_2(g) \rightarrow 2H_2O(l) \quad \Delta H_r^{\ominus} = -572 \text{ kJ mol}^{-1}$$

The standard enthalpy change of this reaction, $\Delta H_r^{\ominus}$ (small r for reaction; or just $\Delta H^{\ominus}$) is twice that for the standard enthalpy change of combustion of hydrogen. This is because $\Delta H_r$ applies to the mole quantities in the equation – 2 moles of hydrogen – whereas the standard enthalpy change of combustion is for burning *one* mole of hydrogen.

## ! SCIENCE FEATURE Biofuels

Biofuels are fuels produced from vegetable matter, and biofuels account for 15 per cent of the total world energy production. A significant proportion is used in developing countries for cooking and heating.

But scientists are also very interested in biofuels for transport. Italy produces some of its diesel from rapeseed oil, and 5 per cent of Austria's diesel is biodiesel. Sugar cane and wheat can be fermented to produce ethanol, which is blended with petrol or used on its own (see the Assignment on page 190 and Fig 8.2(b)).

Growing crops for fuel is the subject of heated debate. Some proponents say biofuels are the answer to increasing levels of carbon dioxide in the atmosphere: the crops remove carbon

**Fig 8.2(b)** Brazilians use Alcool as vehicle fuel. It is made from the fermented and distilled juice of sugar cane

dioxide from the atmosphere to make sugars and oils, and the gas returns into the atmosphere when fuels made from them are burnt, so there is no net carbon dioxide increase. But other analysts point out that this ignores the fossil fuels that must be burnt to provide the fuel for planting the crop, making the fertilizer, harvesting the crop, and transporting and processing it.

This debate will run for years, but as oil extraction becomes more costly, it is likely that by 2025 a significant amount of Europe's fuel will be biofuel.

**Fig 8.2(a)** The oil from this field of oilseed rape can be converted into biodiesel fuel

## ALTERNATIVE FUELS TO PETROL

Table 8.1 gives the standard enthalpy changes of combustion of some fuel alternatives to petrol. There is no standard enthalpy change for petrol because it is a complex mixture of about 100 compounds, mainly hydrocarbons, of which most are alkanes.

The standard enthalpy change of combustion of octane, one of the constituents of petrol, is $-5470$ kJ mol$^{-1}$, which is a good indication of why petrol is used in cars.

| Table 8.1 The standard enthalpy changes of combustion of fuel alternatives to petrol | | | |
|---|---|---|---|
| **Fuel** | **Main constituent** | **Formula and standard state** | **$\Delta H_c^\ominus$ of main constituent (kJ mol$^{-1}$)** |
| Hydrogen | Hydrogen | $H_2(g)$ | $-286$ |
| Compressed natural gas (CNG) | 90% methane | $CH_4(g)$ | $-890$ |
| Liquid petroleum gas (LPG) | 95% propane | $C_3H_8(g)$ | $-2219$ |
| Methanol | Methanol | $CH_3OH(l)$ | $-726$ |

## 3 MEASURING ENTHALPY CHANGES

The process of measuring energy changes is known as **calorimetry**, after the old unit for heat, a calorie. Any vessel that is used to measure enthalpy changes is known as a **calorimeter**.

### ENTHALPY CHANGE OF COMBUSTION

See Table 8.1 for some enthalpy changes of combustion. Being combustion reactions, they are all exothermic. We can find these values by measuring the increase in temperature of a certain mass of water to which energy from the combustion reaction is transferred, making its temperature rise. One gram of water requires 4.2 joules of energy to raise its temperature by 1°C; this is water's **specific heat capacity**.

> **Specific heat capacity ($c$) is the energy required to raise the temperature of 1 gram of substance by 1 K. It is measured in J g$^{-1}$ K$^{-1}$.**

So the specific heat capacity of water is 4.2 J g$^{-1}$ K$^{-1}$ (joules per gram per kelvin).

We can use a simple equation to work out the energy transferred to the water:

$$\text{Energy transferred} = \underset{\substack{\text{mass of}\\ \text{water}\\ \text{g}}}{m} \times \underset{\substack{\text{specific heat}\\ \text{capacity of water}\\ \text{J g}^{-1}\text{ K}^{-1}}}{c} \times \underset{\substack{\text{temperature}\\ \text{change}\\ \text{K}}}{\Delta T}$$

### SIMPLE DETERMINATION OF THE ENTHALPY CHANGE OF COMBUSTION OF A LIQUID FUEL

A simple experimental set-up to measure the enthalpy change of combustion of ethanol is shown in Fig 8.4. The experiment is carried out as follows:

● **Step 1.** Weigh the spirit burner with ethanol in it at the beginning of the experiment and take the water temperature.
● **Step 2.** Allow the spirit burner to heat up a known mass of water by about 20 °C. Stop heating and take the temperature of the water. This gives the temperature rise in Kelvin.
● **Step 3.** Weigh the spirit burner again. This gives the mass of the ethanol used.

**SEE PAGE 218 TO FIND OUT MORE ABOUT THE COMPOSITION OF PETROL.**

**? TEST YOURSELF**

**C** Draw energy level diagrams for the standard enthalpy changes of combustion of hydrogen and propane.

$$\Delta H_c^\ominus (H_2) = -286 \text{ kJ mol}^{-1}$$
$$\Delta H_c^\ominus (C_3H_8) = -2219 \text{ kJ mol}^{-1}$$

**Fig 8.3** This car runs on CNG, which is 90 per cent methane

**✔ REMEMBER THIS**

Sometimes you will see enthalpy change of combustion referred to as just **enthalpy of combustion**. The fact that enthalpy is changing in the reaction is understood.

**✔ REMEMBER THIS**

A temperature rise of 1 K is the same as a temperature rise of 1 °C.

**? TEST YOURSELF**

**D** How many kilojoules of energy will be required to raise the temperature of 1 kg of water by **(a)** 1°C, **(b)** 20 °C?

**Fig 8.4** Measuring the enthalpy change of combustion of a liquid fuel

The next Example explains how this experiment may be used to calculate the enthalpy change of combustion for ethanol.

## ? TEST YOURSELF

**E** Read the method for measuring enthalpy change of combustion and examine Fig 8.4 on the previous page.
**a)** Why is the calorimeter made of metal and not glass?
**b)** Why does a temperature rise in °C give you the temperature rise in Kelvin?
**c)** The draught shield is used to reduce loss of energy to the surroundings. What other things could be done to minimize energy losses in this experiment?

## ? TEST YOURSELF

**F** The experiment in the Example gives an inaccurate result. Compare the value of the enthalpy change of combustion of ethanol in the Example with the value given in Table 8.1, on page 169. What aspects of the experiment lead to such a low calculated value?

## ○ EXAMPLE

**Q** 0.40 g ethanol raises the temperature of 100.00 g water in the metal calorimeter by 21.0 °C. Calculate the enthalpy change of combustion of ethanol.

**A**

**Step 1.** Work out the energy transferred to the water:

Energy transferred to the water $= mc\Delta T$

$$= 100.00 \times 4.2 \times 21.0 = 8820 \, J = 8.82 \, kJ \text{ (ignoring significant figures)}$$

**Step 2.** Calculate the amount of fuel used in moles:

Mass of 1 mole of ethanol ($C_2H_5OH$)

$$= (12 \times 2) + (1 \times 5) + 16 + 1 = 46 \, g$$

Moles of ethanol in 0.40 g $= \dfrac{\text{mass in grams}}{\text{mass of one mole (in grams)}}$

$$= \dfrac{0.40}{46} = 0.0087 \, \text{mol (correct to two sig. figs)}$$

**Step 3.** Work out the enthalpy change of combustion, that is, the energy transferred when 1 mole of ethanol burns:

0.0087 mol ethanol releases 8.82 kJ

1 mole of ethanol releases $8.82 \times \dfrac{1}{0.0087} = 1014 \, kJ$

So the enthalpy change of combustion of ethanol $= -1000 \, kJ \, mol^{-1}$ (correct to two significant figures)

## ▶ EXTENSION The bomb calorimeter

The bomb calorimeter (Fig 8.5) measures enthalpy changes of combustion. The weighed sample is inside a stainless steel container – the bomb – filled with oxygen under pressure, and the fuel is ignited electrically.

The principle is the same as for the simplified experiment on page 169. Energy is transferred from the combusted fuel to the surrounding water, and the temperature rise is measured. However, this apparatus gives a much more accurate value than the simple version, provided the readings are taken quickly, because energy losses to the surroundings are reduced to almost zero. In our simplified experiment on page 169, another potential error arises when the spirit burner is weighed, as some of the ethanol may evaporate between weighings.

Enthalpy change is transfer of energy into or out of the system *at constant pressure*. As the

labels: thermometer; stirrer; water; insulated container holding the water; electrically heated coil to ignite sample; sample: liquid or solid; stainless steel bomb containing oxygen under pressure; crucible

**Fig 8.5** The bomb calorimeter. The sample is ignited electrically and as it burns the energy is transferred to the water with minimal loss. The thermometer measures the rise in temperature

## ▶ EXTENSION The bomb calorimeter

bomb has a constant *volume*, it does not measure enthalpy change exactly. However, because the difference is small, we need not be concerned with the mathematical correction used to adjust the value to constant pressure.

### THE ENERGY VALUES OF FOOD

The bomb calorimeter is used to measure the enthalpy changes of combustion of different foods. Although our bodies don't burn foods in quite the same way as we burn natural gas in a cooker, the outcome is the same. Oxygen is still required and the energy we obtain from compounds in foodstuffs is the same as if they were burnt in a bomb calorimeter. The crucial difference is that when we burn fuels in cooking or in the combustion

engine of a car, most of the released energy is wasted and goes to heat up the surroundings.

When we 'burn' glucose in our bodies, there is not just one high-temperature reaction, but several small steps with small transfers of energy, each step catalysed by complex molecules called enzymes. The human machine is nowhere near 100 per cent efficient, and we do lose energy to our surroundings (hence, our body warmth), but much of the energy of the step-wise reactions is either used to maintain electrical and chemical body functions or stored, and so does not emerge as heat. Of course, when we store energy it is usually as fat, so taking in more energy than we require will make us obese.

**? TEST YOURSELF**

**G** Look back to page 169 and the enthalpy of combustion experiment.
**a)** (i) If some of the ethanol evaporates between weighings, and so is not burnt, would this lead to a higher or a lower value for the calculated enthalpy change of combustion?
(ii) How could the weighings be carried out in the bomb calorimeter experiment to avoid any losses through evaporation?
**b)** (i) Study Fig 8.5, which shows the bomb calorimeter, and explain the features of the apparatus that minimize energy loss to the surroundings.
(ii) What is the purpose of the stirrer?

## ENTHALPY CHANGE OF NEUTRALIZATION

Many reactions take place in aqueous solution and the enthalpy changes of these reactions can be measured using an expanded polystyrene cup fitted with a lid (Fig 8.6). Expanded polystyrene is a very good insulator and also absorbs little energy from the reaction itself.

When an acid reacts completely with an alkali the reaction is called a **neutralization**. The reaction between aqueous sodium hydroxide and aqueous hydrochloric acid can be represented by the equation:

$$HCl(aq) + NaOH(aq) \rightarrow NaCl(aq) + H_2O(l)$$

In this reaction all the reactants are present as ions, so:

$$H^+(aq) + Cl^-(aq) + Na^+(aq) + OH^-(aq) \rightarrow Na^+(aq) + Cl^-(aq) + H_2O(l)$$

You will notice that the ions reacting are only $H^+(aq)$ and $OH^-(aq)$, while $Na^+(aq)$ and $Cl^-(aq)$ take no part in the reaction and are called **spectator ions**. So the neutralization reaction is more simply written as

$$H^+(aq) + OH^-(aq) \rightarrow H_2O(l)$$

So standard enthalpy change of neutralization is defined as:

> **The enthalpy change when an acid and a base react to form one mole of water under standard conditions (298 K, 100 kPa).**

The experiment to measure this enthalpy change is carried out as follows:
● **Step 1.** Measure out accurately a known volume and amount of aqueous acid and take the temperature of the solution.
● **Step 2.** Mix the acid from Step 1 with a known volume and amount of alkali in a polystyrene cup and stir the mixture thoroughly.
● **Step 3.** Calculate the temperature rise. This is usually done by taking the temperature of the alkali in the polystyrene cup each minute for four

— stirring
thermometer accurate to at least 0 °C

— expanded polystyrene cup with lid

**Fig 8.6** A simple calorimeter to measure the enthalpy change of neutralization minutes. On the fifth minute the acid is added, the mixture stirred and then the temperature measured again every minute from the sixth minute until the twelfth minute. A graph is drawn (Fig 8.7) and the temperature rise as the acid is added to the alkali is extrapolated.

**Fig 8.7** Graph of temperature against time plotted in an experiment to measure the enthalpy change of a neutralization reaction

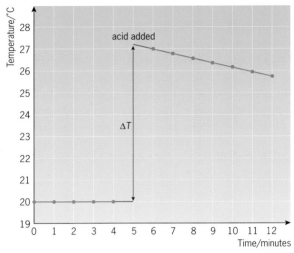

## EXAMPLE

**Q** An experiment to measure the enthalpy change of neutralization is carried out. When 50.0 cm$^3$ HCl(aq) is added to 50.0 cm$^3$ NaOH(aq), both of 1.0 mol dm$^{-3}$ concentration, the temperature rise is 6.0 °C. Calculate the enthalpy change of neutralization.

**A  Step 1.** Work out the energy transferred to the water. The acid and alkali are very dilute so the volume of water is assumed to be:

$$50.0 \text{ cm}^3 \text{ HCl(aq)} + 50.0 \text{ cm}^3 \text{ NaOH(aq)} = 100.0 \text{ cm}^3$$

The mass of 100 cm$^3$ water = 100 g (1.0 cm$^3$ water has a mass of 1.0 g)

Energy transferred to the water = $mc\Delta T$

$$= 100.0 \times 4.2 \times 6.0 = 2520 \text{ J} = 2.52 \text{ kJ (ignoring significant figures)}$$

**Step 2.** Calculate the amount of acid and alkali used in moles:

Number of moles NaOH in 1000 cm$^3$ = 1.0 mol

So number of moles NaOH (and also HCl) in 50.0 cm$^3$ = $1.0 \times \dfrac{50.0}{1000}$

$$= 0.050 \text{ mol}$$

To remind yourself how to do this calculation see page 139.

**Step 3.** Work out the enthalpy change of neutralization, that is, the energy transferred when 1 mol of water is formed from 1.0 mol NaOH and 1.0 mol HCl.

When 0.050 mol of water is formed the energy released = 2.52 kJ

1 mole of water being formed releases $2.52 \times \dfrac{1}{0.050}$ = 50.4 kJ

So enthalpy change of neutralization = −50 kJ mol$^{-1}$ (correct to two significant figures)

The data book value $\Delta H^\ominus_{\text{neut}}$ = −57.2 kJ mol$^{-1}$, so heat energy was lost to the surroundings in this experiment.

**SEE QUESTION 1.**

## ? TEST YOURSELF

**H** Calculate a value for the enthalpy change of neutralization when 25.0 cm$^3$ of 1.0 mol dm$^{-3}$ H$_2$SO$_4$(aq) is neutralized by 50.0 cm$^3$ NaOH(aq). The temperature rise on mixing is 8.2 °C.
[Hint: Remember 1 mol H$_2$SO$_4$ produces 2 mol H$^+$(aq).].

# 4 RELEASING ENERGY FROM FUEL MOLECULES

We know that fuel molecules need oxygen to combust and so release energy. Where does this energy come from? To answer this question we need to look at the energies of the bonds that hold atoms together, and what happens when these bonds are broken and made.

## BREAKING BONDS: BOND ENTHALPY

**Bond enthalpy** or **bond energy** is the energy required to break a bond between two atoms in a gaseous molecule. However, because the energy required to break one bond is so small, we define bond enthalpy as the energy required to break *one mole* of bonds. Thus:

> **Bond enthalpy, *E*, is the energy required to break one mole of bonds of the same type in gaseous molecules under standard conditions (298 K, 100 kPa).**

Consider hydrogen, which has an H—H bond:

$$H-H(g) \rightarrow H(g) + H(g) \quad \Delta H^{\ominus} = +436 \text{ kJ mol}^{-1}$$

Bond enthalpy, *E*, is often written as:

$$E(H-H) = +436 \text{ kJ mol}^{-1}$$

Notice that **bond breaking is an endothermic process**. This is because we are *putting in* energy to overcome the force of electrostatic attraction in the bond.

The bond energy in oxygen molecules is for breaking a *double* bond:

$$O=O(g) \rightarrow O(g) + O(g) \quad \Delta H^{\ominus} = +498 \text{ kJ mol}^{-1}$$

So $E(O=O) = +498 \text{ kJ mol}^{-1}$. This is larger than for the H—H bond, because double bonds are stronger than single bonds and so require more energy to be broken.

Up to now, we have considered bond energies only in *diatomic* molecules (molecules that contain two atoms). There are two O—H bonds in the triatomic water molecule. The energy required to break the first O—H bond is not the same as the energy required to break the second:

$$H-O-H(g) \rightarrow H(g) + O-H(g) \quad \Delta H^{\ominus} = +502 \text{ kJ mol}^{-1}$$

$$O-H(g) \rightarrow O(g) + H(g) \quad \Delta H^{\ominus} = +427 \text{ kJ mol}^{-1}$$

The first bond is in a molecular environment in which two O—H bonds exist, and more energy is required to break this bond. The second O—H bond is in a changed environment, and clearly this has an effect on its bond energy. A total of 929 kJ mol$^{-1}$ is required to break both the O—H bonds in water, so we say the *average* bond enthalpy for the O—H bond in this molecule is +464 kJ mol$^{-1}$.

We also find differences in the bond enthalpy of the O—H bond in other molecules because of the different molecular environments of the bond. For example, the bond enthalpy of the O—H bond in methanol is +437 kJ mol$^{-1}$. Usually, the differences in bond enthalpies are not very great and we average them to give a good approximation which we can use.

Fig 8.8 The force of attraction between the shared pair of electrons and the positive nuclei keep these two hydrogen atoms together. Breaking this covalent bond will require energy

**? TEST YOURSELF**

**I** The nitrogen molecule is very unreactive. Why is this so?
Hint: if you are not sure, look at Fig 4.12, page 73.

**? TEST YOURSELF**

**J** **a)** Two bond energies in Table 8.2 are not average bond energies. Identify these and explain your reasoning.
**b)** Use Table 8.2 to calculate the amount of energy to atomize one mole of gaseous methanol (convert the molecule into its separate, gaseous atoms).
$$CH_3OH(g) \rightarrow C(g) + 4H(g) + O(g)$$
**c)** Methanol is normally a liquid at 298 K. Explain why it must be a gas in this calculation.
Hint: What other energy is involved if methanol is not gaseous?

Table 8.2 Average bond enthalpies of some common bonds

| Bond | $E(X-Y)$/ kJ mol$^{-1}$ |
|---|---|
| C—C | +347 |
| C=C | +612 |
| C≡C | +838 |
| C—H | +413 |
| C—O | +358 |
| C=O (in CO$_2$) | +805 |
| H—H | +436 |
| O—H | +464 |
| O=O | +498 |

## MAKING BONDS

So far, we have only considered *breaking* bonds, which requires energy, but we still have not answered the question: Where does the energy come from that is released when fuels burn in oxygen?

If it takes energy to break bonds, the same amount of energy must be released when bonds are formed. This reflects the **first law of thermodynamics (law of conservation of energy) – energy can neither be created nor destroyed** – and it certainly applies to chemical reactions. Take hydrogen:

$$H–H(g) \rightarrow H(g) + H(g) \quad \Delta H^{\ominus} = +436 \text{ kJ mol}^{-1}$$

$$H(g) + H(g) \rightarrow H–H(g) \quad \Delta H^{\ominus} = -436 \text{ kJ mol}^{-1}$$

**Bond making is exothermic.**

## ENERGY CHANGES IN BOND BREAKING AND MAKING

We are now in a position to look at energy changes from bond breaking and bond making. For example, when hydrogen burns in oxygen:

$$H–H(g) \quad + \tfrac{1}{2}O=O(g) \rightarrow \qquad H–O–H(g)$$

$$E(H–H) \quad + \tfrac{1}{2}E(O=O) \rightarrow \quad 2 \times \ -E(O–H)$$

$$+436 \qquad + \tfrac{1}{2}(498) \qquad\qquad 2(-464)$$

The enthalpy change for this reaction is:

$$\Delta H^{\ominus} = +436 + 249 + (-928) = -243 \text{ kJ mol}^{-1}$$

So:

**The energy given out when fuels burn is the difference between the energy required to break bonds and the energy released when bonds are made.**

We can show this on an energy level diagram (enthalpy profile diagram), such as Fig 8.9.

**Fig 8.9** The energy level diagram (enthalpy profile diagram) for hydrogen burning in oxygen, showing the energy for breaking and making bonds and the enthalpy change of combustion

## ? TEST YOURSELF

**K** **a)** Compare the value given in the text for the enthalpy change for the combustion of hydrogen in oxygen to the enthalpy change of combustion of hydrogen in the Example on page 168. Explain the difference.

**b)** For endothermic reactions, what differences would you expect to see in the energy level diagram?

## USING BOND ENERGIES TO ESTIMATE ENTHALPY CHANGES OF REACTION

Average bond energies can be used to estimate the enthalpy changes for different reactions.

### ● EXAMPLE

**Q** Calculate the energy released when 1 mole of methane burns completely in oxygen using average bond energy data in Table 8.2 on page 173.

**A**

**Step 1.** Write down the equation for the reaction:

$$CH_4(g) + 2O_2(g) \rightarrow CO_2(g) + 2H_2O(g)$$

**Step 2.** Draw out full structural formulas to show all the atoms and bonds:

$$
\begin{array}{c}
\quad\ \ H \\
\quad\ \ | \\
H-C-H + 2O{=}O \rightarrow O{=}C{=}O + 2H-O-H \\
\quad\ \ | \\
\quad\ \ H
\end{array}
$$

**Step 3.** Write down which bond energies you will need to look up, and decide whether bonds are broken (endothermic) or formed (exothermic). Use Table 8.2 on page 173.

| Bond breaking | Bond making |
|---|---|
| $4E(C{-}H) + 2E(O{=}O)$ | $2 \times {-}E(C{=}O) + [4 \times {-}E(O{-}H)]$ |
| $4 \times +413 + 2 \times +498$ | $2 \times -805 + 4 \times -464$ |
| $+2648$ | $-3466$ |

$$\Delta H^\ominus = +2648 + (-3466)$$
$$= -818 \text{ kJ mol}^{-1}$$

### ? TEST YOURSELF

**L** Methanol is the fuel used in Formula One motor racing. Use bond energy values in Table 8.2, page 173, to calculate the energy released when one mole of methanol is burnt.

$$CH_3OH(g) + 1\tfrac{1}{2}O_2(g) \rightarrow CO_2(g) + 2H_2O(g)$$

The standard enthalpy change of combustion of methane is $-890$ kJ mol$^{-1}$. This is different from the value obtained in the Example because the standard state of water is liquid at 298 K. *When dealing with bond energies, all the molecules must be in the gaseous state*, and energy is released when gaseous water condenses.

The standard enthalpy change of combustion of methanol is $-726$ kJ mol$^{-1}$. This is less than the value for methane, so why is there a difference? Fuels release their energy when they form bonds with oxygen. The two compounds have the same number of carbon and hydrogen atoms, but one O–H bond is already formed in the methanol molecule.

$$
\begin{array}{cc}
\quad H & \quad H \\
\quad | & \quad | \\
H-C-H & H-C-O-H \\
\quad | & \quad | \\
\quad H & \quad H \\
\text{methane} & \text{methanol}
\end{array}
$$

When methane reacts, all four O–H bonds must be made, which gives a more exothermic reaction. Fuels that contain oxygen are sometime called oxygenates; they usually have a lower enthalpy change of combustion than fuels with the same number of carbon atoms, but without oxygen.

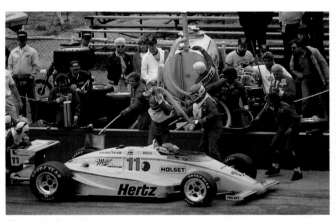
**Fig 8.10** Refuelling a racing car that runs on methanol

### ✔ REMEMBER THIS

Fuels that are oxygenates are partially oxidized (have oxygen added to them). This piece of information should help you with question 3(c) in the Assignment at the end of this chapter.

## SCIENCE FEATURE Methanol as a fuel

Methanol powers the engines of Formula One motor racing – see Test Yourself L. These are some of the advantages that make it an ideal fuel for high performance engines.

● It has a high octane number, meaning that it does not *knock* (prematurely explode) when the gases are compressed by a piston in an engine (see page 217).
● Although flammable, it is less volatile than petrol, and if a fire starts it is easier to put out because methanol is miscible (mixes) with water.
● Methanol is an oxygenate, and so less oxygen is required to fully combust it in a car engine. (In contrast, in cars using petrol, carbon monoxide is a major pollutant, formed because not enough oxygen is taken in for full combustion.)

● It is fairly cheap to produce. It can be made from natural gas or coal, or from methane through the decomposition of rotting waste (making it a biofuel).
● It can be blended with petrol (as is done in several countries).

Too good to be true? Well, there are problems using methanol.

● It is a highly toxic alcohol, so the vapour is a hazard when refuelling.
● It produces methanal as a pollutant from the exhaust, an aldehyde (see page 350) that is carcinogenic (causes cancer).
● It absorbs water from the air and is very corrosive in existing petrol tanks.

Will methanol become more popular as a fuel? A recent study suggests that by 2050 it could provide half of the global liquid fuel requirements, given advances in technology to overcome the above problems.

## ? TEST YOURSELF

**M** **a)** Why is water not used to put out petrol fires?
**b)** Look at Table 8.3 below and the information about petrol just under it. Think of another disadvantage of using methanol as a fuel.

## ? TEST YOURSELF

**N** **a)** Write out the equation for the standard enthalpy change of combustion of octane.
**b)** Work out which bonds are formed, and how many of each one.

## ✔ REMEMBER THIS

$$\text{Energy density} = \Delta H_c^\ominus \times \frac{1000}{\text{mass of 1 mole}}$$

## ? TEST YOURSELF

**O** **a)** What is the energy density of propane, which makes up 95 per cent of liquid petroleum gas (LPG)? You will need information from Table 8.1 on page 169 to answer this question.
**b)** (i) What is the volume of 1 kg of hydrogen at 25 °C and 100 kPa? 1 mole of any gas at this temperature occupies 24 dm$^3$ (see page 120).
(ii) 1 litre of petrol has a mass of 740 g. How many litres of petrol in 1 kg?

## BREAKING AND MAKING MORE BONDS

Octane (molecular formula $C_8H_{18}$), one of the chemicals in petrol, has 18 C–H bonds and seven C–C bonds. Methane has just four C–H bonds. While this means that you need more energy to break the bonds in a mole of octane, far more energy is released when new bonds are formed with oxygen.

The standard enthalpy change of combustion of octane is –5470 kJ mol$^{-1}$.

octane  H–C–C–C–C–C–C–C–C–H

### ENERGY DENSITY

Most means of transport have to carry their own fuel around with them – think of petrol in the tank of a car. So we want to obtain the maximum energy we can from any mass of fuel. With transport fuels we need to know the amount of energy released by 1 kg of fuel. We call this the **energy density** of the fuel. To calculate the energy density we need to know the standard enthalpy change of combustion ($\Delta H_c^\ominus$) and the mass of one mole of the fuel.

**Table 8.3 The energy density values of four fuels**

| Fuel | Formula | $\Delta H_c^\ominus$/kJ mol$^{-1}$ | Mass of 1mole/g | Energy density/kJ kg$^{-1}$ |
|---|---|---|---|---|
| Hydrogen | $H_2(g)$ | –286 | 2 | 143 000 |
| Methane | $CH_4(g)$ | –890 | 16 | 27 800 |
| Methanol | $CH_3OH(l)$ | –726 | 32 | 22 700 |

Petrol has an energy density of approximately 46 000 kJ kg$^{-1}$, which makes it a very concentrated energy source. It is a mixture of many hydrocarbons and you can read more about it in Chapter 10.

From Table 8.3, hydrogen would appear to have excellent prospects as a fuel of the future. But there is still the problem of how to store it on board a vehicle. If you try Test Yourself O(b), you will see why.

**SEE QUESTION 2.**

---

> ! **SCIENCE FEATURE** Metal hydrides: the key to on-board hydrogen storage?

One major disadvantage of hydrogen as a fuel is how to store it. Liquefying it means cooling it to –253°C, which costs four times more than making an equivalent amount of petrol. Then it has to be kept cold. The expense is worth it for the Space Shuttle, page 166, but it is no surprise that alternative storage methods are being sought for the more general use of hydrogen as a fuel.

One possible method is to store hydrogen as a solid. This doesn't mean freezing it, which takes too much energy, but combining it as a metal hydride. Magnesium hydride, $MgH_2$, is 7.7 per cent hydrogen by mass.

A litre of magnesium hydride contains almost as much hydrogen as a litre of liquefied hydrogen, though it is a great deal heavier.

The hydrogen is released by this reaction:

$$MgH_2(s) + H_2O(l) \rightarrow Mg(OH)_2(s) + H_2(g)$$

In magnesium hydride, magnesium is ionically bonded to hydrogen in a small whole-number ratio. But another type of metal hydride – an interstitial hydride – can soak up hydrogen rather like a sponge. The metal is *bathed* in hydrogen. At the metal surface, the hydrogen molecule splits into its atoms and the atoms occupy *holes* in the metal lattice. Very large quantities of hydrogen can be absorbed and released on heating. There remains the problem of the weight of the hydride, but chemists are actively searching for new alloy hydrides of lower weight.

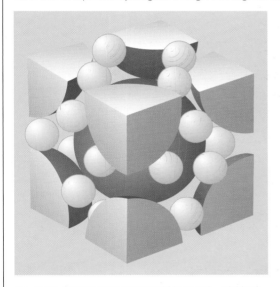

**Fig 8.11** An interstitial metal hydride (metal = niobium). The smaller hydrogen atoms are trapped between the larger metal atoms in the lattice

**Fig 8.12** A prototype coach run on hydrogen in Augusta, Georgia, USA. A petrol engine requires only slight modification and the hydrogen is stored as an interstitial metal hydride

---

# 5 ACTIVATION ENERGY, ENERGY PROFILES AND THE COLLISION THEORY

## HOW A REACTION GETS STARTED

Octane, one of the components of petrol, has an enthalpy change of combustion of –5470 kJ mol⁻¹. So why doesn't it burst into flames as soon as it is exposed to oxygen in the air? The reason is that it takes energy to break bonds. This must happen before oxygen molecules and octane molecules can react together.

To break all the bonds in oxygen and octane vapour would require +7437 kJ mol⁻¹ of energy – a very large amount. So, in fact, it appears difficult for octane to start reacting at all. Yet we know that a match thrown on to petrol causes it to react spectacularly. Clearly, not *all* the bonds need to be broken before other bonds start forming, and once bonds start to form, energy is released to break other bonds. This keeps the reaction going. (Look back at page 175 and calculate the energy required to break all the bonds in octane and oxygen.)

The minimum energy required for a reaction to start is called the **activation energy, $E_a$**. When this energy is supplied to molecules in the system, the bonds begin to stretch and break. Reactions usually occur because the molecules collide with enough energy to make this happen. Sometimes there is sufficient energy in the system for a reaction to occur at room temperature, but for petrol in an engine, energy needs to be supplied to reach the activation energy level; this comes from a spark.

Chemists draw **energy profiles**, or **enthalpy profiles**, for reactions, such as those in Fig 8.13. They show how the energy changes as the reactions proceed.

**REMEMBER THIS**

In an enthalpy profile diagram the vertical axis ($y$ axis) is usually labelled enthalpy rather than energy

(a)  Progress of reaction (reactants → products)

(b)  Progress of reaction (reactants → products)

**Fig 8.13** Energy profile for **(a)** the exothermic reaction of octane with oxygen, and **(b)** an endothermic reaction. Notice that activation energy $E_a$ is always endothermic

**? TEST YOURSELF**

**P** Draw an energy profile diagram for:

**a)** methanol burning in oxygen (see Table 8.1, page 169);

**b)** nitrogen reacting with oxygen to give nitrogen monoxide:

$$\tfrac{1}{2}N_2(g) + \tfrac{1}{2}O_2(g) \rightarrow NO(g)$$
$$\Delta H^{\ominus} = +90 \text{ kJ mol}^{-1}$$

This occurs in the high temperatures of car engines and leads to pollution. See page 225.

**c)** Why is the reaction in **b)** not an enthalpy change of combustion?

**▶ EXTENSION Why do endothermic reactions take place?**

When a spark ignites petrol, there is a **spontaneous reaction** with oxygen, which produces carbon dioxide and water. It does not matter that energy was needed to start this reaction – once started, it is spontaneous (takes place of its own accord).

Let's consider the spontaneous combustion of octane, which is one of the compounds in petrol:

$$2C_8H_{18}(g) + 25O_2(g) \rightarrow 16CO_2(g) + 18H_2O(g)$$

We know that this is an exothermic reaction and that energy is released to the surroundings because the energies of octane and oxygen are higher than the energies of carbon dioxide and water – see Fig 8.13. So we could say that for a reaction to occur it must be exothermic.

However, this does not explain spontaneous reactions that are endothermic, such as the reaction between nitrogen and oxygen to produce nitrogen monoxide:

$$N_2(g) + O_2(g) \rightarrow 2NO(g) \quad \Delta H = +180 \text{ kJ mol}^{-1}$$

## ▶ EXTENSION Why do endothermic reactions take place?

As this reaction is a major cause of pollution from vehicle exhausts, we know that it happens spontaneously in the high temperature of a petrol engine. To understand why spontaneous reactions occur we must seek another explanation that involves entropy.

### ENTROPY CHANGES AND SPONTANEOUS REACTIONS

We first met entropy on page 143. Entropy is a measure of the disorder of a system. The entropy of a gas is higher than the entropy of a solid: there are more ways of arranging molecules in a gas, and so there is more disorder.

> **For a reaction to occur spontaneously there must be an overall increase in entropy.**

This is the **second law of thermodynamics**. Thermodynamics is the study of energy transfer. We have already encountered the **first law of thermodynamics**:

> **Energy can neither be created nor destroyed in physical and chemical processes.**

One of the most important endothermic reactions for life on this planet is photosynthesis. Although it is endothermic, there must be an increase in entropy or it would not be able to occur at all. You can read more about entropy in Chapter 25.

## COLLISION THEORY

The collision theory is a model that chemists use to explain how reactions occur. When two molecules react, bonds are broken or made, and there is usually a rearrangement of the atoms. According to collision theory, for this to happen the molecules first have to collide. The collision model was developed from the kinetic–molecular model of gases discussed on pages 127 to 129.

In 1 cm$^3$ of gas at atmospheric pressure and room temperature, it has been calculated that there are about $10^{27}$ collisions between two molecules every second. If all these collisions were to lead to reactions, then all reactions that involve gases would be over in microseconds. For example, nitrogen and oxygen molecules, which under normal conditions coexist without reacting, would long ago have reacted to fill the atmosphere with polluting nitrogen oxides. This means that only a certain number of collisions result in reactions. These are called **effective collisions** and occur when molecules collide with the minimum energy needed to start a reaction, the **activation energy**.

## ACTIVATION ENERGY AND STABILITY

The higher the activation energy for a reaction, the less likely it is to occur. Even though octane looks very unstable compared with carbon dioxide and water, the high energy of activation acts as an **energy barrier** to the reaction (see Fig 8.13(a)). In petrol, this prevents the reaction from occurring when the tank is filled at the garage, and is the reason why there are 'No smoking' signs!

**Reaction kinetics** is the study of *rates* of reactions and we look at this study in more detail at the end of this Chapter and in Chapter 26. We say petrol is **kinetically stable**, even though it is **energetically unstable**.

 **REMEMBER THIS**

Octane is energetically unstable because it is at a higher energy level than carbon dioxide and water. Sometimes we use the term **thermodynamically unstable** instead of energetically unstable.

# LOWERING THE ENERGY BARRIER: CATALYSIS

In your previous chemistry course you probably learnt that **catalysts** usually increase the rate of chemical reactions. We also mention this in Chapter 1. Catalysts play a crucial role in the lives of everybody. Protein catalysts called enzymes dramatically increase the rates of thousands of different chemical reactions that take place in our bodies every second. And catalysts are involved in the manufacture of most of the chemicals on which people rely. As a definition:

> **Catalysts are substances that alter the rate of a chemical reaction without becoming permanently involved in it.**

But they do become involved temporarily, because:

> **Catalysts work by providing an alternative reaction pathway with a lower energy of activation.**

While we don't use a catalyst in the combustion of petrol, catalysts are now used in catalytic converters to speed up reactions that control emissions from the exhausts of cars (see pages 223 to 225).

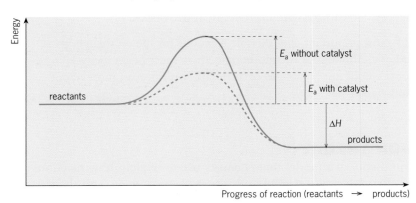

**Fig 8.14** Energy profile of a reaction with a catalyst and without a catalyst. Notice that the enthalpy change is the same, whichever route is followed. Otherwise, it would break the law of conservation of energy (first law of thermodynamics)

Sometimes catalysts slow down the rate of a reaction. When they do, they are called negative catalysts or **inhibitors**. These do not work by raising the activation energy, but usually react with molecules produced in the intermediate stages of a reaction, and remove them from the reaction pathway.

# 6 ANOTHER ENTHALPY CHANGE: THE ENTHALPY CHANGE OF FORMATION

Like the enthalpy change of combustion on page 167, the **enthalpy change of formation** is just one more enthalpy change of reaction. But this time, instead of burning one mole of compound completely in oxygen, we *form one mole of compound from its elements*.

> **The standard enthalpy change of formation $\Delta H^{\ominus}_{f,298}$ is the enthalpy change when one mole of a compound is formed from its elements under standard conditions, that is, 298 K and 100 kPa pressure.**

The standard enthalpy change of formation of water:

$$\Delta H_f^\ominus (H_2O(l)) = -286 \text{ kJ mol}^{-1}$$

applies to this reaction:

$$H_2(g) + \tfrac{1}{2}O_2(g) \rightarrow H_2O(l)$$

Notice that the value above is also the standard enthalpy change of combustion of hydrogen.

The equation for the standard enthalpy change of formation of methanol is:

$$C(s) + 2H_2(g) + \tfrac{1}{2}O_2(g) \rightarrow CH_3OH(l) \quad \Delta H_f^\ominus = -239 \text{ kJ mol}^{-1}$$

## WHY STANDARD ENTHALPY CHANGES OF FORMATION ARE IMPORTANT

You can see the value of knowing enthalpy changes of combustion – because, for example, they allow us to work out the energy we can obtain from a fuel. But why bother with the standard enthalpy changes of formation?

The enthalpy of a substance is its energy content. We cannot measure this and give it an absolute value. It is all the energy associated with the molecules in the substance, which includes the nuclei, the electrons and the movements of whole molecules.

We can only measure enthalpy *changes*, and to do this we must have a reference point. The one that scientists have chosen is the **standard enthalpy change of formation**. It applies to any compound, *while for elements it is zero*. It is a very important piece of information about a compound, and once you know the value for each of the different substances in a reaction, you can calculate the enthalpy change for that whole reaction.

However, before we consider this, how do we obtain values for the standard enthalpy change of formation?

## 7 MEASURING THE STANDARD ENTHALPY CHANGE OF FORMATION

Some standard enthalpy changes of formation can be measured experimentally. For example, we can measure $\Delta H_f^\ominus$ of carbon dioxide by using a bomb calorimeter and combusting graphite. The $\Delta H_f^\ominus$ of magnesium oxide can be measured in the same way.

When a compound is easily synthesized from its elements under normal conditions, we can usually measure $\Delta H_f^\ominus$ directly. However, many standard enthalpy changes of formation cannot be measured directly, which is where we need an indirect approach using **energy cycles**.

## THE INDIRECT METHOD FOR CALCULATING ENTHALPY CHANGES OF FORMATION USING ENERGY CYCLES

This method relies on **Hess's law**, which states:

**If a change can be brought about by more than one route, then the overall enthalpy change for each route must be the same – provided that the starting and finishing conditions are the same for each route.**

This is conservation of energy again – energy cannot be created or destroyed in chemical reactions – so the energy changes for a reaction must be the same, whether it takes place in one step or in a whole series of steps. We can show this by constructing an energy cycle like that shown in Fig 8.15.

**Fig 8.15** An energy cycle illustrating Hess's law. It does not matter whether the reaction takes place by the one-step route or goes through several intermediate steps. The overall enthalpy change for route 2 must be the same as that for route 1, so:

$$\Delta H_1 = \Delta H_2 + \Delta H_3 + \Delta H_4$$

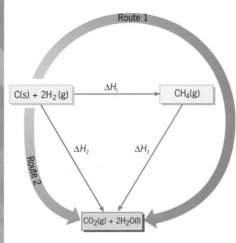

**Fig 8.16** A Hess's law energy cycle being used to calculate $\Delta H_f^\theta$ ($CH_4$(g)) indirectly

**SEE QUESTIONS 3, 4, 5 AND 6.**

Methane cannot be prepared directly from its elements, but we can use Hess's law to calculate its standard enthalpy change of formation. By burning carbon, hydrogen and methane, we can measure their standard enthalpy changes of combustion directly. We can therefore devise an energy cycle with two routes in it, as shown in Fig 8.16.

By Hess's law:

the overall enthalpy change for route 1 = the enthalpy change for route 2

So:
$$\Delta H_1 + \Delta H_3 = \Delta H_2$$

$\Delta H_1$ is what we are trying to find, in this case the standard enthalpy change of formation of methane, $\Delta H_f^\ominus$ ($CH_4$(g)).

For route 1:

$\Delta H_2$ is $\Delta H_c^\ominus$ for carbon (graphite) = $-393$ kJ mol$^{-1}$

**plus** $2 \times \Delta H_c^\ominus$ for hydrogen = $2 \times -286$ kJ mol$^{-1}$

(Remember: $\Delta H_c^\ominus$ for hydrogen is for burning one mole of hydrogen. In this energy cycle there are two moles of hydrogen.)

$\Delta H_3$ = standard enthalpy change of combustion of methane, $\Delta H_c^\ominus$ ($CH_4$(g)) = $-890$ kJ mol$^{-1}$.

So:
$$\Delta H_1 = \Delta H_2 - \Delta H_3 = -393 + 2(-286) - (-890)$$

$$\Delta H_f^\ominus (CH_4(g)) = -75 \text{ kJ mol}^{-1}$$

## EXAMPLE

**Q** Calculate the standard enthalpy change of formation of methanol, given that:
$$\Delta H_c^\ominus (CH_3OH(l)) = -726 \text{ kJ mol}^{-1}$$
$$\Delta H_c^\ominus (C(s)) = -393 \text{ kJ mol}^{-1}$$
$$\Delta H_c^\ominus (H_2(g)) = -286 \text{ kJ mol}^{-1}$$

**A Step 1.** Write out the enthalpy change you have been asked to calculate – in this case $\Delta H_f^\ominus$ ($CH_3OH$(l)):
$$C(s) + 2H_2(g) + \tfrac{1}{2}O_2(g) \rightarrow CH_3OH(l)$$

**Step 2.** Construct an energy cycle with two alternative routes. It helps if you always put the enthalpy change you want to calculate along the top of the cycle and call it $\Delta H_1$.

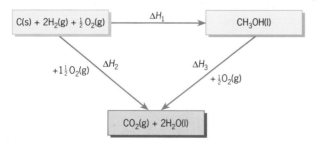

**Fig 8.17** A Hess's law energy cycle to calculate $\Delta H_f^\theta$ ($CH_3OH$(l))

**Step 3.** Write out the enthalpy changes for the two routes:
$$\Delta H_1 + \Delta H_3 = \Delta H_2$$

**Step 4.** Decide what enthalpy changes are represented by $\Delta H_1$, $\Delta H_2$ and $\Delta H_3$. It is a good idea to do this on the energy cycle diagram first.
$$\Delta H_1 = \Delta H_f^\ominus (CH_3OH(l))$$
that is, the enthalpy change we wish to calculate.
$$\Delta H_2 = \Delta H_c^\ominus (C(s)) + [2 \times \Delta H_c^\ominus (H_2(g))]$$
$$\Delta H_3 = \Delta H_c^\ominus (CH_3OH(l))$$

**Step 5.** Rearrange the equation in Step 3 and insert the values of the enthalpy changes that are known:
$$\Delta H_1 = \Delta H_2 - \Delta H_3$$
$$\Delta H_f^\ominus (CH_3OH(l)) = \Delta H_c^\ominus (C(s)) + [2 \times \Delta H_c^\ominus \Delta H_2(g))]$$
$$- \Delta H_c^\ominus (CH_3OH(l))$$

$$\Delta H_f^\ominus (CH_3OH(l)) = -393 + 2(-286) - (-726)$$
$$= -239 \text{kJmol}^{-1}$$

Thus the standard enthalpy change of formation of methanol is $-239$ kJ mol$^{-1}$.

## ? TEST YOURSELF

**S** Work out the standard enthalpy change of formation of:
**a)** propane – see Table 8.1 and the Example above for the values you need.

**b)** carbon monoxide $\Delta H_c^\ominus$ (CO(g)) = $-283$ kJ mol$^{-1}$).
**c)** Why is it impossible to obtain $\Delta H_f^\ominus$ (CO(g)) directly?

# 8 USING STANDARD ENTHALPY CHANGES OF FORMATION TO CALCULATE ENTHALPY CHANGES OF REACTION

Knowing standard enthalpy changes of formation we can calculate enthalpy changes of reaction, and again we apply Hess's law.

---

### ● EXAMPLE

**Q** The Space Shuttle Orbiter uses methylhydrazine as fuel, which is oxidized by dinitrogen tetroxide to provide the energy for propulsion. Calculate the standard enthalpy change $\Delta H_r^\ominus$ for this reaction:

$$4CH_3NHNH_2(l) + 5N_2O_4(l) \rightarrow 4CO_2(g) + 12H_2O(l) + 9N_2(g)$$

These are the standard enthalpy changes of formation:

$$\Delta H_f^\ominus (CH_3NHNH_2(l)) = +54 \text{ kJ mol}^{-1}$$

$$\Delta H_f^\ominus (N_2O_4(l)) = -20 \text{ kJ mol}^{-1}$$

$$\Delta H_f^\ominus (CO_2(g)) = -393 \text{ kJ mol}^{-1}$$

$$\Delta H_f^\ominus (H_2O(l)) = -286 \text{ kJ mol}^{-1}$$

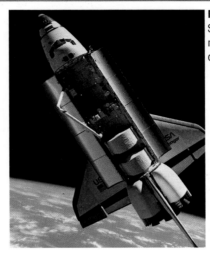

**Fig 8.18** The Space Shuttle Orbiter is powered by methylhydrazine and dinitrogen tetroxide

**A** We can follow a similar procedure to that of the previous Example.

**Step 1.** Write out the enthalpy change you have been asked to calculate. This has already been given in the question, so this step is not required.

**Step 2.** Construct an energy cycle with two alternative routes. Remember that it helps if you always put the enthalpy change you want to calculate along the top and call it $\Delta H_1$.

**Fig 8.19** A Hess's law energy cycle to calculate $\Delta H_r^\ominus$

**Step 3.** Write out the enthalpy changes for the two routes:

$$\Delta H_2 + \Delta H_1 = \Delta H_3$$

Notice that this is different from the previous Example.

**Step 4.** Decide what enthalpy changes are represented by $\Delta H_1$, $\Delta H_2$ and $\Delta H_3$. Remember that it is a good idea to do this on the energy cycle diagram first.

$\Delta H_1 = \Delta H_r^\ominus$, that is, the enthalpy change we wish to calculate:

$$\Delta H_2 = 4\Delta H_f^\ominus (CH_3NHNH_2(l)) + 5\Delta H_f^\ominus (N_2O_4(l))$$

$$\Delta H_3 = 4\Delta H_f^\ominus (CO_2(g)) + 12\Delta H_f^\ominus (H_2O(l))$$

**Step 5.** Rearrange the equation in Step 3 and insert the values of the enthalpy changes that are known:

$$\Delta H_1 = \Delta H_3 - \Delta H_2$$

$$\Delta H_r^\ominus = 4\Delta H_f^\ominus (CO_2(g)) + 12\Delta H_f^\ominus (H_2O(l)) - [4\Delta H_f^\ominus (CH_3NHNH_2(l)) + 5\Delta H_f^\ominus (N_2O_4(l))]$$

$$\Delta H_r^\ominus = 4(-393) + 12(-286) - [4(+54) + 5(-20)]$$

$$= -5004 - [+116] = -5120 \text{ kJ mol}^{-1}$$

So:

$$4CH_3NHNH_2(l) + 5N_2O_4(l) \rightarrow 4CO_2(g) + 12H_2O(l) + 9N_2(g)$$
$$\Delta H_r^\ominus = -5120 \text{ kJ mol}^{-1}$$

The standard state of dinitrogen tetroxide is a gas at 298 K. However, in this case we have used the standard enthalpy change of formation of the liquid dinitrogen tetroxide, because this is how it is carried in the Orbiter. This is the standard enthalpy change of formation in its non-standard state.

**SEE QUESTION 7.**

## ? TEST YOURSELF

**a)** Work through the Example on page 183. Why don't we need to look up the standard enthalpy change of formation of nitrogen?

**b)** Methanol can be produced from coal and the final stage of this process is:

$$CO(g) + 2H_2(g) \rightarrow CH_3OH(l)$$

Calculate the standard enthalpy change for this reaction using the following information:

$$\Delta H_f^{\ominus} (CO(g)) = -110 \text{ kJ mol}^{-1}$$
$$\Delta H_f^{\ominus} (CH_3OH(l)) = -239 \text{ kJ mol}^{-1}$$

## ✔ REMEMBER THIS

The study of rates of reaction is called reaction kinetics (from the Greek word for *to move*). You can read more about reaction kinetics in Chapter 26.

The rate of a reaction can be simply defined as the amount of substance formed per unit of time (e.g. moles per second).

Look closely at Step 5 in the Example on page 183 and you will notice that:

$$\Delta H_r^{\ominus} = \Sigma \Delta H_f^{\ominus} \textbf{ (products)} - \Sigma \Delta H_f^{\ominus} \textbf{ (reactants)}$$

The symbol $\Sigma$ means 'the sum of', and remember that the standard enthalpy changes of formation must be multiplied by the number of moles given in the equation.

The equation above in bold can be used to calculate the enthalpy change for any chemical reaction, which makes it very useful. You do not need to draw an energy cycle, but as you can see, it is still an application of Hess's law.

## 9 RATES OF REACTION AND THE COLLISION THEORY

In Chapter 1, pages 8–9, we considered the importance of being able to control reaction rates in industry. The rate at which a reaction is going to occur and how it can be slowed down (or accelerated), are crucial information. For example, unwanted chemical reactions that lead to food deterioration are slowed down by putting food in a refrigerator or freezer; desired chemical reactions in the cooking of food are accelerated by the energy from an oven. Billions of chemical reactions are happening all around us all the time – from those that occur inside our bodies to those involved in rusting and rotting, and in the life processes of plants and animals. Understanding how reaction rates can be accelerated or slowed down is extremely important and the collision theory introduced on page 179 of this chapter will help us to do this.

In the collision theory model, for a reaction to occur between two particles (atoms, molecules or ions) they must collide with a certain minimum energy called the **activation energy, $E_a$**. Fig 8.22 is an energy profile diagram for an exothermic reaction, showing the activation energy. The higher this

**Fig 8.20** The gas explosion that caused this fire took only a fraction of a second

**Fig 8.21** Divers find that wrecks do not alter much over decades. The rusting of the *Titanic*, which sank in the North Atlantic in 1912, is such a slow reaction that it has been possible to salvage many iron artefacts from the ship more than 90 years later

activation energy is, the fewer the number of molecules that will have sufficient energy to react at any particular temperature, so the rate of reaction will be slow. At the very top of the energy barrier, bonds are forming and being broken, and the molecules are in a **transition state** between reactants and products (Fig 8.23). The idea of a transition state comes from another model of reaction kinetics called the **transition state theory**. This model

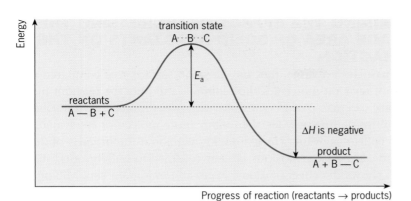

**Fig 8.22** An energy profile diagram for an exothermic reaction

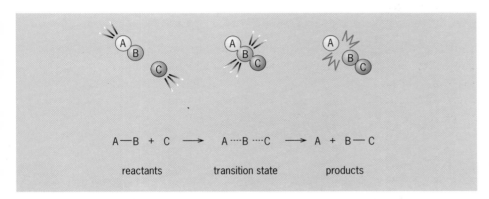

**Fig 8.23** An effective collision between two particles has the required activation energy. The transition state shows bonds breaking and forming as a result of the collision

looks at the collision in detail. While collision theory is a good model for simple gaseous reactions, the transition state theory is applicable more generally. However, all subsequent discussions in this chapter are based on the collision theory.

## EXPLAINING THE EFFECT OF CONCENTRATION AND PRESSURE ON REACTION RATE

Increasing the concentrations of certain reacting species accelerates the reactions in which they are involved. Can collision theory explain this observation?

To react, particles must first collide. So, increasing the number of particles in a given volume must increase the number of collisions and, it would be reasonable to assume, therefore increase the number of effective collisions (Fig 8.24).

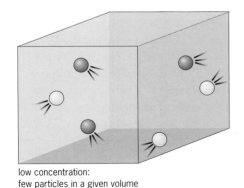

low concentration:
few particles in a given volume

high concentration:
many particles in the same volume

**Fig 8.24** Effect of increasing the concentration of reacting species

Collision theory also explains why increasing the pressure under which gaseous reactions take place increases their rate of reaction. This is illustrated in Fig 8.25. Note that, at a given temperature, increasing the pressure in gases increases the concentration of molecules.

low pressure:
molecules are
spread out

high pressure:
volume has been decreased
(at constant temperature),
forcing the molecules
closer together

**Fig 8.25** The effect of increasing the pressure of reacting gases, held at a constant temperature

### EXPLAINING THE EFFECT OF INCREASING THE SURFACE AREA OF SOLID REACTANTS ON THE RATE OF REACTION

If solid particles are large, they have a small surface area compared to the amount of reactant molecules they contain, and only the reactant molecules at the surface can take part in collisions with other molecules. If a solid particle is ground into a fine powder, then many more molecules are available for effective collisions. Miners have long known about the hazards of dust explosions, but dust explosions are not confined to underground mines. Fig 8.26 shows the effect of a dust explosion in a grain silo. Milling flour also has the same risks.

**Fig 8.26** In 1977, in New Orleans, an explosion of dust killed 37 people. A spark ignited very fine dust in grain silos and the terrifyingly fast rate of combustion with the oxygen in the air caused the explosion

## 10 ENERGY DISTRIBUTION AMONG MOLECULES IN A GAS

Molecules have energy due to their motion. It is called **kinetic energy**. At any instant in time, some molecules have very high energies relative to the rest, because they are moving very fast, while other molecules have very low energies relative to the rest, because they are moving very slowly. Most molecules, however, have a range of energies between these extremes. The distribution of energies among molecules was calculated statistically by the Scots physicist James Clerk Maxwell in 1859, and more generally applied by the Austrian physicist Ludwig Eduard Boltzmann in 1871. The result is the **Maxwell–Boltzmann distribution of molecular energies**, shown in Fig 8.27.

The Maxwell–Boltzmann distribution explains why, at a particular temperature, only a certain number of collisions are effective ones. Fig 8.28 shows the effect of increasing temperature. At the higher temperatures, a greater proportion of molecules has higher energies and so a greater proportion possesses energies equal to, or above, the activation energy, giving more effective collisions.

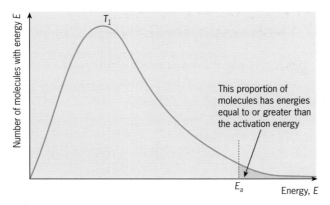

Fig 8.27 Maxwell–Boltzmann distribution curve at a particular temperature

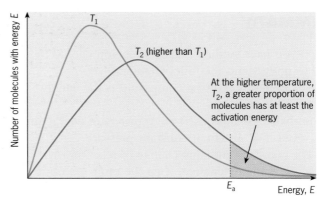

Fig 8.28 Maxwell–Boltzmann distribution curve at temperatures $T_1$ and $T_2$

## MAXWELL–BOLTZMANN DISTRIBUTION CURVE EXPLAINS THE EFFECT OF TEMPERATURE ON REACTION RATE

As Fig 8.28 shows, increasing the temperature increases the proportion of molecules that has the minimum activation energy needed for them to react when they collide. This means that the rate of reaction increases with temperature. In fact, only a small temperature increase leads to a large increase in rate.

**For every 10 °C rise, the rate of most reactions approximately doubles.**

## MAXWELL–BOLTZMANN DISTRIBUTION CURVE EXPLAINS THE EFFECT OF A CATALYST ON REACTION RATE

By having its energy of activation lowered, a reaction occurs at a faster rate; this is what a catalyst does. A catalyst provides an alternative route for a reaction to take place. To explain this in terms of the collision theory, consider again the Maxwell–Boltzmann distribution of molecular energies. If the activation energy is lowered, a greater proportion of molecules has sufficient energy to make effective collisions. This is illustrated in Fig 8.29.

> ✔ **REMEMBER THIS**
>
> The Maxwell–Boltzmann distribution curve is sometimes referred to as the **Boltzmann distribution**

**FOR MORE INFORMATION ABOUT CATALYSTS, SEE PAGES 223–225.**

Fig 8.29 Maxwell–Boltzmann distribution of molecular energies, showing the effect of adding a catalyst. Note that, by lowering the activation energy, there is a large increase in the number of molecules that possess at least the activation energy

**SEE QUESTION 8.**

## SUMMARY

When you have studied this chapter, you should be able to understand the following ideas:

■ The enthalpy ($H$) of a compound is the energy stored in it.

■ Enthalpy changes ($\Delta H$) are transfers of energy into or out of a system at constant pressure.

■ The standard enthalpy change of reaction ($\Delta H^{\ominus}_{r,298}$) is the enthalpy change when the mole quantities expressed in a balanced equation react under standard conditions (298 K, 100 kPa) with all substances in their standard states.

■ The standard enthalpy change of combustion ($\Delta H^{\ominus}_{c,298}$) and the standard enthalpy change of formation ($\Delta H^{\ominus}_{f,298}$) are both standard enthalpy changes of reaction.

■ Some enthalpy changes can be measured directly, using a bomb calorimeter.

■ Other enthalpy changes are measured indirectly by using Hess's law.

■ Average bond energies can give a good estimate of the enthalpy change of a reaction.

■ Activation energy is the minimum energy required for a reaction to occur.

■ High activation energies make reactants kinetically stable, even if they are energetically unstable.

■ Catalysts lower the activation energy by providing an alternative reaction pathway.

■ Collision theory is a model that explains how reacting molecules must collide with sufficient energy, the activation energy, for a reaction to occur

■ The Maxwell–Boltzmann distribution of energies in a sample of gaseous molecules explains why increasing temperature or the use of a catalyst speeds up reactions

## ? EXAM QUESTIONS

1 A student added 50.0 cm³ of hydrochloric acid to 50.0 cm³ of sodium hydroxide solution in a polystyrene cup. The temperature rose by 6.5 °C. The initial concentration of each solution was 1.00 mol dm⁻³.

a) Write an ionic equation for the reaction occurring.

b) Calculate the number of moles of acid used in the reaction.

c) Calculate the heat energy evolved in the reaction. (Assume that the final solution has a specific heat capacity of 4.18 J g⁻¹ K⁻¹ and a density of 1.00 g cm⁻³.)

d) Calculate the molar enthalpy change for the reaction.

AQA Unit 2: Foundation Physical and Inorganic Chemistry June 2001 Q2

2 The enthalpy change of combustion of two fuels is given in Table 8.Q1.

**Table 8.Q1**

| Fuel | Enthalpy of combustion/kJ mol⁻¹ |
|---|---|
| Hydrogen, $H_2$ | −280 |
| Octane, $C_8H_{18}$ | −5510 |

a) Calculate the enthalpy change per unit mass for each of the fuels.

b) Suggest, giving two reasons, which substance is the more useful as a fuel for motor cars and give your reasoning.

c) Suggest one disadvantage of using the fuel chosen in b).

Edexcel 6242/P.01 Unit Test 2 June 2001 Q6

3 The bombardier beetle defends itself from attack by mixing together solutions of hydrogen peroxide and hydroquinone, which react to produce enough energy to bring the mixture to its boiling point. An audible explosion occurs as the beetle fires a hot spray at its attacker.

$$C_6H_4(OH)_2(aq) + H_2O_2(aq) \rightarrow C_6H_4O_2(aq) + 2H_2O(l)$$

hydro-           hydrogen      quinone
quinone         peroxide

**Fig 8.Q3** When under attack, the bombardier beetle uses chemical warfare

a) State Hess's law.

b) Construct an energy cycle based on Hess's law, using data from the following reactions:

$$C_6H_4(OH)_2(aq) \rightarrow C_6H_4O_2(aq) + H_2(g)$$
$$\Delta H^{\ominus} = +177 \text{ kJ mol}^{-1}$$
$$H_2O_2(aq) \rightarrow H_2O(l) + \tfrac{1}{2}O_2(g) \quad \Delta H^{\ominus} = -95 \text{ kJ mol}^{-1}$$
$$H_2(g) + \tfrac{1}{2}O_2(g) \rightarrow H_2O(l) \quad \Delta H^{\ominus} = -286 \text{ kJ mol}^{-1}$$

c) Calculate the enthalpy change of the reaction inside the beetle.

**d)** Enzymes catalyse the hydroquinone–hydrogen peroxide reaction. Define the term catalyst, and explain how a catalyst functions in this reaction.

**4**

**a)** **(i)** Explain what is meant by the term *enthalpy change of reaction*.
**(ii)** Describe an experiment to determine the enthalpy change of the reaction between dilute hydrochloric acid and aqueous sodium hydroxide. Show how the value of $\Delta H$ would be calculated from the data obtained.
**(iii)** Draw an enthalpy level diagram for the neutralization reaction above. Indicate on your diagram the enthalpy change of the reaction and hence compare the relative stabilities of reactants and products.

**b)** Explain, giving one example, the usefulness of Hess's law in determining $\Delta H$ values.

IBO Standard Level Paper 2 May 2000 Q5

**5** Oxyacetylene welding relies on the high energy released from this reaction:

$$C_2H_2(g) + 2O_2(g) \rightarrow 2CO_2(g) + H_2O(l)$$
$$\Delta H^{\ominus} = -1301 \text{ kJ mol}^{-1}$$

**Fig 8.Q5** Oxyacetylene welding or ethyne–oxygen welding

**a)** State Hess's law.
**b)** Construct an energy cycle to calculate $\Delta H^{\ominus}_{f,298}$ ($C_2H_2(g)$), given:

$$\Delta H^{\ominus}_{f,298} (CO_2(g)) = -393 \text{ kJ mol}^{-1}$$
$$\Delta H^{\ominus}_{f,298} (H_2O(l)) = -286 \text{ kJ mol}^{-1}$$

**c)** The standard enthalpy change of combustion of hydrogen $\Delta H^{\ominus}_{c,298}$ ($H_2$) = $-286$ kJ mol$^{-1}$. Explain why this is the same value as the standard enthalpy change of formation of water.

**6**

**a)** State Hess's law.
**b)** Define the term standard enthalpy change of combustion.
**c)** The equation for the combustion of ethanol in air is:

$$C_2H_5OH(l) + 3O_2(g) \rightarrow 2CO_2(g) + 3H_2O(l)$$

and the structural representation of this is given in Fig 8.Q6

H–C($_{}$)($_{}$)($_{}$)–H(g) $\longrightarrow$ 3C(g) + 8H(g)  $\Delta H^{\ominus}$ = +4020 kJ mol$^{-1}$

**Fig 8.Q6**

**(i)** Calculate the enthalpy change for this reaction using the average bond enthalpy values given in Table 8.Q6.

**Table 8.Q6**

| Bond | C–H | C–O | O=O | C–C | O–H | C=O |
|---|---|---|---|---|---|---|
| Average bond enthalpy/kJ mol$^{-1}$ | +412 | +360 | +496 | +348 | +463 | +743 |

**(ii)** Draw and label an enthalpy level diagram to represent this reaction.

Edexcel 6242/P.01 Unit Test 2 June 2001 Q5

**7**

**a)** Define the term *standard enthalpy of combustion*.
**b)** Using the data given below, calculate the standard enthalpy change for the following reaction:

$$CH_4(g) + 2O_2(g) \rightarrow CO_2(g) + 2H_2O(l)$$
$$\Delta H^{\theta}_f \; CO_2(g) = -394 \text{ kJ mol}^{-1}$$
$$\Delta H^{\theta}_f \; H_2O(l) = -286 \text{ kJ mol}^{-1}$$
$$\Delta H^{\theta}_f \; CH_4(g) = -75 \text{ kJ mol}^{-1}$$

**c)** **(i)** State what is meant by the term *mean bond enthalpy*.
**(ii)** Using the standard enthalpy of formation of methane given in **b)** and the data given below, calculate the mean bond enthalpy of the C–H bond in methane.

$$C(s) \rightarrow C(g) \qquad \Delta H^{\theta} = +715 \text{ kJ mol}^{-1}$$
$$H_2(g) \rightarrow 2H(g) \qquad \Delta H^{\theta} = +436 \text{ kJ mol}^{-1}$$

**(iii)** Using the C–H bond enthalpy calculated in **c)(ii)** and the standard enthalpy change for the reaction given in Fig 8.Q3, calculate the mean bond enthalpy of the C–C bond in propane.

H–C($_{}$)($_{}$)–O–H + 3O=O $\longrightarrow$ 2O=C=O + 3H–O–H

**Fig 8.Q7**

NB: If you failed to complete **c)(ii)**, you may assume that the mean bond enthalpy of the C–H bond is +390 kJ mol$^{-1}$. (This is not the correct value.)

AQA Unit 2: Foundation Physical and Inorganic Chemistry June 2001 Q1

**8** Describe, with the aid of a diagram, how the Boltzmann distribution can help to explain the effect of temperature rises and the presence of catalysts on the rates of chemical reactions.

OCR 2813/1 How Far, How Fast? January 2001 Q5

# KEY SKILLS **ASSIGNMENT**

## ETHANOL: A GREEN FUEL OF THE FUTURE?

Brazil is the world's largest user of ethanol as a transport fuel. It is made from the juice of sugar cane, which is fermented and distilled to give a hydrated mixture of 95 per cent ethanol and 5 per cent water. Four million cars, trucks and vans have engines that run on it. Another five million use gasohol, a blend of 22 per cent ethanol and 78 per cent petrol (gasoline).

**1** **a)** Explain what is meant by:
$$\Delta H^{\ominus}_{c,298} (C_2H_5OH) = -1367 \text{ kJ mol}^{-1}.$$
Draw an energy level diagram to represent this enthalpy change.
**b)** Using average bond energies from Table 8.A1, calculate a value for the enthalpy change of combustion of ethanol. Why does this differ from the value given in **a)**?

**Table 8.A1  Some average bond energies**

| Bond (X–Y) | Average bond energy/kJ mol$^{-1}$ | Bond (X–Y) | Average bond energy/kJ mol$^{-1}$ |
|---|---|---|---|
| C—C | +347 | O—H | +464 |
| C—H | +413 | O=O | +498 |
| C—O | +336 | C=O | +805 |

**c)** Although fermentation takes place in aqueous solution, we can estimate the enthalpy change of reaction using:
$$C_6H_{12}O_6(s) \rightarrow 2C_2H_5OH(l) + 2CO_2(g)$$
Calculate the standard enthalpy change of this reaction from the following data:
$$\Delta H^{\ominus}_{f,298} (C_6H_{12}O_6(s)) = -1273 \text{ kJ mol}^{-1},$$
$$\Delta H^{\ominus}_{f,298} (C_2H_5OH(l)) = -277 \text{ kJ mol}^{-1},$$
$$\Delta H^{\ominus}_{f,298} (CO_2(g)) = -393 \text{ kJ mol}^{-1}.$$

During the first world oil crisis in 1973, the price of a barrel of oil rose from $2.5 to $10.5. Until then, Brazil was earning a little more in world trade than it was spending. But with few oil reserves, the country depended entirely on imported oil, which it could no longer afford. In 1975, the Brazilian Pro-Alcool programme was born, aimed at reducing the country's dependence on imported oil. All cars made after 1979 had to run on either gasohol or ethanol fuel.

Economics were the issue, and not the environment, when it was decided to use ethanol as the main transport fuel. Yet, when the fuel is burnt, it releases into the atmosphere just the amount of carbon dioxide that the sugar cane plant took in to make the sugar in the first place. But what of the energy needed to produce ethanol from sugar?

All the energy used to crush the cane, ferment the juice and distil it comes from burning the crushed cane stems and not from fossil fuels, so the carbon dioxide intake and output remain in balance. Also, though artificial fertilizers are used to increase sugar cane production, a variety of cane has been bred that requires only a low nitrogen input.

**Fig 8.A1**  Much of Brazil's sugar cane is cut by hand

**2** **a)** In terms of the carbon dioxide balance in the atmosphere, how do you think producing sugar for fuel compares with i) producing oil for fuel, ii) producing sugar for food?
**b)** In Brazil, most of the sugar cane is harvested by hand. Explain why this reduces extra carbon dioxide emissions significantly.
**c)** How many tonnes of carbon dioxide are produced by one tonne of ethanol?
(Values for $A_r$: C = 12, H = 1, O = 16.) Hint: See pages 7 and 8.
**d)** Suggest why the production of artificial fertilizers releases carbon dioxide into the atmosphere.

Ethanol is an excellent motor fuel. As it is an oxygenate, it burns very efficiently in the car engine, producing almost 60 per cent less carbon monoxide from the exhaust than does petrol. Emissions of nitrogen dioxide are reduced as well. Ethanol has a high octane number, so avoiding 'knocking' in the engine and the need for lead or other compounds to be added to increase its octane rating. (Knocking is covered on page 218.)

**3** **a)** What is meant by the term 'oxygenate'?
**b)** Why is carbon monoxide produced from car exhausts?
**c)** How do you think that ethanol in petrol, or on its own, reduces carbon monoxide emissions? Hint: See page 175.
**d)** Why does it matter if carbon monoxide is produced?

The European Union (EU) over-produces many of the foods it grows, and using land to grow crops to produce fuels is becoming an attractive alternative. This would reduce dependence on fossil fuels and may contribute to the EU's commitment to stabilize and reduce carbon dioxide emissions.

**3** **a)** Will carbon dioxide emissions be reduced by growing fuel crops? Explain your answer.
**b)** What are the advantages and disadvantages of embarking on a Brazilian-type organically derived alcohol programme?

# 09

# Oil and the petrochemical industry

# Oil and the petrochemical industry

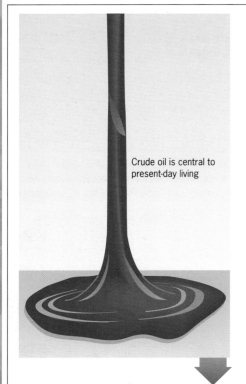

Crude oil is central to present-day living

The Stone Age, the Iron Age and the Bronze Age are periods of history during which humans began to exploit particular substances and so made massive improvements to their lives over a short period of time. In our own time, it is oil that has had the greatest impact on our lives.

From 1859, when the first oil well was sunk by Edwin Drake in the United States, societies have relied increasingly on the production of oil, especially for fuel: 90 per cent of oil is burnt, very much of it used to provide energy for cheap transport. However, if the present were to become known as the Age of the Car, this would ignore what we now do with the remaining 10 per cent of oil. This is transformed through chemical reactions into medicines, paints, insecticides, dyes, detergents, plastics and other chemicals on which we now depend.

Oil is suitably named when called 'black gold', to indicate how precious it is. Maybe future historians will look back and call the age we live in the Oil Age, but it will probably have lasted a brief 200 years from the drilling of the first well to the exhaustion of supplies. In fact, we may well see the end of the Oil Age in our own lifetimes.

# 1 CRUDE OIL

The oil that comes out of the ground is known as crude oil, or petroleum (from the Latin *petra* for rock and *oleum* for oil). It is a thick, dark, smelly liquid. Two thousand years ago, the Chinese occasionally came across it when they dug for brine; then, they used it as a fuel. Inhabitants of the Middle East found it lying on the ground, and because the **volatile** (easily evaporated) components had evaporated, a sticky tar called pitch remained. This they used to waterproof their wooden boats. The ancient Mexicans even used it as chewing gum!

Crude oil is a very complex mixture of hundreds of different compounds. Most are **hydrocarbons**, compounds of carbon and hydrogen only. There are also compounds that contain nitrogen, oxygen and sulphur and even some molecules that have metals in them.

Since crude oil is a mixture, its composition varies. Oil from some wells may contain up to 98 per cent hydrocarbons, while oil from others may have as little as 50 per cent. Hydrocarbons are the most sought-after component, for energy and for making thousands of compounds, from plastics to drugs, that are essential to modern society. Hydrocarbon composition also varies; but more about this on page 200.

## HOW OIL WAS FORMED

The process that formed the oil we find today began up to 400 million years ago, as tiny marine animals and plants died, sank to the bottom of the oceans and became trapped in mud. No oxygen was present in this mud, and bacteria decomposed them anaerobically. An **anaerobic reaction** is one that occurs without oxygen. The decay products stayed in the mud and were acted on by pressure, radioactivity and high temperatures, until finally the mixture we know as oil was created. This took millions of years. As evidence that crude oil came from living organisms, some of the hydrocarbon molecules resemble those found in rubber, cholesterol and vitamin A.

Under pressure, the mud eventually turned into rock, squeezing out the oil into porous rock above. In some areas, the oil seeped to the surface; but fortunately in others it became trapped in the porous rock, which became a reservoir because there was an impervious layer of rock above it.

**Fig 9.1** A microscope picture of sandstone grains. The spaces in between the grains are known as pores. In an oil reservoir, these pores contain crude oil and water

**Fig 9.2** Crude oil trapped by a layer of impervious rock to form an oil reservoir

## ? TEST YOURSELF

**A** Natural gas is associated with oil deposits. What is the principal component of natural gas?

## ► EXTENSION Recovering oil from the ground

There are widely differing estimates of how much oil is left on Earth, and of how much can actually be extracted. So estimates of how long oil supplies will last vary even more widely.

When a new oil reservoir is discovered and wells are sunk, the oil often flows quite freely to the surface, squeezed out by the pressure of rock above it. This is called **primary recovery**. It recovers only 10 to 30 per cent of the oil in the reservoir. **Secondary recovery** methods may involve flooding the reservoir with steam. The energy from the steam makes heavier components of the oil less viscous and drives the oil to the well. Even so, much of the oil still stays in the ground. Whether secondary methods are employed depends on economics. When the price of oil is low, it may cost more to retrieve the extra oil from the ground than it is worth. **Tertiary methods** are being developed for the twenty-first

century, and these require new chemistry and new techniques. Detergents allow oil and water to mix, and may be used to help to extract some of the deposits of heavier oil. The use of polymer solutions and alkalis may also help wrest the oil from the pores of the rock. If these methods succeed, thousands of millions more barrels of oil could be extracted. But the price of oil will have to rise considerably before tertiary extraction becomes cost-effective. Clearly, decisions made in the next few years will decide how long we can enjoy the benefits of crude oil.

**Fig 9.3** Drilling on the Brent Charlie oil rig in the North Sea

# 2 WHY ARE THERE SO MANY CARBON COMPOUNDS?

It was once thought that the chemicals produced in living organisms were different from those found in non-living materials. They were thought to have an associated 'life force' and were therefore known as **organic chemicals**. The rocks and minerals of the Earth, and the chemicals made from them, were called **inorganic chemicals**. But in 1828, a German chemist, Fredrich Wöhler, managed to produce an organic compound, urea (found in urine), from an inorganic compound, ammonium cyanate, just by heating it in his laboratory:

$$\underset{\text{ammonium cyanate}}{NH_4OCN} \xrightarrow{\text{heat}} \underset{\text{urea}}{NH_2CONH_2}$$

This simple experiment helped to dispel the myth of a life force, but we still refer to the chemistry of carbon compounds as **organic chemistry**, and that of all other compounds as inorganic chemistry. During the nineteenth century, many organic chemicals were synthesized (made) and this has continued at an increasing rate, particularly during the past few decades. There are now over eight million recorded organic chemicals, and most of these do not occur in nature. Some are completely new compounds, while others are variations of those found in nature, designed to enhance the properties of particular groups of atoms. The search for new compounds and new ways to synthesize known compounds is a very exciting aspect of chemistry. (You can read about how computers are helping to design drug molecules on page 67.)

# 3 THE SPECIAL NATURE OF CARBON

There are about 100 000 known inorganic substances, a tiny number compared with the vast array of carbon compounds. To explain this, we need to consider just what is so special about carbon.

## CARBON CANNOT EXPAND ITS OCTET

Carbon forms covalent bonds. It is in Group 4, so it has four electrons in its outer shell, which means that it can form four covalent bonds. In common with all the elements in Period 2, it cannot expand its outer shell of eight electrons. This means that its compounds are resistant to chemical attack. We know that methane, $CH_4$, in natural gas does not react with water at normal temperatures and pressures. Silicon is also in Group 4, but it can expand its octet. As a result of this ability, the silane $SiH_4$ reacts vigorously with water. The **mechanism** for this reaction involves a water molecule donating a lone pair of electrons to the silicon atom.

## CARBON CAN FORM STRONG BONDS WITH ITSELF

The ability to form bonds between atoms of the same element in a compound is called **catenation**. While this is not unique to carbon, the C–C bond energy is very high, as you can see from Table 9.1. This means that carbon can form chains of carbon atoms. On page 176 we described octane, a component of petrol, which has a chain of eight carbon atoms:

**Fig 9.4** $SiH_4$ reacting with a water molecule. Notice that when the water molecules form their covalent bonds with this silane, there are 12 outer-shell electrons around the silicon atom

**? TEST YOURSELF**

**B** Why doesn't carbon form ionic bonds?

**C** Draw the dot-and-cross diagrams for methane and water. You can check your diagrams by looking back to pages 72 and 73.

**D** When water donates a lone pair of electrons to silicon, what is the name of the bond formed? See page 74 if you are not sure.

However, in poly(ethene), commonly called polythene, there are thousands of carbon atoms in a chain:

$$
\begin{array}{c}
\text{H} \quad \text{H} \quad \text{H} \quad \text{H} \quad \text{H} \quad \text{H} \quad \text{H} \quad \text{H} \quad \text{H} \quad \text{H} \quad \text{H} \quad \text{H} \\
| \quad\ | \quad\ | \quad\ | \quad\ | \quad\ | \quad\ | \quad\ | \quad\ | \quad\ | \quad\ | \quad\ | \\
\text{\small WV}\ \text{C}-\text{C}-\text{C}-\text{C}-\text{C}-\text{C}-\text{C}-\text{C}-\text{C}-\text{C}-\text{C}-\text{C}\ \text{\small WV} \\
| \quad\ | \quad\ | \quad\ | \quad\ | \quad\ | \quad\ | \quad\ | \quad\ | \quad\ | \quad\ | \quad\ | \\
\text{H} \quad \text{H} \quad \text{H} \quad \text{H} \quad \text{H} \quad \text{H} \quad \text{H} \quad \text{H} \quad \text{H} \quad \text{H} \quad \text{H} \quad \text{H}
\end{array}
$$

Both octane and poly(ethene) are energetically unstable in air under normal conditions, because their products, carbon dioxide and water, are at a much lower energy level. The reaction with oxygen is **spontaneous** at room temperature. This means that once started the reaction takes place of its own accord, but poly(ethene) bags and petrol don't suddenly catch fire when exposed to air. This is because of the large activation energy, which makes them kinetically stable. This activation energy is the energy required to break the C–C and C–H bonds.

Although silicon atoms form short chains with each other, they are not stable under normal conditions. Take, for example, the silane:

$$
\begin{array}{c}
\text{H} \quad \text{H} \quad \text{H} \quad \text{H} \\
| \quad\ | \quad\ | \quad\ | \\
\text{H}-\text{Si}-\text{Si}-\text{Si}-\text{Si}-\text{H} \\
| \quad\ | \quad\ | \quad\ | \\
\text{H} \quad \text{H} \quad \text{H} \quad \text{H}
\end{array}
$$

This compound is unstable and decomposes to silicon and hydrogen at room temperature if no oxygen is present. In air, it bursts into flame for two reasons. First, the reaction with oxygen has a low energy of activation; and second, the Si–O bond is very strong. This means that silicon dioxide, which is sand, is energetically very stable. The stability of the Si–O bond is the reason why so much of the Earth's rocks and soils are made up of silicon dioxide and silicates.

No other element forms chains with itself like carbon. But its uniqueness does not stop at chains – it can form rings as well, such as the example in Fig 9.5.

### CARBON CAN FORM DOUBLE AND TRIPLE BONDS

The strength of the C–H and C–C bonds means that compounds such as octane are fairly unreactive with other chemicals, the reaction with oxygen being a notable exception. However, the inclusion of double and triple carbon–carbon bonds changes all this. Ethene, for example, is the backbone of the organic chemical industry, and it is the π-bond that makes it so reactive (see Fig 9.6).

## 4 REPRESENTING HYDROCARBON MOLECULES

### MOLECULAR AND STRUCTURAL FORMULAS

We first met molecular formulas on page 3. Structural formulas show more information about the arrangement of atoms in a molecule. The molecular formula of methane is $CH_4$ and that of propane is $C_3H_8$. These can be drawn out to show how the atoms are bonded together:

These representations of molecules are often referred to as **displayed formulas**, **graphical formulas** or even **full structural formulas**. We shall use the term *displayed formula* when we draw out all the atoms and

---

**Table 9.1** Some average bond energies between atoms of the same elements near carbon in the Periodic Table. Also included are the bond energies of C–H, Si–H and Si–O

| Bond | Bond energy/kJ mol⁻¹ |
|------|----------------------|
| C—C | 347 |
| Si—Si | 226 |
| N—N | 158 |
| P—P | 198 |
| C—H | 413 |
| Si—H | 318 |
| Si—O | 466 |

✔ **REMEMBER THIS**

**Bond energy** or **bond enthalpy** is the energy required to break a bond between two atoms in a gaseous molecule, leaving two separate atoms. See page 173 for more information

**YOU CAN FIND OUT MORE ABOUT SPONTANEOUS REACTION, ACTIVATION ENERGY AND KINETIC STABILITY BY LOOKING BACK AT CHAPTER 8 AND ALSO BY READING CHAPTER 26. POLY(ETHENE) IS DISCUSSED IN DETAIL ON PAGES 316 AND 431. INFORMATION ABOUT THE STRUCTURE SILICON DIOXIDE IS IN CHAPTER 7.**

**Fig 9.5** Cyclohexane: an example of a ring structure

**Fig 9.6** A molecule of ethene showing the π-bonds

**SEE PAGES 78 AND 82 TO REMIND YOURSELF ABOUT THE BONDING IN ETHENE. THERE IS MUCH MORE ABOUT ITS CHEMISTRY IN CHAPTER 14.**

✔ **REMEMBER THIS**

A molecular formula tells you how many atoms of each element there are in a molecule of the compound (see page 3).

**? TEST YOURSELF**

**E** What is the molecular formula of octane? You can see its structural formula on page 194.

bonds. Structural formulas can also be written in an abbreviated form. For example, the abbreviated structural formula of propane can be shown as:

$$CH_3-CH_2-CH_3 \text{ or even } CH_3CH_2CH_3$$

None of these accurately represents the three-dimensional shape of the molecule. The ball-and-stick models are shown in Fig 9.7.

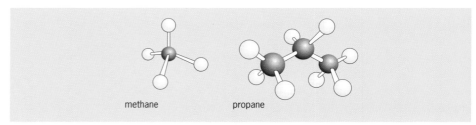

methane          propane

**Fig 9.7** Ball-and-stick models of methane and propane

Notice the tetrahedral arrangement of the hydrogen atoms around the carbon atoms. This can be represented as in Fig 9.8.

H—C bond out of the plane of paper

---- bond into the plane of paper

— bond lies on the plane of paper

**Fig 9.8** Shape diagram of methane and propane

To keep drawing three-dimensional representations of the shapes of molecules would be very inconvenient, which is why we use displayed or structural formulas instead.

**? TEST YOURSELF**

**F** Draw the displayed formulas (showing all atoms and bonds) of ethane, $CH_3CH_3$ and pentane, $CH_3CH_2CH_2CH_2CH_3$.

## COMPOUNDS WITH CARBON–CARBON DOUBLE BONDS

Ethene and propene have carbon–carbon double bonds in their compounds. They are known as **unsaturated hydrocarbons**, because the double bond can allow them to bond with more atoms. Any hydrocarbon with double or triple carbon–carbon bonds in it is called **unsaturated**. Hydrocarbons with no double or triple bonds are called **saturated hydrocarbons**.

The structural formulas of ethene and propene are:

$$CH_2=CH_2 \quad \text{and} \quad CH_3CH=CH_2$$

**SEE PAGE 78 TO REMIND YOURSELF ABOUT THE SHAPE OF ETHENE.**

Their displayed formulas are shown in Fig 9.9. Notice that we show the bonds around a double-bonded carbon at about 120°.

**? TEST YOURSELF**

**G** **a)** Work out which of the following compounds will be unsaturated from the molecular formulas:

$C_4H_{10}$, $C_7H_{14}$ and $C_9H_{18}$.

Hint: Remember that carbon always has four bonds.

**b)** Draw the displayed formulas of the following compounds:

$CH_3CH_2CH=CH_2$, $CH_3CH=CHCH_3$

What do you notice about the molecular formulas of these two compounds?

ethene

propene

**Fig 9.9** The displayed formulas of ethene and propene

# RING COMPOUNDS

As we noted earlier in the chapter, carbon can form rings as well as chains, which adds enormously to the variety of possible organic compounds. Three examples are given in Fig 9.10.

**Fig 9.10** The structural and displayed formulas of cyclopentane, cyclohexene and benzene

Benzene has a ring inside its carbon skeleton, and from the above formula it does not seem to have four bonds around each carbon. The ring represents a **delocalized π-cloud** of electrons, which gives benzene a different set of properties from other ring compounds. **Delocalized electrons** are electrons that do not belong to any one carbon atom.

## SKELETAL FORMULAS

Formulas that show only how the carbon atoms are bonded are called **skeletal formulas**. These depict the carbon skeleton of the molecule without any hydrogen atoms. However, when there are other atoms (such as oxygen or nitrogen) in the molecule, these are shown. Hexane is a hydrocarbon found in

**Fig 9.11** Ball-and-stick model of hexane and its skeletal formula

crude oil and used in petrol. Its skeletal formula, shown in Fig 9.11, is the typical zigzag shape of a hydrocarbon chain.

Oct-3-ene has the structural formula:

$$CH_3CH_2CH_2CH_2CH=CHCH_2CH_3$$

Its skeletal formula is shown in Fig 9.12.

The skeletal formulas you will encounter most are those of ring compounds. Fig 9.13 illustrates the skeletal formulas of the compounds in Fig 9.10.

**Fig 9.13** The skeletal formulas of cyclopentane, cyclohexene and benzene

### TEST YOURSELF

**H**
**a)** Draw the displayed formula of cyclobutane, $C_4H_8$.
**b)** Compare this molecular formula with the molecular formulas in part **b)** of question G. What do you notice?

---

SEE CHAPTER 15 FOR MORE INFORMATION ON BENZENE AND ITS STRUCTURE.

---

### TEST YOURSELF

**I** Draw the skeletal formulas of:
**a)** octane;
**b)** $CH_3CH_2CH=CH_2$;
**c)** $CH_3CH=CHCH_3$.
Note: there are two ways of drawing the last skeletal formula. The Assignment on page 88 tells you why.

---

SEE THE ASSIGNMENT ON PAGE 88, WHICH SHOWS HOW SKELETAL FORMULAS CAN BE USEFUL IN REPRESENTING LARGER HYDROCARBON CHAINS.

---

**Fig 9.12** The skeletal formula of oct-3-ene

---

### TEST YOURSELF

**J** Draw the structural formulas and displayed formulas of the compounds in Fig 9.14.

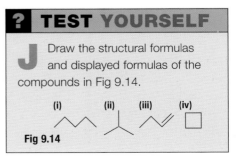

**Fig 9.14**

# 5 ISOMERISM

If you have done the TEST YOURSELF questions in the previous section, you may have noticed that sometimes compounds with the same molecular formula can have different arrangements of their atoms. This is called **isomerism**.

Molecules with the same molecular formula but with different arrangements of their atoms are called isomers.

There are two types of isomerism: **structural isomerism** and **stereoisomerism**. These are summarized in Fig 9.15.

**Isomerism**
Molecules with the same molecular formula but different arrangement of their atoms

**Structural isomerism**
Atoms are bonded in a different order

**Stereoisomerism**
Atoms are bonded in the same order but are arranged differently in space. See Chapters *14* and *19*

**Fig 9.15**

We shall deal in this chapter with structural isomerism. From Fig 9.15 you can see that **structural isomers have the same molecular formula, but different structures**. Their structural formulas are different because the atoms are bonded in a different order.

Consider propane, found in liquefied petroleum gas fuel. It has the molecular formula $C_3H_8$, but only *one* arrangement is possible for its molecule. Even though the carbon atoms can rotate about a single bond, there is still only one structure, which is shown in Fig 9.16.

The displayed formula used to show the rotation of the carbon atom in Fig 9.17 looks exaggerated, so it is tempting to think that it is not the same structure. The carbon atom chain has been shaded so that you see immediately that the three carbon atoms are still in one unbranched chain.

Now we consider the molecular formula $C_4H_{10}$. It is possible to have the two different structures shown in Fig 9.18. These two isomers have different properties, even though they have the same molecular formula. For example, butane boils at a higher temperature than methylpropane.

**Fig 9.16** Ball-and-stick model of propane and its displayed formula

**Fig 9.17** The tinted area shows that the carbon atoms are still part of the same unbranched chain

butane

methylpropane

**Fig 9.18** Ball and stick models of butane and methylpropane and their structural formulas

As the number of carbons in the molecular formula goes up, the number of isomers increases dramatically. For the molecular formula $C_{10}H_{22}$, there are 75 isomers, but when the formula is $C_{30}H_{62}$, the number of possible isomers is more than 400 million! However, hardly any of these have been synthesized and very few occur in nature, but these figures do illustrate how carbon forms so many different compounds.

## ? TEST YOURSELF

**K** Explain why the structures in Fig 9.19 are not additional isomers of butane:

a)

b)

**Fig 9.19**

## ● EXAMPLE

**Q** Draw the structural formulas of the possible isomers of $C_5H_{12}$.

**A** You are unlikely to be asked to find the number of isomers of hydrocarbons with molecular formulas of more than seven carbon atoms. The only way you can approach this problem is by making attempts and then improving them, but try to be systematic.

Always start with the longest hydrocarbon chain you can make from the given formula, as in Fig 9.20.

**Fig 9.20**

Now look for possible branches. But remember, the carbon atoms can rotate about a single bond. In this case, there are only two more possible structures that have branches from the longest chain, as in Fig 9.21.

**Fig 9.21**

A common mistake is to draw a third structure, as shown in Fig 9.22. However, this is actually the structure of 2-methylbutane, and all that has happened is that the molecule has been turned round.

**Fig 9.22**

So far, we have not explained how we name the different hydrocarbons. Don't worry – it is easier than it looks. We deal with naming some of these on pages 212 to 214.

## 6 MORE STRUCTURAL ISOMERS

So far, we have only looked at saturated hydrocarbon chains, but structural isomers occur whenever there are different ways of bonding the atoms together. Two examples are given in Fig 9.23.

**Fig 9.23** Structural isomers

Earlier in this chapter, we said that carbon could bond to atoms other than hydrogen. When this happens, the properties of the molecule change. When we substitute an atom, or a group of atoms, into a hydrocarbon molecule, we call the atom or group a **functional group**.

**? TEST YOURSELF**

**L** Draw the structural formulas of isomers with the molecular formula $C_6H_{14}$.
Hint: If you draw more than five, they will not all be different isomers.

**SEE QUESTION 1.**

**? TEST YOURSELF**

**M** **a)** What is the molecular formula of the two isomers in Fig 9.18?
**b)** Why is the structure in Fig 9.24 not another isomer?

**Fig 9.24**

THERE IS MORE ABOUT FUNCTIONAL
GROUPS ON PAGE 206.

## ? TEST YOURSELF

**N** There are two possible isomers with the formula $C_2H_6O$. One is an ether with a functional group C—O—C. Ethers do not feature much in your advanced course, but the other isomer has a different functional group and has already appeared in Table 8.1 on page 169.
**a)** Draw the displayed formula of both isomers. You can check one of the displayed formulas by looking at page 292.
**b)** Identify the functional group of the isomer of ether. You can check your answer by looking at Table 9.4 on page 206.

SEE QUESTION 2.

## ✔ REMEMBER THIS

Saturated hydrocarbons have no double bonds or triple bonds. The carbon atoms are bonded to the maximum possible number of hydrogen atoms.

## ? TEST YOURSELF

**O** What is the name given to molecules, such as butane and methylpropane, that have the same molecular formula?

A functional group largely determines the set of properties a molecule possesses. The C=C double bond in a molecule is a functional group, because it gives a characteristic set of properties to the molecule. Cl is a halogen functional group and the molecular formula $C_3H_7Cl$ has two isomers, which are shown in Fig 9.25.

**Fig 9.25** The two isomers of $C_3H_7Cl$

# 7 THE HYDROCARBONS IN CRUDE OIL

Crude oil is a mixture, and because of this its composition varies. Crude oil is composed mainly of hydrocarbons, and we have already seen that there can be several hundreds of these because of carbon's ability to form stable chains, branched chains and rings. No two deposits of oil contain exactly the same hydrocarbons in the same proportions. However, we can divide the hydrocarbons found in oil into three major classes of compound: **alkanes**, **cycloalkanes** and **arenes** (also known as aromatic hydrocarbons). We have met representatives of all three classes earlier in this chapter.

## ALKANES

Alkanes are saturated hydrocarbons. They can either be branched chains or straight chains. Table 9.2 summarizes some of the alkanes featured in this chapter.

| **Table 9.2 Some of the alkanes discussed in this chapter** | | | |
|---|---|---|---|
| **Alkane** | **Molecular formula** | **Structural formula** | **Displayed formula** |
| methane | $CH_4$ | $CH_4$ | |
| ethane | $C_2H_6$ | $CH_3CH_3$ | |
| propane | $C_3H_8$ | $CH_3CH_2CH_3$ | |
| butane | $C_4H_{10}$ | $CH_3CH_2CH_2CH_3$ | |
| methylpropane | $C_4H_{10}$ | $CH_3CHCH_3$ <br> $CH_3$ | |

Look at the molecular formulas of the alkanes in Table 9.2 and you should spot a mathematical relationship. For every carbon atom in each formula, there are twice that number of hydrogen atoms, plus another two. This can be represented by the **general formula $C_nH_{2n+2}$**, in which $n$ is the number of carbon atoms. Let's try out this general formula with methane:

$$n = 1 \quad \text{because there is one carbon atom.}$$

So the molecular formula for methane is:

$$C_1H_{(2 \times 1)+2} = CH_4$$

## CYCLOALKANES

Most of the cycloalkanes found in crude oil are based on five- or six-membered carbon rings, as shown in Fig 9.26.

**Fig 9.26** Three examples of cycloalkanes

## ARENES

**Arenes** or **aromatic hydrocarbons** contain at least one benzene ring (see Fig 9.27). The name 'aromatic' originally came from the characteristic smell of some naturally occurring compounds that contained a benzene ring.

**Fig 9.27** Two examples of arenes

SEE CHAPTER 15 FOR MORE INFORMATION ON ARENES.

**? TEST YOURSELF**

**P** What is the molecular formula of alkanes with nine carbon atoms?

### EXTENSION The thirst of the industrialized world for crude oil

Although crude oil deposits are found throughout the world, over half the known oil reserves are concentrated in the Middle East. So, on 2 August 1990, when Iraq invaded Kuwait, a major exporter of crude oil, worldwide supplies were threatened. The invasion highlighted modern society's dependence on oil for energy, and sent shock waves through many countries, including the United States and member states of the European Union, which depend on importing this vital energy supply. The politics of the Middle East are an essential concern to most of the world and will remain so as supplies of crude oil dwindle in the twenty-first century.

**Fig 9.28** A Kuwaiti oil field on fire after the retreat by Iraq in the Gulf War

# 8 MAKING CRUDE OIL USEFUL

Crude oil as it comes out of the ground is of little practical use. To make it useful, it must first be separated into its components, which is done by **fractional distillation**.

## FRACTIONAL DISTILLATION

Fractional distillation is the separation of a mixture of compounds by their different boiling points. When crude oil is separated there are five major fractions, as shown in Table 9.3.

| Table 9.3 The major fractions of crude oil and their uses | | | | |
|---|---|---|---|---|
| Fraction | Boiling point range/°C | Number of carbon atoms | Percentage of crude oil | Uses |
| Refinery gas | −161 to −20 | $C_1$ to $C_4$ | 1 to 2 | Fuel and as a feedstock for petrochemicals |
| Gasoline/ naphtha | 30 to 200 | $C_5$ to $C_{10}$ | 15 to 30 | Petrol for transport and as a feedstock for petrochemicals; the part of the fraction so used is called naphtha |
| Kerosene | 175 to 275 | $C_{10}$ to $C_{16}$ | 10 to 15 | Fuel for jets, paraffin for heating |
| Diesel oil | 275 to 375 | $C_{12}$ to $C_{25}$ | 15 to 25 | Fuel for transport, power plants and heating |
| Residue | 375 | $>C_{25}$ | 40 to 50 | Oil-fired power stations, polishing waxes, lubricating oils, bitumen on roads |

When crude oil reaches a refinery it is heated to about 400 °C and much of it vaporizes. As vapour and liquid, it is fed into a fractionating column, which is about 60 metres high (see Fig 9.29). The liquid collects at the bottom of the column and is called the residue. The vapour passes on up the column and as it cools it turns back to liquid, collecting in trays at various heights. The smaller the vapour molecules, the further the vapour can travel up the column before it condenses. This is because smaller molecules form liquids with lower boiling points (see also pages 214 to 216). At the top of the column, gases that contain very small molecules are collected. These hydrocarbons boil below 20 °C.

**Fig 9.29** The primary fractional distillation of crude oil takes place in a fractionating column. The column is cut away to show some of the trays. Each tray contains many bubble caps, although only three are shown here

## ? TEST YOURSELF

Q Some sulphur remains after refining.

**a)** What compound is formed when sulphur is burnt?

**b)** What environmental consequences are there when this compound is released into the atmosphere?
See page 228 to check your answer.

Vapour rises up the column

Each bubble cap forces the vapour to bubble through the condensed liquid in the tray. This causes the smaller molecules to enter the vapour and continue moving up the column

furnace

400 °C

crude oil

refinery gas - b.p. −161 to +20 °C

petrol

gasoline b.p. 30 to 200 °C

naphtha

kerosene b.p. 175 to 275 °C

diesel oil b.p. 275 to 375 °C

very hot steam

residue

This is the first distillation – the **primary distillation** – and the fractions coming from the column are further separated by a variety of processes. Before a fraction can be used, the sulphur must be removed (this forms a valuable by-product).

## MEETING THE DEMAND: USING ALL THE FRACTIONS

The gasoline fraction is the main source of petrol for car and other vehicle engines, but the amount of gasoline fraction produced is never enough to meet the demand for petrol. Typically, about 40 per cent of the output from a distillation column may be required as petrol. Compare this percentage with that given in Table 9.3 for the gasoline fraction. So, other fractions have to be altered by chemical processes to produce the extra petrol needed.

## CRACKING LARGER MOLECULES

Petrol for use in cars and other vehicles requires alkanes in the range from $C_5$ to $C_{10}$. To meet demand, longer alkanes from other fractions are shortened by being split. This process is known as **cracking**. There are two basic processes for splitting alkane chains:
- using heat, which is called **thermal cracking**;
- using catalysts, which is called **catalytic cracking** (or 'cat cracking' for short).

Whichever process is employed, sufficient energy must be supplied to split the very strong C–C and C–H bonds (look back at Table 9.1, page 195). The advantage of thermal cracking is that molecules in the **residue** can be cracked. Catalytic cracking only works on the **distillate** (the liquid that has been distilled) such as that from the diesel oil fraction. However, catalytic cracking tends to produce more branched-chain alkanes, which are better for use in petrol.

Take, for example, the molecule of dodecane, $C_{12}H_{26}$:

$$CH_3CH_2CH_2CH_2CH_2CH_2CH_2CH_2CH_2CH_2CH_2CH_3$$

$$\xrightarrow[\text{catalyst}]{\text{zeolite}} CH_3\overset{\overset{\displaystyle CH_3}{|}}{C}HCH_2CH_2\overset{\overset{\displaystyle CH_3}{|}}{C}HCH_3 + CH_3CH_2CH{=}CH_2$$

Notice that in this case the molecule has been split and a branched-chain alkane produced. Branched chains are useful in petrol blending because they have higher octane numbers (see page 219). Of course, the molecule can split in almost any way, so a huge variety of hydrocarbons, both straight chains and branched, are produced. These are separated by fractional distillation.

## BOND BREAKING IN THERMAL CRACKING AND CAT CRACKING

The main difference between thermal cracking and catalytic cracking is the way in which covalent bonds in the carbon chains are broken. A covalent bond consists of a pair of electrons that is shared between two atoms (see page 72). The splitting or **fission** of the covalent bond in thermal cracking leaves each species produced with one electron of this pair (Fig 9.31). This type of bond splitting is called **homolytic fission**.

When a species has an unpaired electron it is called a **free radical**. The unpaired electron makes the free radical very reactive and in the reaction mixture it will react again to produce either an alkane or an alkene.

**Fig 9.30** A fractionating column in a refinery

THE COMPOSITION OF PETROL IS DISCUSSED IN GREATER DETAIL IN CHAPTER 10.

**? TEST YOURSELF**

**R** **a)** In which fraction will dodecane be found?
Hint: See Table 9.3.
**b)** Write equations to show the cracking reactions of dodecane to form:
i) ethene and a straight-chain alkane;
ii) propene and a branched-chain alkane.
If you cannot remember the formulas for ethene and propene, look at page 196.

**Fig 9.31** Homolytic fission produces two free radicals

YOU CAN READ MORE ABOUT HOMOLYTIC
FISSION IN CHAPTERS 10 AND 12.

Thermal cracking requires high temperatures (between 400 and 900 °C) and high pressures (up to 7000 kPa), so it is expensive. On completion, it produces a high percentage of alkenes, which are used to produced most of the organic chemicals on which our modern world relies. The Chapter Assignment (page 209) gives more information about thermal cracking and the product molecules formed.

Catalytic cracking involves a different type of bond splitting known as **heterolytic fission**. In this case the shared pair of electrons both go to one of the species produced when the covalent bond is broken. A positive and a negative ion are produced.

Fig 9.32 Heterolytic fission produces a positive and negative ion

The positive ion in Fig 9.32 is called a **carbocation**. This reacts with other alkane molecules to produce a high percentage of branched-chain alkanes, which are ideal for use in petrol (see Chapter 10 page 217).

Catalytic cracking is carried out at about 450 °C with a slight pressure and is the major method of producing petrol and also aromatic hydrocarbons.

SEE QUESTION 4.

## REFORMING AND ISOMERIZATION

Catalytic **reforming** is used to modify molecules to suit demand. Straight-chain alkanes can be converted into cycloalkanes and arenes. Different catalysts are used, depending on the product required. The catalysts used are often bimetallic catalysts such as platinum–rhodium. An example is given in Fig 9.33. Notice that during reforming, whatever happens to the molecule, the number of carbon atoms remains the same.

Fig 9.33 Catalytic reforming to produce cycloalkanes and aromatic compounds

We have already mentioned the need to produce branched-chain alkanes to increase the octane number of petrol. This can be done by taking a straight-chain alkane, heating it to break one of the C—C bonds and allowing the molecule to reform, which often produces a branched chain. This process is called **isomerization**.

$$CH_3CH_2CH_2CH_2CH_3 \xrightarrow{\text{platinum catalyst; heat}} CH_3CH_2CHCH_3$$
$$| \atop CH_3$$

SEE QUESTIONS 4, 5 AND 6.

# ⚠ SCIENCE FEATURE  Zeolites

Zeolite means boiling stone (from the Greek *zeein* for boiling and *lithos* for stone). The name was coined by a Swedish scientist, Baron Cronstedt, in 1756, when he discovered a mineral that released steam and appeared to boil when heated. This name now applies to about 40 naturally occurring minerals that display the same property when heated. Many more zeolites are made synthetically. The steam comes from water trapped in tiny pores and channels in the zeolite. These channels are the key to understanding how this remarkable substance is such an effective catalyst. It took some 200 years from Cronstedt's discovery until synthetic zeolites were first used commercially in 1959 to crack hydrocarbons.

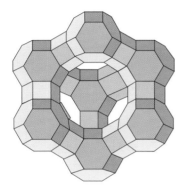

**Fig 9.34** Zeolite Y is used in cracking hydrocarbon molecules

Zeolites are a family of aluminium silicates. Their silicon, oxygen and aluminium atoms provide a regular network of pores and interconnecting channels, rather like a sponge (see Fig 9.34). The size of the pores is critical to the use of zeolites as catalysts for the host chemicals. Pore sizes (in picometres, $10^{-12}$ m) range from 300 pm to 1000 pm, which is why the pores trap water molecules (which have an effective diameter of about 270 pm). If the pore size is large enough, the zeolite can accommodate hydrocarbon molecules, which are cracked inside the structure.

Synthetic rather than natural zeolites are now used for catalytic cracking, and there is much research into ways to alter the sizes of pores and channels by inserting different atoms into the structure. Zeolites have saved the oil industry billions of pounds because they are so efficient and produce more commercially desirable products than the clay catalysts previous used.

The uses for zeolites don't stop with cracking. In New Zealand, which has vast reserves of natural gas, but hardly any oil, a synthetic zeolite, ZSM-5, converts methanol into petrol. Methane is first converted into methanol, then ZSM-5 removes water from the molecules, to leave hydrocarbon chains trapped inside the zeolite. These are then driven out of the zeolite as petrol.

**Fig 9.35** Zeolite ZSM-5 converts methanol produced from natural gas into petrol

## Zeolites in washing powder

Almost certainly you have zeolites in your house. Look at a packet of washing powder and you will probably find more than a quarter of the powder is made from zeolites. The function of zeolites here is to remove calcium ions, which cause water hardness, and replace them with sodium ions, which soften the water. About three-quarters of the world's production of zeolites ends up in detergents.

Zeolites also function as molecular sieves. For example, detergents need to be biodegradable. The largest part of the detergent molecule is a hydrocarbon chain. If this chain is branched, microorganisms at the sewage works cannot break down the detergent.

At the refinery, unbranched hydrocarbons for detergent manufacture are separated from those with branched chains by passing them into the channels of a zeolite. The branched chains are too large to enter the zeolite, so just the unbranched hydrocarbons pass through it.

**Fig 9.36** Zeolite acting as a molecular sieve. Unbranched chains can enter channels, whereas branched chains cannot. The zeolite is removed and the unbranched chains are driven from the zeolite by heating

There is speculation that zeolites may have provided the organizing influence that produced the biologically active molecules that form the basis of life on Earth.

# 9 PETROCHEMICALS

SEE QUESTION 7.

More than 90 per cent of crude oil is burnt; the rest forms the basis of the petrochemical industry, which produces almost all of our organic chemicals. We have already said that alkanes are rather unreactive. However, the cracking process produces small-chain alkenes, of which ethene is the most important. Ethene provides a synthetic route to many organic chemicals. The reactivity of its double bond provides a way to insert many different functional groups into hydrocarbon molecules and thereby allow the full potential of hydrocarbons to be realized.

## FUNCTIONAL GROUPS

A functional group is an atom or group of atoms that determines the properties of a molecule. It is usually the functional group that provides the reaction site of the molecule. Table 9.4 summarizes some of the functional groups you will meet in subsequent chapters.

SEE QUESTION 8.

**Table 9.4**

| Functional group | Name of class of compound | Typical example | Use of example |
|---|---|---|---|
| —C—X<br>X = F, Cl, Br, I | Halogenoalkanes | Chloroethane, $C_2H_5Cl$ | Starting material for the production of time-release capsules for medicines |
| —O—H | Alcohols | Ethanol, $C_2H_5OH$ | Industrial solvent |
| $-C{\underset{O}{\overset{H}{\lessgtr}}}$ | Aldehydes | Ethanal, $CH_3CHO$ | Making synthetic rubber |
| $>C=O$ | Ketones | Propanone, $CH_3COCH_3$ | Industrial and domestic solvents, e.g. nail polish remover |
| $-C{\underset{O}{\overset{OH}{\lessgtr}}}$ | Carboxylic acids | Ethanoic acid, $CH_3COOH$ | Production of pharmaceuticals and in vinegar |
| $-N{\underset{H}{\overset{H}{\big\langle}}}$ | Amines | Ethylamine, $C_2H_5NH_2$ | Important intermediate for many industrial compounds |

SEE QUESTION 9.

# SUMMARY

After studying this chapter, you should know that:

- Carbon is unique in its formation of stable chains, branched chains and rings, giving rise to millions of different chemicals.
- Crude oil is a hydrocarbon mixture of alkanes, cycloalkanes and arenes (aromatic hydrocarbons).
- Components of crude oil can be separated by fractional distillation in a fractionating column. The five major fractions separated in a fractionating column are refinery gases, gasoline (includes petrol and naphtha), kerosene, diesel oil and residue.
- Larger hydrocarbon molecules are broken down (cracked) into smaller ones for petrol and to produce alkenes, such as ethene, for the petrochemical industry.
- Thermal cracking produces a high percentage of alkenes via a free radical mechanism. It takes place at high temperature and pressure.
- Catalytic cracking produces branched-chain alkanes suitable for motor fuel and aromatic hydrocarbons. It takes place via a carbocation mechanism.

- Reforming is a way to produce cycloalkanes and arenes from straight-chain alkanes.
- Isomerization converts straight-chain alkanes into branched-chain isomers.
- Crude oil is a finite resource. Its use as a feedstock for most organic chemicals has changed the world, yet most of it is burnt as fuel.
- Saturated hydrocarbons have no double or triple carbon–carbon bonds in their molecules.
- Unsaturated hydrocarbons possess double or triple carbon–carbon bonds.
- Isomerism occurs when molecules have the same molecular formula, but different ways of arranging their atoms. There are two types of isomerism: structural isomerism and stereoisomerism.
- Structural isomers have the same molecular formula, but different structures because the atoms are bonded in a different order.
- Functional groups give a molecule characteristic properties.

# ? EXAM QUESTIONS

**1** Draw displayed formulas for five structural isomers of $C_7H_{16}$.

**2** There are four structural isomers of molecular formula $C_4H_9Br$. The structural formulas of two of these isomers are given in Fig 9.Q2.

**Isomer 1**          **Isomer 2**

**Fig 9.Q2**

**a)** Draw the structural formulas of the remaining two isomers.

**b)** Name isomer 1. [Refer to Chapter 12.]

AQA Unit 3(a) Introduction to Organic Chemistry June 2001 Q1a(i),(ii)

**3** Catalytic cracking and thermal cracking of petroleum fractions produce different types of products, both of which are valuable. Discuss the **catalytic cracking** of a petroleum fraction. In your answer you should quote the conditions used and the type of mechanism involved. Identify the products formed and state their uses. Give **two** economic reasons for the catalytic cracking of alkanes.

AQA Unit 3(a) Introduction to Organic Chemistry June 2001 Q7

**4** Crude oil is an important source of chemicals that can be obtained by fractional distillation and subsequent processing involving cracking, isomerization and reforming.

**a)** During fractional distillation, explain why hydrocarbons that contain few carbon atoms distil at lower temperatures than hydrocarbons with many carbon atoms.

**b)** **(i)** What is meant by *cracking*?
**(ii)** Suggest an equation that illustrates the cracking of decane, $C_{10}H_{22}$.
**(iii)** Although heat alone can be used to crack hydrocarbons, it is far more common for oil companies to use catalysts also. Suggest **two** reasons why oil companies use catalysts.

**c)** Isomerization produces branched hydrocarbons.
**(i)** Why should oil companies want to make branched hydrocarbons from straight-chain hydrocarbons?
**(ii)** Show the structure of a compound that could be obtained from the **isomerization** of hexane. Name the compound.
**(iii)** One of the important hydrocarbons produced during reforming is benzene. Construct a balanced equation for its formation when hexane is reformed.

OCR Chemistry (A) Chains and Rings Specimen Paper 2812 Q2

# ? EXAM QUESTIONS

**5** Cracking and reforming are essential processes in producing useful substances from oil.

**a)** Why is cracking a useful process?

**b)** Suggest an equation for the cracking of decane, $C_{10}H_{22}$.

**c)** Although heat alone can be used to crack hydrocarbons, oil companies often use catalysts. A catalyst will speed up the rate of a chemical reaction. Suggest another reason why oil companies use catalysts.

**d)** State two ways in which *reforming* changes the structure of the oil fractions.

**6** The fractions from crude oil can be processed further by either cracking, reforming or isomerization.

**a)** Outline, with the aid of suitable examples and equations, each of these processes.

**b)** State clearly the industrial importance of each process.

OCR Chemistry (A) Chains and Rings June 2001 Q8

**7** The formulas of five hydrocarbons associated with the oil industry are:

**A** $C_8H_{16}$; **B** $C_2H_4$; **C** $C_{15}H_{32}$, **D** $C_3H_6$, **E** $C_6H_6$.

**a)** Which of these is an alkane?

**b)** Draw the displayed formulas of molecules **B** and **C**.

**c)** What is the name given to the arene molecule **E**? Draw its displayed formula.

**d)** Molecule **C** is a straight-chain hydrocarbon that can be cracked to produce molecules **A**, **B** and **D**. Write balanced equations to represent three possible cracking reactions that would produce these three hydrocarbons.

**e)** Explain why molecules **B** and **D** are such important molecules in petrochemical manufacture.

**8** Name the functional groups in the following molecules:

**a)** $C_3H_7OH$; **b)** $CH_3NH_2$; **c)** $C_2H_4$;

**d)** $C_2H_5CO_2H$; **e)** $CH_3CHO$.

**9** Explain the meaning of the following terms, giving an example:

**a)** functional group;

**b)** general formula;

**c)** unsaturated hydrocarbon;

**d)** arene;

**e)** cracking;

**f)** reforming.

# KEY SKILLS **ASSIGNMENT**

## A CRACKING WAY TO PRODUCE ETHENE

**One of the fundamental components of a petrochemical plant is the steam cracker. The process of steam cracking involves heating a hydrocarbon feedstock to temperatures of 800–850 °C in the presence of steam. The chief product is ethene, but other alkenes, such as propene, butenes and buta-1,3-diene, are produced as well.**

Steam cracking is thermal cracking and it is used in preference to catalytic cracking, even though the temperatures required when using catalysts are much lower. Catalytic cracking produces high-quality petrol with many branched-chain alkanes, but the yield of ethene tends to be low.

**1** **a)** What is meant by the terms *hydrocarbon* and *cracking*?
**b)** Explain, using the bond energy data in Table 9.1, page 195, which bonds in the hydrocarbon feedstock are most easily broken.

Approximately 0.6 tonnes of steam are used for every tonne of feedstock. The purpose of using steam is to dilute the feedstock, which produces a high yield of ethene. However, an added benefit of steam is that it reduces carbon deposits, which would clog up the cracker, and allows more even heating of the reactants.

The steam–hydrocarbon mixture passes into the furnace where it is heated to the high temperatures required for cracking. However, it stays in the furnace for only a short time; this **residence time** is critical in producing good yields of alkenes. A typical residence time is between 0.1 and 0.5 seconds, depending on the feedstock used.

**Fig 9.A1** A steam cracker furnace in the Joint Venture Olefin 6 complex at Wilton. The complex was commissioned in 1979 and is owned by Hunstman. It is the largest unit in Europe, with the capacity to produce almost 800 000 tonnes of ethene per year

Table 9.A1 gives the product yields for different feedstocks. Once the cracked mixture has left the furnace, it must be cooled rapidly, which is done with cold water. This prevents recombination of the product molecules. The different compounds are then separated, mainly by fractional distillation.

**Table 9.A1 Different feedstocks and the percentage by mass of the different products obtained after cracking**

| Product | Feedstock | | | | |
|---|---|---|---|---|---|
| | Ethane | Propane | Butane | Naphtha | Diesel oil |
| Ethene | 78 | 43 | 39 | 30 | 23 |
| Propene | 3 | 16 | 15 | 16 | 14 |
| $C_4$s (i.e. butenes and buta-1,3-diene) | 3 | 5 | 7 | 10 | 11 |
| Light gases (hydrogen, methane, other) | 14 | 31 | 27 | 17 | 10 |
| Petrol (rich in aromatics) | 2 | 4 | 10 | 23 | 20 |
| Fuel oil | 0 | 1 | 2 | 4 | 22 |

**2** **a)** Which feedstock gives the highest yield of ethene?
**b)** What other factors must be taken into account when deciding on which feedstock to use?

**3** **a)** Butenes have the molecular formula $C_4H_8$. Draw the two straight-chain structural isomers.
**b)** What is meant by the term aromatics?
**c)** Naphtha is part of which main fraction?
**c)** How many carbon atoms are in the diesel oil hydrocarbons?

**4** **a)** Write a balanced equation using displayed formula for the production of ethene and hydrogen from ethane.
**b)** Work out the percentage yield of ethene from ethane.
Hint: Look back at page 8 for details of percentage-yield calculations.
**c)** Steam cracking is a continuous process. What is meant by this statement? (See page 12 if you are not sure.)
The economics of this continuous industrial process are complex. In some years, over-production of ethene in the world has led to dramatic falls in its price, while in other years there has been a worldwide shortage, making ethene a valuable commodity.

$C_4$ compounds are also very useful by-products of ethene production. For example, buta-1,3-diene ($CH_2$=CHCH=$CH_2$) is widely used in the manufacture of synthetic rubber for tyres. However, Europe produces a surplus of buta-1,3-diene while the United States is a net importer; this is related to the

# KEY SKILLS **ASSIGNMENT**

feedstock that goes into the cracker. To maintain the price of buta-1,3-diene, much of it must be converted into butane and its isomer by hydrogenation (reaction with hydrogen). Some of this butane is recycled back into the cracker, while the rest is sold on the LPG (liquefied petroleum gas) market.

**5 a)** In Europe, much of the ethene produced comes from naphtha. Why might this lead to a surplus of buta-1,3-diene?
**b)** Which feedstock might the United States be using?

**c)** Draw the displayed formula of buta-1,3-diene.
**d)** Write a balanced equation for the hydrogenation of buta-1,3-diene to produce butane.
**e)** What advantage is there to the petrochemical manufacturer in recycling butane made from buta-1,3-diene into the cracker?

**6** Find out the main uses of ethene and benzene. While there is plenty of information in this book, also investigate other source material.

# 10

# Alkanes, the car engine and atmospheric pollution

**Smog in 2001** Milan enveloped in mist and smog.

People have become too reliant on the car. In the UK, for example, seven out of ten journeys of a mile or more are now made by car, accounting for 86 per cent of the total distance travelled by any means – and this is likely to increase. Travel by bus and coach, once much more important than travel by car, has been declining steadily for the past 40 years and now accounts for only 6 per cent.

Over four days in December 1991, nitrogen dioxide built up in the London atmosphere to unusually high concentrations and 160 more deaths than expected were recorded. The culprit was almost certainly the car. In August 1994, hot weather led to an increase of ozone in the lower atmosphere across Europe, and in the UK, the Department of the Environment urged drivers to leave their cars at home. On 1 October 1997, the atmosphere in Paris was so polluted that half the cars were banned from entering the city and free public transport was offered.

The amount of carbon dioxide in the atmosphere is increasing and the burning of petrol in cars is a major contributor to this increase. Chemists and engineers can do only so much to reduce car emissions. So people must somehow be weaned from their addiction to the car if the present trend towards serious global warming, with its alarming consequences, is to be stopped.

## ✔ REMEMBER THIS

Remember that **saturated hydrocarbons** have no double or triple bonds. The carbon atoms are bonded to the maximum possible number of hydrogen atoms (see page 196). The term refers to a time when chemists added hydrogen to various organic compounds and those that would not take up extra hydrogen were called saturated.

## 1 ALKANES

Petrol and diesel, the most common fuels for road transportation, are both mixtures that consist mainly of alkane hydrocarbons. Chapter 8 explains where the energy comes from when these fuels are burnt in air, and shows that petrol and diesel are both concentrated energy sources. Chapter 9 shows that both fuels are the products of the fractional distillation of crude oil. In this chapter, we look more closely at these two fuels and the properties of the alkanes that they contain. We also look at the environmental consequences of using petrol and diesel.

### NAMING ALKANES

Alkanes are saturated hydrocarbons. Around each carbon atom are four $\sigma$ bonds (sigma bonds) to hydrogen atoms or other carbon atoms. In Chapter 9, page 200, we found that alkanes could be represented by the general formula $C_nH_{2n+2}$. Table 10.1 lists the first six straight-chain alkanes.

## ? TEST YOURSELF

**A** Look at Table 10.1. The next two straight-chain alkanes in this series are heptane, with seven carbon atoms per molecule, and octane, with eight. What are their molecular formulas and structural formulas?

**Table 10.1 The molecular formulas, structural formulas and names of the first six straight-chain alkanes**

| Molecular formula | Structural formula | Name |
|---|---|---|
| $CH_4$ | $CH_4$ | Methane |
| $C_2H_6$ | $CH_3CH_3$ | Ethane |
| $C_3H_8$ | $CH_3CH_2CH_3$ | Propane |
| $C_4H_{10}$ | $CH_3CH_2CH_2CH_3$ | Butane |
| $C_5H_{12}$ | $CH_3CH_2CH_2CH_2CH_3$ | Pentane |
| $C_6H_{14}$ | $CH_3CH_2CH_2CH_2CH_2CH_3$ | Hexane |

Past four, the number of carbon atoms in a molecule is given by a prefix derived from either the Latin or Greek word for the number. For example, in pentane, *pent* tells you that there are five carbon atoms, and *ane* tells you that the compound is saturated. (The suffix ane was coined by August Hofmann, an eminent German chemist.)

Another word you will see for the system of naming compounds is **nomenclature**.

Try this with nonane, Fig 10.1.

*non* is from the Latin word for nine

*ane* denotes a saturated molecule

**Fig 10.1**

So the molecular formula is $C_9H_{20}$.

Naming straight-chain alkanes is thus straightforward, but what about branched chains? Look at Table 10.2, which shows some **alkyl groups**. Alkyl groups are alkanes with one hydrogen removed so that they can bond with other atoms.

**Table 10.2  The names and structural formulas of some alkyl groups**

| Name | Structural formula |
| --- | --- |
| Methyl | $CH_3-$ |
| Ethyl | $CH_3CH_2-$ |
| Propyl | $CH_3CH_2CH_2-$ |
| Butyl | $CH_3CH_2CH_2CH_2-$ |

The name of this branched-chain alkane is methylpropane:

$$CH_3CHCH_3$$
$$|$$
$$CH_3$$

The longer carbon chain has three carbons (hence propane) and a methyl side group. Sometimes, the molecule is called 2-methylpropane to show that the methyl group is attached to the second carbon in the propane chain. But in the case of this molecule, the 2- is usually left out, as this is the only place the methyl group can go.

Fig 10.2 shows the structural formula of 2-methylbutane. Note that the carbon atoms in the longer chain are numbered so that the position of attachment of the alkyl group is the lowest number possible, 2.

```
     4          3          2          1
    CH3 —— CH2 —— CH —— CH3     longest carbon
                    |            chain has 4 atoms:
                                 hence butane
                   CH3 —— methyl side group attached
                          to 2nd carbon
```

**Fig 10.2**

The molecule in Fig 10.3 is **not** called 2-ethylbutane, because the longest continuous carbon chain has five carbon atoms, and hence its name is 3-methylpentane.

```
              CH2 —— CH3                          CH2 —— CH3
               |                                   |
CH3 —— CH2 —— CH —— CH3        CH3 —— CH2 —— CH —— CH3

    2-ethylbutane  X              3-methylpentane  ✓
```

**Fig 10.3**

**REMEMBER THIS**

You will need to remember the prefixes of the first eight alkanes.

**? TEST YOURSELF**

**B** **a)** Draw the structural formula and the displayed formula of nonane.
Hint: Look back at page 195 if you need to remind yourself about displayed formulas.
**b)** What is the name of the straight-chain alkane with a molecular formula $C_{10}H_{22}$? Look at Table 10.3 on page 214 to check your answer.

**REMEMBER THIS**

Sometimes it is convenient to refer to an alkyl group as R.

**? TEST YOURSELF**

**C** **a)** What is the name of the alkyl group with five carbon atoms?
**b)** Work out the general formula for the alkyl group.

**D** Draw the straight-chain isomer of methylpropane.
Hint: If you are not sure how to do this look back at pages 197 and 198.

**E** **a)** Why is the following not a structural isomer of 2-methylbutane?
To check your answer, see page 199.

$$CH_3CHCH_2CH_3$$
$$|$$
$$CH_3$$

**b)** Sometimes this molecule is just called methylbutane. Why?

**? TEST YOURSELF**

**F** Draw the structural formulas of 3-ethyl-2,2-dimethylhexane and 4-propylheptane.
Note that in the first compound, ethyl comes before methyl: by convention, side groups are put in alphabetical order. See the Example Box on page 214.

## EXAMPLE

**Q** The compound shown in Fig 10.4 used to be called iso-octane. It is important in working out the octane numbers of fuels (see page 218). What is its systematic name?

**Fig 10.4**

**A Step 1.** Look for the longest carbon chain (called the **parent alkane**) and name it. In this case, the chain has five carbon atoms, so it will be called pentane.

**Step 2.** Name every side group. The methyl group is the only side group, but there are three of them. We must ensure the name includes all the side groups, so we use *tri*, and hence trimethylpentane. If there were two groups, we would use *di*, and if four, *tetra*.

**Step 3.** Indicate the position of all of the side groups on the longest carbon chain by numbering the carbon atoms. Remember to use the lowest numbers possible. When

there are two or more different alkyl groups, they are placed in alphabetical order irrespective of their numbered position on the chain. (See the compound named in Test Yourself F).

We have numbered the carbon atoms on the molecule so that two of the methyl groups have the lowest possible number, 2. You will notice that the numbers in a name are separated by commas, that the numbers and letters are separated by hyphens and that every side group has a number.

So the **systematic name** of iso-octane is:

2,2,4-trimethylpentane.

The system we have used to name this compound is recommended by the International Union of Pure and Applied Chemistry (IUPAC) and is accepted by chemists throughout the world. However, if you enter the chemical industry, you may still hear non-systematic names such as iso-octane.

SEE QUESTIONS 1 AND 2.

## ALKANES FORM A HOMOLOGOUS SERIES

We have already seen that alkanes can be represented by a general formula, $C_nH_{2n+2}$. This makes them a **homologous series**.

> **The compounds in a homologous series have similar chemical properties, and gradually changing physical properties as the number of carbon atoms increases. They can be represented by a general formula.**

As you go through this book you will encounter other homologous series. Chemists find it useful to classify organic compounds into homologous series, because of the similar chemical properties of the members. This means they can be studied as a group, rather than studying each individual compound.

### PHYSICAL PROPERTIES OF ALKANES

The term **physical properties** refers to those properties of a substance that can be measured or observed without changing the composition or identity of the substance. Physical properties include melting point, boiling point, colour and density.

## ? TEST YOURSELF

**G** Look at Table 10.3. As the number of carbon atoms in each molecule increases, what is the increase in the number of hydrogen atoms?

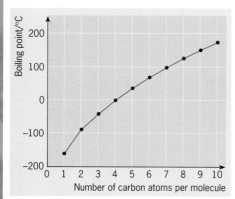

**Fig 10.5** The graph shows the gradual change of boiling points of straight chain alkanes as the number of carbon atoms per molecule increases

SEE QUESTION 3.

**Table 10.3 Three physical properties of the first ten straight chain alkanes. Note the gradual change in these physical properties**

| Alkane | Structural formula | Melting point/°C | Boiling point/°C | Density/g cm$^{-3}$ |
|---|---|---|---|---|
| Methane | $CH_4$ | −182 | −164 | 0.466 |
| ethane | $CH_3CH_3$ | −183 | −89 | 0.572 |
| propane | $CH_3CH_2CH_3$ | −190 | −42 | 0.585 |
| butane | $CH_3(CH_2)_2CH_3$ | −138 | −1 | 0.601 |
| pentane | $CH_3(CH_2)_3CH_3$ | −130 | 36 | 0.626 |
| hexane | $CH_3(CH_2)_4CH_3$ | −95 | 69 | 0.660 |
| heptane | $CH_3(CH_2)_5CH_3$ | −91 | 98 | 0.684 |
| octane | $CH_3(CH_2)_6CH_3$ | −57 | 126 | 0.703 |
| nonane | $CH_3(CH_2)_7CH_3$ | −54 | 151 | 0.718 |
| decane | $CH_3(CH_2)_8CH_3$ | −30 | 174 | 0.730 |

As already stated, the physical properties in a homologous series change gradually as the number of carbon atoms increases. The steady change in the boiling points of the alkanes is used to separate them when crude oil is passed into a fractionating column.

## THE EFFECT OF BRANCHED CHAINS ON THE BOILING POINT

The occurrence of branching on a carbon chain reduces the boiling point. The more branching there is in an alkane, the lower will be its boiling point compared with its straight-chain structural isomer. Pentane and its isomers (Fig 10.6) clearly demonstrate this.

**Fig 10.6** The displayed formulas of isomers with a molecular formula of $C_5H_{12}$. Branching leads to a reduction in boiling point

pentane b.p. 36 °C

2-methylbutane b.p. 28 °C

2,2-dimethylpropane b.p. 10 °C

Branching also affects the **volatility** of an alkane. Volatility describes the tendency of a liquid to evaporate. Branched chain alkanes are very important in blending petrol because of their volatility, as you can see on page 216.

## 2 INTERMOLECULAR FORCES IN ALKANES: INDUCED DIPOLE–INDUCED DIPOLE

We have seen two trends in the boiling points of alkanes. Straight-chain alkanes show an increase in boiling point as the number of carbon atoms increases, and branching reduces the boiling point when the number of carbon atoms remains the same.

The particles in a liquid are held close together by forces. The stronger the forces are, the higher is the boiling point of the liquid. The forces that hold alkane molecules in a liquid are weak **intermolecular forces**, which is why alkanes have relatively low boiling points.

Fig 4.48, page 84, shows that a molecule of hydrogen chloride is **polar** because the centres of negative and positive charge do not coincide. In a polar molecule, there is a **permanent dipole**: one part of the molecule always has a slightly positive charge and another part always has a slightly negative charge. Highly polar molecules can often attract each other quite strongly. Alkanes are sometimes referred to as **non-polar** molecules because they have no permanent dipole. So how do intermolecular forces arise and cause the molecules to attract each other? And how is the increase in carbon atoms related to the trend of increasing boiling point?

The electrons in atoms and molecules are in constant motion. Although alkanes have no permanent dipole, the electron charge cloud around their molecules is not always evenly distributed. At any one instant, there may be a **temporary dipole** when the centres of positive and negative charge do not coincide. This **instantaneous dipole** induces dipoles in neighbouring molecules. A positive charge at one end of a molecule induces a negative charge in a molecule close to it.

**FOR THE SEPARATION OF THE COMPONENTS OF CRUDE OIL, SEE PAGE 202.**

**? TEST YOURSELF**

**a)** In Table 10.3, the density measurements for each alkane were taken at 298 K, except for those that are gases at this temperature. For these compounds, measurements were taken at just below their boiling points. Which density measurements were not taken at 298 K? (See page 124 if you are not sure how to convert Kelvin to degrees Celsius.)

**b)** i) Dodecane has 12 carbon atoms in its molecule. Predict its boiling point and density by drawing a graph similar to that in Fig 10.5 (opposite).
ii) By drawing a graph of melting point against the number of carbon atoms in a molecule, predict how many carbon atoms the first alkane to be a solid at 298 K will have.

**TO REMIND YOURSELF ABOUT PARTICLES IN SOLIDS, LIQUIDS AND GASES, SEE PAGES 127 TO 129.**

**✔ REMEMBER THIS**

Intermolecular means between molecules. Remember that δ+ and δ– show the positive and negative poles in a molecule. The symbol δ tells you that it is a partial charge.

**? TEST YOURSELF**

Why do you think highly polar molecules can often attract each other quite strongly? To check your answer, look at page 264.

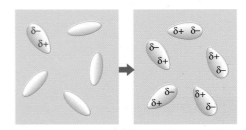

**Fig 10.7 Left:** An instantaneous dipole forms due to the uneven distribution of the electron cloud. **Right:** This induces dipoles in neighbouring molecules, which then attract each other

**? TEST YOURSELF**

**J** Explain why the boiling points of the noble gases increase as you descend Group 0.

**READ MORE ABOUT INDUCED DIPOLE–INDUCED DIPOLE FORCES IN NOBLE GASES AND OTHER ELEMENTS ON PAGE 160.**

**? TEST YOURSELF**

**K** **a)** Explain why the volatility of petrol in Spain is different from that of petrol in the UK.
**b)** Increasing the volatility of petrol can be done by using short straight-chain alkanes. What other alkanes can be used to increase volatility?

**SEE QUESTIONS 4 AND 5.**

✔ **REMEMBER THIS**

Electronegativity is the power of an atom in a molecule to attract electrons to itself.

These opposite charges can now attract. The force between them is known as an **induced dipole–induced dipole force**, or a **van der Waals force** after the Dutch physicist Johannes van der Waals (1837–1923). In the next instant, the movement of electrons in the molecules leads to a collapse of these temporary dipoles and elsewhere in the substance other instantaneous dipoles form. Averaged over time, alkane molecules do not have a permanent dipole, but billions of instantaneous dipoles are continuously occurring, providing a weak intermolecular force.

The more electrons there are in a molecule, the stronger are the induced dipole–induced dipole forces, because there can be a greater distortion of the charge cloud. As the number of carbon atoms increases in the straight-chain alkanes, so do the number of electrons, which in turn increases the induced dipole–induced dipole forces between the molecules, and so leads to increased boiling points.

Although induced dipole–induced dipole forces are relatively weak, they are extremely important. They explain, for example, why noble gases such as helium can be condensed at 4 K. If there were no forces between helium atoms, they could never form a liquid. Induced dipole–induced dipole forces are present in all substances, but they are most important in those that are non-polar, such as alkanes.

We can use the same model of induced dipole–induced dipole forces to explain why branching in alkanes reduces the boiling point. Look back at Fig 10.6. You can see that pentane is a straight chain, which allows the molecules to line up beside each other, thereby giving a greater surface area over which these weak forces can act. Branching reduces this surface area and so the intermolecular forces are correspondingly weaker, which leads to lower boiling points.

**! SCIENCE FEATURE Getting the volatility right**

For petrol to burn in an engine, it must vaporize and be mixed with air. So a crucial property of petrol is high volatility. On extremely cold winter mornings, the petrol must be volatile enough to vaporize and mix with air. If it isn't, the engine won't start. However, on hot summer days, if the petrol is too volatile, it may vaporize in the fuel feed-pipe and so cause a vapour lock that prevents the petrol mixing with air and reaching the engine. Again, the engine won't start. Another drawback of high volatility is that if the petrol vaporizes too readily, too much of it enters the engine and there won't be enough oxygen for it to burn efficiently.

In the UK, the composition or blend of petrol is changed four times a year, according to the season. So, during the winter, petrol contains more of the alkanes with low boiling points, such as butane and pentane, while in the summer, the proportion of volatile components is reduced and more compounds of low volatility are incorporated.

## 3 REACTIONS OF THE ALKANES

Alkanes are quite unreactive because of the nature of the C–C and C–H bonds. First, both are very strong bonds that are difficult to break. Second, both these bonds are non-polar as the electronegativities of carbon and hydrogen are similar. The non-polar nature of the molecules makes them unreactive to polar reagents and ions, and they do not react with acids, bases, metals or oxidizing agents such as potassium manganate(VII). It is their very unreactivity that makes them so useful as lubricants and plastics.

The few reactions that alkanes can undergo are of fundamental importance. Two of these reactions – cracking and reforming – are used to produce petrol with desirable properties (see pages 202 to 204). Another involves halogens (for more about the reaction of halogens with alkanes, see pages 266 to 271). Cracking also produces alkenes, which are the bedrock of the petrochemical industry. For more about the petrochemical industry, see pages 206 and 304.

# 4 COMBUSTION OF ALKANES

We know that a spark produces a spontaneous reaction between petrol vapour and oxygen. The spark provides the activation energy for this reaction. Alkanes burn in oxygen, releasing a great deal of energy, which is one of the reasons why they are such popular fuels. They all produce carbon dioxide and water when there is enough oxygen for them to burn completely.

Butane is used to increase the volatility of petrol. Its reaction with oxygen is:

$$C_4H_{10}(g) + 6\tfrac{1}{2}O_2(g) \rightarrow 4CO_2(g) + 5H_2O(l) \quad \Delta H_c^{\ominus} = -2877 \text{ kJ mol}^{-1}$$

Heptane is another component of petrol. Its reaction with oxygen is:

$$C_7H_{16}(g) + 11O_2(g) \rightarrow 7CO_2(g) + 8H_2O(l) \quad \Delta H_c^{\ominus} = -4817 \text{ kJ mol}^{-1}$$

If there is insufficient oxygen, alkanes are not completely oxidized. They produce carbon monoxide, which is very poisonous, and even carbon (soot). This is a real problem when burning petrol in vehicle engines.

SEE QUESTION 6.

**? TEST YOURSELF**

**L a)** What is the name given to reactions that release energy to the surroundings?
**b)** What are the meanings of the terms spontaneous reaction and activation energy?
**c)** Alkanes and oxygen need a spark before they react. So chemists say that they are kinetically stable but energetically unstable. Explain what they mean by this statement.
**d)** What are the desirable properties of a fuel?
The answers can be found in Chapter 8.

TO CHECK WHAT IS MEANT BY $\Delta H_c^{\ominus}$ SEE

SEE QUESTION 7.

---

## ! SCIENCE FEATURE The petrol engine

The petrol engine is classed as an internal combustion engine because the fuel burns inside the engine and the products of this combustion drive the engine directly. Fig 10.8 illustrates the four-stroke cycle on which vehicle engines work.

The drive to make petrol engines more fuel-efficient centres upon how much the petrol–air mixture can be compressed. In engines of the 1920s, each piston compressed the mixture to a quarter of its original volume. But in today's engines the petrol–air mixture in each cylinder is compressed to one-tenth of its original volume, thereby extracting much more energy from the combustion of the hydrocarbons.

**Fig 10.8** How a four-stroke petrol engine works

inlet valve open
exhaust valve closed
petrol–air mixture

piston moves down, drawing in petrol–air mixture

inlet valve closed
exhaust valve closed

piston moves up to compress petrol–air mixture

spark plug sparks
inlet valve closed
exhaust valve closed

piston moves down by explosion of petrol–air mixture

inlet valve closed
exhaust valve open

piston moves up, expelling products of combustion

### ✔ REMEMBER THIS

2,2,4-Trimethylpentane used to be known as iso-octane, which is why the scale is called octane number or octane rating.

**SEE QUESTION 8.**

## OCTANE NUMBER OF A PETROL BLEND

As the petrol–air mixture is compressed in a cylinder (see Fig 10.8), it becomes hot and sometimes ignites without the aid of a spark from the spark plug **(auto-ignition)**. When this happens, the fuel does not burn smoothly, but explodes in different parts of the cylinder, which generates a rapid knocking noise. **Knocking** can damage the cylinder and piston head and, because the petrol–air mixture does not ignite at the right time, it leads to a loss of power and inefficient use of fuel.

It is the job of the petrol blender to produce a knock-free fuel. So the blender needs a measure of a blended petrol's resistance to knocking. Branched-chain alkanes burn more smoothly in an engine than straight-chain alkanes. 2,2,4-Trimethylpentane has a low tendency to auto-ignite when compressed and is given an octane number of 100. The unbranched-chain alkane heptane knocks readily, even under mild compression. It is given an octane number of 0. Mixtures of these two alkanes are used to assign octane numbers to petrol blends. So if, under test conditions, a particular petrol blend knocks at the same compression as a mixture of 90 per cent 2,2,4-trimethylpentane and 10 per cent heptane, the octane number of that petrol blend is 90.

## ADDING LEAD COMPOUNDS TO PETROL

The advantage of high-octane fuels is that they can be highly compressed, which gives more power per piston stroke and a more efficient use of fuel. In the 1920s, it was discovered that adding small amounts of a certain lead compound to petrol significantly increased the octane number, a practice that was adopted by all the major petrol suppliers.

However, the use of lead in petrol to raise the octane number is now being phased out in most countries for two reasons, both of them connected to the pollution of the environment by lead discharged through exhausts. The first is that lead is a poison that accumulates in the bodies of humans and other animals. Evidence suggests that when children breathe in airborne lead, their IQ level is liable to be lowered, and for adults, their ability to concentrate is reduced. Since the advent of unleaded petrol in the UK in 1986, the concentration of airborne lead in the environment has been reduced by 80 per cent. The second reason why leaded petrol is being phased out is that it inhibits the action of catalysts in catalytic converters, which are essential to reduce pollution from vehicle exhausts.

### ⚠ SCIENCE FEATURE Blending unleaded petrol

If lead is not used to improve the octane number, an alternative way must be found. One solution is to dissolve into the blended petrol small straight-chain alkanes such as butane. The shorter the chain, the less likely is auto-ignition of the mixture. As the alkane chain becomes longer, so the tendency to auto-ignite increases and the octane number decreases. However, too many short-chain alkanes with low boiling points increase the volatility of the petrol too much. This could lead to vapour lock (page 216), and it increases evaporation of hydrocarbons from the fuel tank and engine – yet another source of atmospheric pollution.

Using branched-chain alkanes is another way to raise the octane number of petrol. The more branched the chain, the higher its octane number. Branched-chain alkanes are

**Fig 10.9** This child is at exhaust fume level

produced at the refinery by cracking and reforming reactions; more about this on page 204.

A third way is to add aromatic hydrocarbons. Some 40 per cent of the petrol used in the UK contains such additives. One of them is benzene, a known carcinogen (cancer-causing agent), which can make up to 5 per cent by volume of a petrol. Benzene is now being linked to childhood leukaemia, thought to be the major cause in some cases, and the level of benzene inside a car can be up to ten times higher than outside.

The octane number can also be improved by adding **oxygenates**. We first came across oxygenates on page 175. Oxygenates are partly oxidized fuels that already contain oxygen in their molecules. Methanol ($CH_3OH$) has an octane number of 114 and ethanol ($C_2H_5OH$) 111, so they burn smoothly under high compression. Also, because they have an oxygen atom already in the molecule, when they are combusted they require less oxygen in the petrol–air mixture. So petrols that contain oxygenates produce less carbon monoxide when burnt.

# 5 POLLUTION FROM VEHICLES AND CATALYTIC CONVERTERS

At the outset of this chapter, we pointed to the problem of vehicle emissions. Now we are going to look in more detail at the chemistry behind the pollution associated with the burning of hydrocarbon fuels, such as petrol in vehicle engines.

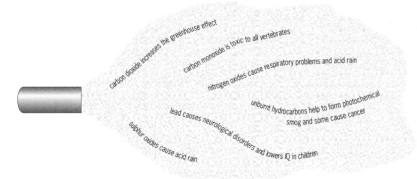

Fig 10.10 Uncontrolled exhaust emissions cause increasing concern

## CARBON DIOXIDE AND THE GREENHOUSE EFFECT

When a hydrocarbon completely combusts, it produces just two products: carbon dioxide and water. Both of these are referred to as greenhouse gases, but what are greenhouse gases and how did they get their name?

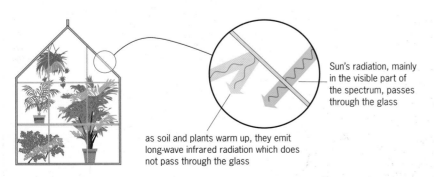

Sun's radiation, mainly in the visible part of the spectrum, passes through the glass

as soil and plants warm up, they emit long-wave infrared radiation which does not pass through the glass

Fig 10.11 Visible light from the Sun enters the greenhouse through the glass. Plants and soil absorb this energy and emit mainly long-wave infrared radiation. The long-wave infrared radiation cannot pass out through the glass, so the greenhouse warms up inside

## ? TEST YOURSELF

**a)** Which radiation transfers the most energy: infrared or visible light?

**b)** Explain why the Earth does not radiate light in the visible region of the electromagnetic spectrum.

Hint: If you are not sure, look back at pages 48 to 49.

## ✔ REMEMBER THIS

A quantum is an indivisible unit or packet of radiation energy (also called a photon). A molecule can exist only in certain fixed vibrational energy levels. When it absorbs a quantum of energy, it becomes excited and jumps to a higher energy level. In this level, it vibrates more. This is similar to electrons absorbing quanta of energy and moving to higher electron energy levels. See page 52 and Chapter 25.

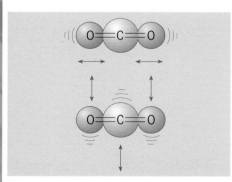

Fig 10.13 Molecules of carbon dioxide absorb infrared radiation, increasing their vibrations

The atmosphere of the Earth acts like glass in a greenhouse. For a start, the atmosphere reflects 30 per cent of the Sun's energy back into space. The remaining 70 per cent, which is mainly visible spectrum light, passes through the atmosphere to strike the Earth. Some of this energy is used by plants for photosynthesis, and some is used to evaporate water from the oceans, lakes and vegetation, but most of it warms the surface of the planet. The warm Earth's surface in turn emits its own radiation, and this is where certain gases in the atmosphere act like greenhouse glass. The energy radiated by the Earth's surface is in the long-wave infrared (IR) region of the spectrum. The greenhouse gases absorb this IR radiation, and their molecules become excited.

Two of the most important greenhouse gases are water and carbon dioxide. Their molecules absorb quanta of IR radiation, which increase their vibration and thus promote them to higher vibrational energy levels. When they fall back to a lower vibrational energy level, they radiate these quanta of energy in all directions. Some of this energy warms the atmosphere, but the rest is transmitted back to the Earth's surface. This is the greenhouse effect.

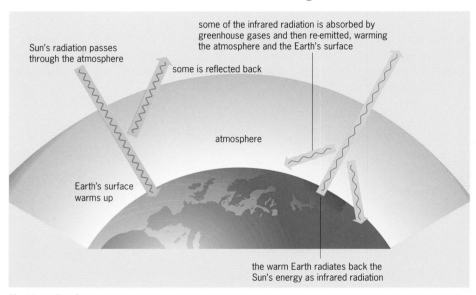

Fig 10.12 The Sun's visible radiation passes through the atmosphere to strike the Earth. The Earth warms up and emits infrared radiation. Some of this is trapped by the greenhouse gases, as shown

## WHY WE NEED THE GREENHOUSE EFFECT IN MODERATION

If carbon dioxide and water were not present in the atmosphere, the Earth would be many degrees cooler. We only have to look at our two closest

Fig 10.14 The extremely dense atmosphere seen here round Venus is mostly carbon dioxide, which keeps its surface roasting hot

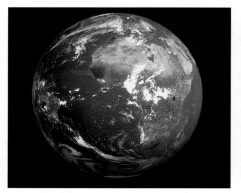

Fig 10.15 Earth enjoys moderate temperatures because greenhouse gases are present in its atmosphere

Fig 10.16 Mars has little atmosphere to prevent infrared radiation from escaping. The temperature of its surface is never greater than 40 °C and can plunge to −80 °C

anetary neighbours to see the effect of greenhouse gases. Venus has an extremely ense atmosphere. It is 96 per cent carbon dioxide and keeps the temperature on Venus a scorching 450 °C. The atmosphere of Mars has about the same proportion of rbon dioxide as the atmosphere on Venus, but is only 1 per cent as dense as the arth's atmosphere. Mars is further from the Sun than Venus so you would expect Mars be cooler, but it has greater temperature fluctuations and cools during its night to 30 °C because there are not enough carbon dioxide molecules to trap the IR radiation.

## HE GREENHOUSE EFFECT AND GLOBAL WARMING

here are about 2750 gigatonnes of carbon dioxide in the Earth's atmosphere, but its oncentration in air is small at 0.0355 per cent, or 355 parts per million (p.p.m.) by olume. We know from analysing the air trapped in polar-ice cores that 250 years ago, at e beginning of the Industrial Revolution, there were about 280 p.p.m. of carbon oxide. Most of the increase has occurred in the past 40 years. This is clearly shown by g 10.17, which is a graph of readings taken by the Mauna Loa Observatory in Hawaii.

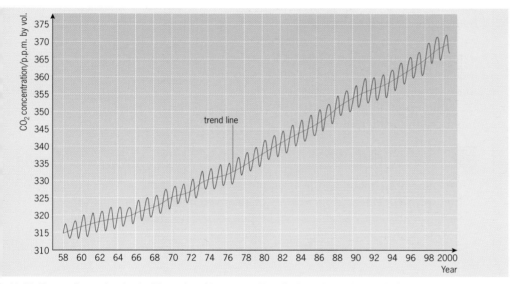

**g 10.17** The readings taken by the Mauna Loa Observatory, Hawaii, show clearly the trend of increasing carbon xide concentration in the atmosphere

At the present time, the rate of increase in carbon dioxide is about 1 p.p.m. per year, hich means that 8 Gt of the gas is added to the atmosphere. However, more than 5 Gt of carbon dioxide is produced each year, much of it from the burning of fossil els for electricity generation or transport. So, where is all the carbon dioxide going? f course, some is removed by plants during photosynthesis, and a lot more dissolves the oceans, lakes and waterways, but there is still much uncertainty about what ppens to the remainder.

Most geoscientists firmly believe that adding such large quantities of carbon dioxide the atmosphere must contribute to the warming of the Earth. If this happens, sea vels will rise and weather patterns throughout the world will be altered. In 1992, an arth Summit was held in Rio de Janeiro in Brazil, and the world's first climate treaty as hammered out by more than 150 nations. This treaty committed the industrialized untries to reduce their greenhouse-gas emissions to 1990 levels by the year 2000.

In 1994, the nations met again to review progress and it was agreed that to stabilize e atmosphere more ambitious targets would have to be set in the years after 2000. So r, there is no agreement as to what these targets should be, or by when they should e achieved. At the same time, carbon dioxide emissions are increasing yearly as the eveloping countries start to industrialize on a large scale, generate more electricity and se more hydrocarbon-driven transport.

## ? TEST YOURSELF

**R** **a)** What carbon dioxide emission targets is the UK currently working towards?

**b)** There are more than 400 million cars in the world today. Assume that petrol consists only of octane, which has a density of 0.7 g cm⁻³:

i) What is the mass of 1 dm³ of petrol?

ii) Write the equation for the complete combustion of octane.

Check your answer on page 176.

iii) What mass of $CO_2$ is produced by burning 1 dm³ of petrol?

(iv) How many tonnes of carbon dioxide are produced daily if 400 million cars use 1 dm³ of fuel a day?

Hint: You may need to look at pages 5 to 6 to remind yourself of work on moles and equations.

## ✔ REMEMBER THIS

Even a 1 °C rise in global temperature would produce unpredictable and possibly devastating climate changes.

The 1992 Earth Summit treaty and the consensus of the 1994 meeting were carried forward in December 1997. After much debate, a further agreement to cut carbon dioxide emissions was put forward, called the Kyoto Protocol, after the city where the agreement was finally made. In 2001, the USA withdrew its support for this agreement, and as this country is the largest emitter of greenhouse gases, accounting for some 22 per cent of the world's total emissions, a further delay to any agreement will be of considerable significance.

## OTHER GREENHOUSE GASES

Carbon dioxide is not the most potent greenhouse gas. If its concentration in the Earth's atmosphere were to double, the effect would probably be an increase in temperature of 1.5–4.5 °C. However, the concentration in the atmosphere of hydrocarbons such as methane, chlorofluorocarbons (CFCs), dinitrogen oxide ($N_2O$) and ozone ($O_3$) are also increasing because of human activity, and these strongly absorb IR radiation. Clearly, the build-up of all greenhouse gases needs careful monitoring and evaluation.

**Table 10.4 Global warming potentials of some atmospheric gases. (Data taken from the Intergovernmental Panel on Climate Change: 1992 Supplement)**

| Gas | Global warming potential |
|-----|--------------------------|
| $CO_2$ | 1 |
| $CH_4$ | 11 |
| $N_2O$ | 270 |
| CFC12 | 7100 |

## ▶ EXTENSION Methane galore

We know from analysing the air in polar-ice cores that the amount of atmospheric methane is increasing. Methane-producing bacteria decompose carbohydrates, such as glucose and cellulose, into methane and carbon dioxide, which are discharged into the atmosphere. From glucose:

$$C_6H_{12}O_6 \rightarrow 3CO_2(g) + 3CH_4(g)$$

The bacterial process is anaerobic: it does not require oxygen, so it happens in bodies of stagnant water. For example, paddy fields produce large amounts of methane because the water and mud that cover the rotting vegetation provide the right conditions for anaerobic bacteria to work. Cattle, too, produce enormous amounts of methane (each cow discharges about 500 dm³ a day in belches from partly digested food in its gut). Even when methane reacts in the atmosphere, it produces mostly ozone, another greenhouse gas.

**Fig 10.18** The growing of rice contributes significantly to atmospheric methane

## EXTENSION Modelling the global climate

Geoscientists do not agree on the contribution made by carbon dioxide to global warming. A few dispute that global warming has occurred at all. This makes it difficult for governments to act and it leaves the public confused. All predictions are based on models of how the climate will react to the increasing quantities of greenhouse gases.

As the Earth warms up, more water vapour will be released into the atmosphere. Water is a potent greenhouse gas, so will this warm the Earth? Or will increased cloud formation lead to global cooling? As carbon dioxide concentration increases, plants will probably photosynthesize more, and thus remove extra carbon dioxide from the atmosphere. Tiny particles in the air, in particular sulphate particles, reflect solar radiation back into space, which leads to global cooling. (Sulphate particles arise mainly from the burning of sulphur in fuels.) The oil fires during the war in Kuwait, and recent volcanic eruptions, produced particles that caused global cooling. Any climate model must take account of such factors. Small wonder, then, that there is a heated debate over global warming.

**Fig 10.19** Forests burning in Indonesia in 1997 set up dust clouds that covered the whole region. In this satellite image, the dust is pink and the fires are red.

## CARBON MONOXIDE EMISSIONS AND CATALYTIC CONVERTERS

In the UK, transport is responsible for 90 per cent of all emissions of carbon monoxide. When petrol does not have enough oxygen for complete combustion, carbon monoxide is produced. It is toxic to all vertebrates (and some invertebrates) and reacts quickly with haemoglobin in red blood cells to produce carboxyhaemoglobin. Once carbon monoxide has reacted with haemoglobin molecules, they do not carry oxygen. Mild symptoms of carbon monoxide poisoning include headaches, dizziness and tiredness. Severe symptoms are fainting and a possibly fatal coma.

The World Health Organization guidelines for carbon monoxide are often exceeded in urban areas all over the world, with health consequences. For example, in New York when carbon monoxide concentrations exceed recommended levels, the incidence of heart attacks increases.

Catalysts can be used to reduce exhaust emissions. Platinum, for example, catalyses the oxidation of carbon monoxide:

> **REMEMBER THIS**
>
> Catalysts are substances that alter the rate of a chemical reaction without becoming permanently involved in it. A catalyst works by providing an alternative reaction pathway that has a lower activation energy. See pages 180 and Chapter 26 for more information.

$$2CO(g) + O_2(g) \xrightarrow{\text{Pt catalyst}} 2CO_2(g)$$

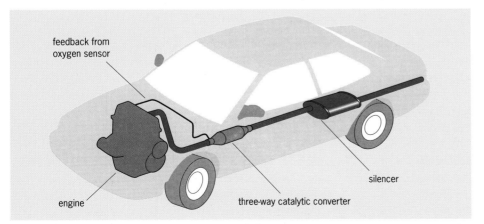

feedback from oxygen sensor

engine

three-way catalytic converter

silencer

**Fig 10.20** The three-way catalytic converter is sited near the engine so that it warms up quickly. An oxygen sensor monitors the oxygen flowing through the exhaust system, and this feeds back to control the fuel–air mixture that enters the engine. Too little oxygen flowing over the catalyst slows down oxidation of carbon monoxide and unburnt hydrocarbons

A catalytic converter is fitted to a vehicle exhaust system to speed up this reaction (see Fig 10.20). It is sited near the engine so that it heats up quickly, because platinum does not start catalysing until its temperature is 240 °C.

**Fig 10.21(a)** The catalytic converter unit

**Fig 10.21(b)** Cutaway of a three-way catalytic converter showing the honeycomb filter. The gases enter the filter, in which a mixture of platinum and rhodium catalyses reactions of the exhaust gases to remove pollutants

support

catalyst coating on honeycomb support

This temperature is reduced to about 150 °C when platinum is mixed with rhodium. The mixture of platinum and rhodium is known as a three-way catalyst, because, in addition to catalysing the oxidation of carbon dioxide, it catalyses two other reactions that involve emission pollutants. Only 1–2 g of each element is used, but because they are coated onto a honeycomb filter of aluminium oxide, the surface area of the catalyst is equivalent to that of two football pitches. The use of **inert** supports to increase the surface area of catalysts reduces the amount of catalyst required and maximizes the surface area available for reaction.

## UNBURNT HYDROCARBONS $C_xH_y$ AND CATALYTIC CONVERTERS

When there is too little oxygen in a petrol engine, unburnt hydrocarbons are present in the exhaust emissions, together with carbon monoxide. Also, volatile hydrocarbons, such as butane, evaporate from the petrol tank on a warm day when a vehicle is stationary. There is increasing concern about **volatile organic compounds (VOCs)**, especially those in the environment. In the UK, almost 40 per cent are from fuel evaporation and exhaust emissions. Long-term exposure to hydrocarbon emissions can impair lung function, while even-short-term exposure can irritate the lung lining. Also, some VOCs, such as benzene, are known carcinogens.

The platinum in a converter catalyses the oxidation of $C_xH_y$ in the exhaust emissions. This is the second way in which pollutants are removed in a three-way catalytic converter:

$$C_xH_y \xrightarrow{\text{O}_2,\ \text{Pt catalyst}} CO_2 + H_2O$$

The catalyst needs to warm up before it becomes effective, so it is during cold starts that most emissions of exhaust hydrocarbons occur. Enough oxygen is also required through the exhaust to oxidize $C_xH_y$ and carbon monoxide. For this reason, an oxygen sensor is fitted just before the catalytic converter to feed back information about oxygen concentration to the vehicle's fuel injection system (see Fig 10.20).

## OXIDES OF NITROGEN AND CATALYTIC CONVERTERS

Air contains mostly nitrogen. Under normal conditions, nitrogen is very unreactive. However, a petrol engine reaches temperatures of 1000 °C, which supplies enough energy to split the very strong triple bond in nitrogen. Nitrogen then reacts with oxygen to form nitrogen oxides ($NO_x$) – mainly nitrogen monoxide (NO).

**Fig 10.22** The flap round this petrol hose nozzle is essential in preventing volatile hydrocarbons from escaping into the atmosphere

Evaporation of hydrocarbons can occur at any time from the blending of petrol at the refinery to the refuelling of a vehicle at a petrol pump. The UK is now well on the way to introducing a closed system for the loading and transport of petrol, so that all hydrocarbon vapours are trapped and recycled before they escape. Hydrocarbon vapours still escape at the petrol pump, almost always because of a poor seal between the tank and the nozzle.

One way to reduce refuelling emissions and evaporation from the petrol tank is to use activated charcoal. The charcoal is contained in a large canister fitted to the air vent of the fuel tank, where it absorbs 90 per cent of hydrocarbon emissions while the vehicle is standing. When the vehicle is moving, air flow over the charcoal returns the hydrocarbons to the tank and reactivates the charcoal.

NO can be further oxidized in the air to give nitrogen dioxide ($NO_2$). While NO is colourless, $NO_2$ is brown; when the atmospheric conditions allow, this can build up as a brown haze in large cities. $NO_2$ contributes to **acid rain**, reacting with water to form nitrous and nitric acids:

$$2NO_2(g) + H_2O(l) \rightarrow HNO_2(aq) + HNO_3(aq)$$

$NO_2$ also catalyses the oxidation of sulphur dioxide in the atmosphere (see pages 226 and 227) and causes respiratory diseases such as bronchitis. The Opener to this chapter mentions the unusual build-up of nitrogen dioxide in London in December 1991. This far exceeded World Health Organization guidelines and was linked to 160 more deaths than were expected at that time of the year.

The third way in which a three-way catalytic converter works is to reduce $NO_x$ back to nitrogen and oxygen, this time using a rhodium (Rh) catalyst:

$$2NO_x(g) \xrightarrow{\text{Rh catalyst}} N_2(g) + xO_2(g)$$

Too much oxygen passing from the engine into the catalytic converter reduces the efficiency of this reaction. So, to achieve the correct fuel–air mixture is critical to the efficient function of a catalytic converter. Hence the need for an oxygen sensor in the exhaust system (Fig 10.20).

## 6 HETEROGENEOUS CATALYSIS AND CATALYST POISONING

The catalysts in converters are always solids, and the reactants are always gases. So the catalysts are said to be **heterogeneous catalysts**, because they are in a different physical state to the reactants. The reactants are **adsorbed** onto the catalyst surface, which means they weakly bond to it. This holds the reactant molecules close together and also allows their

**?** **TEST YOURSELF**

**V** Write a balanced equation for the formation of NO.

**✔** **REMEMBER THIS**

The chemical reactions that lead to the production of acid rain are complicated and beyond the scope of this book. This account is a much simplified one.

**?** **TEST YOURSELF**

**W** Dinitrogen oxide ($N_2O$) is also formed in vehicle engines. This is non-toxic to humans and is used as an anaesthetic; one of its names is laughing gas. Seven per cent of $N_2O$ in the atmosphere comes from vehicle exhausts. What major environmental problem is linked to a build-up of $N_2O$? To check your answer, look at Table 10.4, page 222.

**Fig 10.24** The catalytic oxidation of carbon monoxide. CO molecules and $O_2$ molecules form weak bonds on the catalyst surface. The adsorbed molecules are held close together and their internal covalent bonds weaken, which allows a reaction with lower activation energy to occur. The molecules are then desorbed from the catalyst surface

**Fig 10.25** An energy profile diagram showing the effect of using a platinum catalyst on the oxidation of carbon monoxide

**SEE QUESTION 9.**

covalent bonds to weaken. This provides the alternative route of lower activation energy (discussed on pages 180 and Chapter 26). New bonds form and the product molecules are **desorbed**.

surface of catalyst

The strength of the weak bonds formed at the catalyst surface with the reactant molecules and product molecules is critical to the efficient function of the catalyst. If the bonds formed are too weak the reactant molecules will not be held in place, and if the bonds are too strong the product molecules will not be able to leave. This is why platinum and rhodium are good catalysts to use in catalytic converters. Tungsten, on the other hand, forms bonds with reactants that are too strong, while those formed with silver are too weak.

Just two tanks of leaded petrol are enough to render a catalytic converter useless. This is called **catalyst poisoning**. Lead poisons the catalyst because it is adsorbed more strongly than the reactant molecules and so it blocks the **active sites** at which the reactants bond. (It was for this reason that the US government made it illegal to use leaded petrol in cars fitted with catalytic converters, and required that the opening to the petrol tank in these cars be made too small to take a leaded-petrol nozzle.)

## 7 SULPHUR OXIDES, ACID RAIN AND HOMOGENEOUS CATALYSIS

Most emissions of sulphur oxides, which are toxic, are from the burning of sulphur in fossil fuels. Sulphur dioxide is produced in large quantities by the reaction:

$$S \text{ (in fuel)} + O_2(g) \rightarrow SO_2(g)$$

and can be further oxidized into sulphur trioxide in the atmosphere:

$$2SO_2(g) + O_2(g) \rightarrow 2SO_3$$

Nitrogen dioxide can act as a catalyst in this reaction, which is another reason why nitrogen dioxide emissions should be reduced where possible. The equations for this catalysis are given in Test Yourself X. As the physical state of the nitrogen dioxide is the same as that of the reactants, it is called a **homogeneous catalyst**.

Dilute solutions of sulphuric acid in rainwater are the main cause of acid rain. ($NO_x$ in the atmosphere also contributes to acid rain by forming nitrous and nitric acid solutions – see page 225.)

Acid rain has had serious and far-reaching consequences, killing trees in forests and lowering the pH in lakes so that fish die. Burning petrol and diesel in vehicle engines is not the main contributor to sulphur dioxide in the atmosphere; it accounts for only 2 per cent of emissions in the UK. By far the worst culprit is the burning of coal in power stations (see page 479).

Sulphur dioxide also poisons the catalyst in catalytic converters by forming strong bonds with the active sites at the surface. Fortunately, mixing the catalyst with aluminium oxide alleviates this because the sulphur dioxide is held in preference by the oxide. The normal conditions of engine running are oxidizing, but during acceleration the conditions in the converter allow the sulphur dioxide to be reduced to hydrogen sulphide, which is then expelled and so the catalyst is not permanently poisoned.

## 8 PHOTOCHEMICAL SMOG

The word 'smog' was first used to describe the combination of smoke, fog and sulphur dioxide that used to build up in London and cause hundreds of extra deaths. The Clean Air Act of 1956 made this type of smog a thing of the past in the UK. However, photochemical smog caused by exhaust emissions from vehicles is a problem in many urban areas. Los Angeles and Athens are two infamous examples. The atmosphere around a large city forms a vast mixing bowl for chemical reactions, and sorting out just what causes photochemical smog has not been easy. Even now, it is not fully understood.

### LIGHT AND THE FORMATION OF SECONDARY POLLUTANTS

Any pollutant in an exhaust emission is called a **primary pollutant**. A **secondary pollutant** is formed in air as a result of the chemical reactions of a primary pollutant. Sunlight supplies the energy to initiate the reactions that form the secondary pollutants ozone ($O_3$) and organic nitrates. Reactions in which the energy is supplied by light are called **photochemical reactions**.

$NO_2$ absorbs photons of ultraviolet light which supply the energy ($hf$) to split one of the covalent bonds holding N and O together:

$$NO_2(g) \xrightarrow{hf} NO(g) + O(g)$$

The splitting of the covalent bond is sometimes called **bond fission**. There are two ways in which a covalent bond can split: heterolytic fission and homolytic fission. In **heterolytic fission** (Fig 10.27), both electrons from the bond go to one atom. The atom that gains an electron becomes negatively charged, while the atom that loses an electron becomes positively charged.

$$X \!-\!\!\overset{\bullet}{\underset{\bullet}{}}\!\!- Y \longrightarrow X^+ + \overset{\bullet}{\underset{\bullet}{}}Y^-$$

**Fig 10.27** Heterolytic fission

**Fig 10.26** These statues on the outside of Exeter Cathedral show the effects of acid rain

SEE QUESTION 10.

SEE QUESTION 11.

$$X \overset{\bullet}{\underset{\bullet}{\rule{2em}{0.4pt}}} Y \longrightarrow X\bullet + \bullet Y$$

**Fig 10.28** Homolytic fission

**SEE QUESTION 12.**

**?** **TEST YOURSELF**

**Y** Use s,p,d notation to write the full electron configuration of an oxygen atom.

If you are not sure how to do this, look back at pages 61–62.

Why do you think it has two unpaired electrons?

**FOR MORE INFORMATION ABOUT OZONE IN THE STRATOSPHERE, SEE PAGES 279 AND THE ASSWIGNMENT IN CHAPTER 26.**

**Fig 10.30** The organic nitrate peroxyethanoyl nitrate (often called PAN after its traditional name peroxyacetyl nitrate). PAN causes breathing difficulties and makes eyes water. It also damages plants

In **homolytic fission** (Fig 10.28), when the bond breaks the bonding pair of electrons are equally shared, so that each atom in the bond gains one electron. The atoms are not charged because the number of protons is balanced by the number of electrons.

**Free radicals** result from homolytic fission. Free radicals are species with an unpaired electron, which often makes them highly reactive. So, in our example, X and Y are both free radicals. The unpaired electron on a free radical can be shown by a raised dot, X•.

Take the homolytic fission of $NO_2$ by light, Fig 10.29.

$$N \overset{O}{\underset{\overset{\bullet}{\underset{\bullet}{O}}}{}} \xrightarrow{hf} \bullet N = O + \bullet \overset{\bullet\bullet}{\underset{\bullet\bullet}{O}} \bullet$$

**Fig 10.29**

Both NO and O are free radicals. In fact, atomic oxygen has two unpaired electrons and is called a **diradical**.

Oxygen atoms are very reactive; one of their reactions involves the production of ozone ($O_3$):

$$O(g) + O_2(g) \rightarrow O_3(g)$$

The presence of ozone in the stratosphere is essential to prevent too much ultraviolet light penetrating to the lower atmosphere (troposphere). However, its build-up close to the Earth's surface is dangerous as it causes respiratory problems, and in high concentrations it produces coughing and nausea. It is also involved in a series of other reactions with unburnt hydrocarbons, which produce yet more ozone and a group of very unpleasant molecules called organic nitrates (Fig 10.30).

## SOLUTIONS TO VEHICLE POLLUTION

Particularly through the development of catalytic converters, chemists have reduced exhaust emissions of some pollutants. Catalysts need to heat up before they can start to operate, so emissions on short journeys or in cold weather are particularly high. Because of the tremendous problems with vehicle pollution in Los Angeles, Californian legislation ruled that from 1998, 2 per cent of cars sold had to be zero emission vehicles (ZEVs), rising to 10 per cent by 2003. The USA has led the way in reducing emissions and up to now US manufacturers have responded. ZEV really means an electric car, but is this truly zero emission when fossil fuels are burnt to provide the electricity to charge up the batteries?

People need to change their attitude to the car. Unless personal car use is reduced, we shall have to continue to accept growing deterioration of the environment.

**Fig 10.31** The cycle path and tramway in Amsterdam, which encourage people to leave their cars at home

**Fig 10.32** The Supertram is one way in which Sheffield is making public transport more attractive

# SUMMARY

After studying this chapter, you should know the following:

■ Alkanes are named systematically using IUPAC conventions.

■ A homologous series is a group of compounds that have similar chemical properties, and gradually changing physical properties. They can be represented by a general formula.

■ Trends in melting point, boiling point and volatility can be explained by weak intermolecular forces, called induced dipole–induced dipole forces or van der Waals forces.

■ As the chain length of alkanes increases, so does the number of electrons, which increases the strength of the induced dipole–induced dipole forces.

■ Branching reduces the surface area over which induced dipole–induced dipole forces can act.

■ Alkanes are non-polar molecules, so do not react with polar reagents.

■ Strong C–C and C–H bonds make alkanes relatively unreactive.

■ Complete combustion of alkanes produces $CO_2$ and $H_2O$.

■ Branched-chain alkanes combust smoothly under pressure in a petrol engine and have high octane numbers.

■ Lead compounds can be added to petrol to increase its octane number, but emissions of airborne lead are toxic and also poison the catalysts in a catalytic converter.

■ $CO_2$ in the atmosphere is an important greenhouse gas. Increasing its concentration could lead to global warming.

■ Exhaust emissions include CO, unburnt hydrocarbons and $NO_x$. Three-way catalytic converters reduce the concentrations of these pollutants, but need to heat up before they can start operating.

■ Heterogeneous catalysts are in a different physical state to the reactants.

■ Homogeneous catalysts are in the same physical state as the reactants

■ $SO_2$ and $SO_3$, produced when sulphur-containing fossil fuels are burnt, are the main cause of acid rain and they are toxic.

■ Free radicals are produced by the homolytic fission of covalent bonds, for which light can provide the energy.

■ Ozone is a secondary pollutant.

■ Photochemical smog is produced by the interaction of sunlight with vehicle exhaust emissions.

# ? EXAM QUESTIONS

**1** Name the compounds shown in Fig 10.Q1.

**Fig 10.Q1**

**2** Draw the structural formulas of
**a)** 3-methylhexane;
**b)** 2,2-dimethylpentane;
**c)** 3-ethylheptane;
**d)** 3-ethyl-2-methylpentane;
**e)** 2,2,3-trimethylpentane.

**3** Alkanes are members of a homologous series. Give three characteristics of members of a homologous series.

**4**
**a)** Explain how intermolecular forces arise between alkanes.
**b)** Explain why boiling points increase as carbon chain length increases in straight-chain alkanes.
**c)** Draw the displayed formula of three isomers with the molecular formula $C_5H_{12}$, and name them.
**d)** Explain the effect of branching on the boiling point of the three isomers you have drawn.

# ? EXAM QUESTIONS

**5** [You may need to refer to page 289.] Select the substance with the higher boiling point in each of the following pairs. Explain your reasoning.
**a)** $C_2H_6$ and $C_3H_8$.
**b)** $CH_3CH_2OH$ and $CH_3OCH_3$.
IBO Chemistry Higher Level Paper 2 May 2000 Q2

**6** [See information in Chapter 9 for this question.] The hydrocarbons in crude oil can be separated by fractional distillation.
**a)** Explain what is meant by the terms:
**(i)** hydrocarbons;
**(ii)** fractional distillation.
**b)** Undecane, $C_{11}H_{24}$, can be isolated by fractional distillation. Calculate the percentage composition by mass of carbon in undecane.
**c)** Undecane can be cracked into nonane and compound **A**. One molecule of nonane contains nine carbon atoms.
**(i)** Write a balanced equation for this reaction.
**(ii)** Name compound **A**.
**d)** Hydrocarbons of formula $C_5H_{12}$ can also be isolated from crude oil.
**(i)** Draw the three structural isomers of $C_5H_{12}$.
**(ii)** Isomers **B**, **C** and **D** can be separated by fractional distillation. State the order, lowest boiling point first, in which they would distil.
**(iii)** Justify the order stated in **d)(ii)**.
**(iv)** Write a balanced equation for the complete combustion of pentane, $C_5H_{12}$.
**(v)** Why do oil companies isomerize alkanes such as pentane?
OCR 2812 Chains and Rings Jan 2001 Q2

**7** State which **one** of the equations below is the correctly balanced equation for the combustion of propane in an excess of air.
A $\quad 2C_3H_6 + 3O_2 \rightarrow 6CO_2 + 6H_2O$
B $\quad C_3H_8 + 5O_2 \rightarrow 3CO_2 + 4H_2O$
C $\quad C_3H_8 + 6O_2 \rightarrow 3CO_2 + 4H_2O$
D $\quad 2C_3H_8 + 7O_2 \rightarrow 6CO_2 + 8H_2O$
WJEC Specimen Papers 2001/2002 CH2 Q2

**8** [Information in Chapter 8 may be helpful.]
**a)** State **one** reason why alkanes are described as *unreactive*.
**b)** Give an example of the use of a particular alkane as a fuel and suggest an alternative energy source to alkane fuels, mentioning **one** advantage and **one** disadvantage of your choice.
WJEC Specimen Papers 2001/2002 CH2 Q13a,b

**9** [See Chapter 4 for information required for **c)** and Chapter 8 for **d)**.]
Butane, $C_4H_{10}$, can be used as a fuel. Here are some data about butane:
Boiling point 0 °C
Standard enthalpy change of combustion
$\Delta H^{\ominus}_{c,298} = -2880$ kJ mol$^{-1}$.
**a)** To what homologous series does butane belong?
**b)** Butane has one structural isomer.
**(i)** Say what you understand by the term **structural isomer**.
**(ii)** Draw the full structural formula of the structural isomer of butane and name it.
**c)** All the H—C—H bonds in butane have the same angle in the three-dimensional molecule. State the value of this angle and, with the help of a labelled diagram, describe the shape of the butane molecule.
**d)** 'The standard enthalpy change of combustion of the isomer in part **b)** will be very similar to that of butane itself as all the bonds are the same.'
Say whether or not this statement is correct and explain why.
**e)** Write a balanced equation (with state symbols) for the complete combustion of butane under standard conditions.
**f)** The isomer will have a higher **octane number** than butane itself.
**(i)** What does the octane number of a petrol tell you about its performance in a high-compression engine?
**(ii)** What feature of the isomer would cause it to have a higher octane number than butane?
**g)** Suggest one technical problem (apart from ones associated with octane number) that might arise in using butane as a fuel for cars.
**h)** The structures of two 'oxygenates' that can be added to petrol to improve the octane rating are given in Fig 10.Q10.

H$_3$C—CH$_2$—CH$_2$—OH     H$_3$C—C—O—CH$_3$
                                        |
                                       CH$_3$
(with CH$_3$ above)

**Compound A**          **Compound B**

**Fig 10.Q10**
**(i)** Name the **functional group** in Compound **A**.
**(ii)** Name Compound **A**.
**(iii)** Name the **functional group** in Compound **B**.
**i)** A car running on butane would produce carbon monoxide as a polluting gas.
**(i)** How is carbon monoxide produced in car engines?

**(ii)** Explain why carbon monoxide is a polluting gas.

**j)** A catalytic converter can convert carbon monoxide in the car's exhaust to carbon dioxide.
**(i)** Explain why such a catalyst is called **heterogeneous**.
**(ii)** Complete the sequence below to explain how such a catalyst works:
1. Reactants get adsorbed on to the catalyst surface.
2.
3. New bonds form.
4.
**(iii)** Lead-free petrol must be used with catalytic converters because lead 'poisons' the catalyst. Explain what this means.

OCR Salters Chemistry for Life Specimen Paper 2850 Q3

**10**

**a)** Describe, using **one** example in each case, the different modes of action of homogeneous and heterogeneous catalysis.

**b)** Many catalysts are very expensive, but their use does allow the chemical industry to operate more profitably. Outline why the use of catalysts provides economic benefits to this industry.

OCR How Far, How Fast? Specimen Paper 2813/01 Q6

**11** MTBE is an oxygenate which is added to petrol. Its structural formula is shown in Fig 10.Q11.

**Fig 10.Q11**

**a)** What is the molecular formula of MTBE?
MTBE belongs to the homologous series of ethers and has a functional group C—O—C.

**b)** Explain the meaning of the terms *homologous series* and *functional group*.
Oxygenates are fuels with oxygen atoms in their molecules, which means they require less oxygen to combust completely.

**c)** Write an equation for the complete combustion of MTBE.

**d)** MTBE has a high octane rating. What is the advantage of a high-octane component in petrol?

**e)** Name three major pollutants from petrol-engine exhausts and explain, using balanced equations, how three-way catalytic converters remove these pollutants.
Ozone is a secondary pollutant from exhaust emissions.

**f)** Explain what is meant by the term *secondary pollutant*.

**g)** Show, using equations, how ozone arises in the lower atmosphere.

**h)** Name one other secondary pollutant present in photochemical smog.
Lead compounds in petrol are still used in some petrol blends to increase octane rating.

**i)** What are the two main reasons for phasing out lead in petrol?

**12**

**a)** Draw dot-and-cross diagrams for the following species and state which are free radicals:
**(i)** $Cl$
**(ii)** $OH$
**(iii)** $H_2O$
**(iv)** $NO_2$
**(v)** $O_3$
**(vi)** $CH_3$

**b)** Explain what is meant by:
**(i)** homolytic fission;
**(ii)** photochemical reaction.

# KEY SKILLS **ASSIGNMENT**

## THE DIESEL ENGINE – HOW IS IT DIFFERENT?

The diesel engine was invented by Rudolph Diesel, a German engineer, who patented his idea in 1892. Diesel engines were used first as generators and marine engines, and it was not until 1922 that they appeared in cars.

Diesel fuel is produced from a higher boiling point fraction of crude oil than that used for petrol. It normally contains $C_{16}$ to $C_{20}$ hydrocarbon molecules, but slow-speed diesel engines of the type fitted in large ships burn hydrocarbons up to $C_{25}$. As with petrol, diesel fuel is composed mostly of alkanes.

**1 a)** What is the molecular formula of hexadecane, an alkane molecule that contains 16 carbon atoms?
**b)** What is meant by *higher boiling point fraction*?
**c)** In which crude oil fraction would you find the molecule $C_{20}H_{42}$? (See Table 9.3, page 202.)
**d)** Explain, using the concept of intermolecular forces and temporary dipoles, why straight-chain alkanes in petrol have a lower boiling point than straight-chain alkanes in diesel fuel.

While everything is done to eliminate auto-ignition in the petrol engine, it is fundamental to the operation of the diesel engine, which has no spark plugs. The energy for the combustion of the fuel–air mixture is supplied by compressing air in each cylinder by 20:1 (i.e. 1 $dm^3$ of air is compressed to 50 $cm^3$). At the top of the compression stroke, diesel fuel is injected into the cylinder and the high temperature of the compressed-air charge provides the activation energy to detonate the fuel.

**2 a)** In a petrol engine, what provides the activation energy to detonate the petrol–air mixture?
**b)** Why must diesel engines be built more strongly than petrol engines?
**c)** Why is auto-ignition a major disadvantage in petrol engines?
**d)** What name is given to the scale that measures petrol's ability to auto-ignite?
**e)** Why are lead compounds added to some petrol blends?
**f)** What are the environmental consequences of leaded fuels?

Diesel engines are more efficient than petrol engines, because at least 90 per cent of the fuel is burnt, compared with about 70 per cent in petrol engines. This reduces carbon monoxide emission to less than that of a petrol engine fitted with a catalytic converter, although hydrocarbon emissions tend to be about the same. As combustion is more efficient, the fuel consumption of a diesel engine is lower than that of a comparable petrol engine.

**3 a)** Why is it important to burn fuel efficiently?
**b)** Explain how emissions of CO occur.
**c)** How does a catalytic converter reduce CO emissions?

The sulphur content of diesel fuel is still a problem, although refineries are gradually reducing it. Also, more nitrogen oxides are produced from diesel-driven vehicles than from petrol-driven vehicles with catalytic converters. The other emissions from diesel-engine exhausts are carbon particles and aromatic compounds. The technology to reduce carbon particles is already far advanced, so if the presence of aromatic compounds could be reduced substantially, the diesel engine would have much more to recommend it.

**4 a)** What environmental problem is caused by sulphur in fuels? Include balanced chemical equations in your answer.
**b)** Explain how nitrogen oxides are produced in diesel exhaust emissions.
**c)** What are the environmental consequences of $NO_x$?
**d)** How do catalytic converters reduce emissions of $NO_x$?
**e)** What health hazards are associated with aromatic compounds in exhaust emissions?

Whatever fuel is used, some people are beginning to question whether the disadvantages of the car outweigh the advantages.

**5 a)** What are the advantages of car use?
**b)** Car exhaust emissions are a major problem in urban areas. How might car use in these areas be discouraged? Hint: Positive as well as negative measures should be considered.
**c)** Comment on the statement: Cars produce fewer emissions per mile on long journeys than on short journeys, so they should not be used for very short journeys.

# 11

# Analysing, identifying and separating substances

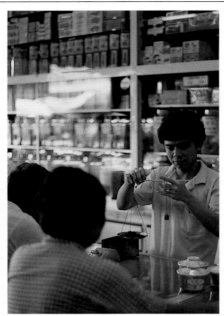

**Chinese chemist** Herbal medicine has always been the norm in China and is gaining greater acceptance in Europe and America

The first recorded use of plants for medicinal purposes was found on Egyptian papyrus dated about 1500 BC. All over the world, from tribal medicine in Africa to folk remedies in Britain, many thousands of plants have been used to cure all sorts of ailments. Often, this knowledge has been handed down by word of mouth, and there is now increasing concern that much of it will be lost for ever as cultures change, tribes disperse and modern medicines of the industrialized world dominate.

The bark of the cinchona tree was used for centuries by the people of Peru to treat malaria, and we now know that it contains the anti-malarial drug quinine. The Chinese use herbal medicines extensively: one of their herbal cures for asthma has been shown to contain ephedrine, which is used in modern medicine to enlarge the air passages of the lungs. Even aspirin has its origins in willow bark.

Interest in tribal and folk remedies has been reawakened since the 1960s. Chemists investigate thousands of plants each year to see if they contain biologically active chemicals that might be developed into medicinal drugs. There is real hope that newly documented remedies will lead to drugs that treat cancer, heart disease and acquired immunodeficiency syndrome (AIDS).

**Fig 11.1** The venom of some snakes immobilizes their prey by dramatically reducing blood pressure. This discovery has already led to greater understanding of how to treat high blood pressure, a potentially fatal condition

## INTRODUCTION

Many organic chemicals come from crude oil, many come from plants and animals and others are made synthetically. Chemists use separation techniques to isolate these substances, followed by analytical techniques to find out their chemical structure. The methods chemists use to identify the formulas of these organic compounds and determine their structures are the subject of this chapter.

### ! SCIENCE FEATURE A cure for cancer?

There is never likely to be one cure for cancer because there are so many different types, but two British scientists, Sir Paul Nurse and Dr Tim Hunt, have made major contributions to our understanding of how cell division is controlled at a molecular level. For their pioneering work they both gained the Nobel Prize for Medicine in 2001.

Cancer occurs when cells divide out of control. To find drugs that may make major contributions to curing some cancers, it is essential to establish how such uncontrolled division happens and to discover the molecules that normally prevent it. The techniques this chapter describes have enabled key molecules

to be identified and the same techniques may eventually lead to the discovery of elusive cures.

**Fig 11.2** Sir Paul Nurse (left) and Dr Tim Hunt celebrate their Nobel Prize for Medicine in Sir Paul's laboratory in London (an award shared with Dr Leland H Hartwell in Seattle)

# 1 WORKING OUT THE FORMULA OF A COMPOUND

We are now going to see how you can work out the formulas of completely unknown compounds, such as those that might be present in a medicinal herb.

## PERCENTAGE COMPOSITIONS AND EMPIRICAL FORMULAS

One very useful piece of information about any compound is its **percentage composition**. This means the percentage by mass of each element in the compound.

If you have just discovered what you think is a new compound, finding the percentage composition would be one of the first investigations you would carry out. You can either decompose a known mass of the compound into its constituent elements, or burn it in oxygen and weigh the products formed, such as carbon dioxide and water (called **combustion analysis**). The masses of any other elements present in the compound, such as nitrogen and chlorine, are found by other means.

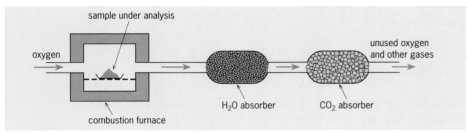

**Fig 11.3** Combustion analysis for carbon and hydrogen. A few milligrams of the compound being analysed are completely combusted. The carbon and hydrogen in the compound form $CO_2$ and $H_2O$, respectively. The increased masses of the two absorbent materials are measured to give the masses of $CO_2$ and $H_2O$ produced. Given that the mass of the original sample is known, the percentage composition can be calculated

Once the percentage composition is known, the **empirical formula** of a compound can be calculated.

> **The empirical formula is the simplest whole-number ratio of the number of atoms of each element in a compound.**

For example, benzene has a molecular formula of $C_6H_6$, but its empirical formula is CH.

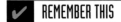
## ● EXAMPLE

**Q** Combustion analysis shows that the compound that gives cinnamon its characteristic aroma has a per centage composition of 81.8 per cent carbon, 6.1 per cent hydrogen and 12.1 per cent oxygen. Calculate its empirical formula. ($A_r$: C = 12.0, H = 1.0, O = 16.0)

|  | **Carbon** | **Hydrogen** | **Oxygen** |
|---|---|---|---|
| Mass in grams: | 81.8 | 6.1 | 12.1 |
| Amount in moles: | $\frac{81.8}{12.0} = 6.82$ | $\frac{6.1}{1.0} = 6.1$ | $\frac{12.1}{16.0} = 0.756$ |
| Simplest ratio: (Divide by the smallest number) | $\frac{6.82}{0.756} = 9.02$ | $\frac{6.1}{0.756} = 8.1$ | $\frac{0.756}{0.756} = 1.00$ |
| Simplest whole-number ratio: | 9 | 8 | 1 |

So its empirical formula is $C_9H_8O$

Another way of calculating the empirical formula from combustion analysis data is to use the actual masses of the different compounds produced, as shown in the next Example.

### EXAMPLE

**Q** 0.100 g of a sugar, known to contain only carbon, hydrogen and oxygen, is completely combusted to give 0.147 g $CO_2$ and 0.0600 g $H_2O$. Calculate its empirical formula.

**A** First, we need to calculate the masses of C, H and O in the compound.

There are 12.0 g of C in 44.0 g $CO_2$
(Remember: 1 mol $CO_2$ = 12.0 + (16.0 × 2) g)
Therefore, the mass of C in 0.147 g = $\frac{0.147}{44.0}$ × 12.0 = 0.0401 g

There are 2.0 g H in 18.0 g $H_2O$ (1 mol $H_2O$ = (1.0 × 2) + 16.0 g)

Therefore, the mass of H in 0.0600 g = 0.0600 × $\frac{2.0}{18.0}$ = 0.0067 g

Since we now know the masses of C and H in 0.100 g of sugar, the rest of the mass must result from O:

$$0.0401 + 0.0067 = 0.0468 \text{ g}$$

Therefore, mass of O in 0.100 g = 0.100 − 0.0468 = 0.053 g

We can proceed now as we did in the previous Example:

|  | **Carbon** | **Hydrogen** | **Oxygen** |
|---|---|---|---|
| Mass in grams | 0.0401 | 0.0067 | 0.053 |
| Amount in moles | $\frac{0.0401}{12.0}$ = 0.00334 | $\frac{0.0067}{1.0}$ = 0.0067 | $\frac{0.053}{16.0}$ = 0.0033 |
| Simplest ratio (Divide by the smallest number) | $\frac{0.00334}{0.0033}$ = 1.0 | $\frac{0.0067}{0.0033}$ = 2.0 | $\frac{0.0033}{0.0033}$ = 1.0 |
| Simplest whole-number ratio: | 1 | 2 | 1 |

So its empirical formula is $CH_2O$

### ✔ REMEMBER THIS

The **relative atomic mass** of an element is the average mass of the atom (taking into account all of its isotopes and their abundance) compared to 1/12 the mass of one atom of carbon-12. See page 37.

**Isotopes** are atoms of the same element with the same atomic number, but different mass numbers. See page 36. **Relative isotopic mass** is defined on page 37.

### ? TEST YOURSELF

**C** **a)** A hydrocarbon is found to contain 84.5 per cent carbon. What is its empirical formula?
**b)** An amino acid contains 0.601 g carbon, 0.799 g oxygen, 0.125 g hydrogen and 0.349 g nitrogen. Calculate its empirical formula.
(The answer is $C_2H_5O_2N$.)

### ? TEST YOURSELF

**D** 0.200 g of a compound found in vinegar on complete combustion gave 0.293 g $CO_2$ and 0.120 g $H_2O$. The compound contained C, H and O only. Work out its empirical formula.

# FINDING THE MOLECULAR FORMULA

The empirical formula tells you the simplest ratio of different atoms in a compound. It does not tell you the actual number of atoms of each element in a molecule. To find the molecular formula, you need to know the **relative molecular mass, $M_r$**, of a compound, which is:

$$M_r = \frac{\text{mass of 1 molecule of the compound}}{\frac{1}{12} \text{ mass of one atom of carbon-12}}$$

The molecular formula may be the same as the empirical formula, or it may be a multiple of it. In Test Yourself D, the empirical formula of the substance found in the vinegar is $CH_2O$. So we have:

Relative mass of $CH_2O$ = 12 + (2 × 1) + 16 = 30

However, $M_r$ ($CH_2O$) = 60, so because 30 × 2 = 60, the empirical formula must be multiplied by 2 to find the molecular formula:

$$(CH_2O) \times 2 = C_2H_4O_2$$

In the second Example, the empirical formula of the sugar is also $CH_2O$, but in this case, $M_r$ = 180. This is 30 × 6, so the molecular formula is:

$$(CH_2O) \times 6 = C_6H_{12}O_6$$

So, when we analyse a compound, we need to know its relative molecular mass. This is where the mass spectrometer comes in.

# 2 MASS SPECTROMETRY

Mass spectrometry is the most accurate method of determining relative atomic and molecular masses, but it has many other applications. The mass spectrometer is used by geologists to date rocks, by anaesthetists to analyse compounds in a patient's breath, by pharmaceutical companies to determine the structure of novel compounds and by the oil industry to work out where samples of crude oil originated. A mass spectrometer has even been taken to Mars to analyse rocks and dust on the planet's surface and gases in its atmosphere. Clearly, the potential of mass spectrometry is enormous.

The mass spectrometer was developed in 1919 by the eminent English physicist Francis Aston, from apparatus used by J.J. Thomson (the discoverer of the electron, see page 31). Aston used his mass spectrometer to show that neon gas was composed of **isotopes**.

Neon atoms were ionized, separated, and then made to hit a photographic plate. Two lines were produced, one much darker than the other, corresponding to relative isotopic masses of 20 and 22. By measuring the relative darkness of the two lines, Aston found that neon-20 atoms were ten times more abundant than neon-22.

From this information, he worked out the average atomic mass of neon to be 20.2. As mass spectrometers became more accurate, a third isotope of neon, neon-21, was discovered. It did not show up in Aston's instrument because only 0.26 per cent of naturally occurring neon is $^{21}$Ne (only 26 atoms in 10 000).

---

**? TEST YOURSELF**

**E** What are the molecular formulas of the following compounds?

**a)** An arene hydrocarbon with empirical formula CH and $M_r$ = 78.

**b)** An acid in ant sting with empirical formula $CH_2O_2$ and $M_r$ = 46.

**c)** Caffeine with empirical formula $C_4H_5N_2O$ and $M_r$ = 194.

---

IN 1922, ASTON RECEIVED THE NOBEL PRIZE FOR HIS WORK IN DEVELOPING THE MASS SPECTROMETER.

## HOW A MASS SPECTROMETER WORKS

All mass spectrometers have five basic operations:
- A vaporized sample is injected into the instrument.
- The sample is ionized to form positive ions.
- The positive ions are accelerated by an electric field.
- The accelerated positive ions are deflected by a magnetic field.
- Positive ions of a particular mass/charge ratio are detected.

### ? TEST YOURSELF

**F** Why is a very low pressure maintained inside the mass spectrometer?

**Fig 11.4** A modern mass spectrometer with digital readout and screen monitor. The sample is injected on the left of the apparatus

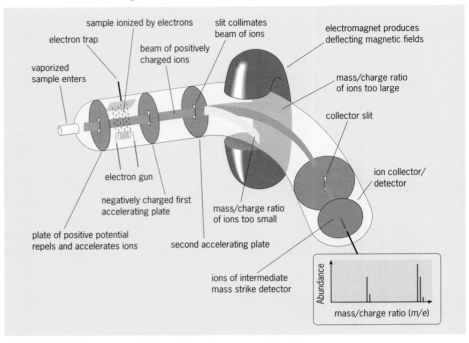

**Fig 11.5** A simplified diagram of the workings of a mass spectrometer

When the sample is not a gas or a volatile liquid, it has to be heated to produce a vapour before it can be introduced into the mass spectrometer. The exceptional sensitivity of the instrument, means that only nanograms ($10^{-9}$ g) of a sample are required. Once inside, the vaporized sample is ionized by bombardment with high-energy electrons. The bombarding electrons knock electrons from the atoms or molecules in the sample, creating positively charged ions.

$$\text{M (g)} + e^- \longrightarrow \text{M}^+ \text{(g)} + e^- + e^-$$
sample     high-energy     ion
       electron

The positive ions formed are accelerated by an electric field of several thousand volts between negatively charged plates. They then enter a strong magnetic field, which deflects them into a series of separate circular paths according to their **mass/charge ratio**.

Positive ions with higher mass/charge ratios are deflected less than those with lower ratios. So, by varying precisely the strength of the magnetic field, ions of a particular mass/charge ratio can be deflected to the collector and recorded. In this way, each separated positive ion beam can be focused on the collector in sequential order of mass/charge ratio, to build up a spectrum.

On the mass spectra in this book, the $x$-axis is labelled: Mass/charge ratio ($m/e$). Although 2+ ions do occur (when two electrons are removed), 1+ ions are far more abundant. This means that the charge $e = +1$, which makes the mass/charge ratios of the peaks equal to the relative masses of the ions. This is why you will sometimes see the $x$-axis labelled simply: Mass.

### ✓ REMEMBER THIS

In the **double-focusing mass spectrometer**, there is a second electric field whose strength can be varied. This focuses the ions according to their kinetic energy before they are deflected by the magnetic field. The resolution of the instrument is therefore improved, allowing a very accurate determination of molecular masses. See page 240.

## ! SCIENCE FEATURE Carbon-14 dating and the Turin Shroud

The Earth is continually bombarded by particles from the cosmos. These 'cosmic rays' crash into atoms in the upper atmosphere and cause them to release high-energy neutrons. They, in turn, smash into the nuclei of nitrogen-14 atoms in the upper atmosphere, and produce carbon-14, which is radioactive.

$$^{14}_{7}N + ^{1}_{0}n \rightarrow ^{14}_{6}C + ^{1}_{1}H$$

Carbon-14 reacts with oxygen to form $^{14}CO_2$, which is used by plants in photosynthesis to make carbohydrates (see Fig 11.6). Through food chains, the isotope becomes incorporated into all living things, including ourselves.

It is important to realize that the amount of carbon-14 produced each year is very small indeed – only about 7.5 kg in total – and that the ratio of carbon-14 to carbon-12 all around us and inside us is also very small. There are twelve million million carbon-12 atoms for every one carbon-14 atom.

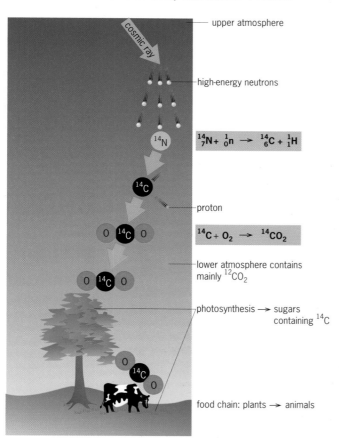

**Fig 11.6** The path taken by carbon-14 as it enters the food chain

As soon as the carbon-14 is produced it starts to decay, but the amount of carbon-14 in the atmosphere stays roughly constant because, over time, the rate of decay of carbon-14 and its rate of production by cosmic rays have become balanced. This ratio of carbon-12 to carbon-14 in any organism stays constant while it is alive, because any carbon-14 that decays is replaced in the turnover of food and nutrients from its environment.

The rate of decay of carbon-14 is measured by its **half-life** (the time taken for half of it to decay), which is 5730 years. When an organism dies it takes in no more carbon-14 and so the amount of carbon-14 starts to decrease. A mass spectrometer is used to determine the ratio of carbon-12 to carbon-14 in dead organic matter, and because we know carbon-14's rate of decay the age of the organic material can be found.

### The Shroud of Turin

This is one of the most famous uses of radiocarbon dating. The Shroud of Turin is a linen cloth, over 4 m long, bearing two images of a man – one of the front and the other of the back of someone who appears to have been crucified. For centuries many people have believed that the cloth wrapped the body of Jesus after his death. According to the first reliable records, the cloth appeared in France in 1350, and was later taken to the Italian city of Turin.

**Fig 11.7** The face on the Shroud of Turin which is thought by many people to have been the burial shroud of Jesus

In recent years, chemical tests were carried out on small samples of the cloth. They revealed that the image was not painted on to the cloth by any known method, and that the blood stains are definitely human.

In 1988, laboratories in Oxford, Zurich and Arizona were each given pieces 2 cm² to date independently. The linen was found to have been made from flax (a plant) grown somewhere between 1260 and 1390. For the cloth to have been the burial shroud of Jesus, the flax would have had to have grown some time before his death. But some people still claim that a burst of energy during the resurrection could have made the flax seem younger than it is. Many questions remain unanswered, including the way that the image was formed.

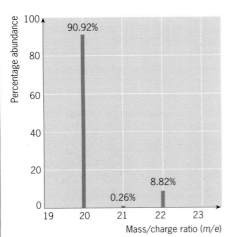

**Fig 11.8** The mass spectrum of naturally occurring neon, showing the percentage abundance of each isotope. It is easy to see why Aston's crude instrument failed to reveal the third isotope

✔ **REMEMBER THIS**

The molecular ion is sometimes called the **parent ion**.

**? TEST YOURSELF**

**G** Calculate germanium's $A_r$ from its mass spectrum, shown in Fig 11.9.

**Fig 11.9** The mass spectrum of germanium

**? TEST YOURSELF**

**H** An acidic substance found in the sting of an ant is analysed using high-resolution mass spectrometry, which gives $M_r$ = 46.0054. A database of compounds is searched and the following molecules are revealed to have values of $M_r$ in this region: $C_2H_6O$, $CH_2O_2$ and $NO_2$. Using accurate values of $A_r$, work out which molecule is present in an ant sting.

## FINDING RELATIVE ATOMIC MASSES

Relative atomic masses can be worked out very accurately using a mass spectrometer. (This was one of the first uses for the instrument.) Take neon, for example, whose mass spectrum is shown in Fig 11.8.

To calculate its $A_r$, we follow the same procedure as that on page 37:

$$A_r(neon) = \frac{(90.92 \times 20) + (0.26 \times 21) + (8.82 \times 22)}{100} = \frac{2018}{100}$$

$$= 20.18$$

Germanium has five naturally occurring isotopes (Fig 11.9) and its relative atomic mass can be calculated in the same way (see Test Yourself G).

## FINDING RELATIVE MOLECULAR MASSES AND MOLECULAR FORMULAS

The mass spectrometer is more commonly used to find relative molecular masses. When a compound such as butane, $C_4H_{10}$, is analysed, its mass spectrum shows a whole series of peaks (Fig 11.10). The peak with the highest mass is the **molecular ion, M⁺**. It is the molecule with one electron removed.

A simple calculation of $M_r$ tells you to expect the molecular ion to have its peak at 58, but Fig 11.10 also shows a peak at 59. This peak results from those molecules of butane that contain an atom of carbon-13 ($^{13}C$). Known as the **M+1 peak**, it is always smaller than the molecular-ion peak because there are far fewer naturally occurring carbon-13 atoms. However, as shown later, the M+1 peak can help you to work out the number of carbon atoms in the molecule.

The other peaks in the mass spectrum are caused by butane molecular ions that have broken into fragments. This fragmentation is important in working out the identity of molecules.

**Fig 11.10** The mass spectrum of butane

**SEE QUESTIONS 1, 2 AND 3.**

## HIGH-RESOLUTION MASS SPECTROMETRY

The mass spectrum in Fig 11.10 is from a spectrometer set for low resolution. At higher resolutions, it may be possible to deduce the molecular formula. We know that butane has an approximate $M_r$ of 58, but $C_4H_{10}$ is not the only possible molecular formula for this value of $M_r$: all the molecules shown in Table 11.1 have an approximate $M_r$ of 58. However, very accurate values for $M_r$ can be obtained using **high-resolution mass spectrometry**. So it is possible to determine which compound is present if accurate values for relative isotopic masses are used.

| Table 11.1  Relative isotopic masses: 12C = 12.0000; 1H = 1.0078; 14N = 14.0031; 16O = 15.9949 | |
| --- | --- |
| **Formulas** | **Accurate $M_r$ (*m/e* values for M⁺ ion)** |
| $C_4H_{10}$ | 58.0780 |
| $C_3H_6O$ | 58.0417 |
| $C_2H_2O_2$ | 58.0054 |
| $C_2H_6N_2$ | 58.0530 |

## USING FRAGMENTATION TO DETERMINE STRUCTURE

The mass spectrum of butane in Fig 11.10 has several peaks because of the way that the butane molecular ion fragments in the mass spectrometer. These fragments can be used to identify parts of the molecule and, in many cases, build up a picture of the structure of the whole molecule. This is illustrated in Fig 11.11, which we can use to identify some of the peaks in Fig 11.10.

**A** is from $CH_3{}^+$, **B** from $C_2H_5{}^+$ and **C** from $C_3H_7{}^+$. The other peaks in the spectrum result from the loss of hydrogen atoms from these fragments. Typical ion fragments produced in the mass spectrometer are shown in Table 11.2.

| Table 11.2  The masses of some typical positive ion fragments | | | |
| --- | --- | --- | --- |
| **Ion** | **Mass/charge ratio (*m/e*)** | **Ion** | **Mass/charge ratio (*m/e*)** |
| $CH_3{}^+$ | 15 | $CH_3CO^+$ | 43 |
| $CO^+$ | 28 | $C_3H_7{}^+$ | 43 |
| $C_2H_4{}^+$ | 28 | $C_2H_5O^+$ | 45 |
| $CHO^+$ | 29 | $C_4H_9{}^+$ | 57 |
| $C_2H_5{}^+$ | 29 | $C_6H_5{}^+$ | 77 |

Apart from deciding the identity of the peaks, the other important thing to look for is the differences in masses between the peaks. A difference of 15 is almost certainly shows the loss of a methyl fragment, and so it is highly likely that the original molecule contains a $CH_3$ group.

When a molecular ion fragments, the chemistry becomes rather complicated, and so we discuss only a small part of it here. In Fig 11.10, the molecular-ion peak (M) is 58. The loss of a methyl fragment produces the peak at 43. The butane molecular ion has fragmented to give $C_3H_7{}^+$ and a free radical $CH_3{}^\bullet$, which is neutral and so is not affected by the electric and magnetic fields of the mass spectrometer.

$$C_4H_{10}{}^+ \rightarrow C_3H_7{}^+ + CH_3{}^\bullet$$

There are other characteristic differences in the masses between peaks that give clues to which other groups have been lost as free radicals. For example, a loss of 17 suggests that an OH group may have been removed, and a loss of 29 suggests a $C_2H_5$ group.

Let's now look at the spectrum of ethanol, $CH_3CH_2OH$ (Fig 11.12).

Fig 11.11 Fragmentation of butane

## ? TEST YOURSELF

**I** **a)** Why are the *m/e* values of the ions in Table 11.2 also their masses?

**b)** Which fragment is responsible for the peak at *m/e* = 28 of the butane mass spectrum?

The most abundant ion gives the strongest signal at the detector. This is called the **base peak** and is given a relative abundance of 100 per cent. All other abundances are percentages of the base peak.

**c)** What is the mass of the base peak in butane?

**J** Free radicals are highly reactive species with unpaired electrons.

**a)** Why is $C_4H_{10}{}^{+\bullet}$ sometimes written $C_4H_{10}{}^+$?

**b)** Why is the particle $CH_3{}^\bullet$ not detected by the mass spectrometer?

**K** What groups may have been removed when there is a mass difference of

**a)** 28;  **b)** 45;  **c)** 77?

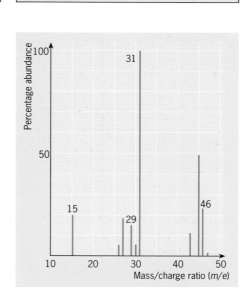

Fig 11.12 The mass spectrum of ethanol

## ? TEST YOURSELF

**L** The compound responsible for the sting of an ant has the mass spectrum shown in Fig 11.13. It has a molecular formula $CH_2O_2$.

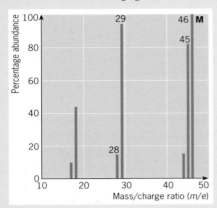

**Fig 11.13**

a) What group of atoms could be lost from the molecular ion to form a peak at $m/e$ 29?

b) Which ions are probably responsible for the peaks at $m/e = 28$, 29 and 45?

c) Suggest a displayed formula for this compound.

**SEE QUESTIONS 4 AND 5.**

The molecular-ion peak is 46, which corresponds to $CH_3CH_2OH^+$. The base peak is at 31. The difference between 46 and 31 is 15, so it is probable that a $CH_3$ group has been lost. Therefore, the ion responsible for the base peak is $CH_2OH^+$. The peak at 29 is from $CH_3CH_2^+$, and that at 15 is $CH_3^+$.

Notice that there is an M+1 peak at 47 because of the isotope carbon-13. In the next section, we see how this peak can be used to determine the number of carbon atoms present in the molecule.

### ISOTOPE PEAKS

The presence of isotopes in a compound will give characteristic peaks. We met one of these peaks when we examined butane and ethanol (see Figs 11.10 and 11.12). The M+1 peak resulted from the isotope carbon-13. Carbon-13 is present naturally as 1.1 per cent of all carbon atoms, information that allows us to work out the number of carbon atoms in a molecule. If there were one carbon atom in a molecule, the M+1 peak height would be 1.1 per cent of the molecular-ion peak, because 1.1 per cent of molecules would contain carbon-13. So, in a molecule that contains five carbon atoms, about 5.5 per cent will be carbon-13. That is, the M+1 peak height will be about 5.5 per cent of the M peak height.

### [M+2] AND [M+4] PEAKS

Chlorine and bromine atoms also give characteristic peaks because of their isotopic compositions. In the case of chlorine, 75 per cent is $^{35}Cl$ and 25 per cent is $^{37}Cl$. This means that when a molecule contains a chlorine atom, there are two peaks at M and **M+2** in the ratio of 75:25, namely 3:1 (see Fig 11.14). Although we refer to these peaks as M and M+2, they both are molecular ions. Similarly M+2 and M+4 peaks arise if there are two chlorine atoms or two bromine atoms in a molecule. (See Test Yourself M(d).)

**Fig 11.14** The mass spectrum of chloroethane

## ? TEST YOURSELF

**M** a) A component of petrol gives an M+1 peak that is 11.1 per cent of the M peak. How many carbon atoms does the molecule contain?

b) In Fig 11.14, the peaks at $m/e = 49$ and 51 are in the ratio 3:1.
(i) Which ions are responsible for these peaks? (ii) Both peaks at 64 and 66 are from molecular ions. Identify these ions. (iii) Which group has been lost from the M peak to give a peak at 49?

c) Bromine has two isotopes, $^{79}Br$ and $^{81}Br$, which are present in equal amounts, that is, 50:50. Sketch the mass spectrum you would expect for bromoethane.

d) When analysed, chlorine gas gives peaks at $m/e = 74$, 72 and 70.
(i) Explain which ions are responsible for these peaks. (ii) Why is the peak at $m/e = 74$ the smallest?

## 3 INFRARED SPECTROSCOPY

Atoms in molecules vibrate about their bonds. These vibrations have definite (quantized) energy levels, in the same way that electrons have quantized energy levels in atoms (see page 52). To increase the vibration in a molecule from one level to the next, a definite amount of energy must be absorbed. These amounts of energy correspond to radiation in the infrared part of the electromagnetic spectrum.

**Fig 11.15** Ways in which the carbon dioxide molecule vibrates

(a) Asymmetric stretch

(b) Symmetric stretch: causes no change in dipole moment

(c) Bending: occurs in two ways (in the plane of the paper and also at right angles to this), so there are two modes caused by bending

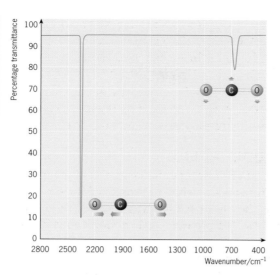

**Fig 11.16** The infrared spectrum of carbon dioxide, showing the two modes of vibration that absorb energy in this region. Note both types of bending occur at the same energy levels and produce just a single peak. When two energy levels coincide they are said to be degenerate

**Fig 11.17** The infrared spectrum of sulphur dioxide, showing the vibrations responsible for the peaks

For infrared radiation to be absorbed, the vibrations must cause a change in the dipole moment of the molecule. Therefore, the symmetrical bonds in $N_2$ and $O_2$ do not absorb infrared radiation. This is why $N_2$ and $O_2$ are not greenhouse gases – they cannot absorb the infrared radiation emitted by the Earth. However, we know from Chapter 10 that carbon dioxide is an important greenhouse gas because it can absorb infrared radiation, because of the ways in which it can vibrate.

Fig 11.15 shows four vibrational modes of carbon dioxide. There is no change in the dipole moment in Fig 11.15(b), and so there is no absorption of infrared radiation in this mode. However, the infrared spectrum of carbon dioxide (Fig 11.16) shows that the modes shown in (a) and (c) absorb in this region.

Notice that in Fig 11.16 the $x$-axis represents **wavenumber**. This is the reciprocal of wavelength $(1/\lambda)$ and is usually measured in $cm^{-1}$. Using it makes the numbers more manageable, since the wavelength of the infrared region lies between $2.5 \times 10^{-5}$ m and $2.5 \times 10^{-6}$ m.

In the case of sulphur dioxide, all three vibrational modes lead to infrared absorption, as shown in Fig 11.17. This is because the V-shaped $SO_2$ molecule has a different shape from the linear $CO_2$ molecule, so all three modes lead to a change in the dipole moment. Notice that the absorptions do not occur at the same wavenumbers. This is because the energy required to excite vibration depends on the strength of the bond. As the C=O bond is stronger than the S=O bond, the corresponding vibrations occur at a higher wavenumber in $CO_2$.

As molecules become more complex, so do their vibrational modes. You may meet terms such as scissoring or rocking. Don't worry about these. When you have to interpret an infrared spectrum, look for one or two peaks that are characteristic of particular bonds.

## ? TEST YOURSELF

**N** **a)** Why is there no absorption peak in the infrared spectrum of hydrogen molecules?
**b)** Why is there only one absorption peak in the infrared spectrum of HCl?
**c)** The absorption peaks of HCl and HI are 2886 $cm^{-1}$ and 2230 $cm^{-1}$, respectively. What does this tell you about the bond energies of H–Cl and H–I?

### ✔ REMEMBER THIS

A **dipole** occurs in a molecule when one part of the molecule has a slightly positive charge and another part has a slightly negative charge. Dipole moment measures the size of the charges and the distance that separates them. HCl is a polar molecule whose vibrations cause fluctuations in the dipole moment as the distance that separates the charges varies, so it absorbs infrared radiation. See pages 82 and 83.

### ✔ REMEMBER THIS

The infrared spectrum in Fig 11.16 appears to be upside down. But notice that the $y$-axis represents transmittance, so the sample is at maximum transmittance (allowing the passage of all infrared radiation) when it is not absorbing radiation.

### ✔ REMEMBER THIS

Equation for the velocity of light:
$$c = \lambda f$$
where $c$ = velocity of light, $\lambda$ = wavelength, $f$ = frequency. So, $1/\lambda$ is directly proportional to $f$. Since energy $E = hf$, the higher the wavenumber (and the higher the frequency), the higher the energy ($h$ = Planck's constant). (See Chapter 3 to remind yourself of this.)

mirrors divide infrared radiation
into two beams

sample
cell

reference
cell

infrared source
(electrically heated filament)

beam chopper
alternates from
sample beam to
reference beam

NaCl prism or
diffraction grating
splits beam into
different wavelengths

detector

recorder

**Fig 11.18** A simplified diagram of a conventional infrared spectrometer

The spectrometer shown in Fig 11.18 is a double-beamed instrument, in which one infrared beam passes through the sample and the other passes through the reference cell. The reference cell may contain air, or the solvent of a solid being analysed as a solution. The purpose of the reference is to eliminate absorptions caused by $CO_2$ and water vapour in the air or by the solvent. The difference in the intensities of the two beams is measured by the detector at each wavenumber and fed to the recorder, which produces a spectrum. When the sample does not absorb, there is no peak at that wavenumber; but when it does absorb radiation, the difference in intensities of the two beams produces a peak at that wavenumber.

Notice that when a prism is fitted instead of a diffracting grating, it is made from sodium chloride rather than glass. This is because sodium chloride is transparent to most infrared radiation frequencies, whereas glass absorbs it strongly.

A more sophisticated instrument is the Fourier transform infrared spectrometer, in which a single beam that contains all the required infrared frequencies is passed through the sample. The emerging beam is fed into a computer, which produces an infrared spectrum by a mathematical technique known as Fourier transformation. This instrument has enabled us to understand how photochemical smog develops by analysing samples taken from a 1.6 km long path of air over Los Angeles and determining concentrations of pollutants down to parts per billion (per $10^9$).

### PREPARING SAMPLES FOR ANALYSIS

Gases are passed into the sample cell (about 10 cm long), which has sodium chloride windows at each end. (Why can't glass windows be used?)
Liquids are smeared into a thin film between two potassium bromide or sodium chloride discs. (Sodium chloride is not transparent to all infrared, so potassium bromide can be used.)

Solids are made either into a solution, or into a disc or a **mull**. To make a disc, the solid is finely ground with potassium bromide and pressed into a mould under very high pressure. The advantage of a disc is that it produces the spectrum of the pure compound. A mull is made by grinding up the sample with a drop of nujol, a long-chain hydrocarbon. The mull is then smeared between discs, as for a liquid sample. The disadvantage of using a mull is that nujol absorbs infrared radiation, which shows up on the spectrum.

**Fig 11.19** Preparing a mull

**Fig 11.20** A potassium bromide disc

# INTERPRETING INFRARED SPECTRA

The spectrum of each and every compound acts like a unique fingerprint and thus makes it possible to identify any compound, often through a computer database. When a totally new compound is discovered, perhaps isolated from a medicinal herb, it will not match any spectrum in the database. But we can still infer a great deal about the compound from the peaks in its spectrum. For example, the functional group C=O absorbs in the range 1680–1750 cm$^{-1}$, so a peak in this region is a strong indication that C=O is present.

Table 11.3 contains some examples of bonds and their characteristic infrared absorption bands. Notice that some bonds give strong absorptions

**REMEMBER THIS**

The fingerprint of a complex molecule arises through the vibrations of the whole molecule; this can be used to check batches of drugs for purity. The identifying peaks of the fingerprint are usually below 1600 cm$^{-1}$.

**REMEMBER THIS**

Arenes are aromatic compounds and have a benzene ring. For more information on arenes, see Chapter 15.

**FOR MORE INFORMATION ON HYDROGEN BONDING, SEE PAGES 133 AND 289.**

**Table 11.3 Bonds and their characteristic infrared absorption bands. Intensity classification: s = strong, m = medium**

| Bond | Location | Wavenumber range/cm$^{-1}$ | Intensity |
|------|----------|----------------------------|-----------|
| O—H | Alcohols and phenols, free (not hydrogen bonded) | 3580–3670 | s |
| N—H | Primary amines | 3350–3500 | m |
| O—H | Alcohols and phenols, hydrogen bonded | 3230–3550 | s (broad) |
| C—H | Alkanes, alkenes, arenes | 2840–3030 | m–s |
| O—H | Carboxylic acids, hydrogen bonded | 2500–3300 | m (broad) |
| C≡N | Nitriles | 2200–2280 | m |
| C=O | Aldehydes, ketones, carboxylic acids, esters | 1680–1750 | s |
| C=C | Alkenes | 1610–1680 | m |
| C—O | Alcohols, ethers, esters | 1000–1300 | s |
| C—Cl | | 700–800 | s |

while others are weaker. The explanation lies with bond polarity. When they vibrate, very polar bonds, such as C=O and O—H, are subject to a greater change in their dipole moments than are non-polar bonds, such as C—H. So polar bonds absorb more energy, and thereby give a stronger peak.

Let's now look at the infrared spectra in Figs 11.21, 11.22 and 11.23. Notice that we have not attempted to identify all the peaks in these three spectra. The more atoms and bonds there are in a molecule, the more complex its spectra becomes.

**Fig 11.21** The infrared spectrum of propanone

**Fig 11.22** The infrared spectrum of ethanoic acid

## ? TEST YOURSELF

**O** Fig 11.23 shows the infrared spectrum of butan-1-ol, $CH_3CH_2CH_2CH_2OH$. Using Table 11.3, identify the peaks marked A and B.

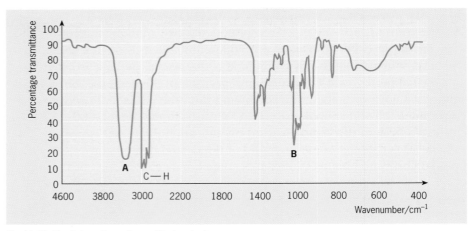

Fig 11.23 The infrared spectrum of butan-1-ol

Infrared spectroscopy is rarely used on its own to identify unknown compounds. However, coupled with data from mass spectrometry, gas–liquid chromatography or combustion analysis, it gives important information about the presence of functional groups.

## ! SCIENCE FEATURE Using infrared spectroscopy

### At the police station

All alcoholic drinks slow down the body's response times. So drinking and driving can be a fatal combination and anyone driving over the legal limit of 35 micrograms per 100 cm³ of breath is likely to be prosecuted.

The roadside breathalyser gives the first indication that a driver may be over the limit. When a breathalyser test proves positive, the next step is to take the driver to the police station for a more accurate analysis, which is now often carried out using infrared absorption.

Fig 11.24 The Intoximeter 3000 shown here has been widely adopted by the British police to analyse breath. It checks the C—H absorption at about 3000 cm⁻¹

### At the roadside and garage

Infrared spectroscopy is very quick and devices that measure pollution from individual cars can be placed at the roadside or used in the annual MOT test to monitor exhaust emissions. In these cases, the C=O bond in carbon monoxide, the bonds in nitrogen oxides and the C—H bonds in unburnt hydrocarbons all give characteristic absorptions.

Fig 11.25 Cars fitted with catalytic converters should emit no more than 0.3% CO and 200 p.p.m. (parts per million) unburnt hydrocarbons

### In the atmosphere

Crucial evidence for the rise of carbon dioxide levels in the atmosphere has been provided by the Mauna Loa observatory in Hawaii (Fig 11.26). The graph of the results from 1958 to the present day is one of the most famous in history and is cited by scientists from around the world when they urge for a reduction in the present output of carbon dioxide emissions. (See page 221.)

Fig 11.26 The rise in atmospheric $CO_2$ levels is monitored here at the Mauna Loa Observatory in Hawaii. (See page 221.)

**P** Read the Feature box before answering these questions.

Look at the infrared spectrum for ethanol in Fig 11.27.

**a)** Identify the bond responsible for the broad peak at about 3400 cm$^{-1}$.

**b)** Breath contains water vapour. Why can't the peak at 3400 cm$^{-1}$ be used to check for alcohol concentration?

**c)** Diabetics may have propanone on their breath. Look at Fig 11.21 and suggest why this could be a problem for the Intoximeter 3000 (see Fig 11.24).

**Fig 11.27**

**d)** The maximum output of unburnt hydrocarbons from car exhausts is given in parts per million (p.p.m.). What is the maximum *percentage* of $C_xH_y$ permissible?

# 4 NUCLEAR MAGNETIC RESONANCE SPECTROSCOPY

SEE QUESTION 6.

In **nuclear magnetic resonance (NMR) spectroscopy**, organic chemists have a very powerful technique for determining the detailed structures of compounds. Like infrared spectroscopy, it uses absorption – but in NMR it is the absorption of radio waves by certain nuclei when they are in a strong magnetic field. When a new organic compound is discovered, its NMR spectrum is often the first to be looked at because it gives detailed information about the hydrogen nuclei (protons) in the molecule. It can also be used to investigate other nuclei, such as $^{13}C$.

A hydrogen nucleus acts like a tiny bar magnet because it spins, generating a minute magnetic field. When a strong external magnetic field is applied to a compound that contains hydrogen atoms, many of their spinning nuclei (protons) line up so that their magnetic fields are in the same direction as that of the external field (like a group of compass needles, each one pointing north). However, some of the nuclei directly oppose the external magnetic field. (Imagine a compass needle pointing south!) This requires more energy, so these nuclei are at a higher energy level than the rest.

### ✔ REMEMBER THIS

Nuclear magnetic resonance spectroscopy is also called magnetic resonance imaging (MRI), especially when its use in medicine is described.

### ✔ REMEMBER THIS

Remember that hydrogen $^1_1H$ has a nucleus that consists of only one proton.

### ? TEST YOURSELF

**Q** As well as $^{13}C$ and $^1H$, the following nuclei also exhibit the property known as spin and generate a magnetic field: $^{19}F$, $^{31}P$, $^{15}N$. Look at the mass numbers and identify a pattern.

### ? TEST YOURSELF

**R** An NMR spectrum is often referred to as a PMR spectrum when the nucleus involved is hydrogen. What does PMR mean?

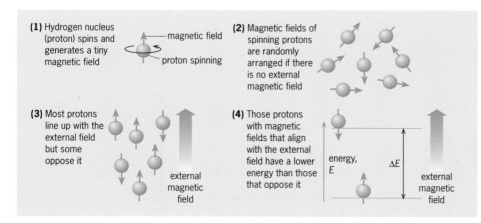

**Fig 11.28** How hydrogen nuclei behave when an external magnetic field is applied to their molecular environment

**Fig 11.29** The hydrogen nuclei (protons) in ethanol are in three different chemical environments. Each gives different proton NMR energy levels

**Fig 11.30** The low-resolution NMR spectrum of ethanol

**Fig 11.31** The low-resolution NMR spectrum of ethanol, showing the integration trace and the ratio of protons

If the compound is now irradiated with a pulse of radio waves of a particular frequency, the lower-energy nuclei absorb these waves and thereby flip to the higher level. This causes a peak in the NMR spectrum.

Both the energy needed to promote a hydrogen nucleus (proton) to a higher energy level and the strength of the applied magnetic field experienced by the nucleus depend on the chemical environment of the nucleus. So, hydrogen atoms in different environments in a molecule absorb radio waves in slightly different parts of the spectrum. Take, for example, the ethanol molecule ($CH_3CH_2OH$) shown in Fig 11.29.

The hydrogen nuclei are in three different chemical environments. This means that the energy gap between the higher and lower energy levels is different for each of these environments. So, when a pulse of radio waves is applied, more of the lower energy nuclei jump to the higher energy level as they absorb radio waves of a particular frequency. In the case of ethanol, the three peaks in the spectrum correspond to the three different frequencies absorbed by the detector.

A compound called tetramethylsilane, $Si(CH_3)_4$, or TMS for short, has all 12 protons (hydrogen nuclei) in the same chemical environment. It therefore gives a very sharp signal. This is used as the standard against which all signals caused by other proton chemical environments are measured, because its protons absorb at a frequency well away from the frequencies of most interest to chemists. TMS is given a value of zero, and the difference between the protons in TMS and the protons in other chemical environments is known as the **chemical shift**, symbol δ.

The NMR spectrum of ethanol is shown in Fig 11.30. The area under each peak is proportional to the number of protons in the particular environment. Many NMR spectrometers give these areas as an integration trace superimposed on the NMR spectrum. By measuring the height of each step on the integration trace, the ratio of the protons in each environment can be worked out. This is shown in Fig 11.31, with the ratio of protons.

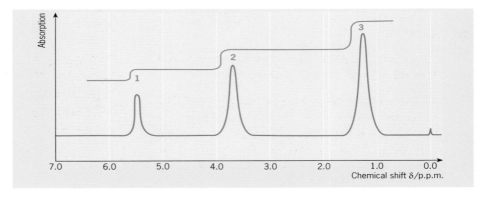

**Table 11.4 Some typical chemical shift values for different types of proton**

| Type of proton | Chemical shift (p.p.m.) | Type of proton | Chemical shift (p.p.m.) |
|---|---|---|---|
| R—CH₃ | 0.9 | R—CH₂—Hal | 3.2–3.7 |
| R—CH₂—R | 1.3 | | |
| R₃CH | 2.0 | R—O—CH₃ | 3.8 |
| | | R—O—H | 4.5* |
| | | RHC=CH₂ | 4.9 |
| $CH_3-C{<}^R_{OR}$ | 2.0 | R**H**C=CH₂ | 5.9 |
| $R-C{<}^{CH_3}_{O}$ | 2.1 | ⬡—OH | 7* |
| | | ⬡—H | 7.3 |
| ⬡—CH₃ | 2.3 | $R-C{<}^O_H$ | 9.7* |
| ⬡—CH₂—R | 2.3—2.7 | | |
| R—C≡C—H | 2.6 | $R-C{<}^O_{O-H}$ | 11.5* |

*The value varies according to the solvent, the nature of R and concentration.

Using Table 11.4, let's work through another NMR spectrum and determine the structure of the compound.

---

## ● EXAMPLE

**Q** A hydrocarbon has a molecular formula $C_8H_{10}$. Work out its displayed formula from the low-resolution NMR spectrum in Fig 11.32.

**Fig 11.32**

**A** As there are three peaks, there must be three types of proton (three different chemical environments).

The peak at 7.2 is probably due to a benzene ring. The integration trace gives a ratio of 5:2:3. This adds up to ten, which in this case is the actual number of hydrogen atoms in the molecule. So there are five hydrogen atoms on the benzene ring. In other words, it is **monosubstituted**.

(Monosubstituted means that only one hydrogen in the benzene ring has been substituted by another atom or group.)

Since there are two more peaks, the peak at 2.4 corresponds to $C_6H_5$—$CH_2$—R, not $C_6H_5$—$CH_3$. From the integration trace, we can see that it does indeed have two hydrogen atoms.

The integration trace shows that three hydrogen atoms are responsible for the peak at 1.0. Even though the value of the peak does not correspond precisely to the value in Table 11.4, it is certainly $CH_3$.

These data give the structural formula in Fig 11.33. The compound is ethylbenzene.

**Fig 11.33**

## ? **TEST YOURSELF**

**T** Using Table 11.4, draw the NMR spectrum you might expect for ethanal (Fig 11.34).

**Fig 11.34**

▶ **EXTENSION Using the NMR spectrometer**

Fig 11.35 A schematic diagram of an NMR spectrometer

The sample is dissolved in a solvent that does not contain hydrogen atoms, such as $CCl_4$, or a solvent in which hydrogen has been replaced by its isotope deuterium. The solution is then placed in a strong magnetic field and a pulse of radio waves passed through it, exciting hydrogen nuclei into their higher spin energy levels. The absorption of radio waves at particular frequencies is detected and recorded.

Either the radio frequency or the magnetic field can be varied to give characteristic peaks.

## HIGH-RESOLUTION NMR SPECTROSCOPY

The NMR spectra of ethanol and ethylbenzene (Figs 11.30 and 11.32, respectively) are at low resolution. High-resolution instruments give more peaks, as shown in the high-resolution NMR spectrum of ethylbenzene in Fig 11.36. These extra peaks yield more information about the structure of molecules.

When different types of proton are next to each other in a molecule, they exert an effect on their respective magnetic fields, known as **spin–spin splitting** or **spin–spin coupling**. Look closely at the displayed formula of ethylbenzene in Fig 11.33. The $CH_3$ protons are next to the $CH_2$ protons, so they have a direct effect on each other. However, the protons in the benzene ring are not next to any other type of proton.

The $CH_3$ protons experience three different magnetic fields because of the way the $CH_2$ protons spin, while the $CH_2$ protons experience four different magnetic fields because of the way the $CH_3$ protons spin, as shown in Fig 11.37.

So, with high-resolution NMR, we can work out the number of nearest-neighbour hydrogen atoms there are in different groups in a molecule. This is made easy to remember by the *n* **+ 1 rule:**

> **If a proton has *n* protons as nearest neighbours then its absorption peak will be split into *n* + 1.**

This helps us to gain greater insight into the structure of unknown molecules.

Fig 11.36 The high-resolution NMR spectrum of ethylbenzene

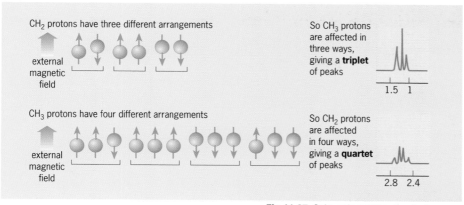

CH₂ protons have three different arrangements

external magnetic field

So CH₃ protons are affected in three ways, giving a **triplet** of peaks

1.5  1

CH₃ protons have four different arrangements

external magnetic field

So CH₂ protons are affected in four ways, giving a **quartet** of peaks

2.8  2.4

**Fig 11.37** Spin–spin splitting in ethylbenzene

## USING D₂O TO IDENTIFY LABILE PROTONS

Certain functional groups, such as OH and $NH_2$, rapidly exchange protons with neighbouring molecules. These protons are said to be **labile**. They usually appear in high-resolution NMR spectra as single peaks, because they do not interact with neighbouring protons.

To identify functional groups with labile protons, heavy water (deuterium oxide, $D_2O$) is used. In the case of ethanol, there is an exchange that produces $CH_3CH_2OD$, and the peak corresponding to the proton in the OH group disappears.

**Fig 11.38** The high-resolution NMR spectrum of ethanol

---

## ! SCIENCE FEATURE Imaging the human body with NMR

Nuclear magnetic resonance is becoming increasingly important in medicine, where it is more usually known as magnetic resonance imaging. The most common nucleus in the body is that of hydrogen, and whole-body scanners investigate tissues by producing images based on the fact that protons in different tissues have different **relaxation times**. When protons are excited and flipped into higher energy levels, they return to the lower energy stage, emitting energy. The time they take to do this is called the relaxation time. This is detected and the data transformed by computer into an image. With NMR, doctors can see an image of the soft tissues in the body, which X-rays cannot distinguish. An added advantage is that, unlike X-rays, no side-effects are known, and therefore scans can be safely taken of a patient at regular intervals.

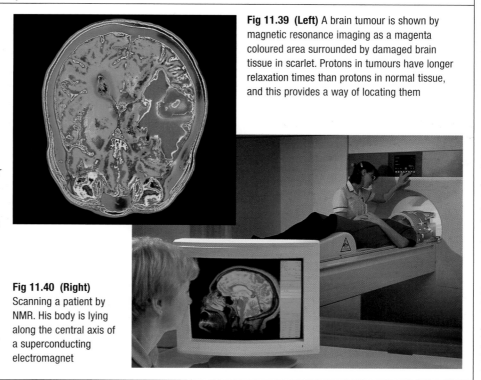

**Fig 11.39 (Left)** A brain tumour is shown by magnetic resonance imaging as a magenta coloured area surrounded by damaged brain tissue in scarlet. Protons in tumours have longer relaxation times than protons in normal tissue, and this provides a way of locating them

**Fig 11.40 (Right)** Scanning a patient by NMR. His body is lying along the central axis of a superconducting electromagnet

# 5 CHROMATOGRAPHY

**Chromatography** is a method of separating and identifying the components of a mixture. (The name is derived from two Greek words, *chroma* for colour and *graphe* for writing.) The technique was discovered and developed by a Russian botanist, Mikhail Tswett, who first reported it in 1903. He was using powdered calcium carbonate packed into a tube to separate coloured plant pigments called chlorophylls. Before this, chemists mainly used crystallization techniques to separate and purify substances. Today, chromatography in one form or another is used in nearly all chemical laboratories, not just to separate and purify, but also to analyse substances, most of which are colourless.

All forms of chromatography operate on the same principle. The mixture is introduced into two different phases, of which one remains stationary while the other flows over it. The phases are known as the **stationary phase** and the **mobile phase**, respectively. The components of the mixture distribute themselves differently between the two phases, according to their affinity for each phase.

There are two main types of chromatography:

**partition chromatography** and **adsorption chromatography**.

## PARTITION CHROMATOGRAPHY

In partition chromatography, the stationary phase is a non-volatile liquid film held on an inert solid surface. The mobile phase is a liquid or gas. The components to be separated distribute themselves between the two phases according to how soluble they are in each.

## ADSORPTION CHROMATOGRAPHY

In adsorption chromatography, the stationary phase is a solid and the mobile phase is a liquid or gas. The components to be separated are **adsorbed** on (bonded to) the solid surface of the stationary phase. Those that are only weakly adsorbed travel faster in the mobile phase than those that are strongly adsorbed.

## PAPER CHROMATOGRAPHY

Chromatography using paper is the one you are probably most familiar with. It is an example of partition chromatography (Fig 11.41). The stationary phase is water adsorbed on the cellulose fibres of the paper. The mobile phase is another liquid solvent.

**Fig 11.41** The principle of paper chromatography

The sample is spotted along a **base-line** drawn in pencil on the paper (Fig 11.41). When the spots are dry, the paper is placed in the solvent, which rises up the paper, taking the components of the sample with it. As they travel up the paper, the components separate according to their different solubilities in the mobile and stationary phases. When the **solvent front** (the leading edge of the solvent) is almost at the top of the paper, the **chromatogram** is taken out and dried (Fig 11.41). If the components are colourless, the chromatogram needs to be **developed** to make them visible. Sometimes, just heating the chromatogram does this. If not, some other method, such as exposure to ultraviolet light, is used.

Components can be identified by the distances they have travelled up the paper compared with the distance travelled by the solvent front. This is known as the **retention factor** or **retardation factor** $R_f$:

$$R_f = \frac{\text{distance moved by component from base-line}}{\text{distance moved by solvent front from base-line}}$$

To obtain the separated components as samples, their spots are cut out and they are dissolved off in solvents.

**SEE QUESTION 11.**

**? TEST YOURSELF**

**W** Look at Fig 11.41.
**a)** Why is the base-line drawn in pencil?
**b)** Why is the air inside the beaker saturated with solvent vapour before the chromatography paper is put in?
**c)** Calculate the $R_f$ values for the two components in sample **B**. (Hint: Measure to the centre of the spot.)
**d)** Explain why one of the components in sample **A** has not risen off the base-line.

## ▶ EXTENSION Column chromatography

This was the first way in which chromatography was carried out, and it is still the simplest. When Tswett was separating plant pigments, he used a column of calcium carbonate powder as the stationary phase, and a hydrocarbon liquid mixture as the mobile phase. He placed the sample on top of the column, where it made contact with the hydrocarbon mixture, which dissolved the pigments in the sample.

Tswett's arrangement is still used today. The column is kept topped up with fresh solvent, which washes the pigment down through the inert solid stationary phase (usually aluminium oxide powder or silica gel). The components that are adsorbed most strongly by the stationary phase take the longest time to flow through the column.

When the components are coloured, they can be identified by eye. But, if colourless, other techniques are used: for example, some components fluoresce in ultraviolet light. Once separation is finished, the solvent is removed by evaporating it off.

Column chromatography in laboratories is usually used to separate minute quantities of mixtures, while in industry it is used in large-scale separation processes that require columns several metres high.

**Fig 11.42** Modern column chromatography in operation

### High-performance liquid chromatography (HPLC)

The performance of column chromatography can be improved by using very fine powder as the stationary phase. In this case, gravity alone is not enough to force the mobile liquid phase through the column, and therefore high pressure must be applied. This is a very efficient process that requires short columns of between 10 and 30 cm in length. Many of the components separated by HPLC absorb ultraviolet light, so it is generally used to identify them.

HPLC has many applications, particularly for the separation and identification of non-volatile substances. Foodstuffs, for example can be checked for additives and contaminants. The detection of steroids in athletes' body fluids is another example.

Some chemist prefer to call this technique high-pressure liquid chromatography, but the abbreviation is the same: HPLC.

**Fig 11.43** An incomplete separation is achieved with solvent A so the chromatogram is turned on its side and placed in a different solvent, B, and the component that appears green is now separated into blue and yellow components

**Fig 11.45** A gas–liquid chromatogram of additives in a soft drink

## TWO-WAY CHROMATOGRAPHY

Sometimes the components of a mixture are not separated completely because they have similar solubilities in a particular solvent. To overcome this problem the paper is rotated 90° and a different solvent is used to ensure a separation. This process is called **two-way chromatography**.

## GAS–LIQUID CHROMATOGRAPHY (GLC)

An unreactive gas, such as nitrogen or helium, is used as the **carrier gas** (the mobile phase). The stationary phase consists of a liquid coating adsorbed on the particles of a finely powdered inert solid. The powder fills a coiled narrow-bore tube called a column. The temperature of the column is controlled by an oven. The arrangement is shown in Fig 11.44. Typically, the tube that forms the column can be 5 to 10 m long, with a bore of 2 to 10 mm.

**Fig 11.44** The main features of a gas–liquid chromatography apparatus

The vaporized sample is injected into the carrier gas as it streams into the column at an even rate. As the components of the sample are carried through the column, those that are more soluble in the stationary phase move at a slower rate. Volatile components are carried more quickly than non-volatile ones, which therefore have a greater opportunity to dissolve in the liquid stationary phase. So the volatile components emerge first from the column.

The time each component remains in the column is known as its **retention time**. This depends on such factors as the flow rate of the carrier gas, the temperature of the oven, and the length and diameter of the column. Each retention time is characteristic of a particular component, which allows it to be separated and identified. The area under each peak displayed on the recorder is the measure of the amount of that component present. The relative amounts of components in a mixture can therefore be calculated, an operation now done by computer.

GLC is very sensitive and is able to detect and measure minute amounts of substances. It has therefore become a standard method for detecting banned

<table>
</table>

> **!** **SCIENCE FEATURE** Revealing the secrets of old masters

When you next go round an art gallery and look at the work of a great artist who perhaps lived four or five centuries ago, give some thought to its chemistry. Paint pigments provide the colour, but they don't stick to the canvas or wooden panel on their own. They need to be bound together and to the supporting surface. For some 5000 years before the Renaissance, artists mainly used egg yolk (called tempera) as the binding medium. Then, in the 1400s, Italian and Flemish artists started to switch to oils that dried, such as linseed, poppy or walnut oil, and this is where GLC comes in.

**Fig 11.46** The Virgin and Child Embracing by Sassoferratto (1609–1685), during restoration of the painting

When a gallery wants to restore a painting, it is essential to know what binding medium the artist used. In the oils used, long chain fatty acids are present as esters (see pages 396 and 407) and these have characteristic retention times in a gas chromatogram. By performing GLC, the different oils used can be identified and then the same oils are mixed with pigments for restoring the picture.

**Fig 11.47** Chromatograms of known oils can be compared with the chromatogram of a tiny sample of paint removed from the picture to be restored to determine which drying oil the artist used. Shown here is the chromatograph of a paint sample from a picture painted in about the year 1500

> **?** **TEST YOURSELF**
>
> **Z** A database of retention times is compiled for each GLC machine using known compounds. Why is it essential to keep the conditions constant after forming this database?

> **✔** **REMEMBER THIS**
>
> GLC analysis of blood or urine samples provides a more accurate result than infrared testing of breath (see Science Feature on page 244) if the breath-alcohol concentration is between 40 and 50 $\mu$g per 100 $cm^3$.

**SEE QUESTIONS 12 AND 13.**

substances in sport, and for determining the amount of alcohol in a motorist's blood when it is close to the legal limit. When connected to a mass spectrometer, it is used to identify substances in foods and medicinal plants.

## THIN-LAYER CHROMATOGRAPHY (TLC)

This is similar to paper chromatography, except that the stationary phase is a fine, inert powder, such as silica gel, which is made into a paste and spread in a thin, even layer over a plastic or glass plate. The layer of paste is then baked on to the plate. Spots of the mixture are applied near the bottom of the plate. As the solvent rises by capillary action up through the spaces between the inert powder particles, the components of the mixture are carried up and separated. When the solvent is almost at the top of the plate, the plate is taken out and dried. $R_f$ values can be obtained in the same way as for paper chromatography.

TLC has three advantages over paper chromatography: it is quicker, the results are more easily reproduced and the separations are more efficient. The technique is used in industry to follow the course of a reaction by studying which components are present. The separated components can be retrieved by scraping off the spots on the stationary phase and dissolving them in a suitable solvent. TLC is also widely used in forensic science to identify explosive residues and to detect cannabis.

**Fig 11.48** Thin-layer chromatography of felt tip pen inks

> **?** **TEST YOURSELF**
>
> **A** Is TLC adsorption or partition chromatography?

# 6 COMBINING TECHNIQUES

## REMEMBER THIS

Further analytical tools are electrophoresis (page 417), ultraviolet/visible absorption spectroscopy (Chapter 27) and atomic emission spectroscopy (Chapter 27).

As you have seen, the chemist has many techniques to call upon when determining the structure of an unknown molecule, and they are hardly ever used in isolation.

Gas–liquid chromatography is an excellent way to separate large numbers of components, whereas mass spectrometry is not the instrument to use if you want to identify substances in complex mixtures because the spectra are too complicated. By combining the two techniques, the separation is first carried out by gas–liquid chromatography, and then the separated components are analysed by the mass spectrometer.

In the same way, infrared spectrometry gives good evidence of the presence of functional groups, but its information needs to be pooled with that from other techniques to gain the complete picture.

Combustion analysis gives information about the percentage composition of elements in a compound by burning a sample in oxygen. The gases produced, such as $CO_2$, $H_2O$ and $SO_2$, can be fed directly into gas–liquid chromatography apparatus, where they are separated and analysed, and their amounts determined.

Another example is the combination of NMR, which determines the different environments of hydrogen nuclei, with mass spectroscopy, which gives fragmentation patterns and an accurate $M_r$. The use of a combination of analytical techniques makes the determination of structural formulas more likely.

## SUMMARY

After studying this chapter, you should know and understand the following:

- Percentage composition by mass of different elements in a compound may be obtained by combustion analysis.
- Empirical formulas can be determined from percentage composition data.
- A mass spectrometer separates streams of positive ions according to their masses and charges.
- Mass spectra of elements can be used to work out accurate values of $A_r$.
- When molecules are ionized in the mass spectrometer, they often break up into fragments that can give information about the structure of a molecule.
- High-resolution mass spectrometry gives highly accurate values of $M_r$, which may lead to the molecular formula of a compound.

- Infrared spectroscopy can be used to identify functional groups that absorb infrared radiation because of their vibrations.
- Nuclear magnetic resonance (NMR) spectroscopy gives information on the location and number of hydrogen atoms in a molecule, because spinning protons create a magnetic field.
- Chemical shift in an NMR spectrum occurs because protons in different chemical environments experience slight variations in the externally applied magnetic field.
- Chromatography involves the distribution of the components of a mixture between a stationary phase and a mobile phase, and provides a way of separating compounds.
- There are several types of chromatography. They include paper chromatography, thin-layer chromatography (TLC), column chromatography and gas–liquid chromatography (GLC).

# ? EXAM QUESTIONS

**1** When a pure, gaseous sample of element **X** is introduced into a mass spectrometer, four mononuclear, singly-charged ions are detected, as shown in the spectrum in Fig 11.Q1.

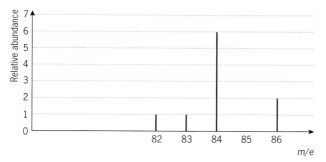

**Fig 11.Q1**

**a)** Describe the process by which the gaseous sample of **X** is converted into ions in a mass spectrometer.
**b)** What adjustment is made to the operating conditions in order to direct the different ions, in turn, onto the detector of a mass spectrometer?
**c)** Use data from the spectrum in Fig 11.Q1 to calculate the relative atomic mass of **X**.
**d)** Identify the element **X**.

AQA CHM1 Unit 1: Atomic Structure, Bonding and Periodicity June 2001 Q1e

**2** Analysis of a sample of bromine in a mass spectrometer showed that it contained a mixture of the $^{79}$Br and $^{81}$Br isotopes in the proportions: $^{79}$Br 55.0%; $^{81}$Br 45.0%.

**a)** Calculate the relative atomic mass of bromine in this sample.
**b)** Outline the basic principles of a mass spectrometer and how mass spectrometry can be used to confirm this isotopic abundance.

OCR 2811 Foundation Chemistry January 2002 Q7

**3**
**a)** When a sample of copper is analysed using a mass spectrometer, its atoms are ionized and then accelerated.
(i) Explain how the atoms of the sample are ionized.
(ii) State how the resulting ions are then accelerated.
**b)** For a particular sample of copper two peaks were obtained in the mass spectrum.

| Peak at *m/e* | 63 | 65 |
|---|---|---|
| Relative abundance | 69.1 | 30.9 |

(i) Give the formula of the species responsible for the peak at *m/e* = 65.

(ii) State why **two** peaks, at *m/e* values of 63 and 65, were obtained in the mass spectrum.
(iii) Calculate the relative atomic mass of this sample of copper, using the results above.

Edexcel 6241/P1 Unit Test 1 June 2001 Q3

**4**
**a)** Fig 11.Q4.1 shows a simple mass spectrometer. For each of the sections listed below, state which lettered part of the spectrometer corresponds to it.

**Fig 11.Q4.1**

Accelerating potential
Gaseous sample
Electron bombardment ionization
Detector
Magnetic field to deflect ions

**b)** Why must the inside of the spectrometer be kept at a high vacuum?
**c)** The spectrometer is set up so that an ion $X^+$ is focussed on the detector. The settings are left unchanged and an ion with a mass less than that of $X^+$ travels through the magnetic field. State and explain whether this ion will be deflected more than, or less than, $X^+$.
**d)** The mass spectrum of propane is shown in Fig 11.Q4.2.

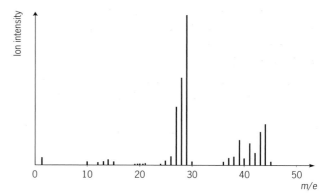

**Fig 11.Q4.2**

(i) Why is there a small peak at *m/e* = 45?
(ii) Why are there no peaks at higher *m/e* values?
(iii) What is the formula of the ion responsible for the peak of greatest intensity?

IBO Chemistry Higher Level Paper 3 Nov 1998 QG1

## ? EXAM QUESTIONS

**5**

**a)** Draw a simple diagram of a mass spectrometer. State briefly how its use could show the existence of isotopes in a gaseous sample of an element.

**b)** Chlorine exists as a mixture of two stable isotopes, $^{35}Cl$ and $^{37}Cl$, present in the approximate ratio 3:1.
**(i)** Calculate the relative atomic mass of chlorine.
**(ii)** Sketch and label a diagram of the mass spectrum of **molecular** chlorine.

IBO Chemistry Higher Level Paper 3 May 2000 QG1

**6**

**a)** Outline **two** methods of preparing a sample of a solid compound in order to obtain an infrared spectrum.

**b)** The two spectra in Fig 11.Q6 were obtained from isomeric compounds **A** and **B** with the formula $C_7H_8O$.

**Fig 11.Q6**

Deduce possible structures for compounds **A** and **B**, explaining in detail how you arrived at your answers.

OCR 9254/4 Paper 4 Option Topics June 2000 Q13

**7** The proton NMR spectrum of a chloroalkyl ketone, **A** ($C_5H_9ClO$), is shown in Fig 11.Q7.

**Fig 11.Q7**

The measured integration trace gives the ratio 1.2 to 1.2 to 1.2 to 1.8 for the peaks at $\delta$ 3.8, 2.8, 2.4 and 1.1 respectively. To answer the following questions, refer to the spectrum in Fig 11.Q7 and to NMR information in data books.

**a)** How many different types of proton are present in compound **A**?

**b)** What is the actual ratio of the numbers of each type of proton?

**c)** The peaks at $\delta$ 2.4 and $\delta$ 1.1 arise from the presence of an alkyl group. Identify the group and explain the splitting pattern.

**d)** What can be deduced from the splitting of the peaks at $\delta$ 3.8 and $\delta$ 2.8?

AQA (NEAB) CH06 March 1998 Q6

**8** Consider the $^1H$ NMR (nuclear magnetic resonance) spectrum that you would expect from methyl propanoate, shown in Fig 11.8.1. **Fig 11.Q8.1**

**a)** Hydrogen atoms within a molecule can exist in different chemical environments. How many different types of hydrogen atoms are there in the molecule?

**b)** Copy Table 11.Q8 and use Table 11.4 and the structure of the molecule to complete Table 11.Q8, about the nature of the spectrum you would expect. (Note, you may not need all of the available spaces.)

| Type of hydrogen | Chemical shift/p.p.m. | Area under peak(s) (area for one H atom = 1) | Splitting (singlet, doublet, etc.) |
|---|---|---|---|
| 1 | | | |
| 2 | | | |
| 3 | | | |
| 4 | | | |
| 5 | | | |

**c)** Methyl propanoate has a number of isomers. For each of the isomers in Fig 11.Q8.2, state **one** difference in the $^1H$ NMR spectrum that would allow you to distinguish it from methyl propanoate.

ethyl ethanoate          propyl methanoate          butanoic acid

**Fig 11.Q8.2**

**d)** Identify which **one** of the four isomers can be distinguished from the other three using infrared spectroscopy, and describe how its spectrum would differ.

IBO Chemistry Higher Level Paper 3 Nov 1999 QG1

**9** Spectroscopy can be used to identify organic molecules. This question is about 3-hydroxy-propanal ($CH_2(OH)CH_2CHO$).

**a)** Draw the displayed formula for 3-hydroxypropanal.

**b)** A sample of 3-hydroxypropanal is analysed using NMR spectroscopy.

# ? EXAM QUESTIONS

**(i)** How many different types of protons will be detected in the low-resolution spectrum?
**(ii)** Describe how the chemical shift of the OH proton can be identified with certainty.
**(iii)** What extra information can be detected from a high-resolution NMR spectrum?

**c)** Predict, with reasons, the absorptions that will be shown on the infrared spectrum of 3-hydroxypropanal.

**10** Two isomers, **A** and **B**, of empirical formula $C_5H_{12}$ give mass spectra that show peaks corresponding to molecular ions and other major fragments with mass-to-charge values as given below.

| **A** | 72 | 57 | 42 | 27 | 12 |
|---|---|---|---|---|---|
| **B** | 72 | 57 | 43 | 29 | 15 |

**a) (i)** Give the molecular formula **A** and **B**.
**(ii)** Give the structural formulas of the **three** isomers of this hydrocarbon.
**(iii)** With respect to the mass spectrum of **A**, identify, with a brief explanation, **one** fragment that would cause the observed mass losses.
**(iv)** With respect to the mass spectrum of **B**, identify **one different** fragment that would cause the observed mass losses.
**(v)** Give the structures of **A** and **B** and name each of them.

**b)** Briefly describe and explain the $^1H$ NMR spectrum of isomer **A**. Indicate the number of different chemical environments of the hydrogen atoms in isomer **B**.

**c)** State why infrared spectroscopy is less useful than mass spectroscopy and $^1H$ NMR spectroscopy in distinguishing between **A** and **B**.

IBO Chemistry Standard Level Paper 3 May 2001 QA1

**11**
**a)** In the use of paper chromatography, a retention factor, $R_f$, is determined. Define $R_f$.

**b)** A drop of green dye is placed 2 cm from the bottom of a strip of filter paper. The filter paper is suspended in a graduated cylinder with 1 cm of the paper immersed in a water–alcohol solvent. After 30 minutes, the green spot is no longer present and there is a yellow spot and a blue spot.
**(i)** Describe how the $R_f$ value of the blue spot could be determined.
**(ii)** Account for the difference in the $R_f$ values of the yellow and blue dyes.
**(iii)** What is the significance of an $R_f$ value of 1.0?

IBO Chemistry Higher Level Paper 3 May 2000 QG1

**12**
**a)** Describe simply and explain the method of gas–liquid chromatography (GLC).
**b)** A 1:1:1:1 mixture of ethanol, propan-1-ol, butan-1-ol and hexan-1-ol was injected into a GLC. The chromatogram shown in Fig 11.Q12 was obtained. The conditions of the run are stated below Fig 11.Q12.

**Fig 11.Q12**

**Conditions:**
The temperature of the oven is 100 °C.
The carrier gas is nitrogen flowing at a rate of 30 $cm^3$ $min^{-1}$.
The stationary phase is carbowax.
The column is 2.0 m long and has an effective cross-section of 6 $mm^2$.
**(i)** Calculate how long the nitrogen takes to pass through the column.
**(ii)** Estimate the volume of nitrogen that passes through the column to carry the ethanol fraction.
**(iii)** Explain why the temperature of the oven (100 °C) does not have to be higher than the boiling points of all the alcohols.
**(iv)** Explain why the carbowax is likely to be hydrophobic (non-polar).
**(v)** Explain the relative heights and shapes of the peaks in the chromatogram.

OCR 9254/4 Paper 4 Option Topics June 2000 Q12

**13** Chromatography is a very powerful tool in each of the following situations. State which chromatographic technique could be used to solve each problem. Describe briefly how the technique could be carried out and the results interpreted.
**a)** It is suspected that the colouring matter used to dye a food product is a mixture, one component of which may be a banned substance. A pure sample of the banned substance is available. You may use a labelled diagram to help if you wish.
**b)** It is believed that an organic liquid may contain traces of ethanol. The quantity of ethanol, if any, is to be determined.

IBO Chemistry Higher Level Paper 3 Nov 1999 QG2

# KEY SKILLS **ASSIGNMENT**

## WONDER CURE FROM WILLOW

In 1763, Edmund Stone sent a letter to the Royal Society (see Fig 11.A1). He had been investigating willow bark to see whether its properties were similar to those of the bark of the Peruvian cinchona tree, which was used to treat malaria. Stone found that, while willow bark did not cure malaria, it did relieve some of the painful symptoms. This was not the first time willow had been used to relieve pain: in 400 BC, Hippocrates, the famous Greek physician, recommended willow leaves to ease the pain of childbirth.

> **[ 195 ]**
>
> XXXII. *An Account of the Succefs of the Bark of the Willow in the Cure of Agues. In a Letter to the Right Honourable* George Earl *of* Macclesfield, *Prefident of R. S. from the Rev. Mr.* Edmund Stone, *of* Chipping-Norton *in* Oxfordfhire.
>
> My Lord,

**Fig 11.A1** Part of the letter sent by Edmund Stone to the Royal Society in 1763

**1** The active chemical in willow bark was isolated in 1829 and called salicin.
**a)** It has the molecular formula $C_{13}H_{18}O_7$. Calculate its percentage composition.
**b)** What technique is used in an analytical laboratory to work out percentage composition?

**2** The pharmacologically active part of salicin is salicylic acid. When this is analysed, it has the following percentage composition: C = 60.86 per cent, H = 4.35 per cent, O = 34.78 per cent. Calculate its empirical formula.

**3** The six most prominent peaks for salicylic acid are shown in a simplified, low-resolution mass spectrum in Fig 11.A2.

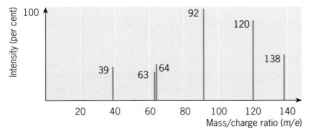

**Fig 11.A2** The mass spectrum of salicylic acid

**a)** Which is the molecular-ion peak?
**b)** Which is the base peak?
**c)** What is the approximate molecular mass of salicylic acid?

**4** The infrared spectrum for salicylic acid is shown in Fig 11.A3. Use Table 11.3, page 245, to identify the groups responsible for some of the major peaks.

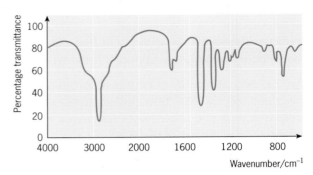

**Fig 11.A3** The infrared spectrum of salicylic acid

**5** What information does NMR spectroscopy give to further determine the structure of salicylic acid?

**Fig 11.A4** The structural formula of 2-hydroxybenzoic acid (salicylic acid)

The structure of salicylic acid is shown in Fig 11.A4. While it was regarded in its day as a wonder drug, salicylic acid (systematic name 2-hydroxybenzoic acid) was unpleasant to take and irritated the mouth and stomach. By modifying the structure of salicylic acid, a German chemist, Felix Hoffman, succeeded in producing a new molecule, which we now know as aspirin. It is still used in many medicines, and is thought to help combat heart disease by suppressing the formation of blood clots.

**Fig 11.A5** The structural formula of aspirin

**6** Aspirin is an ester of salicylic acid. Research the following:
**a)** The functional group of an ester.
**b)** How aspirin might be synthesized in the laboratory from salicylic acid.
**c)** The use of esters as food flavourings.

# 12

# Halogenoalkanes

**A polyfluorohydrocarbon may** well prove to be the right compound for a synthetic blood substitute. It would enable emergency medical workers at accident and injury sites to give life-saving transfusions immediately

After a serious accident, such as a plane crash in a remote area, blood transfusions may have to be carried out under difficult conditions, but there may not be time to identify and match a victim's blood group beforehand. Ideally, the rescue team needs a synthetic blood substitute that will suit any victim and do the vital job of carrying large amounts of oxygen efficiently around the body.

The blood substitute must also be inert and non-toxic. It needs to be chemically stable so that it has a long shelf-life. Some polyfluorohydrocarbons have these properties, and so the search is now on for the polyfluoro carbon compound that is the best synthetic blood substitute.

For over 50 years, it was because of these properties – of being inert, non-toxic and chemically stable – that the related chemicals chlorofluorocarbons (CFCs) were thought to be the best choice for industrial and domestic applications – as refrigerants, electrical insulators and aerosol propellants.

Then, in the 1980s, the same properties were found to make CFCs an environmental hazard: scientists confirmed that CFCs released into the air break down the ozone in the stratosphere. They warned that, even if CFCs were banned, because they are so stable it would be a long time before the ozone could build up to its original level. Since then, CFCs are being replaced in their previous applications by alternative compounds.

## 1 HALOGENO HYDROCARBONS

A **halogeno hydrocarbon** (also called a halogenated hydrocarbon) is a compound that contains carbon, hydrogen and at least one of the halogens. This means that the molecule contains at least one polar covalent bond, namely, the carbon–halogen bond.

A **halogenoalkane** or haloalkane is an alkane in which one or more of the hydrogen atoms have been substituted by halogen atoms.

**SEE PAGES 82–83 FOR INFORMATION ON POLAR COVALENT BONDS.**

### USES OF HALOGENO HYDROCARBONS
Halogenated hydrocarbons are synthesized as useful chemicals in their own right, as well as being intermediates in the synthesis of other chemicals.

#### AS CHEMICALS IN THEIR OWN RIGHT
Chloroalkanes have been used as solvents and anaesthetics for at least a century, but they are poisonous and damage vital organs in the body, including the liver. Synthetic methods have now been developed that successfully introduced several different halogen atoms into a carbon skeleton. This changed the situation immediately. The new chlorofluorocarbons were non-toxic, low-boiling liquids or gases at room temperature and chemically inert. They seemed just the answer for many applications, for example as the fluids for air conditioning systems, refrigerators, aerosols and blowing plastic foams.

**Fig 12.1** The sales of aerosol cans that contain CFCs have been phased out since the damage that CFCs do to the atmosphere became understood

Later, other halogen atoms were introduced and the halons were developed for fire extinguishers. Unfortunately, as the chapter Opener (opposite) points out, the very properties that make halogeno hydrocarbons so useful have caused a global environmental problem – ozone depletion in the upper atmosphere.

Some chlorinated hydrocarbons have been synthesized as insecticides and pesticides. But again, their chemical stability leads to environmental problems.

## AS SYNTHETIC INTERMEDIATES

The manufacture of many pharmaceutical products and polymers requires halogenated hydrocarbons as intermediates in their organic synthesis. A chlorine or bromine atom in an organic molecule often increases its reactivity and allows a change of functional groups or an extension of the carbon skeleton. The polymer poly(chloroethene), polyvinyl chloride PVC, is manufactured from chloroethene; and tetraethyl lead(IV), which is the 'lead' in petrol (being phased out), is synthesized from chloroethane.

## NAMING HALOGENO HYDROCARBONS

Although many halogeno hydrocarbons have traditional names, such as chloroform ($CHCl_3$) or vinyl chloride ($CH_2CHCl$), it is important that you can systematically name halogeno hydrocarbons.

The names of all organic compounds are based on the carbon skeleton and the functional groups present. In halogeno hydrocarbons the functional group is the carbon–halogen bond. The presence of a halogen is indicated by the use of a prefix, namely *fluoro*, *chloro*, *bromo* or *iodo*. The number and location of the halogen atoms must also be specified. The number of each halogen atom is specified by the use of *mono*, *di*, *tri*, *tetra*, *penta*, *hexa*, and so on, before the prefix. For example, the trichloromethane molecule contains three chlorine atoms, and the dibromo-dichloroethane molecule contains two chlorine atoms and two bromine atoms.

**Isomers** are substances that have the same molecular formula, but their atoms are arranged differently. They may have different structural or displayed formulas, or different arrangements about a double bond. This means that, for example, dibromo-dichloroethane is not sufficient to describe fully one particular compound. Fig 12.3 shows the four different structural isomers that could each be called dichloro-dibromoethane.

This means that the *position* of the halogen atom must also be specified in the name. We use the system of numbering each carbon atom which we started to use on page 212. In Fig 12.3, structure **A** is called 2,2-dibromo-1,1-dichloroethane, and structure **B** is called 1,2-dibromo-1,1-dichloroethane. Notice that the order of the halogens in the name is given alphabetically.

**YOU CAN READ MORE ABOUT PESTICIDES AND INSECTICIDES IN CHAPTER 28.**

**SEE PAGE 212 TO FIND OUT HOW TO NAME THE CARBON SKELETON.**

**Fig 12.2** The displayed formula for vinyl chloride shows it has the hydrocarbon skeleton of ethene, so its systematic name is chloroethene

### ? TEST YOURSELF

**A** How many halogen atoms are there in one molecule of 1,2,2,3-tetrachloro-5-iodoheptane?

**B** **a)** What is the systematic name for compounds with structures labelled **C** and **D** in Fig 12.3?
**b)** Draw the displayed formulas for the following halogeno hydrocarbons:
(i) 2,2-dichloropropane;
(ii) 1,2,3,4-tetrachlorocyclohexane;
(iii) 1,2-dichloro-3,3,4-triiodoheptane;
(iv) 1,2-difluoroethene;
(v) tetrafluoroethene.

### ✔ REMEMBER THIS

Sometimes, all the hydrogen atoms in an alkane are replaced by halogen atoms. If the halogen atoms are all identical, the prefix 'per' is often used. So that, for example, $C_2F_6$ is often referred to as perfluoroethane rather than hexafluoroethane.

**SEE PAGE 198 FOR MORE INFORMATION ON ISOMERISM.**

### ? TEST YOURSELF

**C** **a)** What is the name of $C_4F_{10}$?
**b)** What is the name of $CH_3CH_2CHClCH_3$?
**c)** What is the name of $C_6Cl_{14}$?

**Fig 12.3** The displayed formulas for all the isomers of dichloro-dibromoethane

# 2 PHYSICAL PROPERTIES OF HALOGENOALKANES

The uses of halogenoalkanes generally depend on their physical properties. One notable property is their low boiling point, which accounts for many of their applications (see the Feature box opposite).

The boiling point of a substance depends on the nature and strength of the *inter*molecular forces in the liquid.

## INTERMOLECULAR FORCES IN HALOGENOALKANES

Halogenoalkanes are all covalently bonded and form simple molecules. The presence of simple molecules in the liquid phase makes them solvents that are electrically non-conducting.

Halogen atoms tend to withdraw electrons from a carbon–halogen bond. In a monohalogenoalkane, the halogen atom will be slightly negative (δ–) and the carbon atom it is bonded to will be slightly positive (δ+). We say that the molecule has a **permanent dipole** and that it possesses a **dipole moment**, a measurable degree of polarity.

chloromethane          trichloromethane

**Fig 12.4** The high electronegativity of chlorine gives both chloromethane and trichloromethane a dipole moment

Chloromethane and trichloromethane are both molecules with a dipole moment, so they have permanent dipoles. The positive part of one molecule can attract the negative part of another molecule. This intermolecular force is referred to as a **permanent dipole–permanent dipole interaction**.

**Fig 12.5** The permanent dipole causes a weak electrostatic attraction between the negative chlorine atom and the positive carbon atom in a neighbouring trichloromethane molecule

**SEE PAGE 225 FOR INFORMATION ON INDUCED DIPOLE–INDUCED DIPOLE FORCES.**

When a large alkyl group is present as well, part of the intermolecular attraction consists of induced dipole–induced dipole van der Waals attractions.

## ? TEST YOURSELF

**D** **a)** Suggest the molecular formula of (i) CFC13,
(ii) CFC112.
**b)** Suggest the name for CFC112.

temporary induced dipole–induced dipole attraction

permanent dipole–permanent dipole attraction

**Fig 12.6** The intermolecular forces in 1-chlorooctane result from both induced dipole–induced dipole and permanent dipole–permanent dipole interactions

## SCIENCE FEATURE Chlorofluorocarbons and halons

The most well-known and infamous of the halogenoalkanes are the chlorofluorocarbons, known as CFCs. As their name suggests, these are halogenoalkanes in which every hydrogen atom has been replaced by either a chlorine or a fluorine atom. Table 12.1 shows some CFCs.

**Table 12.1 Some typical CFCs and their uses**

| Molecular formula | Name | CFC designation | Uses |
|---|---|---|---|
| $CFCl_3$ | Trichloro-fluoromethane | CFC11 | Blowing foam plastics, refrigeration, air conditioning |
| $CF_2Cl_2$ | Dichloro-difluoromethane | CFC12 | Blowing foam plastics, refrigeration, air conditioning, aerosol propellant, sterilization and food freezing |
| $C_2F_3Cl_3$ | Trichloro-trifluoroethane | CFC113 | Solvent |
| $C_2F_4Cl_2$ | Dichloro-tetrafluoroethane | CFC114 | Blowing foam plastics, refrigeration, air conditioning |
| $C_2F_5Cl$ | Chloro-pentafluoroethane | CFC115 | Refrigeration, air conditioning |

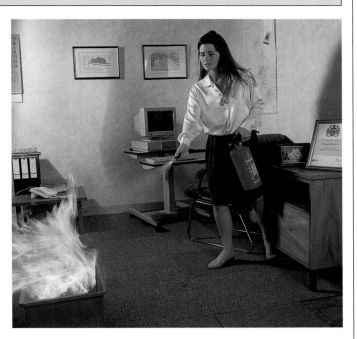

**Fig 12.7** The CFC refrigerant circulates through pipes at the back of the fridge, when a refrigerator is disposed off the CFCs must be extracted carefully to avoid the gases escaping into the atmosphere

**Fig 12.8** Halon fire extinguishers, once used on fires near electrical equipment, are now replaced by carbon dioxide gas extinguishers

When a hydrogen atom is included in a CFC, the compound is called HCFC (hydrogen-chlorofluorocarbon).

Another variety of fully halogenated alkanes is called the **halons**. These contain bromine atoms as well as fluorine and chlorine atoms. Halons are used extensively in fire extinguishers because they are chemically unreactive and non-toxic. Being much denser than air, they effectively blanket a fire, which keeps out the air and so extinguishes the fire.

Typical halons include trifluoro-bromomethane (Halon 1301), and bromo-chloro-difluoromethane (Halon 1211). Unfortunately, these halons also cause the same environmental problems as CFCs, and research is under way to find substitutes.

## BOILING POINTS OF CFCS

Boiling occurs when intermolecular forces are overcome, so the molecules can separate from each other and spread out. Thus, the stronger the intermolecular attraction, the higher the boiling point. The type of intermolecular forces in CFCs are similar to those in other halogenoalkanes, but the dipole moment of the molecules is likely to be much smaller since the individual bond polarities tend to cancel out. This means that the attraction due to permanent dipole–permanent dipole interactions is liable to be quite weak. As a result, even though the relative molecular masses of CFCs are quite large compared with, say, the monohalogenoalkanes, they will have lower boiling points.

**? TEST YOURSELF**

**E** Why is it important that halons are more dense than air?

**Fig 12.9** The bond polarities nearly cancel out one another, so that dichloro-difluoromethane has a small dipole moment

**Table 12.2 Boiling points of some halogenated alkanes**

| Compound | Boiling point/°C |
| --- | --- |
| 1-Fluorobutane | (see question **F (a)** below) |
| 1-Chlorobutane | 77 |
| 1-Bromobutane | 100 |
| 1-Iodobutane | 130 |

**SEE QUESTIONS F(A) AND 1**

**? TEST YOURSELF**

**F** **a)** Predict the boiling point of 1-fluorobutane.

**b)** What does the difference in boiling points of the compounds in Table 12.2 suggest about the strength of intermolecular attraction in 1-chlorobutane and 1-iodobutane?

**c)** The boiling point of 2-chloro-2-methylpropane is 51 °C, and that of 1-chlorobutane is 77 °C, even though they have the same molecular mass. Explain why there is this difference. Hint: read page 216 for some ideas.

**G** Draw in the bond polarities for trichloro-fluoromethane, and hence draw in the dipole moment of the molecule.

**✔ REMEMBER THIS**

The electron shells are shown here as circles, but in Chapter 4 they are not. Either way is acceptable.

**? TEST YOURSELF**

**H** Draw a dot-and-cross diagram of each of the following and use this to decide which particle is a free radical: $Cl^-$, $Cl^+$, I, $BF_3$, $CH_3$.

As CFCs have weak intermolecular attraction, they are volatile liquids and vaporize easily at low temperatures. Many CFCs have two properties that make them efficient refrigerants: they absorb heat energy readily when vaporizing, and they can be easily compressed back into liquids, releasing this energy. Provided a heat exchanger efficiently removes the heat released during liquefaction, net cooling of the refrigerant CFC will occur. If the intermolecular forces were larger, it would be difficult to vaporize the CFC and so little heat would be absorbed.

# 3 PREPARATION OF HALOGENOALKANES BY FREE RADICAL SUBSTITUTION

Even before the wide-scale use of halons and CFCs, the preparation of halogenoalkanes had become an important synthetic reaction because halogenoalkanes are extremely useful in making other chemicals.

The main methods of preparation of halogenoalkanes are described in Fig 12.10. One route involves a direct substitution of a hydrogen atom by a halogen atom. It is called **free radical substitution**.

**Fig 12.10** A summary of the preparation routes for halogenoalkanes. Electrophilic addition is described in Chapter 14, and the conversion of alcohols into halogenoalkanes is described later in this chapter and in Chapter 13

## FREE RADICALS

Alkanes are very unreactive compounds, with the exception of their reaction with oxygen. Their lack of polar covalent bonds is the reason for this behaviour. Nucleophiles are particles that seek out electron-deficient centres and donate pairs of electrons to form a covalent bond, while electrophiles are particles that seek out electron-rich centres and accept pairs of electrons to form a covalent bond. It is difficult for alkanes to react with nucleophiles or electrophiles, because alkanes are non-polar molecules.

Alkanes require a different type of particle with which they can react, that is, one which is highly reactive and does not need to seek out a polar covalent bond before it can react. The particle in question is called a **free radical**. This is an atom or group of atoms that possesses at least one unpaired electron in its outer shell.

**Fig 12.11** Dot-and-cross diagrams of three important free radicals

Most free radicals are extremely reactive because they pair up their unpaired electron with an electron removed from a covalent bond. Alternatively, two free radicals can react with each other, pairing up the unpaired electrons to form a covalent bond.

A free radical is usually indicated by a dot on the right-hand side of the formula of the particle. For example, a chlorine atom Cl is a free radical, which is made clear by writing Cl•.

## HOMOLYTIC FISSION

As they are so reactive, free radicals cannot be stored in reagent bottles. They must be made in the reaction vessel, ready to react immediately. To make a free radical, a covalent bond is broken in such a way that the electrons in the shared pair of the covalent bond become two single electrons, one electron per atom. This is known as **homolytic fission** of a covalent bond.

Fig 12.12 is a representation of this process, here to make a chlorine free radical from a chlorine molecule. A curly half-headed arrow shows the movement of one electron.

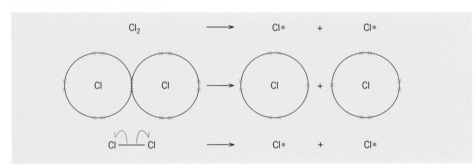

**Fig 12.12** Chlorine free radicals are formed by the homolytic fission of a single bond in a chlorine molecule

The homolytic fission of a covalent bond requires energy (it is an endothermic process). For example, the production of the chlorine free radical requires half of the bond dissociation energy of a chlorine molecule:

$$\tfrac{1}{2}Cl_2(g) \rightarrow Cl•(g)$$

## FREE RADICAL SUBSTITUTION OF METHANE

Methane is a typical alkane, which reacts with chlorine in the presence of ultraviolet light to give a large variety of products.

Fig 12.13 gives you an idea of the variety of **substitution** products that result from the free-radical chlorination of methane. This is the process by which a hydrogen atom is swapped for a chlorine atom.

But there are also other products whose origin is less obvious. The only way to explain this large collection of products is to examine the mechanism of the reaction.

**Fig 12.13** The displayed formulas of six products of the free-radical chlorination of methane

**Fig 12.14** The complete mechanism of the free-radical substitution of methane

## REACTION MECHANISMS

The **mechanism** of a reaction details the smaller steps in the process that results in the formation of the end products from the starting materials. In particular, a mechanism seeks to explain which bonds are broken and formed, and when and how they are broken and formed.

Free-radical substitution is split into three, clearly defined steps:

- **Initiation** This is the formation of the free radical.
- **Propagation** This involves reactions that maintain the presence of the free radicals.
- **Termination** This involves reactions that remove the free radicals.

### INITIATION

We have already described the formation of a chlorine free radical. Ultraviolet radiation is used to supply the energy to break the chlorine molecule into free radicals as follows:

$$Cl_2(g) \xrightarrow{\text{ultraviolet light}} 2Cl\bullet(g)$$

The equation tends to mislead, since it suggests that all the chlorine molecules are changed into free radicals. In fact, only an extremely small percentage of chlorine molecules are dissociated, and most remain as chlorine molecules.

### PROPAGATION

Reactions take place when the appropriate particles collide. Since free radicals are so reactive, they almost invariably react with any particle they collide with. In the chlorine free-radical substitution of methane, once the very first free radical is produced, it is most likely to collide with either a chlorine molecule or a methane molecule. The collision with the chlorine molecule does not lead to any change.

$$Cl\bullet(g) + Cl_2(g) \rightarrow Cl_2(g) + Cl\bullet(g)$$

When the highly reactive free radical collides with a methane molecule, it gains an electron from one of the covalent bonds in the molecule. In this case it is the C—H bond. This gives:

$$CH_4(g) + Cl\bullet(g) \rightarrow CH_3\bullet(g) + HCl(g)$$

This reaction produces a new free radical, $CH_3\bullet$, which is the methyl free radical. The methyl free radical is also very reactive so as soon as it collides with a particle, it reacts. Which particle it collides with is just a question of chance or probability. The most likely will be the particles that are in the greatest concentration. Certainly, at the start of the reaction, these will be either methane or chlorine molecules:

$$CH_3\bullet(g) + Cl_2(g) \rightarrow CH_3Cl(g) + Cl\bullet(g)$$

Notice that the two steps just described regenerate the chlorine free radical, which explains why they are called propagation steps.

**1** $-\overset{|}{\underset{|}{C}}-H + Cl\bullet \rightarrow -\overset{|}{\underset{|}{C}}\bullet + H-Cl$    **2** $-\overset{|}{\underset{|}{C}}\bullet + Cl_2 \rightarrow -\overset{|}{\underset{|}{C}}-Cl + Cl\bullet$

**Fig 12.15** The propagation steps in the free-radical substitution of a C–H bond. Notice that the chlorine free radical is regenerated at the end of the second step

Fig 12.15 generalizes these two steps, explaining the process as a characteristic reaction of a C—H bond associated with a carbon atom with four single bonds.

The product chloromethane still has three hydrogen atoms each bonded to carbon atoms. So, when a chlorine free radical collides with a chloromethane molecule, a reaction will take place:

$$CH_3Cl(g) + Cl\bullet(g) \rightarrow CH_2Cl\bullet(g) + HCl(g)$$

$$CH_2Cl\bullet(g) + Cl_2(g) \rightarrow CH_2Cl_2(g) + Cl\bullet(g)$$

Provided there is sufficient chlorine present, this reaction can continue until all the hydrogen atoms have been substituted. The composition of the reaction mixture varies with the time allowed for the reaction and the mole ratio of methane and chlorine.

## TERMINATION

The propagation steps cannot go on forever, because eventually chance takes a hand and a free radical collides with another free radical. Although this is not very likely, it is not impossible – particularly when there are many millions of collisions per second.

When two free radicals collide, they react and form a covalent bond. As any two free radicals may collide, note that all of the following are possible termination steps.

$$Cl\bullet(g) + Cl\bullet(g) \rightarrow Cl_2(g)$$

$$Cl\bullet(g) + CH_2Cl\bullet(g) \rightarrow CH_2Cl_2(g)$$

$$CH_3\bullet(g) + CH_3\bullet(g) \rightarrow C_2H_6(g)$$

Also note that one of the products is ethane, a molecule that does not even contain chlorine.

## FLUORINATION OF ALKANES

Fluorine is a much more reactive halogen than chlorine, and so it should react much more readily with alkanes. The bond dissociation energy for a fluorine molecule is $158\,kJ\,mol^{-1}$, considerably smaller than that for a chlorine molecule at $244\,kJ\,mol^{-1}$. The reason for this low bond energy is the repulsion experienced by each atom because of the very small bond length of $254\,pm$ in a fluorine molecule. This makes the initiation step for fluorine much easier than that for chlorine.

The reaction of fluorine with alkanes does not need the presence of ultraviolet light and it proceeds much faster than the reaction of chlorine. Consequently, it is difficult to produce monofluoro derivatives of alkanes. The reaction is often explosive and produces the perfluoro derivative (all H atoms are substituted by F atoms). The reaction of fluorine with alkanes is slowed down by diluting the fluorine with nitrogen.

**? TEST YOURSELF**

**K** **a)** How could you ensure that you obtain a large proportion of chloromethane during the free-radical substitution of methane with chlorine? **b)** How could you ensure that you obtain a large proportion of tetrachloromethane during the free-radical substitution of methane with chlorine?

**L** Write down two more possible termination steps during the free-radical chlorination of methane.

**M** Ethane reacts with chlorine in the presence of ultraviolet light to form a mixture of products, including chloroethane, dichloroethane, trichloroethane, tetrachloroethane, hydrogen chloride and ethane. Account for the formation of each of these products by writing down the detailed mechanism for the reaction.

**SEE QUESTIONS 1 AND 2.**

**? TEST YOURSELF**

**N** Write down the mechanism of the reaction of methane with excess fluorine by a free-radical substitution.

## SCIENCE FEATURE Uses of organofluorines

Organofluorine compounds occur rarely in nature, but because of the wide range of their applications – from inert solvents in the electronics industry to highly active pharmaceuticals – their development and production are increasingly important.

The three-dimensional shape of an organofluorine molecule determines its biological activity. Substitution of a fluorine atom for a hydrogen atom rarely changes the shape since the change in bond length is small, being from 120 pm in a C—H bond to 147 pm in a C—F bond. The C—F bond is also very strong and resistant to reaction with water (hydrolysis), which is why pharmaceuticals that contain this bond are liable to remain active in the aqueous medium of a living cell for a long time.

Perfluoro compounds are very unreactive, which has led to yet further applications for which being chemically inert is vital. Perfluorodecalin, for example, dissolves oxygen very efficiently and has been successfully transfused into human volunteers as a blood substitute.

Other perfluorohydrocarbons are being developed to provide a fluid that will preserve donor organs before transplantation. There are still problems with these perfluoro compounds, since to be of use they must remain as an emulsion in water.

**Fig 12.16** The displayed structures of PTFE and perfluorodecalin

Unfortunately, they tend to separate during storage.

Polytetrafluoroethene (PTFE) is a perfluorinated polymer that is used to make non-stick surfaces, such as on saucepans and frying pans.

Coatings are being developed from polymers that include fluoro-substituted alkanes, since these side chains make the surface much easier to clean by their ability to shrug off stains and dirt. Recently, clothes have been developed with such a coating, and it may be possible to make a surface that could remain graffiti free.

**Fig 12.17** Will organofluorines provide surfaces for building materials that will make them easy to wipe free of graffiti?

**SEE QUESTION 3.**

## BROMINATION OF ALKANES

Bromine is less reactive than chlorine, and the free-radical reaction with alkanes is much slower than with chlorine. This is surprising since the bond energy for a bromine molecule is only $151 \text{ kJ mol}^{-1}$. The reason for this is related to the strength of the carbon–halogen bond that is formed, since a carbon–chlorine bond is much stronger than a carbon–bromine bond, and hence more stable.

When a few drops of liquid bromine are added to hexane, a liquid alkane, and the mixture is left in direct sunlight, there is a very slow reaction. Eventually, the orange colour of the bromine disappears, leaving a mixture of all the structural isomers of monobromoalkane as the main products:

$$C_6H_{14}(l) + Br_2(l) \rightarrow C_6H_{13}Br(l) + HBr(g)$$

**? TEST YOURSELF**

**O** Draw the displayed formulas of all the structural isomers of monobromohexane.

## FREE-RADICAL SUBSTITUTION AS A SYNTHETIC REACTION

Although free-radical substitution is a very successful way of introducing halogen atoms into an alkane molecule, it suffers from a major drawback: often a large variety of products is formed.

The aim of synthesis is to make a compound in high yield and purity, using few reactions and involving cheap starting materials. The large number of products formed during free-radical substitution makes it a poor tool in organic synthesis.

The only way to make the reaction useful is to minimize the number of possible products, so that a relatively high yield of the required product is obtained. Simply, this means involving reactants that have very few hydrogen atoms to be substituted.

---

**▶ EXTENSION**

### SUBSTITUTION IN BENZENE

Look at the reaction in Fig 12.18. Methylbenzene has three C–H bonds not attached to the benzene ring; these hydrogen atoms are chemically identical and behave in a similar way to the hydrogen atoms in methane. Therefore, it is quite easy to synthesize chloromethylbenzene by passing chlorine into boiling methylbenzene in the presence of ultraviolet light. By keeping the available chlorine concentration low, the monosubstituted product chloromethylbenzene acid is the major product.

Fig 12.18 The reaction of methylbenzene with chlorine

Ethanoic acid also contains three hydrogen atoms that are similar to those found in methane and so ethanoic acid is chlorinated to give chloroethanoic acid:

$$CH_3COOH(l) + Cl_2(g) \rightarrow CH_2ClCOOH(l) + HCl(g)$$

### SYNTHESIS OF THE AMINO ACID GLYCINE

Amino acids are the building blocks of proteins. Glycine (aminoethanoic acid) is the simplest amino acid. It has a two-carbon skeleton, so an ideal candidate for its synthesis is ethanoic acid. Fig 12.19 shows a synthetic route that could be used to make glycine, employing the free-radical chlorination reaction.

Fig 12.19 The synthesis of the amino acid glycine from ethanoic acid

**? TEST YOURSELF**

**P** How might you modify the reaction conditions to make trichloromethylbenzene by the chlorination of methylbenzene? Write an equation for this reaction.

YOU CAN READ MUCH MORE ABOUT THE MECHANISM OF THE HALOGENATION OF ALCOHOLS ON PAGES 293–294.

# 4 PREPARATION OF HALOGENOALKANES BY THE HALOGENATION OF ALCOHOLS

Monohalogenoalkanes are best prepared by replacing the hydroxyl group in an alcohol with a halogen atom. A variety of reagents can be used for this conversion and some of these are shown in Table 12.3. Note that R stands for an alkyl group such as methyl, ethyl or propyl.

**Table 12.3 Reagents and conditions that can be used to make halogenoalkanes from alcohols. Concentrated hydrochloric acid and concentrated hydrobromic acid are often generated *in situ* by the reaction of the corresponding sodium halide with concentrated sulphuric acid:**

$$NaBr + H_2SO_4 \rightarrow NaHSO_4 + HBr$$

| Halogenoalkane | Reagent | Conditions | Equation |
|---|---|---|---|
| Chloroalkane | Concentrated hydrochloric acid | Heating under reflux | $ROH + HCl \rightarrow RCl + H_2O$ |
| Chloroalkane | Phosphorus(V) chloride | Heating under reflux | $ROH + PCl_5 \rightarrow RCl + HCl + POCl_3$ |
| Bromoalkane | Concentrated hydrobromic acid | Heating under reflux | $ROH + HBr \rightarrow RBr + H_2O$ |
| Bromoalkane | Phosphorus(III) bromide | Heating under reflux | $3ROH + PBr_3 \rightarrow 3RBr + H_3PO_3$ |
| Iodoalkane | Red phosphorus and iodine | Heating under reflux | $3ROH + PI_3 \rightarrow 3RI + H_3PO_3$ |

SEE QUESTION 4.

**? TEST YOURSELF**

**Q** **a)** What is the major product when 3-methylhexan-2-ol is heated with red phosphorus and iodine?
**b)** Which reagents do you need to make (i) iodoethane, (ii) 3-bromohexane?

# 5 REACTIONS OF HALOGENOALKANES

Halogenoalkanes are saturated compounds, which means that they have no double or triple carbon–carbon bonds in their molecules. This limits the reactions available, since saturated compounds cannot undergo addition reactions. Halogenoalkanes undergo two major types of reaction, **substitution** and **elimination**. During substitution, the halogen atom is swapped for one atom or a group of atoms. The elimination reaction involves the loss of a molecule of hydrogen halide to produce an alkene.

**Fig 12.20** The two different reactions of halogenoalkanes

## SUBSTITUTION REACTIONS

### POLAR COVALENT BOND

The carbon–halogen bond is a polar covalent bond. The carbon atom is electron deficient, because the pair of electrons within the covalent bond is drawn closer to the highly electronegative halogen atom. A polar covalent bond, such as a carbon–chlorine bond, is open to attack by a species (an ion or a molecule) that seeks out electron-deficient centres.

**? TEST YOURSELF**

**R** State which one of the following compounds has the most electron-deficient carbon atom: bromomethane, chloromethane, fluoromethane or iodomethane.

**Fig 12.21** The electron cloud in carbon–halogen bonds

## NUCLEOPHILES

A **nucleophile** is a chemical species (ion or molecule) which donates an electron *pair* in order to make a covalent bond with an electron-deficient centre. A nucleophile has a complete outer shell of electrons and at least one lone pair. It is this lone pair of electrons that is donated to form the covalent bond.

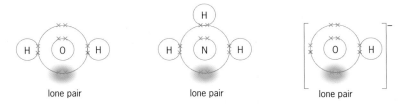

**Fig 12.22** Dot-and-cross diagrams of nucleophiles $H_2O$, $NH_3$, and $OH^-$

When we describe a species as a good nucleophile, the term 'good' refers to the ability of the species to donate an electron pair to form a covalent bond. The halide ions, for example, do not very easily donate the lone pair of electrons in their outer shells, and are therefore classed as poor nucleophiles.

## NUCLEOPHILIC SUBSTITUTION

The electron-deficient, $\delta+$, carbon atom of the polar carbon–halogen bond is attacked by a nucleophile, which donates a pair of electrons to make a covalent bond. Since a carbon atom is normally surrounded by no more than four covalent bonds (four bonding pairs of electrons), this donation must result in the breaking of another bond.

Fig 12.23 shows that when a nucleophile reacts with a halogenoalkane, a substitution reaction takes place. This reaction is called **nucleophilic substitution**, since the species that starts the substitution is a nucleophile.

**Fig 12.23** Nucleophilic substitution of a halogenoalkane by a nucleophile

During the reaction, a carbon–nucleophile bond is made and a carbon–halogen bond is broken. The halogen is sometimes called the **leaving group**, because it is the species that is lost. In this case, the leaving group must be able to gain the pair of bonding electrons. This is not a problem for a halogen atom, since it forms a halide ion, which easily supports a negative charge.

During nucleophilic substitution, **heterolytic fission** of a covalent bond occurs. In heterolytic fission, one of the two atoms joined by the covalent bond that breaks receives both electrons from the bond, so after the fission one atom carries a positive charge and the other a negative charge.

## ENERGY CONSIDERATIONS DURING NUCLEOPHILIC SUBSTITUTION

It is possible to estimate the enthalpy change of reaction by considering the bond energies of the carbon–nucleophile bond and carbon–halogen bond. Although this ignores the energy changes due to the formation of the nucleophile and the enthalpy changes due to interactions with the solvent, it is nevertheless a good way to make some predictions.

## ? TEST YOURSELF

**S** Draw a dot-and-cross diagram for each of the following and use it to decide whether the particle is a nucleophile:

**a)** $NH_2^-$
**b)** $CN^-$
**c)** $BCl_3$

**SEE QUESTION 5.**

On page 173, we describe bond breaking as an endothermic process, which requires the transfer of energy *from* the surroundings, and bond formation as an exothermic process, which involves energy transfer *to* the surroundings.

**Fig 12.24** Bond breaking and bond forming for enthalpy calculations

An estimate for the enthalpy change for the nucleophilic substitution can be calculated using:

$\Delta H$ = (bond energy of carbon–halogen) – (bond energy of carbon–nucleophile)

A useful guide to the feasibility of a reaction can be obtained by looking at the enthalpy change of reaction. An exothermic reaction is often more feasible than an endothermic reaction that has a positive $\Delta H$. In the same way, a reaction that is highly exothermic is often more feasible than one that is less exothermic.

This discussion ignores entropy considerations, which are difficult to predict in the example we are looking at.

This analysis of the energy changes during reactions leads to some simple ideas. Consider the reaction of hydroxide ion with fluoroalkanes, chloroalkanes, bromoalkanes and iodoalkanes. The same bond is made each time, namely the carbon–oxygen bond. But the bond that is broken, the carbon–halogen bond, becomes weaker and weaker, and so the overall enthalpy change is more negative. So, nucleophilic substitution is easier with iodoalkanes than with fluoroalkanes. The strength of the bond to be broken in the reaction, the carbon–halogen bond, plays an important part in determining whether nucleophilic substitution will take place.

This also means the nucleophile must make a strong bond with the carbon atom rather than a weak bond, which releases little energy into the surroundings. Since the carbon–halogen bonds are quite weak, the halide ions do not act as nucleophiles in this reaction.

**Fig 12.25** During the reaction of a chloroalkane and aqueous hydroxide ion, a carbon–chlorine bond is broken and a carbon–oxygen bond is made. This means that the enthalpy change of reaction can be estimated as –20 kJ mol$^{-1}$. This value should be a good estimate of the enthalpy change for the reaction between any chloroalkane and a hydroxide ion

YOU CAN READ MORE ABOUT THE FEASIBILITY OF REACTIONS ON PAGE 178 AND CHAPTER 25.

**? TEST YOURSELF**

Estimate the enthalpy change of reaction for:

**a)** reaction between hydroxide ion and fluoroethane;

**b)** reaction between hydroxide ion and bromoethane.

You will need to look up the appropriate bond energies.

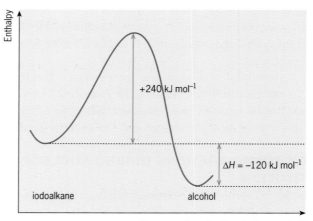

SEE QUESTION 6.

**Fig 12.26** Energy profile diagrams for the reaction between a fluoroalkane or an iodoalkane with a hydroxide ion

# MECHANISM OF NUCLEOPHILIC SUBSTITUTION

So far, we have only discussed which bonds have been broken and which ones have been formed. The next question to ask is: What is the sequence of bond making or bond breaking? One mechanism assumes that the two processes take place together, so that as one bond is forming the other bond is breaking. This mechanism is called $S_N2$, since it involves two particles colliding in the slowest step of the mechanism. This mechanism is shown in Fig 12.27. The curly arrow shows the movement of an electron pair. Primary halogenoalkanes, $RCH_2X$ (where R is an alkyl group and X is Cl, Br and I), react in this way.

The intermediate can be written as

**Fig 12.27** This mechanism is known as $S_N2$, since two species, the nucleophile and the halogenoalkane, are involved in the only step. Primary and many secondary halogenoalkanes show this mechanism during nucleophilic substitution

The nucleophile donates a pair of electrons to the electron-deficient carbon and at the same time the pair of electrons in the carbon–halogen bond move towards the electronegative halogen atom, eventually breaking the covalent bond and forming a halide ion.

# NUCLEOPHILIC SUBSTITUTION REACTIONS OF HALOGENOALKANES

Hydroxide ion, ammonia, amines and cyanide ions are all good nucleophiles and will take part in the nucleophilic substitution reaction with chloroalkanes, bromoalkanes and iodoalkanes. As already described, the carbon–iodine bond is the weakest carbon–halogen bond of the three, so iodoalkanes readily take part in nucleophilic substitution. The nucleophilic substitution reactions of chloroalkanes are much slower.

## HYDROLYSIS

Halogenoalkanes react extremely slowly, if at all, with water. However, as we have already seen, they undergo a nucleophilic substitution reaction with aqueous hydroxide ions. This reaction is known as hydrolysis and involves boiling aqueous sodium hydroxide, for example, with the halogenoalkane. A hydroxyl group is substituted for the halogen atom and so an alcohol is produced. So, 1-bromobutane would be hydrolysed to form butan-1-ol:

$$CH_3CH_2CH_2CH_2Br + NaOH \rightarrow CH_3CH_2CH_2CH_2OH + NaBr$$

In these reactions, the hydroxide ion is behaving as a nucleophile.

Fluoroalkanes are not hydrolysed since the carbon–fluorine bond is too strong. The ease with which halogenoalkanes can be hydrolysed increases as the atomic number of the halogen increases and the carbon–halogen bond becomes weaker.

## ANALYSIS OF HALOGENOALKANES

Aqueous silver nitrate can be used to detect the presence of aqueous halide ions, but it will not detect the halogen in a halogenoalkane since the halogen is present in a covalent bond. To detect the halogen, the halogenoalkane must first be hydrolysed to produce aqueous halide ions.

YOU CAN READ MORE ABOUT THE
REACTION OF AQUEOUS SILVER
NITRATE WITH AQUEOUS HALIDE
IONS ON PAGE 508.

SEE QUESTION 7.

The halogenoalkane is heated with aqueous sodium hydroxide. During hydrolysis, aqueous halide ions are produced. The hydrolysis mixture is acidified with dilute nitric acid, then aqueous silver nitrate is added. The halide ion present in the mixture is precipitated as the silver halide:

$$Ag^+(aq) + X^-(aq) \rightarrow AgX(s), \text{ where } X = Cl, Br \text{ or } I$$

The colours of the precipitate are given in Table 12.4 and shown in Fig 12.28.

| Table 12.4 Test for halide ion in the hydrolysis mixture | |
| --- | --- |
| Halide ion in hydrolysis mixture | Add excess nitric acid followed by aqueous silver nitrate |
| Chloride | White precipitate |
| Bromide | Pale cream precipitate |
| Iodide | Pale yellow precipitate |

Fig 12.28 The resultant precipitates in the halide ion test

## REACTION WITH AMMONIA

Ammonia is another nucleophile (Fig 12.21), since the lone pair on nitrogen can be donated to an electron-deficient centre to make a covalent bond. Ammonia reacts with halogenoalkanes to form an amine. For example, 1-bromobutane will produce butylamine:

$$CH_3CH_2CH_2CH_2Br + NH_3 \rightarrow CH_3CH_2CH_2CH_2NH_2 + HBr$$

There are two serious drawbacks with this reaction: ammonia reacts with one of the products (HBr), and the amine produced is also a nucleophile that can further react with the halogenoalkane, see opposite page.

In section 2 of this chapter, we met a similar problem: a reaction that gives a poor yield and many products is of little use as a synthetic method.

Luckily, the reaction conditions can be manipulated to produce good yields. If excess concentrated ammonia is used, the acidic hydrogen halide product reacts with the excess ammonia to produce an ammonium salt. So, in our example, ammonium bromide would be produced:

$$HBr + NH_3 \rightarrow NH_4Br$$

The excess ammonia also means that there is always a vast excess of ammonia nucleophile compared with the product. This ensures the maximum yield of amine. The best reaction conditions involve heating the alcoholic halogenoalkane with excess ammonia in a sealed tube.

YOU CAN READ MORE ABOUT AMINES
IN CHAPTER 28.

### SCIENCE FEATURE Preparation of amphetamine

The drug amphetamine is a stimulant that gives a sense of well-being and speeds up the metabolism, so that someone who takes it before doing a sport would feel less fatigued. It is prepared using a halogeno compound as an intermediate. Fig 12.29 shows the synthetic route employed.

Fig 12.29 The synthesis of amphetamine

An alcohol, 2-phenylpropan-2-ol, is halogenated to give 2-bromo-2-phenylpropane, which then allows a nucleophilic substitution to take place using excess ammonia. The bromo compound is a synthetic intermediate that allows the conversion of an alcohol into an amine while keeping the same carbon skeleton.

## REACTION WITH AMINES

Amines are nucleophiles because of the lone pair on the nitrogen atom, so they can react with halogenoalkanes to give secondary, tertiary or quaternary ammonium salts.

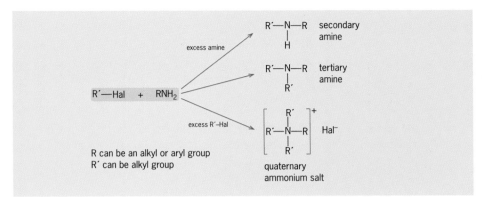

**Fig 12.30** The reactions of amines with halogenoalkanes

## REACTION WITH THE CYANIDE ION

Whenever possible, the starting material chosen for a synthesis already has the correct carbon skeleton. The synthesis of amphetamine in Fig 12.28 illustrates this point, where the nine-carbon skeleton is in place in the starting material. There are times, however, when the chosen starting material does not have the correct carbon skeleton and part of the synthesis is to construct the correct skeleton. An important synthetic reaction is the *extension* of the carbon skeleton by one carbon.

One convenient way involves the nucleophilic substitution of a halogen atom with a cyanide ion. Fig 12.31 shows the dot-and-cross diagram for the cyanide ion. The negative charge resides on the carbon atom, and it is the non-bonding pair on this carbon that is donated to an electron-deficient carbon.

A nucleophilic substitution reaction with the cyanide ion leads to the formation of a carbon–carbon bond rather than a carbon–nitrogen bond. The mechanism of this is clearly shown in Fig 12.32. The net effect is to increase the carbon skeleton by one carbon. The resulting functional group is called a **nitrile**.

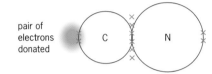

**Fig 12.31** The dot-and-cross diagram for the cyanide ion

**Fig 12.32** The mechanism of the reaction of the cyanide ion with a halogenoalkane

The normal reaction conditions are potassium cyanide or sodium cyanide in ethanol as a solvent. An organic solvent is chosen so that the halogenoalkane dissolves and thus allows more intimate contact with the ionic sodium cyanide. 1-Bromobutane reacts with potassium cyanide in ethanol to form pentanenitrile, so changing a four-carbon skeleton into a five-carbon skeleton.

**Fig 12.33** Changing a four-carbon skeleton into a five-carbon skeleton

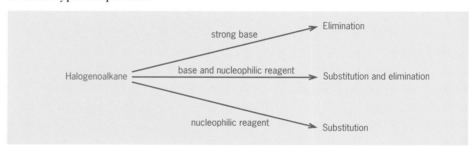

**Fig 12.34** A general example of elimination from a halogenoalkane

**Fig 12.35** The mechanism for the elimination of a hydrogen halide from a halogenoalkane

IN HETEROLYTIC FISSION, A COVALENT BOND IS BROKEN SO THAT BOTH ELECTRONS IN THE BOND MOVE TO ONE OF THE ATOMS (SEE PAGE 311)

**Fig 12.36** A summary of elimination and substitution reactions

SEE QUESTION 11.

## ? TEST YOURSELF

**X** **a)** 1-Iodopropane reacts with hot aqueous sodium hydroxide to give both elimination and substitution products. Draw the displayed formulas of both sets of products, and explain with the aid of equations how they arise.

**b)** 1-Iodopentane is refluxed under heating with ethanolic sodium hydroxide. Predict the major product of the reaction.

SEE CHAPTER 28 FOR INFORMATION ON BASES.

SEE QUESTIONS 12 AND 13.

**Fig 12.37** Elimination products of 3-iodohexane

# ELIMINATION REACTIONS

When halogenoalkanes are heated strongly in the absence of air, an elimination reaction takes place with the formation of an alkene and the hydrogen halide. This reaction is quite impractical to use for manufacture, so the elimination reaction is encouraged by the use of a strong base. Strong bases are proton acceptors and will accept the proton lost during elimination. This, in turn, helps the heterolytic fission of the carbon–halogen bond to give a halide ion. The hydroxide ion is a strong base in ethanol solution. A suitable reagent for the elimination reaction in halogenoalkanes is ethanolic KOH.

Fig 12.35 shows a possible mechanism for the elimination, indicating the help the base gives in the elimination of the proton and the halide ion. Just as in nucleophilic substitution, the actual order of bond making and bond breaking can vary with different halogenoalkanes and different bases. (A full discussion of this is beyond the scope of this book.)

# ELIMINATION VERSUS SUBSTITUTION

Perhaps you may have already thought of this problem: the hydroxide ion is a nucleophile as well as a strong base, which means that in a reaction with a halogenoalkane there is always the possibility of substitution *and* elimination products. It would be more accurate to say that there will always be a mixture of both types of product.

The proportion of elimination and substitution products can be varied by changing both the temperature and the solvent used for the reaction. Table 12.5 summarizes how the proportions can be changed.

**Table 12.5 A summary of changing conditions for competing reactions**

|  | Elimination | Substitution |
|---|---|---|
| Nucleophile | Poor nucleophile | Good nucleophile |
| Basicity | Strong base | Weak base |
| Solvent | Ethanol | Water |
| temperature | High | Low |

There is a further complication: there can often be more than one elimination product because there may be several hydrogen atoms that can be lost.

# 6 REACTIONS OF CHLOROFLUOROCARBONS (CFCS)

The most remarkable thing about CFCs is their complete lack of reactivity towards reagents that react with monohalogenoalkanes. CFCs do not undergo nucleophilic substitution or elimination reactions. This lack of reactivity can be explained by the inability of carbon to expand its octet and by the strength of the C–F bond.

Once CFCs are released into the environment there are no natural ways in which the compounds can be broken down. The normal decomposition processes that involve hydrolysis with water, reaction with oxygen, or bacterial or microbial decay do not take place. This naturally leads to a build-up of CFCs in the environment, and there is a risk that they will increase within the cells of living organisms. In the stratosphere, the CFCs can react, but only because of the presence of ultraviolet light.

## FREE-RADICAL REACTIONS OF CFCS

CFCs contain two types of carbon–halogen bonds: the C–F and the C–Cl bonds. The C–F bond is very strong, as shown by its bond energy of $439 \, kJ \, mol^{-1}$, whereas the C–Cl bond is weaker with a bond energy of $330 \, kJ \, mol^{-1}$. This means that in the presence of ultraviolet light a C–Cl bond, rather than a C–F bond, will undergo homolytic fission to form a chlorine free radical. The equation shows the homolytic fission of a CFC:

$$CF_3Cl(g) \rightarrow CF_3\bullet(g) + Cl\bullet(g)$$

## OZONE DEPLETION

We have seen that the chlorine free radical is very reactive and will react with many other particles. In the stratosphere, it collides with ozone molecules to form chlorine monoxide and oxygen:

$$Cl\bullet(g) + O_3(g) \rightarrow ClO\bullet(g) + O_2$$

Chlorine monoxide is a free radical itself and this reaction is one of the propagation steps. The concentrations of ozone and chlorine monoxide in the stratosphere seem to confirm this reaction. The graphs in Fig 12.38 show that where there is a high chlorine monoxide concentration, there is a low ozone concentration, and vice versa.

Chlorine monoxide reacts with oxygen atoms present in the stratosphere to regenerate the chlorine free radical:

$$ClO\bullet(g) + O(g) \rightarrow Cl\bullet(g) + O_2(g)$$

The regeneration of the chlorine free radical completes the propagation step. The chlorine free radical can then react with more ozone.

The net result of these two reactions is the conversion of an oxygen atom and an ozone molecule into two oxygen molecules:

$$O(g) + O_3(g) \rightarrow 2O_2(g)$$

It is estimated that, on average, one chlorine free radical (hence one CFC molecule) can destroy over 100 000 ozone molecules before the chain reaction stops. It stops when the chlorine free radical collides with other molecules, such as hydrocarbons, to form hydrogen chloride that can leave the stratosphere dissolved in water. The rate of ozone depletion by the presence of chlorine free radicals now exceeds its rate of formation in parts of the stratosphere, especially over the Antarctic.

**YOU CAN READ MORE ABOUT THE LACK OF REACTIVITY OF TETRACHLOROMETHANE IN CHAPTER 21.**

**Fig 12.38** Chlorine monoxide concentration and ozone concentration at different latitudes

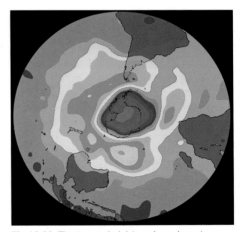

**Fig 12.39** The 'ozone hole' (purple and grey) over the Antarctic. It will increase in size unless the use of chlorofluorocarbons declines. The consequences include damage by excess ultraviolet radiation to skin and the DNA of plant and animal cells, which could lead to mutations

## ? TEST YOURSELF

**Y** Draw the dot-and-cross diagram for chlorine monoxide to show it has one unpaired electron.

**Z** State a function of the chlorine free radical in the process described in the text.

**SEE QUESTION 14.**

## ! SCIENCE FEATURE The nature of ozone

There are two forms of oxygen, $O_2$ and $O_3$. They are called **allotropes**. Ozone, $O_3$, is thermodynamically unstable with respect to oxygen, $O_2$:

$$2O_3(g) \rightarrow 3O_2(g)$$

Ozone is pale blue as a gas and deep blue as a liquid. It has a sharp, irritating odour and produces headaches in humans. Even at one part per million in air, it is poisonous. But the presence of ozone in the stratosphere (a layer 12 to 50 km above the Earth's surface) acts as a protective barrier that prevents much of the high energy UV-B radiation from the Sun

**Fig 12.40** Photochemical smog. In Paris for a period from 1 October 1997, to cut down on pollution cars were restricted from entering the city. A policeman stops a car with a licence plate that ends with an even number and is travelling on an odd-numbered date

reaching the Earth and destroying organisms. It is a strange paradox that we are worried both about an increase in ozone concentration in the lower atmosphere and about a decrease in its concentration in the stratosphere.

Ozone is present in the lower atmosphere, formed there by the reaction between nitrogen oxides and volatile organic compounds in the presence of sunlight. It is produced along with photochemical smogs. In the past 100 years, the levels of ozone concentration have doubled over much of Europe and North America.

The effects of ozone in the lower atmosphere include an estimated 5 to 10 per cent loss in potential crop yield. It damages textiles and makes their colours fade faster. Rubber (such as car tyres) deteriorates faster, and human lungs function less well, as the airways are irritated, which causes coughing.

Ozone in the stratosphere is produced by the reaction of oxygen atoms and oxygen molecules:

$$O(g) + O_2(g) \rightarrow O_3(g)$$

Stratospheric ozone absorbs UV-B radiation by undergoing homolytic fission to form an oxygen atom and an oxygen molecule. If the ozone layer were absent or depleted, much more UV-B would reach the Earth's surface and severely damage plant, marine and human organisms. For example, the incidence of skin cancers and cataracts would increase, crop yields would decrease and fish stocks would plummet.

**YOU CAN READ MORE ABOUT PHOTOCHEMICAL SMOGS ON PAGE 227.**

## SUMMARY

After studying this chapter, you should know that:

- Bromoalkanes, chloroalkanes and iodoalkanes can be prepared by the reaction of alcohols with concentrated hydrobromic acid, concentrated hydrochloric acid and concentrated hydroiodic acid, generated *in situ* by using the sodium halides and concentrated sulphuric acid , respectively, or by the reaction of the appropriate phosphorus halides with an alcohol.
- Alkanes react with chlorine, fluorine and bromine by free-radical substitution to give a mixture of halogeno and polyhalogenoalkanes.
- Free-radical substitution involves three steps: initiation by ultraviolet light, propagation and termination.

- Fluoroalkanes are very unreactive because of the strength of the C—F bond.
- Bromoalkanes, chloroalkanes and iodoalkanes react by nucleophilic substitution with cyanide ion to form nitriles, with ammonia to give amines and with aqueous hydroxide ion to give alcohols.
- Bromoalkanes, chloroalkanes and iodoalkanes can eliminate hydrogen halides to form alkenes when treated with hot ethanolic alkali.
- CFCs and halogenoalkanes have weak permanent dipole–permanent dipole interactions in the liquid state, so often have low boiling points.
- CFCs are non-toxic and inert, and have a low boiling point.
- CFCs are responsible for ozone depletion in the upper atmosphere because they provide chlorine free radicals

# ? EXAM QUESTIONS

**1** Describe the chlorination of methane, stating the conditions and showing the mechanism of the reaction as far as chloromethane.

WJEC Specimen Papers 2001/2002 Paper CH2 Q13c

**2** Ethane and chlorine react in a homolytic **free radical substitution reaction** when exposed to light. The two propagation steps are given in Fig 12.Q2.

**Fig 12.Q2**

**a)** Explain the meaning of the term **free radical**.

**b)** Explain the meaning of the term **substitution reaction**.

**c)** Explain the movement of the C—H bond electron pair in step 1.

**d)** Give an example of a possible termination step in this reaction.

Edexcel A/AS Module Test 2 June 2000 Q2c

**3** Fluoroalkanes and chlorofluoroalkanes are two classes of important industrial chemicals.

**a)** Explain in terms of bond energies why these compounds are relatively inert compared with the corresponding iodo- and bromo-compounds.

**b)** Fluoroalkanes and fluorohalogenoalkanes have many important uses. Give three uses and explain how the chemical and physical properties of these compounds make them suitable for the use given.

**4** [Note: in question 4 you may need to remind yourself about moles and percentage yield by referring to Chapter 1.]

Alcohols can be converted into chloroalkanes by reaction with hydrochloric acid, HCl. 2-Chloro-2-methylpropane can be prepared by shaking together $5.1 \, \text{cm}^3$ (4.0 g) of 2-methylpropan-2-ol with $20 \, \text{cm}^3$ of concentrated HCl. After 10 minutes two separate layers begin to form.

**a) (i)** What is the molecular formula of 2-methylpropan-2-ol, $(CH_3)_3COH$?

**(ii)** Write a balanced equation for the reaction between 2-methylpropan-2-ol and HCl.

Use the data in Table 12.Q4 to answer the questions that follow.

**Table 12.Q4**

| Compound | Relative molecular mass | Density/ g cm$^{-3}$ | Boiling point/°C |
|---|---|---|---|
| 2-Methylpropan-2-ol | 74 | 0.78 | 83 |
| 2-Chloro-2-methylpropane | 92.5 | 0.84 | 51 |
| Water | 18 | 1.0 | 100 |

**b)** Of the two layers formed, one is aqueous and the other contains the organic product. Suggest whether the upper or lower layer is likely to contain the organic product. Explain your reasoning.

**c)** The organic layer was shaken with a dilute solution of sodium hydrogencarbonate, $NaHCO_3$. A gas was given off.
**(i)** Identify the gas that was given off.
**(ii)** Suggest the chemical that could have reacted with the $NaHCO_3$ to form the gas.

**d)** The resulting impure organic liquid was dried with anhydrous calcium chloride and then distilled. 3.75 g of pure 2-chloro-2methylpropane was produced.
**(i)** At what temperature would you expect the pure organic product to distil?
**(ii)** Calculate how many moles of 2-methylpropan-2-ol were used in the experiment.
**(iii)** Calculate how many moles of pure 2-chloro-2-methylpropane were produced.
**(iv)** Calculate the percentage yield of 2-chloro-2-methylpropane in this experiment.

OCR AS Module Chains and Rings Jan 2001 Q5

**5** Classify the type of mechanism that occurs in the reaction represented by:
$$C_2H_5Br + OH^- \rightarrow C_2H_5OH + Br^-$$

WJEC CH2 June 2001 Q13b

**6** Using the appropriate bond enthalpies (energies) suggest why:

**a)** hydroxide ions will react with 1-iodobutane but iodide ions will not react with butan-1-ol,

**b)** fluorocarbons are less susceptible to nucleophilic substitution than iodoalkanes,

**7** Aqueous silver nitrate can be used to monitor the hydrolysis of halogenated compounds. The procedure used is to heat the halogenated compound with aqueous sodium hydroxide and then to test a small sample of the reaction mixture with excess nitric acid, followed by aqueous silver nitrate.

**a)** Describe and explain the results of testing the following substances in this way:

**(i)** 1-iodobutane;
**(ii)** 2-bromopentane;
**(iii)** 3-chlorohexane.

**b)** Rank the following halogenated compounds in order of increasing ease of hydrolysis. Explain how you choose the order.
1-bromopentane, 1-chloropentane, 1-fluoropentane, 1-iodopentane.

**8** Predict the major organic product for the following reactions of 2-bromopentane with:
**a)** warm aqueous potassium hydroxide;
**b)** potassium cyanide in ethanol;
**c)** excess heated ammonia in a sealed tube;
**d)** potassium methoxide, $K^+ \ ^-OCH_3$, in methanol.

**9** The reactions described below refer to compounds with the molecular formula $C_4H_9Br$.
**a)** Write down a general equation to show the hydrolysis of $C_4H_9Br$ using aqueous potassium hydroxide.
**b)** Write down a general equation to show the elimination reaction of $C_4H_9Br$ using ethanolic sodium hydroxide.
**c)** **(i)** Draw the displayed formula and name all the isomers of $C_4H_9Br$.
**(ii)** Which isomer on hydrolysis gives an alcohol with three methyl groups?
**(iii)** Which isomer reacts with concentrated ethanolic sodium hydroxide to give a single hydrocarbon?
**(iv)** Which isomer reacts with excess ammonia to make butylamine?
**(v)** Which isomer reacts with sodium cyanide in ethanol to give 2-methylbutanenitrile?

**10** There are four structural isomers of molecular formula $C_4H_9Br$. The structural formulas of two of these isomers are given in Fig 12.Q10.

**Fig 12.Q10**   isomer 1                     isomer 2

**a)** **(i)** Draw the structural formulas of the remaining two isomers.
**(ii)** Name isomer 1 in Fig 12.Q10.
**b)** **(i)** Give the structural formula and name of the organic product of the reaction between isomer 2 in Fig 12.Q10 and potassium cyanide.

**(ii)** Name and outline the mechanism involved in this reaction.
**(iii)** Suggest why $CH_3CH_2CH_2CH_2F$ reacts less rapidly than $CH_3CH_2CH_2CH_2Br$ with potassium cyanide.

AQA AS Module Unit 3a Introduction to Organic Chemistry June 2001 Q1(a)(b)

**11** The reaction scheme in Fig 12.Q11 shows some of the reactions of methylbenzene and some of its derivatives.

**Fig 12.Q11**

**a)** Give the reagents and essential conditions for each of the three steps.
**b)** Conversion of **B** to **C** is nucleophilic substitution.
**(i)** What is the name of the competing reaction that often takes place along with nucleophilic substitution?
**(ii)** Explain why this competing reaction cannot take place with compound **B**.
**c)** Suggest a reason why methylbenzene is not directly converted into structure **D**.
**d)** Describe how compound **D** can be changed into a compound with eight carbons in its carbon skeleton.

**12**
**a)** A sample of 1-bromopropane is divided into two portions. The first portion is heated with aqueous sodium hydroxide and the second portion is heated with ethanolic sodium hydroxide.
**(i)** Identify the major organic product for each reaction.
**(ii)** Explain the mechanism that leads to each product.
**(iii)** Write a balanced equation for each reaction.
**b)** Starting with 1-bromopropane, outline how you would prepare a sample of
**(i)** propylamine (1-aminopropane),
**(ii)** butanenitrile

**13**
**a)** A compound **L** has the following composition by mass:
C 35.0%; H 6.6%; Br 58.4%
Calculate the empirical formula of **L**. (For help, see page 235.)

**b)** When **L** is treated with dilute aqueous KOH solution, the alcohol 2-methylpropan-1-ol is formed. Draw the structural or displayed formula of this alcohol and hence the displayed formula of **L**.

**c)** When **L** is heated with alcoholic KOH, an alkene **M** is formed. Give the name and displayed formula of **M**.

**14** Some reactions of 1-bromobutane, $CH_3CH_2CH_2CH_2Br$, are shown in Fig 12.Q14.

**Fig 12.Q14**

**a)** For each of the reactions shown, name the reagent and the solvent used.

**b)** Under the same conditions, 1-chlorobutane was used in reaction 1 in place of 1-bromobutane. What difference (if any) would you expect in the rate of reaction? Explain your answer.

**c)** Dichlorodifluoromethane, $CCl_2F_2$, is an example of a chlorofluorocarbon (CFC) that was commonly used as a propellant in aerosols, Nowadays, CFCs have limited use because of the damage they cause to the ozone layer.

**(i)** Draw a diagram to show the shape of a molecule of $CCl_2F_2$.

**(ii)** Predict an approximate value for the bond angles in a molecule of $CCl_2F_2$.

**(iii)** Suggest a property that makes $CCl_2F_2$ suitable as a propellant in an aerosol.

**(iv)** When CFCs are exposed to strong ultraviolet radiation in the upper atmosphere, homolytic fission takes place to produce free radicals. Explain what is meant by the term homolytic fission.

**(v)** Suggest which bond is most likely to be broken when $CCl_2F_2$ is exposed to ultraviolet light. Explain your answer.

**(vi)** Identify two free radicals most likely to be formed when $CCl_2F_2$ is exposed to ultraviolet light.

OCR AS Module 3 Chains and Rings June 2001 Q3

# KEY SKILLS **ASSIGNMENT**

### THE RISE AND FALL OF CFCS

**The first CFC to be discovered was CFC12 in 1892, but it was not until 1928 that Thomas Midgley was able to develop a simple one-step process for making CFCs. He used the reaction of tetrachloromethane with anhydrous hydrogen fluoride and an antimony(V) chloride catalyst:**

$$CCl_4 + xHF \rightarrow CCl_{4-x}F_x + xHCl$$

**1** Write down an equation to show how CFC11 ($CFCl_3$) can be prepared from tetrachloromethane.

**2** **a)** Write down an equation to show how CFC12 ($CF_2Cl_2$) can be prepared from tetrachloromethane
**b)** Use the equation to calculate the minimum mass of tetrachloromethane that is needed to make 1 tonne of CFC12. Hint: You may need to look at page 7.

By 9.55 p.m. on 29 June 1990, the death warrant for CFCs had been signed in London, some 70 years after their first commercial production. During this time, CFCs had been used world-wide in industrial and domestic applications, in the belief that they were inert and harmless to both life and the environment.

The detection of ozone depletion in the stratosphere and the rapid increase in the concentration of CFCs became unquestionably linked. Table 12.A1 shows how the concentrations of CFC have increased up to the first international protocol on the use of CFCs in Montreal in 1987, when the first restrictions on their use came into force.

| Table 12.A1 Atmospheric concentrations of CFCs/parts per trillion | | |
|---|---|---|
| **Year** | **CFC11 ($CCl_3F$)** | **CFC12 ($CCl_2F_2$)** |
| 1976 | 133 | 217 |
| 1978 | 159 | 266 |
| 1980 | 179 | 307 |
| 1982 | 193 | 330 |
| 1984 | 213 | 366 |

**3** Draw a graph of the data in Table 12.A1 and estimate the atmospheric concentration of each CFC in 1994. Assume that there had been no significant changes in production, use and government legislation.

It is not just the concentrations of CFCs that are a cause for concern. It is also the estimated lifetime of these compounds in

# KEY SKILLS **ASSIGNMENT**

the atmosphere and their ozone-depleting potential (ODP). The latter compares the ability of different CFCs to destroy the ozone layer in the stratosphere.

**Table 12.A2 Lifetime and ODP values for some CFCs and related compounds**

| Compound | Lifetime in atmosphere/years | ODP | Per cent contribution to ozone depletion |
|---|---|---|---|
| CFC11 | 74 | 1 | 26 |
| CFC12 | 111 | 1 | 45 |
| CFC113 | 90 | 0.8 | 12 |
| $CCl_4$ | 67 | 1.1 | 8 |
| $CH_3CCl_3$ | 8 | 0.1 | 5 |
| Halon 1301 | 110 | 10 | 4 |

By 1994 the production of CFCs had dropped dramatically, as shown in Fig 12.A1.

**4** Governments agreed by the mid-1990s to cut down and eventually stop the production and use of CFCs within the following few years. Use the data in Fig 12.A1 and Table 12.A3 to estimate the possible changes in atmospheric CFC concentration and ozone depletion.

**5** Halon 1301 ($CBrF_3$) has the highest ODP value.
**a)** Which halogen free radical is most likely to be produced in the stratosphere? Write an equation for this reaction.

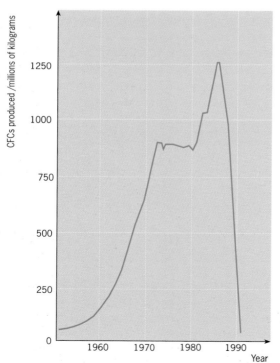

**Fig 12.A1** Change in production of CFCs from 1950 to 1990 (source: DuPont, Worldwatch estimates)

**b)** Use your answer in part a) to suggest why it has the highest ODP of all the values quoted.

Now that the dangers of CFCs are apparent, there is a need to ensure that no more are allowed to escape into the environment. The disposal of CFCs is a problem, since they cannot be dumped at sea or in a land-fill site. One way is to incinerate the waste CFCs with other organic waste materials. This is very difficult because it needs very high temperatures and leads to other dangerous acidic gases polluting the atmosphere. There is also the risk that a CFC will not burn and so will still escape into the atmosphere. Research is underway to find a method of combusting CFCs that does not produce poisonous gases and is 100% efficient.

**6** One way to dispose of CFC is incineration. CFCs burn very poorly except at extremely high temperatures. Suggest why it is very difficult to burn CFCs.

Now that CFCs are being phased out, the chemical industry faces the problem of finding alternatives to them. The obvious replacement is to incorporate hydrogen molecules into CFCs to obtain HCFCs. HCFCs have the same physical properties as CFCs and so offer the same opportunities for use as refrigerants and solvents. But these compounds pose much less of a threat to the ozone layer since the hydrogen atom gives more reactivity to the molecule. However, these products will inevitably still have an effect on the ozone layer, although they have a much reduced ODP value, typically between 0.02 and 0.1.

Unfortunately, there are still several important environmental issues to resolve, since HCFCs react in the atmosphere to produce a cocktail of highly dangerous acidic compounds, such as hydrogen fluoride, hydrogen chloride and trifluoroethanoic acid. One suggestion is that HCFCs should only be used if there is a recycling programme, so that only minute amounts of HCFCs actually escape into the atmosphere.

**7** Describe some of the advantages and disadvantages of using HCFCs rather than CFCs.

**8** Many scientific advances seem to offer a greater quality of life but, often, not many years pass before drawbacks appear. Discuss this statement with respect to CFCs.

**9** The full environmental impact of HCFCs is not yet known and it may take tens of years before it is completely established. Should this stop the introduction of HCFCs now?

The Assignment in Chapter 26 is also concerned with CFCs and ozone depletion

# 13

# Alcohols

# 13 Alcohols

CH₃

CH₃

CH₃

CH₃

CH₃

CH₃

HO

**Cholesterol** The structure is complicated, with four rings and two functional groups

**An artery of the heart (coronary artery)** The wall is orange. The build-up of cholesterol is yellow. The pink between is tissue that the artery wall produces as a reaction to the cholesterol. Blood is confined to the blue area

Cholesterol is a member of the class of chemicals called steroids, which are both naturally occurring and synthetic. The most abundant steroid is cholesterol, which (as the 'ol' at the end of its name suggests) contains an alcohol group. It is found in practically all animal tissues, and forms an important part of cell membranes. Cholesterol is particularly abundant in the brain and spinal cord. It is estimated that the total amount of cholesterol in a 75 kg person is about 250 g.

High levels of cholesterol in the blood are associated with two medical conditions, gallstones and atherosclerosis (hardening and thickening of the arteries when cholesterol from the blood builds up on their inside walls). Blood cholesterol levels can now be measured easily, and people with a high level are advised to reduce their intake of cholesterol-rich foods and saturated fats.

Cholesterol was first isolated in 1770, but it took almost another two centuries before the full structure was established. As the diagram shows, cholesterol has a complicated structure that contains four rings and two functional groups.

## 1 THE HYDROXYL GROUP

Alcohols are organic compounds that contain carbon, hydrogen and oxygen. They contain the hydroxyl functional group, OH, which is an oxygen atom bonded to a hydrogen atom. The hydroxyl group is bonded directly to an alkyl or cycloalkyl group (see Fig 13.1).

There is sometimes confusion in the use of the name alcohol, because it is the name of a class of compounds as well as the name of a particular compound. Alcohol is the popular term for ethanol. But in this book, you can safely assume that when we refer to alcohols we mean the class of compounds – not the single compound.

**PHENOLS ARE COMPOUNDS THAT CONTAIN A HYDROXYL GROUP DIRECTLY ATTACHED TO A BENZENE RING. YOU CAN READ MORE ABOUT PHENOLS ON PAGE 342 IN CHAPTER 15.**

R—O—H

R = alkyl group
e.g. $CH_3CH_2-$

R = cycloalkyl group

e.g.

**Fig 13.1** The generalized structure of alcohols

**MAKE SURE THAT YOU CAN RECOGNIZE AN ALCOHOL WHEN IT IS PART OF A VERY COMPLICATED MOLECULE. LOOK FOR THE HYDROXYL GROUP ATTACHED TO A CARBON ATOM.**

### ▶ EXTENSION Alcohols in nature

Many naturally occurring products are alcohols, including sugars and other carbohydrates, some fragrances, vitamins, pheromones, amino acids and steroids.

Serine is an amino acid used by organisms in the construction of proteins. The structure of serine is shown in Fig 13.2.

H
|
HO—CH₂—C—COOH
|
NH₂

serine: an amino acid that is also an alcohol

**Fig 13.2** The structure of serine

A lack of the alcohol vitamin A in a person's diet causes night blindness and dry skin. And without sufficient vitamin C – another compound that contains alcohol functional groups – a person would suffer from scurvy. The structures of vitamins A and C are shown in Fig 13.4.

Fig 13.3 Citrus fruit, such as the oranges shown here, are rich in vitamin C, which contains the hydroxyl functional group

Vitamin A is a polyene and also a primary alcohol

Vitamin C is a cyclic molecule with more than one alcohol group

Fig 13.4 The structures of vitamin A and vitamin C

The fragrance of scented flowers is the result of a complex mixture of compounds, many of which are alcohols. The structures of three of these are shown in Fig 13.5.

geraniol          nerol          linalool

Fig 13.5 The structures of three of the many compounds responsible for the fragrance of flowers. Geraniol, for example, is a constituent of rose fragrance

Propane-1,2,3-triol is used to make fats and oils that the human body uses as energy stores and insulation, and also in the construction of membranes that surround living cells. Polysaccharides contain many hydroxyl groups, as shown by the structure of the sugar maltose in Fig 13.6.

Fig 13.6 The structure of maltose, a polyhydric alcohol that contains eight hydroxyl groups

## ALCOHOLS IN USE

Alcohols with low relative molecular mass, such as methanol, ethanol, the propanols and the butanols, are liquids and used extensively as solvents, both for chemical reactions and in the production of paints and cosmetics. Many alcohols have the ability to dissolve both polar and non-polar compounds. This property is explained later (see page 289) in terms of the different types of intermolecular attraction between alcohol molecules and other molecules.

Alcohols are also very useful intermediates in the synthesis of compounds, since they have a functional group that can undergo different types of reaction.

ethane-1,2-diol: a component of antifreeze

butan-2-ol: used extensively as a solvent and to make butanone, another solvent

chloramphenicol: a selective bactericide

Fig 13.7 The displayed formulas of three typical alcohols with their main uses.

## NAMING ALCOHOLS

The name of an alcohol is derived from the name of its carbon chain, using the suffix *ol*. The position number for the hydroxyl group must also be specified in the name. The rule is that the carbon atom attached to the hydroxyl group is given the lowest number possible. When there is more than one functional group present, the position numbers are determined by the most important group for naming purposes. (For example, the position of the hydroxyl group is more important than the position of a double bond or of a halogen atom, but less important than the position of a carbonyl group.) The presence of two or more hydroxyl groups is designated by *di*, *tri*, and so on, placed immediately before the suffix *ol*.

**Fig 13.8** Six typical alcohols with their carbon position-numbers

# 2 TYPES OF ALCOHOL

There are three major categories of alcohols: **primary**, **secondary** and **tertiary**. This classification is based on the number of carbon atoms attached to the carbon to which the hydroxyl, OH, group is bonded. A primary alcohol has one carbon atom attached, a secondary alcohol has two carbon atoms attached and a tertiary alcohol has three carbon atoms attached. The significance of this classification is that, although all alcohols have some reactions in common, their other reactions are different and depend on whether they are primary, secondary or tertiary alcohols.

For convenience, methanol is classified as a primary alcohol, even though it does not have a carbon atom attached to the carbon to which the OH group is bonded.

**Fig 13.9** The basic structures of primary, secondary and tertiary alcohols

## ▶ EXTENSION Monohydric, dihydric, trihydric and polyhydric alcohols

Another classification of alcohols is based on the number of hydroxyl groups present per molecule. A monohydric alcohol has one hydroxyl group, a dihydric alcohol has two, a trihydric alcohol has three and a polyhydric alcohol has many hydroxyl groups per molecule.

Ethane-1,2-diol is a dihydric alcohol that is used in car antifreeze. Propane-1,2,3-triol (otherwise known as glycerol) is a trihydric alcohol that is an important component of fats. Glucose, the major source of energy in animals, is a polyhydric alcohol, with five hydroxyl groups per molecule.

## ALCOHOLS AS A HOMOLOGOUS SERIES

Methanol, ethanol, propan-1-ol, butan-1-ol and pentan-1-ol are alcohols that are members of a homologous series (see page 214). They are all primary alcohols with similar chemical properties, and can be represented by the general formula $C_nH_{2n+1}OH$.

As in all homologous series, the physical properties of each member show observable trends as the number of carbon atoms per molecule increases. Table 13.1 summarizes this.

**Table 13.1 Physical properties of the homologous series of primary alcohols**

| Alcohol | Formula | Melting point/°C | Boiling point/°C |
|---|---|---|---|
| Methanol | $CH_3OH$ | −98 | 65 |
| Ethanol | $CH_3CH_2OH$ | −117 | 78.5 |
| Propan-1-ol | $CH_3CH_2CH_2OH$ | −127 | (see Test Yourself D) |
| Butan-1-ol | $CH_3CH_2CH_2CH_2OH$ | −90 | 116 |
| Decan-1-ol | $CH_3(CH_2)_8CH_2OH$ | 6 | 228 |

## BOILING POINTS

Table 13.1 shows that the boiling points of the primary alcohols $C_nH_{2n+1}OH$ increase with the carbon chain length, $n$. For an alcohol to boil, its molecules must overcome the two types of intermolecular force that keep them together. One is the weak induced dipole–induced dipole force that operates between all molecules as a result of the temporary dipoles within the molecules. This force becomes stronger as the molecule becomes larger and contains more electrons. The induced dipole–induced dipole force in alcohols is the attraction between the alkyl part of one molecule and the alkyl part of another molecule.

The other intermolecular force is the much stronger **hydrogen bond**. The hydroxyl group has a polar covalent bond in which the oxygen atom is slightly negative ($\delta-$) and the hydrogen atom is slightly positive ($\delta+$). This is because the oxygen is much more electronegative than hydrogen. This results in the molecule having a permanent dipole. The negative oxygen atom attracts the positive hydrogen atom of the hydroxyl group of another molecule. This is the hydrogen bond.

**Fig 13.11** Hydrogen bonding and induced dipole–induced dipole forces in decan-1-ol. The intermolecular forces of attraction in decan-1-ol are induced dipole–induced dipole forces between alkyl groups and hydrogen bonds between hydroxyl groups. With a long-chain alkyl group, the strength of the induced dipole–induced dipole force is nearly the same as that in the corresponding alkane

## SOLUBILITY IN WATER

The solubility of the primary alcohols $C_nH_{2n+1}OH$ decreases as the number of carbon atoms per molecule increases. Methanol and ethanol mix completely with water in all proportions, whereas nonan-1-ol, in which $n = 9$, is virtually insoluble in water. For a molecule to dissolve in water, it must be able to interact with water molecules. Water is a polar solvent, so the two hydroxyl groups in a water molecule can form intermolecular hydrogen bonds with methanol or ethanol molecules.

In the case of nonan-1-ol, its hydroxyl groups form hydrogen bonds with the water molecules, but its long carbon chains form induced dipole–induced dipole interactions only with other nonan-1-ol molecules. Therefore, the predominant intermolecular attractions are between nonan-1-ol's molecules rather than between nonan-1-ol and water molecules.

**FOR MORE INFORMATION ON INDUCED DIPOLE–INDUCED DIPOLE FORCES, SEE PAGES 159 AND 215.**

**Fig 13.10** Hydrogen bonding in methanol

**FOR INFORMATION ON HYDROGEN BONDING IN WATER, SEE PAGES 133–134.**

## ❓ TEST YOURSELF

**D** Using data from Table 13.1, estimate the boiling point of propan-1-ol.

**E** **a)** Explain why the boiling point of butan-1-ol is higher than that of methanol.
**b)** Explain why the boiling points of decane and decan-1-ol are quite similar, but those of methane and methanol are very different.

**SEE QUESTION 2.**

**Fig 13.12** Intermolecular hydrogen bonds between methanol and water

**THERE IS MORE ABOUT DISSOLVING IN CHAPTERS 21 AND 22.**

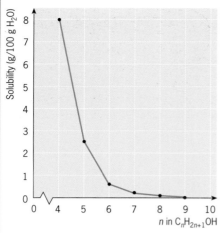

**Fig 13.13** Variation of solubility of the homologous series of primary alcohols in water

## SUGARS

Sugars are a group of naturally occurring polyhydric alcohols that normally have a carbon chain or ring of six or five carbon atoms. They also contain one ketone group or one aldehyde group.

The glucose molecule has five hydroxyl groups. It readily dissolves in water because its hydroxyl groups can form hydrogen bonds with water molecules. It is this solubility that enables us to taste sugar in food or drink. There is more information on sugars and other carbohydrates on pages 358 and 362.

**Fig 13.14** Structures of two simple sugars

**Fig 13.15** Intermolecular forces between a glucose molecule and water molecules. (For the sake of clarity, an interaction is not shown at every hydroxyl group)

---

### ? TEST YOURSELF

**F** Is methanol or nonan-1-ol more soluble in an alkane solvent, such as nonane, than in water? Explain your answer, using ideas about intermolecular forces.

**G** Describe the structural differences between the ring form of glucose in Fig 13.15 and the chain form in Fig 13.14.

---

### ! SCIENCE FEATURE Artificial sweeteners

Diabetics do not produce sufficient insulin to change the excess glucose in their blood into glycogen, a storage carbohydrate. So they must carefully regulate their intake of glucose to stop the blood sugar level from getting too high. But diabetics like sweet-tasting foods. The solution is to use artificial sweeteners, such as sorbitol, which is used to sweeten diabetic chocolate and other confectionery.

**Fig 13.16** The structure of sorbitol

The 3-D shape of the sorbitol molecule is very like that of the glucose molecule. The receptors in the tongue respond to sorbitol in the same way that they respond to glucose, and so sorbitol tastes sweet. However, despite sorbitol's similar taste to glucose, it isn't metabolized by the body, and a diet too rich in sorbitol can cause diarrhoea.

Sorbitol is also used to sweeten dry wines. Adding extra sugar would sweeten the wine, but would also promote unwanted additional fermentation with the production of carbon dioxide, and the pressure could burst the bottle. Since sorbitol is not metabolized by yeast, further fermentation is not a problem.

The market for artificial sweeteners goes beyond helping diabetics and wine-making. Many low-calorie products have now replaced sugar as an ingredient. Many artificial sweeteners chemically decompose with heating, so they cannot be used to sweeten products, such as cake mixtures, which are cooked.

Most artificial sweeteners are much sweeter than sugar, which means that a much smaller mass of the substance will give the same taste. Saccharin, for example, is about 300 times sweeter than sugar, though it has a bitter aftertaste. Another sweetener is aspartame, which is 200 times sweeter than sugar, but some people are seriously affected by it.

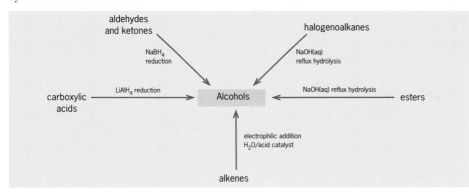

Fig 13.17 The structure of aspartame. Notice that it is completely different from the structure of glucose

These molecules have a different shape from those of the simple sugars, so chemists are not yet sure which molecular shapes are responsible for the sweetness. Recent research has attempted to make sugar-like molecules by substituting some OH groups with chlorine atoms. But there must be extensive testing before such products can be put into foods, in order to ensure that they are completely free of side effects.

? **TEST YOURSELF**

H **a)** Describe the structural differences between sorbitol and glucose.
**b)** Why does sorbitol dissolve in water?

## 3 PREPARATION OF ALCOHOLS

There are five ways to make alcohols, as summarized in Fig 13.18. It follows that alcohols feature in a large number of reactions, which makes them good synthetic intermediates.

**FURTHER INFORMATION ON THE PROCESSES GIVEN IN FIG 13.18 CAN BE FOUND IN THIS BOOK, AS FOLLOWS:**
**HYDROLYSIS OF HALOGENOALKANES, PAGE 275;**
**REDUCTION OF ALDEHYDES AND KETONES, PAGE 356;**
**REDUCTION OF CARBOXYLIC ACIDS, PAGE 383;**
**HYDROLYSIS OF ESTERS, PAGE 396.**

aldehydes and ketones → ($NaBH_4$ reduction) → Alcohols

halogenoalkanes → (NaOH(aq) reflux hydrolysis) → Alcohols

carboxylic acids → ($LiAlH_4$ reduction) → Alcohols

esters → (NaOH(aq) reflux hydrolysis) → Alcohols

alkenes → (electrophilic addition $H_2O$/acid catalyst) → Alcohols

**Fig 13.18** Preparative routes to alcohols

## MANUFACTURE OF ETHANOL

Ethanol is a compound in great demand as a solvent, a fuel and a synthetic intermediate in the production of other chemicals. It is, of course, present in alcoholic drinks, for which it is produced by the fermentation of glucose contained in grapes, for example.

### FERMENTATION OF SUGARS

Ethanol is one of the oldest manufactured chemicals, going back to the time when someone is thought to have first accidentally discovered the fermentation process – possibly 5000 years ago.

During fermentation (anaerobic respiration), aqueous glucose is converted into carbon dioxide and aqueous ethanol. The reaction is catalysed by the enzymes present in yeast:

$$C_6H_{12}O_6(aq) \rightarrow 2CH_3CH_2OH(aq) + 2CO_2(g)$$

The reaction is carried out between temperatures of 25 and 45 °C; any warmer and the enzymes in yeast are denatured and no longer function as a catalyst. It is also important that oxygen is not present, because if so ethanol

**YOU CAN READ MORE ABOUT ENZYMES AND DENATURING ON PAGE 424.**

**Fig 13.19** Yeast floating on the liquid of a brewing fermentation vat

THERE IS MORE INFORMATION ON THE USE OF FERMENTATION TO PRODUCE AN ALTERNATIVE FUEL ON PAGE 190.

FURTHER INFORMATION ON THE ADDITION REACTIONS OF ETHENE IS GIVEN IN CHAPTER 14.

---

### ? TEST YOURSELF

**a)** Write down the equation for the addition reaction between concentrated sulphuric acid and ethene.

**b)** Write down the equation for the hydrolysis of ethyl hydrogensulphate to give ethanol.

**c)** What is the name of the alcohol produced when cyclopentene is hydrated?

SEE QUESTION 3.

---

will be oxidized to produce ethanoic acid. Fermentation does not produce pure ethanol, but instead a dilute solution of ethanol is formed. To obtain higher concentrations of ethanol, the aqueous ethanol has to be fractionally distilled.

### HYDRATION OF ETHENE

We have described the manufacture of ethene from the cracking of larger hydrocarbon molecules on page 208. Large quantities of ethene are used to manufacture ethanol in a reaction known as hydration. This is achieved by reacting ethene and steam together at 300 °C and 6000 kPa (60 atmospheres) pressure in the presence of a phosphoric(V) acid catalyst:

$$C_2H_4(g) + H_2O(g) \rightarrow CH_3CH_2OH(g)$$

Ethene can also be hydrated in the laboratory, but different conditions must be used. Ethene is bubbled through concentrated sulphuric acid and undergoes an addition reaction to form ethyl hydrogensulphate, which is hydrolysed to give ethanol:

$$C_2H_4 + H_2SO_4 \longrightarrow CH_3CH_2OSO_3H$$

$$CH_3CH_2OSO_3H + H_2O \xrightarrow{60\,°C} H_2SO_4 + CH_3CH_2OH$$

**Fig 13.20** Flow diagram showing hydration of ethane

### FERMENTATION OR HYDRATION?

The manufacture of ethanol by fermentation and by hydration each has its advantages and disadvantages. Ethanol made by fermentation is a renewable resource, since the sugar required for the process comes from plants that can be grown again in the following season to make more sugar. However, the hydration process uses ethene that is made by cracking of long-chain hydrocarbons obtained from crude oil, which is a non-renewable resource. Fermentation is a batch process, so the quality of the product cannot be controlled as easily as in the continuous process of hydration. The batch process of fermentation allows a chemical company to match supply with demand fairly easily, since extra batches can be included if there is a need for extra ethanol.

Hydration of ethene produces pure ethanol, whereas fermentation produces a dilute solution of ethanol, which must be purified by distillation.

### ⏵ EXTENSION The manufacture of methanol

Methanol is used in industry and in the laboratory as a solvent and reagent; also as car antifreeze and as an engine fuel. Though an alcohol, methanol is poisonous and can cause blindness if inhaled or drunk.

Until about 1925, methanol was made by heating wood in the absence of air, and collecting and condensing the volatile substances given off. This is why methanol is sometimes called wood spirit. It is now manufactured by

the reaction of carbon monoxide or carbon dioxide with hydrogen. Both reactions are carried out at about 400 °C and under a pressure of some 1500 kPa, using heterogeneous catalysts (for more about heterogeneous catalysts, see Chapter 26):

$$CO(g) + 2H_2(g) \rightarrow CH_3OH(g)$$

The catalysed conversion of either carbon monoxide or carbon dioxide into methanol provides a route (at least in theory) for converting the products of combustion of a fuel into a practical liquid fuel. The methanol itself can be further converted into other useful products.

These synthetic routes provide a way for alkanes and alkenes to be synthesized without the need of a raw material such as crude oil. Perhaps more research will have to be directed in this direction if people are going to continue to rely on hydrocarbons as an energy source and as a raw material for the petrochemical industry.

**Fig 13.21** The catalytic conversion of methanol

# 4 REACTIONS OF ALCOHOLS

## THE NATURE OF THE HYDROXYL FUNCTIONAL GROUP

The hydroxyl functional group in an alcohol contains one covalent bond and is attached to a carbon atom by another covalent bond. This means that the functional group can undergo two types of reaction. The first involves the breaking of the carbon–oxygen, C–O, single bond, and the second involves the breaking of the oxygen–hydrogen, O–H, single bond.

The C–O bond can be broken either by a substitution reaction or by an elimination reaction.

When the O–H bond is broken, an alcohol behaves as an acid. This is because the breaking of the bond results in the transfer of a hydrogen ion. Since the oxygen atom of the hydroxyl group has two lone pairs of electrons, an alcohol can also behave as a nucleophile, reacting with a molecule that has an electron-deficient centre. Finally, an alcohol can take part in redox reactions.

The large range of reactions available to this functional group helps to explain why alcohols are very good synthetic intermediates.

**Fig 13.22** General summary of the reactions of alcohols. $R^1$, $R^2$, $R^3$ and $R^4$ are either alkyl groups or hydrogen atoms

## SUBSTITUTION REACTIONS: HALOGENATION OF ALCOHOLS

The conversion of an alcohol into a halogenoalkane is a substitution reaction that involves the swapping of a halogen atom for a hydroxyl group. However, a halide ion is not a good enough nucleophile to substitute a hydroxyl group directly. The strength of the C–O bond is a great barrier to the substitution reaction.

THE COMPETITION BETWEEN SUBSTITUTION AND ELIMINATION IN HALOGENOALKANES IS DISCUSSED ON PAGE 278.

THE ROLE OF BOND ENERGY IN
NUCLEOPHILIC SUBSTITUTION IS
COVERED ON PAGE 274.

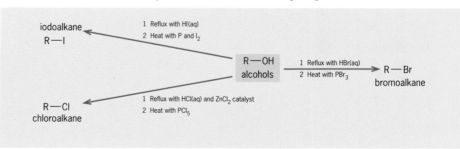

Fig 13.23 The mechanism of the substitution of an alcohol using concentrated hydrobromic acid

The only way that the substitution reaction can succeed is if the C—O bond is substantially weakened. An acidic catalyst can be used to do this.

A lone pair of electrons on the oxygen atom is donated to a proton in an acid–base type interaction. The intermediate formed has a very weak C—O bond, since the oxygen has developed a positive charge. Nucleophilic substitution by a bromide ion is now very easy, with the water molecule as the leaving group. Normally, the reaction is carried out by refluxing together a mixture of alcohol, sodium bromide and concentrated sulphuric acid. The concentrated sulphuric acid and the sodium bromide together generate concentrated hydrobromic acid:

$$H_2SO_4 + NaBr \rightarrow HBr + NaHSO_4$$

For the substitution of the hydroxyl group by a halogen atom, the corresponding acid can be used. Iodo compounds are prepared by refluxing the alcohol with concentrated hydroiodic acid, bromo compounds by refluxing with concentrated hydrobromic acid and a trace of concentrated sulphuric acid as a catalyst, and chloro compounds by refluxing with concentrated hydrochloric acid using zinc chloride as a catalyst.

There are alternative ways to achieve this substitution. All of them involve weakening the C—O bond by a reaction with a phosphorus atom. Typically, phosphorus(V) chloride, phosphorus(III) bromide, or a mixture of phosphorus and iodine (essentially phosphorus(III) iodide) is refluxed with the alcohol to obtain the corresponding halogenoalkane. Representing an alcohol by the formula ROH, where R is an alkyl group, the equations for these three reactions are:

$$PCl_5 + ROH \rightarrow RCl + POCl_3 + HCl$$
$$PBr_3 + 3ROH \rightarrow 3RBr + H_3PO_3$$
$$PI_3 + 3ROH \rightarrow 3RI + H_3PO_3$$

Fig 13.24 Summary of the preparation of halogenoalkanes from alcohols

**TEST YOURSELF**

**K** Predict the products of the following reactions:
**a)** pentan-2-ol refluxed with concentrated hydrobromic acid and a trace of concentrated sulphuric acid;
**b)** cyclohexanol refluxed with concentrated hydroiodic acid;
**c)** propan-1-ol refluxed with concentrated hydrochloric acid and zinc chloride catalyst.

**TEST YOURSELF**

**L** How would you prepare:
**a)** iodoethane;
**b)** 2-chloropentane,
**c)** 2-iodo-3-methylhexane?

**TEST YOURSELF**

**M** A compound has the formula $C_2H_6O$. It does not give fumes of hydrogen chloride when reacted with phosphorus(V) chloride. Suggest an identity for the compound.

### EXTENSION Phosphorus[V] chloride for working out structures

Phosphorus(V) chloride reacts with hydroxyl groups in alcohols and carboxylic acids to produce fumes of hydrogen chloride, which are easily detected. So this reaction is used to analyse compounds to identify hydroxyl groups and establish whether they are alcohols or carboxylic acids.

# ELIMINATION REACTIONS: DEHYDRATION TO FORM ALKENES

As we have already described, alcohols can react by substitution and elimination. The elimination reaction is more frequently referred to as dehydration, since it involves the loss of a molecule of water as well as the formation of an alkene (Fig 13.25). In dehydration, an alcohol is heated with a suitable catalyst. The catalysts are often dehydrating agents, such as concentrated sulphuric acid or concentrated phosphoric(V) acid.

## LABORATORY DEHYDRATION

In the laboratory, it is convenient to take an alcohol and reflux it with a dehydrating agent. Concentrated sulphuric acid can be used, but concentrated phosphoric(V) acid is usually preferred, because it minimizes the number of side products formed. Fig 13.26 shows the dehydration of cyclohexanol to give cyclohexene.

With some alcohols, several alkenes may be produced, but normally the most-substituted alkene is the major product.

**Fig 13.27** The dehydration products of 2-methylpentan-2-ol

# 5 OXIDATION OF ALCOHOLS

As a working definition in organic chemistry, *oxidation* can be considered as either the gain of an oxygen atom by a substance or the loss of two hydrogen atoms from it. (To memorize this, think of water, $H_2O$: one oxygen and two hydrogen atoms.)

Since reduction is the chemical opposite of oxidation, *reduction* can be considered as either the gain of two hydrogen atoms or the loss of one oxygen atom.

These atoms are always denoted by [O] and 2[H]. The square brackets signify that we are not referring to atomic hydrogen or oxygen. The substance that supplies the oxygen, [O], and/or removes the hydrogen, 2[H], is called the **oxidizing agent** or oxidant. The **reducing agent** or reductant removes the oxygen, [O], and/or supplies the hydrogen, 2[H].

**Fig 13.25** The dehydration of an alcohol to form an alkene

THERE IS MORE ON THE DEHYDRATION OF ALCOHOLS ON PAGE 310.

**Fig 13.26** The dehydration of cyclohexanol to give cyclohexene

## ? TEST YOURSELF

**N**
**a)** Draw the dehydration product of cyclohexanol.
**b)** Draw all the alkenes that can be made by the dehydration of butan-2-ol.
**c)** Which of these alkenes is likely to be formed in the greatest amount?

SEE QUESTION 4.

TO READ ABOUT REDOX REACTIONS IN MORE DETAIL, LOOK AT CHAPTER 5.

## ▶ EXTENSION Stoichiometric and half equations

The stoichiometric equation for the oxidation of a primary alcohol is very complicated and tends to hide the processes taking place. For example:

$$3RCH_2OH + K_2Cr_2O_7 + 4H_2SO_4 \rightarrow 3RCHO + K_2SO_4 + Cr_2(SO_4)_3 + 7H_2O$$

Even the slightly simpler ionic equation is still complicated:

$$3RCH_2OH + Cr_2O_7^{2-} + 8H^+ \rightarrow 3RCHO + 2Cr^{3+} + 7H_2O$$

It is easier to look at the oxidation half equation for the reaction (see Chapter 21). During the reaction, the alcohol loses two electrons, as in the half equation:

$$RCH_2OH \rightarrow RCHO + 2H^+ + 2e^-$$

The electrons supplied by the oxidation of the alcohol reduce the dichromate(VI) ion.

$$Cr_2O_7^{2-} + 14H^+ + 6e^- \rightarrow 2Cr^{3+} + 7H_2O$$

Fig 13.28 The formation of butanal from butan-1-ol

K₂Cr₂O₇(aq)
H₂SO₄(aq)
25 °C

geraniol          geranial

Fig 13.29 The oxidation of geraniol to geranial completely changes the odour of the molecule

## OXIDATION OF PRIMARY ALCOHOLS INTO ALDEHYDES

Consider butan-1-ol (Fig 13.28). The removal of two hydrogen atoms produces a new functional group – the aldehyde group, CHO.

In the oxidation of a primary alcohol, the reagent is acidified aqueous potassium dichromate(VI), a mixture of aqueous potassium dichromate(VI) and dilute sulphuric acid.

The conditions must be chosen to prevent any subsequent oxidation of the aldehyde to form a carboxylic acid. Either the reaction is carried out at room temperature and the oxidizing agent is added to the alcohol so that the oxidizing agent is never in excess, or the reagents are heated and the aldehyde product is allowed to distil out from the reaction vessel before it can be oxidized further.

During the oxidation, orange dichromate(VI) ions are reduced to form blue–green chromium(III) ions. The equation can be written as:

$$RCH_2OH + [O] \rightarrow RCHO + H_2O$$

In this equation, the oxidizing agent is not given in full, but is represented by the oxygen [O] that it supplies to remove the two hydrogen atoms. This is a much simpler expression of the reaction than the full stoichiometric or ionic equation for the reaction.

The oxidation of ethanol by acidified potassium dichromate(VI) to give ethanal can be written as:

$$CH_3CH_2OH + [O] \rightarrow CH_3CHO + H_2O$$

The primary alcohols found in rose oil can be oxidized in the same way to give the corresponding aldehydes, as shown in Fig 13.29. It is interesting to note that this change of functional group alters the rose scent to that of a citrus fruit.

## OXIDATION OF PRIMARY ALCOHOLS INTO CARBOXYLIC ACIDS

By changing the conditions of oxidation, it is possible to remove two hydrogen atoms and gain one oxygen atom to produce a carboxylic acid. The oxidation is represented by:

$$RCH_2OH + 2[O] \rightarrow RCO_2H + H_2O$$

The alcohol is refluxed with acidified potassium dichromate(VI). The acid used is dilute sulphuric acid.

Ethanol is oxidized to give ethanoic acid by refluxing with acidified potassium dichromate(VI). The reaction can be represented as:

$$CH_3CH_2OH + 2[O] \rightarrow CH_3COOH + H_2O$$

The half equation for the oxidation of ethanol into the carboxylic acid involves the loss of four electrons from ethanol. This corresponds to the gain of [O] and the loss of 2[H]:

$$H_2O + CH_3CH_2OH \rightarrow CH_3COOH + 4H^+ + 4e^-$$

Ethanol can sometimes be oxidized by atmospheric oxygen. This can happen during fermentation, giving the ethanol a vinegary taste:

$$CH_3CH_2OH + O_2 \rightarrow CH_3COOH + H_2O$$

Fig 13.30 The ethanoic acid in wine vinegars is obtained by the aerial oxidation of ethanol

# OXIDATION OF SECONDARY ALCOHOLS INTO KETONES

Secondary alcohols can be oxidized to give ketones rather than aldehydes. The alcohol is refluxed with the acidified potassium dichromate(VI). Sulphuric acid is used to acidify the reagent.

The reaction involves the removal of two hydrogen atoms. So, for example, propan-2-ol is oxidized to give propanone as follows:

$$3CH_3CHOHCH_3 + Cr_2O_7^{2-} + 8H^+ \rightarrow 3CH_3COCH_3 + 2Cr^{3+} + 7H_2O$$

Notice that we have used the ionic formulas and left out the potassium ions, which remain unchanged during the reaction. The half equation for the oxidation of a secondary alcohol involves two electrons and is similar to the half equation for ethanol being oxidized to give ethanal. Another, simpler way to represent the oxidation of propan-2-ol is:

$$CH_3CHOHCH_3 + [O] \rightarrow CH_3COCH_3 + H_2O$$

**Fig 13.31** Equation describing the oxidation of butan-2-ol

---

**!** **SCIENCE FEATURE  Cutting costs to make propanone**

Propanone is an important industrial and domestic solvent. It is manufactured from propan-2-ol, but acidified potassium dichromate(VI) is not used because the reagents are expensive and the special conditions required are difficult to maintain in a large industrial plant. Instead, propan-2-ol vapour is passed over a copper catalyst at 300 °C, which removes hydrogen molecules in a reaction called dehydrogenation (Fig 13.32).

$$CH_3-\underset{OH}{\underset{|}{CH}}-CH_3 \xrightarrow[\text{heat}]{\substack{Cu \\ \text{catalyst}}} CH_3-\underset{O}{\underset{\|}{C}}-CH_3 + H_2$$

**Fig 13.32** The dehydrogenation of propan-2-ol

---

# OXIDATION OF TERTIARY ALCOHOLS

It is impossible to oxidize a tertiary alcohol without breaking a carbon–carbon bond. Acidified potassium dichromate(VI) has no effect on tertiary alcohols. (Look back to Fig 13.9 for the structure of a tertiary alcohol.)

# DISTINGUISHING PRIMARY, SECONDARY AND TERTIARY ALCOHOLS

Since the three classes of alcohols give different reactions with acidified potassium dichromate(VI) they can be distinguished from one another by the products and colour changes in these reactions, as Table 13.2 shows.

**?** **TEST YOURSELF**

**S** **a)** What is the name of the organic product when cyclohexanol reacts with acidified potassium dichromate(VI)?
**b)** Write an equation for the oxidation of butan-2-ol by acidified dichromate(VI) ions.
Hint: Use the given equation and substitute butan-2-ol for propan-2-ol.

**T** Write down a possible product from the reaction of vitamin C (Fig 13.4) with acidified potassium dichromate(VI).

**?** **TEST YOURSELF**

**U** Name the product, if any, when each one of the following is mixed with acidified potassium dichromate(VI) at room temperature.
**a)** butan-1-ol; **b)** butan-2-ol;
**c)** 2-methylpropan-2-ol

---

**Table 13.2 Distinguishing alcohols**

| Alcohol | Acidified potassium dichromate(VI) at room temperature | | Refluxing under heating with acidified potassium dichromate(VI) | |
|---|---|---|---|---|
| | Product | Observation | Product | Observation |
| Primary | Aldehyde | Orange to green | Carboxylic acid | Orange to green |
| Secondary | Ketone | Orange to green | Ketone | Orange to green |
| Tertiary | No reaction | Stays orange | No reaction | Stays orange |

Propan-1-ol (primary alchohol)  Propan-2-ol (secondary alchohol)  2-methyl-propan-2-ol (tertiary alchohol)

**Fig 13.33** During the oxidation of primary and secondary alcohols using acidified potassium dichromate(VI), the colour changes from **orange** to **green, but tertiary alcohols do not change**

**SEE QUESTIONS 6–11.**

**TESTS TO DISTINGUISH BETWEEN ALDEHYDES AND KETONES ARE COVERED ON PAGES 355–358.**

### ? TEST YOURSELF

**V** A straight-chain alcohol with the formula $C_5H_{11}OH$ reacts with acidified potassium dichromate(VI) to give a ketone and, in a separate test, also gives a yellow precipitate with alkaline aqueous iodine. Suggest an identity for the alcohol that is consistent with this data.

The methyl group is electron releasing compared with a hydrogen atom, so it strengthens the O–H bond

$pK_a = 15.5$

Fluorine atoms are highly electronegative and withdraw electrons from the O–H bond, which weakens it

$pK_a = 12.4$

**Fig 13.34** Structures and $pK_a$ values for two alcohols: the larger the value of $pK_a$ the weaker the acid. (Read about $pK_a$ on page 368.)

### ? TEST YOURSELF

**W** Explain why 2-chloroethanol is a stronger acid than ethanol. Hint: What is the effect of the very electronegative chlorine atom on the strength of the O–H bond?

**READ MORE ABOUT THE STRENGTHS OF ACIDS AND WEAK ACIDS IN CHAPTER 17.**

### ? TEST YOURSELF

**X** Write down the equation for the reaction of methanol with sodium. What is the name of the alkoxide produced?

**SEE QUESTION 12.**

### ✔ REMEMBER THIS

Heterolytic fission involves breaking a covalent bond so that one atom receives both of the shared pair of electrons from the covalent bond.

**THE REACTIONS OF ALCOHOLS WITH CARBOXYLIC ACIDS AND THEIR DERIVATIVES ARE DESCRIBED IN MORE DETAIL IN CHAPTER 18.**

## TRI-IODOMETHANE TEST FOR CH₃CHOH GROUP

Alkaline aqueous iodine reacts with ethanol and secondary alcohols that contain the $CH_3CHOH$ group to give a yellow precipitate of tri-iodomethane. This reaction is used as a chemical test in analysis and identification (see also page 358).

## 6 ALCOHOLS AS ACIDS

### RELATIVE ACIDITY OF ALCOHOLS

An acid is a proton donor. An alcohol is an acid because it can donate a proton if the O–H bond is broken. So, the weaker the O–H bond is, the stronger is the acidity of the alcohol. The strength of the O–H bond can be changed by changing the atoms attached to it. For example, electron-releasing groups strengthen the O–H bond and thereby make the alcohol a weaker acid.

Alkyl groups are electron releasing compared with the hydrogen atom, so it follows that the order of increasing acid strength in alcohols is tertiary, secondary, primary. Water, which also contains the O–H bond, is a stronger acid than primary alcohols.

### REACTION OF SODIUM WITH ALCOHOLS

Sodium reacts violently with water to form hydrogen and aqueous sodium hydroxide. During the reaction, an O–H bond is broken:

$$2H_2O(l) + 2Na(s) \rightarrow 2NaOH(aq) + H_2(g)$$

Ethanol reacts in the same way with sodium, to produce hydrogen and a compound called sodium ethoxide. During the reaction, the sodium fizzes and produces a colourless solution. The reaction is slower than that of sodium with water, because the O–H bond in ethanol is harder to break than the O–H bond in water:

$$2C_2H_5OH + 2Na \rightarrow 2C_2H_5O^-Na^+ + H_2$$

Notice that the two equations show the same pattern.

Sodium reacts with other alcohols, to produce hydrogen and what is known as an alkoxide. The name of the alkoxide is derived from the alkyl chain of the alcohol:

$$2ROH + 2Na \rightarrow 2RO^-Na^+ + H_2$$

The alcohol in this example acts as a very weak acid, since it loses a proton. $RO^-Na^+$ is known as a sodium alkoxide.

## 7 ALCOHOLS AS NUCLEOPHILES

The oxygen atom of the OH group can donate a pair of electrons to make a covalent bond, which allows the group to act as a nucleophile.

Alcohols are poor nucleophiles, but they can react with electron-deficient centres provided the deficiency is sufficiently high. Typically, alcohols behave as nucleophiles in their reactions with carboxylic acids or their derivatives. Probably the most important reaction is that with a carboxylic acid to form an **ester**. Esters are sweet-smelling substances used in perfumes and in food flavourings.

**Fig 13.35** The esterification reaction

**Fig 13.36** Reaction network showing an alcohol behaving as a nucleophile

## SUMMARY

After studying this chapter, you should know:

- Alcohols contain the hydroxyl, OH, functional group attached directly to a carbon atom.
- Short-chain alcohols are soluble in water, but long-chain alcohols are insoluble.
- Primary alcohols can be oxidized by acidified potassium dichromate(VI) to give aldehydes. The aldehyde is often distilled from the reaction mixture as it is being made to prevent further oxidation.
- Primary alcohols can be oxidized by hot acidified potassium dichromate(VI) to give carboxylic acids.
- Secondary alcohols can be oxidized to give ketones, but tertiary alcohols cannot be oxidized.

- Alcohols can be converted by substitution into halogenoalkanes, using phosphorus halides and the appropriate hydrohalic acid, HBr(aq), HCl(aq) or HI(aq).
- Alcohols can be dehydrated to give alkenes.
- Alcohols can behave as nucleophiles and react with carboxylic acids and acid chlorides to form esters.
- Many alcohols are used as solvents.
- Ethanol is manufactured by the anaerobic fermentation of aqueous glucose or by the hydration of ethene in the presence of an acid catalyst.
- Alcohols react with sodium to give sodium alkoxides.

## ? EXAM QUESTIONS

**1** Name the structure of this compound:
$$CH_3-CH_2-CH_2-CH_2OH$$
Edexcel Nuffield Specimen Unit Test 1 January 2000 Q6(a)

**2** Table 13.Q2 shows information about some alcohols that form part of an homologous series.

**Table 13.Q2**

| Name | Formula | Boiling point/°C | Relative molecular mass |
|------|---------|------------------|-------------------------|
| Methanol | $CH_3OH$ | 65 | 32 |
| Ethanol | $C_2H_5OH$ | 78 | 46 |
| Propan-1-ol | $C_3H_7OH$ | 97 | 60 |
| Butano-1-ol | $C_4H_9OH$ | | 74 |
| Pentan-1-ol | $C_5H_{11}OH$ | 138 | |
| Hexan-1-ol | $C_6H_{13}OH$ | 158 | 102 |

**a)** Identify the functional group common to all alcohols.

**b)** What is the general formula for these alcohols?

**c)** What is the formula of the next alcohol in the series?
OCR AS Module Chains and Rings Jan 2001 Q1(a)

**3** There are two different industrial methods for the production of ethanol. Describe and explain, with the aid of equations, the industrial production of ethanol from glucose and from ethene.
OCR AS Module Chains and Rings Jan 2001 Q7

**4** Sulphuric acid is used in the laboratory as a dehydrating agent to convert primary alcohols into alkenes.

**a)** Draw the graphic (full structural) formula for 2-methylbutan-1-ol.

**b)** Write the graphic (full structural) formula and the systematic name for the alkene formed when 2-methyl-butan-1-ol reacts with an excess of sulphuric acid.

**c)** During one stage of the dehydration of a primary alcohol to give an alkene, the protonated alcohol molecule undergoes *nucleophilic* attack by a hydrogensulphate ion. State what is meant by the term *nucleophile*.
WJEC Specimen Paper CH2 2001/2 Q14(d)

**5** Ethane-1,2-diol, $CH_2OHCH_2OH$, has been used illegally to sweeten wine, with serious consequences, as this is oxidized to give toxic ethanedioic acid, $(COOH)_2$, in the body.

## ? EXAM QUESTIONS

**a)** Draw the graphic (full structural) formula of an isomer of ethane-1,2-diol which is **not** an alcohol.

**b)** Describe how ethane-1,2-diol can be oxidized to give ethanedioic acid in the laboratory, assuming that the reaction used is the general method for the oxidation of primary alcohols. Your answer should include the oxidizing agent used, the conditions and relevant observations.

WJEC Specimen Paper CH2 2001/2 Q12(a)

**6** One of the isomers of $C_4H_{10}$ is the alcohol 2-methylpropan-2-ol which has the structural formula given in Fig 13.Q6.

$$CH_3-\underset{\underset{OH}{|}}{\overset{\overset{CH_3}{|}}{C}}-CH_3$$

**Fig 13.Q6**

**a)** There are three other **structural** isomers of $C_4H_{10}O$ that are also alcohols.
**(i)** Draw their structural formulas.
**(ii)** Describe a test to show that each of the isomers in **(i)** contains an OH group.

**b)** Draw the structural formula of the final organic product of the reaction when each of the three alcohols in **a) (i)** is heated under reflux with a solution of potassium dichromate(VI) in dilute sulphuric acid.

Edexcel Module Test 2 Jan 2001 Q1 (not a(ii), c)

**7**

**a)** **(i)** Pentan-1-ol, $CH_3CH_2CH_2CH_2CH_2OH$, was oxidized by adding it dropwise to acidified potassium dichromate(VI) and distilling off the organic product immediately. Write an equation for this reaction, showing clearly the structural formula of the organic product. You may use [O] to represent the oxidizing agent.
**(ii)** Pentan-2-ol, $CH_3CH_2CH_2CH(OH)CH_3$, was oxidized by heating it under reflux with acidified potassium dichromate(VI). Write an equation for this reaction, showing clearly the structural formula of the organic product. You may use [O] to represent the oxidizing agent.

AQA Unit 3(a) Introduction to Organic Chemistry June 2001 Q2(a)

**8** **a)** Draw the displayed formulas and name all the alcohols with the molecular formula $C_4H_{10}O$. Classify each alcohol as primary, secondary or tertiary.

**b)** Explain how you would distinguish experimentally between the primary, secondary and tertiary alcohols of $C_4H_{10}O$.

**9** Cyclohexene can be converted into cyclohexanol.

**a)** Draw the displayed formula for cyclohexene and cyclohexanol.

**b)** **(i)** Describe briefly how cyclohexene can be converted into cyclohexanol.

**(ii)** Write an equation for this reaction
**(iii)** What type of reaction is this?

**c)** Cyclohexene has a boiling point of 83 °C, but cyclohexanol has a boiling point of 161 °C. Explain why cyclohexanol has a much higher boiling point than cyclohexene.

**10** Predict the products for each of the following reactions. For each reaction give a detailed explanation for your choice of product.

**a)** $CH_3CH_2CH_2CH_2CH_2CH_2OH \xrightarrow{\text{HBr(aq), reflux}}$

**b)** $CH_3CH_2CH_2CH_2CH_2CH_2OH \xrightarrow[\text{H}_2\text{SO}_4\text{(aq) 0–5 °C}]{\text{K}_2\text{Cr}_2\text{O}_7\text{(aq)}}$

**c)** $CH_3CH_2CH_2CH_2CH_2CH_2OH \xrightarrow{\text{conc H}_2\text{SO}_4,\ 130\ °\text{C}}$

**d)** $CH_3CH_2CH_2CH_2CH_2CH_2OH \xrightarrow[\text{reflux under heating}]{\text{KMnO}_4\text{(aq)/H}_2\text{SO}_4\text{(aq)}}$

**11** Samples of butan-1-ol, butan-2-ol and 2-methylpropan-2-ol were reacted under reflux with aqueous acidified potassium dichromate(VI).

**a)** Copy and complete Table 13.Q11 for each of the three alcohols. State what you would observe and identify the organic product of each reaction.

**Table 13.Q11**

| Alcohol | Observation | Organic product (if any) |
|---|---|---|
| butan-1-ol | | |
| butan-2-ol | | |
| 2-methylpropan-2-ol | | |

**b)** Describe a simple chemical test by which you could distinguish between the organic products you have given in **a)** for butan-1-ol and for butan-2-ol.

**12 a)** State the type of reaction undergone and give the structural formula of each of the organic products obtained when propan-2-ol reacts with the following reagents:
**(i)** HBr;   **(ii)** sodium metal;
**(iii)** alkaline aqueous iodine.

**b)** Alcohol **A** has esters that are responsible for the flavours of various fruits and has the molecular formula $C_5H_{12}O$. Reaction of **A** with acidified potassium dichromate(VI) produces a compound **B**, $C_5H_{10}O_2$. Heating **A** over $Al_2O_3$ produces compound **C**, $C_5H_{10}$. Vigorous oxidation of **C** forms 2-methylpropanoic acid as one of the products. Suggest structures for **A**, **B** and **C**, and explain the reactions involved.

# KEY SKILLS **ASSIGNMENT**

## FERMENTATION AND LOW-ALCOHOL BEERS

**The fermentation processes involved in bread-making, wine-making and brewing are among the oldest applications of chemistry. But chemists did not begin to understand what happens during fermentation until the 1800s. In 1810, Gay-Lussac published the equation for the conversion of sugar into ethanol and carbon dioxide, but almost 90 years elapsed before Buchner demonstrated that yeast extract without yeast cells would ferment aqueous glucose. We now realize that the reaction is catalysed by many enzymes, collectively called zymase.**

**1** Explain the importance of fermentation in bread-making. Glucose can exist as a chain form and a ring form, shown in Fig 13.A1. In aqueous solution, it is almost entirely in the ring form.

chain structure of glucose

ring structure of glucose

**Fig 13.A1** Displayed formulas of the ring and chain forms of glucose

**2** **a)** The number 1 carbon in the chain structure is part of the aldehyde functional group. Give the systematic name for the chain structure of glucose. Hint: Read the Assignment on page 362.
**b)** Compare the structures of the two forms of glucose. You may want to consider the shape of each molecule, isomerism and the functional groups present.

The fermentation of aqueous glucose requires anaerobic conditions to make ethanol. Also, the temperature must not rise significantly above 32 °C, otherwise the rate of fermentation slows down and the reaction eventually stops.

**3** **a)** Write down an equation to represent the fermentation of glucose to form ethanol and carbon dioxide.
**b)** Suggest the names of the products when the fermentation is carried out in the presence of oxygen.
**c)** Explain why the fermentation of aqueous glucose slows down and eventually stops when the temperature is raised significantly above 32 °C. Hint: You may want to read about enzymes in Chapter 19.
**d)** The fermentation of aqueous glucose is an exothermic process. What precautions must brewers take to ensure efficient fermentation?

The manufacture of alcoholic drinks accounts for much of the ethanol produced by fermentation. Once fermentation is finished, beer, lager and wine need only to be bottled, canned or casked. Spirits such as whisky require a higher concentration of ethanol. This is achieved by distilling the fermentation mixture.

Since the mid-1980s, there has been a significant increase in the consumption of low-alcohol and alcohol-free beers (Table 13.A1). Low-alcohol beer is manufactured in two ways. One method is to ferment partially and then stop the process (Fig 13.A2). The second method is to ferment fully and then remove the alcohol (Fig 13.A3).

**Table 13.A1 Categories of beer**

| Category | Percentage by volume of ethanol |
|---|---|
| Alcohol-free | No more than 0.05 |
| De-alcoholized | 0.05–0.5 |
| Low-alcohol | 0.5–1.2 |
| Typical beer | 4–5 |

**Fig 13.A2** Making low-alcohol beer by partial fermentation

**Fig 13.A3** Making low-alcohol beer by distillation at reduced pressure

# KEY SKILLS **ASSIGNMENT**

Partial fermentation produces low-alcohol beer rather than alcohol-free beer. The method has two drawbacks: the beer tastes sugary, and it is susceptible to microbial contamination. On the other hand, full fermentation followed by removal of the ethanol allows both low-alcohol and alcohol-free beer to be brewed. The ethanol is removed by distillation, which is carried out at reduced pressure so that the esters responsible for taste and aroma can also be removed and stored. An additional benefit of low-pressure distillation is that other flavour components are not subjected to high temperatures that may degrade them.

Further distillation completely removes the ethanol, which is stored for other uses. The beer can be reconstituted by adding the esters extracted during distillation, carbon dioxide and as much ethanol as is required.

**4** **a)** What is fractional distillation, and what property does its success depend on?
**b)** Compare the advantages and disadvantages of the two methods for making low-alcohol and alcohol-free drinks.
**c)** One of the reasons for the increase in the consumption of alcohol-free beer is that it enables people to drink beer without breaking religious observance. Suggest some more reasons for this change of habit.

Most people still prefer to drink normal beer rather than alcohol-free beer, but they should be aware of the dangers of ethanol. Ethanol is a mild depressant that slows down both physical and mental activity. In sufficient quantities it can kill.

**5** **a)** Suggest from the information in Table 13.A2 why it is difficult to keep drinking ethanol until you die.
**b)** What do you notice about the rate of removal of ethanol from the blood shown in Fig 13.A4?
**c)** Explain why it can still be dangerous for you to drive some hours after you have finished drinking.
**d)** A 70 kg moderate drinker has 2 dm³ of beer:
(i) Estimate the time it takes for all the ethanol to be lost from the blood.
(ii) Give reasons why this can only be an approximate answer.
**e)** A person, who weighs 70 kg, drinks 2 dm³ of beer per hour for 4 hours. Estimate the blood ethanol level after 4 hours. Hint: Use a graph.

**Table 13.A2 The relationship between the non-stop consumption of beer and the effect of ethanol for a man of weight 70 kg**

| Volume of typical beer/dm³ | Blood ethanol level/% by volume | Effect |
|---|---|---|
| 0.72 | 0.05 | Mild sedation and tranquillity |
| 1.44 | 0.1 | Lack of coordination |
| 2.16 | 0.15 | Obvious intoxication |
| 3.60 | 0.3 | Unconscious |
| 7.70 | 0.5 | Possible death |

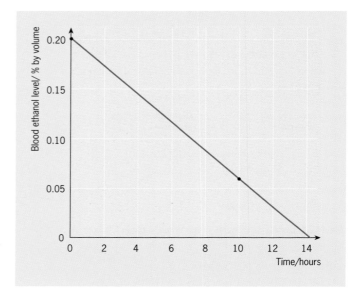

**Fig 13.A4** The rate of removal of ethanol from the blood after drinking has stopped

# 14

# Alkenes

# 14 Alkenes

Chemists are beginning to discover molecules that can be made to act like the memory of a computer, which stores binary information by switching between two states – 'on' or 'off'. The molecular shape of certain molecules that contain double bonds alters when they undergo a reversible photochemical change, and this is the 'switch' that chemists are exploiting.

For when such a molecule is changed from a *trans*-isomer into a *cis*-isomer, the distinct change in shape of the molecule offers a way of storing information. For example, the *trans* form could be designated 'off' (corresponding to binary digit 0) and the *cis* form designated 'on' (corresponding to binary digit 1). Given that the change easily reverses, information storage, retrieval and erasure become feasible. The hope is that, by depositing such memory molecules on an electrode, a hundred million bits of information could be stored on an area the size of a fingernail.

**Memory molecules** A memory molecule changes shape to its *cis* form when it absorbs a photon of ultraviolet light. This form could correspond to the binary digit 1 or 'on'. The *trans* form of the molecule would then correspond to the binary digit 0 or 'off'

## 1 THE PETROCHEMICAL INDUSTRY

The petrochemical industry produces very many organic substances from crude oil and natural gas. About 90 per cent of the world's output of basic organic chemicals is now derived from these two fossil fuels. Petrochemicals include substances as diverse as solvents, detergents, pesticides, insecticides, polymers, plastics, drugs, paints and textiles. Without doubt, improvements in the quality of life since the early 1900s are a direct consequence of the development of the petrochemical industry.

Dmitri Mendeleev, who originated the Periodic Table, wrote that crude oil was too valuable to be burnt as a fuel, but should be used as a source for organic chemicals. That was in 1872, and many people would argue that it is still the case.

### CRACKING AND THE PRODUCTION OF ALKENES

In an earlier chapter we described crude oil as a complex mixture of organic compounds. Even after the fractional distillation of crude oil, the fractions obtained are still complex mixtures (although the components of a fraction do have similar boiling points). Most of the hydrocarbons present in crude-oil fractions are alkanes, which normally are unreactive. However, alkanes can undergo the process called **cracking** to produce a variety of organic molecules that can be used to synthesize other, more generally useful, organic compounds. Cracking converts saturated hydrocarbons into unsaturated hydrocarbons. In particular, a group of versatile and important unsaturated hydrocarbons, called **alkenes**, are made and are much more reactive than alkanes.

**Fig 14.1** Introducing alternative energy sources, such as this wind farm, may reduce the demand for crude oil. But alternative energy sources will have to be better developed and more freely available before they supply a sizeable fraction of the energy we consume

**THE FRACTIONAL DISTILLATION OF CRUDE OIL IS DESCRIBED ON PAGE 202.**

**CRACKING IS DESCRIBED ON PAGES 203 TO 204.**

### IMPORTANCE OF ALKENES

Why are alkenes so important to the petrochemical industry? The main reason is that the carbon–carbon double bond of an alkene molecule makes it much more reactive than an alkane molecule. So, compounds with different carbon chains or with different functional groups can readily be synthesized from alkenes.

The chemistry of the alkene functional group is described in this chapter, which illustrates the wide range of substances that can be made from quite simple alkenes.

Another reason for their importance is that alkenes are readily available in large amounts from the cracking of crude oil.

## ! SCIENCE FEATURE  Alkenes and alkynes as fuels

Surplus alkenes can be used both as fuels and as starting materials for petrochemical processes. Being hydrocarbons, alkenes can burn, and so transfer energy to their surroundings.

Similarly, the alkyne ethyne ($C_2H_2$), which has a triple bond, is mixed with oxygen and burnt to produce an extremely high-temperature flame, which is used to cut iron and steel. The equipment is commonly known as the oxyacetylene torch (acetylene is the traditional name for ethyne).

$$C_2H_2(g) + 2\tfrac{1}{2}O_2(g) \rightarrow 2CO_2(g) + H_2O(g)$$

**Fig 14.2** The temperature of an oxyacetylene flame is high enough to cut easily through steel

## ▶ EXTENSION Natural gas as a source of petrochemicals

Natural gas is an abundant resource. Even if no new gas fields are discovered, present known sources should last until about 2050. So, could natural gas be used as a source of petrochemicals as well as a fuel?

Unfortunately, natural gas is mainly methane, which is a molecule that does not contain any carbon–carbon double bonds. However, much research is in progress to find ways of using methane to make alkenes to provide an alternative supply of raw materials for the petrochemical industry. For example, in a process known as an *oxidative coupling*, methane is reacted with oxygen at high temperatures in the presence of an oxide catalyst. In one such reaction, methane is converted into ethene – a precursor of a large number of petrochemicals:

$$2CH_4 + O_2 \rightarrow C_2H_4 + 2H_2O$$

The coupling reaction is believed to have a free-radical mechanism. As a result, the selectivity is low, so many other products are formed. Oxidative coupling has to be closely controlled since, with the wrong proportions of methane and oxygen, an explosive combustion reaction could occur.

## ? TEST YOURSELF

**A** When decane, $C_{10}H_{22}$, is cracked many products are formed.
**a)** In one cracking reaction decane forms propene, $C_3H_6$. Suggest the molecular formula of the other hydrocarbon produced in this reaction.
**b)** In another cracking reaction decane forms ethene, $C_2H_4$. Write a balanced equation for this cracking reaction.

## ? TEST YOURSELF

**B** Write down the equations for the complete combustion of ethene ($C_2H_4$) and of propene ($C_3H_6$).

## ✔ REMEMBER THIS

Alkynes are a homologous series of hydrocarbons with the general formula $C_nH_{2n-2}$ that have a molecule with a carbon–carbon triple bond.

## ? TEST YOURSELF

**C** **a)** Write down the equation for the complete combustion of methane in air.
**b)** Suggest how the risk of explosion during the oxidative coupling reaction can be reduced.

**Fig 14.3** The displayed formula of ethene

**NAMING A CARBON SKELETON AND NUMBERING ITS CARBON ATOMS ARE COVERED ON PAGES 212 TO 214.**

**Fig 14.4** The displayed formula of 4-methylhex-2-ene

# 2 ALKENES AND ISOMERISM

Alkenes are a homologous series of unsaturated hydrocarbons that have a molecule with one carbon–carbon double bond. This means that their carbon chain has two fewer hydrogen atoms than the same carbon chain in alkanes, and has the general formula $C_nH_{2n}$. As already explained, it is the presence of the double bond that gives alkenes much greater reactivity than alkanes.

## NAMING ALKENES

The presence of the carbon–carbon double bond (C=C) is identified by the suffix *ene*. So, for example, ethene (Fig 14.3) has a two-carbon chain with one C=C bond, and propene has a three-carbon chain with one C=C bond. Notice that the displayed formula of ethene has been drawn with angles of about 120° between the bonds, which is their actual value. It is not essential to draw these bond angles accurately, but it is good practice, and will help you to remember them.

The name of an alkene must also specify the position of the C=C bond and the number of carbon atoms in the longest chain that contains the C=C bond. The use of *eth*, *prop*, *but*, and so on, to indicate the length of the carbon chain is described on page 213.

We therefore proceed as follows:
● First, locate the longest chain that contains the double bond.
● Then, number the carbon atoms in the chain from the end that gives the lowest possible position number to the C=C bond.

Fig 14.4 shows the displayed formula of an alkene. The longest carbon chain that includes the C=C bond is shaded. It contains six carbon atoms, which means that *hex* is used to indicate the chain length. The carbon atoms must be numbered so that the C=C double bond is located between carbon numbers 2 and 3, and so the name of the compound will end with hex-2-ene. Notice that only the number of the first carbon atom of the double bond is included in this part of the name. Finally, the position of the methyl side chain must be specified in the name. In this example, the methyl group is attached to carbon 4. So the full name must be 4-methylhex-2-ene.

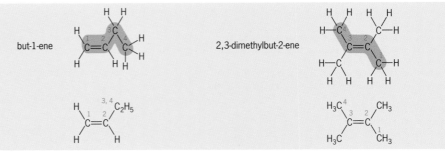

**Fig 14.5** The displayed formulas of but-1-ene and 2,3-dimethylbut-2-ene. The lower formulas are easier to write and give practically as much information as the upper ones. The coloured patches show the longest carbon chain that contains the C=C bond

## STEREOISOMERISM

The C=C bond is stronger than the C–C bond, but it is locked in position and cannot rotate in the way that the C–C bond does. This gives rise to stereoisomerism in compounds with double bonds. Compounds that are stereoisomers of one another have the same make-up of atoms, and their atoms are bonded in the same order, but the arrangements of the atoms in space are different.

Fig 14.7 shows two compounds that have the same name, but-2-ene. But are they really the same? Each molecule has its own molecular shape, defined by the bond length and the bond angles within the molecule. Simple geometry shows that the length from carbon 1 to carbon 4 is not the same in both compounds. So they cannot be identical – they are isomers.

These two isomers have different physical properties. For example, the melting point of *cis*-but-2-ene is –139 °C, but that of the *trans*-but-2-ene is –106 °C. However, the chemical properties are very similar as both molecules contain the same type and number of bonds.

This type of isomerism is known as geometric isomerism, or *cis–trans* isomerism, since the bond angles are all identical, but the geometries and shapes of the molecules are different. The prefix *cis* (Latin) means 'on the same side', and so the *cis* isomer has the two methyl groups on the same side of the double bond. The prefix *trans* (also Latin) means 'on the other side', and so the *trans* isomer has the methyl groups on opposite sides of the double bond. The *trans* isomer has the greater distance between carbon 1 and carbon 4.

Figure 14.8 summarizes the different types of isomerism.

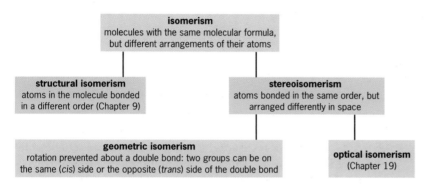

**Fig 14.8** Different types of isomerism

## NATURE OF THE DOUBLE BOND

*Cis–trans* isomers exist because the C=C bond is not free to rotate. This is unlike the behaviour of the C–C carbon bond, which has free rotation about the bond, and therefore (see Fig 14.10) the positions of the two methyl groups in butane are not fixed. In but-2-ene, the carbon–carbon double bond hinders rotation, and so the two methyl groups are locked in their positions.

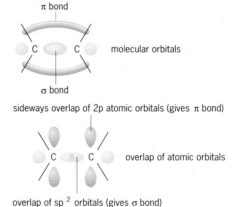

**Fig 14.10** Butane has free rotation at the C—C bond, while in but-2-ene the C=C bond prevents rotation. When the single C—C bond between the carbon 2 and carbon 3 atoms rotates, the methyl groups also rotate. Although the conformations look like *cis* and *trans*, rotation of the C—C bond changes one conformation into the other, and back again. The double bond cannot rotate, so the *cis* form cannot become the *trans* form

The inability of the double bond to rotate is explained by the nature of the bond itself. Any double bond is composed of two different types of bond, the σ bond and the π bond (Fig 14.11).

**Fig 14.7** The displayed formulas of geometric isomers of but-2-ene. The distance between carbon 1 and carbon 4 is different in these two forms, because the shape of the *cis* molecule is different from that of the *trans* molecule

**SEE QUESTION 1.**

sideways overlap of 2p atomic orbitals (gives π bond)

overlap of atomic orbitals

overlap of sp² orbitals (gives σ bond)

**Fig 14.11** The orbital model of the double bond showing the two types of bond

**HYBRID ATOMIC ORBITALS AND THE NATURE OF THE DOUBLE BOND ARE ALSO COVERED ON PAGES 82 AND 196.**

The σ bond is formed by the overlap of two sp² hybrid atomic orbitals, one on each carbon atom. The bond lies mainly in the region directly *between* the two nuclei. The π bond is formed by the side-to-side overlap of two 2p atomic orbitals, one on each carbon atom. As a result, the π bond lies *above and below* the σ bond. The two carbon atoms are thus bonded together in two different ways, with the π bond preventing rotation about the carbon–carbon bond.

Note that the overlap of the atomic orbitals is less for the π bond, so the energy needed to break the π bond is less than that needed to break the σ bond.

## ▶ EXTENSION Interchange of *cis* and *trans* isomers and the photochemistry of vision

The only way to change a *cis* isomer into a *trans* isomer is first to break the π bond, and thereby allow free rotation about the σ bond, and then to re-form the π bond. This can be done by a **photochemical reaction** using ultraviolet light.

When ultraviolet light of the correct frequency is absorbed by the molecule, the π bond is broken, to form a **diradical**. A diradical is a particle that contains two atoms each with an unpaired electron in its outer shell.

During the time that the diradical exists, there is free rotation about the σ bond. As free radicals are highly reactive, the diradical very quickly re-forms the double bond to pair up the electrons. The result is a mixture of the *cis* and the *trans* isomer.

**Fig 14.12** Photochemical *cis–trans* transformations. The π bond can re-form in either structure A or B, to produce both *cis* and *trans* isomers

The property of *cis–trans* isomerism is not limited to compounds with carbon–carbon double bonds. Compounds with carbon–nitrogen and nitrogen–nitrogen double bonds also form geometric isomers. For example, Fig 14.13 shows a candidate for the kind of memory molecule mentioned in the chapter Opener. The completely different shapes of the two isomers are obvious.

**Fig 14.13** A possible memory molecule. The distance between X and Y is much shorter in the *cis* isomer than in the *trans* isomer. The *trans* isomer can be changed into the *cis* isomer by a photochemical reaction

X = C₈H₁₇
Y = C₆H₁₁O₂

### PHOTOCHEMISTRY OF VISION

Light is detected on the inner layer of the eye known as the retina. The retina consists of two types of cell that can respond to light: rods and cones. Rods respond to differences in intensity of light, but not to colour. Cones respond to colour. However, both types of cell function in a similar way.

**Fig 14.14** Detection of light by the retina is a direct result of the photochemical transformation of a *cis* isomer into a *trans* isomer

Light is detected because retinal, a molecule inside the cells of the retina, undergoes a photochemical reaction. When a photon of light reaches a *cis*-retinal molecule, it is changed into the *trans*-retinal form, as in Figs 14.15 and 14.16, and so its shape changes.

Retinal interacts with a protein responsible for producing the nerve impulses sent along the optic nerve to the brain; when the form changes from *cis* to *trans*, a nerve impulse is generated. It is not yet fully understood how the change of shape within the protein molecule produces a nerve impulse, but the energy from the light is in some way transferred, and an electrical signal results.

**Fig 14.15** *Cis*- and *trans*-retinal. The two isomers have completely different overall shapes. The coloured patches highlight the *cis* and *trans* double bonds at which the difference arises

**Fig 14.16** The action of *trans*-retinal at the active site

## PHYSICAL PROPERTIES OF ALKENES

The alkenes form a homologous series. The first three members are gases, but all the higher members are liquids or solids. The increase in the boiling point of compounds as the number of carbon atoms per molecule increases is typical for all homologous series, not just the alkenes.

| Table 14.1 Melting and boiling points of four alkenes | | | |
|---|---|---|---|
| **Alkene** | **Structural formula** | **Melting point/°C** | **Boiling point/°C** |
| Ethene | $CH_2{=}CH_2$ | −169 | −105 |
| Propene | $CH_3CH{=}CH_2$ | −185 | −48 |
| But-1-ene | $CH_3CH_2CH{=}CH_2$ | −185 | −6 |
| Pent-1-ene | $CH_3CH_2CH_2CH{=}CH_3$ | −165 | 30 |

## 3 PREPARATION OF ALKENES

## CRACKING ALKANES

Saturated hydrocarbons can be cracked to give alkenes (see page 203). Though cracking can be carried out in the laboratory, this method is unsuitable because the mixture it produces is difficult to separate into individual alkenes.

Thermal cracking involves heating long-chain alkanes at a very high temperature so that the molecules break down to give hydrocarbons with shorter carbon chains. Since the C—C bond in alkanes is non-polar, the breaking of this bond is believed to involve **homolytic fission** with the formation of alkyl free radicals. Homolytic fission is normally initiated by ultraviolet light, but a very high temperature provides sufficient thermal energy for the process.

Cracking lacks selectivity, since its progress is determined by which C—C bond is first broken, and then by the type of collision in which the free radical is involved. Typically, the cracking of a hydrocarbon such as decane leads to a collection of short-chain alkanes, alkenes and hydrogen. (Free radicals and homolytic fission are described on pages 266 and 267; cracking is described on pages 203–204.)

**Fig 14.17** Cracking liquid paraffin in the laboratory

**THERE IS MORE ABOUT THE DEHYDRATION OF ALCOHOLS ON PAGE 295.**

**Fig 14.18** Butan-1-ol loses a water molecule in an elimination reaction to form but-1-ene

# DEHYDRATION OF ALCOHOLS

A suitable way to prepare alkenes is to dehydrate an alcohol. This reaction is an example of elimination and involves the loss of a molecule of water (Fig 14.18). Normally, an acidic catalyst, such as concentrated phosphoric(V) acid or concentrated sulphuric acid, is used to aid elimination of the water (Fig 14.19). Industrially, the dehydration is usually catalysed by passing the alcohol vapour over heated aluminium oxide.

**Fig 14.19** Dehydration of an alcohol (butan-1-ol) is catalysed by an acid. The reaction starts with protonation of the alcohol. After elimination of the water molecule, the proton is regenerated

Sometimes, the dehydration of an alcohol leads to the formation of two or more alkenes, which can be either structural or geometric isomers (Fig 14.20). Normally, the most stable alkene is produced in a greater proportion than the other isomers. This is the alkene with most alkyl groups attached to the carbon atoms that form the double bond (the most substituted alkene).

**Fig 14.20** Dehydration of butan-2-ol. Three isomeric butenes can be made by the removal of different hydrogen atoms

**? TEST YOURSELF**

**I** **a)** State which alcohol you would need to make (i) ethene, (ii) propene and (iii) cycloheptene.
**b)** (i) Write down all the possible isomers that can be formed by the dehydration of pentan-2-ol.
(ii) Which of these is likely to be produced in the greatest amount?

**? TEST YOURSELF**

**J** Draw the structures of all of the alkenes that can be produced when 2-iodohexane is heated with ethanolic sodium hydroxide.

# ELIMINATION OF HYDROGEN HALIDES FROM HALOGENOALKANES

When a halogenoalkane is heated with a strong base, such as ethanolic sodium hydroxide, the corresponding alkene is produced. Again, the reaction may give more than one alkene and may be complicated by substitution products as well (Fig 14.21).

**Fig 14.21** Elimination of hydrogen iodide from 3-iodohexane

**FOR MORE INFORMATION ABOUT THE ELIMINATION OF HYDROGEN HALIDES FROM HALOGENOALKANES, SEE PAGE 278.**

# 4 REACTIONS OF ALKENES

The double bond in an alkene is non-polar, but it is electron rich because of the π bond. When the π bond is broken, the carbon atoms remain joined together by the σ bond.

## ELECTROPHILES

An **electrophile** is a particle that can accept a pair of electrons to make a covalent bond. Often, an electrophile has only six electrons in its outer shell, so that by accepting two others it achieves a stable octet. A typical electrophile is $Cl^+$, with only six electrons in its outer shell. Metal ions are not electrophiles because, although they are positive, they cannot gain a pair of electrons to make a covalent bond. In fact, a metal ion is often positively charged to preserve a stable octet of electrons.

Electrophiles are generated by the **heterolytic fission** of a covalent bond. In heterolytic fission, the bond is broken as both bonding electrons migrate to the atom that is more electronegative, leaving the less electronegative atom positive. Fig 14.22 shows the heterolytic fission of a covalent bond. Again, we follow the convention of using a curly arrow to show the migration of a pair of electrons.

## ELECTROPHILIC ADDITION

Almost all double bonds will undergo a reaction known as **addition**. During addition, two substances react together to give one product and the original double bond becomes a single bond, as shown in Fig 14.23. In the case of the C=C bond, an electrophile is often needed to break the bond. Since the initial step in the mechanism involves an electrophile, it is called **electrophilic addition**.

### MECHANISM OF ELECTROPHILIC ADDITION

The first stage in electrophilic addition is the formation of the electrophile. (We will return later to discuss this stage in more detail, because it is not always as simple as is shown in Fig 14.24.) The electrophile 'attacks' the double bond. At the same time, a pair of electrons migrates from the double bond towards the electrophile. As seen in Fig 14.24, the effect of this is to break the carbon–carbon double bond and to make a single bond to the electrophile. In the final stage, the nucleophile reacts with the carbocation that has been formed, to complete the addition.

The net effect is to add atoms at the double bond to form a saturated compound. It is this ability to react by addition that accounts for the difference in reactivity between an alkane and an alkene.

**Fig 14.24** A generalized mechanism for electrophilic addition to an alkene. E = electrophile, Nu = nucleophile

## ADDITION OF HYDROGEN HALIDES

Hydrogen bromide reacts with alkenes to produce bromoalkanes. The other hydrogen halides react with alkenes in a similar way. Hydrogen halides act as electrophiles, because the hydrogen end of the molecule is electron deficient as a result of the difference in electronegativity between the hydrogen and halogen atoms. The hydrogen end of the molecule can accept an electron pair,

**? TEST YOURSELF**

**K** Explain why the π bond can be broken more easily than the σ bond.

**? TEST YOURSELF**

**L** State which of the following particles are electrophiles: $Br^+$, $CH_3$, $CH_3^+$, $H_2O$ and $Na^+$.

$$X \overset{\frown}{-} Y \longrightarrow X^+ + Y^-$$

**Fig 14.22** Heterolytic fission of a covalent bond. Atom Y is normally more electronegative than atom X. It therefore accepts the bonding pair of electrons during heterolytic fission

**Fig 14.23** A generalized addition reaction. During addition, two reactants form one product and the double bond becomes a single bond

**THE ABSENCE OF REACTIVITY IN ALKANES IS COVERED ON PAGE 216.**

**✓ REMEMBER THIS**

A carbocation is also known as a carbonium ion. It is a positive ion that has a carbon atom attached to only three other atoms or groups of atoms, $R_3C^+$.

Fig 14.25 The addition of hydrogen bromide to either *cis*- or *trans*-but-2-ene gives the two optical isomers of 2-bromobutane

**OPTICAL ISOMERISM IS ANOTHER FORM OF STEROISOMERISM: SEE PAGE 411.**

**THE REACTIONS OF HALOGENOALKANES ARE COVERED IN DETAIL IN CHAPTER 12.**

while simultaneously the hydrogen–halogen bond is broken, as shown in Fig 14.24 (see page 311). The reaction with hydrogen iodide is particularly favoured because of the relatively low bond energy of the H—I bond.

Ethene reacts with hydrogen chloride to give chloroethane. Likewise, both *cis*- and *trans*-but-2-ene yield 2-bromobutane. It does not matter what the starting isomer is, because after the addition the saturated product has acquired a single bond that can freely rotate, with loss of the *cis–trans* isomerism. In fact, as Fig 14.25 shows, two products are formed, which in this case are **optical isomers**.

The production of many petrochemicals starts with this type of electrophilic addition to give a halogenoalkane. The introduction of the carbon–halogen bond offers further possibilities for organic synthesis.

## ▶ EXTENSION Addition of hydrogen halides to unsymmetric alkenes: Markovnikov's rule

An unsymmetric alkene, such as propene, is an alkene that has different groups attached to the carbon atoms of the C=C bond.

The electrophilic addition of a hydrogen halide to an unsymmetric alkene can give more than one addition product, according to which end of the double bond the electrophile attacks. For example, the electrophilic addition of hydrogen bromide to propene can yield two products, 1-bromopropane and 2-bromopropane, depending on which end of the double bond is attacked by the electrophile (essentially a proton). As Fig 14.27 shows, the two products arise from different carbocations, which have different stabilities.

Fig 14.26 The displayed formulas of four symmetric and four unsymmetric alkenes

Fig 14.27 The mechanism of the electrophilic addition of hydrogen bromide to propene

Fig 14.28 The three types of carbocation

A **primary carbocation** (Fig 14.28) is a species in which a positive carbon atom is attached to just one other carbon atom (of the alkyl or aryl R group). It is much less stable than a **secondary carbocation**, which is much less stable than a **tertiary carbocation**. The source of this stability is the alkyl or aryl group, which tends to donate electron density to the positive carbon atom and thereby stabilizes the positive charge. It follows that two alkyl (or aryl) groups give more stability than one, and three give more stability than two. This action of pushing the electron density is called an **inductive effect**. The major product is formed from the most stable carbocation. Therefore, in the example of the addition of hydrogen bromide to propene, the major product is 2-bromopropane.

**Markovnikov's rule** predicts the way that hydrogen halides are added to unsymmetric alkenes:

**An electrophile adds to an unsymmetric alkene so that the most stable carbocation is formed as an intermediate.**

The term 'electrophile' is used to express the general concept, although the electrophile is often an electron-deficient hydrogen atom or a proton.

## HYDRATION OF ALKENES

The hydration of an alkene is the electrophilic addition of water to give an alcohol. This reaction needs an acidic catalyst, since water itself is not a good electrophile. In the laboratory, sulphuric acid is used as the catalyst. With reactive double bonds, dilute acid is sufficient, but with alkenes, such as propene and ethene, concentrated acid is required.

Fig 14.30 details the use of concentrated sulphuric acid as a catalyst. The proton transferred by the acid is needed to start the reaction as it facilitates the breaking of the double bond. At the end of the reaction, the acid is regenerated.

ethyl hydrogensulphate

ethanol is formed on hydrolysis of the ethyl hydrogensulphate

sulphuric acid transfers a proton

regenerated at end of reaction

**Fig 14.30** The mechanism of the electrophilic addition of water, as shown by the hydration of ethene using concentrated sulphuric acid and subsequent hydrolysis of the intermediate

In the petrochemical industry, alkenes are hydrated to give several important solvents. Hydration is achieved either by the catalytic addition of water as steam to the alkene at 7000 kPa (70 atmospheres) pressure at 300 °C, using $H_3PO_4$ as a catalyst, or by using the concentrated sulphuric acid route just described. Ethene is converted into ethanol, propene into propan-2-ol, and but-1-ene into butan-2-ol (Fig 14.31). All three alcohols are extremely useful industrial solvents.

Since the addition at double bonds follows Markovnikov's rule, the number of alcohols that can be made is limited. In particular, it is difficult to make primary alcohols.

**Fig 14.31** The hydration of ethene, propene and butene. Because of Markovnikov's rule, the hydration of propene and but-1-ene does not produce a large proportion of the primary alcohols propan-1-ol and butan-1-ol. Instead, the secondary alcohols propan-2-ol and butan-2-ol are formed

## ? TEST YOURSELF

**Q** Draw the displayed formulas of the products of the reactions between:
**a)** bromine and ethene;
**b)** chlorine and cyclohexene;
**c)** chlorine and ethene.

## ? TEST YOURSELF

**R** Chlorine will react with ethene to give 1,2-dichloroethane. The mechanism is identical to the bromination of ethene. Write down the mechanism for this reaction.

Fig 14.33 The intermediate positive ion (carbocation) can react with other nucleophiles, such as a chloride ion or water, if they are present

**SEE QUESTIONS 4, 5 AND 6.**

## ? TEST YOURSELF

**S** Explain the formation of two products during the reaction between aqueous bromine and but-2-ene.
Hint: Read about the interception of a carbocation by other nucleophiles.

## REACTION WITH HALOGENS

Halogens can add to double bonds to give dihalogenoalkanes. For example, bromine reacts with propene to give 1,2-dibromopropane:

$$CH_3CH{=}CH_2 + Br_2 \rightarrow CH_3CHBrCH_2Br$$

## MECHANISM OF THE REACTION

Compared with a hydrogen halide, it is much more difficult to visualize a bromine *molecule* as an electrophile, since it is not a polar molecule and both of its atoms have a stable octet of electrons. Nevertheless, a bromine molecule is able to accept an electron pair, provided that, at the same time, the bromine–bromine single bond is broken by *heterolytic* fission.

As the bromine molecule approaches the double bond, the pair of electrons in the Br–Br bond moves to one of the bromine atoms (Fig 14.32). Why should this happen? After all, there is no positive or negative end to the bromine molecule, as there is in the addition of hydrogen bromide. The reason is that a temporary dipole is created in the bromine molecule as it approaches the $\pi$ electrons of the double bond. It is as though the electrons in the bromine–bromine single bond are repelled by the $\pi$ electrons. This temporary dipole produces an electron deficiency at the end of the molecule nearer the double bond, and thereby allows the electrophilic attack. The intermediate carbocation then reacts with a bromide ion to complete the electrophilic addition.

**Fig 14.32** The mechanism of the electrophilic addition of bromine to propene

However, it is possible to intercept the carbocation with other nucleophilic reagents (Fig 14.33). So, if a chloride ion is also present in the reaction mixture, it can react as a nucleophile; and if the reaction is carried out in the presence of water, a water molecule can react with the carbocation.

This means that the choice of solvent used for the bromination is important. If it were aqueous bromine, for example, it would be impossible to stop the interception of the carbocation by a water molecule acting as a nucleophile.

**Fig 14.34** Aqueous bromine reacts with but-2-ene to give two products

## TEST FOR UNSATURATION

Aqueous bromine may be used to test for the presence of a C=C bond, since the reagent changes colour during electrophilic addition. Aqueous bromine is orange, but after reaction with an alkene it forms colourless products. Alternatively, bromine in an inert solvent, such as tetrachloromethane or hexane, can be used. In both of these solvents the same colour change is observed, orange to colourless.

# HYDROGENATION

Alkenes can also undergo other sorts of addition reactions, such as catalytic hydrogenation. In this case, hydrogen is added to the C=C bond to make a saturated compound (Fig 14.35). The reaction mechanism is not electrophilic addition, but often involves **heterogeneous catalysis** (catalyst and reactant in different phases) using a nickel catalyst and high pressure (Fig 14.36).

Hydrogenation is not an important reaction in the petrochemical industry because it leads to the formation of saturated (unreactive) compounds, and petrochemical feedstocks are often already saturated. As a synthetic reaction, it is used when the raw material is derived from a natural source. For example, margarines are produced by the hydrogenation of polyunsaturated compounds derived from plant oils.

**FOR MORE ABOUT HETEROGENEOUS CATALYSIS, SEE PAGES 225–226.**

**Fig 14.35** Hydrogenation converts an alkene to an alkane

**1** Ethene molecule approaches nickel surface

$H_2C=CH_2$

—Ni—Ni—Ni—Ni—

**2** Ethene molecule forms an intermediate on nickel surface

$H_2C---CH_2$

—Ni—Ni—Ni—Ni—

**3** Hydrogen molecule approaches nickel surface

$H_2C---CH_2$

—Ni—Ni—Ni—Ni—

**4** Hydrogen molecule forms an intermediate on nickel surface adjacent to ethene

$H_2C---CH_2$

—Ni—Ni—Ni—Ni—

**5** Hydrogen and ethene form a 'common intermediate' and addition occurs

$H_2C---CH_2$

—Ni—Ni—Ni—Ni—

**6** Ethane molecule leaves nickel surface

$H_3C---CH_3$

—Ni—Ni—Ni—Ni—

**Fig 14.36** Heterogeneous catalysis in the hydrogenation of ethene

**Fig 14.37** The unsaturated compounds in sunflower oil have some of their double bonds hydrogenated to change the oil into a solid

**POLYUNSATURATED FATS ARE THE SUBJECT OF THE ASSIGNMENT ON PAGES 223 AND 224.**

**? TEST YOURSELF**

**T** A sample of a natural product, 0.0100 moles, with molecular formula $C_{22}H_{26}$ needs 0.0502 moles of hydrogen for complete hydrogenation. How many C=C bonds are present in one molecule of the natural product?

**SEE QUESTION 7.**

## WORKING OUT STRUCTURES

Hydrogenation is used to determine the number of C=C bonds in a compound. One mole of C=C bond requires one mole of hydrogen to fully hydrogenate it. Therefore the number of C=C bonds in the compound can be estimated by comparing the number of moles of hydrogen that have reacted with a fixed amount of compound.

## OXIDATIVE ADDITION

Alkenes react with cold acidified dilute potassium manganate(VII) to give a diol product (with two OH groups). The reaction involves both oxidation and addition to the double bond, so it is sometimes called an oxidative addition. During the reaction, the colour of the manganate(VII) ion changes from purple to colourless. When ethene is bubbled into cold acidified dilute potassium manganate(VII), ethane-1,2-diol (a component of antifreeze) is produced:

$$CH_2=CH_2 + H_2O + [O] \rightarrow CH_2OHCH_2OH$$

Notice that the oxidation is represented in the equation in its simplified form using [O] (see page 297).

This reaction is not used to prepare diols, since an excess of potassium manganate(VII) will oxidize the diols further.

When the alkene is treated with acidified hot concentrated potassium manganate(VII), the double bond in the alkene is broken to give two carbonyl compounds. So, for example, hex-2-ene is oxidized to give ethanoic acid and butanoic acid (Fig 14.38).

**Fig 14.38** The oxidative addition of alkenes to give diols. When the alkene is ethene, the product is ethane-1,2-diol

**Fig 14.39** Cleavage of the carbon–carbon double bond of hex-2-ene gives two carboxylic acids

This cleavage of the double bond can be useful during synthesis, particularly when the alkene is a cyclohexene, since it changes the carbon skeleton without changing the number of carbon atoms (Fig 14.40). The dioic acid obtained is an important intermediate in the manufacture of nylon 6,6.

**Fig 14.40** Cleavage of the double bond in a ring compound, such as cyclohexene, produces a chain compound. The cleavage is shown by the changed positions of the asterisked carbon atoms

## ADDITION POLYMERIZATION

Alkene molecules can be joined to give molecules with very long chains, known as polymers. The constituent alkene molecules of polymers are called monomers. In polymerization, thousands of monomers are combined to form a single polymer chain. An alkene forms an addition polymer when a double bond is converted into a single bond and two extra single bonds are created. The resulting polymer consists of repeating units called monomer units.

For example, ethene is converted into poly(ethene), or polythene, by the action of a catalyst, high pressure and a moderate temperature (Fig 14.41). Table 14.2 lists five addition polymers and their main uses.

There is more information on addition polymers and polymerization in Chapter 20.

**Fig 14.41** Poly(ethene) is formed by the repeated catalysed addition of ethene monomers to give a saturated polymer

**Table 14.2 Five common addition polymers and their uses**

| Alkene monomer | | Polymer | Structure of polymer (monomer unit) | Uses |
|---|---|---|---|---|
| Ethene | $\begin{array}{c} H \quad\quad H \\ C=C \\ H \quad\quad H \end{array}$ | Poly(ethene) | $\left[\begin{array}{c} H \;\; H \\ -C-C- \\ H \;\; H \end{array}\right]_n$ | Bags, insulation for wires, squeezy bottles |
| Propene | $\begin{array}{c} H \quad\quad CH_3 \\ C=C \\ H \quad\quad H \end{array}$ | Poly(propene) | $\left[\begin{array}{c} H \;\; CH_3 \\ -C-C- \\ H \;\; H \end{array}\right]_n$ | Bottles, plastic plates, clothing, carpets, crates, ropes and twine |
| Phenylethene | $\begin{array}{c} H \quad\quad C_6H_5 \\ C=C \\ H \quad\quad H \end{array}$ | Poly(phenylethene) | $\left[\begin{array}{c} H \;\; C_6H_5 \\ -C-C- \\ H \;\; H \end{array}\right]_n$ | Insulation, food containers, model kits, flowerpots, housewares |
| Chloroethene | $\begin{array}{c} H \quad\quad Cl \\ C=C \\ H \quad\quad H \end{array}$ | Poly(chloroethene) | $\left[\begin{array}{c} H \;\; Cl \\ -C-C- \\ H \;\; H \end{array}\right]_n$ | Synthetic leather, water pipes, floor covering, guttering, window frames, curtain rails, wall cladding |
| Methyl-2-methylpropenoate | $\begin{array}{c} \quad\quad O \\ \quad\quad \| \\ H \quad C-OCH_3 \\ C=C \\ H \quad CH_3 \end{array}$ | Poly(methyl-2-methylpropenoate) | $\left[\begin{array}{c} \quad O \\ \quad \| \\ H \;\; C-OCH_3 \\ -C-C- \\ H \;\; CH_3 \end{array}\right]_n$ | light fittings, car rear lights, record player lids, tap tops, simple lenses |

## ▶ EXTENSION Manufacture of poly(chloroethene) from ethene

Poly(chloroethene) is an important polymer, as seen from those uses listed in Table 14.2. It is manufactured from chlorine and ethene in a series of simple chemical reactions. First, ethene is reacted with chlorine to form 1,2-dichloroethane in a typical electrophilic addition:

$$Cl_2 + CH_2{=}CH_2 \rightarrow CH_2ClCH_2Cl$$

Next, the 1,2-dichloroethane from this reaction is heated very strongly to eliminate hydrogen chloride to form the monomer chloroethene. This involves a typical reaction of halogenoalkanes:

$$CH_2ClCH_2Cl \rightarrow CH_2{=}CHCl + HCl$$

Hydrogen chloride is a hazardous chemical that must not be allowed to reach the atmosphere. However, it is also a source of valuable chlorine. Therefore, the hydrogen chloride generated in the second reaction is recycled to make more 1,2-dichloroethane:

$$CH_2{=}CH_2 + \tfrac{1}{2}O_2 + 2HCl \xrightarrow[\text{catalyst}]{Cu^{2+}} CH_2ClCH_2Cl + H_2O$$

Chloroethene from the second reaction is a known carcinogen, and so it too is not allowed to escape from the reaction vessel into the atmosphere. Instead, this monomer is kept in a liquid state by increasing the pressure; it is then polymerized in the presence of a free-radical initiator to form poly(chloroethene):

$$nCH_2{=}CHCl \rightarrow {+}CH_2{-}CHCl{\,}_n$$

The polymerization is highly exothermic and has a free-radical chain mechanism. To avoid explosions the rate of reaction is carefully controlled by adding 'scavenger' chemicals to remove any free radicals.

**Fig 14.42** Poly(methyl-2-methylpropenoate) is a transparent polymer used for car lights

**SEE QUESTION 10.**

**Fig 14.43** Houses with unplasticized polyvinyl chloride (uPVC) wall fittings, window frames and doors

Fig 14.44 The displayed formula of epoxyethane

Fig 14.45 The partial oxidation of ethene to make epoxyethane

Fig 14.48 Most cars rely on water circulating round the engine to keep it cool. Antifreeze allows cars to be used when the temperature is below freezing

# 5 EPOXYETHANE

We have already described how cracking of long-chain alkanes produces smaller alkenes such as ethene. About one-quarter of all the ethane produced by cracking is converted into a very reactive synthetic intermediate called epoxyethane. Epoxyethane is a very strained molecule, since it has a ring of three atoms, two carbon and one oxygen atoms.

## MANUFACTURE OF EPOXYETHANE

Epoxyethane is manufactured by a reaction between oxygen and ethene. The reaction is carried out in the presence of a finely divided silver catalyst at a temperature between 250 and 400 °C and a pressure of between 1000 and 3000 kPa (10 to 30 atmospheres).

Epoxyethane is an extremely flammable and explosive gas at room temperature, so great care must be taken during its manufacture to avoid explosions.

## REACTIONS OF EPOXYETHANE

Epoxyethane reacts readily with many nucleophiles. During the reaction the ring is opened and a chain structure is produced.

Fig 14.46 The reaction of nucleophiles with epoxyethane

## HYDRATION WITH WATER

The lone pairs on a water molecule make it a nucleophile. It will react with epoxyethane to produce ethane-1,2-diol. The reaction is quite slow at room temperature, but can be accelerated using an acid catalyst, such as dilute sulphuric acid, at 60 °C.

Fig 14.47 Reaction of epoxyethane with water

Ethane-1,2-diol is used in antifreeze mixtures, which are added to the water cooling systems of cars to stop the water freezing in cold weather.

## HYDRATION WITH LESS WATER

Epoxyethane will also react with a limited amount of steam to produce polymeric ethers called poly(alkoxy) ethanols. These poly(alkoxy) ethanols are used in the manufacture of plasticizers, polyurethane, solvents and detergents.

$$H_2O \ + \ n \left[ \begin{array}{c} H_2C - CH_2 \\ \diagdown O \diagup \end{array} \right] \longrightarrow \ HO(CH_2CH_2O)_nH$$

**Fig 14.49** Reaction of epoxyethane with a limited amount of steam to make poly(alkoxy) ethanols

## REACTION WITH ALCOHOLS

Alcohols can behave as nucleophiles in the same way as water and will react with epoxyethane to give compounds that contain an ether and a hydroxyl functional group. These compounds are called monoalkyl ethers. They are used as solvents in the paint and printing industries.

$$CH_3CH_2CH_2CH_2OH \ + \ CH_2 - CH_2 \ \diagup\!\!\diagdown \ O \ \longrightarrow \ CH_3CH_2CH_2CH_2OCH_2CH_2OH$$

**Fig 14.50** Reaction of butan-1-ol with epoxyethane

## 6 ALKENES AS PETROCHEMICAL STARTING MATERIALS

With their double bonds, alkenes can take part in many useful synthetic reactions. Almost all involve electrophilic additions and the introduction of functional groups to the carbon chain. It is these functional groups that allow further synthetic reactions and provide a route to a large number of organic substances.

## ETHENE

Fig 14.52 features nine substances that can be synthesized directly from ethene. Most of the reactions involve electrophilic addition to the double bond and demonstrate why alkenes are of much more use than alkanes in the manufacture of petrochemicals. Alkanes have such a limited range of reactions that ethane could only be converted easily into one of the substances in Fig 14.52, chloroethane.

**Fig 14.51** Poly(alkoxy) ethanols are added to lipstick to make it feel smooth and easy to apply

**? TEST YOURSELF**

Y Predict the products of the following reactants with epoxyethane:
**a)** methanol; **b)** propan-2-ol;
**c)** ammonia.

**Fig 14.52** Addition and polymerization reactions of ethene lead to a variety of useful petrochemicals

## PROPENE AND BUTENE

**? TEST YOURSELF**

**Z**  **a)** Identify the reactions in Fig 14.53 that are examples of electrophilic addition.

**b)** Identify the reactions in Fig 14.52 that are examples of polymerization.

**SEE QUESTIONS 11 AND 12.**

butan-2-one
CH₃COCH₂CH₃
solvent

↑ oxidation

butan-2-ol
CH₃CHOHCH₂CH₃
solvent

↑ H₂O + catalyst

but-1-ene
CH₃CH₂CH=CH₂

↓

buta-1,3-diene
CH₂=CH—CH=CH₂

polymerization

↓

buta-1,3-diene rubber
polymer: synthetic rubber

**Fig 14.54** But-1-ene is converted into solvents and synthetic rubber

phenol
C₆H₅OH
synthetic intermediate

propan-2-one
CH₃COCH₃
solvent

(i) O₂
(ii) H⁺

oxidation

2-phenylpropane
CH₃
|
C₆H₅—CH
|
CH₃

propan-2-ol
CH₃CHOHCH₃
solvent

C₆H₆ acid catalyst

H₂O + catalyst

**PROPENE**
CH₃CH=CH₂

high pressure, catalyst

poly(propene)

⎡ CH₃ H ⎤
| | |
—C—C—
| | |
⎣ H H ⎦ₙ

polymer, e.g. used in making crates

1,2,3-trichloropropane
CH₂ClCHClCH₂Cl

Cl₂

Cl₂,hf

Cl₂

3-chloroprop-1-ene
CH₂ClCH=CH₂
synthetic intermediate

1,2-dichloropropane
CH₃CHClCH₂Cl

Br₂

2,3-dibromo-1-chloropropane
CH₂ClCHBrCH₂Br

**Fig 14.53** Propene is converted into many useful petrochemicals and synthetic intermediates

**Fig 14.55** Poly(propene) is used to make marine rope

## SUMMARY

After studying this chapter, you should know the following:

■ Alkenes are more reactive than alkanes because of the presence of an electron-rich π bond.

■ Alkenes are manufactured from long-chain alkanes by thermal or catalytic cracking.

■ Alkenes are a major source of petrochemicals, such as polymers and plastics, solvents, detergents and pharmaceuticals.

■ Alkenes burn in excess air to give carbon dioxide and water. Gaseous alkenes are sometimes used as fuels.

■ Alkenes react by electrophilic addition with halogens to give dihalogenoalkanes, hydrogen to give alkanes, hydrogen halides to give halogenoalkanes, and water (catalysed by acid) to give alcohols.

■ The addition of an electrophile to an unsymmetric alkene forms the most stable carbocation.

■ Alkenes can be oxidized by cold dilute potassium manganate(VII) to give diols.

■ Alkenes can be oxidized by hot acidified concentrated potassium manganate(VII) to give carbonyl compounds; the reaction cleaves the carbon–carbon double bond.

■ Alkenes are monomers that form addition polymers.

■ Ethene can be converted into the reactive synthetic intermediate epoxyethane by partial oxidation with oxygen in the presence of a silver catalyst

■ Epoxyethane reacts with nucleophiles, such as water, to make ethane-1,2-diol and poly(alkoxy) ethanols.

■ The double bond involves the overlap of two different types of atomic orbital to form a σ bond and a π bond.

■ Unsymmetric alkenes show geometric (*cis–trans*) isomerism because of the absence of free rotation about the carbon–carbon double bond.

## ? EXAM QUESTIONS

**1** State which *one* of the compounds in Fig 14.Q1 will exhibit geometrical isomerism:

**Fig 14.Q1**

WJEC Chemistry CH2 June 2001 Q2

**2** Table 14.Q2 gives data about the homologous series of alkenes.

**Table 14.Q2**

| Alkene | Molecular formula | $M_r$ | Boiling point/°C | Melting point/°C | Density/ g cm$^{-3}$ |
|---|---|---|---|---|---|
| Ethene | $C_2H_4$ | 28 | −105 | −169 | |
| Propene | $C_3H_6$ | 42 | −48 | −185 | |
| But-1-ene | $C_4H_8$ | 56 | −6 | −185 | |
| Pent-1-ene | $C_5H_{10}$ | 70 | 30 | −165 | 0.641 |
| Hex-1-ene | $C_6H_{12}$ | 84 | 64 | −140 | 0.674 |

**a)** Hept-1-ene is the next member of this homologous series. Predict a value for each of the following for hept-1-ene: **(i)** molecular formula, **(ii)** $M_r$, **(iii)** boiling point, **(iv)** melting point, **(iv)** density.

**b)** Explain the increasing boiling points with the increasing relative molecular mass.

**c)** Which members of this series are gases at room temperature and atmospheric pressure?

**3 a)** Explain what is meant by the terms *structural isomerism* and *cis–trans isomerism*. In your answer refer to isomers of $C_4H_8$ *only*. Clearly identify by formula and by name each isomer and state the key feature required for each type of isomerism.

**b)** Under certain conditions alkenes, such as propene, can react to form alcohols. State the reagents and conditions required. Write a balanced equation and explain why it is possible to produce *two* different alcohols.

OCR AS Module Chains and Rings Jan 2002 Q7

**4** Describe and explain how ethene reacts with bromine. You should include in your answer:
- any observations;
- the name of the organic product;
- a full description of the mechanism.

OCR AS Module Chains and Rings Jan 2001 Q7

**5** Electrophilic addition of hydrogen bromide to compound **A** in Fig 14.Q5 produces compound **B** as the major product. Outline the mechanism for the reaction in which compound **A** is converted into compound **B**. Identify, by name or structure, the isomer of compound **B** that is also formed in this reaction, and explain why compound **B** is the major product.

**Fig 14.Q5**

AQA. AS Module Unit 3a Introduction to Organic Chemistry June 2001 Q6(b)

**6** Ethene is bubbled through aqueous bromine.
**a)** Two products that contain bromine are formed:
**(i)** Draw displayed formulae for the two products.
**(ii)** Describe the mechanism for the formation of one of these two products.

**b)** If the aqueous bromine contains aqueous sodium chloride, a third product is formed, 2-bromo-1-chloroethane. Explain how this product is formed.

**7** Fig 14.Q7 shows a polyunsaturated alkene, Y, found in the secretion of a termite.

**Fig 14.Q7**

**a)** Write down the molecular formula for Y.

**b)** Does Y possess geometric (*cis–trans*) isomers? Explain your answer.

**c)** 0.012 mole of Y was hydrogenated. How many moles of hydrogen were needed to completely hydrogenate Y?

**d)** Y is shaken with aqueous bromine water. Describe and explain the observations you would make.

**8** Explain each of the following observations.
**a)** When hot acidified concentrated potassium manganate(VII) reacts with either *cis*- or *trans*-but-2-ene, only one product is formed. But when it reacts with *cis*- or *trans*-pent-2-ene, two products are formed.

**b)** Bromine in hexane reacts with ethene to give one product, but aqueous bromine reacts to give two products.

**c)** Chlorine in the dark reacts with but-1-ene to give one product, but in the presence of sunlight it gives several products.

## ? EXAM QUESTIONS

**9** Predict the major organic products for the reaction of 2,3-dimethylbut-2-ene with the following reagents. (Look at Fig 14.5 if you need to remind yourself of the displayed formula for 2,3-dimethylbut-2-ene.)

**a)** cold acidified potassium manganate(VII);

**b)** aqueous bromine;

**c)** chlorine in the dark;

**d)** hydrogen in the presence of a nickel catalyst;

**e)** hydrogen bromide.

**10** [Note: You may need to refer back to Chapter 9 to answer part **a)**.]

**a)** Explain, using *one* example in *each* case, together with relevant conditions, the following terms:
  **(i)** homolytic bond fission;
  **(ii)** cracking.

**b)** Ethene can be used as a starting material for the manufacture of a wide range of important polymers.
  **(i)** *Briefly* describe the structure of, and bonding in, a molecule of ethene.
  **(ii)** Give an equation to represent the formation of poly(ethene) from ethene.
  **(iii)** Give the names of *two* important polymers of alkenes or substituted alkenes.

**c)** **(i)** Write down *two* possible products of addition of hydrogen bromide to propene.
  **(ii)** Give the structure of *each* of the two carbocations (carbonium ions) that are formed at each first stage of the reaction in c)(i).
  **(iii)** State which carbocation is responsible for the major product formed. Give a reason to support your answer.

WJEC Chemistry CH2 June 2001 Q11

**11** Table 14.Q11 gives the names and molecular formulas of some alkenes present in crude oil.

**Table 14.Q11**

| Name | Molecular formula |
|------|-------------------|
| Methane | $CH_4$ |
| Ethane | $C2H_6$ |
| Propane | $C_3H_8$ |
| Hexane | $C_6H_{14}$ |
| Octane | $C_8H_{18}$ |
| Decane | $C_{10}H_{22}$ |
| Octadecane | $C_{18}H_{38}$ |

**a)** **(i)** What is the general formula of the alkanes?
  **(ii)** What is the difference in formulas between successive members of the alkanes?

**(iii)** Suggest the molecular formula of the alkane hexadecane?

**b)** **(i)** Explain why the cracking of long-chain alkanes such as octadecane is important.
  **(ii)** Under certain conditions octadecane, $C_{18}H_{38}$, can be cracked into hexane and an alkene. What is the molecular formula of the alkene?

**c)** Hexane can be cracked further into propane and propene, as shown in Fig 14.Q11.

hexane → propane + propene

Fig 14.Q11

**(i)** Predict the value of bond angle **a** in propane and bond angle **b** in propene.
**(ii)** Propene is unsaturated and contains a π bond. Explain, with the aid of a diagram, how p-orbitals are involved in the formation of the π bond in propene.
**(iii)** Describe how propane and propene could be distinguished by a simple chemical test.

**d)** A major commercial use of propene is the production of the polymer poly(propene).
  **(i)** Draw a section of poly(propene) showing *two* repeat units.
  **(ii)** State the type of polymerization.

**e)** Hydrocarbon polymers such as poly(propene) can be disposed of by using land-fill sites or by burning in an incinerator.
  **(i)** State *one* disadvantage of disposal in a land-fill site.
  **(ii)** State *one* possible advantage and one disadvantage if they are disposed of by burning.

OCR AS Module Chains and Rings Jan 2002 Q3

**12**

**a)** Propene, $C_3H_6$, readily undergoes electrophilic addition reactions. Copy the left part of Fig 14.Q12(a) and show, with the aid of curly arrows, the mechanism of the electrophilic addition reaction of propene with bromine.

Fig 14.Q12(a)

**b)** Propene also reacts as shown in Fig 14.Q12(b).

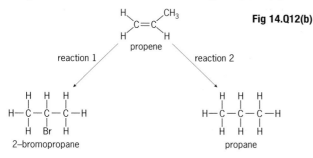

**Fig 14.Q12(b)**

**(i)** State a suitable reagent for reaction 1 in Fig 14.Q13(b).

**(ii)** State a suitable reagent and conditions for reaction 2 in Fig 14.Q12(b).

**(iii)** In the presence of an acid catalyst, propene can react with steam to form a mixture of two alcohols. Draw the structures of the two alcohols.

**c)** The scientists Ziegler and Natta were awarded a Noble Prize for chemistry in 1963 for their work on polymerization. Part of this work involved the polymerization of propene into poly(propene).

**(i)** What type of polymerization forms poly(propene)?

**(ii)** Draw a section of poly(propene) to show *two* repeat units.

**(iii)** State *two* difficulties in the disposal of poly(propene).

OCR AS Module Chains and Rings Jan 2001 Q4

# KEY SKILLS **ASSIGNMENT**

## UNSATURATED AND SATURATED FATTY ACIDS

Between a quarter and a half of the normal energy content of a person's diet comes from fats and oils. (This figure is beginning to decline as people become more aware of the dangers of eating too much fatty food.)

Fats and oils are members of a group of organic compounds called esters. There is no structural difference between a fat and an oil. The two terms merely distinguish between the state of the substances: a fat is a solid and an oil is a liquid at room temperature.

Esters are the result of the reaction between a carboxylic acid and an alcohol called propane-1,2,3-triol. The carboxylic acids in nature are often referred to as fatty acids and normally consist of even-numbered carbon chains (Table 14.A1). The fatty acids were given trivial names after their sources, so palmitic acid was derived from palm oil, lauric acid from the laurel tree, myristic acid from *Myristica fragrans* (the biological name for nutmeg) and stearic acid from stear (the Greek word for tallow or fat).

**Structure of a fat or oil**

Fig 14.A1 Structures of fats and oils and of the alcohol and acids that react to make them

**Table 14.A1 Eight fatty acids**

| Number of C atoms | Systematic name | Trivial name | Structural formula |
|---|---|---|---|
| 12 | Dodecanoic acid | Lauric acid | $CH_3(CH_2)_{10}COOH$ |
| 14 | Tetradecanoic acid | Myristic acid | $CH_3(CH_2)_{12}COOH$ |
| 16 | Hexadecanoic acid | Palmitic acid | $CH_3(CH_2)_{14}COOH$ |
| | *cis*-Hexadec-9-enoic acid | Palmitoleic acid | $CH_3(CH_2)_5CH=CHCH_2(CH_2)_6COOH$ |
| 18 | Octadecanoic acid | Stearic acid | $CH_3(CH_2)_{16}COOH$ |
| | *cis*-Octadec-9-enoic acid | Oleic acid | $CH_3(CH_2)_7CH=CHCH_2(CH_2)_6COOH$ |
| | *cis,cis*-Octadec-9,12-dienoic acid | Linoleic acid | $CH_3(CH_2)_4(CH=CHCH_2)_2(CH_2)_6COOH$ |
| | *cis,cis,cis*-Octadec-9,12,15-trienoic acid | Linolenic acid | $CH_3CH_2(CH=CHCH_2)_3(CH_2)_6COOH$ |

**1 a)** Octadecanoic acid is a solid, and *cis*-octadec-9-enoic acid is a liquid. In simple terms, how can you explain this difference, since they both have almost the same relative molecular mass? Hint: see page 215.

**b)** What is the molecular formula for hexadecanoic acid?

**c)** Suggest a chemical test that could be used to distinguish between octadecanoic acid and *cis*-octadec-9-enoic acid.

**d)** *cis,cis,cis*-Octadec-9,12,15-trienoic acid is a polyunsaturated fatty acid. Explain the meaning of the term polyunsaturated.

**e)** Fatty acids can be synthesized by animals and plants. What evidence is there in Table 14.A1 that the synthesis involves a derivative of ethanoic acid, $CH_3COOH$?

Vegetable oils tend to have a greater proportion of unsaturated fatty acids than animal fats, as shown in Table 14.A2.

Saturated fats are thought to be part of the cause of atherosclerosis, a thickening of the arteries. It is now recognized that unsaturated and, in particular,

# KEY SKILLS **ASSIGNMENT**

**Table 14.A2 The composition of eight fats and oils**

| Fat or oil | Saturated fatty acids | | | | | | Unsaturated (one C=C double bond) | | Polyunsaturated | Unsaturated |
|---|---|---|---|---|---|---|---|---|---|---|
| | $C_4, C_6, C_8, C_{10}$ | $C_{12}$ | $C_{14}$ | $C_{16}$ | $C_{18}$ | | $C_{16}$ | $C_{18}$ | $C_{18}$ | $C_{20}, C_{22}, C_{24}$ |
| tallow (cow) | | | 2–3 | 24–32 | 14–32 | | 1–3 | 35–48 | 2–4 | |
| lard (pork) | | | 1–2 | 28–30 | 12–18 | | 1–3 | 41–48 | 6–7 | 2 |
| butter | 7–10 | 2–3 | 7–9 | 23–26 | 10–13 | | 5 | 30–40 | 4–5 | 2 |
| whale | | | 4–5 | 11–18 | 2–4 | | 13–18 | 33–38 | | 17–31 |
| olive | · | | 0–1 | 5–15 | 1–4 | | 0–1 | 69–84 | 4–12 | |
| coconut | 10–22 | 45–51 | 17–20 | 4–10 | 1–5 | | | 2–10 | 0–2 | |
| soya bean | | | 0–1 | 6–10 | 2–6 | | | 21–29 | 54–67 | |
| linseed | | | | 4–7 | 2–5 | | | 9–38 | 28–91 | |

polyunsaturated fats (those that contain the polyunsaturated fatty acid residues) are much better for health.

**2** **a)** Why is whale oil a healthier source of fat than tallow or lard?

**b)** Suggest why whale oil is no longer considered to be an acceptable alternative to tallow or lard in a human diet.

**c)** What is unusual about the composition of coconut oil?

**d)** Linseed oil is not used as an edible oil. One of its uses is to harden the surface of wood (such as the willow in a cricket bat). On exposure to air, the oil solidifies and hardens. Suggest a reason, in terms of its composition, why this happens to linseed oil, whereas olive oil remains a liquid.

The consumption of low-fat spreads and margarines has greatly increased. These products contain vegetable oils that have been solidified. The only way to solidify polyunsaturated fats is to hydrogenate some of their double bonds. To do this, the vegetable oil is mixed with a nickel catalyst in the presence of hydrogen under pressure.

**3** **a)** Write down an equation to show the hydrogenation of *cis*-hexadec-9-enoic acid.

**b)** What is the name of the product formed?

**c)** Explain how hydrogenation could be used to find out the number of double bonds present in a pure sample of an unknown fatty acid of known relative molecular mass.

Not all the double bonds in the polyunsaturated fatty acid are hydrogenated – just those needed to produce the correct physical properties required of a margarine or low-fat spread. However,

**Fig 14.A2** The apparatus used to hydrogenate unsaturated oils

there is one problem with the hydrogenation: it can change the geometric isomerism present in the fatty acid.

Most naturally occurring fatty acids are of the *cis* variety, but after hydrogenation up to 25 per cent of the *cis* double bonds may be changed into *trans* double bonds. The presence of the *trans* double bond completely alters the three-dimensional shape of a fatty acid. Such *trans* fatty acids are no longer 'essential', and there is unconfirmed evidence that they may be responsible for certain heart diseases. It seems that the promotion of polyunsaturated fats as healthier food may have inadvertently introduced a health risk from *trans* fatty acids themselves.

**4** **a)** During hydrogenation, some *trans* double bonds are produced. What must happen to the double bond during the hydrogenation for a *cis*-isomer to be changed into a *trans*-isomer?

**b)** Draw the displayed formula of *cis*-octadec-9-enoic acid, showing clearly the arrangements of the atoms around the carbon–carbon double bond.

**c)** Draw the displayed formula of *trans*-octadec-9-enoic acid, showing clearly the arrangements of the atoms around the carbon–carbon double bond.

**d)** Suggest why the *trans*-isomer cannot be metabolized in the same way as the *cis* fatty acids.

Fats such as butter turn rancid when left at room temperature for some time. During this change, a complex oxidation reaction occurs that involves the cleavage of double bonds in the fatty acids, with the formation of carboxylic acids with much shorter chains. It is these carboxylic acids that have such an unpleasant odour.

**5** **a)** Butter contains some *cis,cis*-octadec-9,12-dienoic acid. Suggest the identity of some of the short-chain carboxylic acids that can be formed from the oxidation of this unsaturated fatty acid.

**b)** Suggest what can be done to prevent butter from going rancid.

# 15

# Aromatic compounds and arenes

# 15 Aromatic compounds and arenes

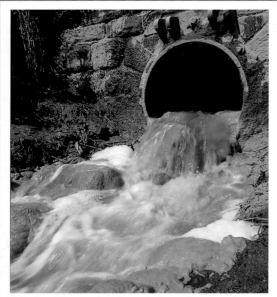

Benzene is one of the most industrially important organic molecules. It features in the manufacture of petrochemicals, which are solvents, detergents, insecticides, dyes and polymers. The benzene molecule is normally unreactive, so to make these compounds requires special conditions such as high pressures and high temperatures. Almost all the reactions of benzene require catalysts, many of them hazardous environmental pollutants that can pollute water supplies and poison the organisms in them.

So there is an urgent need to develop environmentally friendly catalysts – *envirocats* – that can be separated easily by filtration from mixtures after they have reacted. Envirocats are based upon clay materials, which are made acidic by adding metal salts. Since they can be filtered from any aqueous effluent, envirocats can be prevented from harming water supplies and aquatic life. In addition, since envirocats are solid, their controlled disposal is easy. They have the economic benefit of conventional catalysts, too – envirocats can be used again and again, but eventually they lose their catalytic activity.

**Pollution** The use of envirocats will greatly reduce the pollution of waterways by effluents

## 1 AROMATIC COMPOUNDS AND THEIR USES

In 1825, a young scientist at the Royal Institution of London was asked to remove the oily residue that collected in the gas cylinders of gas lamps and analyse it. He found that the residue contained a previously unknown hydrocarbon, for which the molecular formula was later shown to be $C_6H_6$. The scientist was Michael Faraday, one of the greatest scientists of the nineteenth century, though his name is seldom linked with organic chemistry.

The molecule of every aromatic compound contains at least one benzene ring. The name 'aromatic', meaning 'with an odour or smell', was given to these compounds because the first few to be discovered and used do, indeed, have odours. However, now we know that this was just a coincidence. Many aromatic compounds have no odour; whereas others have a particularly nasty smell. Also, some fragrant compounds do not contain a benzene ring (see page 287). Aromatic compounds can contain other functional groups as well as alkyl groups.

The benzene ring is based upon a hexagonal arrangement of six carbon atoms and six delocalized $\pi$ electrons, which give the ring unique stability. Fig 15.1 shows benzene itself. The hydrogen atoms in the benzene molecule can be replaced by alkyl, aryl or other functional groups, as Fig 15.2 shows. Note the way in which the benzene ring is represented in compounds: it is a skeletal formula. The atomic symbols are not included, so remember that the benzene ring consists of six hydrogen atoms and six carbon atoms. The significance of the circle drawn inside the hexagon is explained on pages 327 and 328.

**Fig 15.1** The displayed formula of benzene

## ? TEST YOURSELF

**A** Draw the structure of each of these aromatic compounds:

a) $C_7H_8$;

b) $C_6H_5Cl$;

c) $C_8H_{10}$.

2-phenylethan-1-ol is found in rose oil and is used in perfumes

methyl-2-hydroxybenzoate (or methyl salicylate) is an analgesic

naphthalene is used in mothballs

CS is an active component of tear gas

warfarin is used as a blood anticoagulant and a rodent killer

salbutamol is used as a bronchodilator in asthma inhalers

an azo dye

**Fig 15.2** Seven aromatic compounds and their uses. Note that each compound has at least one benzene ring

## SOME USES OF AROMATIC COMPOUNDS

Fig 15.2 features compounds that have one thing in common: the benzene ring. They are therefore all called aromatic compounds. All but two (2-phenylethan-1-ol and naphthalene) have to be synthesized industrially. The starting point for some of these compounds is benzene, $C_6H_6$, so the benzene part of their carbon skeleton is in place. Chemical changes are then needed to substitute one or more of the hydrogen atoms with one or more of the alkyl or aryl groups.

## 2 STRUCTURE OF BENZENE

Benzene is one of the class of hydrocarbons known as arenes. Arenes are unsaturated hydrocarbons, characterized by great stability and non-reactivity, which distinguishes them from alkenes. All arenes contain a delocalized $\pi$ system of electrons in their benzene rings.

### KEKULÉ AND THE FIRST STRUCTURE OF BENZENE

Following the discovery of benzene, chemists had the challenge to try and determine its structure. In the early nineteenth century, benzene's molecular formula, $C_6H_6$, did not fit the way that chemists thought carbon atoms bonded in organic molecules.

It was 40 years later that August Kekulé proposed the ring structure of benzene, the first time a chemist realized that carbon atoms could form rings as well as chains. His brilliantly original idea was attributed to a dream he had about a snake biting its tail. Kekulé drew the structure of benzene as cyclohexa-1,3,5-triene (Fig 15.5).

Kekulé's structure suggests that benzene has three single bonds and three double bonds, so that it should react in the same way as cyclohexene and be able to undergo electrophilic addition. It further suggests that benzene should have long carbon–carbon bonds, C—C, and short carbon–carbon bonds, C=C.

### FAULTS IN KEKULÉ'S MODEL

Structural investigations using X-ray crystallography and infrared spectroscopy show that the benzene molecule is a regular hexagon of carbon atoms, with

**Fig 15.3** Salbutamol, an aromatic compound that contains alcohol and amine functional groups, is the world's most widely used bronchodilator: a bronchodilator relaxes the airways of people with breathing problems

> ✔ **REMEMBER THIS**
>
> **Arenes** are strictly hydrocarbons.
> **Aromatic** compounds can contain other functional groups.

**Fig 15.4** A snake biting its tail featured in Kekulé's inspirational dream

Sometimes Kekulé's structure is shown without the carbon and the hydrogen atoms being shown

**Fig 15.5** Kekulé's structure of benzene (cyclohexa-1,3,5-triene)

the lengths of all six carbon–carbon bonds equal to 140 pm. This value lies between the average bond length for a single bond, 154 pm, and the average bond length for a double bond, 134 pm.

A further complication is that benzene does not behave like an alkene. It reacts by electrophilic *substitution* rather than by electrophilic addition (see page 311). Table 15.1 shows some of the differences between cyclohexene, an alkene with a six-carbon ring, and benzene. The reactions of the two compounds are very different. This suggests that benzene cannot be an alkene and therefore cannot have Kekulé's structure of cyclohexa-1,3,5-triene.

**Table 15.1 Comparison of the reactions of benzene and cyclohexene**

| Reagent | Cyclohexene (alkene) | | Benzene (arene) | |
|---|---|---|---|---|
| Bromine | Electrophilic addition to give 1,2-dibromocyclohexane | | Electrophilic substitution to give bromobenzene if catalyst used | |
| Chlorine | Electrophilic addition to give 1,2-dichlorocyclohexane | | Electrophilic substitution to give chlorobenzene if catalyst used | |
| Hydrogen | Addition to give cyclohexane | | Addition to give cyclohexane | |
| Cold aqueous acidified potassium permanganate | Oxidative addition to give cyclohexane-1,2-diol | | No reaction | |
| Hydrogen bromide | Electrophilic addition to give bromocyclohexane | | No reaction | |
| Oxygen (combustion) | Carbon dioxide and water | | Carbon dioxide and water | |
| Iodomethane | No reaction | | Electrophilic substitution to give methylbenzene if catalyst used | |

## ENTHALPY CHANGE OF HYDROGENATION

Unsaturated compounds, such as alkenes, react with hydrogen under pressure in the presence of a nickel catalyst to form alkanes (Fig 15.6 and 15.7). This reaction is known as **hydrogenation** (see page 315).

**Fig 15.6** Bond breaking and bond making during hydrogenation. Whatever groups are attached to the double bond, the bonds made and broken during hydrogenation are the same. When a molecule has a C=C bond, 1 mole of molecules will involve the breaking of 1 mole of C=C and 1 mole of H—H to make 1 mole of C—C and 2 moles of C—H

Bonds broken during hydrogenation of 1 mole of alkene:
1 mole of C=C
1 mole of H—H

Bonds made during hydrogenation of 1 mole of alkene:
1 mole of C—C
2 moles of C —H

The **enthalpy change of hydrogenation** is defined as the enthalpy change that occurs when one mole of an alkene reacts with hydrogen to form an alkane. Studies of the hydrogenation of different alkenes show that for a molecule with one C=C bond, the enthalpy change of hydrogenation is about $-120$ kJ mol$^{-1}$. For a molecule with two double bonds, the enthalpy change of hydrogenation is about $-240$ kJ mol$^{-1}$.

Kekulé's structure for benzene has three C=C bonds, so we might expect the enthalpy change of hydrogenation for benzene to be about $-360$ kJ mol$^{-1}$. However, experimental data shows that its value is only $-208$ kJ mol$^{-1}$, which is a considerably smaller enthalpy change than the theoretical value. The only conclusions that can be drawn from the experimental data are that the benzene ring does not contain three C=C bonds, and that benzene is more stable than cyclohexa-1,3,5-triene.

**Fig 15.8** Benzene molecules as seen by a scanning tunnelling electron microscope, which shows the electron distributions in the molecules

## MODERN THEORIES OF THE BENZENE MOLECULE

The value of the enthalpy change of hydrogenation of benzene is conclusive evidence that the benzene ring does not have the C=C bonds that Kekulé predicted. In addition, the enthalpy change transfers much less energy to the surroundings than his model predicts, which indicates that the bonding of the carbon atoms in the benzene ring is much stronger than Kekulé's model predicted, and so benzene is more stable.

## RESONANCE STRUCTURE MODEL OF THE BENZENE MOLECULE

Kekulé's model of benzene is inaccurate in the ways we have seen, and does not match the known physical dimensions of the benzene molecule. Now let's

**Fig 15.7** Hydrogenation of alkenes to form saturated compounds called alkanes

R$^1$, R$^2$, R$^3$ and R$^4$ can be H, alkyl or aryl groups

**? TEST YOURSELF**

**B** **a)** Write down equations to show the hydrogenation of:
(i) cyclohexene;
(ii) cyclohexa-1,3-diene;
(iii) cyclohexa-1,3,5-triene (Kekulé's structure).
**b)** Predict the enthalpy change of hydrogenation for:
(i) cyclohexene;
(ii) cyclohexa-1,3-diene;
(iii) cyclohexa-1,3,5-triene.

**Fig 15.9** Energy level diagram for the enthalpy change of hydrogenation for benzene

**Fig 15.10** Two Kekulé structures of benzene that contribute to the resonance hybrid model of benzene

**Fig 15.11** The displayed formula of benzene, in which the six delocalized π electrons are represented by a circle

**ORBITAL OVERLAP IS COVERED ON PAGE 81, AND BONDING IN ETHENE ON PAGE 307.**

look at Fig 15.10, which shows the structure of a benzene molecule as a hybrid of two Kekulé structures. The structure of benzene is called a **resonance hybrid**. Despite the name 'resonance hybrid', this model does not imply that the hybrid structure is continually changing from one to the other, but rather that it is a cross between the two structures, and possesses some of the characteristics of each. This means that each carbon–carbon bond has some single-bond and some double-bond character. For this model it is often better to draw the structure of the benzene ring as shown in Fig 15.11. The circle represents a bond that contains six π electrons that are delocalized around the ring.

### ORBITAL OVERLAP MODEL

Covalent bonds are formed when atomic orbitals overlap (Fig 15.12). The C–H bond is a σ bond formed by the overlap of a hydrogen 1s atomic orbital with an atomic orbital on carbon. The C–C bond is a σ bond formed by the overlap of two atomic orbitals, one from each carbon atom.

In benzene, there is a third type of bond, which is formed by the side-to-side overlap of six 2p atomic orbitals on carbon. This forms an orbital that spreads, rather like a ring doughnut, above and below the plane of the carbon atoms that possess the six electrons. These electrons are not found between any particular atoms and so are known as delocalized π electrons.

**(a)**

C–H σ bonds are made by the overlap of a hydrogen atomic orbital with a carbon atomic orbital
C–C σ bonds are made by the overlap of two carbon atomic orbitals

**(b)**

π bonds are formed by side-by-side overlap of all six 2p atomic orbitals

**Fig 15.12** Molecular orbital description of benzene: (a) σ bonds; (b) π bonds

### ? TEST YOURSELF

**a)** Write down the balanced equation for the complete combustion of benzene in air.
**b)** Would you expect the enthalpy change of combustion predicted from the bond energies of Kekulé's model to be the same as the experimentally measured value? Explain your answer. Hint: You many need to read page 174 about the enthalpy changes during bond breaking and bond making.

**SEE QUESTION 1.**

### STABILIZATION ENERGY

Earlier in the chapter, we saw that the enthalpy change of hydrogenation of benzene is about 150 kJ mol$^{-1}$ less than expected from the Kekulé model. This suggests that benzene is about 150 kJ mol$^{-1}$ more stable than cyclohexa-1,3,5-triene (Kekulé's structure). This energy difference is sometimes referred to as the stabilization energy of benzene and is associated with the delocalized π electrons.

The idea of stabilization energy means that to break the benzene ring you need *more* energy than is predicted from Kekulé's model.

**Fig 15.13** In combustion, benzene transfers less energy to its surroundings than would be expected from the Kekulé structure. This is because of the stability given by the delocalized electrons in the benzene ring

## STRUCTURAL FORMULA FOR BENZENE

Fig 15.14 shows three ways to represent the displayed formula for benzene. Note that neither the hydrogen atoms nor the carbon atoms are shown. It is a skeletal formula. Every time you see one of these skeletal formulas of the benzene ring, remember that there are six hydrogen atoms attached to the ring. The circle inside the third hexagon represents the delocalized electrons. Each representation in Fig 15.14 has its advantages and disadvantages. Formulas A and B show three discrete double bonds (which do not exist in benzene), and formula C shows that some electrons are delocalized, but not how many. Nowadays formula C is the most accepted way to represent the structure of benzene.

**Fig 15.14** Three representations of the benzene ring. These are all skeletal formulas

**? TEST YOURSELF**

**D** What is the molecular formula for:
**a)** naphthalene;
**b)** pyrene?

## 3 NAMING AROMATIC COMPOUNDS

Arenes is the group name we give to benzene and the hydrocarbons that are just benzene rings joined together (fused). Arenes have similar chemical properties because their benzene rings have delocalized electrons. Fig 15.15 shows three arenes and their traditional names.

We have already described aromatic compounds in general as those that contain at least one benzene ring. Aromatic compounds are not necessarily hydrocarbons, but can contain a wide variety of other functional groups.

| naphthalene | pyrene | anthracene |

**Fig 15.15** Three arenes: they all have fused benzene rings

## SUBSTITUTED ARENES

To name aromatic compounds, we use the name of the parent arene together with a prefix or suffix to indicate the groups of atoms attached to the benzene ring. So, for example, the prefix 'methyl' in the name methylbenzene indicates that the methyl group has replaced one of the hydrogen atoms in the benzene ring. When more than one hydrogen atom is replaced by methyl groups, as in dimethylbenzene, we specify the positions of the methyl groups by numbering the carbon atoms. So there are three structural isomers that could be called dimethylbenzene: 1,2-dimethylbenzene, 1,3-dimethylbenzene and 1,4-dimethylbenzene.

1,3-Dichlorobenzene has a benzene ring with two chlorine atoms in place of hydrogen atoms. The first chlorine is attached to the carbon we number as carbon 1, and the second to carbon 3. A benzene ring is numbered clockwise or anticlockwise, depending on which direction gives the lower position-numbers. Fig 15.17 shows an example of the procedure.

As Fig 15.18 shows, 1-chloro-2-methylbenzene has two different groups attached to the benzene ring. Notice that we number the carbon attached to the chloro group as carbon 1, rather than the carbon attached to the alkyl group. (Halogens are named before alkyl groups.)

**Fig 15.16** Methylbenzene

**? TEST YOURSELF**

**E** **a)** Draw the three structural isomers of dimethylbenzene.
**b)** Explain why there is not a compound called 1,5-dimethylbenzene.

**F** Draw the structures of:
**a)** ethylbenzene;
**b)** 1,2-diethylbenzene;
**c)** 1,3-diethylbenzene.

**G** **a)** Draw the structures of:
(i) 1-chloro-3-iodobenzene;
(ii) 1-chloro-4-iodobenzene.
**b)** A student names a compound 4,5-dichlorobenzene. Explain why this name must be incorrect.

**Fig 15.17** Naming 1,3-dichlorobenzene Note: this is 1,3-dichlorobenzene and not 1,5-dichlorobenzene, to keep the position-numbers as low as possible. The structure could equally well have been drawn as its mirror image, in which case the numbering would have been anticlockwise

**Fig 15.18** 1-Chloro-2-methylbenzene

$C_6H_5-$

two representations of the phenyl group

**Fig 15.19** The phenyl group

**H** **a)** Write down the displayed formula for:
(i) 2-phenylethan-1-ol;
(ii) phenylethene.
**b)** What is the name of each of the four compounds in Fig 15.21?

$C_6H_5CH_2CH_2CH_2OH$

$C_6H_5CH_2CH_2CH$

**Fig 15.21**

**I** Draw the structures of:
**a)** 2,4,6-trichlorophenol;
**b)** 2,4,6-trimethylphenol.

benzoic acid          benzaldehyde

**Fig 15.24** Benzoic acid and benzaldehyde

## PHENYL GROUP

The phenyl group is $C_6H_5$ (Fig 15.19). It is found attached to carbon chains and to other functional groups. The phenyl group is an example of an **aryl** group. An aryl group is the aromatic equivalent of an alkyl group (see page 213) and always contains at least one benzene ring. In phenylamine, for example, the phenyl group is attached to the amine functional group. Phenylamine has the formula $C_6H_5NH_2$.

phenylethanone          or     $C_6H_5COCH_3$

**Fig 15.20** Phenylethanone. The position number of the phenyl group is not given, because the phenyl group has to be on carbon 1 for the compound to be a ketone

Compounds with some substituents (groups) on the benzene ring retain their traditional names. We refer to phenol instead of hydroxybenzene. In phenol, the carbon attached to the OH group is called carbon 1.

**Fig 15.22** Phenol

**Fig 15.23** 2,4,6-Tribromophenol

## BENZOIC ACID AND BENZALDEHYDE

Benzoic acid and benzaldehyde (Fig 15.24) are two aromatic compounds that have retained their traditional names. Note carefully that both of these compounds contain seven carbon atoms rather than six, because the functional groups contain a carbon atom.

**! SCIENCE FEATURE** Arenes in exhaust fumes

Petrol contains aromatic hydrocarbons, which help to improve its octane rating. (See an account of octane rating on page 218.)

In an internal combustion engine, complete combustion of the fuel (petrol or diesel) rarely occurs. Some carbon monoxide, soot and smoke is always produced. Analysis of the soot by gas–liquid chromatography and spectroscopy shows that it contains the arenes benzene, naphthalene, phenanthrene, pyrene, coronene and ovalene (Fig 15.25). The presence of benzene is of particular concern, since it is a known carcinogen.

**Fig 15.25** Gas–liquid chromatography and spectroscopic analysis of the volatile components of the soot from diesel-engine exhaust reveal many different arenes (structurally represented here as hexagons)

**Fig 15.26** There is much concern about the amount of arenes emitted in vehicle exhaust fumes, since many arenes are known carcinogens

**YOU CAN READ ABOUT GAS–LIQUID CHROMATOGRAPHY ON PAGE 254.**

? **TEST YOURSELF**

**J** Deduce the molecular formula for each of the arenes shown in Fig 15.25.

# 4 DEMAND FOR AROMATIC COMPOUNDS

The synthesis of aromatic compounds often starts with one of the arenes. Aromatic compounds are needed for new technologies and new synthetic methods, and so the demand for arenes continues to expand. Until the 1940s, the main source of commercially produced benzene was coal, but today the principal source is crude oil.

## AROMATIC COMPOUNDS FROM COAL

When coal is heated to a high temperature in the absence of air, coke and a complex mixture of compounds are produced. Fig 15.27 shows the variety of aromatic compounds that can be formed. Notice that not only arenes are produced, but also substituted benzenes such as phenols.

**Fig 15.27** Destructive distillation of coal to produce arenes and substituted arenes. The importance of coal in this application is declining

**Fig 15.28** Reforming of cyclohexane to give benzene. Note that the carbon skeleton (the six-membered ring) remains intact

**Fig 15.29** Reforming of heptane

**CRACKING AND REFORMING ARE COVERED ON PAGES 203 TO 204.**

## ? TEST YOURSELF

**K** **a)** Into which aromatic hydrocarbon can 1,2-dimethylcyclopentane be reformed?
**b)** Write a balanced equation for this reaction.

## ✔ REMEMBER THIS

An electrophile is a species that accepts a lone pair of electrons and forms a covalent bond.

# AROMATIC COMPOUNDS FROM CRUDE OIL

Benzene and other aromatic compounds form only a small proportion of crude oil, and fractional distillation does not produce sufficient benzene or other aromatic compounds to match demand. So, as in the case of alkenes, cracking and reforming are processes used to increase the supply of arenes.

## CATALYTIC REFORMING

In catalytic reforming, $C_6$ to $C_{10}$ hydrocarbon vapours are heated in the presence of hydrogen at high pressure and high temperature, typically 900 °C, using a transition-element catalyst.

During the reforming process, the number of carbon atoms per molecule remains the same, but a change occurs in the carbon skeleton and/or the number of hydrogen atoms per molecule. For example, cyclohexane can be reformed to give benzene (Fig 15.28), and heptane can be reformed to give methylbenzene (Fig 15.29). In the second example, note that not only are hydrogen atoms lost, but also the carbon skeleton is rearranged.

## MANUFACTURE OF BENZENE

Naphtha is a light fraction of crude oil that contains liquid alkanes of low relative molecular mass (see page 204). To make benzene, naphtha vapour is passed over a catalyst, such as platinum or molybdenum(VI) oxide, on aluminium oxide at 500 °C and 10–20 atmospheres of pressure. Compounds such as hexane found in naphtha are converted into ring compounds such as cyclohexane, which is then dehydrogenated to make benzene.

$$CH_3CH_2CH_2CH_2CH_2CH_3 \rightarrow C_6H_{12} \rightarrow C_6H_6$$

# 5 ELECTROPHILIC SUBSTITUTION REACTIONS OF BENZENE

In almost all synthetic schemes that start with an aromatic hydrocarbon (such as benzene), the substitution of one or more of the hydrogen atoms occurs. This substitution is normally electrophilic, and so only those groups that can behave as electrophiles are able to replace hydrogen atoms attached to the benzene ring.

## MECHANISM OF ELECTROPHILIC SUBSTITUTION

The presence of the delocalized electrons means that benzene is electron rich. Therefore, an electrophile that is able to accept an electron pair donated by the ring is the most likely type of reagent to react with benzene. This is very similar to the first step in the electrophilic addition of an alkene (see page 311). The resulting positive species is much more stable than a normal carbocation, since it is possible to delocalize the positive charge around the ring (Fig 15.30). A proton is then eliminated from the positive ion to restore that stability of the benzene ring. We saw the unique stability of the benzene ring when looking at the bonding in benzene. This stability is also apparent in the mechanism of electrophilic substitution.

**Fig 15.30** The mechanism of electrophilic substitution of benzene

intermediate has positive charge delocalized over five carbon atoms

Electrophilic substitution clearly needs an electrophile, but what is not obvious is that the electrophile must be an exceptionally good electrophile. This is because it has to react with a molecule made extremely stable by the presence of the π system of delocalized electrons. Benzene has a delocalized π system of electrons over all six atoms of the ring. But Fig 15.30 shows that in the positive intermediate, the π system involves only five of the carbon atoms. In practice, this means that the electrophile normally needs to be a fully positive species that is formed as a result of the heterolytic fission of a covalent bond (Fig 15.32). An induced dipole within a molecule is not normally enough to achieve electrophilic attack in benzene.

It is not easy to form a sufficiently powerful electrophile to attack a benzene molecule, so a catalyst is used in the reaction mixture to generate the electrophile. Typical electrophiles that can be generated in this way include $Br^+$, $Cl^+$, $R^+$, $RCO^+$ and $NO_2^+$. (R is an aryl or alkyl group.) You will notice that the charge carried by the atoms involved is not often what you would have expected by considering their electronic arrangements, which offers an explanation for why catalysts are needed to generate the electrophiles.

## HALOGENATION OF BENZENE

Chlorine and bromine can both be converted into electrophiles that will attack a benzene ring. The resulting substitution reaction makes chlorobenzene or bromobenzene.

$$C_6H_6 + X_2 \rightarrow C_6H_5Cl + HX \text{ (where X is Br or Cl)}$$

Notice that the equation is written with molecular formulas rather than structural or displayed formulas. This is acceptable provided that there is only one structure possible.

## CHLORINATION OF BENZENE

When chlorine is bubbled through a mixture of refluxing benzene in the presence of a catalyst (such as aluminium chloride or iron(III) chloride), electrophilic substitution takes place to produce chlorobenzene:

$$C_6H_6 + Cl_2 \xrightarrow{\text{FeCl}_3, \text{ reflux}} C_6H_5Cl + HCl$$

We already know the mechanism of this reaction: $Cl^+$ is represented by $E^+$ in Fig 15.30. The major question is: How does aluminium chloride help to generate the electrophile?

We might expect aluminium chloride to be ionic, since it is a metal compound. However, in the anhydrous state it has a high degree of covalent character, so it is better when drawing a dot-and-cross diagram to show the electrons as shared (Fig 15.33). This structure has the molecular formula $AlCl_3$, though there is strong evidence to suggest that the formula is actually $Al_2Cl_6$.

The dot-and-cross diagram of Fig 15.33 shows that $AlCl_3$ is an electron-deficient molecule, since the outer shell of the aluminium atom has only six electrons.

**Fig 15.31** The overall reaction of electrophilic substitution

**Fig 15.32** The electrophile is formed by the heterolytic fission of a covalent bond, so that both electrons in the bond go to atom X

**? TEST YOURSELF**

**L** **a)** Write down the equation for the reaction of $Cl^+$ with the benzene.
**b)** Draw the mechanism for the reaction of benzene with $NO_2^+$.

**M** What is the electric charge you would expect to be carried by a chlorine atom when it forms an ion?

**FOR MORE INFORMATION ON ANHYDROUS ALUMINIUM CHLORIDE, SEE CHAPTER 21.**

**Fig 15.33** Dot-and-cross diagram for aluminium chloride

**Fig 15.34** Dot-and-cross diagrams to show the formation of $Cl^+$. The electron deficiency of the aluminium chloride is transferred to the chlorine electrophile $Cl^+$. Note that the aluminium atom has eight electrons in its outer shell after the reaction, while the $Cl^+$ ion has six

R = alkyl group

**Fig 15.35** The alkylation of benzene: the Friedel–Crafts reaction

**Fig 15.36** Generation of carbocations by two routes

**Fig 15.37** Dot-and-cross diagram to show the formation of a methyl carbocation from iodomethane

Anhydrous aluminium chloride transfers its electron deficiency to a chlorine atom, to form the electrophile $Cl^+$ (Fig 15.34) and a tetrachloroaluminate ion, $AlCl_4^-$. The catalyst is often referred to as a halogen carrier. Once the electrophile is generated, the reaction proceeds to give the substitution product. The proton eliminated by the benzene ring at the end of the reaction can then react with the $AlCl_4^-$ ion to give hydrogen chloride.

Anhydrous iron(III) chloride acts as a catalyst in the same way. In fact, iron filings could be used as a catalyst, because iron(III) chloride would be formed in the reaction vessel by the reaction between iron and chlorine.

### BROMINATION OF BENZENE

When a solution of bromine in benzene is refluxed with aluminium chloride, electrophilic substitution takes place to form bromobenzene:

$$C_6H_6 + Br_2 \rightarrow C_6H_5Br + HBr$$

Notice that the catalyst is not included in the overall chemical equation, even though it is intimately involved in the production of the electrophile. We do not have to use aluminium bromide as the catalyst, because aluminium chloride will work instead, in which case the bromo-trichloroaluminate ion, $AlBrCl_3^-$, is formed instead of $AlBr_4^-$.

### ALKYLATION OF BENZENE

This reaction is known as the Friedel–Crafts reaction. In it, a hydrogen atom is substituted by an alkyl group (Fig 15.35). The importance of this reaction is that a carbon–carbon bond is made. This time the problem is to make a carbocation that can act as an electrophile.

There are two main ways in which the electrophile (the carbocation) can be generated. One involves the heterogeneous fission of a halogenoalkane, and the other the electrophilic addition of a proton to an alkene (Fig 15.36).

alkene          carbocation          carbocation

### CARBOCATIONS FROM HALOGENOALKANES

Typically, benzene, a halogenoalkane and anhydrous aluminium chloride are refluxed together to give an alkylbenzene. For example, iodomethane gives methylbenzene:

$$C_6H_6 + CH_3I \xrightarrow{\text{AlCl}_3,\ \text{reflux}} C_6H_5CH_3 + HI$$

Let's look at the formation of the carbocation that permits this reaction. Fig 15.37 shows that it is similar to the way in which the electrophile $Cl^+$ is formed (see Fig 15.34). Again, the aluminium chloride can transfer its electron deficiency, this time to an alkyl group, to form the carbocation.

## CARBOCATIONS FROM ALKENES

Alkenes typically react by electrophilic addition (Fig 15.38). When the electrophile is a proton, a carbocation is formed as an intermediate. When this is formed in the presence of benzene, the carbocation can react with benzene, rather than completing the addition reaction.

ELECTROPHILIC ADDITION TO ALKENES IS COVERED ON PAGES 311 TO 314.

Many different types of acid catalyst can be used in this reaction, including concentrated sulphuric acid, concentrated phosphoric(V) acid and aluminium chloride.

**Fig 15.38** The mechanism of the acid-catalysed reaction of an alkene (in this case ethene) with benzene. Note that the proton H$^+$ reacts to generate the carbocation, but is later regenerated; hence it is a catalyst

### EXTENSION Manufacture of poly(phenylethene)

Poly(phenylethene), common name polystyrene, is familiar to us as the material of hot drinks cups, ceiling tiles and packaging for fragile objects. It is manufactured from two readily available hydrocarbons, ethene and benzene (Fig 15.39).

**Fig 15.39** The manufacture of poly(phenylethene)

**Fig 15.40** Mobile Telephones are made from poly(phenylethene)

Ethene reacts with benzene in the presence of aluminium chloride, with hydrochloric acid as an acid catalyst, to form ethylbenzene. This is an example of electrophilic substitution. The purpose of the acid catalyst is to generate an ethyl carbocation that acts as the electrophile.

The ethylbenzene from the first reaction provides the correct carbon skeleton for the next stage. This involves a dehydrogenation reaction, in which two hydrogen atoms are lost to form phenylethene. Finally, the phenylethene is polymerized normally under the influence of a free-radical initiator.

**Fig 15.41** Expanded poly(phenylethene), also known as polystyrene, is used for packaging material

## ACYLATION OF BENZENE

Acylation is the substitution of a hydrogen atom by an acyl group, RCO (Fig 15.42).

SEE QUESTION 3.

THERE IS MORE INFORMATION ON THIS REACTION IN THE ASSIGNMENT, PAGE 347.

**Fig 15.42** Acylation of the benzene ring. A hydrogen atom is substituted by the acyl group RCO

**Q a)** Write down the structure of the product formed in the reaction between propanoyl bromide, $CH_3CH_2COBr$ and benzene.
**b)** Draw the structure of the reagent used to make 1-phenylbutan-1-one from benzene.

This reaction requires the formation of an acylium ion, $RCO^+$, to act as an electrophile. The acylium ion is produced by the heterolytic fission of a carbon–chlorine bond in an acyl chloride catalysed by anhydrous aluminium chloride (Fig 15.43). For example, ethanoyl chloride, aluminium chloride and benzene react together to give phenylethanone (Fig 15.44).

**Fig 15.43** Generation of an acylium ion

**Fig 15.44** Reaction of ethanoyl chloride with benzene

## NITRATION OF BENZENE

It is possible to introduce the nitro group into the benzene ring (Fig 15.45) by means of a nitrating mixture – in this case, concentrated nitric acid and concentrated sulphuric acid. Nitrobenzene is formed when benzene is heated at less than 60 °C with a nitrating mixture. Keeping the temperature below 60 °C stops the formation of di- or trinitrobenzene.

The electrophile in the reaction (Fig 15.46) is the nitryl or nitronium ion, $NO_2^+$, which is generated by a reaction between the two acids:

$$2H_2SO_4 + HNO_3 \rightarrow NO_2^+ + 2HSO_4^- + H_3O^+$$

This equation represents the overall process that starts with an acid–base reaction using two acids. The concentrated sulphuric acid donates a proton to the nitric acid, and then the protonated nitric acid loses a molecule of water:

$$H_2SO_4 + HNO_3 \rightarrow HSO_4^- + H_2NO_3^+$$

$$H_2NO_3^+ \rightarrow H_2O + NO_2^+$$

**Fig 15.45** Nitration of benzene

**Fig 15.46** Mechanism of the nitration of benzene. The nitryl ion, $NO_2^+$ is the electrophile and accepts an electron pair from the delocalized $\pi$ system of the benzene ring. The positively charged intermediate is stabilized because it has a delocalized $\pi$ system, but only over five carbon atoms. In this way, the positive charge is spread around the five atoms. Finally, a proton is eliminated from the intermediate to form a benzene ring. The proton reacts with $HSO_4^-$ to regenerate the sulphuric acid

The nitration of benzene allows other nitrogen-containing functional groups to be introduced, since, for example, it is easy to reduce the nitro group to an amine, to give phenylamine. It is otherwise impossible to introduce the $NH_2$ group into a benzene ring: $NH_2$ is nucleophilic in nature, not electrophilic, so it cannot react with a benzene ring.

**Fig 15.47** The reduction of nitrobenzene to form phenylamine. This provides a route to a reactive intermediate in the manufacture of dyes (see Chapter 27)

**SEE QUESTIONS 4 AND 5.**

## SULPHONATION OF BENZENE

Sulphonation involves the reaction of concentrated sulphuric acid on its own with benzene (Fig 15.48). It is believed that sulphur trioxide, $SO_3$, is the electrophile. In the reaction, a hydrogen atom in the benzene ring is replaced by the sulphonic acid functional group to produce benzenesulphonic acid. Aromatic sulphonic acids are important components of detergents.

In the presence of cold, dilute, aqueous sodium hydroxide, benzene sulphonic acid is converted into the salt sodium sulphonate:

$$C_6H_5SO_3H + NaOH \rightarrow C_6H_5SO_3^-Na^+ + H_2O$$

However, benzenesulphonic acid is hydrolysed by boiling with aqueous sodium hydroxide to form sodium phenoxide (Fig 15.49). In this way, the OH group can be introduced into a benzene ring, even though it is impossible to generate OH⁺ (the electrophile that would be necessary to produce phenol directly from benzene). So, in two easy steps, phenols can be produced from benzene or substituted benzenes.

**Fig 15.48** The sulphonation of benzene

Fig 15.49 The hydrolysis of benzenesulphonic acid to form sodium phenoxide

### ? TEST YOURSELF

**R** Write down the equation to show the hydrolysis of benzenesulphonic acid to form sodium phenoxide.

## 6 ADDITION REACTIONS OF BENZENE

Although it doesn't contain carbon–carbon double bonds, benzene is clearly an unsaturated compound, and under certain conditions will undergo an addition reaction to give a cyclohexane ring. The reaction is difficult, because the exceptional stability of the benzene ring has to be overcome.

### HYDROGENATION OF BENZENE

Benzene is converted into cyclohexane when it is reacted with hydrogen under pressure in the presence of a nickel catalyst:

$$C_6H_6 + 3H_2 \rightarrow C_6H_{12}$$

It is impossible to stop the reaction to give either cyclohexene or cyclohexadiene, which is further evidence that benzene has a delocalized system of electrons rather than individual C=C bonds.

### ? TEST YOURSELF

**S** What is the molecular formula of the hydrocarbon obtained by the hydrogenation of naphthalene?

### ADDITION OF CHLORINE TO BENZENE

In the presence of chlorine and a catalyst, such as aluminium chloride or iron(III) chloride, benzene undergoes electrophilic substitution to form chlorobenzene. If the conditions are changed, and chlorine is bubbled through refluxing benzene in the presence of ultraviolet light without a catalyst, an addition reaction occurs with chlorine to give isomers of 1,2,3,4,5,6-hexachlorobenzene.

### ? TEST YOURSELF

**T** Write down the equation to show the addition of chlorine to benzene.

**SEE QUESTIONS 6 AND 7.**

## 7 ELECTROPHILIC SUBSTITUTION REACTIONS OF SUBSTITUTED BENZENES

Electrophilic substitutions are characteristic of all compound with a benzene ring, not just arenes and other aromatic hydrocarbons. This raises the question: What happens to a molecule that has two functional groups – for example, in phenol, the benzene ring and the hydroxyl group? The answer is simple: the compound, in this case phenol, undergoes two sets of reactions, one involving the benzene ring and the other the hydroxyl group.

**Fig 15.50** In a substituted benzene, there are three sets of hydrogen atoms: on carbon atoms 2 and 6, on carbon atoms 3 and 5, and on carbon atom 4

Consider a substituted benzene $C_6H_5X$, in which X is an atom or a group of atoms such as a methyl group. As the compound contains the benzene ring, it will undergo electrophilic substitution. It may also undergo another set of reactions because of the presence of X.

Limiting the discussion to reactions of the benzene ring in the substituted benzene, $C_6H_5X$, two questions arise. The first is: Will the reaction be faster or slower than that with benzene? The second is: Which hydrogen atom will be substituted? When the reaction is faster, X is said to activate the benzene ring. When the reaction is slower, X is said to deactivate the benzene ring.

Now look at Fig 15.50, which shows a substituted benzene. Three sets of hydrogen atoms can be substituted. Not all of the hydrogen atoms shown will be substituted. The group X directs the substitution so that only certain hydrogen atoms are substituted. The effect of the substituent X on electrophilic substitution is summarized in Table 15.2.

| Table 15.2 Summary of the effects of substituents on electrophilic substitution of the benzene ring | | |
|---|---|---|
| **Substituent** | **Reactivity compared with benzene** | **Direction of substitution** |
| $CH_3$ and other alkyl groups | Activating | 2, 4 and/or 6 |
| Cl, Br, and I | Deactivating | 2, 4 or 6 |
| $NO_2$ | Deactivating | 3 or 5 |
| OH and $OCH_3$ | Highly activating | 2, 4 and 6 |

## ELECTROPHILIC REACTIONS OF PHENOL

Phenol consists of the benzene ring and the attached hydroxyl group. What effect does the hydroxyl group have on electrophilic substitution? The answer is easy to find out experimentally: aqueous bromine reacts with a solution of phenol in aqueous sodium hydroxide at room temperature to give 2,4,6-tribromophenol, which is a white precipitate. The reaction proceeds without a catalyst – the substitution does not require a catalyst to generate the electrophile. Also, more than one hydrogen atom is substituted.

**Fig 15.51** Reaction of aqueous bromine with phenol

Therefore the hydroxyl group is *highly activating*, since the reaction takes place much faster and more easily than with benzene. The hydroxyl group also *directs* the substitution to positions 2, 4 and 6 in the benzene ring (Fig 15.52).

The reason why phenol reacts much faster than benzene is that the lone pair of electrons on the oxygen atom becomes part of the delocalized $\pi$ system. The $\pi$ system is therefore extended over *seven* atoms rather than six (six carbon atoms and one oxygen atom). This has the effect of *increasing the electron density* in the $\pi$ system, thereby making the benzene ring much more susceptible to electrophilic attack.

Phenol is very susceptible to electrophilic attack and is a very useful synthetic intermediate, as shown in Fig 15.53.

**Fig 15.52** The positions to which the hydroxyl group in phenol directs substitution

**Fig 15.53** Five important electrophilic substitution reactions of phenol. Note that because phenol has an activated benzene ring, the conditions are less extreme than those required for substitution in benzene

# ELECTROPHILIC SUBSTITUTION REACTIONS OF PHENYLAMINE

Phenylamine ($C_6H_5NH_2$) behaves towards electrophilic reagents in the same way as phenol, because the $NH_2$ functional group is very similar to the OH group in terms of its effect on the $\pi$ system of the benzene ring. Both the nitrogen atom in the $NH_2$ group and the oxygen atom in the OH group have lone pairs.

The $\pi$ system is again extended over seven atoms, this time to include the nitrogen atom of the amine group. The net effect is to increase the electron density of the $\pi$ system, so making electrophilic substitution much easier than in benzene. Phenylamine reacts with aqueous chlorine to give 2,4,6-trichlorophenylamine, and with aqueous bromine to give the tribromo derivative.

# ELECTROPHILIC SUBSTITUTION REACTIONS OF METHYLBENZENE

The traditional name for methylbenzene is toluene. As Table 15.2 indicates, the methyl substituent activates the benzene ring and directs the electrophilic substitution to positions 2, 4 and 6. In practice, the electrophilic substitution of methylbenzene always leads to a mixture of products.

## NITRATION OF METHYLBENZENE

The explosive compound trinitrotoluene (TNT), or more correctly 2,4,6-trinitromethylbenzene, is made by the nitration of methylbenzene (Fig 15.54). The reaction has three stages: first, one nitro group is substituted for a hydrogen atom, then a second group and finally a third group.

The mononitro compound is produced by heating concentrated nitric acid and concentrated sulphuric acid at 60 °C. Above this temperature, dinitro compounds are produced, and the trinitro compound is only produced above 120 °C.

$$3HNO_3 \ + \quad \longrightarrow \quad + \ 3H_2O$$

**Fig 15.54** Making TNT by nitrating methylbenzene

**? TEST YOURSELF**

**a)** Write equations to show the reaction of phenylamine with:
(i) aqueous chlorine;
(ii) aqueous bromine.
**b)** Use the information in Table 15.2 to predict the product of the nitration of:
(i) chlorobenzene;
(ii) 1,3-dichlorobenzene;
(iii) nitrobenzene.

THE REACTIONS OF ALKYLBENZENES ALSO APPEAR IN THE ASSIGNMENT TO THIS CHAPTER (PAGE 347).

**Fig 15.55** TNT used in a limestone quarry

**Fig 15.56** The chlorination of methylbenzene in the presence of a halogen carrier.

? **TEST YOURSELF**

**a)** Which mononitromethylbenzenes are formed when methylbenzene is nitrated? Draw their structures.
**b)** Which dinitromethylbenzenes are formed when methylbenzene is nitrated? Draw their structures.

SEE QUESTION 8.

## CHLORINATION OF METHYLBENZENE

When chlorine is bubbled through hot methylbenzene in the presence of aluminium chloride or iron(III) chloride, there is electrophilic substitution of one of the hydrogen atoms bonded to the carbon atoms in the benzene ring, with the formation of a mixture of 1-chloro-2-methylbenzene and 1-chloro-4-methylbenzene.

# 8 OTHER REACTIONS OF SUBSTITUTED BENZENES

We have already described the effect of substituents on the typical reaction of a compound with a benzene ring, electrophilic substitution. The benzene ring can also modify the reactions of functional groups attached to the benzene ring.

### PHENOL

Phenol is the name of the single compound $C_6H_5OH$, whereas the term phenols is reserved for a group of compounds that contain a hydroxyl group attached to a benzene ring. We described the electrophilic substitution reactions of phenol earlier on this chapter, reactions that occur because of the presence of the benzene ring. Phenols also undergo reactions because of the presence of the hydroxyl group. Although alcohols also contain a hydroxyl group, the reactions of phenols and alcohols are quite different. The presence of the benzene ring modifies the reactivity of the hydroxyl group.

### PHENOLS AS ACIDS

Phenols are classified as weak acids because they can act as proton donors when in aqueous solution. In water, the proton is donated to a water molecule:

$$C_6H_5OH(aq) \rightleftharpoons C_6H_5O^-(aq) + H^+(aq)$$

This dissociation lies very much on the left-hand side, but phenols are much more acidic than alcohols. The reason for this is the presence of the benzene

---

! **SCIENCE FEATURE** The phenols in tea

Tea is almost certainly the most popular hot drink in the world. In the United Kingdom alone, each person drinks on average three cups every day. There are several reasons why this simple infusion of dry plant leaves is so popular. It has a characteristic taste and attractive colour derived from the compounds in the leaves.

Every cup of tea contains about 40 mg of caffeine, which is a stimulant to the central nervous system. There are also amino acids, carbohydrates and mineral ions. Complex phenols that are water soluble give the tea the astringent taste and refreshing effect we find appealing. As tea leaves are processed, some compounds are oxidized, which gives the brown colour characteristic of tea.

**Fig 15.57** Tea leaves contain water-soluble phenols that contribute to the taste and colour of a cup of tea

Fig 15.58 shows one of the phenolic components (epigallocatechin gallate) of tea. You can see that it is a much more complicated molecule than the compound phenol itself, with several benzene rings incorporated in the structure.

**Fig 15.58** The structure of one of the phenolic constituents of tea ring. We have already discussed on page 340 that the lone pair on oxygen is donated into the benzene ring to produce an extended system. This not only increases the electron density of the system of the benzene ring, but also reduces the electron density between the oxygen and hydrogen atoms in the hydroxyl group. The O—H bond is thus weakened compared to the O—H bond in an alcohol and is easier to break.

You can also consider the stability of the anion produced when a proton is donated by a phenol molecule. The phenoxide ion, $C_6H_5O^-$, is much more stable than an alkoxide ion, such as $C_2H_5O^-$, because the negative charge is delocalized around the benzene ring. As a result phenols show some reactions that are typical of weak acids. Phenol reacts with the reactive metal sodium and with the alkali aqueous sodium hydroxide, but it is not sufficiently acidic to produce carbon dioxide from carbonates.

## REACTION WITH AQUEOUS SODIUM HYDROXIDE

Phenol reacts with aqueous sodium hydroxide to produce sodium phenoxide, an ionic compound. In this reaction phenol transfers a proton to a hydroxide ion to form water:

$$C_6H_5OH(s) + OH^-(aq) \rightarrow C_6H_5O^-(aq) + H_2O(l)$$

As the state symbols in the equation show, phenol appears to dissolve in aqueous sodium hydroxide, but this must be seen as a chemical reaction rather than the physical process of dissolving. Alcohols do not show this reaction.

## REACTION OF SODIUM WITH PHENOL

Phenol reacts with sodium to produce sodium phenoxide. As in the case of alcohols, the reaction involves the heterolytic fission of the O—H bond. The reaction must be carried out in an inert solvent to ensure that the sodium and the phenol, both solids, do make contact.

$$2C_6H_5OH + 2Na \rightarrow 2C_6H_5O^-Na^+ + H_2$$

## REACTION OF PHENOLS AS NUCLEOPHILES

Alcohols can form esters when they react with carboxylic acids or acid chlorides. Phenols do not react with carboxylic acids to make esters, but will react with acid chlorides to make esters.

Normally, sodium phenoxide is used (aqueous sodium hydroxide and phenol) as the reagent, because the phenoxide ion is a much better nucleophile than a neutral phenol molecule. The sodium hydroxide present also neutralizes the hydrogen chloride formed as a by-product. With ethanoyl chloride, the ester phenyl ethanoate is formed (Fig 15.59):

$$C_6H_5O^-Na^+(aq) + CH_3COCl(l) \rightarrow CH_3COC_6H_5(s) + NaCl(aq)$$

## TESTING FOR PHENOLS

Phenols will react with aqueous iron(III) chloride to form a violet colour.

## METHYLBENZENE

Methylbenzene contains a methyl group attached to the benzene ring. We should expect that the methyl group would react in the same way as the methyl group in other hydrocarbons, such as ethane or methane. This is true for some reactions, but in others the proximity of the benzene ring gives the methyl group added reactivity.

**Fig 15.59** The reaction between ethanoyl chloride and sodium phenoxide

**SEE QUESTION 9.**

**? TEST YOURSELF**

Draw the structure of the organic product of the following reactions:
**a)** sodium phenoxide and propanoyl chloride;
**b)** sodium phenoxide and benzoyl chloride.

**SEE QUESTION 10.**

**FREE-RADICAL CHLORINATION IS COVERED ON PAGE 267.**

## SIDE-CHAIN OXIDATION

Methylbenzene is oxidized by acidified potassium manganate(VII) to form benzoic acid:

$$C_6H_5CH_3 + 3[O] \rightarrow C_6H_5COOH + H_2O$$

This is called side-chain oxidation, since it is the side chain rather than the ring that is oxidized.

Side-chain oxidation occurs with any alkyl group, but always forms benzoic acid, regardless of the chain length of the alkyl group (Fig 15.60). When the benzene ring contains more than one alkyl group, each one is oxidized in the same way. This reaction can be used to determine the substitution pattern within an aromatic hydrocarbon. For example, if the side-chain oxidation of a hydrocarbon of molecular formula $C_8H_{10}$ gives benzoic acid, then the hydrocarbon must be ethylbenzene, whereas if benzene-1,2-dicarboxylic acid is formed then the hydrocarbon would be 1,2-dimethylbenzene.

**Fig 15.60** Side-chain oxidation of ethylbenzene to form benzoic acid

## FREE-RADICAL CHLORINATION

In Chapter 12, we described the free-radical substitution reaction of methane with chlorine. Methylbenzene reacts in a similar way with chlorine in the presence of ultraviolet light. Only the hydrogen atoms in the methyl group are substituted by chlorine atoms; those hydrogen atoms attached to the benzene ring are unaffected.

$$C_6H_5CH_3 + Cl_2 \xrightarrow{hf} C_6H_5CH_2Cl + HCl$$

$$C_6H_5CH_2Cl + Cl_2 \xrightarrow{hf} C_6H_5CHCl_2 + HCl$$

$$C_6H_5CHCl_2 + Cl_2 \xrightarrow{hf} C_6H_5CCl_3 + HCl$$

# 9 BENZENE AND THE PETROCHEMICAL INDUSTRY

Benzene is an important source of many petrochemicals. The electrophilic substitution reaction followed by other synthetic transformations allows a variety of different substituted benzenes to be produced. Some of the important reactions of benzene are shown in Fig 15.61.

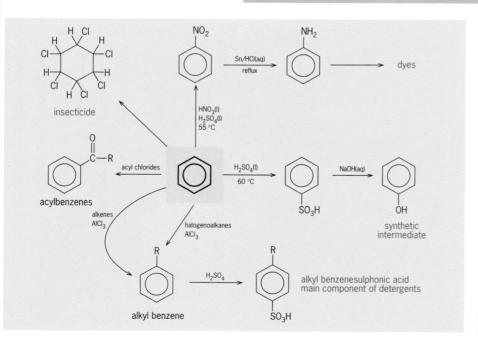

**Fig 15.61** A summary of the reactions of benzene

# SUMMARY

After studying this chapter, you should know the following:

■ The benzene ring is a regular hexagon of carbon atoms joined by six σ bonds. Six π electrons are delocalized in an orbital above and below the ring.

■ Benzene is made by cracking the naphtha fraction and by reforming $C_6$ alkanes or cycloalkanes.

■ Arenes react by electrophilic substitution rather than electrophilic addition: an electrophile is substituted for a hydrogen atom, which is lost as a proton.

■ An arene is alkylated by reaction with a halogenoalkane in the presence of aluminium chloride, and is acylated by reaction with an acyl chloride or acyl bromide in the presence of aluminium chloride.

■ An arene is alkylated by reaction with alkenes in the presence of an acidic catalyst.

■ An arene is chlorinated or brominated by the reaction between chlorine or bromine and the arene in the presence of a halogen carrier, such as iron(III) chloride or aluminium chloride.

■ A nitrating mixture consists of concentrated sulphuric acid and concentrated nitric acid, and produces $NO_2^+$, which can react with arenes to form nitroarenes.

■ Phenol and phenylamine have highly activated benzene rings, and with electrophiles form trisubstituted substitution products at positions 2, 4 and 6.

■ Methylbenzene reacts faster than benzene towards electrophilic reagents and forms substitution products at position 2, 4 or 6.

■ Phenol is a weak acid and reacts with sodium and sodium hydroxide to give sodium phenoxide.

# ? EXAM QUESTIONS

**1 a)** Describe the bonding present in benzene.
**b)** Explain how the shape of the benzene molecule can be explained in terms of its bonding.

**2**
**a)** Describe the reagents and conditions for the conversion of benzene into chlorobenzene.
**b)** Explain why chlorobenzene is resistant to alkaline hydrolysis.

WJEC Chemistry Specimen Paper CH4 2001/2002 Q2b,c

**3** [Note: Part (c) requires information from Chapter 14.] Benzene can be converted into phenylethene (styrene) by the series of reactions shown below.

**a) (i)** Name the mechanism for the reaction in step 1.
**(ii)** Give mechanistic equations for the reaction in step 1, clearly indicating the role of the aluminium bromide.
**b) (i)** What reaction condition is necessary for step 2?
**(ii)** State the name of the reaction mechanism for step 2, and outline the steps involved.
**c) (i)** Name and explain the mechanism of the reaction described in step 3.

**(ii)** In step 3 a different reaction takes place if a warm dilute aqueous solution of potassium hydroxide is used instead of a hot, alkanolic one. Give the structural formula of the major organic product formed under these conditions, and state the name of the mechanism.

IBO Chemistry Higher Level Paper 3 Nov 1998 QH1

**4** Benzene reacts with a mixture of concentrated nitric acid and concentrated sulphuric acid at 50 °C to form nitrobenzene according to the equation in Fig 15.Q4.

**Fig 15.Q4**
**a)** Name the mechanism by which this reaction proceeds.
**b)** What is the initial role of the sulphuric acid when it functions as a catalyst in this reaction?
**c)** The product, nitrobenzene, can be further nitrated with a mixture of concentrated nitric and sulphuric acid, but the temperature needs to be increased to 100 °C. Explain why nitrobenzene is less reactive towards nitration than benzene.

IBO Chemistry Higher Level Paper 3 Nov 2000 QH2(a)(b)(c)

## ? EXAM QUESTIONS

**5 a)** Describe the bonding in benzene and explain why benzene reacts less readily than alkenes with electrophiles.

**b)** Describe, including a mechanism, the nitration of benzene by electrophilic substitution.

OCR Chains, Rings and Spectroscopy Specimen Paper 2000 Q7

**6** The characteristic reaction of benzene is *electrophilic substitution*.

**a)** Select a reaction of benzene that illustrates this type of reaction. Give the reagents, the equation, the conditions under which it occurs and the name of the organic product for the reaction you have chosen.

**b)** For the reaction selected in **a)**:
**(i)** identify the electrophile;
**(ii)** give an equation to show its formulation;
**(iii)** give the mechanism for the substitution reaction.

**c)** Give two *specific* safety precautions you would need to take in *carrying out* the reaction in a).

**d)** Given one reaction of the functional group that was introduced into the benzene ring as a result of the reaction performed in **a)**. Give the reagents, conditions and equation for this reaction.

**e)** The enthalpy of hydrogenation of a single C—C bond is of the order of $-120$ kJ mol$^{-1}$.
**(i)** Assuming that benzene consists of a ring with three separate double bonds, predict the enthalpy change for the reaction shown below.

**(ii)** The enthalpy of hydrogenation of benzene is actually $-205$ kJ mol$^{-1}$. What can you deduce from this and your answer to part **e)(i)** about the stability of the benzene ring? Use an enthalpy level diagram to illustrate your answer.

Edexcel Specimen Unit Test 4 Jan 2000 Q3

**7** Two reactions of benzene are summarized below.

**a)** What is the meaning of the circle in the diagram of both benzene and substrate **E**?

**b)** Name the substrates **A** to **E**.

**c)** Name the type of reaction involved in the formation of **A** and **E**.

Edexcel Specimen Unit Test 4 Jan 2000 Q1

**8**

**a)** 1,4-Dimethylbenzene may be converted into **A** or **B** under different conditions as shown.

**(i)** State the reagents and conditions 1 needed to form A, and suggest a mechanism for the formulation of **A**.
**(ii)** State the reagents and conditions 2 needed to form **B**, and suggest a mechanism for the formation of **B**.

**b)** Methylbenzene can react under conditions 1 or conditions 2 to give isomers of molecular formula $C_7H_7Br$. Draw all products of formula $C_7H_7Br$ formed under
**(i)** conditions 1 (that give **A**);
**(ii)** conditions 2 (that give **B**).

OCR, Further Organic (6 825) March 2001 Q4

**9**

**a)** Explain why it is impossible to make phenol from benzene in a one-step reaction.

**b)** Describe a two-step process that can convert benzene into phenol.

**c)** In the presence of a suitable catalyst it is possible to hydrogenate phenol. Draw the displayed structure of the product of this hydrogenation.

**d)** The OH group in phenol is said to highly activate the benzene ring. Explain what this means.

**10**

**a)** Phenylethene is made by heating ethylbenzene in the presence of a catalyst. Write the graphic (full structural) formula for ethylbenzene and *one* of its isomers.

**b)** State *one* way in which these two isomers could be identified from each other.

WJEC Paper CH4 2001/2002 Q4(c)(ii)

# KEY SKILLS **ASSIGNMENT**

## THE FRIEDEL–CRAFTS REACTIONS OF ARENES

**In 1877, the French chemist Charles Friedel and his American co-worker James Crafts discovered an important type of reaction for arenes. Known as the Friedel–Crafts alkylation and acylation, it enables alkyl groups and acyl groups to be substituted into the benzene ring (Figs 15.A1 and 15.A2). The importance of Friedel–Crafts reactions is that they make carbon–carbon bonds by linking straight or branched chains to arene rings.**

R = alkyl group

**Fig 15.A1** The alkylation of benzene

R = alkyl group

**Fig 15.A2** The acylation of benzene

**1** **a)** Draw the structure of the main organic product of the Friedel–Crafts reaction of butanoyl chloride, $CH_3CH_2CH_2COCl$, with benzene.
**b)** Suggest the structure of one of the main organic products of the Friedel–Crafts reaction of
(i) naphthalene with ethanoyl chloride, $CH_3COCl$, and
(ii) anthracene with iodoethane.

The alkylation and acylation reactions were found to be catalysed by Lewis acids (electron-pair acceptors), such as anhydrous aluminium chloride, anhydrous iron(III) chloride and anhydrous zinc chloride. Later on, the reaction was developed to include the alkylation of arenes using alkenes (Fig 15.A3). In this Friedel–Crafts reaction, acidic catalysts such as phosphoric(V) acid can be used.

AlCl₃ or
phosphoric acid

**Fig 15.A3** The alkylation of benzene using an alkene

**2** Suggest the structure of one of the main organic products of the Friedel–Crafts reaction between pyrene and ethene.

A Friedel–Crafts reaction is normally carried out by mixing the liquid arene with the catalyst and then adding the alkylating or acylating agent. (When the arene is a solid, a solvent such as nitrobenzene or carbon disulphide has to be used.) All the reactants and apparatus must be kept absolutely dry during the reaction.

The temperature of the reaction varies, depending on the reactivity of the particular benzene ring in the arene. So, some Friedel–Crafts reactions need heat, while others take place at room temperature. Once the reaction is finished, the reaction mixture is added to water, which effectively destroys the catalytic properties of aluminium chloride.

On an industrial scale, these reactions produce a large volume of effluent that may contain carcinogenic arenes and a high concentration of metal ions, such as aluminium.

**3** **a)** Suggest why the reaction itself must be carried out in dry conditions.
**b)** What problems do you foresee in carrying out a Friedel–Crafts reaction on an industrial scale?
**c)** Suggest one possible problem in the use of nitrobenzene as a solvent for a Friedel–Crafts reaction of a solid arene.

Envirocats (see page 326) are being developed for Friedel–Crafts reactions. The appropriate clays are made porous and have metal ions inserted to produce insoluble solids with Lewis acid properties. Added to the reaction mixture, they catalyse Friedel–Crafts reactions. The envirocats developed so far cost about the same as the standard Friedel–Crafts catalysts, but the yield and the rate of reaction is lower. At the end of the reaction, the envirocats are filtered off, dried and reused many times. When they finally lose their catalytic activity, they can be safely buried in land-fill sites.

**4** Discuss critically the advantages of using envirocats in an industrial process over the usual Friedel–Crafts catalysts.

The mechanism of the Friedel–Crafts reaction is electrophilic substitution. The purpose of the catalyst is to generate the electrophile (Figs 15.A4 and 15.A5).

$$R—Hal \quad AlCl_3 \longrightarrow R^+ + [Hal—AlCl_3]^-$$

**Fig 15.A4** The generation of a carbocation electrophile from a halogenoalkane and from an alkene

$$R—\overset{\overset{\displaystyle O}{\|}}{C}—Cl \quad AlCl_3 \longrightarrow R—\overset{\overset{\displaystyle O}{\|}}{C}^+ + [AlCl_4]^-$$

**Fig 15.A5** The generation of an acylium ion electrophile from an acyl chloride

**Fig 15.A6** Two ways to represent the structure of an acylium ion

$$R—\overset{\overset{\displaystyle O}{\|}}{C}^+ \qquad R—\overset{\overset{\displaystyle O^+}{\|\|}}{C}$$

# KEY SKILLS **ASSIGNMENT**

**5 a)** What feature of anhydrous aluminium chloride allows it to facilitate the heterolytic fission of a carbon–halogen bond?
**b)** Using a table of bond energies, suggest a reason why iodoalkanes are preferred over chloroalkanes in the Friedel–Crafts reaction.
**c)** Complete the mechanism for the reaction of the carbocation and for the acylium ion (Fig 15.A6) with benzene.

Fig 15.A4 does not fully detail the generation of the electrophile. Very often, the carbocation rearranges its carbon skeleton to make a more stable carbocation. Tertiary carbocations are more stable than secondary carbocations, which are more stable than primary carbocations. This means that there is a limited access to alkylbenzenes because the final electrophilic substitution product comes from the rearranged carbocation (Fig 15.A7). So, reactions of 1-iodobutane, 2-iodobutane and 2-iodo-2-methylpropane with benzene all form the same alkylbenzene via a tertiary carbocation. See page 212 for more information on carbocations.

$$CH_3 - CH_2 - CH_2 - CH_2 - Cl \ + \ AlCl_3$$

primary carbocation $\quad CH_3 - CH_2 - CH_2 - CH_2^+ \ + \ AlCl_3$

rearrangement

tertiary carbocation
$$CH_3 - \underset{\underset{+}{|}}{\overset{\overset{CH_3}{|}}{C}} - CH_3$$

**Fig 15.A7** Rearrangement of the carbon skeleton to form the most stable carbocation, which then reacts with the benzene ring

**6 a)** Explain the order of stability of carbocations, from tertiary (most stable) to primary (least stable).
**b)** Draw the tertiary carbocation that 1-iodobutane, 2-iodobutane and 2-iodo-2-methylpropane form in Friedel–Crafts reactions. Hence draw the product of the reaction of any of these three isomers with benzene.

There is a further complication with alkylation. Once one alkyl group has been substituted, it activates the benzene ring, and so sometimes more than one alkylation per benzene ring takes place.

An alternative two-stage route to alkylbenzenes does not suffer from either a second electrophilic substitution or the rearrangement of the electrophile. It involves acylation to give a ketone, followed by reduction to give an alkylbenzene (Fig 15.A8). However, there is one drawback with this method: it requires much more aluminium chloride catalyst, since the ketone produced after the acylation forms a complex with the catalyst.

So ethylbenzene can be made in the following way:

**Fig 15.A8** The indirect alkylation of benzene

**7 a)** Calculate the overall percentage yield for the missing entry marked ? in Table 15.A1 (see page 332).
**b)** Explain why you cannot make 2-phenylpropane using Route 2.
**c)** Discuss the advantages and disadvantages of making alkylbenzenes by (i) the one-step alkylation method, and (ii) the two-step alkylation–reduction method.

**Table 15.A1 The comparative percentage yields when some alkylbenzenes are made in a one- or two-step process**

| Alkylbenzene to be made | Route 1: direct alkylation using iodoalkane and aluminium chloride | Route 2: direct acylation using acyl chloride and aluminium chloride followed by reduction | | |
|---|---|---|---|---|
| | % yield | % yield acylation | % yield reduction | Overall % yield from benzene to alkylbenzene |
| Ethylbenzene | 76 | 87 | 91 | 79 |
| Propylbenzene | 5 | 85 | 90 | ? |
| 2-Phenylpropane | 78 | Cannot be made this way | | |

# 16

# Aldehydes and ketones

# Aldehydes and ketones

**Steroid structure** All steroids possess the four-ring structure. You can read about cholesterol, another steroid, on page 286.

**Oral contraceptives** These liberated women from the fear of unwanted pregnancies and thereby revolutionized sexual behaviour. However, these chemicals in recycled drinking water have been partly blamed for reducing the sperm count of males

The most publicised hormones must be progesterone and testosterone. The reason is sex.

Progesterone is the female sex hormone, which prepares the uterus for pregnancy after an egg has been fertilized, and which prevents the ovaries from releasing any more eggs. Testosterone is the male sex hormone responsible for sexual development and drive in males, and for muscle growth. Although they have very different functions, these hormones are structurally very similar (they are steroids with a carbonyl functional group), and it takes only a simple chemical reaction in the laboratory to convert progesterone into testosterone.

Once their structures became known in the 1940s, chemists set about synthesizing them, and soon synthetic sex hormones featured in a number of medical applications. However, few people would have predicted the social revolution that was to follow when synthetic progestins, a group of chemicals that mimic the action of progesterone, were developed. The synthetic progestin norethynodrel was the basis of the first oral contraceptive, known simply as 'the pill', introduced in 1960. There is perhaps no more striking example of the influence of chemistry on our lives.

## 1 RECOGNIZING AND NAMING ALDEHYDES AND KETONES

Aldehydes and ketones contain perhaps the most important functional group in organic chemistry – the **carbonyl group**, **C=O** (Fig 16.1). Hence, they are known as **carbonyl compounds**. The carbonyl functional group is found in many important biological molecules, from insect pheromones to human sex hormones. It occurs in the molecules in our eyes that are responsible for vision, and gives lemons their characteristic flavour. It is also involved in the manufacture of many important industrial chemicals, from plastics to solvents.

In aldehydes (Fig 16.2), the carbon atom of the carbonyl group (the carbonyl carbon) is bonded to at least one hydrogen atom, while in ketones (Fig 16.3) it is bonded to two carbon atoms, each from either an alkyl or an aryl group.

**Fig 16.1** The carbonyl functional group

### ✔ REMEMBER THIS

A functional group is an atom or group of atoms that gives a molecule a characteristic set of properties (see page 206).

A pheromone is a chemical that communicates information between members of the same species. (see page 88).

An **aryl** group contains a benzene ring.

**Fig 16.2** An aldehyde. R can be hydrogen, an alkyl group or an aryl group

**Fig 16.3** A ketone. The R′ group may be different from the R group. R cannot be hydrogen, otherwise the compound is an aldehyde

Four naturally occurring carbonyl compounds are shown in Fig 16.4.

**Fig 16.4** Four naturally occurring carbonyl compounds

Menthone: found in mint leaves; gives a peppermint flavour

Citral: an oily liquid that contributes to the flavour and aroma of oranges and lemons

A moth sex pheromone

Oestrone: a female sex hormone

## NAMING ALDEHYDES AND KETONES

The systematic naming of aldehydes and ketones is simple. The names of aldehydes end in *al*, and those of ketones end in *one*. The number of carbon atoms in the chain (including the carbon of the functional group) provides the rest of the name, which is based on the alkane name with the end *e* removed.

**Fig 16.5** Deriving the names of aldehydes and ketones

ethanal
ethan(e) al

propanone
propan(e) one

pentan-2-one
pentan(e) -2-one

Eight of the more important aldehydes and ketones are listed in Table 16.1.

### ▶ EXTENSION Methanal

Formaldehyde, the traditional name for methanal, is still used by most industrial chemists. In fact, it is the most-used industrial aldehyde. Methanal is produced by the air oxidation of methanol, using an iron or silver catalyst:

$$CH_3OH(l) + \tfrac{1}{2}O_2(g) \xrightarrow{\text{Fe or Ag catalyst, 500 °C}} HCHO(g) + H_2O(l)$$

Methanal is used to make plastics such as Bakelite (one of the first plastics) and other phenolic resins (for which it is reacted with phenol), urea–formaldehyde resins (for which it is reacted with urea), melamine resins and Formica.

Your kitchen almost certainly contains surfaces and equipment that are products of reactions with methanal – kitchen work tops and pan handles are just two examples. So does your bathroom, because methanal is used as a preservative in some shampoos and bath foams.

A 40 per cent aqueous solution of methanal – better known as formalin – is used to preserve biological specimens (Fig 16.6) and as a disinfectant and fungicide.

**Fig 16.6** The head end of a dogfish, preserved in formalin, showing blood vessels of the gills

**Table 16.1 Some important aldehydes and ketones**

| Structural formula | Systematic name |
|---|---|
| HCHO | Methanal |
| $CH_3CHO$ | Ethanal |
| $CH_3CH_2CHO$ | Propanal |
| ⬡—CHO | Benzaldehyde |
| $CH_3COCH_3$ | Propanone |
| $CH_3CH_2COCH_3$ | Butanone |
| ⬡—$COCH_3$ | Phenylethanone |
| $CH_3CH_2CH_2COCH_3$ | Pentan-2-one |

## ? TEST YOURSELF

**C** Apart from the aromatic compounds, all the other aldehydes and ketones in Table 16.1 are very soluble in water.

**a)** What type of intermolecular bonding explains this solubility? Hint: If you are not sure of your answer, look back at page 288.

**b)** Draw a diagram to show the intermolecular bonding between an ethanal molecule and a water molecule.

**c)** Why do you think aromatic carbonyl compounds are not very soluble in water?

# 2 THE CARBONYL GROUP AND NUCLEOPHILIC ADDITION

alkene functional group   carbonyl functional group

**Fig 16.7** Alkene and carbonyl functional groups

### ✔ REMEMBER THIS

An **electrophile** is a species that can accept a lone pair of electrons to form a covalent bond. A π bond is formed by the sideways overlap of two p orbitals.

p orbitals overlap to give the π bond

**Fig 16.9** The sideways overlap of two orbitals

**Fig 16.10** Nucleophile Nu: attacks the carbonyl carbon and the π bond breaks

### ✔ REMEMBER THIS

A nucleophile has a lone pair of electrons with which it can form a covalent bond. It is attracted to a centre of positive charge.

**SEE QUESTION 1.**

As the carbonyl functional group has a double bond, it undergoes *addition* reactions – just like the double bond of the functional group in alkenes (see page 311).

Alkenes undergo **electrophilic addition** reactions because the loose electron cloud of the π bond (see page 81) is attractive to electrophiles. There is a π bond in the carbonyl group as well, but it is between two atoms of different electronegativities, C and O. So the density of the electron cloud is greater at the more electronegative oxygen end, thereby making the bond polar.

**Fig 16.8** The difference between the π bonds in the alkene and carbonyl functional groups

Since C=O is polar, the electron-deficient carbonyl carbon is susceptible to attack by nucleophiles. We have already come across nucleophilic reactions with the polar carbon–halogen bond of halogenoalkanes (page 272), and with the polar C–O bond of alcohols (pages 293–294). In both these cases, substitution reactions occur. However, the carbonyl group has a *double* bond, and therefore as the nucleophile attacks and forms a covalent bond with the carbonyl carbon, the π bond splits and forms another single covalent bond, which results in an addition reaction.

## NUCLEOPHILIC ADDITION BY HYDROGEN CYANIDE

The reaction of hydrogen cyanide (HCN) with the carbonyl group is a very important *synthesis* reaction, because it adds another carbon atom to the molecule.

Fig 16.11 shows two nucleophilic addition reactions. The reaction with propanone is used in the production of poly(methyl-2-methylpropenoate), better known by its ICI trade name of Perspex. The first reaction in the synthesis of Perspex is the production of a hydroxynitrile (cyanohydrin).

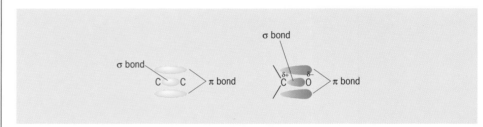

propanone | a hydroxynitrile (sometimes called a cyanohydrin) | ethanal | another hydroxynitrile

**Fig 16.11** Two nucleophilic addition reactions

The nucleophilic addition reaction of hydrogen cyanide was one of the first reaction mechanisms ever to be investigated; this was by a British chemist, Arthur Lapworth, in 1903. A **reaction mechanism** shows the steps by which a reaction takes place. The nucleophile is the cyanide ion ($CN^-$), rather than the HCN molecule.

**Fig 16.12** Nucleophile attacks the ketone group

HCN is a poor nucleophile, and if $CN^-$ ions are not present, the reaction is very slow. Either the addition of potassium cyanide (KCN) can provide the $CN^-$ ions or they can be generated from HCN by adding an alkali:

$$HCN + OH^- \rightarrow H_2O + CN^-$$

A proton then bonds to the oxygen atom. Often this proton is transferred from HCN or $H_2O$:

**Fig 16.13** Proton bonds to the oxygen

The net result is the addition of HCN to the molecule. Notice that in this case the proton was transferred from HCN, and a $CN^-$ ion is regenerated.

### EXTENSION Hydroxynitrile: the millipede's defence

To deter predators, if not finish them off for good, one species of millipede uses the deadly gas HCN – hydrogen cyanide – yet manages not to harm itself. This is because inside the millipede are separate stores of a hydroxynitrile derivative of benzaldehyde, which is harmless, and an enzyme that catalyses the compound's reaction to form benzaldehyde and HCN; the two are not mixed until they leave the animal's body. (Note this is the reverse of the reaction in Test Yourself D.)

When the millipede is attacked, it discharges both compounds. The enzyme becomes mixed with the hydroxynitrile, and the hapless attacker is enveloped in hydrogen cyanide.

**Fig 16.14** *Aphleoria corrigata*. Curled up, this giant millipede looks harmless, but it has a lethal spray for any attacker

## NUCLEOPHILIC ADDITION OF H⁻

The hydride ion ($H^-$) comes from, $NaBH_4$ (sodium tetrahydridoborate(III)). The mechanism (Fig 16.15) is similar to that of the addition of hydrogen cyanide, only in this case the nucleophile is $H^-$. Once the $NaBH_4$ has reacted with the carbonyl compound, water is added to provide the protons.

Notice that an alcohol functional group is produced. This reaction is also called a *reduction*. The equation for the reduction of aldehydes in general is written:

$$RCHO + 2[H] \rightarrow RCH_2OH$$

**? TEST YOURSELF**

D  Write the equation for the reaction of benzaldehyde with HCN.

**SEE QUESTION 2.**

**Fig 16.15** Nucleophilic addition takes place in the first step, and water then transfers the proton to make an alcohol functional group

### ✔ REMEMBER THIS

$LiAlH_4$ (lithium tetrahydridoaluminate(IV) is a more powerful reducing agent than $NaBH_4$. It also reacts with water, which is why it is dissolved in dry ethoxyethane. $NaBH_4$ is a more convenient reagent as it can be used in aqueous or alcoholic solution and the reaction is less vigorous.

**Fig 16.16** 2,4-Dinitrophenylhydrazone is a brightly coloured precipitate and its formation is a test for carbonyl groups, as in this test tube

H is put in brackets to signify that *it comes from a reducing agent*, which in this case is NaBH$_4$ (this is a simplified way to balance the equation and does not represent atomic hydrogen). This reaction is discussed again on page 356, where we look at the reduction of aldehydes and ketones. In the meantime, try Test Yourself E, and then check your answer at the end of the book.

## 3 REACTION WITH 2,4-DINITROPHENYLHYDRAZINE

2, 4-Dinitrophenylhydrazine (also called Brady's reagent) reacts with carbonyl groups to give orange-coloured precipitates, called 2, 4-dinitrophenylhydrazones. This reaction is used to test for the presence of carbonyl group in unknown compounds.

The coloured precipitates from the reaction can be purified by recrystallization to give products that have very precise melting points, which can therefore be used to identify a particular aldehyde or ketone.

**Fig 16.17** Production of ethanal 2,4-dinitrophenylhydrazone

The *ethanal derivative* ethanal 2,4-dinitrophenylhydrazone (Fig 16.17) melts at exactly 168 °C, so its melting point can be used to identify ethanal, a method that used to be important before different types of spectrometers became widely available for identifying substances (see Chapter 11). Before this, the best way to recognize unknown carbonyl compounds was to produce from the liquid a solid derivative with a sharp melting point. The melting point is also a good way to check the purity of a sample, since any impurity lowers the melting point.

**▶ EXTENSION Addition—elimination mechanism of the 2,4-dinitrophenylhydrazine reaction**

The two nucleophilic addition reactions in the previous section produce stable products. The reaction between the carbonyl group and 2,4-dinitrophenylhydrazine gives an unstable product, which spontaneously reacts with the elimination of a water molecule.

The mechanism of the reaction (Fig 16.18) shows that the lone pair of electrons on the first nitrogen atom provides the basis of the nucleophilic addition, and the intermediate compound is then formed by an internal rearrangement. The final product, a 2,4-dinitrophenylhydrazone (Figs 16.17 and 16.18), is formed from the unstable intermediate by the elimination of a water molecule. So this reaction involves both an addition reaction and an elimination reaction – hence addition–elimination.

**Fig 16.18** Addition–elimination reaction of 2,4-dinitrophenylhydrazine

The production of 2,4-dinitrophenylhydrazone derivatives is still important for X-ray crystallography, which requires solid compounds.

A mixture of carbonyl compounds can be separated by reacting the mixture with 2,4-dinitrophenylhydrazine and then using thin-layer chromatography. The various 2,4-dinitrophenylhydrazones separate when a suitable solvent is used.

**FOR INFORMATION ON X-RAY CRYSTALLOGRAPHY, SEE PAGE 76.**

**FOR INFORMATION ON THIN-LAYER CHROMATOGRAPHY, SEE PAGE 255.**

---

## ! SCIENCE FEATURE Sun tan from a bottle

Over-exposure to sunlight is known to cause skin cancer, yet millions of people still work hard to give themselves a suntan because they think it makes them look healthier and more attractive.

**Fig 16.19** Despite grim warnings about the consequences, sunbathing still has millions of devotees

However, suntans haven't always been fashionable. Until about 200 years ago, upper-class women who went on leisurely walks would protect their pale complexions with bonnets and parasols. This distinguished them from the bronzed working men and women who laboured outdoors in the sunshine.

Attitudes gradually began to change with the industrial revolution. While factory and office workers toiled in buildings that shut out the sunshine all day, better off people began to think of a suntan as a sign of wealth and position in society. This trend grew when it was the fashion in the early 1900s to go on touring holidays of continental Europe – the forerunners of today's package holidays.

A natural tan is caused by the dark pigment melanin, which is produced in the skin to absorb harmful ultraviolet rays. A similar tan-like effect can be produced after a few hours using DHA, a colourless ketone that reacts with the protein in the outer skin to produce a brown pigment. DHA, traditionally called dihydroxyacetone, is 1,3-dihydroxypropanone:

$$HO-CH_2-\overset{\overset{\displaystyle O}{\|}}{C}-CH_2-OH$$

There are problems with a 'chemical' tan. Cells of the outer layer of skin are dead and so within a few weeks these have rubbed off – and so has the tan! Also, because only dead cells react with DHA, where these are in a thicker layer, such as at the elbows and knees, the tan is darker than elsewhere. It is also worth noting that a chemical tan does nothing to protect the skin from harmful ultraviolet rays.

---

## ▶ EXTENSION Grignard reagents and aldehydes and ketones

If magnesium turnings are added to a solution of halogenoalkane (RX) in ethoxyethane, an exothermic reaction produces a solution of Grignard reagent:

$$R–X + Mg \rightarrow R–Mg–X$$

R can be either an alkyl or an aryl group and X is a halogen.

The reagents are named after their discoverer Victor Grignard, a French chemist. One of the most important uses of Grignard reagents is to synthesize alcohols from aldehydes and ketones. The mechanism for this reaction is nucleophilic addition. The carbon–magnesium bond is highly polar (Fig 16.20) and attacks the carbonyl group (Fig 16.21) to form an intermediate that is decomposed by adding dilute acid (Fig 16.22).

**Fig 16.20** Highly polar carbon–magnesium bond

$$\overset{\delta-}{\underset{\diagdown}{C}}-\overset{\delta+}{Mg}-X$$

**Fig 16.21** Nucleophilic attack on the carbonyl group

$$\overset{\delta+}{C}=\overset{\delta-}{O}$$
$$\overset{\delta-}{R}:\overset{\delta+}{MgX}$$

$$-\overset{|}{\underset{R}{C}}-OMgX \xrightarrow{\text{dilute acid}} -\overset{|}{\underset{R}{C}}-OH$$

**Fig 16.22** Intermediate is decomposed by dilute acid

Primary, secondary and tertiary alcohols can be made using this reaction (Figs 16.23 to 16.25).

$$\underset{H}{\overset{H}{\diagdown}}C=O + RMgX \rightarrow H-\overset{H}{\underset{R}{C}}-OMgX \xrightarrow{\text{dilute acid}} H-\overset{H}{\underset{R}{C}}-OH$$

**Fig 16.23** Primary alcohols using methanal

$$\underset{H}{\overset{R'}{\diagdown}}C=O + RMgX \rightarrow R'-\overset{H}{\underset{R}{C}}-OMgX \xrightarrow{\text{dilute acid}} R'-\overset{H}{\underset{R}{C}}-OH$$

**Fig 16.24** Secondary alcohols using other aldehydes other than methanal

$$\underset{R'}{\overset{R''}{\diagdown}}C=O + RMgX \rightarrow R'-\overset{R''}{\underset{R}{C}}-OMgX \xrightarrow{\text{dilute acid}} R'-\overset{R''}{\underset{R}{C}}-OH$$

**Fig 16.25** Tertiary alcohols using ketones

### ? TEST YOURSELF

**G** What alcohols would be produced from the reaction of the following reagents?
**a)** $CH_3CH_2MgBr$ + HCHO;
**b)** $CH_3CH_2CH_2CH_2MgBr$ + $CH_3COCH_3$;
**c)** $CH_3CH_2MgBr$ + $CH_3CHO$.

**Fig 16.26** The aldehyde and ketone functional groups

# 4 DIFFERENCES BETWEEN ALDEHYDES AND KETONES

So far, we have treated carbonyl compounds together and looked at the reactions of the C=O functional group in both aldehydes and ketones. They do, however, differ in their reactions with oxidizing and reducing agents. Therefore, we now look at reactions of the aldehyde functional group separately from those of the ketone functional group.

## REDUCTION OF ALDEHYDES AND KETONES

With a reducing agent, such as NaBH$_4$ or LiAlH$_4$, aldehydes give **primary alcohols** and ketones give **secondary alcohols**, as shown in Fig 16.21. Notice that the reducing H in the equation is in brackets (to balance the equation simply – it does not represent atomic hydrogen). The reaction is not that simple, as you can see from its mechanism on pages 352–354.

**Fig 16.27** Reduction of aldehydes and ketones

As an example, benzaldehyde gives the balanced equation shown in Fig 16.28.

**Fig 16.28** Reduction of benzaldehyde

The reduction reactions of aldehydes and ketones are essentially the reverse of the oxidation reactions of primary and secondary alcohols (see page 296).

## OXIDATION OF ALDEHYDES

Aldehydes are prepared by oxidizing primary alcohols, and ketones by oxidizing secondary alcohols. In the case of primary alcohols, if the aldehyde is not removed from the oxidizing agent immediately it is formed, it is further oxidized to give the carboxylic acid (Fig 16.29). The aldehyde is removed from the reacting mixture by distillation.

**Fig 16.29** Oxidation of primary and secondary alcohols to form aldehydes and ketones, respectively

Ketones resist further oxidation, because they have no oxidizable hydrogen bonded to the carbonyl group. However, with prolonged heating they can be oxidized by strong oxidizing agents.

By contrast, the oxidation of aldehydes occurs even in air, so that bottles of opened aldehydes soon contain amounts of carboxylic acids. This difference in the reactivities of aldehydes and ketones to oxidation is one of the main reasons they are considered as separate classes of compounds. It also provides an ideal way to distinguish between them.

## OXIDATION OF ALDEHYDES BY ACIDIFIED POTASSIUM DICHROMATE(VI)

When an acidified solution of the orange dichromate(VI) ion is warmed with an aldehyde, it is reduced to the green chromium(III) ion. The effect of this reaction is shown in Fig 16.30, in which the aldehyde is ethanal.

The equation for the reaction with ethanal is given in Fig 16.31.

$$CH_3-C{\overset{H}{\underset{O}{}}} \quad + \quad [O] \quad \xrightarrow[\text{heat}]{Cr_2O_7{}^{2-}(aq),\, H^+(aq)} \quad CH_3-C{\overset{O}{\underset{OH}{}}}$$

**Fig 16.31** Oxidation of ethanal

▶ **EXTENSION Using half equations**

The half equations for this reaction are:

$$Cr_2O_7{}^{2-}(aq) + 14H^+(aq) + 6e^- \rightarrow 2Cr^{3+}(aq) + 7H_2O(l)$$
$$CH_3CHO(aq) + H_2O(l) \rightarrow CH_3COOH(aq) + 2H^+(aq) + 2e^-$$

Combining these to cancel out the electrons gives the redox equation:

$$3CH_3CHO(aq) + Cr_2O_7{}^{2-}(aq) + 8H^+(aq) \rightarrow$$
$$3CH_3COOH(aq) + 2Cr^{3+}(aq) + 4H_2O(l)$$

## OXIDATION OF ALDEHYDES BY FEHLING'S AND BENEDICT'S SOLUTIONS

Either or both of these reagents may be used in your chemistry course. They are alkaline solutions that contain copper(II) as **complex ions** (see Chapter 24), so the colour of both is blue (Fig 16.32). The copper(II) ions act as a mild oxidizing agent. When the reagent is mixed with an aldehyde and heated, the aldehyde is oxidized to give a carboxylic acid, while the copper(II) ions are reduced to a brick-red precipitate of copper(I) oxide, $Cu_2O$.

A balanced equation for this reaction is:

$$RCHO + 2Cu^{2+}(aq) + 2H_2O \rightarrow RCOOH + Cu_2O(s) + 4H^+(aq)$$

The reaction for ethanal is shown in Fig 16.33.

$$CH_3C{\overset{H}{\underset{O}{}}} \quad \xrightarrow[\text{or Fehling's solution}]{Cu^{2+}(aq) \text{ in Benedict's}} \quad CH_3C{\overset{O}{\underset{OH}{}}} \quad + \quad Cu_2O(s)$$

ethanal          ethanoic acid      brick red precipitate

**Fig 16.33** The oxidation of ethanal by Fehling's and Benedict's solutions

This reaction can be used to distinguish between an aldehyde and a ketone. While this reaction is less used now that we have spectrometers, it is still an important test in some situations. For example, diabetes is easily diagnosed

**Fig 16.30** Left: Acidified potassium dichromate(VI) before ethanal is added and the mixture heated. Right: After the oxidation of ethanal, the green $Cr^{3+}$ is clearly visible

**SEE QUESTION 3.**

THE USE OF HALF EQUATIONS IN REDOX REACTIONS IS COVERED IN CHAPTERS 5 AND 25.

**Fig 16.32** Left: The copper(II) complex ion is responsible for the blue colour of Fehling's solution. Right: After oxidation of the aldehyde, the copper(II) ion is reduced to $Cu_2O$, in the form of a brick-red precipitate

**SEE QUESTION 4.**

using Benedict's or Fehling's reagent, which detects glucose in urine samples, and Benedict's reagent in tablet form is used by diabetics to monitor their blood sugar levels.

### TOLLENS' REAGENT AND THE OXIDATION OF ALDEHYDES

Tollens' reagent provides yet another way to test for aldehydes. This time, the oxidizing agent is a complex of silver(I) ions, which are reduced to silver in the test. Fig 16.34 shows silver coating the inside of a test tube, making a 'silver mirror'.

The complex, which has the formula $[Ag(NH_3)_2]^+$, is made by mixing together aqueous solutions of ammonia and silver nitrate. The resulting solution is called Tollens' reagent, or ammoniacal silver nitrate. When warmed with an aldehyde, it produces the 'silver mirror'.

benzaldehyde                    silver mirror

**Fig 16.35** Producing a silver mirror by reacting benzaldehyde with Tollens' reagent

**Fig 16.34** Left: Ammoniacal silver nitrate before the addition of the aldehyde. Right: The aldehyde reduces the $[Ag(NH_3)_2]^+$ to Ag to give the silver mirror effect

This is a simplified balanced equation for the reaction:

$$RCHO + 2Ag^+(aq) + H_2O \rightarrow RCOOH + 2Ag(s) + 2H^+$$

The reaction provides one of the ways in which mirrors are silvered.

**SEE QUESTIONS 5, 6, 7 AND 8.**

### SUGARS

As mentioned above, glucose gives a positive test with both Fehling's and Benedict's solutions. As it has this reducing property, glucose is known as a **reducing sugar**. In one of its forms, glucose contains an aldehyde group, which accounts for this property.

**MORE INFORMATION ON GLUCOSE AND ANOTHER SUGAR, FRUCTOSE, CAN BE FOUND IN THE ASSIGNMENT ON PAGE 362.**

---

**! SCIENCE FEATURE** Alcoholic drinks and the formation of aldehydes

When you have an alcoholic drink, ethanol passes into your bloodstream. The liver has to break down the ethanol. In the first stage, it is oxidized to ethanal:

$$CH_3CH_2OH + [O] \xrightarrow{\text{enzyme in liver}} CH_3CHO + H_2O$$

Ethanal is then oxidized into other products. However, if you drink a lot of ethanol in a short space of time, the ethanal that enters the bloodstream is distributed throughout the body. Your face goes red, you may feel an unpleasant tingling in the limbs and nausea, and your blood pressure may drop.

Methanol is added to ethanol that is intended for other uses, to make it unfit for drinking. This mixture is commonly known as methylated spirits. Once inside the body, methanol is oxidized into an aldehyde, which rapidly causes liver damage and blindness. So, unfortunately, people determined to drink meths are risking their lives.

---

**? TEST YOURSELF**

**J** What is the name of the aldehyde formed in the bodies of meths drinkers?

**SEE QUESTIONS 9 AND 10.**

### TRIIODOMETHANE TEST

This is a useful test for the $CH_3CO$ group in carbonyl compounds. An alkaline solution of aqueous iodine is warmed with the substance suspected to contain this group (Fig 16.36). (Iodine has a very low solubility in water. Therefore, the aqueous solution is made up by dissolving it in KI solution.) If a yellow precipitate of triiodomethane is produced, it is highly likely that $CH_3CO$ is present.

**Fig 16.36** Triiodomethane test. Note: R could also be hydrogen in this case

As $I_2(aq)/NaOH(aq)$ is an oxidizing agent, the triiodomethane test also gives a positive result with alcohols that contain the $CH_3CH(OH)$ group (Fig 16.37).

**Fig 16.37** Note that R could also be hydrogen in this case

**? TEST YOURSELF**

**K** State which of the following compounds will give a positive result with the triiodomethane test:
**a)** $CH_3CHO$;
**b)** the compound in Fig 16.38;

**Fig 16.38**
**c)** $C_2H_5COCH_3$;
**d)** $C_2H_5CHO$; **e)** $C_2H_5OH$.
Hint: Draw the structures out more fully before you decide. Only one of the five compounds does not give a positive test.

## SUMMARY

After studying this chapter, you should know the following:
- The carbonyl group C=O is found in both aldehydes (RCHO) and ketones (RCOR'), so they are called carbonyl compounds.
- The polar nature of the carbon–oxygen double bond in the carbonyl group makes it susceptible to nucleophilic addition reactions. This contrasts with the electrophilic addition reactions of the alkene carbon–carbon double bond.
- 2,4-Dinitrophenylhydrazine gives characteristic orange precipitates with carbonyl compounds. The resulting 2,4-dinitrophenylhydrazone derivatives can be used to identify specific aldehydes and ketones from their precise melting points.

- Both aldehydes and ketones undergo nucleophilic addition reactions with HCN and with $H^-$ ions (from, for example, $NaBH_4$).
- Aldehydes (but not ketones) are oxidized to give carboxylic acids by mild oxidizing agents, such as $Cr_2O_7^{2-}/H^+$, alkaline solutions of $Cu^{2+}$ (Fehling's solution) and $Ag^+$ (Tollens' reagent). These reactions can be used to distinguish between aldehydes and ketones.
- Alkaline aqueous iodine gives triiodomethane ($CHI_3$) with $CH_3CO$ compounds, and also with alcohols that contain $CH_3CH(OH)$.

## ? EXAM QUESTIONS

**1** A carbonyl group C=O contains a double covalent bond between the carbon and oxygen atoms. This is made up of one σ bond and one π bond.

Explain, using words and diagrams, what is meant by
**a)** a σ bond and
**b)** a π bond.

**2** The unsaturated diketone shown in Fig 16.Q2.1 is excreted by the bombardier beetle.

**Fig 16.Q2.1**

What is formed (Fig 16.Q2.2) when this compound reacts with hydrogen cyanide?

**Fig 16.Q2.2**

OCR (UCLES) 9253/3 November 2000 Q26

**3** **a)** Compound **A** can be converted into three different organic compounds as shown by the reaction sequence shown in Fig 16.Q3. Give the structures of the new compounds **B**, **C** and **D**.

**Fig 16.Q3**

**b)** Outline a mechanism for the formation of **C**.

AQA CH03 Feb 1998 Q6a,b

**4** Two organic compounds, **A** and **B**, both have the molecular formula $C_4H_8O$. Both **A** and **B** give yellow precipitates when added to Brady's reagent (2,4-dinitrophenylhydrazine).
**a)** What can be deduced about **A** and **B**?
**b)** Compound **A** gives a red precipitate when warmed with Benedict's solution, but **B** shows no reaction when treated this way.
**c)** Draw the displayed formula of compound **B** and give its systematic name.

**d)** The compound in Fig 16.Q4 is a product of the reaction between an organic compound and an inorganic reagent.

$$CH_3 \quad OH$$
$$\underset{CH_3}{\overset{|}{C}}\quad CN$$

**Fig 16.Q4**

**(i)** Give the structural formula of the original organic compound.
**(ii)** Give the name or formula of the inorganic reagent.

Edexcel (Nuffield) Specimen Unit Test 4 Q2

**5**
**a)** Three isomers **A**, **B** and **C** have the molecular formula $C_4H_8O$. All three compounds give an orange precipitate with 2,4-dinitrophenylhydrazine reagent. **B** and **C** also give a silver mirror when warmed with ammoniacal silver nitrate solution. Write the structural formulae of **A**, **B** and **C**.
**b)** Substance **A** reacts with the Grignard reagent, $C_2H_5MgBr$.
**(i)** Give the equation for the preparation of this Grignard reagent.
**(ii)** State the conditions for this preparation.
**(iii)** Write the structural formula of the product obtained when this Grignard reagent reacts with substance **A**.

**6** [Information about compound **C** can be found in Chapter 18.]

Several compounds have the molecular formula $C_5H_{10}O_2$ (Fig 16.Q6).

**Fig 16.Q6**

**a)** Name the functional groups present in the three structural isomers **A**, **B** and **C**.
**b)** Draw the structural formulas of the organic products (if any) that would be formed on treatment of **A**, **B** and **C** with the reagents given below. State 'no reaction' if you predict no reaction would occur:
**(i)** with ammoniacal silver nitrate;
**(ii)** with acidified aqueous $K_2Cr_2O_7$.
**c)** Draw the structural formulas of the organic products formed by the reaction:

## ❓ EXAM QUESTIONS

**(i)** of **A** with $NaBH_4$;
**(ii)** of **B** with $NaBH_4$;
**(iii)** of **B** with KCN in the presence of an acid catalyst;
**(iv)** of **C** with aqueous sodium hydroxide.

OCR 6821 General and Organic Chemistry June 2000 Q1

**7**

**a)** **(i)** Pentan-1-ol, $CH_3CH_2CH_2CH_2CH_2OH$, was oxidized by adding it dropwise to acidified potassium dichromate(VI) and distilling off the organic product immediately. Write an equation for this reaction showing clearly the structural formula of the organic product. You may use [O] to represent the oxidizing agent.

**(ii)** Pentan-2-ol, $CH_3CH_2CH_2CH(OH)CH_3$, was oxidized by heating it under reflux with acidified potassium dichromate(VI). Write an equation for this reaction showing clearly the structural formula of the organic product. You may use [O] to represent the oxidizing agent.

**b)** By stating the reagents, conditions and observations, show how you would distinguish between the two oxidation products formed in parts **a)(i)** and **a)(ii)**.

AQA Unit 3(a) Introduction to Organic Chemistry June 2001 Q2

**8** The insecticide DDT is made by heating together chlorobenzene with trichloroethanal (chloral) under suitable conditions.

$$2C_6H_5Cl + CCl_3CHO \longrightarrow$$ DDT

**Fig 16.Q8.1**

**a)** State a test and the result obtained that would identify the presence of the group shown in Fig 16.Q8.2 in an aqueous solution of trichloroethanal.

$$A \xrightarrow[\textbf{Step 1}]{Mg} CH_3MgBr \xrightarrow[\textbf{Step 2}]{C_4H_8O} C_5H_{12}O \xrightarrow[\textbf{Step 3}]{K_2Cr_2O_7/H_2SO_4}$$

$$\textbf{B}$$

$$C_5H_{10}O \xrightarrow[\textbf{Step 4}]{I_2/NaOH} \text{Yellow ppt.} + \textbf{E}$$

$$\textbf{C} \qquad\qquad \textbf{D}$$

**Fig 16.Q8.2**

**b)** Write the graphic (full structural) formula for the organic compound obtained when chloral reacts with acidified potassium dichromate(VI) solution

WJEC 2001/2 Specimen Paper CH4 Q2d(i),(ii)

**9** Which statement about ethanal and propanone is incorrect?

**a)** Both may be prepared by the oxidation of an alcohol.

**b)** Both change the colour of warm acidified potassium dichromate(VI) from orange to green.

**c)** Both react with 2,4-dinitrophenylhydrazine reagent.

**d)** Both give a positive triiodomethane (iodoform) test.

OCR (UCLES) 9253/3 November 2000 Q25

**10** Consider the following reaction scheme in which all of the compounds have straight chains.

**a)** **(i)** Give the structural formula of **A**.
**(ii)** State the conditions under which the reaction in **Step 1** is carried out.

**b)** After giving consideration to the sequence of reactions in **Steps 2, 3** and **4**:
**(i)** write the *structural* formula of $C_4H_8O$;
**(ii)** write the *structural* formula of **B**.

**c)** **(i)** Identify the yellow precipitate **D** formed in **Step 4**.
**(ii)** Give the structural formula of **E**.

**d)** Give the structural formula for **C**.

Edexcel Module Test 4 January 1999 Q1a–d

# KEY SKILLS **ASSIGNMENT**

## SUGARS

Sugars are important and well-known biological molecules. They contain a carbonyl group, so are either aldehydes or ketones. Two examples, glucose and fructose, exist in both open-chain and ring forms, as Figs 16.A1 and 16.A2 show.

**Fig 16.A1** Open-chain forms of glucose and fructose

**Fig 16.A2** Open-chain and ring forms of glucose

**1 a)** What are the molecular formulas of glucose and fructose?
**b)** What is meant by the term carbonyl group?
**c)** Draw the functional group in an aldehyde and a ketone.
**d)** Is fructose an aldehyde or a ketone?
**e)** Name the other functional group present in both molecules.

In aqueous solution, the ring and open-chain forms of glucose are in equilibrium. Since the ring form is at the highest concentration, this is the structure that is usually drawn. Nevertheless, there is enough of the open-chain aldehyde to show characteristic reactions of this group.

**2 a)** What name is given to the mechanism of the reaction that produces the ring form from the open-chain form of glucose?
Hint: Look at the carbonyl functional group.
**b)** Draw the carbonyl group to show the π and σ bonds.
**c)** The alkene C=C also has π and σ bonds, yet its addition reactions occur by a different mechanism. Explain why there is this difference.

**d)** Using the open-chain form of glucose, write the equation for its reaction with HCN.

**3** 2,4-Dinitrophenylhydrazine gives a coloured crystalline precipitate with fructose. Using the open-chain form of fructose, write the equation.

As mentioned on page 358, glucose gives a positive result with the Fehling's test because it has an aldehyde group. However, fructose also gives a brick-red precipitate with Fehling's solution – which is a surprising result. In fact, all monosaccharides, whether ketones or aldehydes, are called reducing sugars because of their ability to act as reducing agents. So, both fructose and glucose give a positive test with Tollens' reagent.

**4 a)** Why is it a 'surprise' that fructose gives a positive Fehling's test?
**b)** How do we know that fructose is reacting as a reducing reagent in this reaction?
**c)** What would you observe in a positive Tollens' test?

Glucose is metabolized in the body in a series of steps. In one step, pyruvic acid is converted into lactic acid (Fig 16.A3).

**Fig 16.A3**

**5 a)** What functional group appears in both lactic acid and pyruvic acid?
**b)** What type of alcohol is lactic acid?
**c)** This reaction could be carried out in the laboratory by the addition of H⁻ ions. Draw this mechanism.

Glucose can be reduced industrially to give sorbitol ($C_6H_{14}O_6$), which is used by diabetics as a sugar substitute.

**6 a)** Write a balanced equation for the production of sorbitol from glucose (open-chain form), using [H] to denote the reducing agent.
**b)** What reagents might be used to make sorbitol from glucose in: (i) the laboratory and (ii) industry?

Sucrose is a disaccharide formed from glucose and fructose. It is the sweetening agent we are probably all most familiar with – the white sugar of the supermarket shelves. It is a **non-reducing sugar** and so does not give a positive test with Tollens' and Fehling's reagents.

**7** Find out about the reaction that forms sucrose.

# 17

# Carboxylic acids and pH

# 17 Carboxylic acids and pH

**Blood chemistry** It is essential for the chemistry of blood that the pH value of the plasma remains fairly constant. Blood contains complex proteins and simple inorganic ions that act as buffers and together control the plasma's acidity

Blood is very intimately connected with the biochemistry of all our tissues, which function correctly only while the blood is maintained at pH 7.4. In fact, death is likely to follow if the blood pH departs from this value for any length of time.

Chemical reactions in the blood itself continually produce acids and alkalis, which would seriously upset the pH level but for the fact that blood also contains buffers. These chemicals resist changes in blood pH and keep its fluctuations within safe limits.

Just as our bodies need a steady pH, so we take steps to control the pH of a wide range of foods and drinks we consume, for their stability as well as for our taste preferences and health. Wines, beers and lagers contain acids, such as ethanoic acid and tartaric acid. On the one hand, acidity helps to preserve them, but on the other too much acid makes them sour and vinegary, so food chemists ensure that the pH strikes the right balance.

When designing new chemical processes, too, industrial chemists often need to find ways to stabilize the pH values of solutions so that they won't significantly change when small amounts of acid or alkali are added. Again, for this buffers are required: often these are carboxylic acids and their soluble salts, the carboxylates, the subjects of this chapter.

**Fig 17.1** Vinegar is a 5 per cent solution of ethanoic acid in water. It is acidic because ethanoic acid donates a proton – a hydrogen ion – to a water molecule. The sour taste is due to the presence of aqueous hydrogen ions

## ✔ REMEMBER THIS

$CH_3COONa$ is an ionic salt and is sometimes written as $CH_3COO^-Na^+$.

## 1 WHAT IS AN ACID?

Carboxylic acids are organic chemicals that contain a group of atoms known as the carboxyl group, COOH. Many common chemicals, such as citric acid in lemon juice and ethanoic acid in vinegar, are carboxylic acids and are described as weak acids. So before we go into the chemistry of carboxylic acids, we need to be clear what is meant by the term 'acid' and why chemists talk of 'weak acids' and 'strong acids'.

Chemists knew about the properties of acids long before they understood what makes an acid. In the late 1700s, they believed that oxygen was the essential element in acids. Then in 1810, Humphry Davy disproved this hypothesis when he discovered that hydrochloric acid consists of only hydrogen and chlorine. So, hydrogen was established as the element common to all acids.

One early definition says that an acid is a compound whose molecule has at least one hydrogen atom that can be replaced by a metal atom. For example, hydrochloric acid, HCl(aq), forms sodium chloride, NaCl, when hydrogen is replaced by sodium. In the same way, ethanoic acid, $CH_3COOH(aq)$, is an acid because it forms sodium ethanoate, $CH_3COONa$. However, as theoretical chemistry developed, more sophisticated and generalized definitions of an acid – and a base – were introduced.

### BRØNSTED–LOWRY THEORY

In 1923, the Swedish chemist Johannes Brønsted and the English chemist Thomas Lowry defined acids and bases:

**An acid is a proton donor and a base is a proton acceptor.**

According to this definition, when an acid reacts with a base, a proton is transferred from the acid to the base.

## 2 ACID–BASE REACTIONS

Put simply, an acid–base reaction is one in which an acid is neutralized by a base, or vice versa. During an acid–base reaction, a proton is transferred from the acid to the base. The Brønsted–Lowry theory of acids and bases really focuses on the behaviour of acids and bases in water. The water molecule is **amphoteric**, which means that it can behave as either an acid or a base (see page 366).

### BEHAVIOUR OF BRØNSTED–LOWRY ACIDS IN WATER

When an acid, HA(aq), is added to water, a proton is transferred to a water molecule:

$$HA(aq) + H_2O(l) \rightleftharpoons H_3O^+(aq) + A^-(aq)$$

In $H_3O^+$(aq), a single water molecule forms a dative covalent bond with a hydrogen ion (see page 74). It is called an **aqueous hydrogen ion** or an **oxonium ion**. (The term hydronium ion is sometimes used.)

So when, for example, pure nitric acid is added to water, there is an immediate reaction in which a proton is donated from the acid to a water molecule:

$$HNO_3(aq) + H_2O(l) \rightarrow H_3O^+(aq) + NO_3^-(aq)$$

Be careful with the use of the formula HCl. The formula HCl(g) is hydrogen chloride, a colourless gas, and HCl(aq) is hydrochloric acid. HCl without (g) or (aq) does not, strictly speaking, refer to either of the two substances.

**Fig 17.3** The colourless gas hydrogen chloride, HCl(g), is made up of covalent molecules. It dissolves in water to form hydrochloric acid, HCl(aq), which consists of ions. When hydrogen chloride dissolves in water it donates a proton to a water molecule

### BEHAVIOUR OF BRØNSTED–LOWRY BASES IN WATER

When a base is added to water, it accepts a proton from the water to form a hydroxide ion. For example, when sodium oxide $Na_2O$ is added to water, it is ionized and the oxide ion $O^{2-}$ accepts a proton from a water molecule to form a hydroxide ion:

$$O^{2-}(s) + H_2O(l) \rightarrow OH^-(aq) + OH^-(aq)$$

The equation has been written with two separate hydroxide ions to emphasize that the oxide ion becomes part of a hydroxide ion when it accepts a proton from a water molecule.

As another example, ammonia, a colourless gas, dissolves in water to form an alkaline solution:

$$NH_3(g) + H_2O(l) \rightleftharpoons NH_4^+(aq) + OH^-(aq)$$

This time, the whole ammonia molecule accepts a proton from a water molecule to form the ammonium ion.

**REMEMBER THIS**

A proton remains when a hydrogen atom loses an electron. So a proton is the same as a hydrogen ion, H+.

**REMEMBER THIS**

An alkali is just a base that dissolves in water.

**Fig 17.2** Dot-and-cross diagram of an oxonium ion, also known as an aqueous hydrogen ion or a hydronium ion

**? TEST YOURSELF**

**A** Write equations to show what happens when:
**a)** hydrogen chloride dissolves in water;
**b)** hydrogen iodide dissolves in water.

**REMEMBER THIS**

Notice that the equation with $NH_3$ does not have an arrow, but has the symbol for a reversible reaction. There is more about the significance of this symbol in relation to bases in Chapter 28.

## AMPHOTERIC NATURE OF WATER

In water, there is always a very small proportion of $OH^-(aq)$ and of $H^+(aq)$. This is the result of an acid–base reaction between two water molecules:

$$H_2O(l) + H_2O(l) \rightleftharpoons H_3O^+(aq) + OH^-(aq)$$

One water molecule donates a proton and the other accepts a proton. This demonstrates that water can behave both as an acid and as a base, so it is an amphoteric compound.

This acid–base reaction can be written in a simpler form as the dissociation of water:

$$H_2O(l) \rightleftharpoons H^+(aq) + OH^-(aq)$$

## DISSOCIATION IN BRØNSTED–LOWRY ACIDS

When a Brønsted–Lowry acid is put into water, a chemical reaction called **dissociation** or **ionization** takes place. A covalent bond between an electronegative atom and a hydrogen atom is broken by heterolytic fission, to leave a proton and a negative ion. So, when the gas hydrogen chloride dissolves in water, it dissociates to form a proton or hydrogen ion, and a chloride ion:

$$HCl(g) \rightarrow H^+(aq) + Cl^-(aq)$$

The arrow in the equation indicates that, in water, every hydrogen chloride molecule dissociates.

In water, sulphuric acid dissociates to form two hydrogen ions per sulphuric acid molecule:

$$H_2SO_4(aq) \rightarrow 2H^+(aq) + SO_4{}^{2-}(aq)$$

The arrow in the equation indicates that, in water, every sulphuric acid molecule dissociates, but in reality many of the sulphuric acid molecules only dissociate to form the hydrogensulphate ion, $HSO_4{}^-$.

$$A{-}H + H_2O \longrightarrow A^- + H_3O^+$$

**Fig 17.4** Dissociation of an acid

**Fig 17.5** Dissociation of sulphuric acid represented by structural formulas

## CONJUGATE BASES

During the dissociation of hydrogen chloride, all the molecules dissociate to form aqueous hydrogen ions and chloride ions. The chloride ion is known as the **conjugate base** of hydrochloric acid. This does not mean that the chloride ion is a base, but that in theory it could accept a proton. That is, it could behave as a base. The conjugate base of any acid is the anion formed after dissociation.

Note that the conjugate base of sulphuric acid is the hydrogensulphate ion, $HSO_4{}^-$, rather than the sulphate ion, $SO_4{}^{2-}$ (see Table 17.1).

**Table 17.1 Five acids and their dissociations**

| Acid | | | | Conjugate base |
|---|---|---|---|---|
| $HCl(aq)$ | $\rightarrow$ | $H^+(aq)$ | $+$ | $Cl^-(aq)$ |
| $HNO_3(aq)$ | $\rightarrow$ | $H^+(aq)$ | $+$ | $NO_3{}^-(aq)$ |
| $H_2SO_4(aq)$ | $\rightarrow$ | $H^+(aq)$ | $+$ | $HSO_4{}^-(aq)$ |
| $HBr(aq)$ | $\rightarrow$ | $H^+(aq)$ | $+$ | $Br^-(aq)$ |
| $HI(aq)$ | $\rightarrow$ | $H^+(aq)$ | $+$ | $I^-(aq)$ |

## CONJUGATE ACIDS

Just as a Brønsted–Lowry acid has a conjugate base, so a Brønsted–Lowry base has a conjugate acid, which is the particle formed once the base has accepted a proton. For example, the conjugate acid of ammonia, $NH_3$, is the ammonium ion, $NH_4^+$, and the conjugate acid of the hydroxide ion, $OH^-$, is the water molecule, $H_2O$.

## STRONG ACIDS

Table 17.1 shows how five common covalently bonded acids dissociate when in water. In each case, almost all the molecules of the acid dissociate to form ions. In fact, we assume that all of the molecules dissociate, and represent the dissociation by an arrow in the equation. Such acids are called **strong acids**, a term meaning that almost all of the acid's molecules dissociate when it is dissolved in water. Do not confuse a strong acid with a concentrated acid. A strong acid is fully dissociated in water, but a concentrated acid has a high number of moles of acid in 1 $dm^3$ of solution.

## WEAK ACIDS

Other acids, such as ethanoic acid, dissolve in water, but do not fully dissociate. Only a very small percentage of their molecules may be dissociated at any one time. They are known as **weak acids**. For example, in a solution containing 0.1 mol $dm^{-3}$ of ethanoic acid, only 1.3 per cent of the molecules dissociate. Ethanoic acid is therefore a weak acid. Do not confuse a weak acid with a dilute acid. The term weak acid means that only a small proportion of molecules dissociate when the acid is dissolved in water:

$$CH_3COOH(aq) \rightleftharpoons CH_3COO^-(aq) + H^+(aq)$$

Note that the dissociation equation includes the symbol for a reversible reaction instead of an arrow. This indicates that the dissociation of ethanoic acid and the association of the ethanoate ion and the hydrogen ion occur at the same time.

As in Figs 17.6 and 17.7, once ethanoic acid has dissolved in water, the rate of the dissociation reaction soon reaches that of the association reaction, so that there is no net change in the concentrations of the undissociated ethanoic acid, the ethanoate ion and the hydrogen ion. We say that the reaction reaches a state of **dynamic equilibrium**, in which the concentration of each species remains the same.

**Fig 17.6** A short time after the ethanoic acid is mixed with water, the concentration of each species remains constant. The concentrations are called equilibrium concentrations. The reaction is in equilibrium because the concentration of each species is constant.

**At start** forward reaction (dissociation) / backward reaction (association)

**Near the start** rate of forward reaction > rate of backward reaction (dissociation) (association)

**At equilibrium** forward reaction / backward reaction

**At equilibrium** rate of forward reaction = rate of backward reaction (dissociation) (association)

**Fig 17.7** Dissociation and association reactions of ethanoic acid in water

This is not the whole story, however. Chemical reactions are still taking place and particles are colliding and reacting, but the rate of the forward reaction is matched by the rate of the backward reaction. Note the y (concentration) axis has no scale between 0 and 1.0. This is because on a uniform y-axis scale, the (equal) concentrations of the aqueous hydrogen ion and the ethanoate ion are so small compared with the concentration of the undissociated ethanoic acid that they would not show as a curve above the x-axis

# ACID DISSOCIATION CONSTANT

Once a weak acid has dissolved in water, it quickly reaches the state of dynamic equilibrium. It is possible to gauge just how weak an acid is by determining the **acid dissociation constant, $K_a$.** The $_a$ in $K_a$ stands for 'acid'.

A weak acid HA dissociates in water and reaches a state of dynamic equilibrium:

$$HA(aq) \rightleftharpoons H^+(aq) + A^-(aq)$$

The acid dissociation constant is given by:

$$K_a = \frac{[H^+(aq)][A^-(aq)]}{[HA(aq)]}$$

The square brackets represent the concentrations of the various particles measured in mol dm$^{-3}$. For example, [H$^+$(aq)] is the concentration of aqueous hydrogen ions in mol dm$^{-3}$. Note that the concentrations are those at the state of dynamic equilibrium.

The acid dissociation constant for ethanoic acid is:

$$K_a = \frac{[H^+(aq)][CH_3COO^-(aq)]}{[CH_3COOH(aq)]} \qquad \text{Units} = \frac{(\cancel{mol\ dm^{-3}})(mol\ dm^{-3})}{(\cancel{mol\ dm^{-3}})}$$

The smaller the value of the acid dissociation constant, the weaker the acid. Ethanoic acid, CH$_3$COOH, has a $K_a$ of $1.8 \times 10^{-5}$ mol dm$^{-3}$, whereas nitrous acid, HNO$_2$, has a $K_a$ of $4.5 \times 10^{-4}$ mol dm$^{-3}$. Notice that the acid dissociation constant includes a unit. It is not just a numerical value.

## p$K_a$

The numerical values of $K_a$ for strong acids normally lie in the range from $10^{-1}$ to $10^2$, but for weak acids they are below $10^{-4}$. To avoid dealing with the small numbers associated with weak acids, we use p$K_a$. This is its definition:

**p$K_a$ is the negative logarithm to base 10 of the acid dissociation constant:**

$$pK_a = -\log_{10} K_a$$

Note that as the acid strength decreases, p$K_a$ increases. So a high p$K_a$ value means a very weak acid.

**Fig 17.8** The connection between p$K_a$ and $K_a$

**○ EXAMPLE**

**Q** The acid dissociation constant for nitrous acid is $4.5 \times 10^{-4}$ mol dm$^{-3}$. What is the p$K_a$ value for nitrous acid?

**A** $pK_a = -\log_{10} K_a$
Substituting the value for $K_a$ gives:
$$pK_a = -\log_{10} (4.5 \times 10^{-4})$$
$$pK_a = 3.3$$

Propanedioic acid has two $pK_a$ values, because one molecule can donate two hydrogen ions. The first is donated more easily than the second:

$$HOOCCH_2COOH(aq) \rightleftharpoons HOOCCH_2COO^-(aq) + H^+(aq) \qquad pK_a = 2.77$$
$$HOOCCH_2COO^-(aq) \rightleftharpoons {}^-OOCCH_2COO^-(aq) + H^+(aq) \qquad pK_a = 5.66$$

## WHY IS SULPHURIC ACID STRONG AND ETHANOIC ACID WEAK?

The ability of a carboxylic acid, such as ethanoic acid, to donate a hydrogen ion depends on the strength of the O–H bond in the carboxyl group. Likewise, the ability of sulphuric acid to donate a hydrogen ion depends on the strength of its O–H bonds. In simple terms, the stronger the O–H bond(s), that is, the greater the bond energy, the weaker the acid. In sulphuric acid the presence of many highly electronegative oxygen atoms withdraws electrons away from the O–H bond and makes it weaker than the O–H bond in ethanoic acid. This is not the whole story, since the stability of the conjugate bases formed is also important. So the sulphate ion and the hydrogensulphate ion are more stable conjugate bases than the ethanoate ion.

**Fig 17.9** The displayed formulas of sulphuric acid and ethanoic acid. The O—H bond in sulphuric acid is very weak compared with the O—H bond in ethanoic acid. The electronegative oxygen atoms in sulphuric acid draw the electron cloud away from the O—H bonds, but this effect is less marked in ethanoic acid. So sulphuric acid is a strong acid and ethanoic acid is a weak acid

**Table 17.2 The relative strengths of some acids. Chloric(VII) acid is the strongest acid of all. It fully dissociates in water. All acids stronger than $H_3O^+$ fully dissociate in water. Notice that they include most of the common laboratory acids. Acids that are weaker than water are not usually considered to be acids. Ammonia, for example, is normally classified as a base**

| | Acid | | | Conjugate base |
|---|---|---|---|---|
| | Chloric(VII) acid | $HClO_4$ | $ClO_4^-$ | Chlorate(VII) ion |
| | Sulphuric acid | $H_2SO_4$ | $HSO_4^-$ | Hydrogensulphate ion |
| | Hydrogen iodide | HI | $I^-$ | Iodide ion |
| | Hydrogen bromide | HBr | $Br^-$ | Bromide ion |
| | Hydrogen chloride | HCl | $Cl^-$ | Chloride ion |
| | Nitric acid | $HNO_3$ | $NO_3^-$ | Nitrate ion |
| increasing acid strength | Oxonium ion | $H_3O^+$ | $H_2O$ | Water | increasing base strength |
| | Hydrogensulphate ion | $HSO_4^-$ | $SO_4^{2-}$ | Sulphate ion |
| | Phosphoric(V) acid | $H_3PO_4$ | $H_2PO_4^-$ | Dihydrogenphosphate(V) ion |
| | Hydrogen fluoride | HF | $F^-$ | Fluoride ion |
| | Nitrous acid | $HNO_2$ | $NO_2^-$ | Nitrous ion |
| | Ethanoic acid | $CH_3COOH$ | $CH_3COO^-$ | Ethanoate ion |
| | Carbonic acid | $H_2CO_3$ | $HCO_3^-$ | Hydrogencarbonate ion |
| | Ammonium ion | $NH_4^+$ | $NH_3$ | Ammonia |
| | Water | $H_2O$ | $OH^-$ | Hydroxide ion |
| | Ethanol | $C_2H_5OH$ | $C_2H_5O^-$ | Ethoxide ion |
| | Ammonia | $NH_3$ | $NH_2^-$ | Amide ion |

**? TEST YOURSELF**

**G a)** What is the connection between the strength of an acid and the strength of its conjugate base?
**b)** Explain the difference in acid strength of the hydrogen halides HBr, HCl, HF and HI.
Hint: Think about the bond energy of the bond that must break for dissociation to take place.

## 3 pH

An acid is a proton donor – when it dissolves in water the $H^+(aq)$ concentration increases. That is, the higher the $H^+(aq)$ concentration, the more acidic a solution becomes.

However, the problem in measuring hydrogen ion concentration is that for many solutions it is extremely small. For example, the hydrogen ion concentration in wine is only about $5 \times 10^{-4}$ mol dm$^{-3}$. In blood, it is even smaller, being about $4 \times 10^{-8}$ mol dm$^{-3}$.

In 1909, the Danish chemist Sören Sörensen, working for the brewers Carlsberg, suggested an alternative way to measure the acidity of a solution – the **pH scale**.

**Table 17.3 The relationship between [H⁺(aq)] and pH**

| [H⁺(aq)]/ mol dm⁻³ | pH | Sample solution |
|---|---|---|
| 10¹ | −1 | |
| 10⁰ or 1 | 0 | ← 1 mol dm⁻³ H₂SO₄ |
| 10⁻¹ | 1 | |
| 10⁻² | 2 | |
| 10⁻³ | 3 | ← vinegar |
| 10⁻⁴ | 4 | ← stomach acid ← wine |
| 10⁻⁵ | 5 | |
| 10⁻⁶ | 6 | ← black coffee |
| 10⁻⁷ | 7 | |
| 10⁻⁸ | 8 | ← pure water ← blood |
| 10⁻⁹ | 9 | |
| 10⁻¹⁰ | 10 | |
| 10⁻¹¹ | 11 | ← milk of magnesia |
| 10⁻¹² | 12 | ← household ammonia |
| 10⁻¹³ | 13 | |
| 10⁻¹⁴ | 14 | ← 1 mol dm⁻³ NaOH |

The pH scale is constructed from the negative logarithms to base 10 of the [H⁺(aq)]. Thus, aqueous hydrogen-ion concentrations that may differ by, say, a hundred million million, are converted into numbers on a scale which runs from just below 0 to just above 14. Accordingly, the pH of a solution is described by the equation:

$$pH = -\log_{10} [H^+(aq)]$$

where the aqueous hydrogen ion concentration is in mol dm⁻³.

### ⚠ SCIENCE FEATURE Measuring the pH of a solution

Our sense of taste is very sensitive to pH, and it provided one of the earliest ways to distinguish between acids and alkalis. We can detect a sour acidic taste in solutions whose pH values are as high as 5, which corresponds to H⁺(aq) concentrations as low as 10⁻⁵ mol dm⁻³. The familiar acidic taste of most fruit juices and soft drinks is caused by the presence of weak acids, which give these drinks a pH value of about 3.

When the pH is below 2, we find the taste distinctly unpleasant and the liquid is sufficiently acid to burn the lining of the mouth. In the same way, we can detect alkalis with a pH value of 9 to 10 as they taste bitter. Liquids with higher pH values than this are dangerous to taste and burn the lining of the mouth.

The pH of a solution is measured accurately using a pH meter (Fig 17.10), which converts the difference in electrical potential between two electrodes into a pH reading. (Electrode potentials and electrochemistry are covered in Chapter 25.)

An indicator, such as universal indicator, which gives a range of colour in solutions of different pH values (Fig 17.11), can be used to estimate the pH value of a solution.

cable to pH meter

plastic cap

filling hole for reference solution

Pt wire

outer electrode

inner electrode

Ag/AgCl reference electrode

sensing area

**Fig 17.10** The pH meter. When the probe is dipped into the solution, an electrical potential difference is established between the two electrodes. This potential difference is amplified and converted into a digital pH reading

**Fig 17.11** Universal indicator gives a colour that depends on pH

**Fig 17.12** A commercial kit that gardeners use to test the pH of soil

# CALCULATING THE pH OF STRONG ACIDS

To calculate the pH of a strong acid, substitute the appropriate $H^+(aq)$ concentration in the equation for pH. This can be deduced from the concentration of the acid itself, since it is assumed that every acid molecule has dissociated.

## ⬤ EXAMPLE

**Q** What is the pH of a solution of 1.00 mol $dm^{-3}$ sulphuric acid?

**A** In water, sulphuric acid is assumed to be fully dissociated because it is a strong acid:

$$H_2SO_4(aq) \rightarrow 2H^+(aq) + SO_4^{2-}(aq)$$

From the equation, every mole of $H_2SO_4$ gives 2 moles of $H^+(aq)$. The concentration of 1.00 mol $dm^{-3}$ refers to moles of the undissociated $H_2SO_4$, so the concentration of $H^+(aq)$ must be twice this. Therefore,

$$[H^+(aq)] = 2 \times 1.00 \text{ mol dm}^{-3} = 2.00 \text{ mol dm}^{-3}$$

Substituting this value in the equation for pH gives:

$$pH = -\log_{10}[2.00] = -0.301$$

 **REMEMBER THIS**

A negative pH value means that the concentration of aqueous hydrogen ions, $[H^+(aq)]$, in mol $dm^{-3}$, is very high.

# CALCULATING THE pH OF WEAK ACIDS

Ethanoic acid has already been described as a weak acid that reaches a state of dynamic equilibrium when dissolved in water. When this state is reached, the concentration of aqueous hydrogen ions is very low compared with the concentration of the undissociated ethanoic acid. We can calculate the pH of aqueous ethanoic acid provided that we know the numerical value of the acid dissociation constant, or the percentage of ethanoic acid molecules that have dissociated.

## ⬤ EXAMPLE

**Q** What is the pH of 0.10 mol $dm^{-3}$ ethanoic acid, given that 1.3 per cent of the ethanoic acid molecules have dissociated?

**A** In water, ethanoic acid forms an equilibrium mixture very quickly:

$$CH_3COOH(aq) \rightleftharpoons H^+(aq) + CH_3COO^-(aq)$$

The 0.10 mol $dm^{-3}$ is the concentration of $CH_3COOH$ before any has dissociated.

Given that only 1.3 per cent of ethanoic acid molecules dissociate, then the concentration of ethanoic acid that has dissociated must be 1.3 per cent of 0.100 mol $dm^{-3}$, which is $1.3 \times 10^{-3}$ mol $dm^{-3}$.

From the equation, every mole of $CH_3COOH(aq)$ that dissociates gives 1 mole of $H^+(aq)$. So, $[H^+(aq)]$ must be $1.3 \times 10^{-3}$ mol $dm^{-3}$.

Substituting this value into the equation for pH gives:

$$pH = -\log_{10}(1.3 \times 10^{-3}) = 2.9$$

**Q** What is the pH of 0.10 mol $dm^{-3}$ ethanoic acid given that the acid dissociation constant, $K_a$, for ethanoic acid is $1.8 \times 10^{-5}$ mol $dm^{-3}$?

**A** Since ethanoic acid is a weak acid, it forms an equilibrium mixture in water:

$$CH_3COOH(aq) \rightleftharpoons CH_3COO^-(aq) + H^+(aq)$$

Therefore, the expression for the acid dissociation constant is:

$$K_a = \frac{[CH_3COO^-(aq)][H^+(aq)]}{[CH_3COOH(aq)]}$$

From the equation, every mole of $CH_3COOH$ that dissociates gives 1 mole of $CH_3COO^-$ and 1 mole of $H^+(aq)$.

So, if $[H^+(aq)]$ at equilibrium is $x$, then it follows that $[CH_3COO^-(aq)]$ must also be $x$.

*(continued on page 372)*

## EXAMPLE (continued from page 371)

It also follows that $[CH_3COOH(aq)]$ must be $0.10 - x$, since some of the ethanoic acid has dissociated.

This can be summarized as follows:

$$CH_3COOH(aq) \;\rightleftharpoons\; CH_3COO^-(aq) + H^+(aq)$$
undissociated
ethanoic acid

| | | | |
|---|---|---|---|
| At start, ie before equilibrium/mol dm$^{-3}$ | 0.10 | 0 | 0 |
| At equilibrium/mol dm$^{-3}$ | $0.10 - x$ | $x$ | $x$ |

These values can be substituted in the equation for the acid dissociation constant. Therefore,

$$1.8 \times 10^{-5} = \frac{x^2}{0.10 - x}$$

(The dissociation of water to form $H^+(aq)$ is ignored since its concentration is so small.)

The equation may be simplified by reference to Fig 17.6 on page 367. The concentration of the $H^+(aq)$ is very small compared with that of the undissociated acid, and therefore the concentration of the undissociated acid is

very nearly 0.10. So we can assume that $(0.10 - x)$ is approximately 0.10. Using this assumption, we can write the equation as:

$$1.8 \times 10^{-5} = \frac{x^2}{0.10}$$

which can be rearranged to give:

$$x^2 = 1.8 \times 10^{-6}$$

Therefore:

$x = \sqrt{(1.8 \times 10^{-6})} = 1.34 \times 10^{-3}$ (ignoring significant figs)

Since $x$ is the $H^+(aq)$ concentration, we can substitute this value of $x$ into the equation for pH:

$$\begin{aligned} pH &= -\log[H^+(aq)] \\ &= -\log (1.34 \times 10^{-3}) \\ &= 2.9 \end{aligned}$$

Notice that even with the assumption made in the second calculation, the value obtained for the pH is the same as that obtained in the Example before this.

## ? TEST YOURSELF

**H** **a)** Calculate the pH of
(i) 1.00 mol dm$^{-3}$ hydrochloric acid
(ii) $1.00 \times 10^{-4}$ mol dm$^{-3}$ hydrochloric acid;
(iii) 0.0520 mol dm$^{-3}$ sulphuric acid.
**b)** Assuming that in 0.100 mol dm$^{-3}$ nitrous acid, 6.5 per cent of the nitrous acid molecules have dissociated, calculate the pH of 0.100 mol dm$^{-3}$ nitrous acid.

**I** What is the $[H^+(aq)]$ and the $[OH^-(aq)]$ in boiling water?

**J** What is the pH of 0.10 mol dm$^{-3}$ methanoic acid, HCOOH, whose $K_a$ is $1.8 \times 10^{-4}$ mol dm$^{-3}$?

SEE QUESTIONS 1 AND 2.

## EXTENSION The pH of boiling water

At 298 K, the pH of pure water is 7.00, which is designated neutral on the pH scale. So the concentration of $H^+(aq)$ in pure water is $1.00 \times 10^{-7}$ mol dm$^{-3}$. An acid has a pH below 7. That is, the concentration of $H^+(aq)$ is greater than $1.00 \times 10^{-7}$ mol dm$^{-3}$.

When a pH meter is placed in boiling pure water, the reading is pH 6.12. So, apparently, boiling water is acidic. But how can it be? In boiling water, the concentration of $H^+(aq)$ has increased, but so has the concentration of the aqueous hydroxide ion. So, there should be no surplus of either type of ion:

$$H_2O(l) \;\rightleftharpoons\; H^+(aq) + OH^-(aq)$$

Now, this is the same situation as that in water at 298 K, which is considered to be neutral. The paradox is easy to explain. In boiling water, a greater proportion of water molecules dissociates at any one time, but there is no surplus of hydrogen ions.

## 4 CARBOXYLIC ACIDS

Carboxylic acids are weak acids. Therefore, they do not fully dissociate when dissolved in water. Many of them will not dissolve in water at all, although they can be shown to be acids since they react with alkalis.

### CARBOXYLIC ACIDS IN NATURE

Many naturally occurring organic compounds are carboxylic acids. Familiar examples already mentioned are citric acid in citrus fruits and ethanoic acid in vinegar. Three more examples are given in Figs 17.13 to 17.15.

Methanoic acid is the simplest carboxylic acid. Discovered in 1670, it was originally called formic acid, after the Latin word formicus for ant, since it is one of the substances responsible for the sting of an ant bite (Fig 17.14).

**Fig 17.13** Two naturally occurring carboxylic acids are ethanedioic acid and lactic acid. Ethanedioic acid is a poisonous white crystalline solid. As a metal salt, ethanedioic acid is found in rhubarb and sorrel, and gives them a sour taste. Lactic acid is formed when lactose, found in milk, ferments. It gives the sour taste of soured milk products such as yoghurt. It is a preservative, deterring the growth of micro-organisms, and is added to many foods, such as confectionery and salad dressings

**Fig 17.14** Methanoic acid, the simplest carboxylic acid, is one of the substances responsible for the irritation caused by an ant bite. It was once prepared by distilling red ants

**Fig 17.15** Benzoic acid is an aromatic carboxylic acid found in raspberries and in some tree barks. It is used as a preservative and an antioxidant in numerous foods, including soft drinks, pickles, salad dressings and fruit products

## THE CARBOXYL GROUP

A carboxylic acid contains the carboxyl group, COOH, which you may see written as $CO_2H$. The carboxyl group contains a carbonyl bond C=O, and a hydroxyl bond O–H. Each bond is attached to the same carbon atom. This arrangement modifies the characteristic behaviour of both the carbonyl bond and the hydroxyl bond, which is why the carboxyl group is considered a functional group in its own right.

**Fig 17.16** The carboxyl group

## NAMING CARBOXYLIC ACIDS

Carboxylic acids are named by adding the suffix oic acid to the name of the carbon chain (Fig 17.17). Unlike many functional groups, the carboxyl group contains one of the carbon atoms of the chain. The oic acid group is almost always on the number 1 carbon atom, so this carbon atom rarely has to be given any other position number.

ethanoic acid

**Fig 17.17** The derivation of the name ethanoic acid

benzoic acid

3–methylbenzoic acid

$CH_3CH_2COOH$
propanoic acid

$CH_3CHOHCOOH$
2-hydroxypropanoic acid

$HOOCCH_2COOH$
propane-1,3-dioic acid

$CICH_2COOH$
chloroethanoic acid

**Fig 17.18** Six carboxylic acids and their systematic names

**L** **a)** Methanoic acid, ethanoic acid and propanoic acid are members of a homologous series. What is the general formula for this homologous series?

**b)** (i) Predict the boiling point of pentanoic acid.

(ii) Explain why it is difficult to predict the melting point of pentanoic acid.

## PHYSICAL PROPERTIES OF CARBOXYLIC ACIDS

Carboxylic acids are either solids or liquids at room temperature (Table 17.4).

The melting points of methanoic acid and ethanoic acid are quite high, and so it is not unusual in cold weather for a bottle of concentrated ethanoic acid to be solid. In methanoic acid and ethanoic acid, the intermolecular forces are predominantly hydrogen bonds. However, as the carbon chain increases in length, there is also a considerable contribution from induced dipole–induced dipole attractions.

**Table 17.4 Melting and boiling points of six carboxylic acids**

| Acid | Formula | Melting point/°C | Boiling point/°C |
|---|---|---|---|
| Methanoic acid | HCOOH | 8.4 | 110.5 |
| Ethanoic acid | $CH_3COOH$ | 16.6 | 118 |
| Propanoic acid | $CH_3CH_2COOH$ | −22 | 141 |
| Butanoic acid | $CH_3CH_2CH_2COOH$ | −5 | 163 |
| Ethanedioic acid | $(COOH)_2$ | 187 | Decomposes |
| Benzoic acid | $C_6H_5COOH$ | 122 | 249 |

## 5 ACIDIC PROPERTIES OF THE CARBOXYL FUNCTIONAL GROUP

As mentioned already, carboxylic acids are weak acids, a property caused by the heterolytic fission of the hydroxyl bond in the COOH group (Fig 17.19).

A carboxylic acid that can dissolve in water forms a weakly acidic solution in water. If a carboxylic acid cannot dissolve in water, it is almost impossible for any of its molecules to donate a hydrogen ion to a water molecule, and so the water remains about neutral. Nevertheless, insoluble carboxylic acids are acids, because they will react with bases, such as aqueous sodium hydroxide.

**Fig 17.19** Dissociation of the carboxyl group involves heterolytic fission of the O–H bond and the subsequent transfer of a hydrogen ion (a proton) to a water molecule

THE EFFECT OF CONCENTRATION ON REACTION RATE IS DESCRIBED IN CHAPTER 26.

## ACID–BASE REACTIONS OF CARBOXYLIC ACIDS

The typical reactions of aqueous acids arise from the presence of the aqueous hydrogen ion. Since carboxylic acids are weak acids, such reactions are much slower than they are with strong acids, such as sulphuric acid.

As Fig 17.20 shows, the acidic reactions of ethanoic acid always form the ethanoate ion.

**Fig 17.20** Acidic reactions of ethanoic acid

# REACTION OF CARBOXYLIC ACIDS WITH ALKALIS AND BASES

Carboxylic acids, such as ethanoic acid, can be neutralized by an alkali, such as aqueous sodium hydroxide:

$$CH_3COOH(aq) + NaOH(aq) \rightarrow CH_3COO^-Na^+(aq) + H_2O(l)$$

This reaction is often carried out in the laboratory as a titration. An aqueous solution of the acid is added from a burette to a known volume of the alkali. At neutralization, an indicator changes colour.

Bases such as magnesium oxide react slowly with aqueous carboxylic acids to form carboxylate salts. For example, dilute ethanoic acid reacts to form aqueous magnesium ethanoate:

$$MgO(s) + 2CH_3COOH(aq) \rightarrow Mg^{2+}(CH_3COO^-)_2(aq) + H_2O(l)$$

This reaction is much slower than the reaction of magnesium oxide with dilute hydrochloric acid because ethanoic acid is a weak acid and so has a much lower concentration of $H^+(aq)$ than does hydrochloric acid, which is a strong acid.

## REACTION WITH SODIUM CARBONATE AND SODIUM HYDROGENCARBONATE

Sodium hydrogencarbonate and sodium carbonate will react with aqueous solutions of carboxylic acids to form carbon dioxide.

$$NaHCO_3(s) + CH_3COOH(aq) \rightarrow CH_3COO^-Na^+(aq) + CO_2(g) + H_2O(l)$$

$$Na_2CO_3(s) + 2CH_3COOH(aq) \rightarrow 2CH_3COO^-Na^+(aq) + CO_2(g) + H_2O(l)$$

## TITRATION CURVES

The changes in pH can be monitored at regular intervals as an alkali is added to an acid. The pH can then be plotted against the volume of alkali added, to obtain a graph known as a **titration curve**.

**Fig 17.21** In titration, an alkali is added dropwise from a graduated burette to an acid of known volume and concentration and that contains an indicator. The addition continues until one drop changes the colour of the indicator, which signals that the alkali has just neutralized the acid. Often, it is preferable to reverse the procedure, and add the acid to the alkali

Fig 17.22 is such a curve for the addition of aqueous sodium hydroxide to ethanoic acid.

Note that, as the alkali is added, the pH at first changes only slightly as the buffer solution is formed (see page 378). Then, near the point of neutralization, it shoots up. With the addition of more alkali, it quickly levels off. The exact neutralization point, or **equivalence point**, can be estimated as the midpoint of the near-vertical portion of the graph, that is, where there is a large change of pH for a very small addition of alkali. This is why you have to add the alkali one drop at a time when you are close to the end-point in a titration.

There are four types of titration curve (Figs 17.22 to 17.25), depending on whether strong or weak acids and bases are used.

**? TEST YOURSELF**

**M** Write down the name of the products of the following reactions:
**a)** calcium oxide and ethanoic acid;
**b)** calcium and propanoic acid;
**c)** magnesium carbonate and ethanoic acid;
**d)** calcium carbonate and ethanoic acid;
**e)** sodium hydrogencarbonate and benzoic acid.

**N** Calcium reacts with 50 cm$^3$ of 1.0 mol dm$^{-3}$ ethanoic acid. The reaction is much slower than the reaction between the same mass of calcium and 50 cm$^3$ of 1.0 mol dm$^{-3}$ hydrochloric acid.
**a)** Write down the balanced symbol equation for the reaction between calcium and ethanoic acid.
**b)** Explain the difference in the rate of reaction with ethanoic acid and hydrochloric acid.

Fig 17.22 The titration curve of a strong base, NaOH(aq), against a weak acid, $CH_3COOH(aq)$. This titration used 0.100 mol $dm^{-3}$ NaOH with 25.0 $cm^3$ of 0.100 mol $dm^{-3}$ $CH_3COOH$

Fig 17.23 The titration curve of a weak base, $NH_3(aq)$, against a weak acid, $CH_3COOH(aq)$. This titration used 0.100 mol $dm^{-3}$ $NH_3(aq)$ with 25.0 $cm^3$ of 0.100 mol $dm^{-3}$ $CH_3COOH$. Note that the pH changes gradually over almost the entire titration with no very sudden change in the pH at neutralization. It is impossible to determine the exact end-point in this type of titration

Fig 17.24 The titration curve of a weak base, $NH_3(aq)$, against a strong acid, HCl(aq). This titration used 0.100 mol $dm^{-3}$ $NH_3(aq)$ with 25.0 $cm^3$ of 0.100 mol $dm^{-3}$ HCl(aq). Note that as the alkali is added, the pH hardly changes until neutralization is close, at which point it shoots up from pH 2 to pH 7.5. Thereafter, the pH gradually levels off

Fig 17.25 The titration curve of a strong base, NaOH(aq), against a strong acid, HCl(aq). This titration used 0.100 mol $dm^{-3}$ NaOH(aq) with 25.0 $cm^3$ of 0.100 mol $dm^{-3}$ HCl(aq). Note again that as the alkali is added, the pH hardly changes until neutralization is close, when it shoots up

An indicator is chosen that changes colour within the pH range of the vertical portion of the titration curve. However, as the graph in Fig 17.23 shows, when a weak acid is titrated with a weak base, there is no sudden change in pH. This makes it very difficult to estimate the equivalence point, since any indicator changes colour within a range of pH values.

### DETERMINING $pK_a$ FROM A TITRATION CURVE

To determine the $K_a$ of a weak acid can be difficult. However, its $pK_a$ can be easily estimated by analysis of its titration curve. The $pK_a$ is the pH value at the half-equivalence point, that is, the point at which only half of the volume of alkali needed to reach the equivalence point has been added.

The reason for this is that, at the half-equivalence point, the concentration of the conjugate base, $A^-$, and that of the undissociated acid, HA, are almost equal. Therefore, they cancel out in the expression for $K_a$:

$$K_a = \frac{[H^+(aq)][A^-(aq)]}{[HA(aq)]}$$

So: $\quad K_a = [H^+(aq)]$ and $pK_a = pH$.

SEE QUESTIONS 2 AND 3.

### ? TEST YOURSELF

**O** Write down an expression for the $K_a$ of an indicator.

**P** **a)** Name an indicator suitable for the titration of a strong acid with a weak base.

**b)** Estimate the $pK_a$ of phenolphthalein and of thymolphthalein (see Table 17.5).

## EXTENSION Indicators

The approximate pH value of a solution can be determined using an indicator solution, or indicator paper, and comparing the colour obtained against a colour chart.

An **indicator** is a substance that changes colour when placed in an alkaline or acidic solution. The change of colour is because of changes in the complex organic structure of the indicator molecule, which affect the absorption of visible light by the molecule (see Chapter 27).

Litmus, a common indicator, is a natural product extracted from a lichen, while methyl orange, another common indicator, is a synthetic organic molecule known as an azo dye.

Indicators are often weak acids, and so in aqueous solutions they dissociate to a hydrogen ion and the conjugate base. It is convenient to use the simplified formula HIn for an indicator, where In⁻ stands for the conjugate base. In aqueous solution, the indicator dissociates:

$$HIn(aq) \rightleftharpoons H^+(aq) + In^-(aq)$$

Normally, the indicator, HIn, has one colour and the conjugate base, In⁻, has another. In the case of litmus, HIn is red and In⁻ is blue.

In an acidic solution (one with an excess of aqueous hydrogen ions), the position of the equilibrium shifts to the left to minimize the effect of the increase of hydrogen ions. So, as the concentration of HIn increases, the solution takes on the colour of the HIn molecule, which for litmus is red. This is an example of Le Chatelier's principle (see Chapters 1, 18 and 28).

When an indicator is added to an alkaline solution, the position of equilibrium moves to the right, because H⁺(aq) reacts with the alkali. Hence, there is a large concentration of In⁻ and the solution takes on the colour of this species.

The pH values for these colour changes vary with different indicators, as shown in Table 17.5.

To be useful in titration, an indicator must change colour within the almost vertical part of the titration curve. Almost all the indicators in Table 17.5 show the end point of a titration between a strong acid and a strong alkali. However, with a weak acid and a strong base, an indicator such as phenolphthalein should be used.

**Fig 17.26** The colour of some flowers depends on the pH of the soil, as shown by these hydrangeas which are pink in acid soil and blue in alkaline soil

**Table 17.5 The main properties of eight widely used indicators**

| Indicator | $pK_a$ | Effective pH range | Colour of acid form | Colour of base form |
|---|---|---|---|---|
| Methyl violet | 1.6 | 0.0–3.0 | Yellow | Violet |
| Methyl orange | 3.2 | 2.1–4.4 | Red | Yellow |
| Bromocresol green | 4.8 | 4.0–5.6 | Yellow | Blue |
| Methyl red | 5.2 | 4.2–6.2 | Red | Yellow |
| Bromothymol blue | 6.9 | 6.0–7.8 | Yellow | Blue |
| Thymol blue | 8.7 | 7.9–9.4 | Yellow | Blue |
| Phenolphthalein | (see P) | 8.3–10.0 | Colourless | Magenta |
| Thymolphthalein | (see P) | 9.3–10.5 | Colourless | Blue |

**SEE QUESTION 6.**

x = volume of alkali needed to reach equivalence or neutralisation point

$\frac{1}{2}x$ = volume of alkali needed to reach half equivalence point

**Fig 17.27** Measuring the $pK_a$ of a weak acid from its titration curve

## TITRATION CURVES FOR DIPROTIC ACIDS

In a **monoprotic acid**, such as ethanoic acid, an acid molecule can donate only one proton to water. In a **diprotic acid**, such as ethanedioic acid, an acid molecule can donate two protons to water. Thus, ethanedioic acid has two $K_a$ values, designated $K_{a1}$ and $K_{a2}$. Their significance is shown in Fig 17.28, in which $K_{a1}$ refers to equilibrium 1 and $K_{a2}$ to equilibrium 2. The titration curve for a diprotic weak acid has two equivalence points (Fig 17.29).

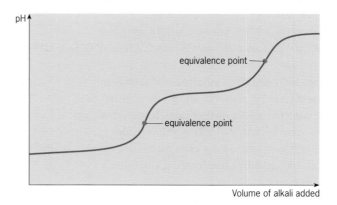

**Fig 17.28** The dissociation of ethanedioic acid

**Fig 17.29** The titration curve of a diprotic weak acid against a strong alkali. Note that there are two equivalence points

---

**▶ EXTENSION Enthalpy of neutralization**

The **enthalpy change of neutralization** is defined as the enthalpy change when one mole of water is produced during the neutralization of an aqueous acid with an aqueous alkali. The neutralization of an acid with an alkali is exothermic and represents the reaction between an aqueous hydrogen ion and an aqueous hydroxide ion:

$$H^+(aq) + OH^-(aq) \rightarrow H_2O(l)$$

This reaction is common to all acid–alkali neutralization reactions, so by Hess's law (see pages 181 and 182) the enthalpy of neutralization should be independent of the acid and the alkali used. When a strong acid is neutralized by a strong alkali the enthalpy of neutralization is 57 kJ mol$^{-1}$, which does not change even if the strong acid or strong base is changed. This value is considerable smaller, however, when a weak acid or a weak base is neutralized. The reason is quite simple – a weak acid is not fully dissociated, so that before the neutralization takes place the acid must dissociate and dissociation is an endothermic process. So the enthalpy of neutralization in this case has the exothermic contribution of the neutralization and an endothermic contribution needed to dissociate the acid or the base.

# BUFFER SOLUTIONS

A buffer solution resists changes in its pH when a small amount of either acid or alkali is added to it. A buffer solution is therefore used to control the pH of another solution to which a small amount of either acid or alkali is to be added. A typical buffer solution is a mixture of either a weak acid and its conjugate base, or a weak base and its conjugate acid. For example, aqueous solutions of a carboxylic acid and its sodium or potassium salt will act as a buffer.

The chemical processes that take place in a solution containing a weak acid, HA, and its sodium salt, NaA, will produce large concentrations of $A^-$ and HA:

$$HA(aq) \rightleftharpoons A^-(aq) + H^+(aq)$$

This equilibrium lies almost entirely on the left. So, we can assume that the concentration of HA at equilibrium is that of undissociated HA. The sodium salt completely ionizes in water, to form $A^-(aq)$. Notice that this process is shown by a right-pointing arrow:

$$NaA(aq) \rightarrow A^-(aq) + Na^+(aq)$$

We can therefore assume that a solution consisting of, for example, 0.100 mol dm$^{-3}$ ethanoic acid and 0.500 mol dm$^{-3}$ sodium ethanoate contains 0.100 mol dm$^{-3}$ of undissociated acid, $CH_3COOH$, and 0.500 mol dm$^{-3}$ of ethanoate ion, $CH_3COO^-$.

FOR MORE ON THE WAY THAT A WEAK BASE AND ITS CONJUGATE ACID TOGETHER ACT AS A BUFFER, SEE CHAPTER 28.

## WHAT HAPPENS WHEN AN ACID IS ADDED TO A BUFFER SOLUTION?

When acid is added to a buffer solution, the extra $H^+$ ions react with the conjugate base present, $A^-$, to produce undissociated acid, HA. This is an example of Le Chatelier's principle, since the equilibrium of the weak acid moves to the left to minimize the effect of the increase in the $H^+(aq)$ concentration. Since most of the extra $H^+(aq)$ added is converted into undissociated HA, the pH hardly decreases.

THERE IS MORE INFORMATION ON LE CHATELIER'S PRINCIPLE ON PAGES 18 AND 392, AND IN CHAPTER 28.

## WHAT HAPPENS WHEN HYDROXIDE IONS ARE ADDED TO A BUFFER SOLUTION?

When extra alkali is added to a buffer solution, the extra $OH^-$ ions react with the $H^+$ present to form water:

$$H^+(aq) + OH^-(aq) \rightarrow H_2O(l)$$

Extra $H^+$ ions are provided by more of the weak acid as it dissociates to maintain the equilibrium between HA(aq), $A^-$(aq) and $H^+$(aq). The overall buffering action can be represented by:

$$HA(aq) + OH^-(aq) \rightarrow H_2O(l) + A^-(aq)$$

**Fig 17.30** A summary of the buffering action of a mixture of sodium ethanoate and ethanoic acid

SEE QUESTIONS 4 AND 5.

## CALCULATING THE pH OF A BUFFER SOLUTION

Buffer solutions with different pH values can be prepared by changing the relative proportions of HA(aq) and A⁻(aq). It is easy to calculate the pH of a buffer solution provided the $K_a$ of the weak acid and the concentrations of the weak acid and its conjugate base are known.

### ? TEST YOURSELF

**R** A buffer solution is 0.750 mol dm⁻³ in propanoic acid and 0.350 mol dm⁻³ in sodium propanoate. What are the approximate concentrations in the buffer solution of undissociated propanoic acid and of propanoate ions?

SEE QUESTIONS 5 AND 6.

### ● EXAMPLE

**Q** A buffer solution is 0.10 mol dm⁻³ in ethanoic acid and 0.20 mol dm⁻³ in sodium ethanoate. Calculate its pH.

**A** Let $x$ be the concentration of the H⁺(aq). We then have:

$$CH_3COOH(aq) \rightleftharpoons CH_3COO^-(aq) + H^+(aq)$$

| | | | |
|---|---|---|---|
| At start/mol dm⁻³ | 0.10 | 0.20 | 0 |
| At equilibrium/mol dm⁻³ | (0.10 − x) | (0.20 + x) | x |

We can assume that $x$ is very small compared with 0.10 and 0.20, since ethanoic acid is a weak acid. Therefore, at equilibrium:

$$[CH_3COOH(aq)] = 0.10 \text{ mol dm}^{-3}$$

$$[CH_3COO^-(aq)] = 0.20 \text{ mol dm}^{-3}$$

Substituting these values into the expression for the $K_a$ for ethanoic acid:

$$K_a = \frac{[CH_3COO^-][H^+]}{[CH_3COOH]} = \frac{0.20 \times x}{0.10} = 2x$$

Now: $K_a = 1.8 \times 10^{-5}$

So: $x = 9.0 \times 10^{-6}$ mol dm⁻³

$$\text{pH} = -\log_{10} [H^+(aq)]$$
$$= -\log_{10} (9.0 \times 10^{-6})$$
$$= 5.0$$

The pH of the buffer solution is therefore 5.0 (to 2 significant figs).

### ! SCIENCE FEATURE Buffering in blood

To remain healthy, the acid–base balance of our blood has to be maintained at a constant pH of 7.4. If, for instance, the blood became acidic and this value dropped (as it does in a medical condition called acidosis), we would have to breathe rapidly to expel more of the acidic gas carbon dioxide. The main mechanism that maintains pH at 7.4 is the buffering action of several conjugate acid–base pairs – including $H_2CO_3$(aq) and $HCO_3^-$(aq), and $H_2PO_4^-$(aq) and $HPO_4^{2-}$(aq) – together with the buffering action of plasma proteins and haemoglobin. The action of the proteins is because of the carboxyl and amino groups in the side-chains of some of the amino acids that make up proteins. The concentrations of $H_2PO_4^-$(aq) and $HPO_4^{2-}$(aq) are too low for this conjugate acid–base pair to have a large buffering effect.

It is the equilibrium between carbon dioxide, carbonic acid, hydrogencarbonate ions and carbonate ions that acts as the most important buffering system in blood plasma:

$$H_2CO_3(aq) \rightleftharpoons HCO_3^-(aq) + H^+(aq)$$

When the pH of blood decreases, the concentration of H⁺(aq) increases and the position of equilibrium shifts to the left to minimize this increase. The carbonic acid concentration does not rise indefinitely, since this too is in equilibrium with carbon dioxide and water, and so the concentration of dissolved carbon dioxide increases:

$$H_2CO_3(aq) \rightleftharpoons CO_2(aq) + H_2O(l)$$

Dissolved carbon dioxide is removed from blood by gas exchange in the lungs:

$$CO_2(aq) \rightleftharpoons CO_2(g)$$

When the pH of blood begins to rise, the $H_2CO_3$(aq)/$HCO_3^-$(aq) equilibrium shifts to the right to produce more H⁺(aq).

▶

SCIENCE FEATURE  Buffering in blood (continued)

Strenuous exercise increases the metabolic rate, which produces more carbon dioxide in tissue respiration. Representing the overall oxidation of glucose:

$$C_6H_{12}O_6(aq) + 6O_2(aq) \rightarrow 6CO_2(aq) + 6H_2O(l)$$

Therefore, the concentration of dissolved carbon dioxide in the blood starts to increase and the net effect tends to raise the blood pH. However, this effect is counteracted by an increase in breathing rate, to speed up gas exchange and thereby reduce the concentration of dissolved carbon dioxide. The blood pH therefore does not increase.

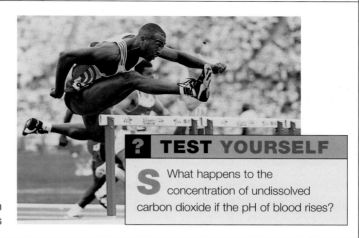

**Fig 17.31** Strenuous sport generates acidic blood, which the body's buffer system counteracts

**? TEST YOURSELF**

**S** What happens to the concentration of undissolved carbon dioxide if the pH of blood rises?

# 6 CARBOXYLATES

**Carboxylates** are the salts of a carboxylic acid. They contain the carboxylate anion (Fig 17.32).

Measurements have established that both the carbon–oxygen bonds in the carboxyl group have the same length, which is longer than the typical C=O bond, but shorter than the typical C–O bond. The reason they are not different lengths is that the negative charge on the anion is delocalized and does not reside on either of the two oxygen atoms. Fig 17.33 shows an alternative way to describe the carboxylate ion.

**DELOCALIZED ELECTRONS ARE COVERED IN MORE DETAIL ON PAGE 330.**

**? TEST YOURSELF**

**T** **a)** Suggest why sodium benzoate has a greater solubility than benzoic acid in cold water.

**b)** Suggest why it is important for food preservatives, such as sodium benzoate, to be soluble in water.

**Fig 17.32** The structure of the carboxylate ion

**Fig 17.33** An alternative way to represent the carboxylate ion, which shows the delocalized nature of the ion. The curved line indicates that the electron is somewhere between the two oxygen atoms

SCIENCE FEATURE  Properties and uses of carboxylates

Most sodium and potassium carboxylates are soluble in water. How soluble they are depends on the length of the carbon chain of the carboxylate ion, and the solubility decreases as the chain becomes longer.

As they are soluble, carboxylates are widely used in the food industry, which adds them to foods as preservatives and as acid regulators (buffers). As well as carboxylates, the food industry uses carboxylic acids; Table 17.6 summarizes these applications. You can probably identify some of them in processed foods by their E numbers.

**Table 17.6 Carboxylic acids and carboxylates as food additives**

| Name | E number | Use |
|---|---|---|
| Sorbic acid | E200 | Occurs naturally in some fruits, but is also made synthetically. Used as a preservative in such foods as soft drinks, cakes and frozen pizzas |
| Sodium sorbate | E201 | Preservatives |
| Potassium sorbate | E202 | |
| Calcium sorbate | E203 | |
| Benzoic acid | E210 | Occurs naturally in cherry bark, raspberries and tea, but normally made synthetically. Used as a preservative and as an antioxidant in fruit products, soft drinks, pickles and salad dressings |
| Sodium benzoate | E212 | Preservatives |
| Potassium benzoate | E213 | |
| Calcium benzoate | E214 | |
| Ethanoic acid | E260 | Vinegar |
| Sodium ethanoate | E281 | Preservatives |
| Calcium ethanoate | E282 | |
| Potassium ethanoate | E283 | |
| Lactic acid | E270 | Found naturally in soured milk and yoghurt, it acts as a preservative and a flavouring. Used in biscuits, confectionery and cakes |
| Propanoic acid | E280 | Preservative in baked foods |
| Potassium lactate | E326 | Acid regulator |

# 7 PREPARATION OF CARBOXYLIC ACIDS

Carboxylic acids are either prepared by an oxidation reaction or by the hydrolysis of a derivative of a carboxylic acid.

## OXIDATION OF PRIMARY ALCOHOLS AND ALDEHYDES

Carboxylic acids can be prepared by the oxidation of primary alcohols. For example, ethanol can be oxidized to give ethanoic acid by heating it under reflux with either acidified potassium dichromate(VI) or acidified potassium manganate(VII):

$$CH_3CH_2OH + 2[O] \rightarrow CH_3COOH + H_2O$$

In a similar way, aldehydes can be oxidized using the same reagents to form carboxylic acids. So phenylmethanal is oxidized to benzoic acid when refluxed with acidified potassium dichromate(VI):

$$C_6H_5CHO + [O] \rightarrow C_6H_5COOH$$

## HYDROLYSIS OF ESTERS

When an ester is refluxed with either dilute acid or dilute alkali, the ester is hydrolysed to give the carboxylic acid or the carboxylate ion. For example, ethyl ethanoate can be hydrolysed to give ethanoic acid (with acid hydrolysis) or sodium ethanoate (with aqueous sodium hydroxide):

$$CH_3COOCH_2CH_3(l) + H_2O(l) \underset{\text{acid hydrolysis}}{\rightleftharpoons} CH_3COOH(aq) + CH_3CH_2OH(aq)$$

$$CH_3COOCH_2CH_3(l) + NaOH(l) \xrightarrow{\substack{\text{alkaline} \\ \text{hydrolysis}}} CH_3COO^-Na^+(aq) + CH_3CH_2OH(aq)$$

The sodium ethanoate can be converted into ethanoic acid by reaction with dilute hydrochloric acid.

## HYDROLYSIS OF AMIDES

Amides can be hydrolysed in the same way as esters. Again, there is the opportunity to use either acid- or base-catalysed hydrolysis. For example, ethanamide, $CH_3CONH_2$, can be hydrolysed to give ethanoic acid and ammonium ion with acid hydrolysis, and ethanoate ion and ammonia with alkaline hydrolysis:

$$CH_3CONH_2(s) + H_2O(l) + HCl(aq) \xrightarrow{\text{acid hydrolysis}} CH_3COOH(aq) + NH_4Cl(aq)$$

$$CH_3CONH_2(s) + NaOH(aq) \xrightarrow{\substack{\text{alkaline} \\ \text{hydrolysis}}} CH_3COO^-Na^+(aq) + NH_3(g)$$

Proteins and polypeptides are polymers that contain the amide linkage. They can be hydrolysed either by refluxing with concentrated hydrochloric acid or by enzymatic action to give amino acids.

## HYDROLYSIS OF NITRILES

Nitriles are hydrolysed by refluxing in either dilute acid or aqueous alkali. Acid hydrolysis gives the ammonium ion and a carboxylic acid. Alkaline hydrolysis gives ammonia and the carboxylate ion. For example, ethanenitrile can be hydrolysed by refluxing with hydrochloric acid to give ethanoic acid, but with aqueous sodium hydroxide it forms sodium ethanoate:

$$CH_3CN(l) + 2H_2O(l) + HCl(aq) \rightarrow CH_3COOH(aq) + NH_4Cl(aq)$$

$$CH_3CN(l) + H_2O(l) + NaOH(aq) \rightarrow CH_3COO^-Na^+(aq) + NH_3(g)$$

---

**DILUTE SULPHURIC ACID IS NORMALLY USED TO ACIDIFY AQUEOUS POTASSIUM DICHROMATE(VI) OR POTASSIUM MANGANATE(VII)**

**THE OXIDATION OF PRIMARY ALCOHOLS IS COVERED IN MORE DETAIL ON PAGE 295.**

**PHENYLMETHANOL IS SOMETIMES CALLED BENZYL ALCOHOL.**

**? TEST YOURSELF**

**U** Write balanced symbol equations for the reactions between:
**a)** propanal and warm acidified potassium dichromate(VI);
**b)** butan-1-ol and hot acidified potassium dichromate(VI).

**MORE INFORMATION ON THE HYDROLYSIS OF ESTERS IS GIVEN ON PAGES 395–397.**

**? TEST YOURSELF**

**V** Write the names of the products of the following hydrolysis reactions:
**a)** heating aqueous sodium hydroxide with propylethanoate;
**b)** refluxing hydrochloric acid with ethylbenzoate;
**c)** heating aqueous sodium hydroxide with $CH_3CH_2CONH_2$;
**d)** refluxing dilute hydrochloric acid with $C_6H_5CONH_2$.

**THE HYDROLYSIS OF PROTEINS AND POLYPEPTIDES IS COVERED IN MORE DETAIL ON PAGE 420.**

# 8 REACTIONS OF CARBOXYLIC ACIDS

The two most important synthetic reactions of carboxylic acids are the production of esters and the production of acyl chlorides.

## ESTERIFICATION

When an alcohol and a carboxylic acid are refluxed together in the presence of an acid catalyst, such as concentrated sulphuric acid, an ester is produced. The reaction does not go to completion, but results in an equilibrium mixture that must be separated to isolate the ester. For example, ethanol, ethanoic acid and a trace of concentrated sulphuric acid give the ester ethyl ethanoate:

$$CH_3CH_2OH(l) + CH_3COOH(l) \rightleftharpoons CH_3COOCH_2CH_3(l) + H_2O(l)$$

## FORMATION OF ACYL CHLORIDES

In acyl chlorides (also known as acid chlorides), the carboxyl's hydroxyl group is replaced by a chlorine atom (Fig 17.34). An acyl chloride is always named after the corresponding carboxylic acid. For example, the acyl chloride of ethanoic acid is called ethanoyl chloride and that of benzoic acid is called benzoyl chloride.

The conversion of carboxylic acids into acyl chlorides is very useful for the synthesis of esters and amides, because acyl chlorides are much more susceptible to nucleophilic attack than are carboxylic acids. Acyl chlorides are therefore used as synthetic intermediates.

Carboxylic acids are converted into acyl chlorides by reaction with sulphuryl(IV) chloride, $SOCl_2$, or phosphorus(V) chloride, $PCl_5$ (Fig 17.35). Normally, the carboxylic acid is refluxed with the $SOCl_2$ or $PCl_5$ and the acyl chloride product is isolated by fractional distillation. For example, ethanoic acid can easily be converted into ethanoyl chloride, and benzoic acid into benzoyl chloride:

$$CH_3COOH(l) + SOCl_2(l) \rightarrow CH_3COCl(l) + SO_2(g) + HCl(g)$$

$$C_6H_5COOH(l) + PCl_5(s) \rightarrow C_6H_5COCl(l) + POCl_3(l) + HCl(g)$$

## REDUCTION OF CARBOXYLIC ACIDS

Powerful reducing agents, such as lithium tetrahydridoaluminate, $LiAlH_4$, reduce carboxylic acids to the corresponding alcohols (this is opposite to the preparation of carboxylic acids by oxidizing primary alcohols):

$$RCOOH + 4[H] \xrightarrow{\text{(i) } LiAlH_4/\text{ether (ii) } HCl(aq)} RCH_2OH + H_2O$$

**Fig 17.34** The acyl chloride functional group, in which the highly electronegative chlorine atom is attached to the carbonyl carbon

**Fig 17.35** The formation of acyl chlorides

### ? TEST YOURSELF

W Write down the symbol equation for the reaction between:
**a)** ethanoic acid and $PCl_5$;
**b)** benzoic acid and $SOCl_2$.

SEE QUESTION 7.

## SUMMARY

After studying this chapter, you should know the following.

■ Brønsted–Lowry acids are proton donors and Brønsted–Lowry bases are proton acceptors.
■ Strong acids are assumed to dissociate fully when dissolved in water. The dissociation of weak acids is an equilibrium process, in which the position of equilibrium lies very much to the side of the undissociated weak acid.

■ $K_a$ is the acid dissociation constant, which for the acid HA is given by the formula:

$$K_a = \frac{[H^+][A^-]}{[HA]}$$

■ pH is the negative logarithm to base 10 of the aqueous hydrogen ion concentration.
■ A buffer solution resists changes in pH when small amounts of acid or alkali are added to it.
■ Many buffer solutions are a mixture of an aqueous weak acid and its conjugate base. ▶

## SUMMARY *(continued from page 383)*

■ An indicator is a compound that changes colour in a solution at a certain pH value. Many indicators are weak acids.

■ Carboxylic acids are typical weak acids; they react slowly with reactive metals such as magnesium to give hydrogen, react with bases to give carboxylate salts and release carbon dioxide from carbonates or hydrogencarbonates.

■ Carboxylic acids can be prepared by the oxidation of primary alcohols or aldehydes by heating under reflux with acidified potassium dichromate(VI), or by the acid- or base-catalysed hydrolysis of esters, amides or nitriles.

■ Carboxylic acids react with alcohols in the presence of an acid catalyst, such as concentrated sulphuric acid, to give esters.

## ? EXAM QUESTIONS

**1**

**a) (i)** Calculate the concentration, in mol $dm^{-3}$, of a solution of hydrochloric acid, HCl, which has a pH of 1.13.
**(ii)** Calculate the concentration, in mol $dm^{-3}$, of a solution of chloric(I) acid, HOCl, which has a pH of 4.23. Chloric acid is a weak acid with $K_a = 3.72 \times 10^{-8}$ mol $dm^{-3}$.

**b)** The pH of 0.100 in mol $dm^{-3}$ sulphuric acid is 0.98.
**(i)** Calculate the concentration of hydrogen ions, $H^+$, in this solution.
**(ii)** Write equations to show the two successive ionizations of sulphuric acid, $H_2SO_4$, in water.
**(iii)** Suggest why the concentration of hydrogen ions is not 0.20 in mol $dm^{-3}$ in 0.100 in mol $dm^{-3}$ sulphuric acid.

Edexcel 6082/P.01 Module Test 2 Jan 2001 Q4a,b

**2** This question relates to an aqueous solution of propanoic acid, $CH_3CH_2COOH$.
**a)** Write an equation to represent the equilibrium reaction in solution.
**b)** Give an expression for the equilibrium constant, $K_a$.
**c)** $K_a$ for propanoic acid is $1.34 \times 10^{-5}$ mol $dm^{-3}$. Calculate **(i)** $pK_a$; **(ii)** the pH of a 0.050 mol $dm^{-3}$ solution of propanoic acid.
**d)** 50.00 $cm^3$ of 0.050 mol $dm^{-3}$ aqueous propanoic acid is titrated with 0.050 mol $dm^{-3}$ aqueous sodium hydroxide, NaOH. Sketch a graph showing the initial pH, the variation in pH with the volume of NaOH added up to and beyond the equivalence point, and the volume of NaOH at the equivalence point.

IBO Chemistry Higher Level Paper 2 Nov 1998 Q7c

**3** The major acidic component of soured milk is lactic acid:

$$CH_3CH(OH)CO_2H$$

When 10.0 $cm^3$ of a solution of lactic acid was titrated against 0.050 mol $dm^{-3}$ sodium hydroxide

the following pH readings were obtained:

| Volume of NaOH added/$cm^3$ | pH | Volume of NaOH added/$cm^3$ | pH |
|---|---|---|---|
| 0 | 2.5 | 12 | 4.4 |
| 2 | 3.1 | 14 | 4.7 |
| 4 | 3.4 | 16 | 9.1 |
| 6 | 3.7 | 18 | 11.6 |
| 8 | 3.9 | 20 | 11.8 |
| 10 | 4.1 | | |

**a)** Plot a graph of these results with pH on the $y$-axis and volume added on the $x$-axis. Comment on the shape of the curve before, at and after neutralization.
**b)** Suggest an indicator, giving a reason for your choice.
**c)** Calculate the concentration of the lactic acid in the solution in **(i)** mol $dm^{-3}$, **(ii)** g $dm^{-3}$.
**d)** deduce the $K_a$ value for lactic acid, giving your reasoning and the units.

**4** Lactic acid and sodium lactate are both used as food additives.
**a)** The systematic name for lactic acid is 2-hydroxypropanoic acid. Draw a displayed formula for lactic acid.
**b)** A solution of 0.100 mol $dm^{-3}$ lactic acid is titrated against 25.0 $cm^3$ 0.100 mol $dm^{-3}$ sodium hydroxide.
**(i)** Suggest an indicator that is suitable to be used for this titration.
**(ii)** Draw the pH titration curve and indicate how it can used to calculate the $pK_a$ of lactic acid.
**c)** An equimolar mixture of lactic acid and sodium lactate is present in a food. These compounds are described as acid regulators. Explain how this mixture can act as a buffer solution.

**5**
**a)** State what is meant by a 'buffer solution'. Using a mixture of ethanoic acid and sodium ethanoate as an example, explain how a buffer functions when small amounts of acid or alkali are added.

**b)** Using the p$K_a$ value of 4.76 for ethanoic acid, calculate the value for $K_a$ for ethanoic acid. What mass of sodium ethanoate needs to be present in 1.00 dm³ of a solution of 0.500 mol dm⁻³ ethanoic acid to give a buffer solution with a pH of 5.0? What would be the resulting pH if twice as much sodium ethanoate was dissolved in 1.00 dm³ of the 0.500 mol dm⁻³ ethanoic acid solution?

IBO Chemistry Higher Level Paper 2 Nov 1998 Q7c

**6** When the indicator bromothymol blue (which can be represented by the formula HIn) dissolves in water, the following dynamic equilibrium is set up.

$$\text{HIn(aq)} \rightleftharpoons \text{In}^-\text{(aq)} + \text{H}^+\text{(aq)}$$
$$\text{yellow} \qquad\qquad \text{blue}$$

The indicator solution appears green because it contains both the yellow HIn and the blue In⁻ forms.

**a)** State **two** features of a *dynamic equilibrium*.

**b)** State Le Chatelier's principle.

**c)** Hydrochloric acid is added to the indicator solution above until no further colour change takes place. Using Le Chatelier's principle, suggest and explain what colour change you might see.

**d)** Aqueous sodium hydroxide is gradually added to the resulting solution in **c)** until no further colour changes take place. Suggest **all** the colour changes you might see. Explain your answer.

OCR 2813/1 How Far, How Fast? June 2001 Q3

**7** The compound shown in Fig 17.Q7 is ibuprofen, a powerful painkiller.

**a)** Write down the molecular formula for ibuprofen.

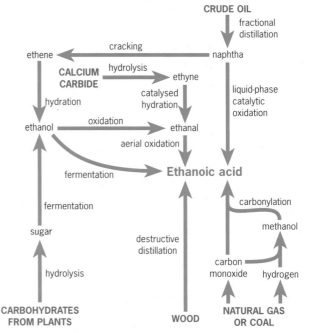

**Fig 17.Q7** Displayed formula of ibuprofen

**b)** Draw the displayed formulas for the organic products of the reaction of ibuprofen with:
**(i)** aqueous sodium hydroxide;
**(ii)** sulphuryl(IV) chloride, $SOCl_2$;
**(iii)** ethanol and a trace of conc. sulphuric acid;
**(iv)** lithium tetrahydridoaluminate(III).

**c)** Predict whether ibuprofen is a soluble or an insoluble painkiller. Explain your prediction in terms of the intermolecular forces that can exist when it is put into water.

**d)** When ibuprofen is heated with acidified potassium manganate(VII), a dicarboxylic acid is isolated. Suggest the structure for this dicarboxylic acid.

# KEY SKILLS **ASSIGNMENT**

## ETHANOIC ACID IN INDUSTRY

**For many years, ethanoic acid was produced on a large scale just as vinegar – an aqueous solution of ethanoic acid used to preserve foods and improve their flavour. More recently, pure ethanoic acid has also been obtained as a synthetic intermediate in the production of a wide range of organic molecules.**

For vinegar, ethanoic acid used to be made by the fermentation of ethanol: when dilute aqueous ethanol is oxidized, an aqueous solution of ethanoic acid is formed.

**1  a)** Write an equation to represent the oxidation of aqueous ethanol by atmospheric oxygen to form ethanoic acid.
**b)** Suggest one problem that arises in the manufacture of concentrated ethanoic acid by this fermentation method.
The destructive distillation of wood is another early method used to manufacture ethanoic acid. The wood is heated in the absence of air and the volatile components produced are condensed. The condensate is a complex mixture of organic products from which ethanoic acid is extracted.

**Fig 17.A1** Five different routes to ethanoic acid

**2** Suggest why the destructive distillation of wood must be carried out in the absence of air.

Fermentation and the destructive distillation of wood have drawbacks: fermentation is slow and yields only a dilute solution of ethanoic acid; the destructive distillation of wood produces a complex mixture that needs to be separated.

So, in the early 1900s, industrial chemists began to devise new processes to make ethanoic acid, using ethanal as an intermediate. There are three principal routes to ethanal. The first involves the conversion of ethanol into ethanal, either by oxidation using atmospheric oxygen, or by dehydrogenation over a heated catalyst such as copper.

**3** **a)** Write down the equation for the oxidation of ethanol by atmospheric oxygen to form ethanal.
**b)** Suggest why the industrial oxidation of ethanol is carried out using atmospheric oxygen rather than an oxidizing agent such as acidified potassium dichromate(VI).
**c)** Dehydrogenation involves the loss of two hydrogen atoms per molecule of ethanol. Write down the equation for the dehydrogenation of ethanol.
The second route to ethanal involves the catalysed hydration of ethyne, $C_2H_2$ (addition of water).

**4** Write down an equation for the hydration of ethyne.
The third route to ethanal involves the catalysed reaction of ethene with carbon monoxide:
$$C_2H_4(g) + CO(g) \rightarrow CH_3CHO(g)$$

**5** A company wants to convert ethene into ethanal. Which of the three synthetic routes described do you think best? Give reasons for your answer.
Ethanal can be oxidized by hot acidified potassium dichromate(VI) in the laboratory to give ethanoic acid. This reagent is too expensive to use industrially and, besides, it would result in large quantities of unwanted chromium(III) salts. Instead, an efficient catalytic oxidation of ethanal is employed. The oxidizing agent is atmospheric oxygen, and using a manganese-containing catalyst ensures that the reaction proceeds rapidly, even at 50 °C.

**6** **a)** Write down the equation for the oxidation of ethanal by oxygen to give ethanoic acid.
**b)** What are the advantages of using atmospheric oxygen over acidified potassium dichromate(VI) in the industrial oxidation of ethanal to give ethanoic acid?
Among more recent developments in the manufacture of ethanoic acid is the direct oxidation of naphtha, one of the distillation products of crude oil, to give a mixture of carboxylic acids. The catalytic oxidation can be controlled to yield much more ethanoic acid than other acids. This reaction means that crude oil can be converted by a two-stage process into ethanoic acid, which avoids the need for cracking and the use of ethene and ethanal as intermediates.

**7** Give some advantages of making ethanoic acid from naphtha by the direct oxidation method rather than using routes via ethanal.
Recently, natural gas, which is mainly methane, has been developed as a source of ethanoic acid. In a series of catalytic reactions, the methane is converted into carbon monoxide and hydrogen, which in turn are converted into methanol. The methanol is then reacted with carbon monoxide to give ethanoic acid:

$$CH_4(g) + H_2O(g) \rightarrow CO(g) + 3H_2(g)$$
$$CO(g) + 2H_2(g) \rightarrow CH_3OH(g)$$
$$CH_3OH(g) + CO(g) \rightarrow CH_3COOH(g)$$

**8** Suggest why the reaction of methanol with carbon monoxide is referred to as carbonylation.
The last series of reactions would suggest that coal might be used as a raw material in the manufacture of ethanoic acid, since the chemistry involved in converting the carbon in coal into carbon monoxide and hydrogen is well-known to industrial chemists. The advantage that coal has over crude oil is that it is not expected to run out for hundreds of years.

**9** Discuss the various advantages of making ethanoic acid from each of the following raw materials:
**a)** crude oil;      **b)** natural gas;
**c)** carbohydrates such as sugar (fermentation);
**d)** wood.
Ethanoic acid has a wide variety of applications: it is used to produce esters (used in food flavourings and as industrial solvents), cellulose ethanoate (better known as cellulose acetate) and benzene-1,4-dicarboxylic acid (a precursor in the manufacture of some polyesters). Chloroethanoic acid is used to make weedkillers. Pharmaceutical products, such as aspirin, are manufactured using ethanoic acid. Ethenyl ethanoate is used in paints.

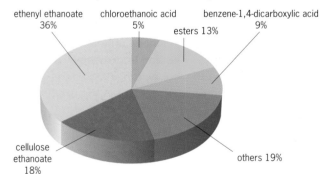

**Fig 17.A2** Applications of ethanoic acid

**10** **a)** Draw the displayed formula for chloroethanoic acid. Suggest how chloroethanoic acid may be prepared from ethanoic acid.
**b)** Pentan-2-ol reacts with ethanoic acid to give an ester used in the food flavouring industry. Draw the displayed formula of the ester produced.
**c)** Discuss the importance of ethanoic acid in the maintenance of today's lifestyle.

# Esters, acyl chlorides, amides and equilibria

**Odours** The characteristic flavour of pear drops is caused by a cocktail of organic compounds, which include carboxylic acid derivatives called esters. The organic compounds have molecules that trigger responses in sensory cells in the nose chamber and the tongue

Our noses can detect hundreds of different odours at unbelievably minute concentrations, perhaps just a few molecules per sniff. Some noses are so sensitive that their owners – humans and dogs included – can learn to identify scores of different substances by their smell alone! How does the nose do it?

The chamber of the nose is lined with thousands of different types of receptor cells that sense the different shapes of molecules. It is likely that a single substance may trigger several types of receptor cell, which in some way together produce the sensation of the odour of that substance. Many natural odours are not just single substances, but complex cocktails of substances, and we don't understand the complicated interaction of the receptor cells involved. Also, we don't know why many substances whose structures differ widely give the same odour as we have yet to discover the connection between molecular structure and odour.

What we do know, however, is that to have an odour a substance must be volatile enough for its vapour to diffuse into the nostrils; it must be water soluble to ensure that it makes its way into the mucus lining the nostrils; and it must be fat soluble so that it can penetrate the receptor cells. Many esters – derivatives of carboxylic acids – possess these properties, which is why they are used a lot to synthesize substances we like to smell, particularly those in food products. The subtle odours of fruits, for instance, are reproduced by blending the appropriate esters and there is an expanding market in attractive smells, such as baked bread and coffee, to distribute in supermarkets and encourage shoppers to buy more food.

## 1 DERIVATIVES OF CARBOXYLIC ACIDS

As explained on page 372, carboxylic acids have a carbonyl group that is directly attached to a hydroxyl group. Derivatives of carboxylic acids still have the carbonyl group, but the hydroxyl group is replaced by the atom of a highly electronegative element. The main derivatives of carboxylic acids are esters, amides (sometimes called acid amides), acid anhydrides and acyl chlorides.

**Fig 18.1** A generalized carboxylic acid and its derivatives. R′ is an alkyl group such as $C_2H_5$, an aryl group such as $C_6H_5$, or a cycloalkyl group

carboxylic acid     ester     acyl chloride     amide     acid anhydride

### POLAR CARBONYL GROUPS

Look at Fig 18.1, which shows the general structure of a carboxylic acid and its derivatives. They all possess a highly polar carbonyl group whose normal polarity is enhanced when the extra atom attached to it is highly electronegative. In carboxylic acids and esters, this atom is an oxygen atom; in amides, it is a nitrogen atom; and in acyl chlorides, it is a chlorine atom.

# USE OF CARBOXYLIC ACID DERIVATIVES

Esters and amides have numerous commercial applications, including perfumes, polymers and pharmaceuticals (Figs 18.3 and 18.4). Acyl chlorides are important synthetic intermediates, but do not have many everyday uses in their own right.

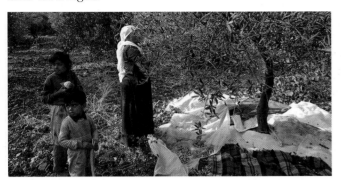

Fig 18.2 Oils from plants, such as olive oil, are esters of long-chain carboxylic acids and the alcohol propane-1,2,3-triol

Fig 18.3 Among the functional groups in vitamin C (ascorbic acid) is a lactone, which is a cyclic ester

Fig 18.4 Penicillins are powerful bactericides whose molecules contain an amide functional group

# 2 ESTERS

An ester is a carboxylic acid derivative in which the hydroxyl group of the acid is replaced by an oxygen atom attached to a carbon chain. Fig 18.5 shows how esters are related to carboxylic acids and to alcohols. An ester has two carbon chains, one related to a carboxylic acid and the other related to an alcohol or a phenol.

## NAMING ESTERS

As esters have two carbon chains, they are more difficult to name than other classes of compounds. For this reason, the main carbon chain of an ester is taken to be the one related to the carboxylic acid, so the main (second) part of the name of an ester is based on the name of this carboxylic acid (like a surname).

Fig 18.5 Esters have a structure that consists of a carboxylic acid portion and an alcohol portion

Fig 18.6 The name of an ester is derived from the name of its alcohol (alkyl) group and the name of its carboxylic acid (alkanoate) group. Note that the structure shown here is the reverse of that given in Fig 18.5

alkyl alkanoate    For example, methyl butanoate

Take, for example, methyl propanoate (Fig 18.7). The carbon chain that is related to the carboxylic acid (the one that contains the carbonyl group) has methyl propanoate     phenyl ethanoate     ethyl benzoate

**Fig 18.7** Structural formulas of methyl propanoate, phenyl ethanoate and ethyl benzoate

---

**? TEST YOURSELF**

**B** Write down the displayed formulas for each of these esters: methyl ethanoate, propyl propanoate, pentyl butanoate and methyl benzoate.

---

**Fig 18.9** The concentration changes that occur during esterification, when 1 mole of alcohol is refluxed with 1 mole of carboxylic acid. Note that each concentration eventually reaches a constant value known as the equilibrium concentration

---

**ANOTHER EQUILIBRIUM CONSTANT – THE ACID DISSOCIATION CONSTANT, $K_a$ – IS DISCUSSED ON PAGE 368.**

---

**✔ REMEMBER THIS**

The term 'stoichiometric equation' is equivalent to 'balanced equation'. That is, the mole ratios of the reactants and the products are stated.

---

three carbon atoms, and the chain related to the alcohol has one carbon atom. Its formula is therefore written as:

$$CH_3CH_2COOCH_3 \quad \text{or} \quad CH_3CH_2CO_2CH_3$$

Note that in these formulas and in the structures of Fig 18.7, the main (carboxylic acid) part is on the left, rather than on the right as in their names.

In the same way, phenyl ethanoate (Fig 18.7) consists of two portions: a two-carbon unit that contains the carbonyl carbon, and the phenyl group (the benzene ring). It has the formula:

$$CH_3COOC_6H_5 \quad \text{or} \quad CH_3CO_2C_6H_5$$

Ethyl benzoate is related to benzoic acid and has the formula:

$$C_6H_5COOC_2H_5 \quad \text{or} \quad C_6H_5CO_2C_2H_5.$$

## 3 ESTERIFICATION AND EQUILIBRIUM CONSTANTS

To make an ester should be simple enough, since when heated in the presence of an acid catalyst, a carboxylic acid and an alcohol react to give an ester and water. This is known as esterification.

**Fig 18.8** The esterification reaction. $R^1$ is an alkyl, an aryl or hydrogen. $R^2$ is an alkyl group

However, the equation in Fig 18.8 does not really indicate the problems with the reaction. To achieve a good yield of ester, the reaction has to be managed carefully since it is reversible. Fig 18.9 shows that when a carboxylic acid is reacted with an alcohol, the reaction does not go to completion. It reaches a balance point at which the concentrations of all four substances (carboxylic acid, alcohol, ester and water) remain constant. The reaction has not stopped, but the concentrations do not change, which means that the rate of the forward reaction then equals the rate of the backward reaction. A situation known as a **dynamic equilibrium** has been reached.

We have already described dynamic equilibria and Le Chatelier's principle in qualitative terms in Chapter 1. In this chapter equilibria are explored further using a quantitative approach.

### EQUILIBRIUM CONSTANTS

Any dynamic equilibrium can be described in terms of its **equilibrium constant, $K_c$**. The equilibrium constant relates the molar concentrations at equilibrium of each species in the reaction:

**The equilibrium constant is the (mathematical) product of the molar concentrations of the products each raised to the power of its coefficient in the stoichiometric equation, divided by the product of the molar concentrations of the reactants, each raised to the power of its coefficient in the stoichiometric equation.**

An equilibrium constant is independent of the concentrations of the reactants and the products, but will change its numerical value as the temperature of a reaction changes.

So, for the reaction $w\text{A} + x\text{B} \rightleftharpoons y\text{C} + z\text{D}$, the equilibrium constant is given by:

$$K_c = \frac{[\text{C}]^y[\text{D}]^z}{[\text{A}]^w[\text{B}]^x}$$

where [ ] stands for the concentration of the species in mol dm$^{-3}$.

For esterification, the general reaction is shown by the equation:

$$\text{RCOOH} + \text{R}'\text{OH} \rightleftharpoons \text{RCOOR}' + \text{H}_2\text{O}$$

where R and R' represent carbon skeletons, such as alkyl or aryl groups.

The equilibrium constant is thus given by:

$$K_c = \frac{[\text{RCOOR}'][\text{H}_2\text{O}]}{[\text{RCOOH}][\text{R}'\text{OH}]}$$

This equilibrium constant does not have a unit, since the units for concentration cancel out.

Notice that, although esterification normally requires an acid catalyst, the catalyst is not shown in the expression for the equilibrium constant. This is because it is not shown in the stoichiometric equation. The catalyst decreases the time it takes to reach equilibrium, but *it does not change the concentrations* obtained at equilibrium.

A reaction whose equilibrium constant has a large numerical value is one in which, at equilibrium, the concentrations of the products are much higher than the concentrations of the reactants.

## CALCULATIONS THAT INVOLVE THE EQUILIBRIUM CONSTANT

Despite the problem of having to extract the required product from an equilibrium mixture, esters are made via the acid-catalysed reaction of an alcohol and a carboxylic acid. The reaction is carried out in a batch process, so the ester is not made continuously but only as needed. When using this equilibrium process on an industrial scale, the chemists need to estimate the yield of ester, and the amounts of starting materials left unreacted, and for this they use the equilibrium constant.

### EXAMPLE

**Q** The $K_c$ for the esterification reaction between ethanol and ethanoic acid is 4.0.

Calculate the yield of ester when 1.0 mole of ethanol and 1.0 mole of ethanoic acid are left to reach equilibrium.

**A** First, write down the stoichiometric equation. At the start of the reaction there is no product. However, once equilibrium has been reached, the reaction vessel contains ethanoic acid, ethanol, ethyl ethanoate and water.

From the equation, every mole of ethanoic acid requires 1 mole of ethanol and makes 1 mole of ethyl ethanoate and 1 mole of water. This means that if the yield at equilibrium is $x$ moles of ethyl ethanoate, then $x$ moles of each of ethanol and ethanoic acid must have reacted. This leaves behind $(1.0 - x)$ moles of each of ethanoic acid and ethanol. This part of the calculation is summarized thus:

| | CH$_3$COOH | + | C$_2$H$_5$OH | $\rightleftharpoons$ | CH$_3$COOC$_2$H$_5$ | + | H$_2$O |
|---|---|---|---|---|---|---|---|
| At start of reaction/moles | 1.0 | | 1.0 | | 0 | | 0 |
| At equilibrium/moles | 1.0 − $x$ | | 1.0 − $x$ | | $x$ | | $x$ |

(continued)

SEE QUESTION 1.

### REMEMBER THIS

The concentrations used to calculate $K_c$ must be those after equilibrium has been reached.

### ? TEST YOURSELF

**C** **a)** Write down expressions for the equilibrium constant for each of the following reactions:
(i) (COOH)$_2$ + 2C$_2$H$_5$OH $\rightleftharpoons$ (COOC$_2$H$_5$)$_2$ + 2H$_2$O;  (ii) 2NO$_2$ $\rightleftharpoons$ N$_2$O$_4$.
Show that in  (i) $K_c$ has no unit and in (ii) the unit is dm$^3$ mol$^{-1}$.

**b)** Ethanoic acid and pentan-1-ol are refluxed together to obtain an equilibrium mixture. The concentration of each substance at equilibrium is shown in Table 18.1.

**Table 18.1**

| Substance | Molar concentration at equilibrium/mol dm$^{-3}$ |
|---|---|
| Ethanoic acid | 2.12 |
| Pentan-1-ol | 2.12 |
| Pentyl ethanoate | 4.32 |
| Water | 4.32 |

(i) Write down a balanced equation for the esterification reaction.
(ii) Write down an expression for the equilibrium constant.
(iii) Calculate the numerical value for the equilibrium constant.

**D** The equilibrium constant, $K_c$, for the reaction between propanoic acid and ethanol is 4.3 at a particular temperature. 1.0 mole of water and 1.0 mole of ethyl propanoate are refluxed together with an acid catalyst. Calculate the number of moles at equilibrium of:

**a)** water;

**b)** ethyl propanoate;

**c)** propanoic acid;

**d)** ethanol.

**● EXAMPLE [continued]**

The equilibrium constant is: $K_c = \dfrac{[CH_3COOC_2H_5][H_2O]}{[CH_3COOH][C_2H_5OH]}$

The concentration of each substance must be substituted into the above expression, so the moles at equilibrium must be converted into concentrations.

Remember (page 140) that
concentration is measured in mol dm$^{-3}$ = $\dfrac{\text{moles}}{\text{volume (dm}^3)}$

So be careful to substitute concentrations into the expression for $K_c$, *not* the number of moles.

Assuming that the volume of the esterification mixture is $V$ dm$^3$, this gives:

$$K_c = 4.0 = \dfrac{\left(\dfrac{x}{V}\right)\left(\dfrac{x}{V}\right)}{\dfrac{1.0-x}{V}\dfrac{1.0-x}{V}}$$

Cancelling $V$ gives: $4.0 = \dfrac{x^2}{(1.0-x)^2}$

Taking the square root of both sides gives: $2.0 = \dfrac{x}{1.0-x}$

So: $2.0 - 2.0x = x$

That is, $x = 0.67$ moles

This means that the exact composition of the equilibrium mixture can be estimated.

The Example shows that when an equimolar mixture of ethanol and ethanoic acid are refluxed together, only about 67 per cent of the starting ethanol and ethanoic acid are converted into ethyl ethanoate. In the laboratory, the ester can be separated from the starting materials, but industrially this represents a large loss of valuable reactants, unless they can be recycled. Therefore, an industrial chemist needs to find a way for this reaction to give a maximum yield, especially when the ester is derived from an expensive starting material.

## LE CHATELIER'S PRINCIPLE

Le Chatelier's principle is described in Chapter 1, and states:

> **The position of the equilibrium of a system changes to minimize the effect of any imposed change in conditions.**

Le Chatelier's principle describes the effect that a change in the pressure, temperature or concentration of a substance in an equilibrium mixture will have on the position of equilibrium. Le Chatelier's principle applies to any reaction that is in equilibrium and allows chemists to manage and manipulate equilibrium reactions. We describe here the consequences of Le Chatelier's principle on the esterification reaction.

### EFFECT OF CONCENTRATION CHANGES ON THE ESTERIFICATION EQUILIBRIUM

Changing the **concentration** of a reactant or a product **does not change the numerical value of the equilibrium constant, $K_c$.** However, it does change the position of equilibrium. As an example, let's look at what happens to the equilibrium mixture formed by the reaction of an alcohol with a carboxylic acid when extra alcohol is added. Temporarily, the reaction is not in equilibrium, but equilibrium is quickly restored by reaction of the alcohol to form the ester. A new equilibrium mixture is produced that has the same equilibrium constant. Fig 18.10 shows what happens to the concentrations of each substance during this process.

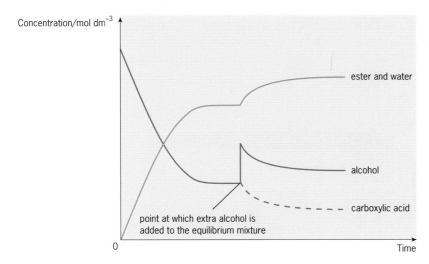

**Fig 18.10** The effect of increasing the concentration of alcohol on the esterification equilibrium. Note that even higher equilibrium concentrations can be achieved if even more alcohol is added to the mixture

This effect is an example of how Le Chatelier's principle operates in practice. The change is the increase of alcohol concentration, which disturbs the equilibrium of the mixture. However, once a new state of equilibrium is established, some of the extra alcohol is changed into ester.

Industrially, this means that if an ester is to be made starting from an expensive alcohol, it is more economical to use an excess of the cheaper carboxylic acid. This ensures that the equilibrium is shifted to the right side and that as much as possible of the expensive alcohol is converted into the ester.

Another way to manage the reversible reaction is to *remove* one of the products the moment it has formed. This prevents the reaction from achieving equilibrium, but it will nevertheless 'keep trying' to, as long as the product continues to be removed from the reaction mixture. In esterification, water is the easier product to remove.

**SEE QUESTION 2.**

**THE EFFECT OF PRESSURE CHANGE ON EQUILIBRIUM IS DISCUSSED IN CHAPTER 28.**

**IF YOU WANT TO READ MORE ABOUT ENERGY CHANGES THAT OCCUR DURING CHEMICAL REACTIONS, SEE PAGES 10 AND 166.**

## EFFECT OF TEMPERATURE CHANGES ON THE ESTERIFICATION EQUILIBRIUM AND $K_c$

Although the equilibrium constant does not change when concentrations of the reactants and products change, it will change as the temperature of the equilibrium is changed. For an **exothermic** reaction the $K_c$ **becomes numerically smaller** as the temperature increases, and for an **endothermic** process the $K_c$ **becomes numerically larger** as the temperature increases. Esterification is an exothermic process: when a carboxylic acid and an alcohol react, energy is transferred to the

**Fig 18.11** A summary of the effect of Le Chatelier's principle on esterification

surroundings. So, if the temperature of the esterification is raised, the equilibrium shifts to absorb energy, that is, in the direction of the endothermic reaction (the left side). As the temperature is increased, the yield of ester decreases, though the rate of production of ester increases. For esterification in industry, a compromise temperature is used so that the rate and the yield are both acceptable, and it is normal for the alcohol and the carboxylic acid to be refluxed together in the presence of concentrated sulphuric acid as a catalyst. The sulphuric acid catalyst has no effect on the position of equilibrium.

**SEE QUESTIONS 1 AND 3.**

Fig 18.12 In the laboratory, esterification is usually carried out by refluxing an alcohol and a carboxylic acid in the presence of an acid catalyst, such as concentrated sulphuric acid

**SEE QUESTION 4.**

**? TEST YOURSELF**

**G** Write down the names of the carboxylic acid and alcohol used to make each of the following esters:
a) ethyl ethanoate;
b) ethyl butanoate;
c) propenyl ethanoate.

## APPLICATIONS OF THE ESTERIFICATION REACTION

The importance of esters as a source of pleasant-smelling substances and food flavourings has been mentioned earlier (see page 388). Almost all these esters are manufactured using the esterification reaction just described with the formation of the ester linkage.

There are several examples of the esterification reaction used to link two carbon chains to form a molecule that has a particular shape associated with biological activity. Two such molecules, benzocaine and procaine, are anaesthetics based on 4-aminobenzoic acid. Among other uses, there are ester solvents (including ethyl ethanoate), ester plasticizers in PVC and ester synthetic fibres (polyesters).

Fig 18.13 The preparation of methyl cinnamate, an ester responsible for the spicy aroma of the matsutake mushroom. It can be prepared from cinnamic acid (3-phenylpropenoic acid), the corresponding carboxylic acid. Note that the esterification occurs via the acyl chloride

methyl cinnamate

## SCIENCE FEATURE  Esters and the food flavourings industry

The simple esters (that is, those with only a few carbon atoms per molecule) tend to have pleasant odours. The characteristic flavours of fruits and the fragrances of flowers often result from esters – commonly a subtle blend of esters and other odoriferous compounds.

When we talk about 'flavour' we mean a combination of taste and odour by receptors on the tongue and in the nose. Almost always, it is a combination of the substances that our various receptors detect that allows us to recognize a particular flavour.

Table 18.2 shows the 11 chemicals, and the amount of each, that chemists have put together to imitate the flavour of pineapple. Notice that many of the esters listed are still known by their

traditional names – the result of long-established use in the food industry. Almost all the esters in this formulation are made by the reaction of the appropriate alcohol and carboxylic acid.

### Table 18.2 Components of imitation pineapple flavouring

| Compound | % in formulation | Compound | % in formulation |
|---|---|---|---|
| Allyl caproate | 5 | Ethyl ethanoate | 15 |
| Butanoic acid | 12 | Isoamyl acetate | 3 |
| Caproic acid | 8 | Isoamyl isovalerate | 3 |
| Ethanoic acid | 5 | Terpinyl propanoate | 3 |
| Ethyl butanoate | 22 | Other essential oils | 19 |
| Ethyl crotonoate | 5 | | |

## ! SCIENCE FEATURE Local anaesthetics

**Fig 18.14** A local anaesthetic is used to numb just the area being treated

Benzocaine and procaine are local anaesthetics (Fig 18.14), which means that they make only small areas of the body insensitive to touch and pain. They were used a lot until recently – benzocaine as an ointment, drug or aerosol to relieve painful conditions of the skin, mouth and respiratory tract, and procaine for dental injections. Both compounds are usually prepared by esterification reactions.

Benzocaine has the systematic name of ethyl 4-aminobenzoate. It is synthesized from 4-aminobenzoic acid (Fig 18.15). The acid is esterified with ethanol in the presence of a little concentrated sulphuric acid to speed up the attainment of equilibrium. An excess of ethanol is used to drive the equilibrium to the right, according to Le Chatelier's principle.

**Fig 18.15** The preparation of benzocaine by the esterification of 4-amino-benzoic acid with ethanol

**Fig 18.16** The preparation of procaine by the esterification 4-aminobenzoic acid

In a very similar process, procaine is also synthesized from 4-aminobenzoic acid (Fig 18.17). This time the alcohol has a much more complicated structure, but the chemistry behind the reaction is the same.

? **TEST YOURSELF**

**H** Explain how an excess of ethanol drives the equilibrium reaction between ethanol and 4-aminobenzoic acid to the right-hand side.

## 4 REACTIONS OF ESTERS

The carbonyl carbon in the ester group is susceptible to nucleophilic attack. The principal nucleophiles that react with it are the water molecule, the hydroxide ion and the hydride ion.

### HYDROLYSIS OF ESTERS

In the hydrolysis of esters, water reacts with the carbonyl carbon and an alcohol is eliminated. The reaction is very slow because the water molecule is a poor nucleophile. It is also a reversible reaction, because it is essentially the reverse of the esterification reaction:

$$RCOOR' + H_2O \rightleftharpoons RCOOH + R'OH$$

It is this reaction with water that prevents esters from being used in some perfumes and deodorants. Pleasant-smelling esters are liable to be hydrolysed by chemicals in perspiration to form carboxylic acids, many of which have an unpleasant smell. They include butanoic acid, which gives the odour we detect in rancid butter. These carboxylic acids are among the components of body odour and are in the scent that dogs pick up when tracking humans (Fig 18.17).

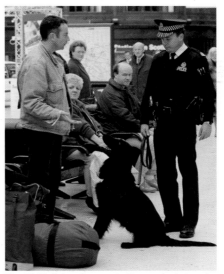

**Fig 18.17** Esters in perspiration are hydrolysed to form unpleasant-smelling carboxylic acids. These constitute the scent that a sniffer dog picks up when tracing someone

# BASE-CATALYSED HYDROLYSIS OF ESTERS

Also known as saponification, this reaction involves boiling an ester with aqueous sodium hydroxide to form the sodium salt of the acid and the corresponding alcohol. Fats and oils are natural esters and their alkaline hydrolysis is the basis of making soap.

During saponification of a natural fat or oil, propane-1,2,3-triol (glycerol) and the sodium salt of a long-chain fatty acid are formed (Figs 18.18 to 18.20). The sodium salt is a major constituent of soaps. Its carboxylate ion is negatively charged at one (hydrophilic) end and non-polar at the other (hydrophobic) end.

**Fig 18.18** Soaps are mainly the sodium salts of carboxylic acids known as fatty acids

**Fig 18.19** Soaps are made by the hydrolysis of fats and oils during boiling with aqueous sodium hydroxide. The reaction is called saponification

**Fig 18.20** Saponification. The fats and oils used are esters in which the alcohol part is propane-1,2,3-triol and the carboxylic acid part is a so-called fatty acid. Hydrolysis of such an ester forms a sodium carboxylate, the main constituent of soap.

polar negative end of carboxylate

non-polar long-chain alkyl group

An ester such as an ethyl benzoate is hydrolysed to form sodium benzoate and ethanol (Fig 18.21).

**Fig 18.21** The hydrolysis of ethyl benzoate to form sodium benzoate and ethanol

**YOU CAN READ MORE ABOUT THE ACIDIC PROPERTIES OF PHENOL ON PAGE 343.**

## ? TEST YOURSELF

Write down the names of the products of the reactions between the following esters and hot, aqueous sodium hydroxide:

**a)** ethyl ethanoate;

**b)** propyl propanoate;

**c)** ethyl benzoate.

**SEE QUESTIONS 4 AND 10.**

Phenyl benzoate is hydrolysed to give two ionic products, since the phenol formed is sufficiently acidic to react with sodium hydroxide to give sodium phenoxide:

$$C_6H_5COOC_6H_5 + 2NaOH \rightarrow C_6H_5COO^-Na^+ + C_6H_5O^-Na^+ + H_2O$$

## ACID-CATALYSED HYDROLYSIS OF ESTERS

It is also possible to hydrolyse an ester by refluxing with a dilute acid. Normally, dilute sulphuric acid or concentrated hydrochloric acid is used as the acid catalyst. This is really the reverse reaction of esterification and, in theory, should lead to an equilibrium mixture that contains carboxylic acid, alcohol, ester and water. However, the presence of excess water in the dilute acid drives the reaction to completion and so the ester is virtually all hydrolysed. Notice that in acidic hydrolysis, the carboxylic acid is obtained rather than the carboxylate salt.

Ethyl methanoate is hydrolysed to give methanoic acid and ethanol:

$$HCOOC_2H_5(l) + H_2O(l) \underset{}{\overset{H^+(aq)}{\rightleftharpoons}} HCOOH(l) + C_2H_5OH(l)$$

## REDUCTION OF ESTERS

Powerful reducing agents, such as lithium tetrahydridoaluminate(III), $LiAlH_4$, reduce esters to two alcohols. During the reaction, the ester linkage is broken and the carbonyl group is reduced to a primary alcohol. This is not a useful reaction industrially, because water and air have to be excluded and lithium tetrahydridoaluminate(III) is very expensive.

$$RCOOR' + 4[H] \rightarrow RCH_2OH + R'OH$$

where R and R′ are alkyl or aryl groups.

### SCIENCE FEATURE Fuels from oils and fats

It is impossible to oxidize esters without breaking their carbon–carbon bonds: they burn in excess oxygen to give carbon dioxide and water. This is the reaction that supplies energy when vegetable oils are used as alternative fuel sources.

Both fats and vegetable oils are used by living systems to transfer energy. They are converted into the corresponding carboxylic acids, which are then oxidized in a complicated series of reactions to form carbon dioxide and water.

Vegetable oils and animal fats are made up of molecules called triglycerides (Fig 18.22). These are esters of the alcohol propane-1,2,3-triol and long-chain carboxylic acids. Vegetable oils are easily extracted from the seeds of plants, and several of them are used in cooking, the most popular ones being olive oil, sunflower oil, soya oil and palm oil.

**Fig 18.22** The structure of triglycerides, the constituents of vegetable oils. When the carbon chain contains a double bond, it is an unsaturated oil or fat

The long carbon chain of the carboxylic acid portion of the triglyceride resembles a long-chain alkane, and hence it can be used to make a renewable source of fuel. But the oil as extracted can't be burnt in an engine because the propane-1,2,3-triol residues that form during combustion would soon clog it up. So the vegetable oil is hydrolysed with an alkali and then acidified, and the carboxylic acid products are isolated. They are then reacted with methanol to form methyl esters (Fig 18.23). The methyl esters are much more volatile than the original oil and it is these that are used as a substitute for diesel fuel.

**Fig 18.23** The synthesis of methyl esters of fatty acids from vegetable oils. The triglycerides in vegetable oils can be hydrolysed to form sodium salts of fatty acids. These salts can be converted into the fatty acids, and finally esterification occurs to form a methyl ester. Methyl esters of fatty acids burn easily and therefore look promising as substitutes for diesel fuel

In the United Kingdom, over one million tonnes of rapeseed is produced each year (Fig 18.24). The oil from rapeseed is easily converted into methyl esters known as rape methyl ester. Rape methyl ester can be used as a diesel substitute and offers several environmental advantages over conventional diesel fuel. It doesn't form sulphur dioxide and emits fewer sooty particles during combustion.

**Fig 18.24** Rape is easily grown in the United Kingdom. Its seed, commonly called rapeseed, contains the oil that is converted into the diesel fuel known commercially as RME (rape methyl ester)

**Fig 18.25** A Reading bus that runs on RME biodiesel

# 5 ACYL CHLORIDES

**Fig 18.26** The structure of an acyl chloride. As a result of the highly electronegative chlorine atom, its carbonyl group is highly polar, too

Nu⁻ is a nucleophile
such as $H_2O$, $NH_3$
or $CH_3CH_2OH$

**Fig 18.27** The mechanism of the addition–elimination reaction of an acyl chloride

An acyl chloride is a carboxylic acid in which the hydroxyl group has been replaced with a chlorine atom (Fig 18.26).

Acyl chlorides are highly reactive compounds that make very useful synthetic intermediates. The preparation of acyl chlorides by reacting sulphuryl(IV) chloride, $SOCl_2$, or phosphorus(V) chloride, $PCl_5$, with a carboxylic acid is described on page 383.

Acyl chlorides are named after the corresponding carboxylic acid, using the suffix *-oyl* followed by chloride. For example, $CH_3COCl$ is ethanoyl chloride, and $C_6H_5COCl$ is benzoyl chloride.

## ADDITION–ELIMINATION REACTIONS OF ACYL CHLORIDES

An acyl chloride has an extremely electron-deficient carbonyl carbon because of the electron-withdrawing effect of the highly electronegative chlorine atom. This enhances the reaction of a nucleophile with the carbonyl carbon. Nucleophilic addition occurs and the carbonyl double bond is broken. The addition product then eliminates the chloride as an ion to regenerate the carbonyl double bond. This mechanism is shown in Fig 18.27. The curly arrows show the movement of an electron pair.

The addition–elimination reaction is an example of a **condensation reaction**, in which two molecules react together with the elimination of a *simple* molecule, such as water, hydrogen chloride or an alcohol. With an acyl chloride, hydrogen chloride is eliminated.

## HYDROLYSIS OF ACYL CHLORIDES

An acyl chloride reacts readily with water to form the corresponding carboxylic acid. Most acyl chlorides fume in air because of this reaction with water vapour: the hydrogen chloride gas formed dissolves in atmospheric moisture to form tiny droplets of hydrochloric acid.

For instance, ethanoyl chloride fumes in moist air to form ethanoic acid and hydrogen chloride:

$$CH_3COCl(l) + H_2O(g) \rightarrow CH_3COOH(l) + HCl(g)$$

In water, hydrochloric acid would be formed instead.

This reaction of the carbon–chlorine bond is much more rapid than reactions involving other carbon–chlorine bonds. The highly electron-deficient carbonyl carbon encourages nucleophilic addition, to be followed later by elimination.

Compare the hydrolysis of acyl chlorides with the reaction of chloroalkanes with water described on page 274.

When an acyl chloride is hydrolysed in alkaline solution, the carboxylate ion is formed instead of the acid:

$$RCOCl + 2NaOH \rightarrow RCOO^-Na^+ + H_2O + NaCl$$

## PREPARATION OF ESTERS FROM ACYL CHLORIDES

As already noted on page 393, the reaction of a carboxylic acid with an alcohol is an equilibrium process, which makes it difficult to obtain good yields of esters. An alternative way to make an ester from a carboxylic acid involves the formation of an acyl chloride and the subsequent reaction of the acyl chloride with an alcohol (Fig 18.28). Acyl chlorides react completely with alcohols and do not form an equilibrium mixture. This provides, in the laboratory, a powerful synthetic route to make esters. Both steps of the process can be achieved with yields over 90 per cent.

**Fig 18.28** Synthesis of an ester via an acyl chloride. An ester can be prepared from a carboxylic acid in a two-stage reaction. First, the acid is converted into an acyl chloride by reaction with either phosphorus(V) chloride or sulphuryl(IV) chloride. Next, the acyl chloride is used in a condensation reaction with an alcohol to give an ester. The advantage of this method is that there are no equilibrium reactions

## MECHANISM OF THE ADDITION–ELIMINATION REACTION BETWEEN ACYL CHLORIDES AND ALCOHOLS

The first stage is nucleophilic addition of the alcohol. An alcohol molecule can act as a nucleophile by donating a lone pair of electrons from the oxygen atom to the electron-deficient carbonyl carbon atom, thereby forming a covalent bond (Fig 18.29).

SEE QUESTION 5.

**Fig 18.29** The mechanism of the reaction of an alcohol with an acyl chloride

The mechanism's next stage involves the transfer of a proton from one highly electronegative atom to another highly electronegative atom. Following proton transfer, hydrogen chloride is eliminated. When the reaction is carried out in the presence of an alkali, the hydrogen chloride reacts to form a chloride ion and it becomes impossible for the reaction to be reversed.

Phenol reacts with acyl chlorides in the same way, and the presence of an alkali can assist the reaction in a second way (Fig 18.30). Phenol is sufficiently acidic for a reaction with an alkali, such as aqueous sodium hydroxide, to form the phenoxide ion, $C_6H_5O^-$. This ion is a much better nucleophile than the neutral phenol molecule, and so the first stage of the reaction, nucleophilic addition, occurs easily.

**Fig 18.30** The mechanism of the reaction of phenol with benzoyl chloride

## REACTIONS OF ACYL CHLORIDES WITH AMMONIA AND AMINES

Ammonia and other amines are nucleophiles because they have a lone pair of electrons on the nitrogen atom. They can react with acyl chlorides in the same way as alcohols. This time, the products formed are amides. Amides are difficult to make from carboxylic acids, since they react with ammonia in an acid–base reaction to form an ammonium salt instead of an amide:

$$RCOOH + NH_3 \rightarrow RCOO^-NH_4^+$$

So the use of an acyl chloride as a synthetic intermediate solves this problem.

And it is easy to produce amides by the reaction of an amine or ammonia with an acyl chloride (Fig 18.31). It is normal to use an excess of the amine or ammonia, so that the excess is available to remove the hydrogen chloride eliminated at the end in an acid–base reaction.

**Fig 18.31** The mechanism of the reaction of ammonia with acyl chlorides

Ethanoyl chloride reacts with ammonia to make ethanamide, whereas with methylamine it will make *N*-methylethanamide:

$$CH_3COCl + NH_3 \rightarrow CH_3CONH_2 + NH_4Cl$$

Notice from this equation that the presence of an excess of ammonia leads to the formation of ammonium chloride:

$$CH_3COCl + CH_3NH_2 \rightarrow CH_3CONHCH_3 + HCl$$

**SEE QUESTIONS 6, 7, 8 AND 9.**

### ? TEST YOURSELF

**a)** Draw the mechanism of the reaction of phenylamine, $C_6H_5NH_2$, with benzoyl chloride to give N-phenyl benzamide, $C_6H_5CONHC_6H_5$.

**b)** Predict the products of each of the following reactions:
(i) ethanoyl chloride and aqueous sodium hydroxide;
(ii) ethanoyl chloride and excess ammonia;
(iii) benzoyl chloride and ammonia;
(iv) benzoyl chloride and methanol;
(v) benzoyl chloride and sodium benzoate.

**Fig 18.32** Six reactions of ethanoyl chloride

**Fig 18.33** Six reactions of benzoyl chloride

# 6 ACID ANHYDRIDES

Acid anhydrides are very similar to acyl chlorides and are also very useful synthetic intermediates. Acid anhydrides can be prepared by distilling a sodium carboxylate with an acid chloride. So sodium ethanoate and ethanoyl chloride will form ethanoic anhydride and leave behind sodium chloride as an involatile salt.

$$CH_3COO^-Na^+ + CH_3COCl \rightarrow (CH_3CO)_2O + NaCl$$

Acid anhydrides are normally named after the corresponding acid, so that $(CH_3CO)_2CO$ is ethanoic anhydride.

## REACTIONS OF ACID ANHYDRIDES

Nucleophiles such as water, hydroxide ion, ammonia, alcohols and amines react with acid anhydrides by an addition–elimination mechanism with the elimination of a carboxylic acid.

A typical reaction is that of ethanoic anhydride with phenol to make phenyl ethanoate:

$$(CH_3CO)_2O + C_6H_5OH \rightarrow CH_3COOC_6H_5 + CH_3COOH$$

Figure 18.36 shows the range of compounds that can be made from ethanoic anhydride.

**Fig 18.34** Acid anhydrides can either have R = R′ or be a mixed anhydride where R does not equal R′

**Fig 18.35** Structure of ethanoic anhydride

**SEE QUESTIONS 9, 10 AND 11.**

**Fig 18.36** Some reactions of ethanoic anhydride

---

**! SCIENCE FEATURE Manufacture of aspirin**

Aspirin is a well-known painkiller. It is made from phenol in a two-step process, shown in Fig 18.37. The formation of the ester linkage in stage two of the synthesis uses ethanoic anhydride. 2-Hydroxybenzoic acid and ethanoic anhydride are refluxed together and the aspirin is precipitated by placing the reaction mixture in water. The aspirin can then be filtered and

purified before it is ready to be sold commercially. Ethanoic anhydride is used to synthesize the ester rather than ethanoyl chloride, since ethanoic acid (the co-product formed) is less volatile than hydrogen chloride (the co-product with ethanoyl chloride). Another advantage with ethanoic anhydride is that it is much cheaper to manufacture than ethanoyl chloride.

**Fig 18.37** The synthesis of aspirin

Fig 18.38 This mountaineer's rope is made from nylon, which is a polyamide. This is a polymer molecule that contains repeating units linked by amide bonds. It is made by the reaction of an acyl chloride (hexanedioyl dichloride) and an amine (1,6-diaminohexane)

**YOU CAN READ MORE ABOUT SECONDARY AMIDES IN PROTEINS AND POLYMERS ON PAGES 420 AND 438.**

## ? TEST YOURSELF

**S** Write down the formula for pentanamide.

**T** What is the structure of N-ethyl ethanamide?

**THE HYDROLYSIS OF SECONDARY AMIDES IN PEPTIDES IS DISCUSSED ON PAGE 420.**

## 7 AMIDES

Amides have an amino group directly attached to the carbonyl carbon (Fig 18.39). Proteins, polypeptides and synthetic polymers such as nylon contain the amide linkage.

Fig 18.39 The displayed formula of amides. Note that they contain a polar carbonyl bond attached to a nitrogen atom

### NAMING OF AMIDES

Amides are named by adding the suffix *amide* to the name of the carbon skeleton. Take, for example, ethanamide, pentanamide and benzamide. Ethanamide is a two-carbon chain in which the number 1 carbon is part of the amide functional group. Its formula can be written as $CH_3CONH_2$. Pentanamide is a five-carbon chain in which the number 1 carbon is part of the amide group. Benzamide consists of a benzene ring attached to an amide functional group (Fig 18.40). It has the formula $C_6H_5CONH_2$.

Fig 18.40 The structure of benzamide

These amides are known as primary amides, because they have only one carbon atom attached to the nitrogen atom of the amide. Secondary amides have two carbon atoms attached, and tertiary amides have three carbon atoms attached, as Fig 18.41 shows.

Fig 18.41 The structure of the three types of amide

Note that the extra carbon atoms are not part of the carbon chain that contains the carbonyl group, so they are considered as substituents. Their position is indicated by *N-*, which signifies they are bonded to the nitrogen atom. For example, *N*-methyl ethanamide has the formula $CH_3CONHCH_3$. It has a two-carbon chain, which includes the carbonyl carbon, and a methyl group, which is attached to the nitrogen atom.

### REACTIONS OF AMIDES

Amides show the same types of reaction as the other derivatives of carboxylic acids, but they are rather less reactive.

#### HYDROLYSIS OF AMIDES

Just as with esters, the reaction of an amide with water is extremely slow, but the reaction can be catalysed by an acid or a base (Fig 18.42). For example, when hydrolysed using boiling hydrochloric acid, ethanamide forms ethanoic acid and ammonium chloride, but when refluxed with aqueous sodium hydroxide, it forms sodium ethanoate and ammonia:

$$CH_3CONH_2 + H_2O + HCl \rightarrow CH_3COOH + NH_4Cl$$

$$CH_3CONH_2 + NaOH \rightarrow CH_3COO^-Na^+ + NH_3$$

**Fig 18.42** The hydrolysis of amides. These reactions are much faster than with water alone

## REACTION WITH PHOSPHORUS(V) OXIDE TO FORM NITRILES

Amides can be dehydrated to form a new class of compounds called nitriles. A nitrile contains the C≡N group:

$$RCONH_2 \xrightarrow[\text{heat}]{P_4O_{10}} RCN + H_2O$$

### ▶ EXTENSION Nitriles

Nitriles are useful synthetic intermediates, since they can be hydrolysed in either acidic or basic conditions, and they can be reduced to form primary amines:

$$RCN + H_2O + NaOH \rightarrow RCOO^-Na^+ + NH_3$$
$$RCN + 2H_2O + HCl \rightarrow RCOOH + NH_4Cl$$

$$RCN + 4[H] \xrightarrow[\text{or } H_2]{LiAlH_4} RCH_2NH_2$$

Nitriles can be prepared easily from halogenoalkanes by a nucleophilic substitution reaction with cyanide ions supplied by potassium cyanide:

$$CH_3CH_2Br + CN^- \rightarrow CH_3CH_2CN + Br^-$$

Alternatively, the nucleophilic addition of hydrogen cyanide to carbonyl compounds provides access to nitriles.

$$CH_3CH_2CHO + HCN \rightarrow CH_3CH_2CH(OH)CN$$

## REACTION WITH ALKALINE AQUEOUS BROMINE

Primary amides react with alkaline aqueous bromine to form amines. The amine has one less carbon atom than the amide. This reaction is sometimes referred to as the Hofmann degradation, because a carbon–carbon bond is broken during the reaction. Normally, the amide is heated with concentrated aqueous sodium hydroxide and aqueous bromine:

$$RCONH_2 \xrightarrow[\text{heat}]{Br_2(aq)/KOH} RNH_2$$

**SEE QUESTION 12.**

## SUMMARY

After studying this chapter, you should know that:

■ Esters, acyl chlorides, acid anhydrides and amides are all derivatives of carboxylic acids.

■ Esters can be prepared by refluxing a carboxylic acid and an alcohol in the presence of a small amount of concentrated sulphuric acid as a catalyst.

■ Esterification is a reversible reaction that reaches a dynamic equilibrium.

■ At equilibrium, the concentrations of all the substances involved remain constant, and the rate of the forward reaction is equal to the rate of the backward reaction.

■ Le Chatelier's principle states that the position of equilibrium changes to minimize any change imposed on the conditions at equilibrium.

■ The equilibrium constant for a reversible reaction is the product of the molar concentrations of the products, each raised to the power of its coefficient in

the stoichiometric equation, divided by the product of the molar concentrations of the reactants, each raised to the power of its coefficient in the stoichiometric equation.

■ The numerical value of the equilibrium constant only changes if the temperature of the equilibrium process changes.

■ Acyl chlorides and acid anhydrides react with alcohols to form esters in reactions that go to completion.

■ All derivatives of carboxylic acids react by nucleophilic addition–elimination (i.e. by condensation reactions).

■ Acyl chlorides and acid anhydrides can be hydrolysed by water to give the corresponding carboxylic acid, whereas amides and esters need to be hydrolysed using an acid or a base catalyst.

■ Amides can be made by the reaction between acid anhydrides or acyl chlorides and ammonia or amines.

## ? EXAM QUESTIONS

**1** A student was given the following instructions for the preparation of ethyl ethanoate.
Place 20 cm³ (15.8 g) of ethanol in a 250 cm³ round-bottomed flask and add 30 cm³ (31.5 g) of glacial ethanoic acid. Add carefully, with stirring, 2 cm³ of concentrated sulphuric acid. Attach a condenser to the flask and reflux under heating for 30 minutes. Allow to cool and then arrange the flask for fractional distillation and collect the distillate boiling up to 90 °C in a receiver. When the fraction has been collected, carefully add 25 cm³ of aqueous sodium carbonate to it. Shake the mixture and separate the organic layer. Allow this to stand overnight in contact with a suitable drying agent. Remove the drying agent and redistill the liquid, collecting the fraction that boils at about 75 °C.

**a)** **i)** Write a balanced equation for the reaction.
**ii)** Draw the structural formula of the organic product.

**b)** Explain the function of the sulphuric acid in this preparation.

**c)** Calculate which of the reagents is present in excess.

**d)** The reaction described usually produces a yield of around 20 per cent. Suggest two reasons why the yield is so low.

**2** Consider the esterification reaction:
$$CH_3CO_2H + CH_3CH_2OH \rightleftharpoons CH_3CO_2CH_2CH_3 + H_2O$$
**a)** **(i)** Give the expression for $K_c$.
**(ii)** An equilibrium mixture from the above reaction

contains 0.90 mol of ethyl ethanoate, 0.90 mol of water, 0.10 mol of ethanoic acid and 2.1 mol of ethanol. It has a volume of 235 cm³. Calculate $K_c$ to two significant figures, having regard for the units.
**(iii)** State, giving a reason, what would happen to the equilibrium composition if more ethanoic acid were to be added to the equilibrium mixture.

**b)** The reaction employs an acid catalyst, usually a little sulphuric acid. State, with a reason, the effect on the equilibrium position if the concentration of the acid catalyst were to be increased.

**c)** **(i)** Using the concept of mean bond enthalpies, show that $\Delta H$ for the esterification reaction is zero.
**(ii)** State, with a reason, the effect of an increase in temperature on the equilibrium composition of this reaction.
**(iii)** Explain why, when esters are hydrolysed in the laboratory, they are heated under reflux with dilute sodium hydroxide solution rather than with dilute sulphuric acid.

Edexcel 6082/P.01 Module test 2 January 1999 Q1

**3** Carefully read the passage below and answer the questions that follow.
*Preparation of pentyl ethanoate*
'Add 5.00 of pentan-1-ol to 5.00 g of ethanoic acid and 1 cm³ concentrated sulphuric acid. Reflux this mixture for one hour and then allow to cool. Fractionally distil the mixture to obtain an impure

# ? EXAM QUESTIONS

sample of the ester. Wash the impure ester with aqueous sodium hydrogencarbonate, followed by water, and then dry the ester over calcium chloride granules. Filter off the ester and finally distil to obtain the pure ester.'
A student carried out this preparation and obtained 2.12 g of pentyl ethanoate.

**a) (i)** Draw the structure of pentyl ethanoate.
**(ii)** What is the molecular formula for pentan-1-ol?
**(iii)** Write the balanced equation for this esterification reaction.
**(iv)** State the reason for adding concentrated sulphuric acid.
**(v)** What is fractional distillation?
**(vi)** Why was the impure ester washed with aqueous sodium hydrogencarbonate?

**b) (i)** Calculate the number of moles in 5.00 g of pentan-1-ol.
**(ii)** Calculate the number of moles in 5.00 g of ethanoic acid.
**(iii)** Which reagent is in excess, pentan-1-ol or ethanoic acid? Explain your answer.
**(iv)** Calculate the theoretical yield of pentyl ethanoate.
**(v)** Calculate the percentage yield of pentyl ethanoate.

**4** Esters are a class of organic compound.
**a)** Describe the base-catalysed hydrolysis of a named ester.
**b)** Describe the manufacture of a named synthetic polyester.

**5**
**a)** Describe how you can make benzoyl chloride starting from benzoic acid.
**b)** Describe with the aid of equations how benzoyl chloride reacts with: **i)** water, **ii)** aqueous sodium hydroxide, **iii)** phenol in aqueous sodium hydroxide.

**6**
**a)** Acyl halides such as ethanoyl chloride, $CH_3COCl$, are useful synthetic intermediates because they are quite reactive.
**(i)** Suggest synthetic steps, giving reagents and conditions, by which ethanol could be converted into ethanoyl chloride.
**(ii)** Ethanoyl chloride, $CH_3COCl$, reacts violently with water at room temperature, whereas chloroethane, $CH_3CH_2Cl$, reacts extremely slowly with water under the same conditions. Suggest a reason why $CH_3COCl$ is **more** susceptible to nucleophilic attack than $CH_3CH_2Cl$.

**b)** Ethyl ethanoate, $CH_3COOCH_2CH_3$, can be made either from ethanol and ethanoic acid or from ethanol and ethanoyl chloride.
**(i)** Give the equations for the two reactions.
**(ii)** Suggest why the reaction with ethanoyl chloride gives a better yield of the ester.
Edexcel 6084/P.01 Module Test 4 June 1999 Q2

**7** Look at the following reaction flowchart involving butanoyl chloride.

**Fig 18.07**

**a)** Draw the structures of the organic compounds **A–F**.
**b)** Write the equation for the reaction to make **D** from butanoyl chloride.
**c)** Describe the mechanism of the reaction between water and butanoyl chloride. Include 'curly arrows' and any relevant lone pairs and dipoles.

**8** Describe the reaction of ethanoyl chloride with:
**a)** ethanol;
**b)** ammonia;
**c)** water.
Give the mechanism for one of these reactions.

**9** The reaction between ammonia, $NH_3$, and ethanoyl chloride, $CH_3COCl$, is an example of an addition–elimination reaction.
**a) (i)** Draw the structure and name the product of this reaction.
**(ii)** Draw the mechanism of this reaction. Include 'curly arrows' and any relevant dipoles and lone pairs.
**(iii)** Use the mechanism to explain what is meant by an addition–elimination reaction.
**b)** The reaction between methylamine and ethanoic anhydride is an example of an addition–elimination reaction.
**(i)** Draw the structure of ethanoic anhydride.
**(ii)** Draw the product of the reaction between methylamine and ethanoic anhydride.

# ? EXAM QUESTIONS

**10** Protoanemomin is a naturally occurring compound with the structure shown in fig 18.Q10.

**Fig 18.Q10**

**a)** Write down the molecular formula for the compound.

**b)** Identify the functional groups present in the compound.

**c)** Protoanemomin is refluxed with aqueous sodium hydroxide. Draw the structure of the product formed.

**d)** Protoanemomin is refluxed with concentrated hydrochloric acid. Draw the structure of the product formed.

**11** Esters and amides can be made from carboxylic acids, acyl chlorides and acid anhydrides.

**a)** State the alcohol and carboxylic acid needed to make propyl butanoate.

**b)** Draw the structure of the acid anhydride needed to make butyl propanoate.

**c)** Suggest some advantages of making the esters in a laboratory from an acyl chloride and an alcohol rather than a carboxylic acid and an alcohol.

**d)** The manufacture of aspirin uses enthanoic anhydride rather than ethanoyl chloride to make an ethanoate ester. Suggest why.

**e)** State the amine and the acyl chloride needed to make propanamide.

**f)** Explain why it is difficult to make amides by the reaction between amides and carboxylic acids.

**12** Ethanamide is a white solid.

**a)** Give the full displayed formula for ethanamide and explain why it is a primary amide.

**b)** Draw the structure of one secondary amide.

**c)** Hydrolysis with hot aqueous sodium hydroxide ethanamide produces an alkaline gas, but this gas is not produced when the hydrolysis is carried out in refluxing concentrated hydrochloric acid. Explain these results with the aid of equations.

**d)** When ethanamide is heated with phosphorus(V) oxide, a colourless liquid **A** can be distilled from the mixture. **A** will react with hot aqueous sodium hydroxide to give an alkaline gas **B** and a solution **C**. **A** will also react with refluxing hydrochloric acid to form a solution that contains two compounds **D** and **E**. Liquid **A** can be reduced to a compound **F** with formula $C_2H_7N$. Identify all the substances **A** to **F**. Give reasons for your answers and include equations for all reactions involved.

# KEY SKILLS **ASSIGNMENT**

## CARBOXYLIC ACID DERIVATIVES AND FOOD

Many foods contain fats or oils. These fats and oils are esters, and they include triglycerides. Triglycerides are formed between propan-1,2,3-triol and carboxylic acids with a long carbon chain.

**1** Many fats and oils are triglycerides. Draw the general structure of a triglyceride. Use R to represent a long-chain alkyl group.

The annual world production of natural oils and fats, such as soya bean oil, palm oil and sunflower oil, is about 50 million tonnes. Natural oils vary in colour and flavour and are refined to produce oils suited for particular uses. For example, in a butter substitute the fat needs a particular texture, flavour and melting range.

Many manufactured foods contain both an oil and water, which must remain mixed and not separate during storage. Additives control this separation and also the flavour.

**2** **a)** Make a list of four food products that contain fats. Then list the fat content of each product and the additives present.
**b)** Use a data booklet to identify the different types of food additive found in the foods you listed in part **a)**.

Although the texture and melting range of fats can be controlled, their flavour poses a problem. Fats and oils can deteriorate by aerial oxidation into products that have unpleasant odours and flavours. For example, oxygen reacts with triglycerides to form hydroperoxides. They contain an oxygen–oxygen single bond, which makes them particularly reactive, because the bond can break homolytically to form free radicals. These are then involved in a chain reaction shown in Fig 18.A1.

**3** **a)** The initiation step (see page 268) involves the homolytic fission of a carbon–hydrogen bond. What is meant by homolytic fission?
**b)** Explain why the oxidation process is referred to as a chain reaction.

The hydroperoxides themselves don't have unpleasant odours or flavours, but they tend to decompose and form short-chain aldehydes, ketones or alcohols that do smell and taste worse.

Since, in aerial oxidation, a hydroperoxide is formed first, which then catalyses further oxidation, there is a time delay before the rate of this auto-oxidation becomes significant. During the delay, there is not enough hydroperoxide to maintain the free radical chain reaction. Antioxidants added to food interfere with the propagation stage in the oxidation process to slow it down.

Metal ions present in the food, such as copper(II) ions, speed up the initial formation of hydroperoxides. To stop this, antioxidant additives, such as citric acid, form a complex with the metal ions and thus slow down the initial formation of hydroperoxides. Two common antioxidants added to fatty foods are BHA (E310) and BHT (321), whose structures are shown in Fig 18.A2. These substances remove free radicals, but are consumed in the process.

**4** **a)** Suggest ways of storing foods that contain fats and oils so as to slow down the initial formation of hydroperoxides.
**b)** Suggest ways of storing foods that would slow down the chain reaction leading to auto-oxidation.

**Fig 18.A1** The oxidation of triglycerides

**Fig 18.A2** Antioxidants, such as BHA and BHT, prevent the oxidation of fats by reacting with free radicals, which stops the chain reaction

**5** Explain why the auto-oxidation of fats starts slowly and then progresses rapidly, even when antioxidants are present.

# KEY SKILLS **ASSIGNMENT**

The separation of the oil from the water in foods is a constant problem. Ideally, an emulsion should be formed that will not separate. Emulsifiers such as lecithin (Fig 18.A3) are added to the oil–water mix to prevent the separation process.

**Fig 18.A3** The structure of lecithin

**6** **a)** Name the functional groups found in lecithin.
**b)** Explain the differences between lecithin and a triglyceride.

The lecithin molecule has an extremely polar end, which is able to form intermolecular attractions with water molecules. At the other end is a non-polar alkyl chain that can form induced dipole–induced dipole interactions with either fat or oil molecules. Lecithin behaves very similarly to household detergents.

**7** Explain how the structure of lecithin enables it to function as an emulsifier.

# 19

# Amino acids and proteins

# Amino acids and proteins

**Extremoenzymes** Organisms such as the algae that colour this volcanic spring in Yellowstone National Park, USA, survive because their enzymes can tolerate high temperatures

Without enzymes – biochemical catalysts – the chemical processes of life that take place in each cell of our bodies would be impossibly slow. When in action, each of these special proteins catalyses its specific chemical reaction with a turnover of thousands – even millions – of molecules per minute.

The suggestion is that enzymes could possibly revolutionize large-scale industrial processes. They are thousands of times faster than inorganic catalysts and most work best at body temperature. Industry uses vast amounts of energy to speed up many reactions, so enzymes could save precious energy. Moreover, they are environmentally friendly because it is easy to dispose of them safely.

But there is a major snag. Enzymes lack stability over the wide range of reaction conditions encountered in industry. So the race is on to find 'extremozymes' – enzymes that can tolerate higher temperatures and a wider range of other conditions. As their enzymes must be able to withstand high temperatures, organisms that survive in volcanic springs are a promising starting point. By understanding the structure of these enzymes, biochemists can work out ways to synthesize extremozymes that will make industrial reactions more efficient – and greener, too.

## 1 WHY PROTEINS ARE IMPORTANT

### REMEMBER THIS

Not all hormones are proteins. Some, including the sex hormones, are steroids (see page 349).

Proteins make up about 15 per cent of our body weight. They are the major components of skin, muscle, nails and hair, giving structural support and holding cells together. While these may be the obvious proteins, others are just as essential:

● enzymes that catalyse the chemical reactions going on inside our bodies;
● many protein hormones that act as chemical messengers within and between cells;
● protein antibodies that protect us from disease;
● protein haemoglobin, which transports oxygen through our arteries.

So, proteins are crucial biological molecules (biomolecules). Yet, despite being so diverse, they all are made up from just 20 small molecules called amino acids.

## 2 NATURALLY OCCURRING AMINO ACIDS

**Fig 19.1** Functional groups in an amino acid

As their name implies, amino acid molecules have two functional groups, an amine group and a carboxylic acid group (Fig 19.1). All naturally occurring amino acids have the amine group on the second carbon atom of the molecule (next to the carboxylic acid group). This is often called the **α-carbon**, see Fig 19.2.

**Fig 19.2** The Greek lettering of the carbon atoms. 2-aminobutanoic acid is an α-amino acid

**Fig 19.3** General formula for α-amino acids

2-Aminobutanoic acid is an α-**amino acid**, or a **2-aminocarboxylic acid**. The more commonly used term is α-amino acid. In this type of acid, the amine and the carboxylic acid functional groups are both bonded to the same carbon atom. α-Amino acids are represented by the *general formula* given in Fig 19.3.

The side chain R varies considerably, as Fig 19.4 shows. The composition of the R group confers an individual set of properties to each amino acid, and this, of course, affects the properties of the proteins in which they are found.

Of the 20 amino acids needed to make up our proteins, eight cannot be synthesized in our bodies. These eight are called essential amino acids and must be part of our diet.

> ✔ **REMEMBER THIS**
>
> The reactions of the amine functional group can be found on pages 277, 438 and 630.
> The reactions of carboxylic acids can be found on pages 383 and 388.

> **?** **TEST YOURSELF**
>
> **A** **a)** Draw the α-amino acid based on butanoic acid.
> **b)** 4-Aminobutanoic acid is also known as gamma-aminobutanoic acid (GABA for short) and is involved in transmitting nerve impulses. Draw this amino acid.
>
> **B** **a)** Which amino acids in Fig 19.4 have polar side chains?
> **b)** Look at leucine and isoleucine. Why is isoleucine so named?
> **c)** (i) The R group of threonine (Thr), another amino acid found in proteins, is CH$_3$CH(OH)–. Draw the displayed formula of this amino acid.
> Hint: Remember that the displayed formula shows all the atoms and bonds (including those in OH groups).
> (ii) What type of alcohol side chain is this?
> (iii) How does the R group in Thr differ from that in serine?

**Fig 19.4** Shown are 12 of the 20 naturally occurring amino acids that make up most proteins. In each case, the R group is in colour

# 3 OPTICAL ISOMERISM AND CHIRAL CARBONS

As discussed on page 198, there are two types of isomerism: **structural isomerism** and **stereoisomerism** (Fig 19.5). The amino acids leucine and isoleucine (see Fig 19.4) are structural isomers because they both have the same molecular formula, but different structural formulas (which answers part **(b)** of Test Yourself B). However, with the exception of glycine, all the α-amino acids can exhibit stereoisomerism, which means that their atoms are bonded in the same order, but arranged differently in space. One form of stereoisomerism, known as geometric (*cis–trans*) isomerism, is described on page 306. The other form of isomerism that is shown by these α-amino acids is **optical isomerism**.

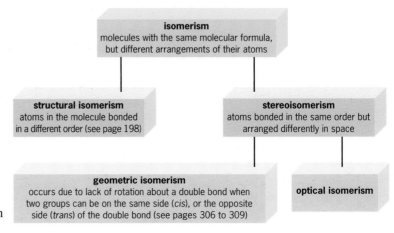

**Fig 19.5** Different types of isomerism

Fig 19.6 Since your hand is not symmetrical, it cannot be superimposed on its mirror image

## CHIRALITY

Look at both your hands together with their palms facing you. One is a mirror image of the other. Now put one hand palm up on this page and the other on top of it, also palm up. You will see that you cannot superimpose them – the thumbs stick out in opposite directions. This property is known as **chirality** and exists because hands are not symmetrical. Chiral objects cannot be superimposed on their mirror images – they are **non-superimposable**. The term chirality is derived from *kheir*, the Greek word for hand.

Molecules, too, can be chiral and so have non-superimposable mirror images called **enantiomers**. In organic compounds, the usual reason for a molecule being chiral is that it has a carbon atom bonded to *four different groups*. When this occurs, the molecule cannot be symmetrical and the carbon atom is called an **asymmetric carbon atom** or a **chiral carbon**.

Fig 19.7 A common way for a molecule to be chiral is when it contains a carbon atom bonded to four different groups. The mirror image of this molecule is non-superimposable and is called an enantiomer

Fig 19.8 The enantiomers of alanine are shown using an imaginary mirror. Note that, although the central carbon atom is bonded to two carbon atoms, these atoms are part of two different groups

Take, for example, alanine. It has a chiral carbon because this atom is bonded to four different groups. It is therefore a chiral molecule, having a pair of enantiomers that cannot be superimposed.

Now look at glycine in Fig 19.9. Note that its carbon is no longer asymmetric because it has two identical groups attached (the hydrogen atoms). The glycine molecule and its mirror image can be superimposed, so they cannot be enantiomers of each other.

The enantiomers of chiral molecules have the same chemical properties in ordinary test-tube reactions, which might be expected given that their atoms are bonded in the same order. However, because their atoms are arranged differently in space, we might expect differences in their physical properties such as melting point, boiling point and solubility. However, these too, are identical, except in one unusual way which gives rise to optical isomerism.

Fig 19.9 Glycine

| ✔ | REMEMBER THIS |
|---|---|

Although enantiomers have identical chemical properties, they do interact differently with the enantiomers of other chiral molecules. Hence, some enantiomers do not smell and taste the same, because they interact differently with the chiral taste and smell receptors.

## WHAT IS OPTICAL ISOMERISM?

With any pair of enantiomers, one enantiomer rotates plane-polarized light in one direction, and the other enantiomer rotates it in the opposite direction. But the angles of rotation are equal. In 1815, Jean-Baptiste Biot, a French physicist, was the first to discover that the crystals of certain substances could rotate a beam of plane-polarized light either to the right (clockwise) or to the left (anticlockwise), looking at the incoming light. At the time, the reason for this rotation was a mystery. It was later discovered that solutions of certain compounds could also exhibit this **optical activity**.

Molecules that rotate plane-polarized light to the left (anticlockwise) are said to be **laevorotatory**. Those that rotate it to the right (clockwise) are called **dextrorotatory**.

**SCIENCE FEATURE** Light and optical activity

Normal light consists of electromagnetic waves that vibrate in all directions perpendicular to the direction of travel, as shown in Fig 19.10. (The electromagnetic nature of light is covered on page 48.) Certain crystals allow light with vibrations in one plane only to pass through them. Such crystals are known as polarizers. The light that emerges from a polarizer is called **plane-polarized light**, or just **polarized light**. The lenses of Polaroid sunglasses provide a good example of polarizers.

When certain substances are placed between two aligned polarizers through which plane-polarized light is passing, the second polarizer stops transmitting light. This is because the substance has rotated the plane of polarization of the light. The substance is said to be **optically active**. Rotating the second polarizer through a certain angle restores its transmission of light. The angle of rotation caused by a substance is measured in this way, which is the principle of the polarimeter.

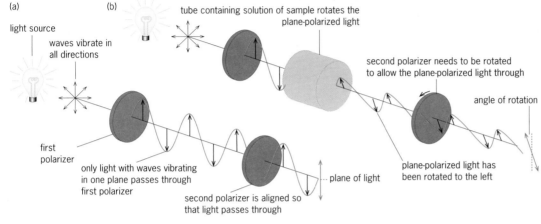

(a)

light source

waves vibrate in all directions

first polarizer

only light with waves vibrating in one plane passes through first polarizer

second polarizer is aligned so that light passes through

plane of light

(b)

tube containing solution of sample rotates the plane-polarized light

second polarizer needs to be rotated to allow the plane-polarized light through

angle of rotation

plane-polarized light has been rotated to the left

**Fig 19.10(a)** Unpolarized light from a source passes through the first polarizer and the light emerges from it in one plane only. This plane-polarized light passes through the second polarizer (analyser) only when both polarizers are aligned. **(b)** A solution of an optically active sample placed in the beam of plane-polarized light rotates the beam. For light to pass through the second polarizer (analyser), the second polarizer too must be rotated until it aligns with the plane of polarization of the light

In 1848, the French scientist Louis Pasteur discovered the important connection between the structure of crystals and their optical activity. He was examining salts of tartaric acid when he noticed that samples of the sodium salt included two sorts of crystal, which were mirror images of each other. Pasteur separated the two sorts of crystal and made a solution of each. He then examined both solutions in a polarimeter, an instrument used to measure optical activity. He found that one sort of crystal rotated plane-polarized light clockwise, while the other rotated it anticlockwise. He also found that, when the concentrations were the same, the two angles of rotation were equal, but in opposite directions.

Pasteur proposed that the molecules making up the crystals must also be mirror images of one another. This proved to be an advance of fundamental importance. Less than 20 years later, the tetrahedral model was proposed for those carbon-containing molecules in which each carbon atom has four single bonds (see the description of methane on page 78). It was then recognized that when a carbon atom is bonded to four different groups, it is asymmetric or chiral.

**EXTENSION** Louis Pasteur

Most people have heard of Louis Pasteur as the inventor of the so-called pasteurization of milk, heat treatment to destroy the bacteria that are naturally in milk and that would otherwise make it 'go off'. This process was just one aspect of his revolutionary work on microorganisms and how they cause food to spoilage and fermentation.

**Fig 19.11** Louis Pasteur (1822–1895)

Among Pasteur's other achievements was his intense research on silkworm disease, which led to his pioneering germ theory. Putting his theory into practice, he produced vaccines for anthrax and rabies.

Pasteur's achievements outside scientific medicine were just as significant. For instance, he solved the puzzle of why certain crystals and

*(continued on page 414)*

## ▶ EXTENSION Louis Pasteur [continued from page 413]

solutions are able to rotate plane-polarized light. He observed that the sodium ammonium salt of tartaric acid (crystals of tartaric acid are found on wine casks) rotated plane-polarized light to the right, while the sodium ammonium salt of racemic acid (another acid found on wine casks during fermentation) had no effect. Apart from this observation, the salts of racemic acid and tartaric acid appeared to be the same, having identical chemical composition and properties. The shapes of the crystals apparently looked the same. However, under the microscope he noticed right- and left-handed crystals in racemic acid.

Pasteur pressed on, painstakingly sorting the two sorts of crystals with the aid of a microscope and tweezers. He made a solution of each sort and examined both solutions in his polarimeter.

Pasteur's discovery of right- and left-handed crystals in racemic acid owed much to luck. He was doing his experiments in winter. Above 26 °C, racemic acid would not have crystallized out into two different forms. Also, the salt of racemic acid he had chosen to investigate is the only salt whose mirror-image crystals have faces clear enough to allow separation with the aid of a microscope (which would not have been as powerful as today's instruments). However, as Pasteur himself remarked, 'Chance favours the prepared mind.'

Racemic acid contains a 50:50 mixture of dextrorotatory and laevorotatory crystals of tartaric acid. Until Pasteur's separation of racemic acid, the laevorotatory form of tartaric acid was unknown. Any 50:50 mixture of left- and right-handed molecules is called a **racemic mixture** after the salt of racemic acid that Pasteur separated. **In a racemic mixture, the optical activity of one isomer cancels out the optical activity of the other isomer.**

**Fig 19.12** Mirror-image crystals of the sodium

ammonium salt of tartaric acid

---

### ? TEST YOURSELF

**C** What is the systematic name of lactic acid?

### ✔ REMEMBER THIS

To depict a pair of optical isomers, draw the 3-D structure of one molecule and then swap the positions of two of the groups in the other molecule.

### ◉ EXAMPLE

**Q** Lactic acid, $CH_3CH(OH)COOH$, is isolated from milk. It is an optical isomer of the lactic acid present in muscles.
**a)** Explain what is meant by the term optical isomerism.
**b)** Draw the full structural formulas of the two enantiomers, showing clearly the feature that causes optical isomerism.

**A a)** Optical isomerism occurs when a molecule has a non-superimposable mirror image (Fig 19.13). The molecules of such a pair are called enantiomers. The two enantiomers rotate plane-polarized light in opposite directions.
**b)** It is often helpful to include an imaginary mirror when drawing optical isomers. In Fig 19.13 the groups around the chiral carbon atom are shown as mirror images, but this is not essential.
The feature that causes optical isomerism in lactic acid is a chiral carbon, which is often indicated by an asterisk ($C^*$).

**Fig 19.13 Enantiomers of the same compound, lactic acid**

imaginary mirror

Many molecules have more than one chiral carbon. For example, the chain form of the glucose molecule has four chiral carbon atoms (Figs 19.14 and 19.15).

**Fig 19.14** (left) The structural formula of the chain form of glucose showing the four chiral carbon atoms, each identified by an asterisk

**Fig 19.15** (right) Although each chiral carbon is bonded to two other carbon atoms, to decide if a carbon atom is chiral you must consider the whole of each group attached to it. This has been done for one carbon atom, showing clearly that there are four different groups around it. Can you see why the other three carbon atoms, indicated in Fig 19.14, are chiral?

### ▶ EXTENSION L- and D-Enantiomers of α-amino acids

Nearly all the amino acids in every organism are made up from one type of enantiomer, known as the L-enantiomer. In the L-enantiomer, the arrangement of the four groups spell CORN (**CO**OH, **R**, **N**H₂) when you look directly down on the H atom and go clockwise round the molecule (Fig 19.18).

**Fig 19.18** To identify an L-amino acid, see whether the groups spell CORN in a clockwise direction when you look down on the hydrogen atom

The proteins in the cell wall of a bacterium are made up of D-amino acids, which is unusual in organisms. The antibiotic penicillin kills bacteria because it interferes with the building of new cell walls of the bacteria that are dividing. Penicillin does not destroy our cell walls when they divide because our protein is made of L-amino acids.

### ⚠ SCIENCE FEATURE Ageing a skull by its smile

When we are born, the dentine inside our teeth contains L-aspartic acid (see Fig 19.4). As we grow older, this starts to change to the D-form. When we die, this process continues at the same rate for hundreds of years. This means that we can accurately date a skull that has been removed from the earth just by knowing the proportions of D- to L-aspartic acid in its teeth – provided that the skull is not too old. This technique is reliable back to the tenth century AD.

**Fig 19.19** As a skull ages, there is an increase in the ratio of the L- to D-isomers of aspartic acid in its teeth. Exhumed skulls have been dated by determining this ratio.
This is the skull of a mediaeval Ilford man

### ? TEST YOURSELF

**D** Optical isomers of the molecule carvone produce different taste sensations. One enantiomer gives the taste of spearmint and is used in spearmint chewing gum. The other enantiomer gives the taste of caraway seeds, which are used to flavour seed cakes.
**a)** Explain what is meant by the term enantiomer.
**b)** Copy the displayed formula of carvone and mark the chiral carbon. If you are in doubt, look at Fig 19.15 and its caption.

**Fig 19.16** Carvone

**E** Copy the fructose molecule shown in Fig 19.17 and identify the chiral carbons using asterisks.

**Fig 19.17** Fructose

**SEE QUESTIONS 1, 2 AND 3.**

COOH
|
CHNH₂
|
CH₂
|
(benzene ring with HO and OH substituents)

**Fig 19.20** The dopa molecule

### ? TEST YOURSELF

**F** Identify the chiral centre on the dopa molecule.

**SEE QUESTION 4.**

$$H_3\overset{+}{N}-\underset{|}{\overset{R}{C}}H-COO^-$$

**Fig 19.21** An amino acid zwitterion

### ✔ REMEMBER THIS

Bases are proton acceptors and acids are proton donors (see page 365).

**THE REACTION OF COOH WITH ALKALIS IS COVERED ON PAGE 375, AND THE REACTION OF NH₂ WITH ACIDS ON PAGE 624.**

### ? TEST YOURSELF

**G** Write balanced equations for the reactions of:
**a)** glycine with sodium hydroxide solution;
**b)** leucine with dilute hydrochloric acid.
Salts are produced in both cases, so remember to show the charges on the ions.
Hint: See Fig 19.4 on page 411 for the formulas of amino acids.

### ? TEST YOURSELF

**H** Look at Fig 19.4, page 411. Which amino acids have an acidic side chain?

## CHIRAL DRUGS

Pharmaceutical drugs work on complex chemical sites around the body because they have the correct shape to fit the receptor sites. L-Dopa is just such a drug (Fig 19.20). It is used to treat Parkinson's disease, which is characterized by uncontrollable shaking in the hands and loss of balance. The L-dopa is absorbed by the brain, where it is converted into dopamine. In the brain, high concentrations of dopamine improve conduction by nerves and thus help to control the shaking. The D-form of the chiral dopa molecule causes very unpleasant side-effects.

Many drugs are chiral. Another example is the drug thalidomide. Between 1956 and 1961 it was prescribed to women who suffered badly from morning sickness during the early stages of pregnancy. Unfortunately, it exists as optical isomers. While one form of the drug is an effective treatment, the other form caused the most dreadful birth defects.

A positive aspect of understanding the chiral nature of some drugs is that smaller doses can be given, since only half as much is required (the other half being the other optical isomer). Also, side effects caused by the wrong optical isomer are eliminated and the pharmacological activity is much improved.

## ZWITTERIONS

Amino acids have two functional groups. The chemical properties of compounds with two functional groups are often the sum of the chemical properties of each individual group. However, with amino acids, the basic amine group and the acidic carboxylic acid group can react with each other. A proton from COOH can be donated to the NH₂ group of the same molecule to give a **zwitterion** (Fig 19.21), a molecule that carries both a positive and a negative charge. (*Zwitter* is the German term for hybrid.)

Their relatively high melting point and high solubility in water indicate that amino acids exist as zwitterions, both in the solid state and in solution. When a dilute acid is added to an aqueous solution of an amino acid (Fig 19.22), the COO⁻ group accepts a proton to form COOH. This leaves a positive ion (a cation). When the pH of the aqueous solution is raised by adding OH⁻ (Fig 19.23), a proton from NH₃⁺ is removed to form NH₂ and a negative ion (an anion).

$$H_3\overset{+}{N}-\underset{|}{\overset{R}{C}}H-COO^-(aq) \; + \; H_3O^+(aq) \longrightarrow H_3\overset{+}{N}-\underset{|}{\overset{R}{C}}H-COOH(aq) \; + \; H_2O(l)$$
$$\text{cation}$$

**Fig 19.22** Dilute acid added to an aqueous solution of an amino acid results in a positive ion and COOH

$$H_3\overset{+}{N}-\underset{|}{\overset{R}{C}}H-COO^-(aq) \; + \; OH^-(aq) \longrightarrow H_2N-\underset{|}{\overset{R}{C}}H-COO^-(aq) \; + \; H_2O(l)$$
$$\text{anion}$$

**Fig 19.23** Adding OH⁻ raises the pH of the aqueous solution and results in a proton being removed from NH₃⁺ to form NH₂ and a negative ion (an anion)

The pH at which zwitterions have the highest concentration in solution in equal amounts is known as the **isoelectric point**. For those amino acids with a neutral R group, the isoelectric point is about pH 6. By contrast, lysine has a basic side chain that contains an extra NH₂ group (Fig 19.24). This means that a solution of lysine is alkaline and its isoelectric point is about 9.5. Those amino acids with an additional COOH in the side chain are acidic and have low isoelectric points.

NH₂
|
(CH₂)₄
|
H₂N—CH—COOH

**Fig 19.24** Lysine has two NH₂ groups, making it a basic amino acid

## ! SCIENCE FEATURE Electrophoresis

Electrophoresis can be used to separate a mixture of amino acids according to the ionic charge at a particular pH. The mixture is placed in a gel or on filter paper (Fig 19.25) in an electric field.

fragments are negatively charged and can be separated into bands using gel electrophoresis. The bands are made visible by tagging them with molecules that contain radioactive phosphorus, $^{32}P$,

chances that two samples of DNA from different people will actually match is estimated to be of the order of one in 25 million. In this way, innocent people can be eliminated from criminal

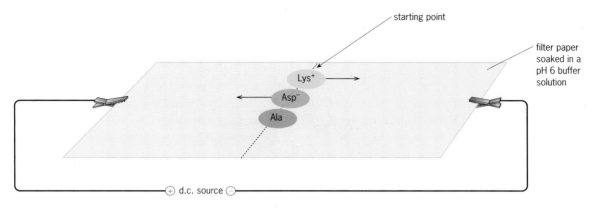

**Fig 19.25** Electrophoresis of a mixture of three amino acids. The pH used in this example is 6, so alanine exists as a zwitterion. At this pH, aspartic acid is negatively charged and lysine is positively charged

Positively charged ions migrate to the cathode, and negatively charged ions to the anode. When the pH is that at which zwitterions are formed with equal positive and negative charges, the zwitterions cannot move in either direction. So, the pH value must be carefully chosen before the mixture of amino acids can be separated and analysed successfully.

Electrophoresis can also be used to separate and identify different proteins in a gel. The large number of different amino acids that make up a protein give it an overall charge at a particular pH. This, combined with the protein's mass and shape, enables it to be separated from other proteins by electrophoresis. (The higher the mass and more irregular their shape, the slower they move.)

### Genetic fingerprinting
Genes, too, can be analysed by electrophoresis. DNA, the molecule that carries genetic information, is broken up into small fragments by enzymes. These

and allow them to fog a photographic film. This forms the basis of genetic fingerprinting used by forensic scientists to identify criminals from blood, saliva or other samples that contain cells. The

**Fig 19.26** A researcher with autoradiograms of DNA fragments separated by gel electrophoresis

investigations and compelling evidence can be provided against the guilty.

The technique is not only used to catch criminals. It can be used to match parents to their children (see Fig 19.27), and also to identify cancerous cells in bone marrow.

**Fig 19.27** Genetic fingerprints to confirm parenthood

## 4 FORMING PEPTIDES

When two amino acids react together, the compound formed is known as a **dipeptide**. The reaction is between the COOH of one amino acid and the $NH_2$ of the other (Fig 19.29), to give a dipeptide. By convention, the amino acid that contains the free $NH_2$ group is always placed on the left. A molecule of water is eliminated and so the reaction is called a **condensation reaction**.

**REMEMBER THIS**

A condensation reaction is also called an addition–elimination reaction. The small molecule eliminated can be $H_2O$, $NH_3$ or HCl. (See page 398.)

peptide linkage

Fig 19.29 The formation of a peptide linkage in a condensation reaction between two amino acids

The shorthand notation for a dipeptide made from alanine and glycine is Ala–Gly. The peptide link is shown by the dash. Three-letter codes or abbreviations are used for the names of the amino acids, as in Fig 19.4.

The CO–NH group is a **secondary amide** functional group, often referred to as just an **amide** group. The **peptide group** or **peptide link** is the amide functional group that joins two amino acids. Either end of the dipeptide can react with another amino acid to form a **tripeptide**. Amino

**TEST YOURSELF**

When alanine reacts with glycine, two dipeptides can be formed: Ala–Gly and Gly–Ala. Draw the structural formulas of both dipeptides.

SEE QUESTION 5.

acids can keep reacting to form longer and longer amino acid chains, known as **polypeptides**. The amino acids in peptides and polypeptides are called **amino acid residues**. **Proteins** are made up of one or more long polypeptide chains. Thus, polypeptides and proteins are condensation polymers of amino acids. In our bodies, the condensation reaction is catalysed by enzymes.

### ✔ REMEMBER THIS

Secondary amides are usually made in the laboratory by reacting a primary amine (RNH$_2$) with an acyl chloride (R'COCl), as in Fig 19.27. This is also called a condensation reaction because a small molecule is eliminated. (See page 398.)

**Fig 19.30** A primary amide reacts with an acyl chloride to form a secondary amide

### ! SCIENCE FEATURE  Peptides don't have to be large

Many biologically important peptides contain just a few amino acid residues. For example, the human brain produces a peptide called leucine enkephalin that contains just five amino acid residues. This peptide was first discovered when the pain-killing actions of morphine and codeine were being investigated in the 1970s. Both morphine and codeine fit into a brain receptor site, which was found to be the start of their pain-killing action.

The mystery scientists set out to unravel was how these two drugs, both obtained from the dried sap of poppies (called opium), could fit into brain receptor sites. They looked for compounds made in the body that might fit into these sites and discovered that the brain produces its own pain-killers, called enkephalins, of which leucine enkephalin is one (Fig 19.31).

**Fig 19.31** Leucine enkephalin. By convention, the amino acid with the free NH$_2$ group is shown at the left end of the molecule, while the amino acid with the COOH group is shown at the right end

Oxytocin, which contains nine amino acid residues, is another small peptide. It is a hormone secreted by the human pituitary gland and is responsible for inducing the uterus to contract at the end of pregnancy. It was the first natural peptide to be synthesized in a laboratory. Vincent du Vigneaud, the biochemist who made this important breakthrough in 1954, was awarded the Nobel Prize for Chemistry for his achievement.

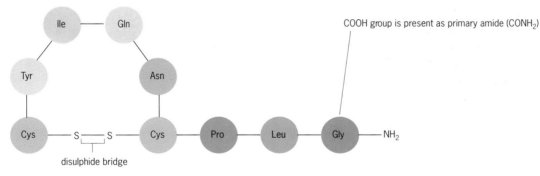

**Fig 19.32** The two sulphur atoms form a bridge holding six of the amino acids in a ring. (Disulphide bridges are covered on page 420)

## 5 PROTEIN STRUCTURE

Proteins are very large molecules, with values of $M_r$ that range from 5000 to several million. When proteins are formed inside a cell, they are too large to pass through the cell membrane and so stay trapped inside. That is, unless tissues are damaged. For example, certain enzymes (which we have seen are proteins) can be found outside cells after a heart attack. Normally, these enzymes would be confined inside the cells of the heart, but during a heart attack some of the cells rupture, which allows these enzymes to escape. The more massive the heart attack, the more cells split open and the higher the concentration of the enzymes found in the blood. The presence of these enzymes in the blood is used by doctors to diagnose the seriousness of a heart attack. In the same way, certain proteins in urine indicate that a kidney has been damaged.

## PRIMARY STRUCTURE OF PROTEINS

Although only 20 amino acids, usually, make up our proteins, the number of ways in which they are combined is vast. Take, for example, lysozyme, an enzyme found in tears, saliva, nasal mucous and milk, and that destroys the cell walls of bacteria by breaking certain bonds through hydrolysis. It has 129 amino acid residues and uses all 20 amino acids (Fig 19.33). The number of different arrangements of the amino acids in the protein chain is $20^{129}$ ($7.0 \times 10^{167}$), which is estimated to be more than twice the number of atoms in our galaxy.

The sequence of the amino acids in a protein is called its **primary structure** and this forms the backbone of the protein.

### ? TEST YOURSELF

**J** Human insulin, which consists of 51 amino acid residues, is a relatively small protein. It has a molecular formula of $C_{254}H_{377}N_{65}S_6$. What is its relative molecular mass?

**Fig 19.33** The primary structure of the enzyme lysozyme. Notice the disulphide bridges

### ✔ REMEMBER THIS

Disulphide bridges are formed between two cysteines by the oxidation of two S–H groups to S–S.

**Fig 19.34** Formation of disuphide bridges

## HYDROLYSIS OF PEPTIDE BONDS

To find the sequence of amino acids in a protein, the peptide links are **hydrolysed** (split by the action of water). This is done by boiling the protein with an acidic or alkaline solution (Fig 19.35), or by using certain enzymes. The protein chain may be hydrolysed completely, to leave a mixture of individual amino acids that can be separated using chromatography. This gives information as to how many different amino acids are present.

### ? TEST YOURSELF

**K** Write an equation similar to that in Fig 19.35 for the alkaline hydrolysis of the peptide link. Hint: To remind yourself about hydrolysis, see page 402.

**Fig 19.35** The acid hydrolysis of one peptide link in a protein chain is often carried out with hot HCl(aq)

If different hydrolysing agents are used, the protein chain is split in different places. Some of the amino acid sequences overlap, which enables the overall sequence of amino acids to be worked out.

**SEE QUESTIONS 6 AND 7.**

## ! SCIENCE FEATURE  Primary structure of insulin

Insulin was the first protein to have its amino acid sequence (primary structure) worked out. This was done by the English biochemist Frederick Sanger, who began in 1944 and spent ten years on the task. One of the techniques he used was to completely hydrolyse the insulin into its component amino acids by heating a mixture of 6 mol $dm^{-3}$ hydrochloric acid with insulin in a sealed tube for 24 hours.

Sanger then separated the amino acids, using paper chromatography (Fig 19.36). Instead of using just one solvent to separate the amino acids, he used two, one after the other, in a technique called two-way chromatography. A chromatogram is made using the first solvent and is left to dry. Then it is turned on its side and the second solvent is applied.

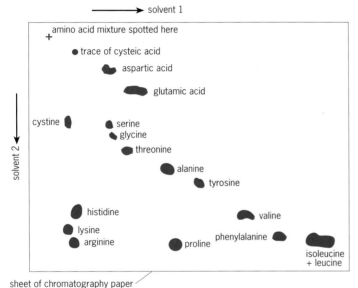

**Fig 19.36** Insulin contains 17 amino acids. A spot of the amino acid mixture obtained by hydrolysing insulin was placed at the cross. The first solvent was allowed to run almost to the top of the paper, causing some of the amino acids to separate out. To complete the separation, after drying, the chromatography paper was turned through 90° and a second solvent was run up the paper. The amino acids are made visible as purple spots using ninhydrin solution

For his painstaking research, Sanger received the Nobel Prize for Chemistry in 1958. In 1980, he was awarded a second Nobel Prize jointly with two American scientists for work on DNA, and so he became only the third person in history to win two Nobel prizes in scientific disciplines.

? **TEST YOURSELF**

**L** Work out the sequence of this simple peptide, which contains five amino acid residues. It is an enkephalin, similar to the one on page 419. The chain has been partially hydrolysed by two different reagents:
Reagent A produces Gly–Phe–Met and Tyr–Gly
Reagent B produces Tyr–Gly–Gly and Phe–Met
Hint: Remember the convention that the amino acid with a free $NH_2$ group is always placed on the left of the molecule, while the amino acid with the COOH group is at the right end.

FOR MORE INFORMATION ON TWO-WAY CHROMATOGRAPHY SEE PAGE 254.

## SECONDARY STRUCTURE OF PROTEINS AND HYDROGEN BONDING

The shape of a protein molecule is what makes it able to perform its enzyme function.

Linus Pauling, the eminent American scientist, first turned his attention to protein structure in the 1930s (see also page 83). After 15 years, he and another American, Robert Corey, proposed that the primary structure of proteins could have one of two orderly arrangements, both held together by hydrogen bonds. The first arrangement is a regular coiling of part of the polypeptide chain, called an α-**helix**. The second arrangement is a folding of the polypeptide chain to make sheets, called a β-**pleated sheet**. These are

## ✔ REMEMBER THIS

The hydrogen bond is a weak electrostatic attraction between an H atom bonded to a highly electronegative atom (N, O or F), which gives it a partial positive charge, and a lone pair of electrons on a neighbouring highly electronegative atom (N, O or F).

**Fig 19.38** Hydrogen bonding

For more information, see page 133.

known as the **secondary structures** of a protein. Both were subsequently discovered through X-ray crystallography.

The hydrogen bonding that holds a secondary structure together is between the N—H of one peptide link and the C=O of another. Since this is within the molecule, it is known as **intramolecular** hydrogen bonding. The α-helix is a right-handed spiral (Fig 19.37) and hydrogen bonds form between every fourth amino acid residue. It is found extensively in wool fibres and allows wool to stretch. When it is pulled, the α-helix elongates, which breaks the hydrogen bonds. When the α-helix is released, the hydrogen bonds re-form as the α-helix returns to its usual shape.

The β-pleated sheet occurs when the amino-acid backbone of the primary structure folds to give sections of parallel chains of amino acids, which again form hydrogen bonds through their peptide links (Fig 19.39). Although the β-pleated sheet is fairly flexible, it cannot be stretched, as the chains of amino acids are already extended. Silk, for example, is composed of the protein silk fibroin. This is almost entirely a β-pleated sheet, which is the reason why silk cannot be stretched like wool.

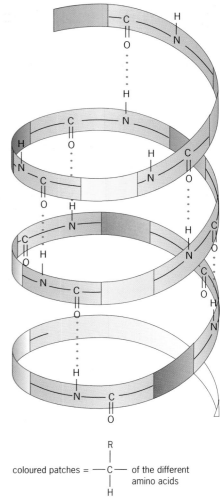

coloured patches = $-\overset{\displaystyle R}{\underset{\displaystyle H}{C}}-$ of the different amino acids

**Fig 19.37** The shape of the α-helix is maintained by hydrogen bonds

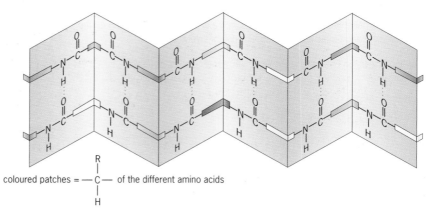

coloured patches = $-\overset{\displaystyle R}{\underset{\displaystyle H}{C}}-$ of the different amino acids

**Fig 19.39** The β-pleated sheet results from hydrogen bonding between parallel chains of amino acids

## EXTENSION Tertiary structure of proteins

The overall three-dimensional shape of a protein is the result of yet more folding and bending to give what looks like a random structure, but most definitely is not. This is the **tertiary structure** and is specific to a particular polypeptide chain. It is this shape that determines how each protein will function.

Proteins can be divided into two sorts: **globular proteins** and **fibrous proteins**.

Globular proteins, as their name implies, are roughly spherical (Fig 19.40). Our enzymes and protein hormones are globular proteins, and their polypeptide chains are folded extensively to give a very compact structure. These proteins are usually soluble because they are arranged in a way that leaves the hydrophilic R-groups of individual amino acid residues on the outside of their structures.

The α-keratin of hair, wool and claws is a fibrous protein in which the α-helix is twisted so that several strands can intertwine. The silk fibroin protein is also fibrous, only this time β-pleated sheets become bound together.

**Fig 19.41** Collagen is a triple helix of polypeptide chains

polypeptide chain

The tertiary structure of a protein is held together by four different types of interaction between the various R side chains of the different amino acids. These are shown in Fig 19.42.

**Fig 19.40** The tertiary structure of the protein myoglobin. Note that sections of the secondary α-helix structure have been folded to give the overall three-dimensional shape. In the middle of the protein is a haem group that contains an iron atom. Myoglobin is found in muscles and is responsible for binding oxygen and releasing it as needed. It was the first protein for which the three-dimensional structure was worked out

Fibrous proteins have their polypeptide chains arranged in bundles to form fibres. They have a structural function. Collagen is a fibrous protein in tendons and muscles that consists of intertwining polypeptide chains (Fig 19.41).

**Fig 19.42** The four types of interaction that hold the tertiary structure of a protein together:
**A** Van der Waals forces (induced dipole–induced dipole bonding) exist when non-polar R groups come close together. They are usually found on the inside of globular proteins where, because they are hydrophobic, they do not interfere with solubility
**B** Ionic linkages occur between $COO^-$ and $NH_3^+$
**C** Hydrogen bonding occurs between polar R groups, such as those that contain OH and $NH_2$
**D** Disulphide bridges result from two cysteine residues coming close together and the two S–H groups oxidizing to S–S

**SEE QUESTION 8.**

## ? TEST YOURSELF

**M** Hydrophilic means water-loving, whereas hydrophobic means water-hating. Hydrophilic groups can interact with water through, for example, hydrogen bonding.

**a)** Look at Fig 19.4 and work out which amino acids have R groups that could form hydrogen bonds with water.

**b)** Non-polar R groups tend to be hydrophobic. Which amino acids in Fig 19.4 have hydrophobic R groups?

## ! SCIENCE FEATURE It makes your hair curl!

Our hair is made up of strands of the protein α-keratin. Its shape is maintained by hydrogen bonds, ionic linkages and disulphide bridges between the keratin strands. The very act of washing our hair disrupts these forces because water molecules can get in between the strands and affect all three interactions – in particular, the hydrogen bonds.

As hair dries, the water molecules leave the keratin and the original forces reassert themselves. So, if we do not immediately comb and set our hair in the shape we want, hydrogen bonds may re-form to curl our hair in a way we don't want.

**Fig 19.43** The very act of washing hair disrupts some of the weaker forces holding the protein molecules of the hair together

## ► EXTENSION Enzymes

An enzyme is a biomolecule catalyst. Almost all enzymes are proteins, and most are found inside cells. Like all catalysts, enzymes alter the energy of activation of a reaction by providing an alternative route for reactants to interact without becoming permanently involved in the reaction. They differ from other catalysts in their amazingly high activity, being between a million and 20 million times more efficient than inorganic catalysts.

Sometimes, each enzyme molecule can catalyse the reaction of thousands of molecules per second. An example is the enzyme catalase, which catalyses the decomposition of hydrogen peroxide:

$$H_2O_2(l) \xrightarrow{\text{catalase}} H_2O(l) + O_2(g)$$

One molecule of catalase catalyses about 50 000 molecules of $H_2O_2$ per second at 0 °C, whereas manganese(IV) oxide, an inorganic catalyst for this reaction, works very much less efficiently at 0 °C.

Enzymes can also be very specific in the reactions they catalyse. For example, out of the vast number of chemicals in the body, an enzyme may catalyse just one particular reaction of one chemical. The three-dimensional shape of an enzyme is critical to its catalytic ability. This shape can be denatured (disrupted) by changes of temperature, by altering the pH of the environment of the enzyme or by exposing it to heavy metal ions, such as mercury(II) ions ($Hg^{2+}$) and lead(II) ions ($Pb^{2+}$).

Enzymes have an optimum temperature and pH at which they work best. Their catalytic activity decreases markedly outside a narrow range, which is why it is dangerous to have a very high temperature for too long during an illness. Most enzymes work efficiently in the range of 25–40 °C, and their shape is irreversibly changed at about 50–60 °C. The optimum pH for most of our body enzymes is about pH 7, but pepsin, a digestive enzyme, has an optimum pH of 2, with its maximum activity in the acid conditions of the stomach.

(You can read more about catalysis on pages 180 and 531.)

## ! SCIENCE FEATURE Immobilizing enzymes

At the start of this chapter, you read about the advantages of using enzymes in industry. Unfortunately, if an enzyme is simply mixed with the reactants to be catalysed, it is difficult to separate at the end of the reaction and is usually destroyed.

However, if the enzyme is attached to an inert surface across which the reactants pass, the enzyme functions as a heterogeneous catalyst (see page 531) and can be used several times over.

Various surfaces can hold the enzyme – from polystyrene to glass beads – and immobilized enzymes are used increasingly in food and chemical manufacture. They are also important in medical diagnosis: clinical tests for glucose and cholesterol rely on immobilized enzymes that react with these compounds.

# SUMMARY

After studying this chapter, you should know the following:

- Amino acids have two functional groups: COOH and $NH_2$.
- Naturally occurring $\alpha$-amino acids have the general formula in Fig 19.44.

**Fig 19.44** General formula for amino acids

- Optical isomerism occurs when two molecules, called enantiomers, exist as non-superimposable mirror images of each other. These molecules are said to be chiral.
- A carbon atom bonded to four different groups is called a chiral carbon. Such carbon atoms give rise to optical isomers.
- A pair of enantiomers rotates plane-polarized light in equal but opposite directions.
- A zwitterion is formed when a COOH group donates a proton to an $NH_2$ group on the same molecule, to give an ion that is both positively and negatively charged.

- CO–NH is called a peptide group or peptide link when it joins two amino acids together.
- Polypeptides are long-chain molecules that contain many amino acids joined by peptide links. Proteins are very large molecules that have $M_r$ values of from 5000 to several million.
- The primary structure of a protein (either an $\alpha$-helix or a $\beta$-pleated sheet) is the sequence of its constituent amino acids. It forms the protein backbone.
- The secondary structure of a protein (either an $\alpha$-helix or a $\beta$-pleated sheet) is held together by hydrogen bonding between C=O and N–H in different peptide groups.
- The tertiary structure is the overall three-dimensional shape of a polypeptide in a protein. It is the result of the secondary structure folding and bending.
- Enzymes are biomolecule catalysts and nearly all are proteins. They have a very high activity and usually a high specificity. They are also very sensitive to changes in pH and temperature.

# ? EXAM QUESTIONS

**1** When compound **A** reacts with water in the presence of an acid catalyst, products **B** and **C** can be obtained (Fig 19.Q1).

**Fig 19.Q1**

**a)** Explain what is meant by *optical activity*. Describe the structural characteristic of an optically active molecule.

**b)** Identify which compound, **B** or **C**, can show optical activity. Draw structures of the two enantiomers to illustrate clearly the relationship between them. How do the two enantiomers differ in their optical activity?

IBO Chemistry Standard Level Paper 3 November 2000 QH2c (part), d

**2** An optically active bromoalkane, **A**, contains 35.0% carbon and 6.57% hydrogen by mass.

**a)** Calculate the empirical formula of bromoalkane **A**.

**b)** Given that the molecular formula of **A** is the same as its empirical formula, draw the structure of **A**.

AQA CH03 Kinetic and Organic Chemistry June 2000 Q5a

**3** The structures of the two isomers, **A** and **B**, are shown in Fig 19.Q3.

**Fig 19.Q3**

Only one of the isomers is chiral. Identify which isomer is chiral. State the physical property caused by the chirality that would allow you to distinguish the isomers.

Edexcel Nuffield Specimen Unit Test 6 Q5b

## ❓ EXAM QUESTIONS

**4** Ibuprofen (Fig 19.Q4) is sold for the alleviation of pain. It is a mixture of optical isomers, only one of which is active as the drug.

ibuprofen

**Fig 19.Q4**

**a)** Explain, using clear diagrams, the term *optical isomers* in relation to the structure of Ibuprofen.

**b)** The optical activity of Ibuprofen is zero. What does this tell you about how much of each of the stereoisomers is present in the mixture?

**c)** What might the advantages and disadvantages be of using a mixture of stereoisomers, rather than just the active stereoisomer, as a drug?

OCR 6821 General and Organic Chemistry June 2000 Q2

**5** Diphenylurea, $C_6H_5NHCONHC_6H_5$, can be used as a plant hormone. State the formula of the compound that is used to produce diphenylurea from phenylamine and state the name of another type of compound that also contains the linkage in Fig 19.Q5.

**Fig 19.Q5**

WJEC Specimen Paper 2001/2 CH4 Q5d

**6** **Aspartame** is an artificial sweetener that is about 200 times sweeter than sucrose. It is a methyl ester of a dipeptide formed from two α-amino acids: aspartic acid and phenylalanine (Fig 19.Q6.1).

the aspartame dipeptide

**Fig 19.Q6.1**

**a)** Explain the meaning of the term **dipeptide**.

**b)** Draw a skeletal formula for the aspartame dipeptide.

**c)** One way to show the composition of aspartame is to hydrolyse it and then separate and identify the substances formed.

**(i)** Briefly describe the procedure that is used to hydrolyse a protein or a dipeptide.

**(ii)** Three compounds are formed when aspartame is hydrolysed. Name these three compounds and explain what further information you would need to establish the structure of aspartame.

**d)** Both aspartic acid and phenylalanine exist in two optical isomer forms, labelled the L-form and the D-form. Only the L-amino acids occur naturally in proteins. Many of the chemical reactions of L and D amino acids are identical, but in the body they can behave differently. For example, D-phenylalanine tastes sweet, but L-phenylalanine is bitter.

**(i)** Copy the structure of phenylalanine in Fig 19.Q6.2, and place an asterisk next to the atom that indicates to you that there are two optical isomers.

**Fig 19.Q6.2**

**(ii)** Draw a vertical line to represent a mirror and either side draw diagrams to illustrate the two optical isomers of phenylalanine.

**e)** A solution of phenylalanine in water will contain few molecules like the one shown in **d)(i)**. Draw a structure for the form of phenylalanine most likely to be present in an aqueous solution.

**f)** A disadvantage of using aspartame as a sweetener is that it breaks down much faster than sucrose. In soft drinks stored in cans or bottles, about 10% of the aspartame is destroyed each month. Suggest a reason why aspartame breaks down in this way.

OCR Salters: Polymers, Proteins and Steel Specimen Paper Q2

**7**

**a)** How many different tripeptides can be formed using three α-amino acids, glycine, alanine and valine, if each amino acid is used only once in each tripeptide?

**b)** **(i)** Name two methods by which an unknown tripeptide can be analysed.

**(ii)** For one of these methods outline the experimental procedures and give the information that would be needed to identify the individual amino acids.

IBO Chemistry Standard Level Paper 3 May 2000 QC2

# ? EXAM QUESTIONS

**8** Three naturally occurring amino acids are valine, glutamic acid and cystine (Fig 19.Q8).

CH(CH₃)₂
H₂N—C—COOH
H
valine

CH₂CH₂COOH
H₂N—C—COOH
H
glutamic acid

CH₂SH
H₂N—C—COOH
H
cystine

**Fig 19.Q8**

**a)** Draw a tripeptide composed from each of these three amino acids.

**b)** Describe how a section of a protein containing these three amino acids can contribute to ordered secondary and tertiary structures of a protein with an α-helix. Your answer should include diagrams and should discuss the relevant bonds and forces that stabilize each structure.

OCR 2815/02 Biochemistry Specimen Paper Q4

# KEY SKILLS **ASSIGNMENT**

## ASPARTAME – A MULTIMILLION POUND SWEETENER

Many of us are sweet-toothed and particularly enjoy the sweetness provided by carbohydrates – in particular, sucrose. Sadly, over-indulgence usually makes us obese. Carbohydrates, such as the sucrose of table sugar, can provide too much energy for the body's immediate use, so it is stored as fat. Hence the attraction of artificial sweeteners, such as aspartame, which provide the sweetness without the excess energy.

Aspartame was discovered in 1965 by James Schlatter, while he was doing research on anti-ulcer drugs. It is a methyl ester of a dipeptide and weight for weight is about 160 times as sweet as sucrose. Aspartame is one of the biggest-selling artificial sweeteners on the market today.

**Fig 19.A1** Aspartame

**1** **a)** Use Fig 19.4 on page 413 to identify the two amino acids that make up this dipeptide.
**b)** Copy the aspartame molecule in Fig 19.A1. On your diagram:
(i) circle the peptide group and the ester functional group;
(ii) identify the two chiral carbon atoms using an asterisk.
**c)** Draw the structural formulas of the ions that are likely to exist in:
(i) an acidic solution;
(ii) an alkaline solution;
(iii) a neutral solution.

**2** Another dipeptide molecule is possible if the two amino acids you identified in question **1a)** react differently. Draw the structural formula of this dipeptide.

It is the shape of aspartame that gives it its sweetness. Our taste buds contain protein receptors into which aspartame can fit and form hydrogen bonds. It is the formation of the hydrogen bonds that is thought to create the sensation of sweetness.

**3** **a)** Which groups on the aspartame molecule could form hydrogen bonds?
**b)** In order to form a receptor site, the protein must have a particular shape:
(i) Explain what is meant by primary structure and secondary structure.

(ii) State the types of force that maintain each structure.
Aspartame has a limited shelf life and its molecules decompose at room temperature at the rate of about 10 per cent per month. It is therefore used to sweeten foods with a quick turnover, such as soft drinks and yoghurts. At higher temperatures, the rate of decomposition is much faster, so aspartame cannot be used in cooking.

**Fig 19.A2** Some products that contain aspartame

Aspartame is believed to be a safe food additive. Unfortunately, about 1 in 15 000 people have a genetic condition called phenylketonuria, which means they cannot metabolize the essential amino acid phenylalanine contained in aspartame. Most people have an enzyme called phenylalanine hydroxylase, which converts phenylalanine into another amino acid, tyrosine, which is harmless.

However, in people who suffer from phenylketonuria, the enzyme is defective and so phenylalanine cannot be converted into tyrosine. Though it is eventually excreted, it accumulates in the blood. If not diagnosed and treated, the condition can cause mental retardation.

**4** **a)** Draw the general formula for an $\alpha$-amino acid.
**b)** The R group of tyrosine is shown in Fig 19.A3. Draw this amino acid.

$$-CH_2-\bigcirc-OH$$

**Fig 19.A3** The R group in tyrosine

**c)** Phenylalanine is an essential amino acid. What is meant by this term?

# 20

# Polymers

# Polymers

**Uses for Biopol** Biopol bottles at stages of increasing biodegradation

**Uses for Bioplol** A credit card made from Biopol

Most polymers that revolutionize our everyday living are manufactured from chemicals that come from non-renewable crude oil. As the reserves of this precious resource are used up, the hunt is on for other sources of polymers, and growing crops that provide them is one promising option. We are all familiar with the natural polymers starch, cellulose and protein; certain bacteria produce other polymers that are the subject of intense research throughout the world.

Biopol, a polyester, is the world's first bacterial polymer to be produced commercially. The polymer makes up 80 per cent of the dry weight of the bacterium *Alcaligenes eutrophus*, which uses it as an energy store in the same way that we use fat.

The first articles made of Biopol, bottles for hair-care products, appeared in 1990 and it is now finding uses in coating paper bags and in producing disposable cutlery. Greenpeace, the environmental group, has used it for its affinity credit card rather than the commonly used polyvinyl chloride (PVC). At present, bacterial polymers are much more costly to produce than synthetic polymers, but they have one outstanding advantage – they are totally biodegradable. In a world in which the disposal of synthetic polymers is reaching crisis levels, as landfill sites become scarce and the incineration of plastic waste adds to atmospheric pollution, this is a very important virtue. Even if it is incinerated the only combustion products are carbon dioxide and water and it releases more energy in the process than its synthetic counterparts.

The genes that produce Biopol have been isolated from the bacterium, and bioengineers are hoping to implant them in potatoes and other crops. So, perhaps fields that produce the polymers essential to modern living will be a familiar sight by the mid twenty-first century.

## 1 WHAT IS A POLYMER?

FOR MORE INFORMATION ON PROTEINS, SEE PAGES 418 TO 424.

More chemists work with polymers than with any other type of material. Polymers enter every aspect of our lives. The DNA that carries the genetic code is a natural polymer, as are proteins, starch and cellulose. Chemists started to make synthetic polymers in the early 1900s and now we are dependent on them. They are used in clothing, packaging, furniture, adhesives, inks, coatings and electrical equipment, to mention just a few.

**Polymers** are very large molecules with very high molecular masses. It took chemists some time to realize that they were dealing with such large molecules. The German chemist Hermann Staudinger, working in the early 1900s, was the first to realize that compounds such as rubber were not just collections of tiny molecules held together by intermolecular forces, but were made up of huge molecules that contained many thousands of atoms covalently bonded together. It was thought at the time that a molecular mass of 5000 was the limit for any molecule, so it took some time for Staudinger's model of a polymer to become accepted.

IN 1953, HERMANN STAUDINGER (1881–1963) WAS AWARDED THE NOBEL PRIZE IN CHEMISTRY FOR HIS PIONEERING WORK WITH POLYMERS.

The word polymer is derived from two Greek words: *polys*, which means many, and *meros*, which means a part. Polymers, then, are large molecules made up of many small molecules called **monomers**. The process of forming polymers from monomers is called **polymerization**. **Plastics** – derived from the Greek word *plastikos*, which means shaped – is the general name for materials that, when heated under pressure, can be shaped or moulded or extruded (extruded means forced through a hole, for example to make rods or tubes). Not all polymer materials possess this property, so to avoid confusion the word 'plastics' is not used until environmental issues are considered on page 442.

## 2 ETHENE AND ADDITION POLYMERIZATION

Poly(ethene) – or polythene as it is popularly known – was discovered by accident in 1933. Two British chemists working at ICI, Eric Fawcett and Reginald Gibson, were trying to produce a ketone using benzaldehyde and ethene under very high pressure (Fig 20.1). They decided to leave the reaction mixture at high pressure over a weekend, even adding more ethene when the reaction vessel sprang a leak.

benzaldehyde          ethene

**Fig 20.1** The reaction Fawcett and Gibson were expecting

By Monday morning, the hoped-for ketone had not been produced. Instead, there was a minute amount of a white waxy solid. They tried to repeat the experiment, but the reaction mixture exploded. The unpredictability of the reaction meant that making this unusual substance was very hazardous, so they stopped.

However, it was eventually realized that this was a polymer. Interest grew, and in 1935 Fawcett and Gibson made another attempt to produce it, this time with reactor vessels that could withstand very high pressures. By December, they had made 8 grams. They continued their experiments to find the optimum conditions. When they added cold ethene at a controlled rate, the reaction stopped going out of control. Soon they had enough solid polymer to establish that it could be melted and moulded.

Another British chemist, Michael Perrin, who took charge of the project in 1935, realized that oxygen was essential to produce this white waxy solid, polythene. (It is probable that oxygen had entered the original reaction vessel when it sprang its leak.) Perrin also showed that polythene was produced without the benzaldehyde of the 1933 experiment.

### ? TEST YOURSELF

**A** If the reaction between benzaldehyde and ethene had worked, it would have been an addition reaction. Explain what is meant by addition reaction.
Hint: To check your answer, see page 311.

### ! SCIENCE FEATURE Polythene and the Second World War

The claim is often made that penicillin, the first antibiotic, was one of the chemicals that won the Second World War because it prevented terrible suffering and needless death from wound infections for thousands of soldiers. Polythene, too, had a valued role in the war. It was commercially produced in 1939,

three years before penicillin. Unlike other insulators, it was impervious to rainwater and sea water, and one of its first uses was to insulate the electrical wiring of aircraft radar sets – radar being crucial to Britain's survival in the war's early years.

**? TEST YOURSELF**

**B** **a)** Draw the displayed formula of part of a poly(ethene) chain that shows three ethene units.
**b)** Explain why poly(ethene) may be classed as an alkane.
Hint: See page 212 to remind yourself about alkanes.

**THERE IS MORE ABOUT ADDITION REACTIONS AND ADDITION POLYMERIZATIONS ON PAGES 311 AND 316.**

**SEE QUESTION 1.**

**REMEMBER THIS**

Free radicals are reactive species with an unpaired electron (see page 266).

**? TEST YOURSELF**

**C** The formation of the free radical from organic peroxides involves homolytic fission. What is meant by this term?
Hint: If you are not sure, refer to page 267.

**REMEMBER THIS**

A half-headed curly (fish-hook) arrow is used to show the direction of shift of a single electron. The direction of shift of an electron pair is shown by a full-headed curly arrow.

**Fig 20.4** Sometimes a chain folds back on itself and shifts the position of the unpaired electron. This allows branching to occur

**Fig 20.5** Letters $m$ and $n$ refer to the number of $CH_2$ groups in the chain which may be very large

Poly(ethene) is called an **addition polymer** because, when the ethene monomers join together in the reaction, no small molecules are eliminated: the polymer is the only product, as Fig 20.2 shows.

**Fig 20.2** Polymerizing ethene

In the equation of Fig 20.2, $n$ represents a very large number. Poly(ethene) chains can be made up of thousands of ethene monomers, hence its systematic name. In the formula for the poly(ethene), the ethene unit is enclosed in brackets which cut the covalent bonds to show that we are representing, the **repeating unit** of a very long chain.

## MECHANISM OF ADDITION POLYMERIZATION

Addition polymerization is a free-radical chain reaction and, as for any such reaction, there must be an **initiation stage**. In this stage, a free radical is formed. Free radicals are often produced from organic peroxides because the covalent bond in the peroxide (O–O) group is weak:

$$R–O–O–R \rightarrow 2R–O\bullet$$

The R-group could be an alkyl group, but a common reagent used to produce free radicals is benzoyl peroxide, $(C_6H_5CO)–O–O–(COC_6H_5)$.

Once formed, this free radical goes on to form another free radical, which starts a chain reaction. This is called the **propagation stage** (Fig 20.3).

**Fig 20.3** The propagation stage in this free-radical addition polymerization reaction

Very long chains that contain thousands of ethene units are produced in less than a second, which is why the reaction can easily go out of control and explode.

Sometimes a large chain folds back on itself and shifts the position of the unpaired electron to the middle portion of the chain by removing a hydrogen atom from a $CH_2$, as shown in Fig 20.4. This means that branched chains can form, which is what happened in Gibson and Fawcett's experiment. Branching gives polythene particular properties: in particular, it affects the density.

The **termination stage** of this chain reaction occurs when two free radicals combine without producing new free radicals (Fig 20.5).

**EXTENSION** Types of poly(ethene) and poly(propene)

## OXYGEN MOLECULES ARE ALSO FREE RADICALS

Oxygen might have initiated the poly(ethene) chain reaction in Gibson and Fawcett's original experiments because it can also exist as a free radical. The model used for bonding in $O_2$ is based on the production of octets of electrons (page 73). In the case of oxygen, this gives a double bond (Fig 20.6).

**Fig 20.6** Model of bonding in oxygen as octets of electrons

The bonding in $O_2$ is known to be more complicated than this. Each O atom in the molecule has an unpaired electron, which makes oxygen a free radical, sometimes known as a diradical (Fig 20.7)

**Fig 20.7** Oxygen is a diradical

However the bond between the O atoms is more complicated than just a single covalent bond.

## LOW-DENSITY POLY(ETHENE)

We have seen that the process invented at ICI to manufacture polythene produces branched chains. This means that the polymer chains do not all lie in parallel, as unbranched molecules could. Instead, they form a tangled mass (Fig 20.8). As the chains cannot lie close together, the density of this polythene is low. It is therefore called **low-density poly(ethene)** or **LDPE**.

**Fig 20.8** Chains of low-density poly(ethene)

The intermolecular forces that hold the poly(ethene) chains together are induced dipole–induced dipole forces (see page 215). As the chains get longer, these weak forces act over a larger surface area. And as they become tangled, the tensile strength increases because when one chain is pulled it drags other chains along with it. However, because the branching keeps the chains further apart, LDPE has a lower tensile strength than polymers with no branched chains. Branching also keeps the melting point fairly low, at about 130 °C.

Some 6 million tonnes of LDPE are manufactured worldwide each year under very high pressures and with a trace of oxygen or an organic peroxide. Among the uses of LDPE are electrical insulation, tough transparent film in packaging, dustbin liners and carrier bags (Fig 20.9).

**Fig 20.9** Hot low-density poly(ethene) being blown to form a bag

## HIGH-DENSITY POLY(ETHENE)

Linear chains of poly(ethene) – chains with no branching – can lie closer together than branched chains. This increases the density. **High-density poly(ethene), HDPE**, was first produced in 1953 in Germany by Karl Ziegler while he was experimenting with an organometallic catalyst. This catalyst consisted of ethyl groups covalently bonded to aluminium atoms (Fig 20.10).

**Fig 20.10** Triethylaluminium is an organometallic catalyst

Ziegler found that a little titanium(IV) chloride, $TiCl_4$, mixed with triethylaluminium in an alkane solvent caused ethene to polymerize at room temperature and atmospheric pressure. Long linear chains of about 100 000 ethene units were formed. Today, the world production of HDPE is 7.5 million tonnes. The conditions for its production have changed little.

$$n C_2H_4 \xrightarrow[50\,°C,\ 1.5\ atm]{(C_2H_5)_3Al/TiCl_4} \ \text{---}CH_2\text{--}CH_2\text{---}_n$$

The mechanism for the addition polymerization of HDPE is ionic, not free radical.

Being linear, HDPE chains can form ordered structures by aligning themselves. This leads to large regions of **crystallinity** within HDPE (Fig 20.11), which give it a higher density than LDPE. The melting point and tensile strength are also higher, and HDPE is harder than LDPE. The molecular regions that are not ordered are known as **amorphous** regions.

areas of crystallinity

**Fig 20.11** The crystalline regions in HDPE result from linear chains that become aligned

*(Continued on page 434)*

▶ **EXTENSION Types of poly(ethene) and poly(propene)** *(continued from page 433)*

Almost half of the HDPE produced is blow-moulded to make bottles and other containers for a range of chemicals from shampoos to bleaches. Another quarter is injected into moulds (injection moulding) to make food storage containers, car petrol tanks, buckets and crates. Medical appliances, such as bedpans, are made from HDPE because its higher melting point allows high-temperature sterilization without loss of shape.

**Fig 20.12(b)** Two bottles made from poly(ethene). The one on the right was made from LDPE, the other from HDPE

**Fig 20.12(a)** Blow-moulding a bottle

### POLY(PROPENE): ANOTHER ADDITION POLYMER

The equation for the addition polymerization of propene is given in Fig 20.13.

**Fig 20.13** The polymerization of propene

If all the methyl groups are oriented in the same direction, the polymer chains can get very close together and be crystalline. A year after Ziegler had used $(C_2H_5)_3Al/TiCl_4$ to catalyse the low-pressure polymerization of ethene, the Italian chemist Giulio Natta used the same catalyst to produce crystalline poly(propene) in which all the methyl groups pointed in the same direction (Fig 20.14).

**Fig 20.14** Isotactic poly(propene). Notice that the methyl groups are all pointing in the same direction

At the suggestion of his wife, he called this **isotactic** (poly)propene, after the Greek words *isos*, which means equal, and *taktos*, which means ordered. This was the first time catalysts had been used to produce a **stereoregular polymer**. Catalysts used to produce stereoregular polymers are called **Ziegler–Natta catalysts**. In 1963, Natta and Ziegler jointly received the Nobel Prize for Chemistry for their contributions to polymer chemistry. **Atactic** poly(propene) has no ordered orientation of its $CH_3$ groups, which means that the polymer chains cannot lie very close together. As would be expected, atactic poly(propene) has a lower melting point and is softer than the isotactic form. Isotactic poly(propene) is moulded into objects such as car bumpers and battery cases, and drawn into fibres for carpets and clothing. It is also made in sheet form for packaging. Isotactic poly(propene) is particularly useful for athletics wear because it does not absorb perspiration, but allows it to evaporate, unlike cotton clothing which absorbs and holds moisture. The atactic form is used in weather-proofing materials and sealants.

**Fig 20.15** Atactic poly(propene): the random orientation of the methyl groups prevents the polymer chains lying close together

---

### ? TEST YOURSELF

**D** After reading the Extension box on Types of poly(ethene), explain why linear polymer chains have higher melting points than branched chains that contain similar numbers of carbon atoms.

**E** There is another stereoregular form of poly(propene) called **syndiotactic** poly(propene). It is used for medical tubing. In this form, the $CH_3$ groups are on alternate sides:
**a)** Draw a portion of its chain structure using similar diagrams to Figs 20.14 and 20.15.
**b)** Predict which properties might be different from atactic poly(propene).

**SEE QUESTIONS 2, 3 AND 4.**

## PVC AND ADDITIVES

Look at Table 14.2 (page 317) in which examples are given of other addition polymers whose monomers are based on the ethene structure and add together in a similar way. PVC or poly(chloroethene) is made from chloroethene monomers, as shown in Fig 20.16.

**Fig 20.16** The polymerization of chloroethene

The non-systematic name for chloroethene is vinyl chloride, hence the name by which the polymer was first known: polyvinyl chloride or PVC. When first made in 1912 in Germany, PVC was hard and brittle. These properties come from the chlorine atom.

Being highly electronegative, the chlorine atom forms a dipole with the carbon atom and the partial negative charge on the chlorine atom attracts a slightly positive hydrogen atom, which also forms a dipole with the carbon atom. This gives rise to permanent dipole–permanent dipole forces which are stronger than induced dipole–induced dipole forces, making PVC stronger than poly(ethene).

The brittleness arises because the larger chlorine atoms tend to catch on each other when the polymer chains are pulled apart. Add to this that PVC often decomposed before it could be moulded, it is small wonder that the German company allowed its patent to lapse in 1926.

**SEE QUESTION 5.**

However, in the same year it was discovered that certain additives change the properties of PVC, making it more flexible and easier to mould. As a result, PVC is now the world's most versatile polymer and second only to poly(ethene) in the amount produced.

The additives that make PVC more flexible and softer are called plasticizers. Plasticizers are molecules that get in between the chains, and enable them to slide over one another more easily. There is, however, concern that plasticizers in PVC food wrapping film may migrate into fatty foods and be harmful.

Another additive prevents the decomposition of PVC by ultraviolet light. Hence, the polymer used in door and window frames has this additive and is called uPVC.

**Fig 20.17** The non-glass surfaces of these flats are made of uPVC

The use of additives to alter its properties means that PVC is found everywhere. Credit cards are made from it, as are precision engineering items, flooring and footwear. If your house has polymer gutters, down-pipes and plumbing, the material is almost certainly PVC.

### ? TEST YOURSELF

**F** Poly(phenylethene) is the systematic name of polystyrene. Part of the polymer chain is shown in Fig 20.18. Draw the structural formula of the monomer.
Hint: Check your answer by looking at page 317. There is also more information about this important polymer on page 337.

**Fig 20.18** Poly(phenylethene)

**SEE QUESTIONS 6 AND 7.**

### ▶ EXTENSION Glass transition temperature

Most polymers contain some crystalline regions and some amorphous regions. Below a certain temperature (called the glass transition temperature, $T_g$) the long polymer chains are fixed in position and cannot move over one another. As the polymer warms up, the chains in the amorphous regions are able to slide over one another and so the polymer softens and becomes more flexible. But the crystalline regions stay rigid and do not start to move until the polymer melts.

Poly(propene) bumpers on cars become brittle in very cold winters when the temperature falls below its $T_g$ of –10 °C. Even a slight knock can cause a bumper to shatter. Another example is poly(propene) food containers kept in a refrigerator. They, too, can become brittle and so crack when someone tries to open them.

**?** **TEST YOURSELF**

**G** The displayed formula of tetrafluoroethene is shown in Fig 20.20. Write an equation for the production of this polymer.

$$\begin{matrix} F & & F \\ & \diagdown & \diagup \\ & C=C \\ & \diagup & \diagdown \\ F & & F \end{matrix}$$

**Fig 20.20** Tetrafluoroethene

**!** SCIENCE FEATURE  Teflon: the non-stick polymer

Teflon was discovered by accident in 1938 when Roy Plunkett, a chemist working for the Du Pont company on tetrafluoroethene gas, could not get any of the gas from one of the steel storage cylinders he wanted to use. Instead of jumping to the conclusion that the gas must have escaped, he weighed the cylinder and found that it weighed the same as if it were full. He sawed it open and found a white powder, which turned out to be poly(tetrafluoroethene) or PTFE – better known as Teflon. Ten years later it was in commercial production.

Teflon has a virtually friction-free surface, is very resistant to heat and chemicals, and is an excellent electrical insulator. Apart from its use as a coating on non-stick frying pans (Fig 20.19), Teflon is used to coat bearings (even those that support loads as heavy as bridges) and in human joint replacements. Unlike other polymers, its properties remain constant over a wide temperature range (–70 to 350 °C).

**Fig 20.19** Teflon makes this frying pan non-stick

## 3 CONDENSATION POLYMERIZATION

In 1928 a brilliant young American chemist, Wallace Carothers, was invited to join the Du Pont company to head a team researching into polymers. Here he made an immense contribution to our understanding of polymer science, which was probably his most valuable work. However, this is not what posterity remembers him for, because in 1935 he produced the first nylon – a wholly synthetic fibre that mimics the protein silk.

**!** SCIENCE FEATURE  Wallace Carothers

Working at the Du Pont company in Delaware, Wallace Carothers led the research team that produced the world's first nylon. He was an internationally renowned expert on polymers, and his theoretical ideas are the foundation of today's polymer science. He was the first industrial chemist to become a member of the prestigious National Academy of Sciences in the USA.

However, his success masked the depression that had plagued him since childhood and, in 1937, feeling that his life's work had been a failure, he committed suicide by drinking a solution of cyanide. Yet, only three years later, nylon was proclaimed an outstanding commercial success.

**Fig 20.21** Wallace Carothers (1896–1937)

Within three years of Carothers joining Du Pont, his group produced the first commercial synthetic rubber, Neoprene, an addition polymer (*neo* means new). Then they switched their attention to **condensation polymerization**. (When monomers form condensation polymers, small molecules, such as water, are eliminated.) Carothers' group investigated two different types of polymer: **polyesters** and **polyamides**. Their first success was the production of a polyester fibre – the world's first wholly synthetic fibre.

# POLYESTERS

Carothers realized that for condensation polymers to be formed, the monomers that make them up need two reactive ends. So monomer A (Fig 20.22) could contain two OH groups, while monomer B could contain two COOH groups, to give polyesters with monomer units joined by ester functional groups.

BEFORE READING ABOUT POLYESTERS YOU WILL FIND IT HELPFUL TO TURN TO PAGE 394 AND REVISE HOW ESTERS ARE MADE.

$$H-O-\overset{\overset{O}{\|}}{C}-A-\overset{\overset{O}{\|}}{C}-O-H \ \ H-O-B-O-H \ \ H-O-\overset{\overset{O}{\|}}{C}-A-\overset{\overset{O}{\|}}{C}-O-H \ \ H-O-B-O-H$$

$$\cdots-\overset{\overset{O}{\|}}{C}-A-\overset{\overset{O}{\|}}{C}-O-B-O-\overset{\overset{O}{\|}}{C}-A-\overset{\overset{O}{\|}}{C}-O-B-O-\cdots$$

$$H_2O \qquad H_2O \qquad H_2O$$

**Fig 20.22** Forming ester links between monomers in a polyester

**REMEMBER THIS**

| carboxylic acid | + | alcohol | ⇌ | ester | + | water |
|---|---|---|---|---|---|---|

$$R-\overset{\overset{O}{\|}}{C}-OH \ + \ R'-OH \ \rightleftharpoons \ R-\overset{\overset{O}{\|}}{C}-O-R' \ + \ H_2O$$

**? TEST YOURSELF**

**H** Write an equation for the formation of ethyl ethanoate from ethanoic acid and ethanol.

**REMEMBER THIS**

The $M_r$ of polymers is an average value, since not all the polymer chains have the same $M_r$.

The polyesters that the Du Pont team produced were thought to be of theoretical interest, but of little practical use. The water eliminated in the condensation reaction appeared to prevent the formation of very long chains.

This problem was solved by Carothers, who invented a 'molecular still' that evaporated the water molecules as they were produced. This enabled very long polymer chains to be produced with $M_r$ values of about 10 000. However, the polyesters made of these were just sticky masses when hot, and tough, opaque solids when cold. Then, in a happy accident, one of the team, Julian Hill, pulled a stirring rod out of a hot, sticky ball of polyester. The result was a long, thin fibre of the polymer that, when it cooled, could be stretched considerably and was very strong.

X-ray analysis revealed what had occurred. Pulling the polymer into a long filament had aligned the polymer chains, which increased the tensile strength of the material (Fig 18.24). The pulling process is known as cold drawing.

**Fig 20.23** Julian Hill demonstrates the birth of the first completely synthetic fibre

crystalline areas          cold drawing aligns the polymer chains, increasing crystallinity and tensile strength

**Fig 20.24** Cold drawing of a polymer aligns its chains in the fibre produced

The polyester fibres that the Du Pont group led by Carothers produced were not destined for commercial success. Although the fibres were strong and pliable, they melted at too low a temperature to be of practical use in clothing, which would have to be ironed. They were also slightly water soluble, another considerable disadvantage!

**REMEMBER THIS**

Examples of happy accidents occur throughout the development of science. Another name for this is **serendipity**. However, it is recognizing the implications of serendipity that marks out great scientists. It was Louis Pasteur who said, 'Chance favours the prepared mind.' Pasteur's serendipitous discovery is featured on page 414.

**I a)** The dimethyl ester of benzene-1,4-dicarboxylic acid also produces Terylene.
(i) Draw the structural formula of this molecule.
(ii) Predict the small molecule that is lost during the condensation polymerization reaction with ethane-1,2-diol.
(iii) Predict the products of the hydrolysis of this type of Terylene by referring to pages 395–397.
**b)** Another polyester can be produced using benzene-1,4-dicarboxylic acid and propane-1,3-diol. Draw the structural formula of the repeating unit of this polymer.

**J** Terylene and similar polyesters are water repellent. Why is it an advantage that, when blended with cotton fibres, the resulting fabric should absorb moisture?

**K** Polyesters also find many uses in medicine. For example, as mesh tubes they can replace blood vessels, with human tissue growing into and around the mesh. While sometimes a permanent mesh is required, it is an advantage at other times to insert a mesh that can be degraded by enzymes in the body. A polymer made from 2-hydroxypropanoic acid (lactic acid, Fig 20.27) forms a degradable mesh.

CH₃
|
HO—CH—COOH
**Fig 20.27** 2-Hydroxypropanoic acid

Write an equation for the production of this polyester.

**Fig 20.28** The secondary amide linkage

O
||
—C—N—
|
H

The polyester story now switches to the laboratories of the Calico Printers' Association in England, where in 1941 J Whinfield and J Dickson invented Terylene, building on the foundations laid by Wallace Carothers. This polyester is produced from benzene-1,4-dicarboxylic acid and the alcohol ethane-1,2-diol (Fig 20.25). The systematic name for this polyester need not concern us. In industry, it is usually referred to simply as polyester or PET.

**Fig 20.25** Condensation reaction for the polyester Terylene

The names Terylene and PET are derived from the traditional names for the two monomers and the polymer, respectively. Benzene-1,4-dicarboxylic acid is also called **tere**phthalic acid and ethane-1,2-diol is also called ethyl**ene** glycol – hence Terylene. The traditional name for the polymer is **p**oly(**e**thylene **t**erephthalate), which gives the acronym PET.

Polyester is the leading synthetic fibre with worldwide production of about 10 million tonnes. Its principal use is in clothing, because it is crease-resistant. Often, it is mixed with other synthetic or natural fibres. For example, a mixture of cotton and polyester is very popular because the cotton absorbs moisture. Since polyester is a good thermal insulator, duvets and anoraks are filled with its fibres. Apart from fibres, it can be produced as an extremely tough yet very thin film, which provides the backing for videotapes and computer discs. It is also becoming the dominant material in packaging and is used in some carbonated drinks bottles.

**Fig 20.26** PET is an ideal material for carbonated drinks bottles because it is light, tough and won't shatter when dropped

## NYLON: A POLYAMIDE

After working on polyesters, Carothers and his team focused their efforts on another type of polymer, in which the monomer units are joined through secondary amide linkages (Fig 20.28), instead of ester linkages. These secondary amide linkages are the same linkages as those found in proteins such as silk.

In 1935, the team produced a polymer, now called nylon 6,6, from a diamine and a dicarboxylic acid (Fig 20.29). Notice the similarity of this condensation reaction to that which produces a polyester. Both use a dicarboxylic acid monomer, but the alcohol groups at each end of the other monomer are replaced with primary amine functional groups ($NH_2$).

**Fig 20.29** The reaction of Carothers and his team that gave rise to nylon 6,6

Nylon 6,6 is not the only polyamide that Carothers' team produced, and so each is distinguished by a pair of numbers unique to that nylon. The first number after the word nylon is the number of carbon atoms in the diamine. The second number is the number of carbon atoms in the dicarboxylic acid.

When nylon 6,6 is cold drawn, it is stretched to four times its original length, which gives the fibres so produced a high tensile strength and elasticity. As a fibre is drawn, the polymer chains align parallel to one another and the tensile strength comes from the hydrogen bonding between the CO and NH groups of adjacent chains (Fig 20.30). The cold drawing also increases the lustre (shininess) of nylon. In a commercial nylon plant, molten nylon is forced through tiny holes to produce the fibres, which are then cold drawn (Fig 20.31).

**Fig 20.30** Part of two linear chains of nylon, showing intermolecular hydrogen bonding

Du Pont went from the laboratory preparation of nylon 6,6 to its manufacture in the remarkably short time of less than five years. By 1937 a pilot plant was operational, and by the end of 1939 a full-scale plant had been built and was in production – all this when the techniques of large-scale condensation polymerization were unknown and there were no bulk supplies of either monomer. Add to this the different technologies required to spin nylon fibres, and we have a remarkable achievement.

The first article to contain nylon was Dr West's Miracle Toothbrush, in which nylon bristles replaced animal bristles. Meanwhile, Du Pont had test-marketed women's nylon stockings and recognized their enormous sales potential. Skirt lengths had become shorter and silk stockings were highly fashionable, but were very expensive. When nylon stockings became widely

**? TEST YOURSELF**

**L** Nylon 5,10 is another polyamide produced by Carothers and his team. He favoured this as the nylon to be developed commercially, but he was overruled and nylon 6,6 became the first nylon to be manufactured. Write the structural formulas of the monomers used for nylon 5,10.

**? TEST YOURSELF**

**M** With reference to page 420, what are the products of the complete hydrolysis of nylon 6,6? What reagents and conditions could be used to carry out this reaction in the laboratory?

**SEE QUESTIONS 8 AND 9.**

granules of monomer

molten monomer undergoes condensation

tiny holes in spinneret

molten polymer forced through spinneret forms fine fibres

**Fig 20.31** The spinneret through which molten nylon is forced before being cold drawn

▶ **EXTENSION Thermoplastic and thermosetting polymers**

All the polymers discussed so far have one property in common – they are **thermoplastic**. Thermoplastic polymers soften when heated and can be moulded. The process of heating and moulding can be repeated many times. At a molecular level, the forces between the polymer chains are weak, such as induced dipole–induced dipole forces or hydrogen bonds (Fig 20.32(a)). These are sometimes referred to as **secondary bonds**. When a thermoplastic polymer is warmed, these forces are broken and the polymer chains can slide over one another, which allows the polymer to be shaped. When the polymer cools, the forces reform with the chains in their new positions.

This ability to be remoulded makes these polymers very versatile and therefore in much demand. Consequently, the annual world production of thermoplastic polymers is seven times that of thermosetting polymers.

**Fig 20.32(a)** Diagrammatic representation of a thermoplastic showing the weak intermolecular forces between the polymer chains, which can be overcome on heating to allow the chains to slide over one another. On cooling, the intermolecular forces reform

**Fig 20.32(b)** Diagrammatic representation of a thermosetting polymer. The covalent bonds form a three-dimensional network that, once formed, cannot be melted again

**Thermosetting polymers** can be moulded only when they are first produced. With strong heating, they decompose and cannot be softened and remoulded. A thermosetting polymer has covalent cross-links between its chains, so the polymer is a three-dimensional network of strong covalent bonds (Fig 20.32(b)). On heating, these break, which causes the polymer to decompose before melting can occur.

## BAKELITE: THE FIRST SYNTHETIC POLYMER

Leo Baekeland, a Belgian who had emigrated to the United States, invented the first wholly synthetic polymer, which he called Bakelite. He made it from the reaction of phenol with methanal. Bakelite was first manufactured in 1910.

The initial reaction involves a substitution in the benzene ring of phenol, which can be at either position 2

**Fig 20.33** 78 r.p.m. records were made from Bakelite

or position 4 (Fig 20.34(a)). Once formed, the product can react with another molecule of phenol with the elimination of a water molecule (Fig 20.34(b)). The three-dimensional network is then built up through a series of similar reactions (Fig 20.34(c)).

**Fig 20.34** Reactions in the formation of Bakelite

As it has a high number of cross-links, Bakelite forms a hard, rigid structure. Like most polymers, it is a good electrical and heat insulator. It was widely used in all types of articles, among them electric sockets and plugs, pan handles, and even music records, before PVC superseded it. But phenol–methanal resins are still important.

## ❓ TEST YOURSELF

**N** **a)** Is Bakelite: (i) a condensation polymer or an addition polymer?
(ii) a thermoplastic or a thermosetting polymer? **b)** What properties of Bakelite would make it a suitable material for pan handles?

available in 1940, the commercial success of nylon was assured. However, there was not enough production capacity to supply the consumer market and the United States war requirements, for nylon was in great demand for ropes and, in particular, for parachutes, which had previously been made from silk.

Nylon stockings were rationed in the US until the end of the Second World War, and it took until the early 1950s before production capacity in Western Europe matched the insatiable demand for nylon stockings and other nylon goods. Today, nylon goods account for 95 per cent of the women's hosiery market.

A variety of machine parts are now made from nylon instead of metals. For these, its properties of toughness, strength and abrasion resistance are much in demand, and specific properties are enhanced by the use of **fillers**, such as glass fibre. Fillers are widely used to tailor the properties of polymers to specific functions. Fillers are also added to provide bulk and make a cheaper product.

# 4 TWO NATURAL POLYMERS: STARCH AND CELLULOSE

**YOU CAN LEARN MORE ABOUT GLUCOSE BY READING THE CHAPTER 16 ASSIGNMENT, PAGE 362.**

Natural polymers are mentioned in the chapter opener, and proteins, which are polymers of amino acids, are discussed in some detail on pages 419 to 424. Two other important polymers are the polysaccharides, starch and cellulose, both made from glucose monomers. Glucose can exist in two distinct forms, as Fig 20.35 shows.

Note the difference between these two glucose molecules by looking at carbon 1. This seemingly minor difference has a marked effect on the properties of starch and cellulose.

Starch is a polymer of α-glucose monomers. It is linked by α-1,4 linkages (Fig 20.36), also called glycosidic linkages. This form of starch is called amylose and is a straight-chain polymer.

**Fig 20.35** The two ring forms of glucose

**Fig 20.36** Starch

**Fig 20.37** Cellulose

We digest starch by hydrolysing it with enzymes to give glucose. Starch is therefore an energy-storage compound, since once we have broken it down to glucose it can be metabolized in the body to give energy. Starch can also be hydrolysed by acids in the laboratory to give α-glucose:

$$(C_6H_{10}O_5)_n + nH_2O \xrightarrow{\text{H}^+(aq)} nC_6H_{12}O_6$$

Cellulose, a linear polymer formed from β-glucose (Fig 20.37), is probably the world's most abundant organic chemical. It has a structural role in plants and is the source of cotton fibre.

In common with most animals, we cannot digest cellulose because we do not have the necessary enzymes to hydrolyse the β-1,4 linkage. However, cellulose forms the fibre we require in our diet.

**? TEST YOURSELF**

**O** Another form of starch, called amylopectin, has branches formed by reaction between the OH groups of carbon 1 and carbon 6. Draw part of this polymer, showing the carbon 1,6 linkage.

**P** Cellulose has considerable strength because its linear chains are able to pack closely together. What intermolecular forces give cellulose its strength?

**SEE QUESTION 10.**

# 5 WHAT DO WE DO ABOUT PLASTIC WASTE?

This problem won't go away, mainly because the polymers we throw away do not degrade (break down). In the UK, most plastic waste goes into landfill (Fig 20.38) and that is where it will stay. So, what are the options?

Most polymers come from non-renewable crude oil, and it takes energy to produce them. In fact, of the energy that goes into making a plastic article, most is used in producing the polymer. So, why not recycle the polymer? The majority of polymers thrown out are thermoplastics, which could be melted down and reused. However, with some exceptions, all kinds of rubbish are thrown away together. It would take much energy and be fairly costly to sort it. Even if plastic waste were kept separate from other household rubbish, the several different types of polymer involved would have to be further separated before they could be reprocessed.

However, research is underway into melting polymers down to form a chemical feedstock from which new polymers can be produced. A project in Japan holds out the prospect of producing petrol from the waste. But to be economically viable and environmentally friendly, any recycling scheme must not use up more energy than it takes to make the polymer in the first place.

Some articles themselves can be reused, which is achieved successfully for PET (polyester) bottles in some countries in Europe. Again, though, the energy consumption of recycling must also be taken into account.

## BIODEGRADABLE PLASTICS

The world's first totally biodegradable polymer in commercial production was Biopol (see page 430). Biopol is made by living organisms, so it is called a **biopolymer**. Starch is a biopolymer that is being developed as a biodegradable packaging material. However, it is brittle and requires the addition of plasticizers. Though water and alcohols can fulfil this function through hydrogen bonding with the polymer chains, they are too volatile. Propane-1,2,3-triol (glycerol) could provide a promising, non-toxic alternative plasticizer.

Apart from biopolymers, there are other ways of making synthetic polymers degradable. For example, cellulose, starch or protein can be impregnated into plastic articles such as poly(ethene) bags. When such an article is buried in a landfill site, microorganisms break down the cellulose, starch or protein. The article then disintegrates. The synthetic polymer chains that remain have an increased surface area, which speeds up their decomposition.

Poly(alkenes) do not biodegrade in landfill sites if they are not impregnated with a substance that makes them biodegrade. However, polyesters and polyamides are hydrolysable at their ester and amide linkages. Enzymes from bacteria or acid catalysts will break the long polymer chains into shorter and shorter fragments, and so eventually they will biodegrade in landfill sites.

Another method with some polymers is to impregnate them with chemical activators so that on exposure to ultraviolet light the polymer chains break down into much shorter chains, which can then be biodegraded. Such polymers are called photodegradable.

There are environmental issues even with degradable plastics. If polymers degrade, then the articles made of them may not be recyclable. Also, there is some concern over the unknown nature of the decomposition products, which may cause more long-term damage to the environment than the original plastics themselves.

**Fig 20.38** Most of the plastic waste in the UK goes into landfill sites

## ? TEST YOURSELF

**Q** Explain why the molecular structure of thermoplastics allows them to be melted and remoulded.

**R** Epoxy resin is a common component of household waste, but it is a major obstacle to plastics recycling because it is a thermosetting polymer. Why is this a problem?

**S** What processes cost energy in recycling a PET coke bottle?

## ? TEST YOURSELF

**T** If plastics biodegrade, the energy of their bonds is not reused. An alternative is to burn plastic waste to give steam, which is used to generate electricity. What are the environmental and energy issues of incinerating the plastics in household rubbish?

**SEE QUESTION 11.**

# SUMMARY

After studying this chapter, you should know:
- Polymers are very large molecules made up of many small molecules called monomers.
- Addition polymers form when monomers react together without the elimination of small molecules.
- Many addition polymers form through a free-radical mechanism.
- Poly(propene) can be manufactured using Ziegler–Natta catalysts to give stereoregular polymers, called isotactic and syndiotactic polymers, which are denser, harder and have higher melting points than atactic poly(propene), which has no ordered orientation of its $CH_3$ groups.

- Condensation polymers are formed from monomers with the elimination of small molecules such as water.
- Polyesters such as Terylene (PET) and polyamides such as nylon 6,6 are synthetic condensation polymers.
- Starch, cellulose and proteins are natural polymers.
- Many synthetic polymers are not biodegradable and so environmental issues surround the disposal or recycling of plastic waste.
- Poly(ethene), PTFE, PVC (poly(chloroethene)) and poly(phenylethene) are all synthetic addition polymers.

# ? EXAM QUESTIONS

**1** The Perspex monomer is shown in Fig 20Q1.

**Fig 20.Q1** The Perspex monomer

It undergoes addition polymerization to form Perspex.
**a)** What is meant by the term addition polymerization?
**b)** Draw the polymer repeat unit for Perspex.

**2** [You may need to refer to Chapter 14 to answer **b)**].
A manufacturer makes poly(propene) from propene monomer, $CH_3CH=CH_3$. The polymer produced has an average relative molecular mass of $1.0 \times 10^6$.
**a)** Draw the full structural formula for propene.
**b)** The reactions of alkenes, such as propene, with bromine can be used as a simple test for the alkene functional group.
**(i)** Describe how you would carry out this simple test on a liquid alkene and say what you would expect to see.
**(ii)** Which structural feature of propene is responsible for its reaction with bromine? Explain how the structural feature is involved in the first stage of the reaction mechanism.
**(iii)** Circle **two** words in the list below that describe the mechanism of the reaction between propene and bromine:

electrophilic radical nucleophilic addition substitution.

**c)** **(i)** Draw a full structural formula that indicates clearly the structure of the repeating unit in poly(propene).

**(ii)** How many monomer units are joined together to give poly(propene) with a relative molecular mass of $1.0 \times 10^6$? [$A_r$ C 12; H 1.]
**d)** Why is the mass referred to as an **average** relative molecular mass?
**e)** Poly(propene) can exist in two forms (Fig 20.Q2):

**Fig 20.Q2**
**(i)** Explain with reasons which of the two forms of poly(propene) you would expect to be more crystalline.
**(ii)** Explain why the more crystalline form can be used to make strong fibres for carpets.
OCR Salters Specimen Paper Minerals to Medicines Q3

**3** **a)** Ethene, $C_2H_4$, can be converted into a polymer called poly(ethene). Draw the structure of part of the poly(ethene) molecule clearly showing the repeating unit.
**b)** Give one use of poly(ethene).
Edexcel Chemistry Specimen Unit Test 2 2000 Q1e(i)

**4** Poly(propene) can be made from propene either by using a Ziegler–Natta catalyst, or by a free-radical mechanism.
**a)** Draw a diagram to show the structure of poly(propene).

## ? EXAM QUESTIONS

**b)** Explain the differences in the properties of the products that result from the two ways of polymerizing propene.

**5** Chloroethene is the feedstock for the manufacture of the polymer poly(chloroethene).
**a)** Draw the displayed formula of chloroethene
**b)** Draw the displayed structure of the polymer poly(chloroethene) showing three repeating units.
**c)** What type of polymer is poly(chloroethene)?
**d)** Poly(chloroethene) can exist in isotactic and atactic forms.
(i) Draw four units of the polymer chain to show these two forms
(ii) What will be the differences in properties of the two forms?

**6**
**a)** Chloroethene, $C_2H_3Cl$, is the monomer from which the important plastic poly(chloroethene), PVC, is made. Draw the displayed formula of chloroethene.
**b)** (i) Complete the **balanced** equation for the formation of poly(chloroethene) from chloroethene, starting: $nC_2H_3Cl \rightarrow$ .
(ii) Poly(chloroethene) can be produced with any degree of flexibility from rigid to pliable. Suggest **two** uses of PVC that demonstrate this.

Edexcel Nuffield Specimen Unit Test 2 Q4a(i),c

**7**
**a)** Phenylethene (styrene) can undergo polymerization to form poly(phenylethene). This polymer is also known as polystyrene.
(i) State the name of this type of polymerization and draw a small section of the polymer.
(ii) Outline the difficulties in the disposal of waste containing this polymer.

**Fig 20.Q7**

**b)** Compound **A** in Fig 20.Q7 can be polymerized. Draw a section of its polymer, showing two repeat units. Suggest a possible problem in its disposal that is additional to the problems of disposing of poly(phenylethene).

**8** Nylon 66 is a polyamide in common use. Two repeat units of this polymer are shown.

$-CO(CH_2)_4CONH(CH_2)_6NHCO(CH_2)_4CONH(CH_2)_6NH-$

It is hydrolysed by boiling with aqueous HCl, followed by careful neutralization.

Which two products from **A** to **D** are formed on this hydrolysis?
**A** $H_2N(CH_2)_6NH_2$ and $CH_3(CH_2)_3COOH$
**B** $H_2N(CH_2)_6NHCOOH$ and $CH_3(CH_2)_3COOH$
**C** $H_2N(CH_2)_6NH_2$ and $HOOC(CH_2)_4COOH$
**D** $CH_3(CH_2)_5NH_2$ and $HOOC(CH_2)_4CONH_2$

UCLES 9253 Multiple Choice Paper 3 Nov 2000 Q30

**9** The repeating units of three polymers are shown in Fig 20.Q9.

**Fig 20.Q9**

**a)** Identify the functional groups that contain an O atom present in **B** and **C**.
**b)** Draw the formulas of the monomers used to prepare **A** and **B**.
**c)** Classify the type of polymerization used to make **A**.
**d)** What holds the repeating units together in polymers?
**e)** (i) Draw a diagram of two sections of **B** to show the major intermolecular force between the strands.
(ii) State the requirements for a strong interaction of this type.
**f)** Identify the major types of intermolecular forces that exist between polymer molecules of **A** and of **C**.
**g)** Polymer **A** is used for laboratory equipment, but polymers **B** and **C** are not suitable for this purpose. Suggest and explain a reason for this difference.

OCR 6826 Polymers and Plastics March 2001 Q1

**10** Describe, using an example in each case, the formation of the following polymers:
**a)** **one** synthetic polymer by addition polymerization;
**b)** **one** *synthetic* polymer and **one** *natural* polymer by condensation polymerization.

OCR 2814 Specimen Paper Chains, Rings and Spectroscopy Q6

**11** Increasing use of plastics is accompanied by issues associated with waste disposal.
**a)** State **two** methods of waste disposal and in each case identify issues, other than cost, associated with its use.
**b)** Give an example of a synthetic polymer that is biodegradable.

OCR 6826 Polymers and Plastics March 2001 Q3

# KEY SKILLS **ASSIGNMENT**

## MORE ABOUT POLYAMIDES

Following their invention of nylon 6,6, the Du Pont team led by Wallace Carothers found that the material had two important properties. One was the high tensile strength of its cold-drawn fibres. The other was its high melting point of about 263 °C.

Shortly after the invention of nylon 6,6, another nylon with similar properties, called nylon 6, was invented by Paul Schlack, working for IG Farbenindustrie in Germany. For nearly 60 years these two nylons dominated the polyamide market, in both fibre and sheet form. However, this dominance is now being challenged by another nylon, bearing the trade name Stanyl. Stanyl has greater tensile strength than both nylon 6,6 and nylon 6, and softens and melts at a higher temperature.

Stanyl is made from 1,4-diaminobutane and hexanedioic acid. It was first prepared by Carothers, but was rejected for further development because it had a low $M_r$ and its properties did not seem at all promising. However, a Dutch company has devised a process whereby Stanyl's $M_r$ is 30 000.

**1** The repeating unit of nylon 6 is shown in Fig 20.A1. Predict the structural formula of its monomer.

**Fig 20.A1** The nylon 6 repeating unit

**2 a)** Write the equation for the production of Stanyl from its monomers.
**b)** Stanyl is a type of nylon. Use the nylon naming system to identify it.
**c)** The $M_r$ of Stanyl is 30 000. Approximately how many monomer units are in the chain.
**d)** Why is the $M_r$ of a polymer an average value?

**3** What is meant by the term cold drawing? Why does this process increase the tensile strength of nylon fibres?

Even though nylons have such high softening and melting points, the temperatures at which articles made of them can be in continuous use is much lower. The maximum working temperature of nylon 6,6 is 120 °C, while that of Stanyl is 150 °C. These temperatures can be increased by using a glass filler. However, the higher the working temperature, the quicker nylons deteriorate and lose the properties on which their industrial and domestic use depends.

Du Pont have been in the forefront of research on aromatic polyamides, now called aramids, for a number of years. Their first commercial success came with an aramid fibre they called Nomex. It is made using the reaction in Fig 20.A2.

**Fig 20.A2** The reaction for making Nomex

Nomex has a melting point above 400 °C. It can be made into fibres, but cannot be moulded like nylon. Its main application is as a fire-resistant material, particularly in lining racing-drivers' suits and firefighters' tunics. Also, it is used in hot-gas filters and is made into a honeycomb paper that, when bonded to other materials, forms flame-resistant building materials.

**Fig 20.A3** Being fire resistant, Nomex is used to line the suits of racing drivers

**4 a)** Name these functional groups: (i) COCl and (ii) $NH_2$.
**b)** What is the name given to the functional group that links the benzene rings in Nomex?
**c)** Explain why Nomex is a condensation polymer.

**5** Another monomer that Du Pont has investigated is given in Fig 20.A4. Draw the repeating unit of this polymer.

**Fig 20.A4** 3-Aminobenzoic acid

The tensile strength of Nomex is an improvement on that of nylon. However, another aramid, called Kevlar, has been

# KEY SKILLS **ASSIGNMENT**

developed from monomers with functional groups in the 1,4 position on the benzene ring (Fig 20.A5).

**Fig 20.A5** Kevlar

Unlike Nomex, Kevlar polymer chains can be aligned to make very strong fibres. A 7 cm diameter steel cable has a breaking strain of about 40 tonnes. An identical Kevlar cable has the same breaking strain, but is five times lighter than the steel cable. Kevlar is embedded in tyres in place of steel reinforcement, which has reduced the weight of a typical truck tyre by 9 kg. However, the market for Kevlar-reinforced tyres has been slow to take off because tyre manufacturers have invested in equipment to produce steel reinforcement, believing that customers would prefer the strength that the word steel implies.

Kevlar is renowned as the padding in bullet-resistant vests (Fig 20.A7). It is also incorporated into aircraft wings, in which strength combined with lightness is essential.

**7 a)** Draw the repeating unit for Kevlar and predict its monomers.
**b)** What is the intermolecular force that holds Kevlar fibres together?
**c)** Why is Kevlar so much stronger than Nomex?
**d)** Why is Kevlar less dense than steel? Hint: Think about what density means.

**8** Establishing the right chemistry to produce fibres with the remarkable properties of Kevlar is only one aspect of its commercial development. After the laboratory stage, what other aspects did Du Pont have to consider to ensure the commercial success of their product?

**Fig 20.A7** A bullet-resistant vest made from Kevlar

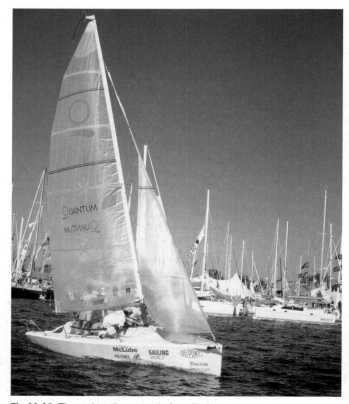

**Fig 20.A8** The yacht sails are made from Kevlar

**Fig 20.A6** Kevlar fibre being twisted into cord for radial tyres at a Goodyear plant

# 21

# Patterns across the Periodic Table

**Angina** When the coronary arteries become narrowed, glyceryl trinitrate can relax them, avoiding the pain of angina

Glyceryl trinitrate tablets have been used for many years to relieve the gripping pain of angina, a heart disease caused by the narrowing of the coronary arteries in the heart.

How glyceryl trinitrate works has been unclear. Now chemists believe they have the answer.

The muscles in the walls of arteries are partly responsible for controlling blood pressure. When the muscles are relaxed, the arteries enlarge and blood pressure drops. Instructions to the artery muscles to relax are carried from the brain by a number of messenger molecules.

One messenger, only recently discovered, is the gaseous oxide of nitrogen called nitrogen monoxide. This is a very small molecule and so is able to diffuse quickly into muscle cells, where it binds with an iron atom in the enzyme responsible for muscle relaxation. The enzyme is activated by the binding, and so the muscle relaxes.

Research evidence suggests that when someone takes glyceryl trinitrate, it is converted into nitrogen monoxide, which is carried to the muscles in the coronary artery and relaxes them. Blood pressure is reduced, thereby relieving the pain of angina.

## 1 PERIODICITY OF PROPERTIES OF THE ELEMENTS

On pages 97 to 99, the periodic classification of the elements is described in relation to some of the properties of the atoms of elements. The similarities between the elements within a group are explained in terms of the number of electrons in the outer electron shell. Similarities in the structure of the outer electron shell lead to a periodicity in the properties of elements, so that the pattern of properties shown by the elements in one period is repeated in the next period. The idea of periodicity is extended in this chapter to include properties of the *compounds* of elements, rather than the elements themselves.

Oxides and chlorides are two of the most common classes of inorganic compounds. Within a single period in the Periodic Table, a bewildering variety of oxides and chlorides occurs – with crystalline solids, materials that look solid but are really supercooled liquids, and colourless gases. Behind this variety, however, is a pattern that we can explain in terms of the structure and bonding of the oxides and chlorides themselves.

This chapter describes and explains the change in the properties of oxides and chlorides of the elements in Period 3 as the atomic (proton) number of the element increases. It relates these properties to both structure and bonding, and to some technological and environmental aspects of chemistry.

### REACTIONS OF ELEMENTS

Almost all the known elements, even some of the noble gases, react to form compounds. When an element reacts, it forms one of the following types of bond: ionic, covalent or dative covalent. This means that there must be a redistribution of electrons around the atom of the element. This redistribution often leads to a particle that is isoelectronic with a noble gas. Some atoms lose electrons to form cations, others gain electrons to form anions, and others share electrons.

When an element reacts, a type of reaction called **redox**, occurs since the oxidation number of an element will necessarily change from zero to either a positive number (with metals and some non-metals) or a negative number (with non-metals). We describe redox reactions in Chapter 5 as a reaction that involves electron transfer and a change in the oxidation number of an element.

## OXIDATION AND REDUCTION

Redox reactions necessarily involve both an oxidation and a reduction. The oxidation number is a most useful way to explain whether a reaction is an example of a redox reaction. During oxidation, the oxidation number of an element increases and electrons are lost. During reduction, the oxidation number decreases and electrons are gained. Do not confuse this with what happens to the oxidizing agent and to the reducing agent. The oxidizing agent can be identified because it contains an element whose oxidation number decreases during the reaction. The reducing agent can be identified because it contains an element whose oxidation number increases during the reaction.

Metals are reducing agents and non-metals are often oxidizing agents. Since metals are positioned on the left-hand side of the periodic table and non-metals on the right-hand side as the atomic (proton) number increases across a period, the elements change from reducing agents to oxidizing agents. Within a period, the best reducing agent is always in Group 1 and the best oxidizing agent is in Group 7.

## REACTION OF ELEMENTS WITH OXYGEN

Many elements react with oxygen in a reaction often called combustion since it involves elements burning. Oxygen is the oxidizing agent and the element will be oxidized, and at the same time oxygen is reduced.

For example, magnesium will burn to form magnesium oxide.

$$2Mg(s) + O_2(g) \rightarrow 2MgO(s)$$

During this reaction the oxidation state of magnesium increases as magnesium atoms lose electrons to form $Mg^{2+}$. These electrons are gained by oxygen to form the oxide ion, $O^{2-}$.

## REACTION OF ELEMENTS WITH CHLORINE

Chlorine is also an oxidizing agent and it will react with many elements to form chlorides. For example, phosphorus will react with excess chlorine to form phosphorus(V) chloride:

$$P_4(s) + 10Cl_2(g) \rightarrow 4PCl_5(s)$$

Phosphorus is oxidized, as its oxidation number increase from 0 to +5; at the same time the oxidation number of chlorine changes from 0 to –1.

**Fig 21.1** Magnesium ribbon burning in air

# 2 PERIODICITY IN CHLORIDES

On page 99, the first ionization energy is described as a periodic function of the atomic (proton) number, with each noble gas having the highest ionization energy within a period. Periodicity is not restricted to the properties of the elements themselves. It is also reflected in the properties of compounds of the elements. By comparing the properties of a range of similar compounds, we can see that the properties are a periodic function of the atomic number of the element in the compound. The two classes of compounds most studied are the oxides and the chlorides of elements. We concentrate here on the chlorides formed by the elements in Period 3, but a similar pattern of properties is shown by the chlorides of elements in other periods.

## CHLORIDES OF ELEMENTS IN PERIOD 3

Chlorine reacts with almost all the elements of this period. The formulas of the chlorides and the oxidation numbers of the elements in the chlorides are given in Table 21.1.

**Table 21.1**

| Element | Formula of chloride | Oxidation number |
|---|---|---|
| Sodium | NaCl | +1 |
| Magnesium | $MgCl_2$ | +2 |
| Aluminium | $Al_2Cl_6$ | +3 |
| Silicon | $SiCl_4$ | +4 |
| Phosphorus | $PCl_3$ | +3 |
|  | $PCl_5$ | +5 |
| Sulphur | $S_2Cl_2$ | +1 |

Table 21.1 clearly shows that there is a predictable change in the oxidation number of each element as the atomic (proton) number increases. It is also evident that the highest oxidation number corresponds to the *group number* of the element involved. This is easy to explain, because each successive element in a period has one more electron available in its outer shell to form a bond with chlorine. Notice that the oxidation state of sulphur in $S_2Cl_2$ does not fit the pattern, but sulphur does form the halide $SF_6$, in which the oxidation number of sulphur is +6.

## MELTING POINTS OF CHLORIDES

Fig 21.3 shows the wide range of melting points of chlorides. There is only one interpretation of this. Namely, the internal structures of the solids are notably different, and the magnitude of the force of attraction between the particles in the different crystal lattices varies widely. We describe here the different crystal lattices in some detail, as well as how each lattice can explain the melting point of the chloride.

**Fig 21.2** Chlorine reacting with aluminium to form white aluminium chloride

**SEE QUESTION 1.**

 **REMEMBER THIS**

The terms 'oxidation number' and 'oxidation state' are interchangeable. Both terms are used in this book.

**? TEST YOURSELF**

**E** The equation for the reaction between sodium and chlorine is:

$$2Na(s) + Cl_2(g) \rightarrow 2NaCl(s)$$

Write down the equations for the formation of all the other chlorides shown in Table 21.1.

**THE DIFFERENT TYPES OF SOLID AND HOW THE SIZE OF THE FORCE OF ATTRACTION BETWEEN THEIR PARTICLES CHANGES THE VALUE OF THE MELTING POINT ARE COVERED ON PAGE 149.**

**Fig 21.3** Melting points of the chlorides of five elements in Period 3

# GIANT IONIC LATTICES

Metals react with chlorine to give ionic chlorides. Metal atoms lose electrons, and chlorine molecules accept them to become chloride ions. There is a strong electrostatic force of attraction between positive ions and negative ions. This force is not in any particular direction, as in the case of a covalent bond. The result is that positive ions attract negative ions to produce a regular pattern of ions in which each positive ion has several negative ions as its nearest neighbours, and each negative ion has several positive ions as its nearest neighbours. Eventually, a giant structure is produced that consists of the regular repetition of a unit cell in three dimensions.

The exact nature of the unit cell, and the coordination number of the positive and negative ions, depend on certain geometric considerations, the ionic radii, the formula of the ionic compound (the ratio of the number of positive to negative ions) and the magnitude of the forces of attraction, which is related to the charge on the ions.

In Period 3 both magnesium chloride and sodium chloride form a giant ionic lattice. The properties of these two compounds are typical of compounds that have a giant ionic structure.

These properties include:
- high melting and boiling points;
- solubility in water;
- electrical non-conductors as solids;
- electrical conductors when an aqueous solution or a molten liquid.

The properties of ionic compounds can be explained by their giant ionic lattice structure. We use the giant ionic lattice of sodium chloride to illustrate how these properties may be explained.

## IONIC LATTICE OF SODIUM CHLORIDE

Composed of sodium ions and chloride ions, sodium chloride is a typical ionic chloride. Each sodium ion has a coordination number of six. That is, it has six chloride ions as its nearest neighbours. Each chloride ion also has a coordination number of six. Since the force of attraction between the sodium ions and the chloride ions is very strong, separating them is difficult and therefore needs lots of energy. Hence, sodium chloride has a high melting point.

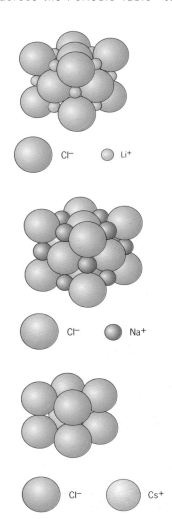

Fig 21.4 Lithium chloride and sodium chloride are face-centred cubic structures, whereas caesium chloride is a body-centred cubic structure. The caesium ion is too large to fit in a face-centred cubic structure

**SEE QUESTION 2.**

**? TEST YOURSELF**

**F** What are the differences in the structure of lithium chloride, sodium chloride and caesium chloride? Suggest why you think there are differences.

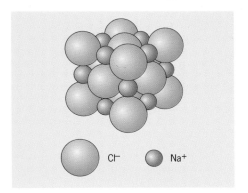

Fig 21.6 Sodium chloride crystal lattice

Fig 21.5 The sodium chloride unit cell

## SOLUBILITY IN WATER

Many ionic compounds dissolve in water to form solutions. The regular pattern of the cations and anions in the lattice is broken down as aqueous or hydrated ions are formed. The energy required to break the strong electrostatic attraction between the cation and the anion is compensated by the energy released when polar water molecules are attracted to the ions. We discuss this energy of dissolution in Chapter 22.

**HOW AND WHY IONIC SUBSTANCES DISSOLVE IN WATER ARE EXPLAINED ON PAGE 483.**

The pH of aqueous sodium chloride is 7.0. Magnesium chloride also dissolves in water, but it forms a solution with a pH slightly less than 7.0. This is because of the polarization of water molecules by the magnesium ion (which is highly charged and has a small ionic radius) forms a small concentration of aqueous hydrogen ions. In solution, aluminium chloride behaves as an ionic chloride, but its aqueous solution is very acidic since an aluminium ion, $Al^{3+}$, has a higher positive charge and a smaller ionic radius than $Mg^{2+}$. The chemistry of aqueous aluminium compounds is described in more detail later in this chapter and in Chapter 22.

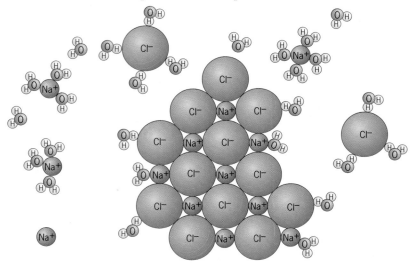

**Fig 21.7** The dissolution of a sodium chloride lattice. Water is a polar molecule and the slightly negative oxygen end is attracted to the positive ions, while the slightly positive hydrogen part is attracted to the negative ions.

**SEE QUESTIONS 2, 3 AND 4.**

## ELECTRICAL CONDUCTIVITY

Solid ionic substances such as sodium chloride cannot conduct electricity because their ions are not free to move. When an ionic substance is melted the ions are free to move, so the molten ionic compound is able to conduct electricity. In the same way, in an aqueous solution of sodium chloride the ions are free to move so the solution conducts electricity.

Solutions of ionic chlorides, or molten ionic chlorides, can be electrolysed because their ions are free to move and act as charge carriers.

## SIMPLE MOLECULAR CHLORIDES

The melting points of all the other chlorides of the elements of Period 3 are mostly quite low. This indicates that the forces of attraction between the particles in their crystal lattices are not particularly strong. The lattices are composed of molecules that are bonded to one another by intermolecular forces. These chlorides have a simple molecular structure.

The properties of simple molecular chlorides such as silicon(IV) chloride is typical of all compounds that have a simple molecular structure. These properties are:

● low melting point, because the intermolecular forces are weak and so easily broken;

● non-conductors of electricity, since there are no free electrons to carry the charge.

We now describe the structure of several of the chlorides to illustrate the formation of simple molecular lattices and weak intermolecular forces.

**SEE QUESTIONS 2, 3 AND 5.**

## ALUMINIUM CHLORIDE

As a metal, aluminium might be expected to form an ionic chloride. Indeed, a bottle of aluminium chloride picked from the chemical store cupboard will probably be ionic. This is because it is hydrated aluminium chloride rather than anhydrous aluminium chloride.

An anhydrous crystalline solid does not have any molecules of water as part of the structure and bonding of the crystal. A hydrated crystal, however, has a lattice that incorporates water molecules. It is important to realize that hydrated crystals are not wet crystals. They are completely dry, since the water is *chemically* bound into the crystal lattice.

Anhydrous aluminium chloride is a simple molecular solid. The formula of the molecule is $Al_2Cl_6$ and it contains two bridging chlorine atoms (Fig 21.8). The intermolecular forces between each $Al_2Cl_6$ unit are fairly weak and, once broken, individual $Al_2Cl_6$ molecules can escape from the lattice. Anhydrous aluminium chloride **sublimes**. That is, when heated, it changes directly from a solid to a gas without having a liquid phase.

## SILICON(IV) CHLORIDE

Often referred to as silicon tetrachloride, silicon(IV) chloride is a colourless liquid with a low boiling point. Its molecules have the formula $SiCl_4$ (Fig 21.9). Each molecule has a tetrahedral shape. The silicon–chlorine bond is polar because the two elements have different electronegativities. This means that the chlorine end of the bond is slightly negative and the carbon atom is slightly positive. Since $SiCl_4$ is a symmetrical molecule, the individual bond dipoles cancel out and the molecule is non-polar. As a result, $SiCl_4$ molecules are attracted to each other by weak intermolecular forces. These are van der Waals forces of attraction or induced dipole–induced dipole attractions. To melt or boil $SiCl_4$ only the weak intermolecular forces need to be broken so that the melting point and the boiling point will be low. The atoms within the molecule remain strongly covalently bonded to each other.

## PHOSPHORUS(III) CHLORIDE

Also known as phosphorus trichloride, phosphorus(III) chloride (Fig 21.10) is a liquid at room temperature with a low boiling point. Its molecule is polar with an overall dipole moment. That is, one end of the molecule is slightly positive and the other end is slightly negative. This happens because the molecule has polar covalent bonds and is not symmetrical, so the individual bond dipoles do not cancel out.

Since one end of the molecule is slightly positive, it can form a weak electrostatic attraction to the negative end of another phosphorus(III) chloride molecule. The electrostatic force of attraction in the phosphorus(III) chloride lattice is not nearly as strong as that in an ionic lattice, since the magnitude of the positive and negative charges is quite small. The molecule has a permanent dipole and it is this that causes the intermolecular attraction. This type of weak intermolecular attraction is known as **permanent dipole–permanent dipole attraction**. Permanent dipole–permanent dipole attractions are stronger than induced dipole–induced dipole interactions.

## PHOSPHORUS(V) CHLORIDE

Also known as phosphorus pentachloride, phosphorus(V) chloride is a white crystalline solid at room temperature. This means that the forces of attraction between particles within the crystal lattice must be stronger than those in the liquid phosphorus(III) chloride.

Phosphorus(V) chloride is unusual in that the particles present in the solid and gaseous phase are different. In the gaseous phase, it is molecular phosphorus(V) chloride, $PCl_5$, whereas in the crystalline solid there is an interesting ionic type of interaction involving $PCl_4^+$ and $PCl_6^-$ (Fig 21.14).

**THE STRUCTURE OF ALUMINIUM CHLORIDE IS DISCUSSED ON PAGE 75.**

**Fig 21.8** The displayed formula of aluminium chloride

**G** Draw the dot and cross diagram for silicon(IV) chloride.

polarity of individual bonds

**Fig 21.9** The displayed formula of silicon(IV) chloride. This molecule is symmetrical so, even though it has four polar covalent bonds, it has no overall dipole moment

Polarity of individual bonds — overall dipole moment

**Fig 21.10** The displayed formula of phosphorus(III) chloride

**Fig 21.11** The polar nature of the phosphorus(III) chloride molecule

intermolecular force

**Fig 21.12** Intermolecular forces in phosphorus(III) chloride

**H** Boron trichloride, $BCl_3$, does not have a dipole moment. Why?

**Fig 21.13** The structure of the $BCl_3$ molecule

**SEE QUESTIONS 1, 2, 5 AND 6.**

THE DOT-AND-CROSS DIAGRAM FOR MOLECULAR PHOSPHORUS(V) CHLORIDE IS GIVEN IN FIG 4.20, PAGE 75.

FOR FURTHER DETAILS ON IONIC BONDING, READ PAGES 69–71.

**? TEST YOURSELF**

Why does phosphorus(V) chloride have a higher melting point than phosphorus(III) chloride?

**Fig 21.15** $S_2Cl_2$ has a similar shape to that of $H_2O_2$

**✔ REMEMBER THIS**

Carbon has no 2d subshells.

**Fig 21.16** The reaction of silicon(IV) chloride with water

YOU MAY WANT TO READ ABOUT EXPANDING THE OCTET ON PAGE 75.

These positive and negative ions are arranged in an ordered pattern. As the positive and negative charges are spread over a large particle, the attraction between the $PCl_4^+$ and $PCl_6^-$ ions is much less than that in an ionic solid, such as sodium chloride, with ions that have small radii.

**Fig 21.14** The structure of the particles present in solid phosphorus(V) chloride

## CHLORIDES OF SULPHUR

We have already noted in Table 5.6 (page 118) that the oxidation state of the most stable chloride of sulphur is +1 rather than the oxidation state that would be predicted from looking at the other chlorides of the elements in Period 3. Sulphur forms $S_2Cl_2$ and the much less stable chlorides $SCl_2$ and $SCl_4$. $S_2Cl_2$ is a pale yellow, fuming liquid formed when chlorine reacts with molten sulphur:

$$2S + Cl_2 \rightarrow S_2Cl_2$$

$S_2Cl_2$ is a simple molecule with weak intermolecular attractions, and so it has a low melting point and does not conduct electricity.

## HYDROLYSIS OF MOLECULAR CHLORIDES

Covalently bonded chlorides behave in a completely different way from ionic chlorides when they are added to water. Ionic chlorides dissolve in water, but covalently bonded chlorides are hydrolysed. Hydrolysis involves the reaction of a compound with water, which results in the decomposition or splitting up of the water.

Often an acidic solution is formed after hydrolysis of a molecular chloride. For example, when silicon(IV) chloride is dropped into water, an immediate reaction takes place with the formation of a strongly acidic solution and a white precipitate of silicon dioxide (sand):

$$SiCl_4(l) + 2H_2O(l) \rightarrow SiO_2(s) + 4HCl(aq)$$

Since carbon is the same group as silicon, you might expect tetrachloromethane (carbon tetrachloride) to behave in the same way and be hydrolysed to give carbon dioxide and hydrochloric acid. This is not the case. In fact, tetrachloromethane is inert towards water or steam. What causes this difference in reactivity?

The mechanism of the hydrolysis of silicon(IV) chloride is believed to involve water molecules forming temporary bonds with the central silicon atom. Electrons from the lone pair on oxygen can be donated into vacant 3d subshells, which are sufficiently low in energy to be available. In the case of carbon, there are no energy levels of sufficiently low energy available. This means that the water molecule cannot temporarily form a bond with the carbon atom, and so no reaction takes place.

Water forms a temporary bond with silicon because silicon expands its octet

Hydrochloric acid is formed when silicon–chlorine bonds break and oxygen–hydrogen bonds break

**Fig 21.17** Possible mechanism of hydrolysis

Both phosphorus chlorides are hydrolysed by cold water to form highly acidic solutions, since hydrochloric acid and either phosphoric(III) acid or phosphoric(V) acid are formed:

$$PCl_3(l) + 3H_2O(l) \rightarrow H_3PO_3(aq) + 3HCl(aq)$$

$$PCl_5(s) + 4H_2O(l) \rightarrow H_3PO_4(aq) + 5HCl(aq)$$

---

## ! ■ SCIENCE FEATURE Tetrachloromethane

Many household 'spot' cleaners used to be composed of chlorinated hydrocarbons such as tetrachloromethane. Its use has recently been banned as part of the London revision of the Montreal Protocol of 1987. This international agreement was one of the first pieces of global environmental legislation to result in the banning of chemicals believed to be responsible for environmental damage, in this case ozone depletion. In fact, in the United Kingdom, even the manufacture of tetrachloromethane has been banned.

Tetrachloromethane is a tetrahedral molecule with the same shape as silicon(IV) chloride and methane. It is non-polar, although it does contain polar covalent bonds because of the presence of the very electronegative chlorine atoms. Tetrachloromethane is able to dissolve other non-polar materials, such as fats and greases (Fig 21.18). The tetrachloromethane molecule forms intermolecular bonds with the fat molecule of the induced dipole–induced dipole type. This helps the dissolving process.

Water-based solvents often cannot dissolve such stains. Water is a polar molecule and so forms strong intermolecular attractions, often hydrogen bonds, with other molecules that are polar. But fat molecules are non-polar, so cannot form strong intermolecular forces of attraction with water and are therefore not dispersed throughout water.

The use of tetrachloromethane or similar non-aqueous solvents to clean fat and grease off clothes is called dry cleaning, since it does not involve water. The use of tetrachloromethane was widespread because it is not hydrolysed by water. However, with the realization of its implication in ozone depletion, coupled with the fact that it is a carcinogen, there has been much research into finding a safer but just as effective solvent.

= fat particle

= tetrachloromethane

**Fig 21.18** The dissolution of fat with tetrachloromethane

---

## HYDROLYSIS OF ALUMINIUM CHLORIDE

Anhydrous aluminium chloride is described on pages 452–453 as a simple molecular substance, so it would be expected to hydrolyse when added to water to give an acidic solution. When aluminium chloride is added to water, the molecular lattice is broken and the bonding changes to ionic, which produces aqueous aluminium ions $Al^{3+}(aq)$. These ions are surrounded by water molecules, with the polar negative end of the water molecule being attracted to the positive aluminium ion. So, it is more accurate to use the formula $[Al(H_2O)_6]^{3+}(aq)$ for the ions.

The aluminium ion has a high charge density, which distorts the electron clouds in a water molecule and so weakens one of the O–H bonds. This distortion is called polarization (Fig 21.20 overleaf). The polarization is so great that the O–H bond breaks, to form an aqueous hydrogen ion:

$$[Al(H_2O)_6]^{3+}(aq) \rightleftharpoons [Al(H_2O)_5(OH)]^{2+}(aq) + H^+(aq)$$

So the solution becomes acidic. In fact, it is sufficiently acidic to fizz immediately and give off carbon dioxide when sodium carbonate is added to it.

**Fig 21.19** The reaction when sodium carbonate is put into aluminium chloride solution: bubbles of carbon dioxide appear

**SEE QUESTIONS 4 AND 5.**

**Fig 21.20** The polarization of water molecules by the aluminium ion. Aqueous aluminium ions are surrounded by six water molecules. The high charge density breaks one of the O–H bonds, releasing a proton

| | Li | Be | B | C | N | O | F |
|---|---|---|---|---|---|---|---|
| **Bonding** | covalent character increases → | | | | | | |
| **Oxidation number** | oxidation number increases → | | | | | | |
| **Structure** | giant → simple molecular | | | | | | |
| **Effect of water (pH)** | dissolves 7 → hydrolysed pH decreases | | | | | | |
| | Na | Mg | Al | Si | P | S | Cl |

**Fig 21.21** A summary of the behaviour of chlorides across Periods 2 and 3

# 3 PERIODICITY IN OXIDES

**Fig 21.22** Sodium burning in oxygen

**Fig 21.23** Sodium oxide is a white solid powder

**Fig 21.24** (Top) Magnesium burning in oxygen. (Bottom) Magnesium oxide is a white solid

**Fig 21.25** Aluminium burns in oxygen to form aluminium oxide, a white solid powder

**Fig 21.26** Phosphorus burns in air to give phosphorus(V) oxide, $P_4O_{10}$, a white solid powder

**Fig 21.27** Sulphur burns in oxygen with a blue flame to give a colourless gas

**Table 21.2**

| Element | Formula of oxide | Oxidation number of element in oxide | Appearance of oxide |
|---|---|---|---|
| Sodium | $Na_2O$ | +1 | White solid |
| Magnesium | $MgO$ | +2 | White solid |
| Aluminium | $Al_2O_3$ | +3 | White solid |
| Silicon | $SiO_2$ | +4 | White solid |
| Phosphorus | $P_4O_6$ | +3 | White solid |
| | $P_4O_{10}$ | +5 | White solid |
| Sulphur | $SO_2$ | +4 | Colourless gas |
| | $SO_3$ | +6 | Colourless gas |
| Chlorine | $Cl_2O_7$ | +7 | Colourless liquid |

Oxygen reacts with almost all the elements of Period 3. Normally, oxides have oxygen with an oxidation number of –2, but there are oxides in which oxygen has an oxidation number of –1. Such oxides are called peroxides.

Over the next few pages, the oxides are discussed in terms of the change in their properties as one goes from one element to another across the period.

As in the case of the chlorides, there is a clear pattern in the maximum oxidation number of each element. Remember that elements in Period 3 can expand their octets, so that the highest oxidation state occurs when all of the outer electrons are used in bonding. Once the oxidation state of the element reaches 4 or above, the bonding between the atoms in the oxide becomes covalent. It is impossible to supply the energy during a chemical reaction for an atom to lose four or more electrons, so that sharing of electrons is the only option left.

The wide range of melting points (Fig 21.28) indicates that there are several types of structure and bonding within these oxides. In fact, they can have giant ionic structures, giant molecular structures and simple molecular structures. The structure with the highest melting point clearly has the strongest attraction between its particles.

## IONIC OXIDES

The metals sodium, magnesium and aluminium form ionic oxides. The metal atoms lose electrons, which are accepted to form oxide ions. As in the case of sodium chloride, the ions are packed in a giant lattice that consists of positive ions surrounded by negative ions, and vice versa.

Magnesium oxide has the same structure as sodium chloride. However, the magnesium ion has a smaller ionic radius than a sodium ion, because of its increased nuclear charge. Also, it has a +2 charge rather than a +1 charge. This means that the positive charge is much more concentrated around a magnesium ion than around a sodium ion. It is said to have a higher charge density. In exactly the same way, the oxide ion has a –2 charge rather than a –1 charge, and its radius is smaller than that of the chloride ion. So, the oxide ion has a greater charge density than the chloride ion.

With the magnesium ions and the oxide ions both having high charge densities, they have an extremely strong attraction for each other. Hence, it is difficult to separate the magnesium ions from the oxide ions in the crystal lattice of magnesium oxide. So this oxide has a very high melting point. Magnesium oxide is therefore widely used as a refractory material, such as for the linings of high-temperature furnaces.

**Fig 21.28** Graph of melting points of the oxides with the highest oxidation numbers

**? TEST YOURSELF**

**a)** Sodium burns in air to form sodium oxide:

$$4Na(s) + O_2(g) \rightarrow 2Na_2O(s)$$

Write down the equations for the formation of the following oxides by burning the element in oxygen:

$MgO$, $Al_2O_3$, $SiO_2$, $P_4O_{10}$ and $SO_2$.

**b)** Which of the oxides in Table 21.2 involve bonding in which an atom has had to expand its octet?

**c)** Give the systematic names for each of the oxides in Table 21.2.

**IONIC RADII ARE DEALT WITH ON PAGE 98.**

**SEE QUESTIONS 5 AND 8.**

**Fig 21.29** A section of the ionic lattice of magnesium oxide

$Mg^{2+}$

$O^{2-}$

**Fig 21.30** Relative sizes and charges of the sodium ion and the magnesium ion

$Na^+$  $Mg^{2+}$

**Fig 21.31** Relative sizes of the chloride ion and the oxide ion

$Cl^-$  $O^{2-}$

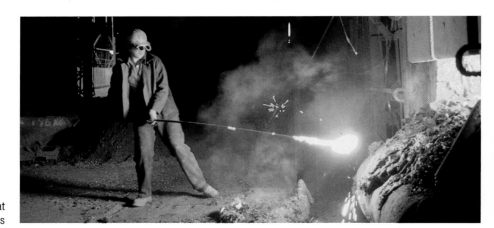

**Fig 21.32** Its heat resistance means that magnesium oxide is used to line furnaces

## ALUMINIUM OXIDE

Aluminium oxide is another ionic lattice, although there is some degree of covalent character to the bonds. This happens because the aluminium ion has a large positive charge and a very small radius. It can therefore polarize oxide ions. Nevertheless, aluminium oxide has a very high melting point.

---

**! SCIENCE FEATURE Aluminium oxide and anodizing**

Over 3000 articles in daily use are made from aluminium. Among these are cooking utensils, kitchen appliances, kitchen foil, electrical conductors, and engineering and building components. The metal has found so many applications because of its low density coupled with a remarkable resistance to corrosion. The plaque carried on board Pioneer 10 – the first constructed object to escape from the Solar System – was made from gold and anodized aluminium because of these two properties.

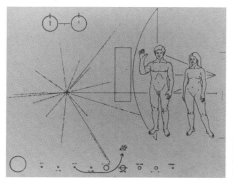

**Fig 21.33** The plaque carried on board Pioneer 10

Aluminium normally has a dull lustre because it is coated in a thin, transparent surface film of aluminium oxide that forms on exposure to air. This layer of oxide protects the underlying metal

from further reaction and oxidation. It is in complete contrast with the formation of rust on iron, which leads to further rusting. Aluminium can be made even more corrosion-resistant by increasing the thickness of the layer of aluminium oxide. This process is called anodizing, in which the aluminium is the anode during the electrolytic decomposition of sulphuric acid.

Another advantage of anodizing is that the layer of aluminium oxide formed can act as a mordant and adsorb (bond to) coloured dyes. In this way, the aluminium surface can be given a more attractive, coloured finish.

**Fig 21.34** Eros, at Piccadilly Circus, London, is made of aluminium

**Fig 21.35** Doors and window frames are often made of aluminium

## BASIC CHARACTER OF IONIC OXIDES

Ionic oxides behave as bases. That is, they are able to accept a proton. This is because they contain the oxide ion, which reacts with two protons to give water:

$$O^{2-} + 2H^+ \rightarrow H_2O$$

Metals tend to form ionic oxides, which explains why metal oxides are basic.

If a metal oxide dissolves in water, it forms an alkaline solution. Most metal oxides are insoluble in water, or at the most are sparingly soluble. Sodium oxide, however, not only dissolves in water, but reacts with it to form sodium hydroxide:

$$Na_2O(s) + H_2O(l) \rightarrow 2NaOH(aq)$$

The ionic equation shows the formation of hydroxide ions:

$$O^{2-}(s) + H_2O(l) \rightarrow 2OH^-(aq)$$

## AMPHOTERIC NATURE OF ALUMINIUM OXIDE

Aluminium oxide does not dissolve in or react with water, but it does show basic properties in its reactions with acids. Aluminium oxide also shows acidic properties, since it reacts with alkalis.

An oxide that shows both basic and acidic properties is called an **amphoteric oxide**. Often, there is some degree of covalent character in the bonding in amphoteric oxides. When aluminium oxide is heated with aqueous sodium hydroxide, it forms a salt called sodium aluminate:

$$2NaOH(aq) + Al_2O_3(s) \rightarrow 2NaAlO_2(aq) + H_2O(l)$$

The formula of sodium aluminate is open to speculation. A common formula used is $Na_3Al(OH)_6$, as is $NaAlO_2$. Note that in this compound, aluminium is found in the *anion*.

## SILICON DIOXIDE: A GIANT MOLECULAR OXIDE

Fig 21.36 A quartz crystal in electronic watches ensures accurate time

Fig 21.37 The displayed formula of the unit cell of silicon dioxide

Fig 21.38 The geometric shape of quartz crystals indicates the regular pattern of their particles

YOU CAN READ MORE ABOUT THE GIANT
MOLECULAR STRUCTURES OF GRAPHITE
AND DIAMOND ON PAGES 154–155.

## ? TEST YOURSELF

**L** What are the differences between the structure of diamond and the structure of silicon dioxide?

$$\overset{\delta-}{O}=\overset{\delta+}{C}=\overset{\delta-}{O}$$
← + + →
bond dipoles cancel out

**Fig 21.39** The displayed formula of carbon dioxide showing the bond dipoles

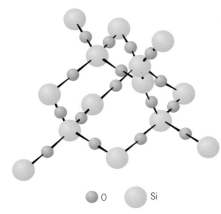

**Fig 21.40** The structure of silicon dioxide

O •    Si ○

Silicon dioxide is also known as sand and quartz. Quartz is a hard, brittle, clear, colourless solid. Among its many applications are architectural decorations, semi-precious jewels, optical components and frequency controllers in radio transmitters. It melts to form a viscous liquid. When this liquid is cooled, the particles have difficulty in taking up a regular pattern, and it supercools to form a glass called silica. Silica glass is also a useful substance, being inert towards most acids. A mixture of boron oxide and silicon dioxide, heated into a liquid and then cooled, forms borosilicate glass, which is heat resistant.

The high melting point of quartz is attributed to its giant molecular structure. Each silicon atom is covalently bonded to four oxygen atoms in a structure like that of diamond (Fig 21.37). Like diamond, silicon dioxide is hard, insoluble in water and a non-conductor of electricity.

The structure of silicon dioxide is in complete contrast to that of carbon dioxide, which is a simple molecular lattice held together by weak intermolecular forces. Theoretically, carbon could form four C–O bonds or two C=O bonds.

More energy is released in making two C=O bonds than four C–O bonds, so carbon dioxide has the displayed formula shown in Fig 21.39.

More energy is released in making four Si–O bonds than two Si=O bonds, so silicon(IV) oxide has the structure shown in Fig 21.40.

## SIMPLE MOLECULAR OXIDES

A simple molecular oxide has simple molecules that are held together in a solid by weak intermolecular forces. Just as for the simple molecular chlorides in the preceding section, the physical properties of these oxides will be similar; they have relatively low melting and boiling points and are non-conductors of electricity.

## OXIDES OF SULPHUR, PHOSPHORUS AND CHLORINE

Sulphur dioxide, $SO_2$ (Fig 21.41), sulphur trioxide, $SO_3$, phosphorus(III) oxide, $P_4O_6$, and phosphorus(V) oxide, $P_4O_{10}$ (Fig 21.42), and the oxides of chlorine are all simple molecules and so form simple molecular lattices in the solid state. All these molecules have a permanent dipole. The negative end of one molecule can attract the positive end of another, which results in the formation of a weak intermolecular force. This force is a permanent dipole–permanent dipole attraction.

Since only weak intermolecular forces exist in all of these oxides, they almost all have relatively low melting points. $P_4O_{10}$ has quite a high melting point because it is a large molecule. Therefore, in addition to the permanent dipole–permanent dipole attraction, there are significant forces of attraction through the induced dipole–induced dipole interaction.

SEE QUESTION 8.

$$\overset{\delta+}{\underset{\delta-}{S}}\diagup\diagdown\overset{}{\underset{\delta-}{}}$$
bond dipoles do not cancel out    overall dipole moment

**Fig 21.41** The structure of sulphur dioxide and its dipole moments

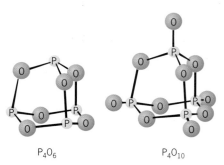

$P_4O_6$       $P_4O_{10}$

**Fig 21.42** The structures of phosphorus oxides

## ACIDIC BEHAVIOUR OF COVALENT OXIDES

In Period 3 the non-metals form covalent oxides, and these oxides are acidic. If an acidic oxide dissolves in water it will form an acidic solution.

For example, sulphur trioxide reacts with water to make sulphuric acid:

$$SO_3(g) + H_2O(l) \rightarrow H_2SO_4(aq)$$

Chlorine(VII) oxide reacts with water to form chloric(VII) acid:

$$Cl_2O_7 + H_2O \rightarrow 2HClO_4$$

This is the strongest acid that has so far been discovered.

When an oxide does not dissolve in or react with water, its acidic properties are less obvious. For example, silicon dioxide does not react with water – if it did, there would be no sandy beaches. But it does react when heated with a basic oxide. So, when sodium oxide and silicon dioxide are heated together, sodium silicate is formed:

$$Na_2O(s) + SiO_2(s) \rightarrow Na_2SiO_3(s)$$

Strong alkalis should not be left for a long time in glass burettes and in bottles with glass stoppers, because the stoppers and stopcocks become fused. This happens because of the slow reaction of the alkali with the silica in the glass, which makes a silicate.

> **ON PAGE 12, THERE IS A DESCRIPTION OF THE REACTION OF CALCIUM OXIDE WITH SAND (SILICON DIOXIDE) TO MAKE CALCIUM SILICATE OR SLAG.**

> **SOME NON-METALS FORM NEUTRAL OXIDES, FOR EXAMPLE CARBON MONOXIDE. METAL OXIDES WITH THE METAL IN A HIGH OXIDATION STATE, E.G. CHROMIUM(VI) OXIDE, ARE ALSO ACIDIC BECAUSE THE BONDING IN THE OXIDE IS COVALENT.**

> **SEE QUESTIONS 1 AND 5.**

**Fig 21.43** A summary of the properties of the oxides of the elements of Periods 2 and 3

**? TEST YOURSELF**

M What acid do you get when the following oxides are reacted with water?
**a)** Sulphur dioxide;
**b)** Phosphorus(III) oxide;
**c)** Phosphorus(V) oxide.
Write equations for the reactions.

# SUMMARY

After studying this chapter, you should know the following:

- Non-metals often react by electron gain and are oxidizing agents, whereas metals react by electron loss and are reducing agents.
- When any element reacts in a chemical reaction or is formed in a chemical reaction, a redox or electron transfer reaction has taken place.
- The maximum oxidation number attained by elements when combined with chlorine increases across Period 3. To obtain an oxidation number above +4, the atom has to expand its stable set of outer electrons beyond an octet.
- When a metal and a non-metal react together, the compound formed is normally ionic. When two non-metals react the compound formed is normally covalent.
- Ionic compounds are composed of a giant lattice of positive and negative ions held in place by strong electrostatic attraction. This results in ionic crystals that have high melting points.
- Ionic solids do not conduct electricity, since they have no mobile charge carriers. However, when molten or in solution, they can be electrolysed and conduct electricity because the ions are free to move.

- Giant molecular compounds are composed of atoms held together by strong covalent bonds. Giant molecular compounds have high melting and boiling points, do not dissolve in water and do not conduct electricity.
- Simple molecular compounds are composed of simple molecules held together by weak intermolecular forces.
- Simple molecular compounds have low melting and boiling points, and are non-conductors of electricity.
- The structures of the chlorides of the elements in Period 3 vary from giant ionic to simple molecular.
- Ionic chlorides usually dissolve in water to form neutral solutions, whereas molecular chlorides are hydrolysed in water to form hydrogen chloride or hydrochloric acid.
- The structures of the oxides of the elements in Period 3 vary from giant ionic to giant molecular to simple molecular.
- Metal oxides are basic or amphoteric in nature, non-metal oxides are acidic or neutral.
- Ionic oxides are basic and covalent oxides are acidic or neutral. An ionic oxide with a high degree of covalent character is amphoteric.

## ❓ EXAM QUESTIONS

**1**

**a)** List the formulas of the chlorides of Na, Mg, Al, Si and P. Why is there no chloride of argon? Give the name of the bonding in the chloride of silicon both within and between molecules.

**b)** Classify the acid–base character of one oxide of each of the elements in the period for Na to S. Illustrate your answer by writing balanced chemical equations for the reaction of magnesium oxide and of a phosphorus oxide with water. Explain why 'pure' rain water is slightly acidic (pH 5.7).

**c)** **(i)** Write balanced equations to show how aluminium oxide reacts with hydrochloric acid and with sodium hydroxide.
**(ii)** Write a balanced equation to show what happens when $FeCl_3$ is added to water.

IBO Chemistry Higher Level Paper 2 Nov 2000 Q5(b,c,d)

**2** How are the physical properties of ionic and simple molecular chlorides related to their structures? Include in your answer references to the change of state, electrical conductivity and solubility in polar and non-polar solvents.

**3** Look at the two diagrams in Fig 21.Q3. They show the unit cells of lithium chloride and caesium chloride.

lithium chloride, LiCl

caesium chloride

**Fig 21.Q3**

**a)** Describe the differences between the two unit cells.

**b)** Suggest factors that could account for the difference in the unit cells?

**c)** Explain why caesium chloride cannot conduct electricity when it is a solid, but can when it is molten.

**d)** Which of the two chlorides will have the higher melting point? Give reasons for your answer.

**4**

**a)** **(i)** Write equations, including state symbols, to show the changes that take place when each of the NaCl(s) and $SiCl_4$(l) is added to water.
**(ii)** Relate these changes to the type of bonding in the chlorides.

**c)** Aluminium oxide is amphoteric.
**(i)** What is meant by the term *amphoteric*?
**(ii)** Write equations that illustrate this property of aluminium oxide.

Modified from Edexcel Chemistry Module Test 1 6081/P.01 Jan 2001 Q3(a,c)

**5**

**a)** **(i)** Use the data in Table 21.Q5 to identify the type of bonding in each of the chlorides listed and state how the given properties depend on the type of bonding. Support your answer with appropriate diagrams.

**Table 21.Q5**

| Chloride | Melting point/K | Boiling point/K |
|----------|-----------------|-----------------|
| Sodium chloride | 1074 | 1686 |
| Aluminium chloride | 451 | Sublimes |
| Phosphorus(III) chloride | 161 | 349 |

**(ii)** Describe how the chlorides behave when added to water. Give equations for any reactions that occur.

**b)** **(i)** State the bonding in the oxides of sodium, magnesium, silicon and phosphorus.
**(ii)** What happens to the pH of pure water when these oxides are added to separate samples of the water? Give equations for any reactions which occur.

IBO Chemistry Higher Level Paper 2 May 1999 Q8

**6**

**a)** Silicon(IV) oxide, $SiO_2$, and phosphorus(III) oxide, $P_4O_6$, are both covalent compounds. The melting temperature of $SiO_2$ is around 1700 °C while that of $P_4O_6$ is 24 °C.
Explain how the melting temperatures of the oxides are related to their structures.

**c)** Gaseous phosphorus pentachloride, $PCl_5$, contains molecules, but at room temperature it is an ionic solid that contains the ions $[PCl_4]^+$ and $[PCl_6]^-$. Draw clear diagrams to show the shapes of the molecule and of the ions. Give a brief explanation as to why each has the shape you give.

London Examinations AS Chemistry 6085/P.01 June 1999 Q3(a,c)

**7**

**a)** Write the formula or formulae of the chlorides that each of the elements Na, Mg, Al, Si and P form.

**b)** Write equations for the changes that occur when sodium chloride and phosphorus(III) chloride are added to water.

**c)** When an aqueous solution of magnesium chloride was evaporated to dryness, a product of composition Mg 35.5%, Cl 52.6 %, O 11.9% by mass was obtained.

(i) Calculate the empirical formula of the product.

(ii) Suggest why anhydrous magnesium chloride was not formed.

**d)** (i) Draw the dot-and-cross diagram for phosphorus(V) chloride in the gaseous state.

(ii) Sketch this molecule to show its shape, indicating the values of the bond angles.

Adapted from ULEAC 1996 Specimen paper 9081, CH1, Q.1

**8** This question refers to the oxides of elements in Period 3.

**a)** State the formulae of the oxides of all the elements of Period 3.

**b)** How does the maximum oxidation number of the element in its oxide change across the period? Account for this change in terms of the electronic structure of the elements involved.

**c)** Magnesium oxide is a basic oxide. What does this indicate about the bonding in magnesium oxide?

**d)** What does the term amphoteric mean ? Illustrate your answer with reference to aluminium oxide.

**e)** Explain why an oxide with a giant molecular structure will have a much higher melting point than one with a simple molecular structure. Illustrate your answer by using examples with each containing a different element from Period 3.

# KEY SKILLS **ASSIGNMENT**

## PATTERNS ACROSS PERIOD 2

**The patterns in the properties of the oxides and chlorides of the elements in Period 3 are also evident in Period 2. Lithium is the only true metal in this period, so it is the only element to form an ionic oxide and an ionic chloride.**

**1** **a)** Lithium chloride has a giant ionic lattice that has the same structure as sodium chloride. Predict the pH of an aqueous solution of lithium chloride.

**b)** Lithium oxide, $Li_2O$, has a giant ionic lattice. Write down an equation to show what happens when lithium oxide reacts with water.

The next element in the period is beryllium. Although it is in Group 2, it cannot really be described as a metal. Beryllium atoms do not easily lose electrons and do not form the beryllium ion, $Be^{2+}$, in compounds. Any ionic beryllium compounds contain hydrated beryllium ions, $[Be(H_2O)_4]^{2+}$. The outer electrons are very tightly held to the nucleus, since there is only one shielding inner shell of electrons. If it were formed, the beryllium ion, $Be^{2+}$, would have such a small ionic radius that its charge density would be extremely large, and so it would polarize negative ions.

So, how does a beryllium atom achieve the stability of a noble-gas electron configuration? If it cannot lose electrons, then it has to share them and share sufficient to achieve the desired electron configuration. Solid beryllium chloride forms a layered lattice (Fig 21.A1) reminiscent of aluminium chloride in that it has bridging chlorine atoms, but different in that the structure is continuous.

**Fig 21.A1** The layered lattice of beryllium chloride

Beryllium oxide has a very high melting point. It is an amphoteric oxide and reacts with both acids and alkalis:

$$BeO(s) + 2H^+(aq) \rightarrow Be^{2+}(aq) + H_2O(l)$$
$$BeO(s) + 2OH^-(aq) + H_2O(l) \rightarrow Be(OH)_4{}^{2-}(aq)$$

**2** **a)** The hydrated beryllium ion has very similar bonding to that of the hydrated aluminium ion shown on page 493. Describe the bonding in the hydrated beryllium ion and suggest a shape for the ion.

**b)** By counting up the number of electrons round the beryllium atom in the hydrated beryllium ion, suggest why this ion is stable.

**c)** What evidence is there in this assignment that beryllium oxide has a giant structure?

**d)** Explain, in terms of its electronic structure, why a simple molecule of beryllium chloride is electron deficient.

**e)** How does the formation of the layered lattice help a beryllium atom obtain a noble-gas electron configuration?

**f)** Predict, with reasons, what would happen when beryllium chloride is added to water.

Boron, the third element in the period, forms boron trichloride, which is a colourless gas at room temperature. Boron trichloride is a Lewis acid, which means that it can accept a pair of electrons. This makes boron trichloride a very useful catalyst in, for example, certain electrophilic substitution reactions of arenes. Boron oxide, on the other hand, is a white solid that has a giant structure.

**3** **a)** Explain, in terms of its electronic structure, why boron trichloride can accept a pair of electrons.

**b)** Write down the equation to show the hydrolysis of boron trichloride.

# KEY SKILLS **ASSIGNMENT**

**c)** Describe the intermolecular forces involved in a lattice of boron trichloride.

Carbon is the fourth element in Period 2. Its more useful oxide is carbon dioxide, which is a colourless gas at room temperature, and dissolves in water to form a weakly acidic solution. Carbon also forms a neutral oxide called carbon monoxide, which is poisonous because of its ability to bind with the iron(II) ion in haemoglobin, the red pigment in blood. Tetrachloromethane is something of an anomaly, because it is a simple covalent chloride that is not hydrolysed by water.

**4** Compare and contrast the oxides and chlorides of silicon and carbon.

Nitrogen is the fifth element in Period 2. It forms a chloride, $NCl_3$, that is a very explosive yellow oil. This chloride is slowly hydrolysed by water to form ammonia:

$$NCl_3(l) + 3H_2O(l) \rightarrow NH_3(aq) + 3HClO(aq)$$

**5** What is unusual about the hydrolysis of $NCl_3$?

Nitrogen forms not just two oxides, like its fellow group member phosphorus, but several simple molecular oxides, ranging from $N_2O$ to $N_2O_5$. The acidity of these oxides increases as the oxidation number of the nitrogen increases, so that dinitrogen(I) oxide, $N_2O$, is neutral and $N_2O_5$ is acidic.

**6** **a)** Table 21.A1 shows the formula of some nitrogen oxides. Copy out and complete the table by working out the oxidation number of nitrogen for each oxide.

**Table 21.A1**

| Formula of oxide | $N_2O$ | NO | $N_2O_3$ | $NO_2$ | $N_2O_4$ | $N_2O_5$ |
|---|---|---|---|---|---|---|
| Oxidation no. of nitrogen | | | | | | |

**b)** Dinitrogen(V) oxide, $N_2O_5$, reacts with water to form an acid. Write an equation for this reaction and hence deduce the name of the acid formed.
**c)** Nitrogen dioxide, $NO_2$, reacts with water to give two acids, one of which is nitrous acid, $HNO_2$. Write an equation for this reaction and hence deduce the name of the other acid.

Although oxygen is the next element in the period, we normally refer to oxides of chlorine rather than chlorides of oxygen. There are several chlorine oxides, all of which are simple molecules. Lastly, there is the noble gas neon, which forms neither a chloride nor an oxide.

**7** One of the oxides of chlorine has the formula $Cl_2O_7$.

**a)** What is the oxidation state of chlorine in this oxide?
**b)** Suggest a name for this oxide.
**c)** This oxide is a simple molecule that contains a bridging oxygen atom between the two chlorine atoms. Suggest a possible displayed formula for the oxide.
**d)** Write down the formula for the acid formed when $Cl_2O_7$ reacts with water.

# The reactive metals

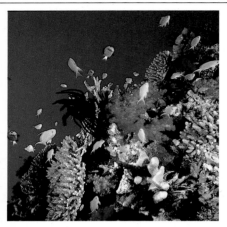

**Calcium carbonate** Coral reefs are composed mainly of the insoluble compound calcium carbonate

**Calcium carbonate** Chalk cliffs are composed of calcium carbonate

Some people think of carbon dioxide as just the greenhouse gas that causes global warming problems, although of course carbon dioxide is essential for life as well. From earliest times on Earth and continuously since then, vast amounts of carbon dioxide from the atmosphere have been locked up in rocks of the Earth's crust as carbonates.

Several different carbonates occur in rocks, but by far the most abundant is that of the highly reactive metal calcium, calcium carbonate, which we see in the enormous chalk cliffs and widespread chalklands in the UK, and also in limestone deposits and coral reefs around the world.

Chalk, limestone and coral reefs exist only because calcium carbonate is an insoluble material, while other calcium salts are soluble. The calcium carbonate comes from the activity of marine animals and microscopic organisms that synthesize the carbonate as part of their life processes, but they can only do this by using calcium salts that are soluble in the waters around them. The calcium carbonate forms their shells, bones or coral protective structures, and as the organisms die and reach the ocean floor, these deposits feed into the rock-forming cycle.

Calcium carbonate formed at the bottom of seas and oceans is therefore a 'sink' for carbon dioxide. It has been suggested that as atmospheric carbon dioxide increases, this will to some extent be offset by an increase in the carbon dioxide content of the oceans and, in turn, by an increase in the conversion of it into calcium carbonate by marine creatures.

## 1 REACTIVE METALS

The classification of elements into metals and non-metals is based upon the physical and chemical properties of the elements and their compounds. To most people, metals are hard, strong and shiny and good thermal and electrical conductors. These properties describe the metals we are familiar with outside the chemistry laboratory, such as iron, lead, silver, gold, zinc, copper and nickel.

Other elements are also classified as metals, but they are less recognizable as such, in that they are soft and have a low melting point. It is only when their chemical properties are described that they are clearly seen to be metals. These include lithium, sodium, potassium, calcium and barium, which are found in the s block of the Periodic Table and are often collectively called the reactive metals.

Aluminium, in Group 3, is sometimes also referred to as a reactive metal. So aluminium is included in this chapter (see page 490).

### CHEMICAL PROPERTIES OF METALS

The main chemical property of a metal atom is its ability to lose one, two or sometimes three electrons to form a positive ion. When an atom loses electrons, it often attains a stable, noble-gas electron configuration. It is this single property that can be used to explain the reactivity of metals. Atoms of reactive metals easily lose electrons; atoms of unreactive metals lose them with difficulty.

The ability of a metal atom to lose electrons to form a positive ion explains why:
- metals are reducing agents;
- metals form ionic compounds with non-metals;
- metals do not normally form compounds with other metals.

The most reactive of all metals are those in the s block of the Periodic Table, since they have an electron configuration that contains only 1 or 2 electrons more than the nearest noble gas. This means that atoms of these elements lose electrons easily to form cations.

**REMEMBER THIS**

OIL RIG: oxidation involves the loss of electrons and reduction the gain of electrons. So metals are reducing agents because they give away electrons easily to an oxidizing agent.

## 2 GROUP 1: THE ALKALI METALS

**Table 22.1 Some properties of the first five alkali metals**

| Metal | Atomic (proton) number | Common oxidation state | Atomic radius/pm | Ionic radius (M⁺)/pm | Density/kg m⁻³ | Melting point/°C | Boiling point/°C | Electron configuration |
|---|---|---|---|---|---|---|---|---|
| Lithium | 3 | +1 | 152 | 60 | 534 | 180 | 1326 | $1s^2\,2s^1$ |
| Sodium | 11 | +1 | 186 | 95 | 968 | 98 | 889 | $1s^2\,2s^2\,2p^6\,3s^1$ |
| Potassium | 19 | +1 | 231 | 133 | 856 | 63 | 757 | $1s^2\,2s^2\,2p^6\,3s^2\,3p^6\,4s^1$ |
| Rubidium | 37 | +1 | 244 | 148 | 1532 | 39 | 679 | $[Kr]5s^1$ |
| Caesium | 55 | +1 | 262 | 169 | 1870 | 29 | 690 | $[Xe]6s^1$ |

The metals in Group 1 are collectively known as the alkali metals. There are six of them: lithium, sodium, potassium, rubidium, caesium and francium. They are the most reactive metals in their appropriate period of the Periodic Table.

These six elements are all remarkably similar in terms of both their chemical and physical properties. Being in the same group, we would expect the elements to have similar chemical properties (see page 93). However, that many of their physical properties are also similar and with, in some cases, an observable and predictable trend is unusual for most groups within the Periodic Table. For example, the melting and boiling points decrease with increasing atomic (proton) number.

It is interesting that the melting points of these elements are very low for metals, which makes them easy to melt. Advantage is taken of this property in some nuclear reactors, in which liquid sodium is used as the primary coolant because it also has a relatively high specific heat capacity and can be pumped easily through the pipes of the cooling system.

**REMEMBER THIS**

Specific heat capacity measures the energy needed to raise the temperature of 1 kg of a substance by 1 °C.

## OCCURRENCE AND EXTRACTION

**Table 22.2 The natural abundances of the alkali metals in the Earth's crust**

| Element | Abundance (% by mass) | Common mineral |
|---|---|---|
| Lithium | $1.8 \times 10^{-3}$ | Lepidolite |
| Sodium | 2.63 | Rock salt and Chile saltpetre |
| Potassium | 2.40 | Sylvite and carnallite |
| Rubidium | $7.8 \times 10^{-3}$ | Lepidolite |
| Caesium | $3 \times 10^{-4}$ | Pollucite |

The physical properties of the alkali metals, coupled with their high reactivity, limit the number of large-scale applications. Nevertheless, there is significant demand for lithium, sodium and potassium.

**Fig 22.1** Sodium chloride is the principal source of the reactive metal sodium. Most sodium chloride is mined

The alkali metals are so reactive that they occur naturally only as compounds. It is impossible to reduce alkali metal compounds using a chemical process such as heating with carbon. The only successful method of reduction involves electrolytic decomposition. Francium, the last of the alkali metals to be discovered, is highly radioactive and, although its chemistry is easily predicted, it has not been fully determined experimentally. This is hardly surprising, since it is estimated that only 15 grams of francium occur in the whole of the Earth's crust.

## MANUFACTURE OF SODIUM

Sodium is manufactured by the electrolysis of molten sodium chloride. The melting point of sodium chloride is 801 °C and so calcium chloride is added to lower the melting point to about 600 °C. The overall reaction is represented by the equation:

$$2NaCl(l) \rightarrow 2Na(l) + Cl_2(g)$$

Electrolysis is the only practical way to achieve this decomposition. Sodium ions are reduced at the cathode and chloride ions are oxidized at the anode:

$$\text{Cathode: } Na^+ + e^- \rightarrow Na$$
$$\text{Anode: } 2Cl^- \rightarrow Cl_2 + 2e^-$$

Since sodium is a highly reactive metal, it is important to ensure that the sodium and chlorine formed are not allowed to recombine. They must be produced in different parts of the electrolytic cell. Fig 22.2 shows the Down's cell used to electrolyse molten sodium chloride. A fine screen prevents the chlorine from diffusing to the cathode.

All the alkali metals are costly to make because of the amount of electrical energy needed.

**Fig 22.2** A Down's cell is used to manufacture sodium. An electrolyte of molten sodium chloride and calcium chloride is electrolysed using a graphite anode and an iron cathode. Molten sodium is collected as it floats on top of the molten electrolyte. A fine screen or gauze prevents the chlorine and sodium from recombining. The chlorine produced is collected and stored

## 3 REACTIONS OF THE ALKALI METALS

The alkali metals are good reducing agents because their atoms can easily transfer the outer electrons to non-metal atoms. Once electrons are lost, the resulting positive ions have noble gas electron configurations.

For example, the sodium atom ($1s^2\ 2s^2\ 2p^6\ 3s^1$) loses an electron to form the sodium ion ($1s^2\ 2s^2\ 2p^6$) with the same electron configuration as argon. This means that in all compounds the elements of Group 1 have an oxidation state of +1.

# FIRST IONIZATION ENERGY AND ELECTRODE POTENTIALS

Lithium has the highest of the first ionization energy values, so it is expected to be the least reactive of the alkali metals.

**Table 22.3 First ionization energies of the Group 1 elements**

| Element | First ionization energy/kJ mol$^{-1}$ |
| --- | --- |
| Lithium | 519 |
| Sodium | 494 |
| Potassium | 418 |
| Rubidium | 400 |
| Caesium | 380 |

The first ionization energy refers to the loss of an electron from a gaseous atom, which does not represent the situation for most reactions. A more appropriate way to describe the ease of electron loss from an alkali metal atom is to refer to its oxidation potential, $E_{oxid}$, since this refers to the reaction of a metal atom to form an aqueous metal ion. For example, the oxidation potential for lithium refers to the following half-equation:

$$Li(s) \rightarrow Li^+(aq) + e^-$$

This is precisely the reaction that occurs when lithium metal reacts with water or dilute acid.

The values of the oxidation potentials show that all the elements in Group 1 are highly reactive in that they can lose electrons easily.

**Table 22.4 Oxidation potentials**

| Half-equation | Oxidation potential, $E_{oxid}$/volts |
| --- | --- |
| $Li(s) \rightarrow Li^+(aq) + e^-$ | + 3.05 |
| $Na(s) \rightarrow Na^+(aq) + e^-$ | +2.71 |
| $K(s) \rightarrow K^+(aq) + e^-$ | +2.93 |
| $Rb(s) \rightarrow Rb^+(aq) + e^-$ | +2.92 |
| $Cs(s) \rightarrow Cs^+(aq) + e^-$ | +2.92 |

The more positive an oxidation potential, the more feasible the reaction. (More information on oxidation potentials is given on page 545.)

## REACTION WITH WATER

The alkali metals are so-called because of their reaction with water. All the metals react vigorously with water to form hydrogen and an alkaline solution.

Take lithium as an example (Fig 22.3). During the reaction, the metal reduces the water by losing electrons to form lithium cations and hydrogen. Lithium reacts with water to form aqueous lithium hydroxide and hydrogen:

$$2Li(s) + 2H_2O(l) \rightarrow 2LiOH(aq) + H_2(g)$$

The ionic equation more clearly describes why an alkali is formed, since aqueous hydroxide ions are formed:

$$2Li(s) + 2H_2O(l) \rightarrow 2Li^+(aq) + 2OH^-(aq) + H_2(g)$$

**Fig 22.3** Lithium reacts with water to form aqueous lithium hydroxide and hydrogen. The water has turned pink because it contains phenolphthalein indicator

**? TEST YOURSELF**

**D** **a)** Write down the electron configuration for each of the following: Li$^+$, K$^+$ and Rb$^+$.
**b)** Why is sodium more reactive than lithium? Use Table 22.3 to help you.

**✔ REMEMBER THIS**

As the atomic (proton) number increases down the group, the number of inner-shell shielding electrons increases and the atomic radius increases. The outer electron is therefore attracted less strongly to the nucleus and so less energy is needed to remove it from the atom. This means that the first ionization energy decreases within a group as the atomic (proton) number increases.

YOU CAN READ ABOUT FIRST IONIZATION ENERGY AND ELECTRON CONFIGURATION ON PAGE 57 AND PAGE 99.

**? TEST YOURSELF**

**E** Predict the first ionization energy for francium.

## ? TEST YOURSELF

**F**  **a)** Write down the balanced equations for the reaction of sodium and of potassium with water.
**b)** Predict what would happen if a piece of francium were added to cold water.

The reactivity of the alkali metals increases with increasing atomic (proton) number, so that the reaction of potassium with water is often accompanied by a lilac flame as the hydrogen produced burns (Fig 22.5).

**Fig 22.4** Sodium reacts with water to form aqueous sodium hydroxide and hydrogen. The yellow flame occurs when the hydrogen produced burns. Notice that sufficient energy is transferred to the surroundings to melt the sodium into a sphere

**Fig 22.5** Potassium is more reactive than sodium or lithium. It reacts violently with water, to produce hydrogen and aqueous potassium hydroxide. The reaction is exothermic and the energy transferred is sufficient to melt the potassium and to ignite the hydrogen formed

## REACTIONS WITH ACIDS

The alkali metals react explosively with dilute acids, such as hydrochloric acid and sulphuric acid:

$$2Li(s) + 2HCl(aq) \rightarrow 2LiCl(aq) + H_2(g)$$

$$2Na(s) + 2H^+(aq) \rightarrow H_2(g) + 2Na^+(aq)$$

They will even displace hydrogen from very weak acids such as alcohols, to form compounds called alkoxides. (The reactions of alkali metals with alcohols are covered on page 298.)

## REACTION WITH AIR

Unlike most metals, the alkali metals are very soft and easy to cut with a knife. The surface obtained after cutting is shiny, but almost immediately tarnishes by reaction with moisture and/or oxygen from the air (Fig 22.6).

All the alkali metals react with air to form a complex mixture of compounds, including the corresponding carbonate. The following reaction scheme shows one way in which sodium may react with air:

$$4Na(s) + O_2(g) \rightarrow 2Na_2O(s)$$

$$Na_2O(s) + H_2O(g) \rightarrow 2NaOH(s)$$

$$2NaOH(s) + CO_2(s) \rightarrow Na_2CO_3.H_2O(s)$$

**Fig 22.6** Sodium, potassium or lithium can be easily cut by a knife to reveal a shiny, silver metal. This surface tarnishes quickly, with the formation of oxide, hydroxide and eventually carbonate

## REMEMBER THIS

The colour of a flame is a result of electron excitation and the consequent release of energy as the electron falls back to a lower energy level. You can read more about electron excitation on pages 52 and 599.

## REACTION OF ALKALI METALS WITH OXYGEN

All alkali metals burn when heated in oxygen, to produce oxides. The combustion is always accompanied by a coloured flame characteristic of the element. Lithium burns with a red flame, sodium with a yellow flame and potassium with a lilac flame.

The oxides formed are not always the predicted $M_2O$, containing the $M^+$ and $O^{2-}$ ions. Peroxides that contain the $O_2^{2-}$ ion and superoxides that contain the ion $O_2^-$ are formed with the more reactive alkali metals. The stability of the peroxides and the superoxides increases with the size of the cation. So, caesium oxide forms $CsO_2$ when burnt in excess oxygen, whereas lithium forms $Li_2O$:

$$4Li(s) + O_2(g) \rightarrow 2Li_2O(s)$$

$$Cs(s) + O_2(g) \rightarrow CsO_2(s)$$

Superoxides are solids. They are very powerful oxidizing agents, and can oxidize water to form oxygen. This is the reaction in some types of breathing mask used in mine rescue, where potassium superoxide is the source of the emergency oxygen supply to the wearer. Moisture reacts with the superoxide to provide this oxygen:

$$4KO_2(s) + 2H_2O(l) \rightarrow 4KOH(s) + 3O_2(g)$$

Note that 4 moles of potassium superoxide provide 3 moles of oxygen. This means that one gram of potassium superoxide can provide approximately 250 $cm^3$ of oxygen at room temperature and atmospheric pressure.

The attraction of this reaction is that as the superoxide is used up, it makes potassium hydroxide, which removes carbon dioxide. This prevents the wearer from breathing in large quantities of carbon dioxide:

$$2KOH(s) + CO_2(g) \rightarrow K_2CO_3(s) + H_2O(l)$$

The overall effect is the removal of carbon dioxide and formation of oxygen. Potassium superoxide can also be used in submarines to provide oxygen, while at the same time it prevents the build-up of dangerous levels of carbon dioxide.

**Fig 22.7** Gas masks protect miners and potholers from excess carbon dioxide in the air

**G** Write down equations to show the reaction of:
**a)** sodium to form sodium peroxide, $Na_2O_2$;
**b)** potassium to form potassium peroxide;
**c)** rubidium to form rubidium superoxide.

# 4 COMPOUNDS OF THE ALKALI METALS

The common feature of almost all compounds of the alkali metals is their high solubility in water. All the compounds of alkali metals are ionic and so they have a giant ionic lattice. As a result they have high melting points and can be electrolysed both in aqueous solution and as molten liquids.

YOU CAN READ ABOUT THE CONDUCTIVITY OF IONIC COMPOUNDS ON PAGE 452 IN CHAPTER 21.

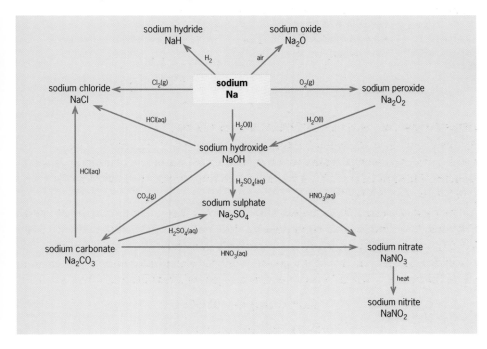

**Fig 22.8** Some important reactions of sodium and its compounds

**? TEST YOURSELF**

**H** Write an equation to show the reaction of rubidium oxide with cold water. Include state symbols in your answer.

**✓ REMEMBER THIS**

A soluble base forms an alkaline solution.

**YOU CAN READ ABOUT STRONG AND WEAK ACIDS ON PAGE 367 IN CHAPTER 17 AND ABOUT STRONG AND WEAK BASES ON PAGE 626 IN CHAPTER 28.**

**THE REACTION OF SODIUM HYDROXIDE WITH CHLORINE IS DESCRIBED ON PAGE 502 IN CHAPTER 23.**

## ALKALI METAL OXIDES

The normal oxides of the alkali metals contain the $O^{2-}$ ion, and so are basic and react with water to give aqueous hydroxides – alkalis. Bases are proton acceptors, so the oxide ion accepts a proton from water to form the hydroxide ion:

$$O^{2-}(aq) + H_2O(l) \rightarrow 2OH^-(aq)$$

Take sodium oxide, $Na_2O$, as an example:

$$Na_2O(s) + H_2O(l) \rightarrow 2NaOH(aq)$$

## HYDROXIDES

Sodium hydroxide – commonly called caustic soda – is a strong alkali. In water, it is fully dissociated into aqueous sodium ions and aqueous hydroxide ions and is referred to as a strong base. All the alkali metal hydroxides are strong bases and react with both weak and strong acids to give a salt and water:

$$OH^-(aq) + H^+(aq) \rightarrow H_2O(l)$$

$$KOH(aq) + HNO_3(aq) \rightarrow KNO_3(aq) + H_2O(l)$$

## SODIUM HYDROXIDE IN QUALITATIVE ANALYSIS

In the laboratory aqueous sodium hydroxide is used as a reagent to test for metal ions in solutions. Aqueous sodium hydroxide precipitates metal ions as their insoluble hydroxides. Very often, the hydroxides have characteristic colours that aid identification of the metal ion. Table 22.5 shows the reactions of aqueous hydroxide ions with various aqueous metal ions.

**Table 22.5 Reactions of aqueous sodium hydroxide with metal ions in solutions**

| Metal ion in solution | Reaction with aqueous sodium hydroxide | Equation |
|---|---|---|
| Calcium | White precipitate | $Ca^{2+}(aq) + 2OH^-(aq) \rightarrow Ca(OH)_2(s)$ |
| Magnesium | White precipitate | $Mg^{2+}(aq) + 2OH^-(aq) \rightarrow Mg(OH)_2(s)$ |
| Copper(II) | Pale blue precipitate | $Cu^{2+}(aq) + 2OH^-(aq) \rightarrow Cu(OH)_2(s)$ |
| Iron(II) | Green precipitate that slowly darkens | $Fe^{2+}(aq) + 2OH^-(aq) \rightarrow Fe(OH)_2(s)$ |
| Iron(III) | Rust-red precipitate | $Fe^{3+}(aq) + 3OH^-(aq) \rightarrow Fe(OH)_3(s)$ |
| Cobalt(II) | Blue precipitate that turns pink in excess aqueous sodium hydroxide with warming | $Co^{2+}(aq) + 2OH^-(aq) \rightarrow Co(OH)_2(s)$ |
| Nickel(II) | Apple-green precipitate | $Ni^{2+}(aq) + 2OH^-(aq) \rightarrow Ni(OH)_2(s)$ |
| Manganese(II) | White precipitate that rapidly darkens in air | $Mn^{2+}(aq) + 2OH^-(aq) \rightarrow Mn(OH)_2(s)$ followed by $2Mn(OH)_2(s) + O_2(g) \rightarrow 2MnO_2.H_2O(s)$ |
| Chromium(III) | Green precipitate that redissolves in excess to give a green solution of aqueous chromate(III) ion | $Cr^{3+}(aq) + 2OH^-(aq) \rightarrow Cr(OH)_3(s)$ followed by $Cr(OH)_3(aq) + 3OH^-(aq) \rightarrow [Cr(OH)_6]^{3-}(aq)$ |
| Silver(I) | Dark-brown precipitate | $2Ag^+(aq) + 2OH^-(aq) \rightarrow Ag_2O(s) + H_2O(l)$ |
| Zinc | White precipitate that redissolves into a colourless solution in excess aqueous sodium hydroxide because of the formation of a soluble zincate ion | $Zn^{2+}(aq) + 2OH^-(aq) \rightarrow Zn(OH)_2(s)$ followed by $Zn(OH)_2(s) + 2OH^-(aq) \rightarrow [Zn(OH)_4]^{2-}(aq)$ |
| Lead(II) | White precipitate that redissolves into a colourless solution in excess sodium hydroxide because of the formation of soluble plumbate(II) ion | $Pb^{2+}(aq) + 2OH^-(aq) \rightarrow PbO(s) + H_2O(l)$ followed by $PbO(s) + 2OH^-(aq) \rightarrow PbO_2^{2-}(aq) + H_2O(l)$ |
| Aluminium | White precipitate that redissolves into a colourless solution in excess sodium hydroxide because of the formation of soluble aluminate ion | $Al^{3+}(aq) + 3OH^-(aq) \rightarrow Al(OH)_3(s)$ followed by $Al(OH)_3(s) + OH^-(aq) \rightarrow [Al(OH)_4]^-(aq)$ |

Ca²⁺ Mg²⁺ Cu²⁺ Fe²⁺ Fe³⁺ Co²⁺ Ni²⁺ Mn²⁺ Cr³⁺ Ag⁺ Zn²⁺ Pb²⁺ Al³⁺

**Fig 22.9** Aqueous sodium hydroxide can be used in qualitative analysis, because it gives characteristic coloured solutions and precipitates (lower left) with aqueous metal ions

## OXY-SALTS

The nitrates, sulphate and carbonates are much more thermally stable than those of other metals. The sulphates and carbonates are thermally stable at Bunsen-burner temperatures, with the exception of lithium carbonate. Even the nitrates decompose to form nitrites and oxygen rather than the oxide and oxygen. This contrasts with most other metal nitrates, which thermally decompose to give the oxide or the metal itself. Again, the exception is provided by a lithium compound, since lithium nitrate forms lithium oxide:

$$4LiNO_3(s) \rightarrow 2Li_2O(s) + 4NO_2(g) + O_2(g)$$

Potassium nitrate thermally decomposes to form potassium nitrite:

$$2KNO_3(s) \rightarrow 2KNO_2(s) + O_2(g)$$

The alkali metals are the only group of metals that form stable, solid hydrogencarbonates. Sodium hydrogencarbonate is an important constituent in baking powder. It decomposes when heated to form sodium carbonate, carbon dioxide and water:

$$2NaHCO_3(s) \rightarrow Na_2CO_3(s) + H_2O(l) + CO_2(g)$$

The thermal stabilities of some of the carbonates and nitrates are described in more detail later in this chapter on pages 482 and 483.

**Fig 22.10** Baking powder contains sodium hydrogencarbonate. When heated in an oven, sodium hydrogencarbonate decomposes to form carbon dioxide, which helps the sponge cake to rise

# 5 ALKALINE EARTH METALS

The elements in Group 2 are collectively called the alkaline earth metals. With the exception of beryllium, these elements are all closely similar. The metallic character (ease of loss of electrons) of the elements increases with increasing atomic (proton) number and, with the exception of beryllium, their compounds are almost all ionic.

## OCCURRENCE AND EXTRACTION

Their compounds, particularly those of calcium and magnesium, are found extensively in the Earth's crust.

Electrolysis of the molten chloride of the element is the normal method of manufacture. The process is similar to that used to manufacture sodium. Magnesium is produced from molten magnesium chloride, much of it obtained from sea-water. It is estimated that 800 tonnes of sea-water are processed to obtain just 1 tonne of magnesium.

**Table 22.6 Occurrence of four Group 2 elements in the Earth's crust**

| Element | Abundance (% by mass) | Mineral |
|---|---|---|
| Magnesium | 1.93 | Magnesite ($MgCO_3$) |
| | | Dolomite ($CaCO_3.MgCO_3$) |
| | | Sea-water and brines |
| Calcium | 3.39 | Dolomite ($CaCO_3.MgCO_3$) |
| | | Marble and limestone ($CaCO_3$) |
| Strontium | 0.02 | Celestite ($SrSO_4$) |
| Barium | 0.04 | Barite ($BaSO_4$) |

## ? TEST YOURSELF

Use the trends in the physical properties of the Group 2 elements to predict the density, the melting point and the boiling point of radium.

**Table 22.7 Properties of Group 2 elements**

| Metal | Atomic number | Oxidation number | Atomic radius/pm | Ionic radius ($M^{2+}$)/pm | Density/ kg m$^{-3}$ | Melting point/°C | Boiling point/°C | Electron configuration |
|---|---|---|---|---|---|---|---|---|
| Beryllium | 4 | +2 | 111 | 31 | 1850 | 1 278 | 2970 | $1s^2 2s^2$ |
| Magnesium | 12 | +2 | 160 | 65 | 1740 | 651 | 1107 | $1s^2 2s^2 2p^6 3s^2$ |
| Calcium | 20 | +2 | 197 | 99 | 1550 | 850 | 1490 | $1s^2 2s^2 2p^6 3s^2 3p^6 4s^2$ |
| Strontium | 38 | +2 | 215 | 113 | 2540 | 770 | 1384 | $[Kr]5s^2$ |
| Barium | 56 | +2 | 217 | 135 | 3500 | 704 | 1638 | $[Xe]6s^2$ |
| Radium | 88 | +2 | 220 | 152 | See l | See l | See l | $[Rn]7s^2$ |

## ! SCIENCE FEATURE  Teeth and bones

'Calcium is good for your bones and teeth.' Is this popular saying correct?

Teeth and bone are composed of two main constituents. One is a protein called collagen, and the other is a complex calcium phosphate compound with the approximate formula $Ca_{10}(PO_4)_6.(OH)_2$. This complex calcium phosphate is called hydroxyapatite. Bone acts as the body's calcium store, but calcium can be removed from it. Unless bones are kept under load, they begin to lose their calcium, which happens to astronauts who stay in space stations for a long time.

In teeth, the complex calcium phosphate undergoes changes in the presence of fluoride ions, since there is a partial replacement of the hydroxide ion in hydroxyapatite with fluoride ions to form fluoroapatite. The fluoroapatite is much more resistant to acids in the mouth and so its presence in teeth reduces decay. This is the reason why fluoride ions are added to toothpastes and some water supplies. It is also known that fluoride ions promote bone growth, but the mechanism is not yet fully understood.

**Fig 22.11** Prolonged space travel can lead to a drastic reduction in bone mass, because the calcium ions within the bone are reabsorbed

# 6 REACTIONS OF THE GROUP 2 ELEMENTS

✓ **REMEMBER THIS**

$Be^{2+}$ is a very small, highly charged ion. It will strongly polarize negative ions, see page 85

Atoms of Group 2 elements lose two electrons to form a stable electron configuration. Their reactivity increases with increasing atomic (proton) number. This is explained by the increasing ease with which electrons can be lost from the atoms. Beryllium is unusual in that it does not easily lose two electrons to form $Be^{2+}$, and many compounds of beryllium are covalent. So, with the exception of beryllium, Group 2 elements are reactive metals because they can very easily lose two electrons per atom. This means that the oxidation number of Group 2 elements in their compounds is always +2.

**Table 22.8 Ionization energy and oxidation potential of five alkaline earth metals.**
The sum of the first and second ionization energies represents the energy transferred when two electrons are lost from each atom in a mole of gaseous atoms:
$$M(g) \rightarrow M^{2+}(g) + 2e^-$$
The oxidation potential is a measure of the ease of electron loss in the reaction:
$$M(s) \rightarrow M^{2+}(aq) + 2e^-$$

| Metal | First ionization energy/kJ mol$^{-1}$ | Second ionization energy/kJ mol$^{-1}$ | Sum of first and second ionization energies/kJ mol$^{-1}$ | Oxidation potential, $E_{oxid}$/V |
|---|---|---|---|---|
| Beryllium | 900 | 1760 | 2660 | +1.85 |
| Magnesium | 736 | 1450 | 2186 | +2.37 |
| Calcium | 590 | 1150 | 1740 | +2.87 |
| Strontium | 548 | 1060 | 1608 | +2.89 |
| Barium | 502 | 966 | 1468 | +2.91 |

The data in Table 22.8 demonstrate the increasing ease with which electrons can be lost. The reactivity of each Group 2 element is less than that of the Group 1 element in the same period because two electrons are lost per atom rather than one, which requires more energy. For example, potassium is considerably more reactive than calcium.

Since Group 2 elements lose two electrons per atom in reactions, it is the sum of the first and second ionization energies that must be considered when comparing reactivities. The ionization energy data and the electrode potential data both indicate that the reactivity of the Group 2 elements increases with atomic number.

**YOU CAN READ ABOUT IONIZATION ENERGY AND ELECTRON LOSS ON PAGE 56, AND ABOUT OXIDATION POTENTIALS ON PAGE 545.**

## REACTION WITH WATER

As already noted, the reactivity of each Group 2 metal is lower than that of the Group 1 metal in the same period. This is exemplified by the reaction of the metals with water. There is virtually no reaction between magnesium and cold water, but with hot water it slowly forms magnesium hydroxide and hydrogen whereas sodium reacts violently with cold water. When hot steam is passed over heated magnesium (Fig 22.12), an exothermic reaction occurs with the formation of hydrogen:

$$Mg(s) + H_2O(g) \rightarrow MgO(s) + H_2(g)$$

Note that magnesium oxide is produced rather than magnesium hydroxide.

Calcium, the next metal in Group 2, is more reactive than magnesium. It reduces cold water to form an alkaline solution, aqueous calcium hydroxide, commonly known as lime-water:

$$Ca(s) + 2H_2O(l) \rightarrow Ca(OH)_2(aq) + H_2(g)$$

**Fig 22.12** The reaction of magnesium with steam

## ? TEST YOURSELF

**J** Write down equations to show the reaction of barium and strontium with water.

**SEE QUESTION 1.**

The ionic equation shows the formation of an alkaline solution with calcium, strontium, barium or radium.

$$M(s) + 2H_2O(l) \rightarrow M^{2+}(aq) + 2OH^-(aq) + H_2(g)$$

This means that the pH of water increases as the reaction proceeds and, since the basicity and solubility of the hydroxides increases down the group, the reaction with barium and water will give a higher pH value than that with calcium (approximately pH 9–11 for calcium).

## REACTION WITH AIR AND OXYGEN

All the Group 2 elements tarnish in air to form a coating of the oxide. This reaction is rapid with those elements having a high atomic number. When these metals are heated in air or oxygen, they burn vigorously to produce the white ionic oxide. Magnesium burns in air and produces a brilliant white light (Fig 22.13):

$$2Mg(s) + O_2(g) \rightarrow 2MgO(s)$$

## ? TEST YOURSELF

**K** Explain why the reactivity of the Group 2 elements increases down the group. Use ideas about electron loss.

**L** Explain why the reaction between magnesium and sulphuric acid is a redox reaction in terms of:
**a)** electron transfer;
**b)** changes in oxidation number.

**M** Write balanced equations for the reactions between:
**a)** calcium and hydrochloric acid;
**b)** barium and hydrochloric acid;
**c)** magnesium and dilute ethanoic acid, $CH_3COOH$.

**Fig 22.13** Magnesium burns with a brilliant white flame

**Fig 22.14** Calcium oxide emits white light when it is heated strongly. Blocks of calcium oxide heated by gas burners were used as stage lights in theatres before electricity was available. It was from this that the phrase 'being in the limelight' originated

## REACTIONS WITH ACIDS

The Group 2 elements react with dilute acids to release hydrogen. The reactivity of the metals increases down the group, as it does with water. The reactions are an example of redox reactions, since the metal atom loses electrons (oxidation) and hydrogen ions gain electrons (reduction). This is represented by the overall ionic equation:

$$M(s) + 2H^+(aq) \rightarrow M^{2+}(aq) + H_2(g)$$

where M = Mg, Ca, Sr, Ba or Ra.

The redox reaction can also be explained in terms of the change in oxidation number. So when magnesium reacts with dilute hydrochloric acid, the oxidation number of magnesium changes from 0 to +2 (oxidation) and that of hydrogen from +1 to 0 (reduction):

$$Mg(s) + 2HCl(aq) \rightarrow MgCl_2(aq) + H_2(g)$$

# 7 COMPOUNDS OF THE GROUP 2 METALS

With the exception of beryllium, most of the compounds of the Group 2 metals are ionic. Although it is a property of ionic compounds that they are soluble in water, some compounds are insoluble and there are distinct trends in solubility with similar compounds. For example, the solubilities of the carbonates and sulphates decrease, but the solubilities of the hydroxides increase as the atomic number of the Group 2 element increases.

## OXIDES AND HYDROXIDES

The oxides and hydroxides of Group 2 elements are basic and neutralize acids. The oxide ion or the hydroxide ion reacts with aqueous protons to form water:

$$O^{2-} + 2H^+ \rightarrow H_2O$$

$$OH^- + H^+ \rightarrow H_2O$$

The solubility of the hydroxides in water increases with increasing atomic (proton) number. Magnesium hydroxide is insoluble in water, whereas barium hydroxide is soluble. The reason for this change in solubility is explained on page 483. When water is added to the oxides, a comparable difference in solubility is observed, but this is because a chemical reaction takes place. For example, calcium oxide (quicklime) reacts with water to form aqueous calcium hydroxide or lime-water, which is sparingly soluble in water:

$$CaO(s) + H_2O(l) \rightarrow Ca(OH)_2(aq)$$

When water is dropped slowly onto the oxide, slaked lime (solid calcium hydroxide) is formed:

$$CaO(s) + H_2O(l) \rightarrow Ca(OH)_2(s)$$

Notice the importance of the state symbols in these two equations – without the symbols the equations would be identical, but they describe different processes.

## USES OF THE OXIDES AND HYDROXIDES

The basic properties of the oxides and hydroxides are used extensively to neutralize acids in a variety of applications. Slaked lime is used to neutralize acid soils and lakes. A suspension of magnesium hydroxide in milk of magnesia is used to cure indigestion.

**Fig 22.15** Milk of magnesia can be in tablet form or a suspension of magnesium hydroxide in water. It is used to neutralize excess acid in the stomach

**Fig 22.16** Farmers use slaked lime to neutralize acid soils

**Fig 22.17** Most lakes in Scandinavia are highly acidic due to acid rain. The acid can be neutralized by spreading slaked lime

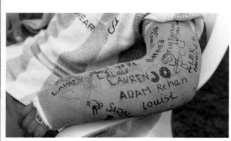

**Fig 22.18** Plaster of Paris is a hydrate of calcium sulphate $(CaSO_4)_2.H_2O$. When water is added to this hydrate, it forms gypsum – a different hydrate of calcium sulphate, $CaSO_4.2H_2O$ – which sets solid. Plaster of Paris is used to make moulds

**SEE QUESTIONS 1, 2 AND 3.**

## LIME-WATER TEST FOR CARBON DIOXIDE

Aqueous calcium hydroxide (lime-water) is used to test for carbon dioxide. It gives a characteristic white precipitate often described as 'milky'. This is a further example of the basic behaviour of the hydroxides, since carbon dioxide is an acidic gas and reacts to form insoluble calcium carbonate:

$$Ca(OH)_2(aq) + CO_2(g) \rightarrow CaCO_3(s) + H_2O(l)$$

When more carbon dioxide is bubbled into the white precipitate, it redissolves to form a colourless solution of aqueous calcium hydrogencarbonate:

$$CaCO_3(s) + CO_2(g) + H_2O(l) \rightarrow Ca(HCO_3)_2(aq)$$

## OXY-SALTS

The oxy-salts (such as the nitrates and sulphates) of the Group 2 elements are white ionic substances. They show distinct trends in their solubility and thermal decomposition, which are explained on pages 482 to 483 in terms of the bonding between ions.

## SULPHATES

The solubility of the sulphates of the Group 2 elements decreases as the atomic (proton) number of the element increases. The reason for this trend is described on pages 485 to 486.

The aqueous barium chloride or barium nitrate test for sulphate ions rests on the insolubility of barium sulphate. A solution of the test chemical is mixed with aqueous barium ions and dilute nitric acid. The sulphate ion is present in the test solution if a white precipitate is formed and no sulphur dioxide is formed:

$$Ba^{2+}(aq) + SO_4^{2-}(aq) \rightarrow BaSO_4(s)$$

## CARBONATES

The carbonates of the Group 2 elements are all considered to be insoluble in water. Therefore, each can be prepared through precipitation by reacting a soluble carbonate, such as aqueous sodium carbonate, with a soluble salt of the element:

$$CO_3^{2-}(aq) + M^{2+}(aq) \rightarrow MCO_3(s)$$

The thermal stability of the carbonates increases with increasing atomic number of the Group 2 elements. Beryllium carbonate decomposes so easily that it cannot be isolated at room temperature. Magnesium carbonate decomposes at 540 °C and barium carbonate at about 1360 °C. The reason for this variation in thermal stability is described on page 482.

$$MCO_3(s) \rightarrow MO(s) + CO_2(g)$$

where M = Be, Mg, Ca, Sr, Ba or Ra.
Calcium oxide (quicklime) is made in large quantities by the thermal decomposition of calcium carbonate (limestone). A temperature of between 900 and 1200 °C is needed, and it is important to remove the carbon dioxide produced, since the decomposition is reversible:

$$CaCO_3(s) \rightleftharpoons CaO(s) + CO_2(g)$$

**Fig 22.19** Cement is made from calcium carbonate. Calcium carbonate and clay are heated strongly together to form a mixture of calcium silicate and aluminium silicate. When cement is mixed, first with a ballast of sand and gravel and then with water, it makes concrete

## SCIENCE FEATURE Calcium carbonate, acid rain and FGD plants

Acid rain results from the presence in the air of sulphur dioxide and oxides of nitrogen. These gases react with the water and oxygen in the air to produce a mixture of acids, including dilute sulphuric acid and dilute nitric acid. A major source of these gases is the power station that uses fossil fuels contaminated with sulphur.

Coal and orimulsion (a mixture of a tar and water imported from South America) are particularly rich in sulphur and cause the most problems at power stations. Oxides of nitrogen are formed in the very high temperatures reached when the fossil fuels burn. They are sufficiently high to allow nitrogen and oxygen from the air to react directly to form nitrogen monoxide. When cool, the nitrogen monoxide reacts with more oxygen to give acidic nitrogen dioxide.

It has now become a high priority to remove these acidic gases from the waste or flue gases before they are emitted. A process called flue-gas desulphurization (FGD) is being introduced, in which cold waste gases are treated with powdered calcium carbonate. The acidic gases react with the calcium carbonate to give calcium nitrate, calcium nitrite,

$Ca(NO_2)_2$, and, mostly, calcium sulphite, $CaSO_3$:

$$SO_2(g) + CaCO_3(s) \rightarrow CaSO_3(s) + CO_2(g)$$

$$4NO_2(g) + 2CaCO_3(s) \rightarrow Ca(NO_2)_2(s) + Ca(NO_3)_2(s) + 2CO_2(g)$$

Almost 90 per cent of the sulphur dioxide and nitrogen dioxide produced by a power station can be removed by this process. The calcium sulphite produced is converted by reaction with oxygen into calcium sulphate, which is sold as gypsum and used in the building industry. The downside of the process is that large quantities of limestone have to be quarried to provide the calcium carbonate.

**Fig 22.20** Power stations remove acidic gases, such as sulphur dioxide, by reacting them with powdered calcium carbonate

## NITRATES

The nitrates of the elements of Group 2 are all white soluble solids that thermally decompose to produce the metal oxide, oxygen and nitrogen dioxide:

$$2M(NO_3)_2(s) \rightarrow 2MO(s) + 4NO_2(g) + O_2(g)$$

where M = Mg, Ca, Sr, Ba and Ra.

The thermal stability of the nitrates increase as the atomic number of the Group 2 elements increases. This trend is explained later on page 483.

## REMEMBER THIS

The more thermally stable a compound, the higher the temperature required to decompose it.

## ▶ EXTENSION The atypical behaviour of beryllium compounds

Beryllium is not a typical Group 2 metal; in fact, it resembles aluminium in Group 3 more than magnesium in Group 2. The reason for this is the very small ionic radius of $Be^{2+}$, which gives the ion a similar charge density to that of $Al^{3+}$.

The beryllium ion, $Be^{2+}$, is very polarizing and distorts the electron cloud of any anion to which it is bonded. This results in either decomposition of the compound, as in beryllium carbonate (which decomposes even at room temperature) or the compound being covalent rather than ionic. Beryllium chloride is a covalent chloride with a layer lattice as shown in Fig 22.21. This structure has each beryllium atom surrounded by four chlorine atoms, two bonded by dative bonds and two by covalent bonds. As a result the beryllium atom has a stable octet of electrons.

Beryllium hydroxide is also atypical in that it is amphoteric, whereas all the other hydroxides of Group 2 metals are basic, so it reacts with both acids and alkalis:

$$Be(OH)_2(s) + 2H_2O(l) + H_2SO_4(aq) \rightarrow$$
$$[Be(H_2O)_4]^{2+}SO_4^{2-}(aq)$$
$$Be(OH)_2(s) + 2NaOH(aq) \rightarrow Na_2Be(OH)_4(aq)$$

Notice that in both the reactions a product is obtained in which beryllium has a coordination number of four, namely four groups around it that donate a pair of electrons to give a stable octet of electrons.

**Fig 22.21** The structure of beryllium chloride

**SEE QUESTION 1.**

**REMEMBER THIS**

The enthalpy change for the breakdown of an ionic lattice into gaseous ions is numerically the same as the lattice energy, but positive. It is endothermic.

**? TEST YOURSELF**

Write down the reaction that corresponds to the lattice energy for each of the following ionic solids:

a) MgO;
b) $MgSO_4$;
c) $Fe_2(SO_4)_3$;
d) $Mg_3N_2$.

✔ **REMEMBER THIS**

A third factor that affects the magnitude of the lattice energy is the type of ionic lattice formed. A discussion of this factor lies beyond the scope of this book.

small ionic radius and high charge on ion produce a strong attraction for ion of opposite charge

large ionic radius and low charge on ion produce a weaker attraction for ion of opposite charge

**Fig 22.22** The magnitude of the attractive force between ions in a crystal lattice depends on both the ionic radius and the charge on the ion

**Table 22.9 Charge density of 15 ions**

| Ion | Charge | Ionic radius/nm | Charge density |
|---|---|---|---|
| Lithium | +1 | 0.060 | 277 |
| Sodium | +1 | 0.095 | 111 |
| Potassium | +1 | 0.133 | 57 |
| Rubidium | +1 | 0.148 | 46 |
| Beryllium | +2 | 0.031 | 2081 |
| Magnesium | +2 | 0.065 | 473 |
| Calcium | +2 | 0.099 | 204 |
| Strontium | +2 | 0.113 | 78 |
| Aluminium | +3 | 0.050 | 1200 |
| Oxide | −2 | 0.140 | 102 |
| Sulphide | −2 | 0.184 | 59 |
| Fluoride | −1 | 0.136 | 54 |
| Chloride | −1 | 0.181 | 31 |
| Bromide | −1 | 0.195 | 26 |
| Iodide | −1 | 0.216 | 21 |

# 8 LATTICE ENERGY

To explain the thermal decomposition and solubility of an ionic substance, the forces that exist between positive and negative ions in its ionic lattice must be understood. Chemists estimate the strength of attraction between a positive ion and a negative ion in an ionic lattice from the enthalpy change called **lattice energy** or **lattice enthalpy**. **Lattice energy is the energy released into the surroundings when one mole of an ionic lattice is made from its constituent gaseous ions**. Therefore, the lattice energy is always negative. The higher the magnitude of the lattice enthalpy, the greater the attraction between the positive and the negative ions.

For sodium chloride, the lattice enthalpy corresponds to the following reaction:

$$Na^+(g) + Cl^-(g) \rightarrow NaCl(s) \qquad \Delta H_{le}^{\ominus} = \text{lattice energy}$$

## FACTORS THAT AFFECT THE MAGNITUDE OF THE LATTICE ENERGY

The lattice energy of each ionic compound is different and depends on the degree of attraction between the negative and the positive ions. This attraction depends on two factors: the ionic radius and the charge on the ion.

The smaller the ionic radius, the greater the density of the positive or negative charge. This gives a greater numerical value for the lattice energy.

The greater the charge on the ion the greater the attraction for an ion of the opposite charge. This also gives a greater numerical value for the lattice energy.

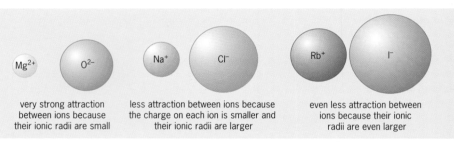

very strong attraction between ions because their ionic radii are small

less attraction between ions because the charge on each ion is smaller and their ionic radii are larger

even less attraction between ions because their ionic radii are even larger

**Fig 22.23** Relative sizes of, and attractions between, three pairs of ions. MgO has a greater lattice energy than NaCl, which has a greater lattice energy than RbI (scale: 1 mm to 20 nm)

## CHARGE DENSITY

The ionic radius and the charge on the ion can be brought together and treated as a single property of the ion called the **charge density**. The charge density is based on the assumption that the charge is spread over the outer surface of the ion, and defined as the charge per unit surface area. Although ions are not solid, we can still think of ions as having a boundary surface.

The surface area is proportional to the square of the ionic radius, and so the charge density of an ion is given by:

$$\text{charge density} = \frac{\text{charge on ion}}{(\text{ionic radius})^2}$$

## BORN–HABER CYCLES

Experimentally, it is impossible to use a direct method to determine the lattice energy of an ionic solid. So an indirect method that involves Hess's law has to be used. Hess's law states that if a change can be brought about by more than one route, then the enthalpy change for each route must be the same, provided that the starting and finishing conditions are the same for each route. (There is more about Hess's law on page 181.)

Applying Hess's law to the determination of lattice energy, there is more than one route from the gaseous ions to the ionic lattice. The direct route and the indirect route shown in Fig 22.24 is known as a Born–Haber cycle. All the enthalpy changes in the indirect route can be measured experimentally.

There are many ways to draw Born–Haber cycles, of which two are shown in Figs 22.24 and 22.25. Others appear elsewhere in this book.

One way to draw a Born–Haber cycle for sodium chloride is first to put down the equation for the formation of the lattice from its gaseous ions, and then to construct an alternative pathway. The pathway will always involve the energy changes shown for sodium chloride.

**Key**

$\Delta H_{le}^{\ominus}$ (NaCl) = lattice energy of sodium chloride

$\Delta H_i^{\ominus}$(Na) = first ionization energy for sodium

$\Delta H_{at}^{\ominus}$(Na) = enthalpy change of atomization for sodium. This is the energy needed to make 1 mole of gaseous atoms from the element in its standard state

$\Delta H_{at}^{\ominus}$(Cl$_2$) = enthalpy change of atomization of chlorine

$\Delta H_{ea}^{\ominus}$(Cl) = first electron affinity for chlorine. This is the enthalpy change to make 1 mole of gaseous negative ions X$^-$(g) from 1 mole of gaseous atoms X(g)

$\Delta H_f^{\ominus}$(NaCl) = enthalpy change of formation of sodium chloride

By Hess's Law:

$\Delta H_{le}^{\ominus}$ (NaCl) = $-\Delta H_i^{\ominus}$(Na) $- \Delta H_{at}^{\ominus}$(Na) $- \Delta H_{ea}^{\ominus}$(Cl) $- \Delta H_{at}^{\ominus}$(Cl$_2$) $+ \Delta H_f^{\ominus}$ (NaCl)

= −496 − 109 − 122 + 368 − 411

= −770 kJ mol$^{-1}$

An alternative way to draw a Born–Haber cycle for sodium chloride is to use a vertical energy axis.

The elements sodium and chlorine in their standard states have an enthalpy of formation of 0 kJ mol$^{-1}$. All endothermic reactions are shown by an arrow pointing upwards, and all exothermic reactions by an arrow pointing downwards. Notice that the lattice energy is exothermic.

**Fig 22.24** Born–Haber cycles for working out the lattice energy of the ionic solid NaCl

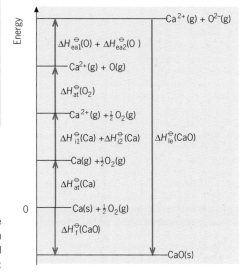

**Fig 22.25** Born–Haber cycle for the formation of calcium oxide. Notice that the ionization energy term for calcium involves both the first and second ionization energies. In the same way, the electron affinity term involves both the first and second electron affinities:

$$O(g) + e^- \rightarrow O^-(g) \quad \Delta H_{ea1}^{\ominus}(O)$$
$$O^-(g) + e^- \rightarrow O^{2-}(g) \quad \Delta H_{ea2}^{\ominus}(O)$$

**SEE QUESTION 4.**

# 9 THERMAL DECOMPOSITION OF IONIC SALTS

The thermal decomposition of an ionic salt must necessarily involve the breakdown of one ionic lattice and the formation of another. For example, when magnesium carbonate is decomposed, it forms magnesium oxide and carbon dioxide. This requires that the lattice of magnesium carbonate is completely broken and the formation of the lattice of magnesium oxide. It therefore follows that the enthalpy change that occurs during this decomposition must be related to the lattice energies of the two ionic substances involved.

## DECOMPOSITION OF THE CARBONATES

The carbonates of Group 2 elements decompose as follows:

$$MCO_3(s) \rightarrow MO(s) + CO_2(g)$$

where M = Be, Mg, Ca, Sr, Ba or Ra.

As already stated, the thermal stability of the carbonates increases as the atomic (proton) numbers of the Group 2 elements increase. This trend is best explained in terms of the ability of the metal ion to **polarize** the carbonate ion and of the magnitude of the lattice energy for the oxide formed.

### POLARIZATION

Small, highly charged, positive ions can polarize negative ions. That is, they are able to distort the electron cloud of the negative ion. Negative ions that have a large ionic radius, such as the carbonate ion, are much more easily polarized than ions that have a small ionic radius, such as an oxide ion. Since the cations $M^{2+}$ (where M is one of the Group 2 elements) all have the same charge, it is the ion with the smallest ionic radius that polarizes the carbonate ion the most. $Be^{2+}$ has the smallest ionic radius and so it causes most polarization of the carbonate ion. Once the carbonate ion is highly polarized, it forms an oxide ion and carbon dioxide. This means that $BeCO_3$ is the least thermally stable and $RaCO_3$ the most thermally stable, since $Ra^{2+}$ has the largest ionic radius.

**THERE IS MORE INFORMATION ON POLARIZATION ON PAGE 85.**

### LATTICE ENERGY

An additional factor is the magnitude of the lattice energy of the oxide produced. This lattice energy is more exothermic with a small cation than with a large cation. So, in the decomposition of magnesium carbonate, the formation of magnesium oxide with a highly exothermic lattice energy can be considered as the driving force for the thermal decomposition.

**SEE QUESTIONS 3, 5 AND 6.**

**Fig 22.26** Magnesium oxide is often used to line furnaces. The exceptionally high lattice energy of magnesium oxide explains why it has a very high melting point

## DECOMPOSITION OF THE CARBONATES OF GROUP 1

The carbonates of Group 1 metals are much more thermally stable than those of Group 2 metals, because the cation has only a single rather than a double positive charge. Lithium carbonate is the least thermally stable, since $Li^+$ has a very small ionic radius, and hence a large charge density, and so can strongly polarize the carbonate ion:

$$Li_2CO_3(s) \rightarrow Li_2O(s) + CO_2(g)$$

# THERMAL DECOMPOSITION OF NITRATES OF GROUPS 1 AND 2

The thermal stabilities of the nitrates of Group 2 elements show the same trend as the carbonates: namely, the stability increases with increasing atomic (proton) number. The reason for this trend in thermal stability is again that the polarization power of the $M^{2+}$ decreases as the ionic radius increases:

$$2M(NO_3)_2(s) \rightarrow 2MO(s) + 4NO_2(g) + O_2(g)$$

where M = Mg, Ca, Sr, Ba and Ra.

Since the Group 1 nitrates have ions that normally are much less polarizing, the decomposition products are the nitrites rather than the oxides:

$$2MNO_3(s) \rightarrow 2MNO_2(s) + O_2(g)$$

where M = Na, K, Rb and Cs.

# 10 SOLUBILITY OF IONIC SALTS

Most ionic salts dissolve in polar solvents such as water. A **solvent** is a liquid that can dissolve other substances. The most common solvent is water. A **solute** is the substance that is dissolved in the solvent. A **solution** is the name given to the mixture of the solute and the solvent. For example, a sodium chloride solution is simply a mixture of water, the solvent, and sodium chloride, the solute.

## SOLUBILITY

The solubility of a substance is defined as the maximum concentration that can be obtained when the substance (the solute) is dissolved in a solvent. This concentration is usually measured either in grams of solute per 100 grams of solvent or in mol dm$^{-3}$. Throughout this section of the book, the chosen unit of solubility is mol dm$^{-3}$.

Solubility changes with temperature, so it should always be quoted at a particular temperature, usually 298 K.

## THE DISSOLVING PROCESS

This section considers what happens when an ionic salt dissolves in water.

In the regular arrangement of an ionic lattice, the cations are electrostatically attracted to the anions (see page 451). When an ionic salt dissolves, these electrostatic forces of attraction are broken and the anions and cations are free to move throughout the solution. In a polar solvent, such as water, the ions are surrounded by water molecules. Since the electronegativity of oxygen is much higher than that of hydrogen, the oxygen atom of a water molecule carries a small negative charge and so is weakly attracted to the cations. The hydrogen atoms of a water molecule carry a small positive charge and so they are weakly attracted to the anions. As a result, both cations and anions are surrounded by many water molecules.

When sodium chloride dissolves, aqueous sodium ions and aqueous chloride ions are formed:

$$NaCl(s) \rightarrow Na^+(aq) + Cl^-(aq)$$

The symbols $Na^+(aq)$ and $Cl^-(aq)$ represent the respective ion surrounded by many water molecules.

Overall, the dissolving process can be considered to be two processes:
● the breaking of the electrostatic attraction between ions in the ionic lattice;
● the formation of aqueous ions with attractive forces being established between the water molecules and these ions.

**Fig 22.27** The dissolving process

## ENERGY TRANSFERS DURING DISSOLVING

When an ionic compounds dissolves in water, there is always an energy transfer. Normally, the energy transfer process is modest and therefore there is only a small change in the temperature of the water. Energy is required to break the electrostatic attraction between ions, and energy is released when the attractive forces between water molecules and the ions are established. The numerical value for these two energy transfers (one endothermic and the other exothermic) is often very similar so that the overall energy change when making a solution is small. Occasionally, a dramatic change occurs in water temperature, but this is nearly always because a chemical reaction is taking place during the formation of the solution.

The **enthalpy change of solution, $\Delta H_{soln}$**, is defined as the enthalpy change when one mole of solute is dissolved in a solvent and extra dilution causes no further change in enthalpy.

## ENTROPY AND DISSOLVING

On page 179, entropy is described as a measure of the disorder of a system. When an ionic solid dissolves to make a solution, the system becomes more disordered (increases in entropy), because all the particles in the solution are free to move, whereas in the solid the particles were in fixed positions (lower entropy). Processes in which the entropy – the amount of disorder – increases are favoured over those in which the entropy does not increase. This means that all substances should dissolve in a solvent. But this is not the case because entropy changes in the surroundings must also be taken into account. The energy transfer processes offer a simpler explanation.

## BORN–HABER CYCLES FOR DISSOLVING

Fig 22.28 shows that the enthalpy change of solution is equal to the sum of three different energy transfer processes. The first of these processes is the lattice energy, the second is the enthalpy change of hydration of the cation and the third the enthalpy change of hydration of the anion. The **enthalpy change of hydration** $\Delta H_{\text{hyd}}$ is the enthalpy change that occurs when 1 mole of gaseous ion is dissolved in excess water to make one mole of aqueous ion.

Although the enthalpy change of hydration is very much a theoretical energy change, it does give a clear indication of the strength of the interaction of the water molecules with the ions. The enthalpy change of hydration is always exothermic, since it necessarily involves bond making between water molecules and the ions. The lattice energy contribution to the enthalpy change of solution is always endothermic, since it involves the breaking of the strong electrostatic interactions between the positive and the negative ions. The lattice energy and the sum of two enthalpy changes of hydration have roughly the same magnitude (but are of opposite sign), so that the enthalpy change of solution is always very small.

**Fig 22.28** Two ways of showing energy cycles for dissolving. The overall enthalpy change of solution is often small, since the exothermic contributions of the enthalpy changes of hydration are often cancelled out by the endothermic contribution of the lattice energy

## ENTHALPY CHANGE OF HYDRATION AND CHARGE DENSITY

The magnitude of the enthalpy change of hydration depends on the ionic radius and the charge on the ion. The smaller the ionic radius and the higher the charge on the ion, the more exothermic the enthalpy change of hydration. This is because an ion with a small radius and a high charge has a high charge density, and therefore water molecules are strongly attracted to it. Conversely, an ion with a large radius and a small charge has a low charge density, and therefore water molecules are weakly attracted to it.

Within Group 2, the enthalpy change of hydration of $M^{2+}$ becomes less exothermic as the atomic (proton) number increases:

$$M^{2+}(g) \xrightarrow{\text{H}_2\text{O}(l)} M^{2+}(aq) \quad \Delta H_{\text{hyd}}^{\ominus} = \text{enthalpy change of hydration}$$

## SOLUBILITY TRENDS OF THE GROUP 2 SULPHATES

On page 477, it is stated that the solubility of the sulphates decreases as the atomic (proton) number of the Group 2 elements increases. This trend can be explained in terms of the energy transfer processes that take place during dissolving.

As Fig 22.29 shows, the enthalpy change of hydration of the $SO_4^{2-}(g)$ is common to all Born–Haber cycles for all of the Group 2 sulphates, so it cannot be responsible for the solubility trend. The lattice energy of the Group 2 sulphates, $MSO_4(s)$, becomes less exothermic as the atomic number of M increases, but the change in lattice energy from $MgSO_4$ to $BaSO_4$ is quite small because of the large size of the sulphate ion.

**Table 22.10 Solubility of the Group 2 sulphates**

| Compound | Solubility at 298 K/mol dm$^{-3}$ |
|---|---|
| Magnesium sulphate | 2.8 |
| Calcium sulphate | $1.4 \times 10^{-3}$ |
| Strontium sulphate | $7.6 \times 10^{-5}$ |
| Barium sulphate | $1.1 \times 10^{-6}$ |

However, the difference in the enthalpy change of hydration of $M^{2+}(g)$ from $Mg^{2+}$ to $Ba^{2+}$ is large because of the big difference in ionic radii between the very small ion $Mg^{2+}$ and the large ion $Ba^{2+}$. Referring to Fig 22.29, it is the enthalpy change of hydration of $M^{2+}(g)$ that determines the solubility trend. The more exothermic this process, the more likely $MSO_4(s)$ is to dissolve.

To sum up: the enthalpy change of solution of $MSO_4(s)$ becomes less exothermic as the atomic number increases, and hence the solubility decreases.

**Fig 22.29** Born–Haber cycle for the determination of the enthalpy of solution, $\Delta H_{soln}$, for the sulphates of Group 2. Note that the enthalpy change of hydration of $SO_4^{2-}(g)$ is common to all the sulphates of Group 2

For the same reason, all salts of Group 2 elements that have an anion with a large ionic radius show the same solubility trend as the sulphates.

## SOLUBILITY TREND OF THE GROUP 2 HYDROXIDES

The solubility of hydroxides, $M(OH)_2(s)$, shows the opposite trend to that of the sulphates. Namely, as the atomic (proton) number increases, the solubility increases. This time, the key factor that determines solubility is the lattice energy of the $M(OH)_2(s)$ rather than the enthalpy change of hydration. Since the hydroxide ion is small, there is a considerable difference between the very high lattice energy of magnesium hydroxide and the relatively low lattice energy of barium hydroxide.

As Fig 22.30 shows, a large lattice energy makes a large endothermic contribution to the enthalpy change of solution and so reduces solubility. Other salts of Group 2 elements with small anions, such as the fluorides, display a solubility trend similar to that of the hydroxides.

**Fig 22.30** The key factor in determining the solubility of the hydroxides of Group 2 is the change in the lattice energy from $Mg(OH)_2$ to $Ba(OH)_2$. $Mg(OH)_2$ has a much larger lattice energy than $Ba(OH)_2$, so the enthalpy change of solution of $Mg(OH)_2$ is less exothermic than that of $Ba(OH)_2$

| Table 22.11 Summary of solubility trends | | |
|---|---|---|
| **Anion** | **Magnesium salt** | **Barium salt** |
| Large, e.g. carbonate and sulphate | High solubility | Very low solubility |
| Small, e.g. hydroxide and fluoride | Low solubility | High solubility |

**SEE QUESTION 6.**

# 11 SOLUBILITY PRODUCT

The term 'insoluble' is not very suitable, since no ionic substance is completely insoluble in water. The dissociation process takes place even with substances referred to as insoluble, such as barium sulphate and calcium carbonate. Sometimes, the term 'sparingly soluble' is used to express the fact that the solubility of a substance is low. The term 'insoluble' should be taken to mean that the solubility of a substance is extremely low.

## SOLUBILITY PRODUCT

When calcium carbonate is shaken up with distilled water and left to settle, the water contains aqueous calcium ions and aqueous carbonate ions in very small concentrations. A dynamic equilibrium is set up that is best represented by the following equation:

$$CaCO_3(s) \rightleftharpoons Ca^{2+}(aq) + CO_3^{2-}(aq)$$

An expression can be written for the equilibrium constant for this process:

$$K_c = \frac{[Ca^{2+}(aq)][CO_3^{2-}]}{[CaCO_3(s)]}$$

**SEE PAGE 390 FOR AN EXPLANATION OF THE EQUILIBRIUM CONSTANT.**

The concentration of a solid is a constant and therefore can be included in a modified equilibrium constant known as the **solubility product, $K_{sp}$**:

$$K_{sp} = [Ca^{2+}(aq)][CO_3^{2-}(aq)]$$

The solubility product is the product of the concentrations of the aqueous ions formed when an insoluble or sparingly soluble ionic substance dissolves in water. Each concentration is raised to the power shown in the dissociation equation. For example, the dissociation equation for barium chromate(VI) is:

$$BaCrO_4(s) \rightleftharpoons Ba^{2+}(aq) + CrO_4^{2-}(aq)$$

Therefore:       $K_{sp}(BaCrO_4) = [Ba^{2+}(aq)][CrO_4^{-}(aq)]$

For lead(II) iodide, the dissociation equation is:

$$PbI_2(s) \rightleftharpoons Pb^{2+}(aq) + 2I^{-}(aq)$$

Therefore:       $K_{sp}(PbI_2) = [Pb^{2+}(aq)][I^{-}(aq)]^2$

> **? TEST YOURSELF**
>
> **V** Write down expressions for the solubility product of the following sparingly soluble salts:
> **a)** calcium sulphate, $CaSO_4$;
> **b)** barium sulphate, $BaSO_4$;
> **c)** silver chromate(VI), $Ag_2CrO_4$;
> **d)** lead(II) chloride, $PbCl_2$.

## CALCULATING SOLUBILITY PRODUCTS

The solubility product of a salt can be calculated by substituting the appropriate concentrations into the expression for the solubility product.

Notice in the first example on page 488 that the solubility product has a unit that depends on the expression for the solubility product. In this case, it is concentration squared. The unit can be worked out by substituting the initial units into the expression for the solubility product:

$$K_{sp}(CaCO_3) = [Ca^{2+}(aq)][CO_3^{2-}(aq)]$$
$$= (mol\ dm^{-3})(mol\ dm^{-3})$$
$$= mol^2\ dm^{-6}$$

> **? TEST YOURSELF**
>
> **W**   **a)** The solubility of barium chromate(VI), $BaCrO_4$, at 298 K is $1.4 \times 10^{-5}$ mol dm$^{-3}$. Calculate the solubility product for barium chromate(VI) at this temperature.
>
> **b)** The solubility of calcium hydroxide at 298 K is $1.25 \times 10^{-2}$ mol dm$^{-3}$. Show that the solubility product for calcium hydroxide at this temperature is $7.81 \times 10^{-6}$ mol$^3$ dm$^{-9}$

## EXAMPLE

**Q** What is the solubility product for calcium carbonate, given that the solubility of calcium carbonate at 298 K is $6.9 \times 10^{-5}$ mol dm$^{-3}$?

**A** Assume that excess calcium carbonate is shaken with distilled water until equilibrium is attained. The solubility of calcium carbonate corresponds to the concentration of the aqueous calcium ions (or the aqueous carbonate ions) in the equilibrium solution:

$$CaCO_3(s) \rightleftharpoons Ca^{2+}(aq) + CO_3^{2-}(aq)$$

At start (before shaking)/mol dm$^{-3}$      0      0

At equilibrium/mol dm$^{-3}$      $6.9 \times 10^{-5}$    $6.9 \times 10^{-5}$

$$K_{sp}(CaCO_3) = [Ca^{2+}(aq)][CO_3^{2-}(aq)]$$

$$= (6.9 \times 10^{-5})(6.9 \times 10^{-5})$$

$$= 4.8 \times 10^{-9} \text{ mol}^2 \text{ dm}^{-6}$$

## CALCULATING SOLUBILITY

When the solubility product is known, it is possible to work out the solubility of the solute.

## EXAMPLE

**Q** The solubility product of magnesium fluoride at 298 K is $6.4 \times 10^{-9}$ mol$^3$ dm$^{-9}$. What is the solubility of magnesium fluoride at 298 K?

**A** Let $s$ be the solubility of magnesium fluoride:

$$MgF_2(s) \rightleftharpoons Mg^{2+}(aq) + 2F^{-}(aq)$$

At equilibrium:      $s$      $2s$

$$K_{sp}(MgF_2) = [Mg^{2+}(aq)][F^{-}(aq)]^2$$

Therefore:      $6.4 \times 10^{-9} = s(2s)^2 = 4s^3$

Therefore:      $s = 1.2 \times 10^{-3}$ mol dm$^{-3}$ (to 2 significant figures).

**? TEST YOURSELF**

**X** Calculate the solubility at 298 K of each of the following:

**a)** barium fluoride, BaF$_2$, $K_{sp} = 1.7 \times 10^{-6}$ mol$^3$ dm$^{-9}$;

**b)** calcium fluoride, CaF$_2$, $K_{sp} = 3.9 \times 10^{-11}$ mol$^3$ dm$^{-9}$;

**c)** strontium sulphate, SrSO$_4$, $K_{sp} = 2.8 \times 10^{-7}$ mol$^2$ dm$^{-6}$.

**SEE QUESTIONS 3 AND 7.**

## COMMON-ION EFFECT

Since the process involved in the solubility product is a dynamic equilibrium, it is possible to change the position of equilibrium according to Le Chatelier's principle (see page 18 and 392).

Consider the solubility of barium carbonate. Aqueous barium ions and aqueous carbonate ions are in equilibrium with solid barium carbonate:

$$BaCO_3(s) \rightleftharpoons Ba^{2+}(aq) + CO_3^{2-}(aq)$$

When extra carbonate ions are added to the equilibrium mixture, the equilibrium shifts to the left to remove the extra carbonate ions. This causes a reduction in the solubility of barium carbonate, the concentration of aqueous barium ions having decreased. This is an example of the common-ion effect. The solubility of barium carbonate is also reduced by the addition of extra aqueous barium ions.

The common-ion effect can be demonstrated quantitatively by means of the solubility product.

## ● EXAMPLE

**Q** Calculate the solubility of barium chromate(VI) in **a)** water and **b)** 1.0 mol dm$^{-3}$ potassium chromate(VI), given that the solubility product of barium chromate is $2 \times 10^{-10}$ mol$^2$ dm$^{-6}$ at 298 K.

**A** **a)** Let $s$ be the solubility of barium chromate(VI).

$$BaCrO_4(s) \rightleftharpoons Ba^{2+}(aq) + CrO_4^{2-}(aq)$$

At equilibrium:                    $s$              $s$

Therefore:     $K_{sp} = [Ba^{2+}(aq)][CrO_4^{2-}(aq)]$

$$= s^2$$

This gives:     $s = 1.4 \times 10^{-5}$ mol dm$^{-3}$.

**b)** In this calculation, it is necessary to consider dissolving barium chromate(VI) in aqueous potassium chromate(VI):

$$BaCrO_4(s) \rightleftharpoons Ba^{2+}(aq) + CrO_4^{2-}(aq)$$

At start/mol dm$^{-3}$:                    0                1.0
At equilibrium/mol dm$^{-3}$:           $s$            $(1.0 + s)$

Barium chromate(VI) is sparingly soluble in water. Therefore $s$ is small compared with 1.0, and so $(s + 1.0)$ is approximately equal to 1.0.

Now:     $K_{sp} = [Ba^{2+}(aq)][CrO_4^{2-}(aq)]$
$$2 \times 10^{-10} = s(1.0)$$
This gives:     $s = 2 \times 10^{-10}$ mol dm$^{-3}$.

Note that the solubility of barium chromate(VI) is very much smaller in potassium chromate(VI) than in water alone. This is the common-ion effect in operation.

---

## ! SCIENCE FEATURE The barium meal

Doctors often need to use X-rays to look inside the intestine and the stomach for cancers, ulcers and blockages. However, using X-rays to diagnose such conditions does pose a problem, because the stomach and the intestine are transparent to X-rays (they pass through), so no image is detected – unlike bone, which is opaque to X-rays.

To overcome this, doctors give their patients a barium meal that contains insoluble barium sulphate and water. The stomach and intestine become coated with the barium sulphate, which is opaque to X-rays, and so a shadow of these organs is cast on an X-ray photographic film.

Using barium sulphate has one drawback: aqueous barium ions are highly toxic. Even though barium sulphate is said to be insoluble, there is still a small concentration of aqueous barium ions in the stomach, which could lead to poisoning. The equilibrium equation shows the production of barium ions:

$$BaSO_4(s) \rightleftharpoons Ba^{2+}(aq) + SO_4^{2-}(aq)$$

The concentration of the aqueous barium ions is reduced by the common-ion effect, brought about by magnesium sulphate added to the barium meal. The aqueous sulphate ions from the soluble magnesium sulphate shift the equilibrium to the left to lower the concentration of aqueous barium ions.

**Fig 22.31** The image of someone who has had a barium meal shows the stomach on the right and intestine on the left

---

## PRECIPITATION

Precipitation is the formation of an insoluble solid when two solutions react together. For example, when a solution that contains aqueous barium ions is mixed with a solution that contains aqueous carbonate ions, a white precipitate of insoluble barium carbonate is formed:

$$Ba^{2+}(aq) + CO_3^{2-}(aq) \rightarrow BaCO_3(s)$$

Note that this is the opposite process to the dissolving of sparingly soluble barium carbonate in water. This means that the solubility product predicts the conditions for a precipitation reaction. In fact, precipitation occurs when the ionic product for the compound exceeds its solubility product. The ionic product of barium carbonate is the product of the concentration of the aqueous barium ion and of the aqueous carbonate ion.

## ? TEST YOURSELF

**Y** Calculate the solubility of barium chromate in 0.1 mol dm$^{-3}$ barium chloride.

## ✔ REMEMBER THIS

Ionic product is the same expression as for $K_{sp}$, but uses the actual concentrations, not those at equilibrium.

**Z** Will a precipitate of magnesium fluoride, $K_{sp}(MgF_2) = 6.4 \times 10^{-9}$ $mol^3 dm^{-9}$, be formed from a solution initially containing $1.0 \times 10^{-3}$ mol dm$^{-3}$ aqueous magnesium ions and $1.0 \times 10^{-3}$ mol dm$^{-3}$ aqueous fluoride ions? Explain your answer fully.

**● EXAMPLE**

**Q** Will a precipitate of barium carbonate ($K_{sp}(BaCO_3) = 8.1 \times 10^{-9}$ mol$^2$ dm$^{-6}$) be formed from a solution that is both 0.1 mol dm$^{-3}$ in aqueous barium ion and 1.0 mol dm$^{-3}$ in aqueous carbonate ion?

**A** Ionic product $(BaCO_3) = [Ba^{2+}(aq)][CO_3{}^{2-}(aq)]$
$$= 0.1 \times 1.0$$
$$= 0.1 \text{ mol}^2 \text{ dm}^{-6}$$
$$K_{sp}(BaCO_3) = 8.1 \times 10^{-9} \text{ mol}^2 \text{ dm}^{-6}$$

Since the ionic product is larger, a precipitate will be formed.

## 12 ALUMINIUM

So far, this chapter has concentrated on the behaviour of the Group 1 and 2 elements and their compounds. But there is one other metal that can reasonably be called reactive and that is aluminium. The surface of aluminium is covered by an impermeable layer of aluminium oxide that masks the true reactivity of the metal. Once this layer has been removed, the metal resembles magnesium in its reactivity.

### OCCURRENCE AND MANUFACTURE OF ALUMINIUM
Aluminium is estimated to make up 7.5 per cent of the Earth's crust, most of it as complex aluminates and aluminosilicates in clays, and also as bauxite, which is a form of aluminium oxide and hydroxide. It is not possible commercially to extract aluminium from clays and so bauxite is the major source of all aluminium metal used today.

### ELECTROLYTIC MANUFACTURE OF ALUMINIUM
The process dates back to 1886, when Charles Hall and Paul Heroult independently discovered a way to extract aluminium from bauxite using electrolysis. The bauxite mineral is mined and then purified to form pure aluminium hydroxide, which is in turn converted into pure aluminium oxide. This is the substance that is electrolysed to form aluminium.

The aluminium oxide is first dissolved in molten sodium aluminium fluoride (cryolite) and some calcium fluoride, and is then electrolysed. Electrolysis of pure aluminium oxide would be too expensive since very high temperatures are needed to melt it. Introducing the two fluorides allows a much lower temperature to be used. As a result of shortages of sodium aluminium fluoride, some plants use sodium fluoride and aluminium fluoride instead.

A carbon (graphite) anode is used. Oxygen is produced by the discharge of oxide ions:

$$2O^{2-} \rightarrow O_2 + 4e^-$$

The high temperature of production is a nuisance, because the oxygen reacts with the carbon anode to form carbon dioxide. So the carbon anode is continually being replaced.

The carbon lining of the cell acts as the cathode, and aluminium is formed by the discharge of aluminium ions. The temperature in the cell is above the melting point of aluminium, so molten aluminium is formed. This sinks to the bottom of the cell, where it is run off and allowed to cool. The aluminium obtained is exceptionally pure, being at least 99.9 per cent aluminium.

carbon anodes (+)

Al₂O₃ dissolved in molten Na₃AlF₆

molten aluminium

bubbles of O₂, CO and CO₂

carbon cathodes (–)

aluminium tapped off

**Fig 22.32** The cell used for the manufacture of aluminium by electrolysis

## APPLICATIONS OF ALUMINIUM

As soon as it was discovered, aluminium found many applications associated with its low density and its resistance to corrosion. More recent applications include foil (often incorrectly called tin foil) for use in the home and in packaging, double-glazed window frames and now car engines and bodies.

**Fig 22.33** A car with aluminium bodywork

**Fig 22.34** Aluminium and its alloys are used in the aviation industry because of their high strength and low density

| Table 22.13 Some properties of aluminium | |
|---|---|
| Electron configuration | $1s^2\ 2s^2\ 2p^6\ 3s^2\ 3p^1$ |
| Melting point | 660 °C |
| Boiling point | 2300 °C |
| Oxidation potential, $E_{oxid}$ | +1.67 V |
| Density | 2.70 g cm⁻³ |

## CHEMICAL PROPERTIES OF ALUMINIUM

Table 22.13 suggests that an aluminium atom should lose three electrons from an atom to form an aluminium ion. The loss of three electrons is always energetically more difficult than the loss of one or two electrons, and the cation formed will always be small and have a large charge density. This means that the small and highly charged cation is able to attract back the electrons it has lost in forming an ionic bond. An aluminium ion is said to polarize the anions, and so many aluminium compounds have a covalent character.

In aqueous solution, six water molecules surround an aluminium ion to form a hydrated ion, $[Al(H_2O)_6]^{3+}(aq)$. For simplicity, this is often written as $Al^{3+}(aq)$.

### REACTION WITH AIR AND OXYGEN

As already noted, aluminium forms an oxide layer on its surface which protects the rest of the metal from further corrosion. When heated in air or oxygen, aluminium powder burns to form aluminium oxide, a white ionic solid:

$$4Al(s) + 3O_2(g) \rightarrow 2Al_2O_3(s)$$

If the oxide layer is removed, for example by wiping the surface with mercury, then aluminium will react with water to form hydrogen.

### REACTION WITH ACIDS

Aluminium powder reacts with dilute sulphuric acid and dilute hydrochloric acid to give the corresponding aqueous aluminium salts. These reactions are initially very slow and need heat to accelerate the reaction rate. But, once

**Fig 22.35** The thermite process involves the reaction of aluminium powder and iron(III) oxide. Aluminium is sufficiently reactive to displace iron from its oxide

started, they become quite violent. The initial slow rate of reaction arises from the time it takes for the acid to react with the protective oxide layer before it exposes the active metal:

$$2Al(s) + 6H^+(aq) \rightarrow 2Al^{3+}(aq) + 3H_2(g)$$

Surprisingly, aluminium does not react with pure nitric acid. It is therefore used as the lining of vessels to transport this acid.

### REACTION WITH ALKALIS

Aluminium is unusual in that it reacts with both acids and alkalis. This property makes aluminium an amphoteric metal. Aluminium powder reacts vigorously – after a slow start – with warm aqueous sodium hydroxide to form hydrogen and aqueous sodium aluminate:

$$OH^-(aq) + 3H_2O(l) + Al(s) \rightarrow [Al(OH)_4]^-(aq) + H_2(g)$$

### REACTION WITH HALOGENS

When one of the halogens is passed over heated aluminium, the corresponding halide is formed by direct combination. The halides formed have a large degree of covalent character:

$$2Al(s) + 3X_2(g) \rightarrow Al_2X_6(s)$$

where X = a halogen atom.

## COMPOUNDS OF ALUMINIUM

Most aluminium compounds have some covalent character, but when dissolved in water or as hydrated crystals, the compounds are ionic, and contain the hydrated aluminium ion.

### ALUMINIUM OXIDE AND ALUMINIUM HYDROXIDE

Both these are white ionic amphoteric solids. With acids, they form the hydrated aluminium ion $[Al(H_2O)_6]^{3+}$; with alkalis they form the aluminate ion $[Al(OH)_4]^-$ in which the aluminium atom forms part of the anion.

Aluminium oxide is used as an abrasive and as a dehydrating agent, particularly in the formation of alkenes from alcohols. Aluminium hydroxide is used in the dyeing industry, since it can adsorb coloured materials onto its surface. This property is also used to remove particulate matter from water during its purification. In this application, a solution of aluminium sulphate is added to the water and forms aluminium hydroxide, which, as it settles, collects particulate matter and removes it from the water. The amount of aluminium sulphate used in this process has to be carefully controlled to prevent excess aluminium ions reaching domestic water supplies, since aluminium ions may be linked to Alzheimer's disease.

### SOLUTIONS OF ALUMINIUM SALTS

Aqueous solutions of aluminium salts are highly acidic, so much so that they can release carbon dioxide from sodium hydrogencarbonate. This acidity is caused by hydrolysis and involves one of the water molecules in the aqueous aluminium cation being polarized to form an aqueous proton:

$$[Al(H_2O)_6]^{3+}(aq) \rightleftharpoons [Al(H_2O)_5OH]^{2+}(aq) + H^+(aq)$$

**REMEMBER THIS**

The formula for the aluminate ion is sometimes written as $[Al(OH)_6]^{3-}$ or as $[Al(H_2O)_2(OH)_4]^-$.

**Fig 22.36** Aluminium sulphate is used in water purification plants to remove solid particles, and its level has to be carefully monitored so as not to reach toxic levels

**? TEST YOURSELF**

**B** Would you expect aluminium carbonate to exist? Explain your answer.

**SEE QUESTION 8.**

All aqueous solutions of aluminium salts react with aqueous hydroxide ions to give first a white precipitate of hydrated aluminium hydroxide and then, with excess hydroxide ions, a colourless solution of the aluminate ion. This reaction involves **deprotonation**, in which each reaction step removes one more proton from the aluminium species:

$$[Al(H_2O)_6]^{3+}(aq) + OH^-(aq) \rightleftharpoons [Al(H_2O)_5OH]^{2+}(aq) + H_2O(l)$$

$$[Al(H_2O)_5OH]^{2+}(aq) + OH^-(aq) \rightleftharpoons [Al(H_2O)_4(OH)_2]^+(aq) + H_2O(l)$$

$$[Al(H_2O)_4(OH)_2]^+(aq) + OH^-(aq) \rightleftharpoons [Al(H_2O)_3(OH)_3](s) + H_2O(l)$$

$$[Al(H_2O)_3(OH)_3](s) + OH^-(aq) \rightleftharpoons [Al(H_2O)_2(OH)_4]^-(aq) + H_2O(l)$$

 **REMEMBER THIS**

Remember that the $H^+$ ion is really hydrated and could be written as $H_3O^+(aq)$. It is this ion that makes an aqueous solution acidic. This can be a severe problem in the environment when aluminium ions are released from the soil. This makes the groundwater very acidic and can lead to toxic metal ions being released into the soil, followed by the destruction of plants and trees.

## SUMMARY

- Group 1 metals are highly reactive metals. Each atom loses one electron when it reacts. In compounds, the oxidation state of Group 1 metals is always +1.

- Group 2 elements are less reactive than the Group 1 element in the same period. In compounds, the Group 2 elements always have an oxidation number of +2.

- The ease with which an atom loses electrons increases with increasing atomic (proton) number in both Group 1 and Group 2.

- The oxides and hydroxides of Group 1 and Group 2 elements are basic and when they dissolve in water, they form strongly alkaline solutions.

- The nitrates of Group 1 (except lithium nitrate) thermally decompose to form the corresponding nitrites and oxygen, whereas the nitrates of Group 2 thermally decompose to give nitrogen dioxide, oxygen and the corresponding metal oxides. The thermal stability of the nitrates increases with increasing atomic (proton) number of the metal.

- The carbonates of Group 1 (other than lithium carbonate) do not decompose at Bunsen temperatures, but most carbonates of Group 2 do to give carbon dioxide and the metal oxides. The thermal stability of the carbonates increases with increasing atomic (proton) number of the metal.

- Lattice energy is the enthalpy change when 1 mole of an ionic lattice is formed from its constituent ions in the gas phase. Lattice energy depends on the charge densities of the ions involved.

- The solubility of the Group 2 sulphates, $MSO_4$, and the Group 2 carbonates, $MCO_3$, decreases with increasing atomic (proton) number of M. This trend is explained by the decrease in the magnitude of the enthalpy change of hydration of $M^{2+}$.

- The solubility of the Group 2 hydroxides, $M(OH)_2$, increases with increasing atomic (proton) number of M. This trend is explained by the decrease in magnitude of the lattice energy of $M(OH)_2$.

- The thermal decomposition of nitrates and carbonates is determined by the ability of the cation present to polarize the large carbonate and nitrate ions.

- Aluminium is manufactured by the electrolytic decomposition of molten aluminium oxide dissolved in sodium aluminium fluoride.

- Aqueous aluminium salts are acidic because of the polarization of water molecules by the very small and highly charged aluminium ion.

- The solubility product can be used to explain the common-ion effect and precipitation.

# ? EXAM QUESTIONS

**1**

**a)** Write an equation for the reaction of strontium with water, and suggest an approximate value for the pH of the resulting solution.

**b)** Describe what is seen when an aqueous solution of barium chloride is added to dilute sulphuric acid. Write an equation for the reaction that occurs.

**c)** Give two examples that illustrate the atypical properties of beryllium compounds in Group II.

**d)** Give one feature of the beryllium ion that causes the atypical properties of beryllium compounds.

AQA Unit 1 Atomic Structure, Bonding and Periodicity, Jan 2001 Q5

**2** In the UK, over 60 million tonnes of limestone are quarried each year. Much of this limestone is used to produce cement. The main chemical in limestone is calcium carbonate, $CaCO_3$.

**a)** Complete the flow-chart in Fig 22.Q2 for reactions starting from calcium carbonate. You should identify each of the substances **A** to **F** by name or formula.

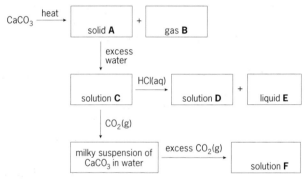

**Fig 22.Q2**

**b)** Cement is a mixture of calcium and aluminium silicates, formed by heating limestone with clay.

$$4CaCO_3(s) + Al_2Si_2O_7(s) \rightarrow$$
$$\underbrace{2CaSiO_3(s) + Ca_2Al_2O_5(s)}_{cement} + 4CO_2(g)$$

A typical bag of cement has a mass of 25 kg. Calculate the mass of limestone (taken as calcium carbonate) required to make 25 kg of cement. The molar mass of cement, taken as $(2CaSiO_3 + Ca_2Al_2O_5)$, is 446.6 g mol$^{-1}$

$A_r$ values: Al, 27.0; C, 12.0; Ca, 40.1; O, 16.0; Si, 28.1.

**c)** Lime mortar is a thick paste made by adding water to a mixture of slaked lime, $Ca(OH)_2$, and sand. As mortar dries out the slaked lime reacts with carbon dioxide in the air, forming calcium carbonate, which causes the mortar to harden.

**(i)** Write an equation to represent the hardening of mortar. Assume the sand does not react.

**(ii)** In time, lime mortar crumbles and needs to be replaced. Suggest why this happens more quickly when the mortar is exposed to air contaminated with acid pollution.

OCR Chemistry 2811 Jan 2002 Q5

**3** The four most abundant salts in sea-water are given in Table 22.Q3.

**Table 22.Q3**

| Salt | Tonnes km$^{-3}$ |
|---|---|
| Sodium chloride | 27 500 000 |
| Magnesium chloride | 6 750 000 |
| Magnesium sulphate | 5 625 000 |
| Calcium sulphate | 1 800 000 |

Magnesium oxide is obtained from sea-water by the steps shown in Fig 22Q.3.

**Fig 22.Q3**

**a)** Explain why the addition of carbonate ions in Step 1 has to be controlled. The numerical values of the relevant solubility products are given below:

Magnesium carbonate      $1.0 \times 10^{-5}$
Calcium carbonate      $5.0 \times 10^{-9}$

**b)** **(i)** Write an equation for Step 3.

**(ii)** Suggest a reagent for Step 4.

**(iii)** Suggest how Step 5 could be carried out.

**(iv)** Write an equation for Step 5.

**c)** State a use for magnesium or magnesium oxide. State which substance you choose.

**d)** **(i)** Write an expression for the *solubility product of magnesium carbonate.*

**(ii)** Calculate the concentration, in mol dm$^{-3}$, of each of the two cations in saturated solutions of the following:

magnesium carbonate
calcium carbonate

**e)** The solid calcium carbonate produced by this process can be decomposed by heating.

# ? EXAM QUESTIONS

**(i)** Write an equation for the thermal decomposition of calcium carbonate.
**(ii)** Calcium carbonate decomposes at a lower temperature than barium carbonate. Explain why this is so.

**f)** Explain why lime is used in agriculture.

UCLES Chemistry 9255 Nov 2001 Q3

**4**

**a)** Draw and label carefully a Born–Haber cycle for the formation of calcium oxide.

**b)** Use the data below to calculate the value of the lattice energies of calcium oxide and iron(II) oxide.

| | |
|---|---|
| Standard enthalpy change of formation of calcium oxide | $-635$ kJ mol$^{-1}$ |
| Standard enthalpy change of formation of iron(II) oxide | $-278$ kJ mol$^{-1}$ |
| Standard enthalpy change of atomization of calcium | $+178$ kJ mol$^{-1}$ |
| Standard enthalpy change of atomization of iron | $+416$ kJ mol$^{-1}$ |
| Standard molar first ionization energy of calcium and standard molar second ionization energy of calcium | $+1735$ kJ mol$^{-1}$ |
| Standard molar first ionization energy of iron and the standard molar second ionization energy of iron | $+2320$ kJ mol$^{-1}$ |
| Standard enthalpy change of atomization of oxygen | $+249$ kJ mol$^{-1}$ of oxygen atoms |
| Standard molar enthalpy of first electron affinity and second electron affinity of oxygen | $+657$ kJ mol$^{-1}$ |

**c)** Suggest why the order of lattice energies is as given by your calculation.

**d)** Calcium(III) oxide has never been formed, and its enthalpy change of formation is thought to be positive.
**(i)** Which terms in the Born–Haber cycle for calcium(III) oxide would be most different from those in the cycle for calcium(II) oxide, and how would they differ?
**(ii)** What is the largest factor that causes the enthalpy change of formation of calcium(III) oxide to be endothermic?

AEB Specimen paper 0654/3 1996 Q5

**5**

**a)** Write the equation for the action of heat on barium nitrate.

**b) (i)** What is meant by the term polarizing power as applied to cations?
**(ii)** Give two factors that affect the polarizing power of cations.
**(iii)** Use this information to explain why it is easier to decompose magnesium nitrate than barium nitrate by heating.

Edexcel 6241 Jan 2002 Q8

**6** Explain the following:

**a)** Magnesium fluoride is much less soluble in water than barium fluoride, but magnesium sulphate is much more soluble than barium sulphate.

**b)** Lithium carbonate will decompose when heated in a Bunsen flame, but sodium carbonate does not.

**7** Chrome yellow, the pigment used for yellow road markings, is lead(II) chromate(VI), $PbCrO_4$.

**a)** Write an equation, with state symbols, for the formation by precipitation of $PbCrO_4$.

**b)** The solubility product of $PbCrO_4$ at 15 °C is $1.69 \times 10^{-14}$.
**(i)** Write an expression for the solubility product, $K_{sp}$, of $PbCrO_4$.
**(ii)** What is the solubility, in mol dm$^{-3}$, of $PbCrO_4$?
**(iii)** Concentrated aqueous lead(II) nitrate is added dropwise to 0.010 mol dm$^{-3}$ potassium chromate(VI).
What is the concentration, in mol dm$^{-3}$ of lead(II) ions, when the first trace of precipitate appears?

UCLES Modular Chemistry 4826 March 1995 Q2

**8** Aluminium is used to make drink cans.

**a)** Describe, giving essential details, how aluminium is manufactured from alumina.

**b)** Explain why aluminium does not corrode easily in moist air.

**c)** Describe the effect of adding aqueous sodium hydroxide drop-by-drop, until it is in excess, to aqueous aluminium sulphate. Write equations to describe the reactions that take place.

# KEY SKILLS **ASSIGNMENT**

## THERMAL STABILITY OF SALTS OF GROUP 1 AND 2 ELEMENTS

The anhydrous carbonates, chlorides, nitrates and sulphates of lithium, sodium, potassium, magnesium, calcium, strontium and barium were separately heated in a thick-walled Pyrex test-tube, under similar conditions, by using the hottest flame of a Bunsen burner. The apparatus used is shown in Fig 22.A1.

**Fig 22.A1**

No gas was produced when the chlorides and the sulphates were heated.

**1** Suggest a reason why no gases were produced when the chlorides and the sulphates were heated.

The soda lime (a mixture of solid sodium hydroxide and calcium hydroxide) was present to remove any acidic gases produced when the nitrates were heated, so that only neutral gases were collected in the gas syringe. The volume of gas obtained was measured every 10 seconds and results are shown in Tables 22.A1 and 22.A2.

**2** **a)** Write down equations, including state symbols, to show the thermal decomposition of all the nitrates used in the experiments.
**b)** Write down the equation to show the decomposition of magnesium carbonate.

**Table 22.A1 Volumes of gas from heated carbonates**

| Name of salt | Volume of gas collected in syringe after time/s cm$^{-3}$ | | | | | | | | | | |
|---|---|---|---|---|---|---|---|---|---|---|---|
| | 0s | 10s | 20s | 30s | 40s | 50s | 60s | 70s | 80s | 90s | 100s |
| Lithium carbonate | 0 | 1 | 4 | 20 | 47 | 76 | | | | | |
| Sodium carbonate | 0 | 0 | 0 | 0 | 0 | 1 | 4 | 7 | 10 | 13 | 16 |
| Potassium carbonate | 0 | 0 | 0 | 0 | 0 | 0 | 0 | 0 | 1 | 2 | 3 |
| Magnesium carbonate | 0 | 4 | 15 | 35 | 61 | 91 | | | | | |
| Calcium carbonate | 0 | 0 | 0 | 0 | 0 | 0 | 1 | 2 | 3 | 4 | 5 |
| Strontium carbonate | 0 | 0 | 0 | 0 | 0 | 0 | 0 | 0 | 0 | 0 | 0 |
| Barium carbonate | 0 | 0 | 0 | 0 | 0 | 0 | 0 | 0 | 0 | 0 | 0 |

**Table 22.A2 Volumes of gas from heated nitrates**

| Name of salt | Volume of gas collected in syringe after time/s cm$^{-3}$ | | | | | | | | | | |
|---|---|---|---|---|---|---|---|---|---|---|---|
| | 0s | 10s | 20s | 30s | 40s | 50s | 60s | 70s | 80s | 90s | 100s |
| Lithium nitrate | 0 | 0 | 2.5 | 11 | 24 | 38 | 52 | 65 | 79 | | |
| Sodium nitrate | 0 | 0 | 1 | 4 | 13 | 22 | 31 | 40 | 49 | 58 | 67 |
| Potassium nitrate | 0 | 0 | 0 | 1.5 | 4 | 8 | 12 | 16 | 20 | 24 | 28 |
| Magnesium nitrate | 0 | 3 | 8 | 29 | 72 | | | | | | |
| Calcium nitrate | 0 | 0 | 3.5 | 10 | 25 | 47 | 79 | | | | |
| Strontium nitrate | 0 | 0 | 0 | 2 | 7 | 16 | 32 | 52 | 72 | 92 | |
| Barium nitrate | 0 | 0 | 0 | 1 | 2.5 | 6.5 | 18 | 30 | 42 | 54 | 66 |

**3** A gas syringe was used to collect the gases produced. Look at the results for calcium nitrate. Suggest why there are no readings taken after 60 seconds.

**4** **a)** Plot the results shown in Table 22.A1 on graph paper. Draw a line of best fit between the points for each salt. For each line, calculate the gradient of the straight-line portion. This is the rate of thermal decomposition of the carbonate.
**b)** Plot the results shown in Table 22.A2 on graph paper. Draw a curve of best fit between the points for each salt. This is the rate of thermal decomposition of the nitrate.

**5** By referring to your graphs or the tables:
**a)** deduce which salts, the carbonates or the nitrates, are less stable towards heat;
**b)** arrange the nitrates in decreasing order of thermal stability, starting with the most stable nitrate;
**c)** arrange the carbonates in decreasing order of thermal stability, starting with the most stable carbonate.

**6** **a)** Copy out and complete the following table. [You will need to refer to Table 22.9 on page 480.]

**Table 22.A3**

| Ion | Charge density | Rate of thermal decomposition of nitrate |
|---|---|---|
| Lithium | | |
| Sodium | | |
| Potassium | | |
| Magnesium | | |
| Calcium | | |
| Strontium | | |
| Barium | | |

**b)** Plot charge density against the rate of thermal decomposition of each nitrate. Comment on the curve that you get.

# 23

# The halogens

We find it quite uncomfortable on the eyes to move rapidly in and out of brilliant sunlight and shadow, even if we are wearing sunglasses. Our eyes cannot adjust quickly enough to sudden changes of light intensity – that is, unless the sunglasses are made from photochromic glass, which compensates for changes in brightness.

Lens glass is usually made photochromic by adding tiny amounts of silver chloride and copper(I) chloride to the molten glass as it cools, which traps the crystals within the structure of the glass. When bright sunlight strikes the glass, the silver chloride decomposes to form metallic silver, which darkens the glass. This reduces the intensity of the light that reaches the eyes. At the same time, chlorine atoms are formed and they react with copper(I) ions to form copper(II) ions and chloride ions. As soon as the exposure to bright light ends, the copper(II) ions are reduced by silver atoms to reform silver chloride and copper(I) chloride. This lightens the glass and allows all the available light to reach the eyes.

In this way, spectacle lenses can be made to darken or lighten according to the intensity of the light, and so make it comfortable for people to keep their sunglasses on, whether they are in bright light or shadow.

**Sunglasses** Photochromic spectacles are easy on the eyes

## 1 WHAT ARE THE HALOGENS?

Halogens is the collective name given to the non-metal elements in Group 7 of the Periodic Table. The name is derived from *hals* (the Greek for salt) and the suffix *gen* (meaning producer). It was first used by the Swedish scientist Jöns Berzelius (1779–1884) to indicate that chlorine, bromine and iodine occurred in sea-water. Chlorine was first prepared in 1774 by the Swedish chemist Carl Wilhelm Scheele, but he did not recognize it as an element. It was Humphry Davy who, in 1810, recognized chlorine as an element and named it after the Greek word for green, *chloros*. In the same way, iodine was derived from *ioides* (the Greed word for violet-like) and bromine from *bromos* (the Greek word for stench). It was not until 1940 that the final member of the halogens, astatine, a radioactive element, was prepared artificially.

**Fig 23.1** Bromomethane is used as a soil fumigator for high value crops such as strawberries

### SOME USES OF THE HALOGENS AND THEIR COMPOUNDS

The halogens are commercially of great importance, although the use of many chlorine-containing compounds is controversial for environmental reasons. Chlorine and fluorine are both used in the production of polymers, such as polyvinyl chloride (PVC) and polytetrafluoroethene (PTFE). Many insecticides contain chlorine, but there is concern over the use of these compounds, despite their obvious ability to kill disease-carrying and crop-eating insects. Bromine has a variety of applications, including the manufacture of fuel additives and of soil fumigators such as 1,2-dibromoethane and bromomethane, which kill pests found in the soil. Silver bromide and silver iodide both have applications in photography.

**Fig 23.2** We take it for granted that the water we drink is sterile. Chlorine is used to remove bacteria from our water supply

# 2 PHYSICAL PROPERTIES OF THE HALOGENS

The halogens have physical properties that are typical of non-metals with a simple molecular structure. The halogens have relatively low melting and boiling points and are very poor conductors of heat and electricity.

**Table 23.1 Properties of four halogens**

| Halogen | Melting point/°C | Boiling point/°C | Atomic (covalent) radius/pm | Electronegativity | Electron configuration |
|---------|---------|---------|---------|---------|---------|
| Fluorine | −223 | −188 | 64 | 4 | $1s^2\ 2s^2\ 2p^5$ |
| Chlorine | −101 | −34 | 99 | 3 | $1s^2\ 2s^2\ 2p^6\ 3s^2\ 3p^5$ |
| Bromine | −7 | 59 | 114 | 2.8 | $[Ar]3d^{10}\ 4s^2\ 4p^5$ |
| Iodine | 114 | 187 | 133 | 2.4 | $[Kr]4d^{10}\ 5s^2\ 5p^5$ |

## SIMPLE MOLECULAR STRUCTURE OF THE HALOGENS

All the halogens have diatomic molecules, which in the solid state are arranged in a simple molecular lattice. Halogen molecules are held in place by weak intermolecular forces known as **induced dipole–induced dipole attraction**. These forces result from an asymmetric distribution of electrons within each halogen molecule. This produces an instantaneous dipole, which induces dipoles in neighbouring molecules. Such a weak attractive force between molecules is easily overcome, so the elements have relatively low melting points and boiling points. The simple molecular structure has no free electrons, so a halogen, either as a solid or as a liquid, cannot conduct electricity.

electron density distorted by asymmetric distribution of electrons within molecule

induced dipole–induced dipole attraction

**Fig 23.4** A temporary dipole is set up in an iodine molecule when its electrons move such that they become distributed asymmetrically, with the electron density at one end of the molecule greater than that at the other end. The top diagram shows how the electron density is temporarily distorted. The temporary dipole induces dipoles in neighbouring molecules and an intermolecular force is set up (bottom diagram). This force is quite weak.

**Fig 23.5** Solid iodine has a cubic structure. The lines of short strokes represent the weak intermolecular forces that hold the iodine molecules in place within the lattice. The other solid halogens have a similar simple molecular lattice

All the halogens form solids with this type of intermolecular force. The larger the halogen molecule (and the more electrons), the easier it is to distort the electron cloud and increase the intermolecular forces. Hence, the melting point of the halogens increases with increasing atomic number.

**YOU CAN READ MORE ABOUT THE PHYSICAL PROPERTIES OF SUBSTANCES WITH A SIMPLE MOLECULAR STRUCTURE IN CHAPTER 7 AND 21.**

**Fig 23.3** At room temperature and atmospheric pressure, chlorine is a pale green gas, bromine an orange liquid and iodine a dark purple solid

## ? TEST YOURSELF

**A** The appearance of the halogens has an observable trend at room temperature and atmospheric pressure. Predict the colour and physical state of fluorine and of astatine under these conditions.

**B** Astatine is the last member of the halogens.
**a)** (i) What is the atomic (proton) number for astatine?
(ii) How many electrons does astatine have in the outer shell of its atom?
(iii) Predict the melting point, boiling point, covalent radius and electronegativity of astatine.
**b)** What is meant by the term electronegativity? Why does the electronegativity of the halogens decrease with increasing atomic (proton) number?
Check your answers on page 83.

**SEE QUESTIONS 1, 2 AND 3.**

**WEAK INTERMOLECULAR FORCES IN ELEMENTS ARE DISCUSSED ON PAGE 215.**

**?** **TEST YOURSELF**

**C** Write down the electron configurations for
**a)** chloride ion, $Cl^-$;
**b)** bromide ion, $Br^-$.

**Table 23.2 Reduction potentials ($E_{red}^{\ominus}$) for**

| Half reaction | Reduction potential $E_{red}^{\ominus}$/V |
|---|---|
| $F_2 + 2e^- \rightarrow 2F^-$ | +2.87 |
| $Cl_2 + 2e^- \rightarrow 2Cl^-$ | +1.36 |
| $Br_2 + 2e^- \rightarrow 2Br^-$ | +1.07 |
| $I_2 + 2e^- \rightarrow 2I^-$ | +0.54 |

**EXPERIMENTALLY, THE REACTION BETWEEN FLUORINE AND AQUEOUS POTASSIUM HALIDES, SUCH AS POTASSIUM BROMIDE, IS COMPLICATED BECAUSE FLUORINE ALSO REACTS RAPIDLY WITH WATER.**

**?** **TEST YOURSELF**

**D** State whether a reaction takes place when the following substances are mixed. If a reaction takes place, write a balanced ionic equation for the reaction:
**a)** aqueous iodine and aqueous potassium bromide;
**b)** aqueous chlorine and aqueous sodium iodide;
**c)** aqueous chlorine and aqueous calcium fluoride;
**d)** aqueous iodine and aqueous sodium astatide.

# 3 CHEMICAL PROPERTIES OF THE HALOGENS

Halogens react by gaining electrons to form an anion in ionic halides, or by sharing electrons to form a covalent bond in molecular halides. In both cases the halogen atom attains a noble gas electron configuration. Most metals form ionic halides and most non-metals form molecular halides. The ability to gain an electron is typical of a non-metal atom.

## THE HALOGENS AS OXIDIZING AGENTS

Since their atoms accept electrons, the halogens are oxidizing agents and in a reaction they are reduced. Of the halogens, fluorine is the most powerful oxidizing agent and astatine the least. This can be explained by the relative size of their atoms and their ability to capture an electron. The fluorine atom is the smallest, with fewer inner-shell shielding electrons, so its nucleus can have a greater attraction or ability to attract an electron. The reduction potentials, $E_{red}^{\ominus}$, of the halogens also illustrate this trend. The reaction of a fluorine molecule to give a fluoride ion has a very positive reduction potential, which indicates that this is a highly feasible process. (The more positive an electrode potential the more feasible the reaction. More details on electrode potentials can be found in Chapter 25.)

Fluorine always has an oxidation number of –1 in its compounds, since a fluorine atom either gains one electron to form an ion or contributes one electron to a shared pair of electrons in a covalent bond. Fluorine oxidizes other substances by gaining electrons from them.

## DISPLACEMENT REACTIONS

Since fluorine is the most reactive halogen, in theory it can react with the halide ion of any of the other halogens. Fluorine becomes the fluoride ion and the free halogen (chlorine, bromine or iodine) is formed from the halide ion. This is called a **displacement reaction**. Experiments are not normally carried out with fluorine because its extraordinary reactivity makes it extremely dangerous. For example, if inhaled, fluorine can seriously damage the respiratory tract.

In aqueous solution, chlorine displaces bromide ions to form bromine:

$$Cl_2(aq) + 2KBr(aq) \rightarrow 2KCl(aq) + Br_2(aq)$$

During this reaction, the colour of the aqueous becomes orange, which indicates that the element bromine has been produced.

Chlorine will also displace iodide ions to form iodine, and again an orange–brown solution is formed.

$$Cl_2(aq) + 2I^-(aq) \rightarrow 2Cl^-(aq) + I_2(aq)$$

In the same way, aqueous bromine displaces aqueous iodide ion to form iodine. (The presence of iodine in a solution can be confirmed by the addition of starch solution. A dark blue coloration is produced.)

In general, the presence of the displaced halogen can be demonstrated easily if hexane is added after the displacement reaction, since bromine will preferentially dissolve in the hexane to give an orange–red solution, while iodine will give a purple solution.

The displacement reaction is an example of a redox reaction. This is because the halogen that reacts is reduced because it gains electrons to form halide ions, and halide ions are oxidized by the loss of electrons to form the halogen.

► **EXTENSION Manufacture of bromine**

The concentration of bromide ion in normal sea water is between 65 and 70 ppm. Therefore, a large quantity of sea-water has to be processed to make significant quantities of bromine. Inland seas, such as the Dead Sea, have considerably higher concentrations of bromide ion and provide a much better feedstock (Fig 23.6).

Chlorine is used to liberate bromine by a displacement reaction. It is bubbled through the acidified sea-water and the bromine formed is removed as a gas from the water by blowing air through it:

$$Cl_2(g) + 2Br^-(aq) \rightarrow Br_2(aq) + 2Cl^-(aq)$$

The bromine vapour formed is difficult to handle and so is converted back into bromide ion by reaction with sulphur dioxide and water:

$$Br_2(g) + SO_2(g) + H_2O(l) \rightarrow H_2SO_4(aq) + 2HBr(aq)$$

This makes a much more concentrated solution of bromide ion, from which it is easier to produce high-purity bromine. A second displacement reaction with chlorine yields bromine vapour, which can be condensed and then purified by distillation.

**Fig 23.6** The Dead Sea is rich in dissolved minerals. It is about 60 times more concentrated in bromide ion than normal sea-water

## REACTIONS OF HALOGENS WITH ELEMENTS

We have already described some of the reactions of metals and non-metals with chlorine in Chapter 21. In every case the halogen oxidizes the element. Fluorine will react the most violently, and astatine the least.

### REACTION OF HALOGENS WITH NON-METALS

Almost all non-metals react with fluorine and chlorine to form covalent halides. The halogen atom shares electrons to make a single covalent bond (Fig 23.7). Very often, the halide contains the element in its highest possible oxidation state. In the reaction between phosphorus and chlorine, the reaction can be controlled to make either phosphorus(III) chloride or phosphorus(V) chloride:

$$P_4(s) + 6Cl_2(s) \rightarrow 4PCl_3(l)$$

$$P_4(s) + 10Cl_2(g) \rightarrow 4PCl_5(s)$$

### REACTIONS OF THE HALOGENS WITH METALS

Most metals react with the halogens to form ionic halides. This reaction is necessarily a redox reaction, in which the metal is oxidized and the non-metal is reduced. Electrons are transferred to halogen atoms from metal atoms. The reaction of a metal with the halogens varies with the halogen. For instance, the reaction between a metal and fluorine is faster, more exothermic and more violent than that between the same metal and iodine.

Since halogens are good oxidizing agents, if the metal has more than one oxidation state then in the halide formed the metal is often in one of its higher oxidation states. So, when chlorine is passed over hot iron wire it forms iron(III) chloride, $FeCl_3$, rather than iron(II) chloride. Similarly, copper and chlorine form copper(II) chloride, $CuCl_2$, rather than copper(I) chloride.

**SEE QUESTIONS 2 TO 6.**

**Fig 23.7** Phosphorus(III) chloride is a covalent compound that forms a simple molecule, $PCl_3$. Electrons are shared so that all the atoms attain a stable octet of electrons. All non-metal halides are covalent compounds, and the halogen atom attains a stable octet. Sometimes, the other atom involved expands its octet. Note: only the outer electrons are shown in the dot-and-cross diagram

### ? TEST YOURSELF

**E**  **a)** Write equations for the reactions between:
(i) iron and chlorine;
(ii) copper and chlorine;
(iii) iron and fluorine.
**b)** What is the name of the product formed in the reactions between:
(i) chromium and chlorine;
(ii) zinc and chlorine;
(iii) barium and iodine?

**THE REACTION OF THE HALOGENS WITH HYDROCARBONS TO FORM HALOGENOALKANES AND HYDROGEN HALIDE IS DESCRIBED ON PAGE 267.**

## ? TEST YOURSELF

**F** **a)** Determine the oxidation number of each of the chlorine species for the following reaction:
$Cl_2(aq) + H_2O(l) \rightarrow HCl(aq) + HOCl(aq)$
**b)** Use the oxidation numbers to show that the reaction is an example of disproportionation

**FOR SOME MORE EXAMPLES OF DISPROPORTIONATION, READ PAGE 111.**

**SEE QUESTION 5.**

## ? TEST YOURSELF

**G** **a)** Construct the ionic equation for the reaction between aqueous chlorine and aqueous sulphur dioxide and water to form a chloride ion and a sulphate ion.
**b)** Suggest why astatine does not oxidize aqueous iron(II) ions into iron(III) ions.

## ? TEST YOURSELF

**H** In a determination to find the amount of iodine liberated by an oxidizing agent, it was found that the number of moles of sodium thiosulphate needed to react with the iodine in a titration was $2.1 \times 10^{-3}$. What was the number of moles of iodine liberated by the oxidizing agent?

## REACTIONS OF CHLORINE WITH COMPOUNDS

Chlorine and the other halogens have an affinity for hydrogen. Chlorine reacts with compounds that contain hydrogen to form hydrogen chloride, or hydrochloric acid when the reaction is carried out in aqueous solution. Chlorine behaves as an oxidizing agent.

### REACTION OF CHLORINE WITH WATER

Chlorine reacts with water to form an acidic solution that is a mixture of two acids, hydrochloric acid and chloric(I) acid. This is an example of a **disproportionation** reaction, in which chlorine is both oxidized and reduced during the reaction:

$$Cl_2(aq) + H_2O(l) \rightarrow HCl(aq) + HOCl(aq)$$

This reaction is the basis of one of the chemical tests for chlorine: namely, it turns moist blue litmus paper first red and then white. Chlorine reacts with the moisture to form the acidic solution which turns the litmus paper red. The chloric(I) acid acts as a bleach and turns the paper white.

### REACTION OF CHLORINE WITH COLD DILUTE ALKALI

Cold aqueous sodium hydroxide can react with chlorine to produce chlorate(I) ions, $ClO^-$, and chloride ions, $Cl^-$:

$$Cl_2(g) + 2OH^-(aq) \rightarrow ClO^-(aq) + Cl^-(aq) + H_2O(l)$$
$$Cl_2(g) + 2NaOH(aq) \rightarrow NaClO(aq) + NaCl(aq) + H_2O(l)$$

A solution of aqueous sodium chlorate(I) is the liquid bleach found in most homes.

### REACTION OF CHLORINE WITH HOT CONCENTRATED ALKALI

When chlorine is added to hot concentrated aqueous sodium hydroxide, the overall reaction is:

$$3Cl_2(g) + 6OH^-(aq) \rightarrow ClO_3^-(aq) + 5Cl^-(aq) + 3H_2O(l)$$

This reaction occurs because of the **disproportionation** of chlorate(I) ions. Chlorate(I) ions lose electrons to give chlorate(V) ions, and gain electrons to form chloride ions:

$$3ClO^-(aq) \rightarrow ClO_3^-(aq) + 2Cl^-(aq)$$

Sodium chlorate(V), $NaClO_3$, is used as a weedkiller. It is also reduced to form $ClO_2$, chlorine dioxide, which is a safer bleach than elemental chlorine for paper and textiles, though it is less efficient.

### OXIDATION OF IRON(II) IONS

Chlorine oxidizes aqueous iron(II) chloride to form aqueous iron(III) chloride:

$$Cl_2(g) + 2Fe^{2+}(aq) \rightarrow 2Fe^{3+}(aq) + 2Cl^-(aq)$$

### REACTION OF HALOGENS WITH AQUEOUS SODIUM THIOSULPHATE

Aqueous chlorine and aqueous bromine oxidize aqueous thiosulphate ions into sulphate ions:

$$4Cl_2(aq) + S_2O_3^{2-}(aq) + 5H_2O(l) \rightarrow 10H^+(aq) + 8Cl^-(aq) + 2SO_4^{2-}(aq)$$

Iodine is a less powerful oxidizing agent than either chlorine or bromine, and so the reaction does not give sulphate ions. Instead, it gives $S_4O_6^{2-}(aq)$:

$$I_2(aq) + 2S_2O_3^{2-}(aq) \rightarrow 2I^-(aq) + S_4O_6^{2-}(aq)$$

## CHEMICAL TEST FOR AN OXIDIZING AGENT

One chemical test for an oxidizing agent uses aqueous potassium iodide. An oxidizing agent liberates iodine from aqueous potassium iodide, and the amount of iodine can be determined using volumetric analysis with a titration against aqueous sodium thiosulphate. The presence of iodine is shown by a blue–black coloration with starch solution. This is a chemical test for iodine.

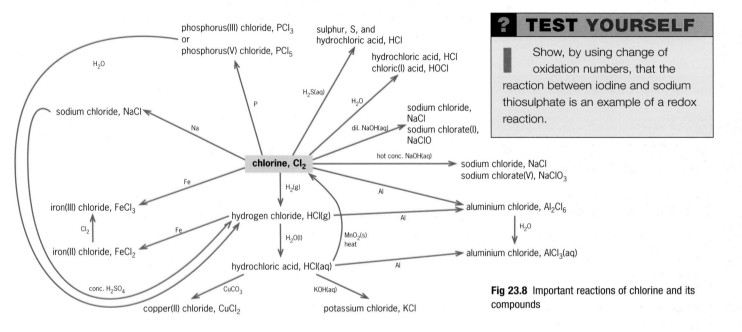

**Fig 23.8** Important reactions of chlorine and its compounds

**? TEST YOURSELF**

**I** Show, by using change of oxidation numbers, that the reaction between iodine and sodium thiosulphate is an example of a redox reaction.

## 4 COVALENT HALIDES

All non-metal halides are covalent and have a simple molecular structure. Typically, non-metal halides are liquids at room temperature with a low boiling point. Some metals will also form covalent halides, which is the case when the oxidation state of the metal is +3 or above. In such cases the metal ion that could be formed, e.g. $Al^{3+}$, has a very small ionic radius and is highly charged, and consequently distorts the electron cloud around the anion so much that the resulting bond has sufficient covalent character to be considered as being covalent.

**? TEST YOURSELF**

**J** **a)** What is the oxidation number of xenon in xenon tetrafluoride and in xenon hexafluoride?
**b)** Write equations to show the reaction of fluorine with xenon to make xenon tetrafluoride and to make xenon hexafluoride.

---

### ► EXTENSION Noble gas compounds

Until 1962, chemists believed that the noble gases, such as xenon, did not react. Since then, several compounds that contain xenon and fluorine have been prepared. Xenon difluoride, for example, is formed when excess xenon is heated with fluorine gas at 400 °C:

$$Xe(g) + F_2(g) \rightarrow XeF_2(g)$$

Xenon difluoride forms colourless crystals, which are stable at room temperature in a dry atmosphere. It is quite surprising that this compound should be so stable. Xenon has a stable octet of electrons, so by reacting with fluorine and attaining an oxidation number of +2 in the difluoride, it should lose its stability.

By changing the mole ratios of xenon to fluorine, it is also possible to prepare xenon tetrafluoride, $XeF_4$, and xenon hexafluoride, $XeF_6$. $XeF_4$ crystals are shown on page 75.

All the xenon fluorides are very powerful oxidizing agents, since xenon in a positive oxidation state is less stable than elemental xenon with an oxidation number of 0. Xenon difluoride oxidizes water to form xenon, hydrofluoric acid and oxygen:

$$2XeF_2(s) + 2H_2O(l) \rightarrow 2Xe(g) + 4HF(aq) + O_2(g)$$

**READ MORE ABOUT COMPOUNDS OF NOBLE GASES ON PAGE 68.**

**YOU CAN READ ABOUT THE HYDROLYSIS OF COVALENT CHLORIDES OF PERIOD 2 AND PERIOD 3 ELEMENTS AND THE RESISTANCE TO HYDROLYSIS OF TETRACHLOROMETHANE IN CHAPTER 21.**

### ? TEST YOURSELF

**K** Construct the equation to show the hydrolysis of phosphorus(III) bromide, $PBr_3$, to make phosphoric(III) acid, $H_3PO_3$, and hydrogen bromide.

**SEE QUESTIONS 7 AND 8.**

### ? TEST YOURSELF

**L** Explain why all the hydrogen halides are polar molecules. Hint: Look at the electronegativities of the atoms involved on page 84.

# HYDROLYSIS OF COVALENT HALIDES

Almost all covalent chlorides can be hydrolysed to give an acidic solution. It is a typical property of covalently bonded chlorides that they can be hydrolysed to form hydrochloric acid or hydrogen chloride if only a limited supply of water is available. One exception to this rule is tetrachloromethane, which does not react with water at all.

The hydrolysis of covalent halides, such as phosphorus(III) bromide and phosphorus(III) iodide, provides a suitable way to prepare hydrogen bromide and hydrogen iodide.

# 5 HYDROGEN HALIDES

All the hydrogen halides have the formula HX, where X is F, Cl, Br, I or At. They all are colourless acidic gases that are highly soluble in water, in which they form an acidic solution. The bonding in a hydrogen halide is covalent, but the molecule is polar with its halogen end being slightly negative and its hydrogen end slightly positive.

## REACTION OF HALOGENS WITH HYDROGEN

The halogens react with hydrogen to give hydrogen halides, but the rate of reaction decreases with increasing atomic (proton) number of the halogen. The reaction between hydrogen and fluorine takes place very rapidly, even at low temperatures, whereas the reaction between hydrogen and iodine is reversible and needs elevated temperatures:

$$H_2(g) + F_2(g) \rightarrow 2HF(g)$$

When heated, hydrogen reacts with chlorine, and when a mixture of the two gases is subjected to ultraviolet light the reaction can be explosive.

**Fig 23.9** The reaction between hydrogen and chlorine in the presence of sunlight has a free-radical mechanism. The initiation reaction involves the homolytic fission of a chlorine–chlorine single bond to form highly reactive chlorine atoms. The chlorine atoms collide with hydrogen molecules to form hydrogen chloride and hydrogen atoms. The hydrogen atoms collide with chlorine molecules to form hydrogen chloride molecules and regenerate chlorine atoms. These two stages are propagation steps, since there is no net loss of chlorine atoms. In the termination steps, pairs of atoms collide to form molecules. (Initiation, propagation and termination are covered on pages 268 to 269.)

### ? TEST YOURSELF

**M** **a)** Write down an expression for the equilibrium constant for the thermal decomposition of hydrogen iodide.

**b)** At 600 K, hydrogen iodide is 19.1 per cent dissociated. Calculate the mole ratios of hydrogen, iodine and hydrogen iodide when a sample of hydrogen iodide is allowed to reach equilibrium at 600 K.

## THERMAL DECOMPOSITION OF HYDROGEN HALIDES

The thermal stability of the hydrogen halides decreases with increasing relative molecular mass, in a similar manner to the decrease in the bond energy from H–F to H–I. In the decomposition of hydrogen iodide, it is better to describe the reaction as an equilibrium process:

$$2HI(g) \rightleftharpoons H_2(g) + I_2(g) \quad \Delta H \text{ is positive}$$

Le Chatelier's principle (see pages 18 and 392) states that the position of equilibrium of a system shifts to minimize the effect of any change in the external conditions, such as increasing the temperature or the pressure. In the equilibrium process above, pressure has no effect whatsoever, because there is no volume change during the reaction – the number of moles of gas is the same on both sides of the equation. Since the decomposition is endothermic, an increase in temperature favours the formation of the two elements, because the reaction from left to right absorbs energy.

## CHEMICAL TEST FOR HYDROGEN HALIDES

Hydrogen halides are colourless acidic gases that are able to react with bases such as ammonia gas. This reaction is interesting in that two gases react together to make a white solid dispersed in a gas, correctly described as a smoke (Fig 23.10). This can be used as a chemical test for the hydrogen halides:

$$NH_3(g) + HCl(g) \rightarrow NH_4Cl(s)$$

## AQUEOUS SOLUTIONS OF THE HYDROGEN HALIDES

All the hydrogen halides dissolve in water to form acidic solutions. Hydrogen chloride forms hydrochloric acid, hydrogen bromide forms hydrobromic acid and hydrogen iodide forms hydroiodic acid:

$$HX(g) + H_2O(l) \rightarrow H_3O^+(aq) + X^-(aq) \qquad \text{where X is Cl, Br or I.}$$

The acid strength increases from hydrochloric acid to hydroiodic acid. This is because the H–I bond is weaker than the H–Cl bond (see Table 23.3). The bonding changes from covalent in the hydrogen halide to ionic in the corresponding acid.

All three acids are strong and display the typical reactions of strong acids (see page 374):

- They form hydrogen with metals above hydrogen in the electrochemical series.
- They form carbon dioxide with carbonates and hydrogencarbonates.
- They form salts with bases, such as metal oxides and hydroxides.
- They fully dissociate to form aqueous hydrogen ions.

Concentrated hydrochloric, hydrobromic and hydroiodic acids contain a large percentage of water, since if an attempt is made to concentrate them further, the gaseous hydrogen halide is evolved.

**Table 23.3 Bond energies of hydrogen halides**

| Hydrogen halide | Bond energy (H–X)/kJ mol$^{-1}$ |
|---|---|
| HF | 562 |
| HCl | 431 |
| HBr | 366 |
| HI | 299 |

**Fig 23.10** Ammonia gas and hydrogen chloride gas react to form a white smoke of ammonium chloride (a test either for ammonia or for hydrogen chloride)

**SEE QUESTIONS 9 AND 10.**

### ? TEST YOURSELF

**N** Predict the products of the reaction of:
**a)** hydrochloric acid with sodium carbonate;
**b)** hydroiodic acid with magnesium oxide;
**c)** hydrobromic acid with sodium hydroxide.

---

### ! SCIENCE FEATURE Hydrofluoric acid and hydrogen fluoride

Hydrofluoric acid is unique among acids in that it reacts with glass. Fortunately, it does not react with plastics; if it did, it would be difficult to find a suitable container in which to store it. During the reaction with glass, a volatile silicon compound, $SiF_4$, is produced:

$$SiO_2 + 4HF \rightarrow SiF_4 + 2H_2O$$
$$CaSiO_3 + 6HF \rightarrow CaF_2 + SiF_4 + 3H_2O$$

Hydrofluoric acid is extremely dangerous to handle, because it acts as a local anaesthetic while burning into the skin and flesh. So if it comes into contact with your skin, you are not aware of the severity of the damage until it is too late. You are left with acutely painful and slow-healing burns.

Dilute hydrofluoric acid is a weak acid. Only about 10 per cent of the HF molecules are dissociated in a 0.1 mol dm$^{-3}$ solution. This is because the H–F bond is very strong and the presence of strong intermolecular hydrogen bonds hinders dissociation.

Hydrogen fluoride is used mainly to produce fluorocarbons and the sodium aluminium fluoride required to manufacture

aluminium. It is also used to prepare a variety of important synthetic chemicals and catalysts.

Hydrogen fluoride also attacks glass, and so it is used to etch glassware.

**Fig 23.11** Hydrogen fluoride is used to etch the markings on thermometers and burettes

**? TEST YOURSELF**

**O** Construct the equations to show the reaction of potassium manganate(VII) with concentrated hydrochloric acid. The products are manganese(II) chloride, potassium chloride, water and chlorine.

**SEE QUESTION 10.**

**Fig 23.12** Calcium chloride is an ionic compound. Two electrons are lost by the calcium atom, one to each chlorine atom. This transfer of electrons means that both chlorine atoms and the calcium atom have a stable octet of electrons. The $Ca^{2+}$ and two $Cl^-$ ions are held together by strong electrostatic attractions, called ionic bonds. Most metal halides are ionic compounds and have metal atoms that transfer electrons to halogen atoms. Note: only the outer electrons are shown in the dot-and-cross diagram

**THE ACTION OF WATER ON ALUMINIUM CHLORIDE IS DISCUSSED ON PAGE 455.**

**SEE QUESTION 4.**

## OXIDATION OF HYDROHALIC ACIDS

Hydrochloric acid, hydrobromic acid and hydroiodic acid can all be oxidized to form the elemental halogen. The reaction involves the removal of hydrogen and an increase in the oxidation number of the halogen atom. For example:

$$PbO_2(s) + 4HCl(aq) \rightarrow PbCl_2(aq) + 2H_2O(l) + Cl_2(g)$$

The ease of oxidation increases as the atomic (proton) number of the halogen increases. So it is quite easy to reduce hydroiodic acid, but much more difficult to reduce hydrochloric acid.

Hydrochloric acid can be reduced with lead(IV) oxide, manganese(IV) oxide or potassium manganate(VII) to make chlorine, but less powerful oxidizing agents can be used to convert hydrobromic acid and hydroiodic acid into bromine and iodine, respectively. In other words, the reducing power decreases from hydroiodic to hydrochloric acid.

## 6 IONIC HALIDES

Ionic halides are made when halogens react with metals. Most metals react when the hot or burning metal is plunged into an atmosphere of the halogen. Ionic halides show the properties typical of compounds with a giant ionic structure. They have high melting and boiling points, do not conduct electricity as solids, often dissolve in water and the solution will conduct electricity. However, as mentioned on page 501, some metals can form ionic halides with covalent character. These have much lower melting points than those of the other ionic halides, and often sublime instead of melting.

### ACTION OF WATER ON IONIC CHLORIDES

Ionic chlorides, such as sodium chloride, dissolve in water to form a neutral solution. However, ionic chlorides with a covalent character dissolve to form an acidic solution because of hydrolysis. Metal halides in which the metal has an oxidation state of +3 have a high degree of covalent character. These halides have a metal cation with a charge of 3+, which is highly polarizing because it has a small ionic radius and a high charge. That is, it has a high charge density (see page 480).

When metal chlorides with covalent character are added to water, the ionic lattice breaks up and the chloride ion and the metal ion are surrounded by water molecules. The highly charged metal ion polarizes one of the water molecules to form a proton, so the solution becomes acidic. Chromium(III) chloride, iron(III) chloride and aluminium chloride all form acidic solutions when they are added to water, because of the polarization of a water molecule by an $M^{3+}$ ion:

$$[Cr(H_2O)_6]^{3+}(aq) \rightleftharpoons [Cr(H_2O)_5OH]^{2+}(aq) + H^+(aq)$$

### REACTION OF SOLID HALIDES WITH CONCENTRATED SULPHURIC ACID

Concentrated sulphuric acid reacts with solid halides to form the corresponding hydrogen halide. Since concentrated sulphuric acid is an oxidizing agent, the reaction is often complicated by oxidation of the hydrogen halide.

Solid sodium chloride reacts to form hydrogen chloride. This reaction is often used to prepare hydrogen chloride in the laboratory:

$$NaCl(s) + H_2SO_4(l) \rightarrow NaHSO_4(s) + HCl(g)$$

The same reaction with sodium bromide yields a collection of products, which include sulphur dioxide, hydrogen bromide and bromine. This is a result of the oxidation of hydrogen bromide formed initially in the reaction by concentrated sulphuric acid:

$$NaBr(s) + H_2SO_4(l) \rightarrow NaHSO_4(s) + HBr(g)$$

$$2HBr(g) + H_2SO_4(l) \rightarrow 2H_2O(l) + SO_2(g) + Br_2(g)$$

When the reaction with concentrated sulphuric acid is repeated with sodium iodide, an even more complicated set of reactions takes place, with the formation of hydrogen iodide, sulphur dioxide, iodine and hydrogen sulphide. This time, violet fumes of iodine are produced (Fig 23.14):

$$8HI(g) + H_2SO_4(l) \rightarrow H_2S(g) + 4H_2O(l) + 4I_2(g)$$

These three reactions provide a way to distinguish between solid ionic chlorides, bromides and iodides.

The reactions of solid ionic halides with concentrated sulphuric acid illustrate the relative ease with which hydrogen halides and halide ions can be oxidized. Hydrogen iodide and iodide ions are the easiest to oxidize, hydrogen fluoride and fluoride ions the most difficult. The reason for this trend in reducing power of the halide ions lies in the change in the ionic radius of the halide. It is easier to remove an electron from the large iodide ion than from the much smaller fluoride ion.

**Fig 23.13** Fumes of bromine are produced when concentrated sulphuric acid is added to sodium bromide

**? TEST YOURSELF**

**P** **a)** Write an equation to show the reaction between concentrated sulphuric acid and sodium iodide to form hydrogen iodide.
**b)** Write an equation to show the reaction between concentrated sulphuric acid and hydrogen iodide to form iodine and sulphur dioxide.
**c)** Predict the reaction products for the reaction between sodium astatide and concentrated sulphuric acid.
**d)** Predict the reaction products for the reaction between sodium fluoride and concentrated sulphuric acid.

**SEE QUESTIONS 9, 10 AND 11.**

**Fig 23.14** Fumes of iodine are produced when concentrated sulphuric acid is added to sodium iodide

## REACTIONS OF THE AQUEOUS HALIDE IONS

In addition to the displacement reactions described on page 500, the aqueous halide ions also take part in precipitation reactions. Most halides are soluble, but silver halides are insoluble. Therefore, silver halides can be precipitated by mixing together solutions of the appropriate aqueous solutions:

$$Ag^+(aq) + X^-(aq) \rightarrow AgX(s)$$

where X is Cl, Br or I.

These precipitation reactions are very useful in qualitative analysis, and are summarized in Table 23.4.

**Fig 23.15** Aqueous silver nitrate forms characteristic coloured precipitates with aqueous chlorides, bromides and iodides. Silver chloride is white, silver bromide is cream and silver iodide is pale yellow

| Table 23.4 Action of dilute nitric acid followed by aqueous silver nitrate and aqueous ammonia on aqueous halide ions | | | | |
|---|---|---|---|---|
| | Fluoride | Chloride | Bromide | Iodide |
| Action of dilute nitric acid followed by aqueous silver nitrate followed by addition of aqueous ammonia | No precipitate | White precipitate of silver chloride that turns violet in sunlight, redissolves in dilute aqueous ammonia to form a colourless solution through the formation of diamminesilver(I) ion [Ag(NH₃)₂]⁺(aq) | Cream precipitate of silver bromide redissolves in concentrated aqueous ammonia to form a colourless solution through the formation of [Ag(NH₃)₂]⁺(aq) | Yellow precipitate of silver iodide that does not redissolve in concentrated aqueous ammonia, but turns white |

**Fig 23.16** In order to avoid drought, rain is made to fall in Australia by seeding clouds with silver iodide crystals

## SCIENCE FEATURE Black and white photography

Silver halides are sensitive to light. For example, in the precipitation reaction given in Table 23.4, the silver chloride formed turns violet in the presence of light. This is because of the decomposition of silver chloride to give silver and chlorine:

$$2AgCl(s) \rightarrow 2Ag(s) + Cl_2(g)$$

The longer the exposure to light, the greater is the amount of silver produced.

Black and white photography works on a similar principle. The film consists of a very thin layer of gelatine and silver bromide crystals deposited on clear plastic. As light shines on the film, an image is captured by the silver bromide crystals. This image is developed by immersing the film in a developer that reacts only with those silver bromide crystals that have been exposed to the light. The developer reduces the silver bromide to metallic silver:

$$2AgBr(s) + C_6H_4(OH)_2(aq) \rightarrow 2Ag(s) + C_6H_4O_2(aq) + 2HBr(aq)$$

Next, the film is placed in a fixer, where the silver bromide crystals not exposed to light, and hence unaffected by the developer, are removed from the film. This leaves the film as a negative with black areas of silver where the film was exposed to light.

The negative is then converted into a permanent picture on photographic paper. Light is shone through the negative onto the photographic paper, which is coated with silver bromide. The action of light on the silver bromide produces a true image that has to be developed to give a permanent photograph.

**Fig 23.17** The negative image

**Fig 23.18** The positive image

# SUMMARY

After studying this chapter, you should know that:

- The reactivity of the halogens decreases with increasing atomic (proton) number.

- The oxidizing power of the halogens decreases with increasing atomic (proton) number and that this can be explained in terms of the ability to capture electrons.

- The oxidation state of a halogen in a compound is normally −1.

- A halogen atom can gain an electron through ionic bonding or share electrons through covalent bonding to attain a noble-gas electron configuration.

- Metals react with halogens to form ionic halides and non-metals react with them to form covalent halides.

- The hydrogen halides are colourless acidic gases that dissolve in water to form acid solutions. Hydrogen chloride forms hydrochloric acid, hydrogen fluoride forms hydrofluoric acid, hydrogen bromide forms hydrobromic acid and hydrogen iodide forms hydroiodic acid.

- Covalent halides or ionic halides with covalent character are hydrolysed to form acidic solutions.

- The thermal stability of the hydrogen halides decreases with increasing atomic (proton) number of the halogen, which can be explained in terms of the bond energy of the hydrogen–halogen bond.

- The ease of oxidation of the hydrogen halides increases with increasing atomic (proton) number of the halogen.

- In aqueous solution, the halogen with the lower atomic (proton) number can displace the halide ion with the higher atomic number.

- Acidified silver nitrate can be used to distinguish between aqueous halide ions. The silver halides formed can be distinguished by their differing solubilities in aqueous ammonia.

# ? EXAM QUESTIONS

**1**

**a)** State and explain the trend in electronegativity of the halogens down Group 7.

**b)** State and explain the trend in boiling points of the halogens down Group 7.

AQA CHM2 Unit 2 Foundation Physical and Inorganic Chemistry Jan 2002 Q6

**2**

**a)** State and explain the trend in volatility in the halogens chlorine, bromine and iodine.

**b)** Describe and explain the thermal stability of the hydrogen halides.

**c)** Explain why hydroiodic acid is stronger than hydrochloric acid.

**d)** Astatine is a member of the halogen family.
(i) Predict two physical properties of astatine.
(ii) Predict three chemical properties of astatine and explain your reasoning by reference to the properties of the other halogens.
(iii) Suggest why astatine is not normally studied in schools or college laboratories.

**3** The halogens are the elements in Group 7 of the Periodic Table.

**a)** State and explain four ways in which the properties of the halogens are similar. Use ideas about the structure and bonding of the elements.

**b)** How is the trend in reactivity down Group 7 shown by displacement reactions. Illustrate your answer with ionic equations.

**4**

**a)** Sea water contains bromide ions and iodide ions.
(i) State how bromine is extracted from sea water.
(ii) Suggest how iodine may be extracted from sea water.

**b)** Explain why fluorine cannot be prepared by a displacement reaction.

**5**

**a)** State the trend in oxidizing power of the halogens chlorine, bromine and iodine.

**b)** State what would be observed if aqueous bromine were to be added separately to samples of aqueous potassium chloride and aqueous potassium iodide. Write an ionic equation for any reaction that occurs.

**c)** When chlorine is dissolved in cold water, a pale green solution, chlorine water, is formed. A piece of universal indicator paper, dipped into chlorine water, first turns red and then becomes white.
(i) Give the formula of the species responsible for the green colour of chlorine water.
(ii) Write an equation for the reaction between chlorine and cold water.
(iii) Explain the colour changes observed when universal indicator paper is dipped into chlorine water.

**d)** Write an equation for the reaction that occurs when chlorine is bubbled into cold, dilute aqueous sodium hydroxide.

AQA CHM2 Unit 2 Foundation Physical and Inorganic Chemistry Jan 2001 Q1

**6** The halogens have different reactivities.

**a)** Describe how displacement reactions can be used to show the different reactivities of chlorine, bromine and iodine. In your answer, you should include equations and observations.

**b)** Explain the trend in reactivity of the halogens.

OCR 2811 Foundation Chemistry Jan 2002 Q6

**7** Phosphorus(III) chloride is hydrolysed in cold water to form hydrochloric acid and phosphoric(III) acid.

**a)** Write an equation, including state symbols, for this reaction.

**b)** Describe how you could confirm the presence of chloride ions in the reaction mixture.

**8**

**a)** Explain in terms of their electron configurations why metals tend to form ionic chlorides and non-metals tend to form molecular (covalent) chlorides.

**b)** Draw dot-and-cross diagrams to show the bonding in:
(i) magnesium chloride;
(ii) aluminium chloride.

**c)** Metal chlorides can be prepared by the reaction between chlorine and the metal. Name the products of the following reactions:
(i) iron with chlorine;
(ii) aluminium with chlorine.

**d)** One of the characteristic properties of molecular (covalent) chlorides is that they are hydrolysed rapidly by water to form hydrogen chloride or hydrochloric acid.
(i) Write an equation to show the hydrolysis of phosphorus(III) chloride.
(ii) One mole of boron trichloride reacts with three moles of water during its hydrolysis. Construct the equation to show the hydrolysis of boron trichloride.
(iii) Aluminium chloride and iron(III) chloride are both hydrolysed by water to form hydrochloric acid. What does this suggest about the bonding in these two compounds?

# ? EXAM QUESTIONS

**9**

**a)** Hydrogen chloride can be made from sodium chloride and concentrated sulphuric acid. Write a balanced chemical equation to represent this reaction.

**b)** **(i)** How would you confirm that a solution said to be HCl(aq) contained chloride ions?
**(ii)** Hydrogen chloride is soluble in water. Explain why the solution is acidic.

**c)** **(i)** Give a chemical test for chlorine, stating what you would do and what you would see.
**(ii)** Hydrogen chloride can be oxidized to chlorine by lead(IV) oxide, $PbO_2$. Write down the equation below and write below it the oxidation numbers of lead and chlorine. [Five oxidation numbers required]

$$PbO_2 + 4HCl \rightarrow PbCl_2 + Cl_2 + H_2O$$

**d)** Sodium iodide reacts with concentrated sulphuric acid to give iodine, not hydrogen iodide. Explain why iodides react differently from chlorides in this case.

Edexcel 6241/P.01 A/AS Unit Test 1 Jan 2002 Q4

**10** Hydrogen chloride can be prepared by the reaction of concentrated sulphuric acid on solid sodium chloride.

**a)** Write an equation for the reaction.

**b)** Describe a chemical test for hydrogen chloride.

**c)** Explain why hydrogen iodide cannot be prepared by the reaction of sodium iodide with concentrated sulphuric acid.

**11**

**a)** State what you would observe on adding aqueous chlorine to separate aqueous solutions of sodium bromide and sodium iodide. Write equations for the reactions that occur.

**b)** State what you would observe on adding concentrated sulphuric acid to separate solid samples of sodium bromide and sodium iodide. In each case, identify all the reduction products. Using half-equations, construct an overall ionic equation for the oxidation of bromide ions by concentrated sulphuric acid.

AQA CHM2 Unit 2 Foundation Physical and Inorganic Chemistry Jan 2002 Q7

**12** Describe how you could distinguish between separate solutions of sodium fluoride, sodium chloride, sodium bromide and sodium iodide using separate solutions of silver nitrate and ammonia.
Predict the effect of concentrated ammonia solution on silver astatide, AgAt, and explain your answer.

AQA CHM2 Unit 2 Foundation Physical and Inorganic Chemistry Jan 2001 Q7

# KEY SKILLS **ASSIGNMENT**

## THE INTERHALOGENS

**Compounds formed between two different halogens are called interhalogens. All interhalogens consist of an atom of the halogen with the higher atomic number bonded to a number of atoms of the halogen with the smaller atomic number. All of the interhalogens are covalently bonded. The interhalogens of the formula XY, where X and Y are two different halogens, closely resemble the halogens $X_2$ and $Y_2$.**

| $IF_3$ T-shaped molecule | $IF_5$ square pyramidal molecule | $IF_7$ pentagonal bipyramidal molecule |

**Fig 23.A1** Structures of some interhalogen molecules

**Table 23.A1 The interhalogens. The state symbols refer to the state at room temperature and atmospheric pressure**

| | | | |
|---|---|---|---|
| $ClF(g)$ | $ClF_3(g)$ | $ClF_5(g)$ | |
| $BrF(g)$ | $BrF_3(l)$ | $BrF_5(l)$ | |
| $BrCl(g)$ | | | |
| $IF(g)$ | $IF_3(s)$ | $IF_5(l)$ | $IF_7(g)$ |
| $ICl(l)$ | $ICl_3(s)$ | | |

**1** Predict the shape of the following interhalogen molecules: **a)** $ClF_3$; **b)** $ClF_5$; **c)** $BF_3$.
Hint: You may want to read about the shapes of molecules on page 77.

**2** Draw a dot-and-cross diagram to show the bonding in IF, $IF_3$, $IF_5$ and $IF_7$.

**3** **a)** What type of intermolecular force is present in solid ICl?
**b)** Predict the structure of solid ICl based upon your knowledge of the structures of solid chlorine and solid iodine.

**4** Predict the physical state of the interhalogen IBr. Give reasons for your answer.
If the molecule has a central halogen atom, it always has a positive oxidation number, because it is bonded to a more electronegative halogen. So the naming of the interhalogens is quite easy.

Start with the central halogen and its oxidation number, and then follow that with the appropriate halide. So, $ICl_3$ is iodine(III) chloride and $IF_7$ is iodine(VII) fluoride.

**5** Name each of the following interhalogens:
**a)** $BrF_3$; **b)** BrF; **c)** ICl.
An interhalogen can be prepared by direct combination of the two halogens.

**6** **a)** Write an equation to show the preparation of iodine(VII) fluoride.
**b)** Suggest the mole ratio of fluorine to bromine that you would need to make bromine(III) fluoride.
**c)** Write an equation to show the preparation of bromine(V) fluoride. Calculate the mass of fluorine needed to make 10 g of bromine(V) fluoride.
Hint: You may want to read page 7.

The interhalogens react in the same way as their constituent halogens. The halogen fluorides are more reactive than halogen chlorides.

**7** Burning sodium is placed in a gas jar of bromine chloride. Suggest the names of the products formed in this reaction.

**8** Chlorine and bromine will add to a molecule of ethene, $C_2H_4$, to give dichloroethane and dibromoethane.
Suggest the product of the reaction between bromine chloride and ethene. Write an equation for the reaction.
Hint: You may want to read about addition reactions of ethene on page **XXX**.

**9** A mixture of chlorine fluoride and hydrogen is placed in sunlight.
**a)** Write an equation for the reaction that occurs.
**b)** Suggest a mechanism for the reaction.

**10** Chlorine reacts with benzene, $C_6H_6$, in the presence of aluminium chloride to give chlorobenzene, $C_6H_5Cl$. The reaction is electrophilic substitution and the aluminium chloride is used to make a chlorine electrophile, $Cl^+$. The reaction is repeated using bromine chloride.
**a)** Draw the dipole in bromine chloride.
**b)** Suggest a product of the reaction between bromine chloride and benzene.
Hint: You may want to read about electrophilic substitution on page 336.

# 24

# The transition elements

**Artery repair** Arteries can collapse or become constricted. This image shows a severe constriction in a coronary artery, which restricts blood flow from the red to the orange part of the artery. To overcome this, a coil of Nitinol wire can be inserted to hold it open and allow the blood to flow

Nitinol is a remarkable alloy that can remember shapes. Deform a piece of the alloy and it will return to its original shape after being heated.

Nitinol is an alloy of nickel and titanium, both transition elements. It was discovered in the USA in 1962 quite by chance, when researchers in the defence industry were seeking ways to make titanium less brittle by adding nickel to it. They found that the temperature at which Nitinol returns to its original shape can be set anywhere between −100 °C and 100 °C by altering the amount of nickel in the alloy.

Nitinol wire can be used to repair damaged or blocked arteries. The wire is wound round a tube of the same diameter as the inside of an artery. It is heated so that this coiled shape is remembered. The wire is then cooled and straightened. It is passed into the artery to be repaired. The temperature of the blood warms up the wire, which returns to its coiled shape. So an artery with a weak wall can be reinforced and a constriction in an artery can be unblocked. Nitinol does not react with body chemicals, and the arterial wall soon grows round it.

## 1 WHAT IS A TRANSITION ELEMENT?

**MORE INFORMATION ON IRON AND STEEL CAN BE FOUND ON PAGES 4 AND 13.**

Ask someone to name the first metal that comes to mind, and the chances are that a transition element will be named. It may be iron, essential to the construction of many buildings, most vehicles and almost all machinery. It may be gold, a precious metal used in jewellery and other forms of ornamentation, and as a protective coating. It may be copper, used in electric wiring and for indoor water pipes, or it may be silver or platinum. It could be nickel or chromium. Everyone is familiar with the transition elements through their use, not in compounds, but on their own or in alloys, such as brass, bronze and steel.

The transition metal compound with the largest annual production (2 600 000 tonnes) is titanium(IV) oxide, $TiO_2$. This compound is bright white when pure, and so is used as the pigment in white paint, white paper and white plastics, as well as in sunscreens.

Transition metals are essential to life. Cobalt is found in vitamin B1, where it acts as a catalyst. The iron in haemoglobin is involved in transporting oxygen round the body. And nickel and copper are essential components of several enzymes.

### REMEMBER THIS

The terms transition element and transition metal are used interchangeably.

**Fig 24.1** The pigment in white paint is $TiO_2$

So what is a transition element? The central block of the Periodic Table is the **d block**. This chapter deals with the ten d block elements in Period 4, sometimes called the 'first row transition elements' as they are the top row of the d block. These elements have a closely similar set of properties, so it is tempting to call them all transition elements. However, on further examination, scandium at one end of the first row and zinc at the other are obviously different from the rest. This has led chemists to define transition elements as **elements that form one or more ions with a partially filled d subshell**. This definition excludes zinc and scandium.

## ELECTRON CONFIGURATION

Table 24.1 shows the electron configurations of the first-row transition elements. Notice that the 4s subshell has already been filled because this is at a lower energy than the 3d subshell. The full electron configuration of titanium (atomic number 22) is $1s^2\,2s^2\,2p^6\,3s^2\,3p^6\,3d^2\,4s^2$.

**Table 24.1** Arrangement of electrons in the outer 4s and inner 3d subshells of the first row of the d block. Each atomic orbital is represented by a box, and each electron by an arrow

| Element | Symbol | Atomic (proton) number | Electron configuration 3d | 4s |
|---|---|---|---|---|
| Scandium | Sc | 21 | [Ar] ↑ ☐ ☐ ☐ ☐ | ↑↓ |
| Titanium | Ti | 22 | [Ar] ↑ ↑ ☐ ☐ ☐ | ↑↓ |
| Vanadium | V | 23 | [Ar] ↑ ↑ ↑ ☐ ☐ | ↑↓ |
| Chromium | Cr | 24 | [Ar] ↑ ↑ ↑ ↑ ↑ | ↑ |
| Manganese | Mn | 25 | [Ar] ↑ ↑ ↑ ↑ ↑ | ↑↓ |
| Iron | Fe | 26 | [Ar] ↑↓ ↑ ↑ ↑ ↑ | ↑↓ |
| Cobalt | Co | 27 | [Ar] ↑↓ ↑↓ ↑ ↑ ↑ | ↑↓ |
| Nickel | Ni | 28 | [Ar] ↑↓ ↑↓ ↑↓ ↑ ↑ | ↑↓ |
| Copper | Cu | 29 | [Ar] ↑↓ ↑↓ ↑↓ ↑↓ ↑↓ | ↑ |
| Zinc | Zn | 30 | [Ar] ↑↓ ↑↓ ↑↓ ↑↓ ↑↓ | ↑↓ |

Since the noble gas core is argon ($1s^2\,2s^2\,2p^6\,3s^2\,3p^6$), titanium can be abbreviated to $[Ar]3d^2 4s^2$.

For each element from scandium to zinc, the number of protons increases by one. This increases the positive charge on the nucleus. However, because the electrons are added to the inner 3d subshell, the outer 4s electrons tend to be shielded from the increasing charge. This partly explains the similarity in physical and chemical properties. Also, because the 3d and 4s subshells have similar energies, the electrons from both can take part in bonding, which gives rise to some characteristic transition metal properties.

Electrons occupy first the orbitals of lowest energy. The orbitals in a subshell are first occupied singly by electrons that spin in the same direction, which helps to minimize the interelectron repulsion. Only when all the d orbitals in the d subshell are singly occupied do electrons start to pair up. For the build-up of electron configurations, see page 62.

## ELECTRON CONFIGURATIONS OF CHROMIUM AND COPPER

Look at Table 24.1. The chromium atom and the copper atom each have only one electron in the 4s subshell. As the number of protons increases, the number of electrons in the d subshell also increases by one each time, until chromium is reached. At chromium, there is a jump of two electrons, one of which is from the 4s subshell. Why is this? The half-filled subshell has five

**? TEST YOURSELF**

**A** Why is the central block of elements called the d block?
Hint: If you are not sure, look at page 58.

**TO REVIEW ELECTRON CONFIGURATIONS, SEE PAGE 61.**

**? TEST YOURSELF**

**B** **a)** Write down the full electron configuration for manganese and nickel.
**b)** Look at Table 24.1. Why are atomic orbitals in the d subshell occupied by single, unpaired electrons first?
Hint: See page 62 if you are not sure.

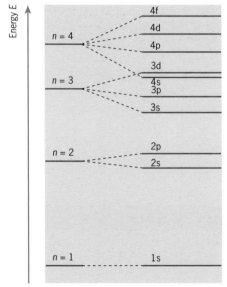

**Fig 24.2** Energy levels of the various subshells in a many-electron atom

**Fig 24.3** Shapes of d orbitals. Four of the five d orbitals have the shape shown on the left and one has the shape shown on the right. Remember that an orbital's shape represents the boundary surface and that there is a 90 per cent probability of finding an electron within it

**? TEST YOURSELF**

**C** What is the electron configuration of:
**a)** $Co^{2+}$ and $Co^{3+}$;
**b)** $Cu^+$ and $Cu^{2+}$;
**c)** $Zn^{2+}$?
Hint: See Table 24.1.

**? TEST YOURSELF**

**D** Scandium forms only 3+ ions. What is the electron configuration of $Sc^{3+}$ and why is scandium not considered to be a transition element?

**? TEST YOURSELF**

**E** What other typical properties of metals are there?
To check your answer, look at page 101.

singly occupied orbitals. So, it is a lower energy arrangement to have $3d^5$ and only one 4s electron, as this removes the paired electron in the 4s orbital and thereby reduces interelectron repulsion. At copper, the most stable arrangement is to have $3d^{10}4s^1$, since a full 3d subshell is a more stable arrangement.

## FORMATION OF TRANSITION METAL IONS

Metal ions are always formed by the loss of electrons, which requires energy. In the case of transition metal ions, the first electrons to be lost are in the 4s subshell. This is because, once the 3d subshell starts to fill, the 4s electrons are repelled further from the positive nucleus. This can be seen with $Mn^{2+}$:

Mn [Ar] $\boxed{\uparrow}\,\boxed{\uparrow}\,\boxed{\uparrow}\,\boxed{\uparrow}\,\boxed{\uparrow}\quad\boxed{\uparrow\downarrow}\ \xrightarrow{-2e}\ $ Mn$^{2+}$ [Ar] $\boxed{\uparrow}\,\boxed{\uparrow}\,\boxed{\uparrow}\,\boxed{\uparrow}\,\boxed{\uparrow}\quad\boxed{\ }$

Thus, the electron configuration of $Mn^{2+}$ is $[Ar]3d^5$.

Fe loses its 4s electrons to form $Fe^{2+}$:

Fe [Ar] $\boxed{\uparrow\downarrow}\,\boxed{\uparrow}\,\boxed{\uparrow}\,\boxed{\uparrow}\,\boxed{\uparrow}\quad\boxed{\uparrow\downarrow}\ \xrightarrow{-2e}\ $ Fe$^{2+}$ [Ar] $\boxed{\uparrow\downarrow}\,\boxed{\uparrow}\,\boxed{\uparrow}\,\boxed{\uparrow}\,\boxed{\uparrow}\quad\boxed{\ }$

which has an electron configuration of $[Ar]3d^6$.

As the energies of the 4s and 3d subshells are still fairly close together, 3d electrons may also be lost to form ions with charges of 3+. In forming $Fe^{3+}$ a 3d electron is lost together with the two 4s electrons to give an electron configuration of $[Ar]3d^5$:

Fe [Ar] $\boxed{\uparrow\downarrow}\,\boxed{\uparrow}\,\boxed{\uparrow}\,\boxed{\uparrow}\,\boxed{\uparrow}\quad\boxed{\uparrow\downarrow}\ \xrightarrow{-3e}\ $ Fe$^{3+}$ [Ar] $\boxed{\uparrow}\,\boxed{\uparrow}\,\boxed{\uparrow}\,\boxed{\uparrow}\,\boxed{\uparrow}\quad\boxed{\ }$

As already noted, scandium and zinc do not show typical transition metal properties. Zinc has an electron configuration $[Ar]3d^{10}4s^2$. The stable full d subshell means that zinc forms only $Zn^{2+}$ ions, $[Ar]3d^{10}$. Recalling the definition of transition metals as those that form one or more ions with a partially filled d subshell, zinc does not qualify as a transition element.

## 2 PHYSICAL PROPERTIES OF TRANSITION ELEMENTS

In terms of their physical properties, transition elements are typical metals. They are hard and dense, have high melting points and are good conductors of heat and electricity. They tend to be strong and durable, and have high tensile strengths, as well as other useful mechanical properties. So they find many applications, from bridges (see the Humber bridge, page 2) to cooking utensils. Contrast this with the s block metals considered on page 467, some of which are soft and have low melting points.

A more detailed examination of the physical properties reveals their close similarity. This is again explained by the electron configurations.

## MELTING POINTS AND HARDNESS

The high melting points and the hardness of transition metals are caused by their strong metallic bonds. As described on pages 152 and 154, metallic bonding is caused by the delocalization of outer electrons, leaving positive metal ions surrounded by a sea of electrons. The more electrons that are delocalized, the stronger the metallic bond. In the case of transition metals, some 3d electrons can be delocalized along with the 4s. This explains why they have higher melting points compared with s block elements, such as potassium and calcium.

— metal cation

— sea of delocalized electrons

**Fig 24.4** Metallic bonding in transition metals involves 3d as well as 4s electrons

## ▶ EXTENSION Ionization energies

Ionization energy is discussed in several chapters of this book. Although it is a physical property, it is fundamental in determining the chemical properties of elements.

Fig 24.5 shows the first four successive ionization energies of the elements of the first transition series. Note how similar all the first ionization energies are across this series. The same similarity is seen with the second ionization energies.

This is because the 4s electrons are the first to be removed, even though it is the 3d subshell that is being filled along the series. This shows that the inner 3d electrons shield the 4s electrons from the increasing nuclear charge. In the case of the third and fourth ionization energies, the 3d electrons are being removed from the same subshell. So, the effective nuclear charge is increasing significantly along the period.

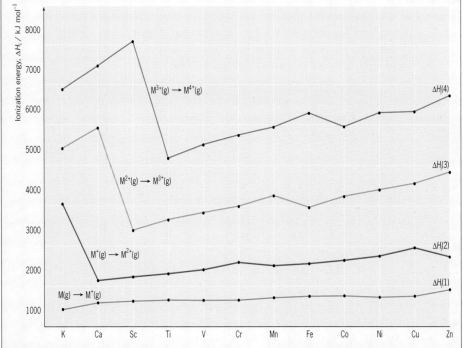

**Fig 24.5** First four successive ionization energies of the elements of the first transition series, with K and Ca included for comparison

## CONDUCTION OF ELECTRICITY

As explained on page 153, the conductivity of metals results from delocalization of the outer electrons in the metallic bond, which makes the electrons free to move under the influence of a potential difference. Transition metal atoms also have the 3d electrons as part of their metallic bond, which makes them better conductors than calcium, which can only use its 4s electrons. In fact, copper and silver are the best metallic conductors at room temperature.

# 3 CHEMICAL PROPERTIES

There is a close similarity of the physical properties of the transition elements, so what about their chemical properties? There are four that characterize transition elements:

- variable oxidation states in their compounds;
- formation of complexes;
- formation of coloured ions;
- catalysis of reactions both as elements and compounds.

Each of these is now examined.

## VARIABLE OXIDATION STATES

The terms oxidation state and oxidation number tend to be used interchangeably (see pages 105 and 110). However, when considering the transition elements, the term oxidation state is usually preferred.

### LOWER OXIDATION STATES IN IONIC COMPOUNDS

There is a wide range of oxidation states in the transition metals. Contrast this with Group 1 and Group 2 elements, each of which has only one oxidation state (+1 and +2 for Groups 1 and 2, respectively). The difference is explained by the closeness of the 4s and 3d energy levels. In calcium, the two 4s electrons are easily removed, but to remove another electron means breaking into the 3p subshell, which requires much more energy. This is not the case with the transition metals. Remember that the first electrons to be lost in forming a transition metal ion are the 4s electrons, which makes the +2 oxidation state common, as in $Fe^{2+}$ and $Co^{2+}$. However, the 3d electrons may also be lost, which means that ions such as $V^{3+}$, $Fe^{3+}$ and $Cr^{3+}$ are also common. Table 24.2 shows this.

| Table 24.2·Comparison of the first four ionization energies of calcium and iron | | | | | |
|---|---|---|---|---|---|
| Element | Electron configuration of element | $\Delta H_i(1)$ | $\Delta H_i(2)$ | $\Delta H_i(3)$ | $\Delta H_i(4)$ |
| Ca | $1s^2\,2s^2\,2p^6\,3s^2\,3p^6\,4s^2$ | 590 | 1145 | 4912 | 6474 |
| Fe | $1s^2\,2s^2\,2p^6\,3s^2\,3p^6\,3d^6\,4s^2$ | 759 | 1561 | 2958 | 5290 |

Notice the big jump in the third ionization energy of calcium as the 3p electron is removed. In the case of iron, the increase is more gradual and the simple ions $Fe^{2+}$ and $Fe^{3+}$ are both formed. The electron configurations of $Fe^{2+}$ and $Fe^{3+}$ are:

The electron configurations of manganese in the +2 and +3 oxidation states are:

### ? TEST YOURSELF

**G** **a)** Solutions of $Fe^{3+}(aq)$ are more stable than $Fe^{2+}(aq)$. Suggest why.
**b)** Why is $Mn^{2+}$ more stable than $Mn^{3+}$?
**c)** Using box diagrams, write down the electrons for $Cu^+$ and $Cu^{2+}$. Although you might predict $Cu^+$ to be the more stable, in fact the +2 oxidation state of copper is the more stable. The reasons for the stability of copper(II) are explained on page 566.

**Fig 24.6** Oxidation states of the first transition series. The most common oxidation states are ringed

Fig 24.6 shows all the common oxidation states of the first transition series. Note that both scandium and zinc have only one oxidation state, which is yet another reason why they should not be classified as transition metals. From titanium to copper, both the +2 and +3 oxidation states exist. At first, the +3 oxidation state is common, which shows the greater stability of the $M^{3+}$ ion. But from manganese onwards the +2 oxidation state becomes more common. The removal of a third electron from the d subshell is energetically more difficult after manganese, because the effective nuclear charge becomes so much greater than it is up to manganese. The reason for the relative stabilities of the +2 and +3 oxidation states is explored further when electrode potentials are considered (see Chapter 25).

### HIGHER OXIDATION STATES

From Fig 24.6, the maximum oxidation state increases across the first transition series, reaching +7 at manganese. Note that this corresponds to the total number of electrons in the 4s and 3d subshells. However, a $Mn^{7+}$ ion or a $Cr^{6+}$ ion would be too polarizing to exist. Also, too much energy would be needed to remove so many electrons. The higher oxidation states are found covalently bonded either in simple compounds, such as $TiO_2$, $V_2O_5$, $CrO_3$ and $Mn_2O_7$, or as ions, such as $VO_2^+$, $Cr_2O_7^{2-}$ and $MnO_4^-$. The availability of partially filled or empty orbitals, particularly in the 4s and 3d subshells, allows the higher oxidation states to be reached.

After manganese, the maximum oxidation state of each element decreases because of the increasing energy needed to involve 3d electrons in bonding, and because singly filled orbitals are required in order to form a covalent bond.

### NAMING THE TRANSITION METAL COMPOUNDS AND IONS

Ionic NaCl is easy to name. Sodium has only one oxidation state, so the compound can be called sodium chloride, and $Na^+$ can be called the sodium ion. However, in the case of, for example, $FeCl_3$ and $FeCl_2$ the name iron chloride does not distinguish the two compounds, so they have to be given systematic names that include the oxidation state of the transition element:

$FeCl_3$, iron(III) chloride   which contains the ion $Fe^{3+}$   iron(III) ion

$FeCl_2$, iron(II) chloride    which contains the ion $Fe^{2+}$   iron(II) ion

Note that the oxidation state is shown by a roman numeral in brackets and that there is no space between the metal name and the oxidation number.

### REMEMBER THIS

In $MnO_4^-$, since O has an oxidation number of –2, the total oxidation number due to O is $-2 \times 4 = -8$. Therefore, the oxidation state of manganese is +7, because this leaves –1 as the charge on the anion.

### REMEMBER THIS

Redox means oxidation (oxidation number increases) and reduction (oxidation number decreases). To find out more about redox reactions and half equations, read Chapter 25.

### REMEMBER THIS

A dative covalent bond is a covalent bond in which the shared pair of electrons come from the same atom. It is also called a coordinate bond. See page 74.

**Fig 24.7** Two complex ions

**SEE QUESTION 1.**

### ? TEST YOURSELF

**L** What is the charge on the central metal ion in the following complexes?

**a)** $[Ag(NH_3)_2]^+$; **b)** $[NiCl_4]^{2-}$;
**c)** $[Fe(H_2O)_6]^{3+}$; **d)** $[Ni(CN)_4]^{2-}$;
**e)** $[Fe(CN)_6]^{3-}$.

Hint: The cyanide ligand is $CN^-$.

---

Here are two more examples of compounds with systematic names:

CuO, copper(II) oxide    $MnO_2$, manganese(IV) oxide

Ions such as $MnO_4^-$ and $Cr_2O_7^{2-}$ are called **oxyanions**, because they are negative ions that contain oxygen. $MnO_4^-$ is called the manganate(VII) ion. The (VII) signifies that the oxidation state of manganese in this oxyanion is +7. $Cr_2O_7^{2-}$ is the dichromate(VI) ion, so the oxidation state of chromium in this oxyanion is +6. Thus, $KMnO_4$ is called potassium manganate(VII) and $Na_2Cr_2O_7$ is called sodium dichromate(VI).

## REDOX REACTIONS AND THE STABILITY OF OXIDATION STATES

Since the transition elements have variable oxidation states in their compounds, they can undergo **redox** reactions. Compounds that contain transition elements in high oxidation states, such as potassium manganate(VII), tend to be oxidizing agents. This is because they are usually less stable than compounds with lower oxidation states. In acid solution, the $MnO_4^-$ ion is reduced to $Mn^{2+}$ by gaining electrons, as can be seen in this **half equation**:

$$MnO_4^-(aq) + 8H^+(aq) + 5e^- \rightarrow Mn^{2+}(aq) + 4H_2O(l)$$

Look again at Fig 24.6 on page 579, which shows that, across the first transition series, the +3 oxidation state is more stable than the +2 state until manganese is reached. This means that $Ti^{2+}$, $V^{2+}$ and $Cr^{2+}$ are highly reducing, because they are easily oxidized to the +3 state. For example:

$$Cr^{2+}(aq) \rightarrow Cr^{3+}(aq) + e^-$$

Traces of oxygen can be removed from other gases by bubbling the gases through a solution of $Cr^{2+}(aq)$.

From manganese to copper, the +2 state becomes more stable. So, compounds with oxidation states of +3 are highly oxidizing. For example:

$$Co^{3+}(aq) + e^- \rightarrow Co^{2+}(aq)$$

The relative stabilities of the different oxidation states are predicted from **standard electrode potentials**, $E^{\ominus}$ (see Chapter 25).

## COMPLEXES

A **complex** is formed when a central metal atom or ion is surrounded by species that donate lone pairs of electrons. The actual bonding in complexes is complicated and beyond the scope of this book. However, it is assumed that the lone pairs from the ligands form dative covalent (coordinate) bonds. **A species that donates a lone pair of electrons is called a ligand**. Ligands are usually molecules such as water or negative ions such as $Cl^-$. Fig 24.7 shows two complex ions.

Many complexes are positively or negatively charged, but some are neutral. In the formula, the central metal atom or ion is written first followed by the ligands. The charge on a complex is the charge of the central metal ion and the charges on the surrounding ligands added together. The overall charge of $[Cu(H_2O)_6]^{2+}$ results from the 2+ charge of $Cu^{2+}$, since the water molecules carry no charge. In the case of $[CuCl_4]^{2-}$, there are four $Cl^-$ ions, to give a charge of 4–, which when added to the 2+ charge of $Cu^{2+}$ gives an overall charge of 2– [i.e., $4(1-) + (2+) = 2-$]. The charge on a complex ion is delocalized over the whole ion and is usually shown outside square brackets. However, complex ions are sometimes drawn with a charged central metal ion.

## COORDINATION NUMBER AND SHAPE

The number of dative covalent (coordinate) bonds from the ligands to the central metal ion is called the **coordination number**. The coordination number determines the shape of the complex. The most common coordination numbers are 4 and 6. Table 24.3 shows the shapes that are often associated with these two numbers, together with the shape of coordination number 2. Note that a coordination number of 4 gives two possible shapes. The more common is the tetrahedral structure. On page 77, the electron pair repulsion theory is discussed, which predicts the tetrahedral, octahedral and linear shapes in Table 24.3. However, d orbital electrons affect the shape in a different way to s and p orbital electrons, and many shapes are not as predicted by this theory. For example, the shape of $[Cu(H_2O)_6]^{2+}$ is not a regular octahedron because four of the copper–oxygen bonds are short and the other two are longer, which gives a distorted octahedron.

**Table 24.3 Shapes of four complexes**

| Coordination number | Shape | Example | Structure |
|---|---|---|---|
| 2 | Linear | $[Ag(NH_3)_2]^+$ | $\left[ H_3N \longrightarrow Ag \longleftarrow NH_3 \right]^+$ |
| 4 | Tetrahedral | $[CuCl_4]^{2-}$ | (tetrahedral structure with Cu centre and 4 Cl ligands) $^{2-}$ |
| 4 | Square planar | $[Ni(CN)_4]^{2-}$ | (square planar structure with Ni centre and 4 CN ligands) $^{2-}$ |
| 6 | Octahedral | $[Co(NH_3)_6]^{2+}$ | (octahedral structure with Co centre and 6 $NH_3$ ligands) $^{2+}$ |

**SEE QUESTION 2.**

### ? TEST YOURSELF

**M** For each of the complex ions in the Science Feature box work out:
**a)** the charge on the central silver ion;
**b)** the shapes of the complex ions.

---

### ! SCIENCE FEATURE The usefulness of three silver complexes

**$[Ag(CN)_2]^-$ in electroplating**
Silver is an unreactive precious metal. If aqueous potassium cyanide is added to solutions of $Ag^+(aq)$ ions then the soluble $[Ag(CN)_2]^-$ complex is formed. This complex is used to electroplate a layer of silver onto objects (Fig 24.8). The object to be plated is made the cathode. During electrolysis $[Ag(CN)_2]^-$ decomposes to release $Ag^+$ ions, which are attracted to the negatively charged cathode where they accept electrons to form the layer of silver required.

**$[Ag(NH_3)_2]^+$ in testing for aldehydes.**
We first met this reaction in Chapter 16 (see page 358). Tollens reagent contains

**Fig 24.8** This unusual vinegar and oil set has been electroplated with silver

the complex ion $[Ag(NH_3)_2]^+$, made by adding an excess of aqueous ammonia to aqueous silver nitrate. It is this complex that is reduced to silver in the silver mirror test for the aldehyde functional group. An example of this reaction is shown in the equation and in Fig 24.9.

$$CH_3CHO + 2Ag^+(aq) + H_2O \rightarrow$$
$$CH_3CO_2H + 2Ag(s) + 2H^+$$

**Fig 24.9** The silver mirror produced when $[Ag(NH_3)_2]^+$ is reduced to Ag is used as a test for aldehydes

(continued on page 522)

! **SCIENCE FEATURE** The usefulness of three silver complexes [continued from page 521]

### $[Ag(S_2O_3)_2]^{3-}$ in photography

If you have ever processed your own black-and-white film you will be familiar with the term **fixer**, which is the name for an aqueous solution of sodium thiosulphate ($Na_2S_2O_3(aq)$). It was only when scientists realized that the $S_2O_3^{2-}$ ion could form a soluble complex with silver ions, $[Ag(S_2O_3)]^{3-}$, that photography became possible.

The black-and-white film is coated with an emulsion that contains insoluble silver bromide. When this compound is exposed to light, silver atoms form on the film's surface. The greater the amount of light that hits the film the more the number of silver atoms deposited. When the film is processed the first stage involves a solution called **developer**. This enhances the image to the correct level by depositing more silver atoms from the silver bromide. The image must then be **fixed**, which is the stage at which the soluble complex ion $[Ag(S_2O_3)]^{3-}$ is formed by adding $S_2O_3^{2-}$ ions from the **fixer** solution and dissolving any remaining light sensitive silver bromide away from the film.

$$AgBr(s) + 2S_2O_3^{2-}(aq) \rightarrow [Ag(S_2O_3)]^{3-}(aq) + Br^-(aq)$$

**Fig 24.10** Cross-section of a black-and-white film

**Fig 24.11(a)** The negative image is produced on the film. The areas where most light has struck appear dark. **(b)** Light is then shone through the negative image to produce the positive image on photographic paper, which is also coated with an emulsion that contains silver bromide

## LIGANDS

All four ligands discussed so far ($H_2O$, $NH_3$, $Cl^-$ and $CN^-$) donate one lone pair of electrons to the central metal atom, and so are called **monodentate**. This means 'single toothed' and is derived from *monos*, the Greek for one, and *dens*, the Latin for tooth. **Bidentate** ligands can form two dative covalent bonds to the metal ion, so they are 'two toothed', as shown in Fig 24.12. The lone pairs come from the nitrogen and oxygen, respectively. When these ligands attach to the central metal ion, they form five-membered rings and the complexes are called **chelates** (from *chele*, the Greek for claw).

1,2-diaminoethane
called en for short

ethanedioate
traditional
name: oxalate

**Fig 24.12** Two examples of bidentate ligands

**Fig 24.13** An example of chelation with 1,2-diaminoethane. Shown on the right is the simplified version of this bidentate ligand, 'en'. The complex is written using 'en': $[Ni(en)_3]^{2+}$

**E**thylene**d**iamine**t**etra**a**cetate ion (EDTA) is the traditional name for a very important chelating agent. As Fig 24.14 shows, six lone pairs of electrons are available to bond with a metal ion, so it is called a hexadentate ligand. This ligand surrounds and encloses metal ions to form a very stable complex ion, which explains why it is so useful.

**Fig 24.14** EDTA$^{4-}$ ion

**Fig 24.15** Lead–EDTA complex ion. Since the charge on the lead ion is 2+ and EDTA has a charge of 4–, the overall charge of the complex ion is 2–

EDTA is added to human blood when it is stored for transfusions because it keeps the calcium ions trapped in the EDTA complex and so prevents the blood from clotting. It is also used during operations to prevent clotting.

Lead(II) ions (Pb$^{2+}$) are toxic in the human body; if they are not removed quickly they can affect the enzymes that catalyse the formation of haemoglobin. Lead(II) ions can cause permanent brain damage in children and young people. EDTA is administered to remove lead(II) ions from the blood and tissue as a stable lead–EDTA complex ion (Fig 24.15), which is excreted in urine. It used equally effectively to treat mercury poisoning. EDTA is so effective at trapping and removing metal ions that it also removes calcium and important trace metal ions that are essential to normal body functioning. These must be replaced during the treatment.

However, EDTA's applications extend far beyond the medical. It is added to some foods to remove traces of metal ions that catalyse air oxidation. In salad dressings, for example, EDTA prevents rancidity by removing the ions that catalyse the oxidation of oil. (For more about this see the Assignment on page 362.) EDTA is even used in liquid fertilizers because it forms a chelate complex with the Fe$^{3+}$ ions. Thus, OH$^-$ ions are prevented from reacting with the Fe$^{3+}$ to form insoluble Fe(OH)$_3$. This allows the soluble iron complex to reach and enter the plant. EDTA is also added to bathroom products – particularly shampoos – to remove calcium ions from the water, and thereby prevent the formation of scum.

## NAMING COMPLEX IONS

To name a complex ion, proceed as follows.

- First, understand how the ligands are named (Table 24.4). Note: ligands that are anions are named by adding *o* to the stem of the anion names.
- The prefixes *mono*, *di*, *tri*, *tetra*, *penta* and *hexa* are used to indicate how many of a particular ligand that is present.
- When more than one type of ligand is present, the names are written in alphabetical order.
- The central metal ion is named with its oxidation state. When the complex is negatively charged, *ate* is added, as shown in Table 24.5. Also included in Table 24.5 are two common non-transition metals. Note: Latinized names are used for iron and copper – ferrate and cuprate, respectively.

**Table 24.4 The names of monodentate ligands**

| | Ligand | Name |
|---|---|---|
| Neutral molecules | H$_2$O | Aqua |
| | NH$_3$ | Ammine |
| | CO | Carbonyl |
| Anions | Cl$^-$ | Chloro |
| | CN$^-$ | Cyano |
| | F$^-$ | Fluoro |
| | OH$^-$ | Hydroxo |

**Table 24.5 The names of nine metals in complex anions**

| Metal | Name of metal in complex anion |
|---|---|
| Aluminium | Aluminate |
| Chromium | Chromate |
| Cobalt | Cobaltate |
| Copper | Cuprate |
| Iron | Ferrate |
| Lead | Plumbate |
| Manganese | Manganate |
| Nickel | Nickelate |
| Vanadium | Vanadate |

## ◯ EXAMPLE

**Q** Name the following complex ions:
$Cu(H_2O)_6]^{2+}$, $[Fe(CN)_6]^{4-}$ and $[CrCl_2(H_2O)_4]^+$.

**A** **$[Cu(H_2O)_6]^{2+}$**

**Step 1** Consider the ligands and their number.

Since there are six water molecules, *hexaaqua* is used.

**Step 2** Work out the oxidation state of the central metal ion.

Water is a neutral molecule, so the overall charge of the complex ion must come from copper, which has a charge of 2+. That is, its oxidation state is +2.

**Step 3** Write the name of the complex ion.

$[Cu(H_2O)_6]^{2+}$ is called hexaaquacopper(II).

**$[Fe(CN)_6]^{4-}$**

**Step 1** Consider the ligands and their number.

The ligands are cyanide ions, hence *cyano*, and there are six, hence *hexa*. So, the name starts with *hexacyano*.

**Step 2** Work out the oxidation state of the central metal ion.

Six $CN^-$ ions have a total charge of 6–. Since the overall charge on the complex is 4–, iron must have a charge of 2+, so its oxidation number is +2.

**Step 3** Write the name of the complex ion.

In this Example, we are dealing with an anion (negatively charged ion), which means the Latinized name *ferrate* must be used. So $[Fe(CN)_6]^{4-}$ is called hexacyanoferrate(II).

**$[CrCl_2(H_2O)_4]^+$**

**Step 1** Consider the ligands and their number.

In this case, there are two types of ligand: anions and neutral molecules.

Two chloride ions are *dichloro* and four water molecules are *tetraaqua*.

Anions are always named before neutral molecules, hence *dichlorotetraaqua*.

**Step 2** Work out the oxidation state of the central metal ion.

Two $Cl^-$ ions contribute a charge of 2–, while water is neutral and so contributes no charge. Since the overall charge is +1, the charge on Cr is 3+, giving it an oxidation number of +3.

**Step 3** Write the name of the complex ion.

$[CrCl_2(H_2O)_4]^+$ is called dichlorotetraaquachromium(III).

## NAMING NEUTRAL COMPLEXES

Neutral complexes are named in exactly the same way as positively charged complexes. For example, $[Ni(CO)_4]$ is called tetracarbonylnickel(0). Note that the oxidation state of nickel in this complex is zero.

## ! SCIENCE FEATURE Haemoglobin

Haemoglobin, a protein molecule, is a vital component of blood. It transports molecular oxygen from the lungs around the body via the arteries. This giant of a molecule is made up of four protein groups (called globulin groups). In the centre of each globulin group is the ion $Fe^{2+}$ bound in a complex. Four sites of attachment around the $Fe^{2+}$ are occupied by a planar ring structure, called a porphyrin (Fig 24.16). The

**Fig 24.16** Porphyrin ring

iron–porphyrin part of the complex is called the haem group (Fig 24.17). The fifth ligand site is occupied by a dative covalent bond to the protein globin, while molecular oxygen loosely attaches to the sixth site.

**Fig 24.17** Haem group

When oxygen is bonded to the $Fe^{2+}$, the whole complex, called oxyhaemoglobin, takes on a red colour. Once the oxygen is removed, it is replaced by a water ligand, which changes the colour of the complex to blue. The complex is now called deoxyhaemoglobin. Deoxyhaemoglobin gives the blood in our veins its blue colour as it returns to the lungs to pick up more oxygen.

**Fig 24.18** Schematic drawing of haemoglobin

**Fig 24.19** When oxygen is needed, it is replaced by water

oxyhaemoglobin          deoxyhaemoglobin

**Fig 24.20** Computer graphic of the haemoglobin molecule. The four globulin groups (coloured yellow and blue) each contain a haem group (white)

Carbon monoxide gas is sometimes referred to as the 'silent killer'. The CO molecule has a lone pair and can form dative covalent bonds with $Fe^{2+}$ in the same place as oxygen bonds with haemoglobin, to form carboxyhaemoglobin. As the bonds formed by CO are stronger than those formed by oxygen the uptake of oxygen is prevented, which is why carbon monoxide is so deadly.

**Fig 24.21** Carboxyhaemoglobin forms when CO bonds with $Fe^{2+}$, which prevents oxygen uptake

**SEE QUESTION 3.**

## ? TEST YOURSELF

**a)** What is the coordination number of iron in the complex shown in Fig 24.18?

**b)** What type of ligand is the porphyrin ring? Hint: How many points of attachment does it have?

**c)** Explain why blood in most arteries is red.

# ISOMERISM IN COMPLEXES

**Isomerism** occurs when molecules have the same molecular formula, but different ways of arranging their atoms. As explained on pages 198, 306 and 411, there are two principal types of isomerism: **structural isomerism** and **stereoisomerism**.

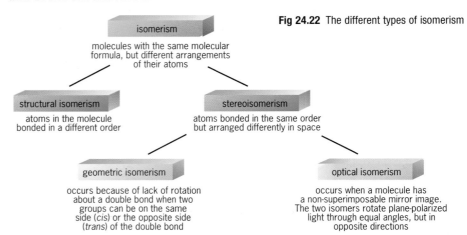

**Fig 24.22** The different types of isomerism

The discussions on pages 198, 306 and 411 are confined to organic molecules, but transition metal complexes can also exhibit isomerism. Both types of stereoisomerism – **geometric isomerism** and **optical isomerism** – occur in transition metal complexes, as the next two sections show.

## GEOMETRIC (*CIS–TRANS*) ISOMERISM

Stereoisomers have their atoms bonded in the same order, but they are arranged differently in space. Geometric isomers (or *cis–trans* isomers) occur when two different kinds of ligands are bonded in different positions around the central metal cation. The octahedral complex ion $[Cr(NH_3)_4Cl_2]^+$ shows geometric isomerism. In the *cis* form, the chloride ligands are on the same side, while in the *trans* form they lie on opposite sides. (*Trans* is the Latin word for 'across'.) The *cis* and *trans* isomers shown in Fig 24.23 have different colours.

## OPTICAL ISOMERISM

When a stereoisomer has a non-superimposable mirror image, it exhibits optical isomerism. Optical isomers have identical physical and chemical properties, but they rotate plane-polarized light in opposite directions (see page 411). Bidentate ligands, such as 1,2-diaminoethane (called en for short), can give rise to optical isomerism in octahedral complexes, such as tri-1,2-diaminoethanecobalt(III) shown in Fig 24.24.

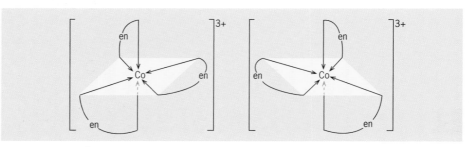

**Fig 24.24** $[Co(en)_3]^{3+}$ exists as two mirror image forms or enantiomers

**Fig 24.23(a)** The *cis* isomer of $[Cr(NH_3)_4Cl_2]^+$ is violet

**Fig 24.23(b)** The *trans* isomer of $[Cr(NH_3)_4Cl_2]^+$ is green

## LIGAND EXCHANGE

In aqueous solutions, transition metal ions exist as aqua-complexes. The water around the central metal ion can be substituted by other ligands, which is called **ligand exchange** or **ligand substitution**. Fig 24.25(a) shows copper(II) sulphate dissolved in water, so the complex ion present is $[Cu(H_2O)_6]^{2+}$. When concentrated hydrochloric acid is added to this solution, it turns green (Fig 24.25(b)). This occurs because the water ligands of the copper(II) complex have been exchanged for chloride ions. $[CuCl_4]^{2-}$ is a more stable complex because chloride ligands have replaced the water ligands:

$$[Cu(H_2O)_6]^{2+}(aq) + 4Cl^-(aq) \rightleftharpoons [CuCl_4]^{2-}(aq) + 6H_2O(l)$$
$$\text{blue} \qquad\qquad\qquad\qquad \text{yellow}$$

The reason for the green colour that is observed is explained in the Extension box on page 528.

$[Cu(NH_3)_4(H_2O)_2]^{2+}$ is more stable than the chloro-complex (see Table 24.7 in the Extension box on stability constants). So, when concentrated ammonia solution is added, ammonia ligands replace chloride ligands to produce a deep blue solution. This can be seen in Fig 24.25(c). Fig 24.25(d) shows the lightening of the solution that occurs when EDTA is added to the solution of ammine complex, which is again as predicted from Table 24.6.

**Table 24.6 The colours and stability of some copper complexes**

| Complex Ion | Colour | Stability |
|---|---|---|
| $[Cu(H_2O)_6]^{2+}$ | Blue | |
| $[Cu(Cl_4)]^{2-}$ | Yellow | |
| $[Cu(NH_3)_4(H_2O)_2]^{2+}$ | Deep blue | Increases |
| $[Cu(EDTA)(H_2O)_2]^{2-}$ | Pale blue | |

**Fig 24.25(a)** Aqueous solution of copper(II) sulphate containing $[Cu(H_2O)_6]^{2+}$(aq) ions. **(b)** $Cu^{2+}$ ions in concentrated HCl, giving the complex ion $[CuCl_4]^{2-}$(aq). **(c)** An excess of ammonia has now been added, producing $[Cu(NH_3)_4(H_2O)_2]^{2+}$. **(d)** EDTA produces a much more stable complex, $[Cu(EDTA)(H_2O)_2]^{2-}$

**SEE QUESTIONS 4 AND 5.**

**FOR AN EXPLANATION OF WHY LIGAND EXCHANGE OFTEN RESULTS IN A CHANGE IN COLOUR, SEE PAGE 607 OF CHAPTER 27.**

Another example of ligand exchange is the reaction that occurs when thiocyanate ions, $SCN^-$, are added to an aqueous solution of iron(III) ions. A blood red solution of $[Fe(SCN)(H_2O)_5]^{2+}$ replaces the yellow $[Fe(H_2O)_6]^{2+}$, as is seen in Fig 24.33 on page 531. This is a very sensitive test for the presence of $Fe^{3+}$(aq) ions.

$$[Fe(H_2O)_6]^{2+}(aq) + SCN^-(aq) \rightleftharpoons [Fe(SCN)(H_2O)_5]^{2+}(aq) + H_2O(l)$$
$$\text{yellow} \qquad\qquad\qquad\qquad\qquad \text{blood-red}$$

If we examine the ligand substitution reaction that occurs when concentrated hydrochloric acid is added to aqueous copper(II) sulphate, we do not obtain the expected yellow colour.

$$[Cu(H_2O)_6]^{2+}(aq) + 4Cl^-(aq) \rightleftharpoons$$
blue

$$[CuCl_4]^{2-}(aq) + 6H_2O(l)$$
yellow

This is because the reaction has not gone to completion, since the solution is green – a combination of the blue and yellow complexes. As this is an equilibrium reaction, the equilibrium constant, $K_c$, can be calculated:

$$K_c = \frac{[CuCl_4{}^{2-}(aq)]}{[Cu(H_2O)_6{}^{2+}(aq)]\,[Cl^-(aq)]^4}$$

Notice in the expression for $K_c$ that the complex ions are shown without square brackets to avoid confusion with the square brackets that indicate concentration. Also, notice that water is not included in the $K_c$ equation because, as the solvent, its concentration is considered to be constant. (For more information on $K_c$, see page 390.).

The equilibrium constant is a measure of how far the position of the equilibrium is over to the right. The further the position of the equilibrium is to the right, the higher the value of $K_c$ and the more stable the complex ion compared with the aqua-complex. $K_c$ is called the **stability constant**. As the numbers

for $K_c$ are often very large, they are made more manageable by using a logarithm scale. Five values of log $K_c$ are shown in Table 24.7.

**Table 24.7 Stability constants**

| Complex ion | Colour | Stability constant, log $K_c$ |
|---|---|---|
| $[Cu(Cl_4)]^{2-}$ | Yellow | 5.6 |
| $[Cu(NH_3)_4(H_2O)_2]^{2+}$ | Deep blue | 13.1 |
| $[Cu(EDTA)(H_2O)_2]^{2-}$ | Pale blue | 18.8 |
| $[Ni(NH_3)_6]^{2+}$ | Purple | 8.0 |
| $[Fe(SCN)(H_2O)_5]^{2+}$ | Blood-red | 3.0 |
| $[Fe(CN)_6]^{3-}$ | Yellow | 31.0 |

## AQUEOUS COMPLEXES AND ACIDITY

As already stated, in aqueous solution, transition metal ions form aqua-complexes. When the pH values of various aqueous solutions of transition metal salts are measured, a pattern emerges. There is an increase in acidity as the charge of the transition metal ion increases. So, for example, $[Fe(H_2O)_6]^{3+}(aq)$ is more acidic than $[Fe(H_2O)_6]^{2+}(aq)$. Why is this?

**Fig 24.26** The highly charged cation pulls electrons from the water molecule, which weakens the O–H bond. A proton is thereby released, to produce $H_3O^+$ with water. The $H_3O^+$ causes acidity

The more highly charged the central ion, the more highly polarizing it is (see pages 84–85). This attracts the electrons of the surrounding water molecules, which allows the release of protons and thus makes the solution acidic. This process is called **deprotonation** of the cation. A number of equilibria exist in solution, two of which are:

Equilibrium 1:

$$[Fe(H_2O)_6]^{3+}(aq) \rightleftharpoons [Fe(OH)(H_2O)_5]^{2+}(aq) + H^+(aq)$$

Equilibrium 2:

$$[Fe(OH)(H_2O)_5]^{2+}(aq) \rightleftharpoons [Fe(OH)_2(H_2O)_4]^{+}(aq) + H^+(aq)$$

The polarizing power of ions with charges greater than 3+ explains why oxyanions are formed. For example, $[Cr(H_2O)_6]^{6+}$ and $[Mn(H_2O)_6]^{7+}$ are theoretically possible, but the central ions are too polarizing. This causes the loss of both protons from some of the surrounding ligand water molecules to produce $O^{2-}$ ions, which form dative covalent bonds to give $CrO_4{}^{2-}$ and $MnO_4{}^-$.

**R** Name the three different complexes shown in equilibria 1 and 2.
Remember: Because water is a neutral ligand, the charge on the aqua-complex is the same as the charge on the central metal cation. For example, Fe has a charge of 3+ in $[Fe(H_2O)_6]^{3+}(aq)$.

**S** Non-transition metal ions in solution can also be acidic. For example, $[Al(H_2O)_6]^{3+}(aq)$ releases protons into an aqueous solution. Write an equation to show the equilibrium that exists when one proton is released. You can check your answer on page 493.

## ADDITION OF AQUEOUS SOLUTIONS OF SODIUM HYDROXIDE OR AMMONIA

Aqueous sodium hydroxide contains hydroxide ions, which remove $H^+(aq)$ ions by forming water:

$$H^+(aq) + OH^-(aq) \rightarrow H_2O(l)$$

This shifts equilibria 1 and 2 (above) to the right and causes the formation of the neutral complex $[Fe(OH)_3(H_2O)_3]$ from a third equilibrium. This is an example of Le Chatelier's principle (see pages 18 and 392). The complex forms as a rusty brown, gelatinous precipitate:

Equilibrium 3:
$$[Fe(OH)_2(H_2O)_4]^+(aq) \rightleftharpoons [Fe(OH)_3(H_2O)_3](s) + H^+(aq)$$
rusty brown precipitate

Similar reactions occur between hydroxide ions and other aqueous transition metal ions, to produce coloured, gelatinous precipitates (Figs 24.28 to 24.30). In the equations below, the aqua-complexes are represented by (aq) and the hydroxoaqua-complexes by the simple metal hydroxide formulas:

$$Cr^{3+}(aq) + 3OH^-(aq) \rightarrow Cr(OH)_3(s)$$
grey–green

$$Mn^{2+}(aq) + 2OH^-(aq) \rightarrow Mn(OH)_2(s)$$
cream

$$Fe^{2+}(aq) + 2OH^-(aq) \rightarrow Fe(OH)_2(s)$$
green

**Fig 24.27** The rusty brown precipitate produced when aqueous sodium hydroxide is added to $Fe^{3+}(aq)$ ions

FOR THE COLOUR OF OTHER METAL HYDROXIDES PRODUCED WHEN AQUEOUS SODIUM HYDROXIDE IS ADDED TO SOLUTIONS OF METAL IONS, SEE PAGES 472 AND 631.

**Fig 24.28** $Cr(OH)_3(s)$ is produced when $Cr^{3+}(aq)$ reacts with NaOH(aq)

**Fig 24.29** $Mn(OH)_2(s)$ is precipitated when $OH^-(aq)$ ions react with $Mn^{2+}(aq)$ ions

**Fig 24.30** The addition of $OH^-(aq)$ to $Fe^{2+}(aq)$ precipitates $Fe(OH)_2(s)$

When excess $OH^-(aq)$ is added to chromium hydroxide, the precipitate redissolves because further deprotonation takes place, as shown in equilibrium 4, to form a soluble complex:

Equilibrium 4:
$$[Cr(OH)_3(H_2O)_3](s) \rightleftharpoons [Cr(OH)_4(H_2O)_2]^-(aq) + H^+(aq)$$

Aqueous ammonia also produces coloured gelatinous precipitates because hydroxide ions are present:

$$NH_3(aq) + H_2O(l) \rightleftharpoons NH_4^+(aq) + OH^-(aq)$$

**Fig 24.31(a)** When concentrated ammonia solution is added dropwise to aqueous copper(II) ions, a precipitate of copper(II) hydroxide forms

**Fig 24.31(b)** On the further addition of excess ammonia solution, the precipitate dissolves as the soluble ammine complex $[Cu(NH_3)_4(H_2O)_2]^{2+}(aq)$ is formed

**FOR MORE EXAMPLES OF REACTIONS WITH AQUEOUS AMMONIA, SEE PAGE 631.**

However, ammonia is also a ligand, and in some cases the precipitates dissolve because a soluble ammine complex is formed. This occurs when aqueous ammonia is added to $Cu^{2+}(aq)$:

$$[Cu(H_2O)_6]^{2+}(aq) + 2OH^-(aq) \rightarrow [Cu(OH)_2(H_2O)_4](s)$$
blue solution                                    pale blue precipitate

On addition of an excess of ammonia solution, the precipitate dissolve to form the deep blue $[Cu(NH_3)_4(H_2O)_2]^{2+}(aq)$:

$$[Cu(OH)_2(H_2O)_4]^{2+}(s) + 4NH_3(aq) \rightarrow [Cu(NH_3)_4(H_2O)_2]^{2+}(aq) + 2H_2O(l) + 2OH^-(aq)$$

# 4 THE FORMATION OF COLOURED IONS

The formation of coloured ions is another characteristic property of transition metals. Most of their compounds are coloured and they form coloured solutions. The colours of nine transition metal ions in aqueous solution are given in Table 24.8. Scandium and zinc ions are also included.

Being colourless is another reason why $Sc^{3+}(aq)$ and $Zn^{2+}(aq)$ are not usually regarded as transition metal ions. A coloured ion results from an incomplete d subsconvl, which neither of these ions has.

The type of ligand also affects the colour, as already noted in the case of $[Cu(H_2O)_6]^{2+}(aq)$, which is a paler blue than the ammine complex $[Cu(NH_3)_4(H_2O)_2]^{2+}(aq)$, which is deep blue. The cobalt chloride test for water is probably a test you have done yourself in the laboratory. Cobalt chloride paper is blue because it contains the complex ion $[CoCl_4]^{2-}$. When water is added, $[Co(H_2O)_6]^{2+}$ is formed, which is pink.

**Fig 24.32** Cobalt chloride paper before and after the addition of water

**? TEST YOURSELF**

**a)** Write the equation for the formation of $[Co(H_2O)_6]^{2+}(aq)$ from $[CoCl_4]^{2-}(aq)$. These are in equilibrium, so remember to include the equilibrium sign.

**b)** Heating moist cobalt chloride paper turns it blue again. Why does this happen? Hint: You will need to have understood Le Chatelier's principle on page 392.

**SEE QUESTION 6.**

**Table 24.8 Colours of some common aqueous ions in the first transition series**

| Ion | Colour | Outer 3d electrons |
|---|---|---|
| $Sc^{3+}(aq)$ | Colourless | $3d^0$ |
| $Ti^{3+}(aq)$ | Purple | $3d^1$ |
| $V^{3+}(aq)$ | Green | $3d^2$ |
| $Cr^{3+}(aq)$ | Violet | $3d^3$ |
| $Mn^{2+}(aq)$ | Pink | $3d^5$ |
| $Fe^{3+}(aq)$ | Yellow* | $3d^5$ |
| $Fe^{2+}(aq)$ | Green | $3d^6$ |
| $Co^{2+}(aq)$ | Pink | $3d^7$ |
| $Ni^{2+}(aq)$ | Green | $3d^8$ |
| $Cu^{2+}(aq)$ | Blue | $3d^9$ |
| $Zn^{2+}(aq)$ | Colourless | $3d^{10}$ |

*$[Fe(H_2O)_6]^{3+}(aq)$ is violet, but because of the deprotonation (see page 529) the yellow ion $[Fe(OH)(H_2O)_5]^{2+}(aq)$ predominates.

The colours of ions are affected too by the number of different types of ligand, as shown by these chromium(III) complexes:

$$[Cr(H_2O)_6]^{3+}(aq) \qquad [CrCl(H_2O)_5]^{2+}(aq) \qquad [CrCl_2(H_2O)_4]^{+}(aq)$$

violet           green           dark green

Many transition metal complexes have characteristic colours that are so distinctive they can be used to test for these particular ions. For example, $Fe^{3+}(aq)$ forms a blood-red complex with thiocyanate ions ($SCN^-$). The addition of aqueous potassium thiocyanate, KSCN, to a solution of $Fe^{3+}(aq)$ is a very sensitive test, since tiny amounts give the typical blood-red coloration:

$$[Fe(H_2O)_6]^{3+}(aq) + SCN^-(aq) \rightarrow [Fe(H_2O)_5SCN]^{2+}(aq)$$

blood-red

$Fe^{2+}(aq)$ ions, however, do not give a coloured complex with thiocyanate ions.

Another way to distinguish between $Fe^{2+}(aq)$ and $Fe^{3+}(aq)$ is to use the complex ion $[Fe(CN)_6]^{3-}(aq)$, which produces a deep blue precipitate, called Turnbull's blue, with $Fe^{2+}(aq)$.

# 5 HETEROGENEOUS AND HOMOGENEOUS CATALYSIS

Catalysis is the final characteristic property of transition elements to be considered here. A catalyst is a substance that alters the rate of a chemical reaction without becoming permanently involved in the reaction. A catalyst works by providing an alternative reaction pathway with a lower energy of activation. (See pages 178 and 185 for more information.)

There are two types of catalyst. **Homogenous catalysts** are in the same phase as the reactants they catalyse, while **heterogeneous catalysts** are in a different phase from the reactants they catalyse. Transition metals and their compounds can act as either of these two types.

## HETEROGENEOUS CATALYSIS

Platinum, for example, plays a crucial role in reducing emissions from car exhausts through its use in catalytic converters. Here it acts as a heterogeneous catalyst because it is in a different phase from the gaseous reactants. One reaction that is catalysed by platinum is the oxidation of carbon monoxide:

$$2CO(g) + O_2(g) \xrightarrow{\text{Pt catalyst}} 2CO_2(g)$$

The Assignment on page 323 shows how nickel catalyses the hydrogenation of carbon–carbon double bonds, an important reaction in the production of fats from oils.

**Fig 24.33** The blood-red colour of $[Fe(H_2O)_5SCN]^{2+}(aq)$ is a very sensitive test for the presence of $Fe^{3+}(aq)$ ions

**Fig 24.34** Turnbull's blue precipitate

WE DISCUSS THE ORIGIN OF COLOUR IN MORE DETAIL IN CHAPTER 27.

EVEN THE WAY THE LIGANDS ARE ARRANGED AROUND THE CENTRAL CATION CAN AFFECT THE COLOUR. TURN TO PAGE 526 TO SEE THE DIFFERENT COLOURS OF THE *CIS* AND *TRANS* ISOMERS OF [Cr(NH$_3$)$_4$Cl$_2$]$^+$.

SEE QUESTIONS 7 AND 8.

> ✔ **REMEMBER THIS**
>
> A phase is defined as a homogeneous (uniform) portion of matter separated from other portions of matter by a boundary surface (on page 132).

double bond in an unsaturated oil

**Fig 24.35** Hydrogenation of carbon–carbon double bonds

**Fig 24.36** Proposed mechanism for the functioning of the iron catalyst in the Haber process. The nitrogen and hydrogen molecules are adsorbed on the surface of the catalyst. The bonds within the molecules break, to leave separate nitrogen and hydrogen atoms. The atoms then rearrange and form new bonds. The ammonia molecules fly off. This mechanism offers an alternative reaction pathway of lower activation energy

**YOU CAN FIND OUT MORE ABOUT HETEROGENEOUS CATALYSTS IN CHAPTER 10, PAGE 223 (CATALYTIC CONVERTERS) AND THE CHAPTER 14 ASSIGNMENT ON PAGE 323 (NICKEL IN THE HYDROGENATION OF UNSATURATED FATS)**

### ✓ REMEMBER THIS

**Adsorb** is the word used when reactants are weakly bonded to a surface. The bonds are formed by a similar mechanism to that when ligands bond to a central metal cation. Desorbtion is the opposite of adsorption.

**SEE QUESTIONS 9 AND 10.**

### ❓ TEST YOURSELF

**U** Give reasons why:
**a)** Catalysts should have rough surfaces rather than smooth ones.
**b)** Platinum is applied as a powder on a cheap solid support in a catalytic converter.
**c)** In some catalysed reactions gases are blown through a *fluidized bed* of finely powdered catalyst so that the catalyst particles float on the gases being blown in.

The Haber process for the production of ammonia (featured on page 19) uses finely divided iron to catalyse the reaction between nitrogen and ammonia:

$$N_2(g) + 3H_2(g) \underset{\text{Fe catalyst}}{\rightleftharpoons} 2NH_3(g)$$

This ranks among the most important industrial reactions because ammonia is used to produce nitrogenous fertilizers. Fig 24.36 shows a proposed mechanism for its heterogeneous catalysis.

The contact process, which leads to the production of sulphuric acid (see pages 655–657), is another important catalysed reaction:

$$2SO_2(g) + O_2(g) \underset{\text{V}_2\text{O}_5 \text{ catalyst}}{\rightleftharpoons} 2SO_3(g)$$

A heterogeneous catalyst works by providing a surface on which the reactants can bond. These are called **active sites** (Fig 24.37).

low magnification

at greater magnification the surface looks very irregular

= active sites

**Fig 24.37** The active sites (shown by the red dots) occur on the surface of the catalyst

However not all transition metals make good heterogeneous catalysts. For example, tungsten bonds strongly to molecules adsorbed on its surface and will not release product molecules, while silver's bonds with adsorbed molecules are too weak to hold them on the surface for long enough for a reaction to occur.

It is the availability of the 3d and 4s electrons coupled with the ability to use variable oxidation states that make transition metals and their compounds such good catalysts.

**Fig 24.38** Gauze of platinum and rhodium provides a huge surface area on which to catalyse the oxidation of ammonia in the production of nitric acid

Let's examine how variable oxidation states allow vanadium(V) oxide to catalyse the reaction that produces sulphur trioxide from sulphur dioxide.

The first reaction that occurs is the oxidation of $SO_2$ into $SO_3$:

$$V_2O_5(s) + SO_2(g) \rightarrow V_2O_4(s) + SO_3(g)$$

Oxidation state of vanadium      +5                  +4

Vanadium is reduced from oxidation state +5 to +4 and at the same time the sulphur is oxidized.

A catalyst must remain chemically unchanged at the end of the reaction, which occurs through a second reaction in which oxygen oxidizes the vanadium back to its +5 oxidation state.

$$V_2O_4(s) + \tfrac{1}{2}O_2(g) \rightarrow V_2O_5(s)$$

Oxidation state of vanadium      +5                  +4

## HOMOGENEOUS CATALYSIS

Transition metal compounds also make excellent **homogeneous catalysts**. Homogeneous catalysts are in the same physical state as the reactants, which usually means in solution. While these reactions are not as commercially significant as heterogeneous catalysis, they are fundamental to many biological reactions. It is the ability of transition metals to exist in more than one oxidation state that is the key to these reactions, since transition metal ions can take part in electron-transfer reactions that provide an alternative pathway of lower activation energy.

The reaction between iodide ions, $I^-(aq)$, and peroxodisulphate(VI) ions, $S_2O_8^{2-}(aq)$, is catalysed by $Fe^{2+}(aq)$ ions:

$$S_2O_8^{2-}(aq) + 2I^-(aq) \rightarrow 2SO_4^{2-}(aq) + I_2(aq)$$

$Fe^{2+}(aq)$ is first oxidized into $Fe^{3+}(aq)$:

$$S_2O_8^{2-}(aq) + 2Fe^{2+}(aq) \rightarrow 2SO_4^{2-}(aq) + 2Fe^{3+}(aq)$$

$Fe^{3+}(aq)$ is then reduced back to $Fe^{2+}(aq)$ by $I^-(aq)$:

$$2I^-(aq) + 2Fe^{3+}(aq) \rightarrow 2Fe^{2+}(aq) + I_2(aq)$$

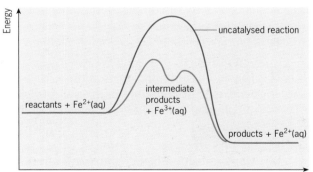

Another example of homogeneous catalysis is the addition of carbon monoxide to methanol, using a rhodium/iodine-based catalyst:

$$CH_3OH + CO \rightarrow CH_3COOH$$

The reaction and the catalyst are all in the liquid phase and there is more than a 99 per cent yield of ethanoic acid. This process was introduced in 1970 and now accounts for a large slice of the world's production of this important chemical. A major use of ethanoic acid is in the production of

? **TEST YOURSELF**

**V** In the catalysis of sulphur trioxide from sulphur dioxide with vanadium (see text), what are the oxidation states of sulphur in $SO_3$ and $SO_2$ and why is this an oxidation of sulphur? Hint: If you need to remind yourself about oxidation states, look back at Chapter 5, pages 106–109.

**Fig 24.39** The reaction between $S_2O_8^{2-}(aq)$ and $2I^-(aq)$ is slow without the addition of an $Fe^{2+}(aq)$ catalyst, which provides an alternative route of lower activation energy ethenyl ethanoate (vinyl acetate), which is the monomer of poly(vinyl acetate), or PVA, found in adhesives and certain paints.

Exciting research is continuing to discover new homogeneous catalysts. One promising area involves organometallic catalysts, which could build up complex organic molecules from carbon monoxide or carbon dioxide. Compared with heterogeneous catalysts, which often involve the use of precious metals such as platinum, homogeneous catalysts generally cost less and their catalytic action often takes place at lower temperatures and pressures.

## AUTOCATALYSIS

Sometimes, catalysts are produced by the reaction itself. This is known as **autocatalysis**. One example is on page 582 of Chapter 26. Another is the reaction of $MnO_4^-(aq)$ with $C_2O_4^{2-}(aq)$:

$$2MnO_4^-(aq) + 5C_2O_4^{2-}(aq) + 16H^+(aq) \rightarrow 2Mn^{2+}(aq) + 10CO_2(g) + 8H_2O(l)$$

The reaction is slow to begin with, but speeds up once $Mn^{2+}(aq)$ ions are formed (see Fig 24.40). The $Mn^{2+}(aq)$ is acting as a homogeneous catalyst.

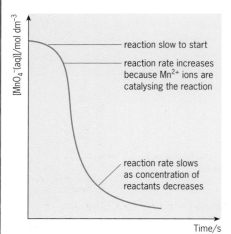

Fig 24.40 The change of concentration of $MnO_4^-(aq)$ when it reacts with $C_2O_4^{2-}(aq)$. This graph is typical of autocatalysed reactions

reaction slow to start

reaction rate increases because $Mn^{2+}$ ions are catalysing the reaction

reaction rate slows as concentration of reactants decreases

$[MnO_4^-(aq)]/mol\ dm^{-3}$

Time/s

### ? TEST YOURSELF

**W** Sketch a graph of the volume of $CO_2$ produced against time for the reaction of $MnO_4^-(aq)$ with $C_2O_4^{2-}(aq)$.

## SUMMARY

After studying this chapter, you should know that:
- A transition element is an element that forms one or more ions with a partially filled d subshell.
- When a transition metal atom forms an ion, the first electrons to be lost are in the 4s subshell.
- The 3d and 4s subshells have closely similar energies, which results in ions with 3+ charges.
- Transition elements are typical metals. They are hard and dense, have high melting points and are good conductors of heat and electricity.
- The chemical properties that characterize transition elements are: variable oxidation states in their compounds, the formation of complexes, the formation of coloured ions and the catalysis of reactions, both as elements and as compounds.
- A complex is formed when a central metal atom or ion is surrounded by species that donate lone pairs of electrons.
- A ligand is a species that donates a lone pair of electrons to a central metal atom or ion.
- The coordination number is the number of dative covalent (coordinate) bonds from ligands to the central metal ion. It determines the shape of the ligands.
- A monodentate ligand forms a single dative covalent bond with the central metal cation.

- A bidentate ligand can form two dative covalent bonds with the central metal ion. When the ligands form five-membered rings with the metal ion, they are called chelates.
- There is a system for naming complex ions.
- Some complex ions show isomerism.
- Ligand exchange may occur if the complex ion produced is more stable.
- The stability of complex ions is measured using the stability constant.
- Hydrated cations (aqua-complexes) may become deprotonated, which causes acidity.
- The addition of aqueous solutions of sodium hydroxide or ammonia may result in the production of coloured hydroxide precipitates. Some of these precipitates dissolve in excess alkali, while others dissolve in excess ammonia.
- Many transition metal complexes have characteristic colours that are so distinctive they can be used to test for these particular ions.
- There are two types of catalyst: homogeneous, which is in the same phase as the reactants, and heterogeneous, which is in a different phase.
- Variable oxidation states are important in both heterogeneous and homogeneous catalysis.
- Some reactions are autocatalysed.

## EXAM QUESTIONS

**1** Explain the following terms, and give an example in each case.

**a)** transition element;

**b)** ligand;

**c)** complex ion.

**2**

**a)** Give the electronic configuration of a cobalt atom and a $Co^{2+}$ ion. [Use (Ar) to begin your answer.]

**b)** Cobalt can be described both as a *d-block element* and as a *transition element*. State what is meant by each of the terms.

**c)** In aqueous solution, water combines with the $Co^{2+}$ ion to form the complex ion $[Co(H_2O)_6]^{2+}$, which gives a pink colour to the solution.
**(i)** What feature of the water molecule allows it to form a complex ion with $Co^{2+}$?
**(ii)** What types of bond are present in the complex ion $[Co(H_2O)_6]^{2+}$?
**(iii)** Suggest the shape of the ion $[Co(H_2O)_6]^{2+}$.

Edexcel 6081/P.01 Module Test 1 Jan 1999 Q1b–d

**3**

**a)** The ethanedioate ion, $C_2O_4^{2-}$, acts as a bidentate ligand.
**(i)** What is meant by the term *bidentate ligand*?
**(ii)** This ligand forms an octahedral complex with iron(III) ions. Deduce the formula of this complex and draw its structure, showing all the atoms present.

**b)** **(i)** Give the name of a naturally occurring complex compound that contains iron.
**(ii)** What is an important function of this complex compound?

AQA CH05 Further Inorganic Chemistry June 2000 Q4a,b

**4** [You may need to refer to Chapter 27 for part **c)** and Chapter 6 for part **d)**]

**a)** Explain what is meant by ligand exchange.

**b)** Describe all the colour changes and observations that take place when an aqueous solution of ammonia is gradually added to a solution of $Cu^{2+}(aq)$, until the ammonia is in excess. Write equations for these transformations.

**c)** Blood gets its colour from oxygen-carrying molecules with organic groups that surround a transition metal ion. In humans this transition metal is iron, and the blood is red. In horseshoe crabs, the metal is copper and the blood is blue, and in sea squirts the metal is vanadium and the blood is green.

The graph in Fig 24.Q4 shows the major absorption peak for human blood. Copy the graph and on it show and label the corresponding absorption peaks for the blood of horseshoe crabs and sea squirts.

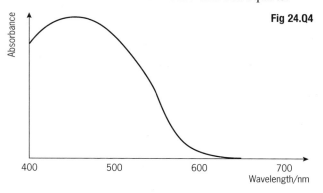

Fig 24.Q4

**d)** A 0.0100 mol sample of an oxochloride of vanadium, $VOCl_x$, required 20.0 cm$^3$ of 0.100 mol dm$^3$ acidified potassium manganate(VII) for oxidation of the vanadium to its +5 oxidation state.
**(i)** Calculate how many moles of potassium manganate(VII) were reacted.
**(ii)** How many moles of electrons were removed by the $MnO_4^-$ ions?
**(iii)** Determine the change in oxidation state of vanadium.
**(iv)** Deduce the value of $x$ in the formula $VOCl_x$.

OCR 2815/06 Transition Elements Specimen Paper Q4

**5** Read the passage below. Identify each of **A**, **B**, **C**, **D**, **E** and **F**, and write equations for all the reactions that occur.
**A** is a black solid that dissolves in dilute sulphuric acid to form a blue solution that contains a cation **B** and an anion **C**.
The addition of aqueous ammonia to the blue solution gives, initially, a blue precipitate **D**, which dissolves when an excess of aqueous ammonia is added, to give a deep blue solution that contains species **E**.
The addition of concentrated hydrochloric acid to the blue solution of **A** gives a yellow–green solution that contains species **F**.

AQA CH05 Further Inorganic Chemistry June 2000 Q6(part)

**6** Cobalt(II) forms the following coloured complexes with water molecules and chloride ions: $[Co(H_2O)_6]^{2+}$ and $[CoCl_4]^{2-}$. Describe how the different ligands, $H_2O$ and $Cl^-$, affect the stereochemistry and colour of these complexes.

OCR 2815/06 Transition Elements Specimen Paper Q4

## ? EXAM QUESTIONS

**7**

This question concerns the atomic structure of iron.

**a) (i)** Write down the 4s and 3d electron configurations (indicating electron spins) for iron, $Fe^0$, and its positive ions, $Fe^{2+}$ and $Fe^{2+}$.

**(ii)** What is the oxidation state of iron in $[Fe(CN)_6]^{4-}$?

**(iii)** Iron can also exist in an oxidation state of +6. Give the formula of a species that contains only iron and oxygen in which Fe(VI) might exist.

**b) (i)** Define the term *ligand*.

**(ii)** Explain why the two iron complex ions $[Fe(H_2O)_6]^{2+}$ and $[Fe(CN)_6]^{4-}$ are different colours.

IBO Chemistry Higher Level Paper 2 Nov 2000 Q2(not b(ii))

**8**

**a)** Hydrated iron(III) chloride is ionic and its solution in water is acidic.

**(i)** Write an ionic equation to show why its solution is acidic, and identify the acid–base conjugate pairs in this reaction.

**(ii)** Describe what you would see when a solution of sodium hydroxide is added until in excess to a solution of iron(III) chloride.

**(iii)** A test for $Fe^{3+}$ ions is to add a solution of thiocyanate ions, $SCN^-(aq)$, which reacts to form a blood-red ion of formula $[FeSCN(H_2O)_5]^{2+}$. State the type of reaction taking place. What two properties associated with transition elements are shown in this reaction?

**b)** Calculate the oxidation number of $FeO_4^{2-}$.

Edexcel Module Test 6 (Synoptic) Jan 2001 Q6c, d (in part).

**9** Iron shows the characteristic properties of a transition mental.

**a)** Why does zinc not show the characteristic properties of a transition metal?

**b)** Give the electronic configuration of the $Fe^{3+}$ ion.

**c)** Iron can form $Fe^{2+}$ ions as well as $Fe^{3+}$ ions. Give an example of another transition element that can show variable oxidation states, showing clearly two of its oxidation states.

**d)** Give the formula and describe the shape of the complex ion formed between $Fe^{3+}$ and water.

**e)** Explain why complexes of $Sc^{3+}$ are colourless, whereas complexes that contain $Fe^{3+}$ are coloured.

**f)** Explain why iron is used in the Haber process to manufacture ammonia, and why it is added in finely divided form rather than in large pieces.

IBO Chemistry Higher Level Paper 2 Nov 1999 Q3

**10**

**a)** Aqueous copper(II) sulphate has a pale blue colour.

**(i)** Give the formula for the species responsible for the colour of the solution.

**(ii)** When aqueous sodium hydroxide is added to such a solution, a pale blue precipitate is formed. Write an ionic equation for this reaction, There is no need to show the copper in a complex form.

**(iii)** When aqueous ammonia is added to aqueous copper(II) sulphate, a pale blue precipitate is formed, but this then dissolves to give a deep blue solution. Give the formula for the species responsible for the deep blue colour.

**(iv)** If aqueous ammonium chloride is added to aqueous copper(II) sulphate, the deep blue solution is not formed. Explain.

**(v)** On the addition of a large excess of ammonium chloride the solution turns green. Explain.

**b)** Give an example of the use of a d-block element or one of its compounds as a catalyst in an industrial process. Supply an equation for the process and name the catalyst.

OCR 6822 General and Inorganic Chemistry Mar 2000 Q4

# KEY SKILLS **ASSIGNMENT**

## PLATINUM COMPLEXES FIGHT CANCER

**Barnett Rosenberg's first degree and his PhD were in physics, and he earned his living as a physicist. So, why is he being featured in a chemistry textbook? The answer is that he and a group of colleagues set up a biophysics department at Michigan State University, and it was in that department that a momentous discovery in chemistry was made in 1964. Like so many discoveries in science, it was made by accident. But, as Louis Pasteur said, 'Chance favours the prepared mind.'**

Rosenberg was investigating the effect of an alternating current on mammalian cell division. To test the apparatus, he first passed a current through a culture of bacterial cells growing and dividing in a nutrient medium. To his amazement, the cells stopped dividing and started growing into long filaments.

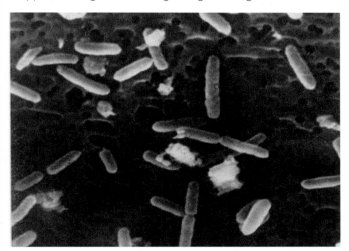

**Fig 24.A1** Bacterial cells before the passage of an electric current

**Fig 24.A2** Long filaments seen after the electric current had passed

Upon further investigation, Rosenberg and his team discovered that the platinum electrodes inserted in the nutrient medium – chosen because they were believed to be unreactive – were forming a product by electrolysis. It was this product that was inhibiting cell division, but not cell growth. After much research, it appeared that platinum was forming complex ions with ammonia and chloride ions. Two likely candidates were $[PtCl_2(NH_3)_2]$ and $[PtCl_4(NH_3)_2]$.

**1** Platinum is a transition element.
 **a)** What is meant by transition element?
**b)** What are the characteristic chemical properties of transition elements?

**2** **a)** Explain what is meant by (i) complex ion and (ii) ligand.
 **b)** Ammonia molecules and chloride ions are ligands in the two platinum complexes mentioned. Draw diagrams to help explain how these species act as ligands.

**3** The platinum complexes formed in the nutrient medium are both neutral.
**a)** What is the oxidation state of platinum in:
(i) $[PtCl_2(NH_3)_2]$ and (ii) $[PtCl_4(NH_3)_2]$?
**b)** Predict the shape of $[PtCl_4(NH_3)_2]$.

An English chemist, Andy Thompson, working with the Rosenberg team, synthesized *trans*-$[Pt(NH_3)_2Cl_2]$ and dissolved some into the bacterial solution. To everyone's intense disappointment, nothing happened and the bacterial cells continued to divide. He subsequently made the *cis*-isomer and this did inhibit cell division. The active chemical had been identified.

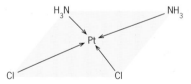

**Fig 24.A3** *cis*-$[Pt(NH_3)_2Cl_2]$, now commonly known as cisplatin

**4** **a)** Geometric isomerism is a form of stereoisomerism. What is meant by stereoisomerism?
**b)** Draw the *trans* isomer of $[Pt(NH_3)_2Cl_2]$.
**c)** $[PtCl_4(NH_3)_2]$ can also exist as *cis* and *trans* isomers. Draw these two isomers and label them.

Cancer cells divide in an uncontrolled way, and often give rise to tumours. Rosenberg approached the USA National Cancer Institute (NCI) and suggested that *cis*-$[Pt(NH_3)_2Cl_2]$, now called cisplatin, might have the same effect on cancer cells as it had on bacterial cell division. At first the NCI were sceptical. After all, many platinum compounds were known to be toxic to the liver. However, Rosenberg demonstrated that cisplatin could destroy tumours in mice, and so cisplatin began the long

# KEY SKILLS **ASSIGNMENT**

journey to becoming accepted as an anti-cancer drug. One of the major problems to overcome was the predicted toxicity to the kidneys. By giving the patient large amounts of water before and after the injection of cisplatin, the compound was diluted in the kidneys and so the toxic effects were reduced. It is now among the most widely used anti-cancer drugs, being particularly effective in the treatment of tumours on the ovaries and testes.

However, the toxicity and its side effects remain. This has stimulated the search for other complexes that may be equally effective but safer. The way cisplatin works has recently been determined by X-ray crystallography. The chloride ligands are exchanged for dative covalent bonds with nitrogen atoms of DNA. This bends the DNA, which causes a loss of function and cell death. However, the ease with which chloride ligands can be exchanged causes the toxicity of cisplatin and some of the unpleasant side effects. Carboplatin replaces the chloride ligands with a bidentate ligand called 1,1-cyclobutanedicarboxylic acid. It, too, is licensed for use in Britain.

**Fig 24.A4** 1,1-Cyclobutanedicarboxylic acid

**Fig 24.A5** Carboplatin

**5 a)** Why is 1,1-cyclobutanedicarboxylic acid called a bidentate ligand?

**b)** The complex formed is called a chelate. What is meant by this term?

**6** Another platinum complex that shows anti-tumour activity replaces the two $NH_3$ ligands with 1,2-diaminoethane $(NH_2CH_2CH_2NH_2)$.

**a)** Draw this square planar complex

**b)** Why do you think this complex is likely to be as toxic as cisplatin?

**7** Before a drug is licensed, extensive clinical trials must take place.

**a)** Why do you think this is important?

**b)** Is there ever a justification for speeding up the process?

**c)** Clinical trials use placebos. Find out what this word means and why the use of placebos is considered essential.

# 25

# Electrode potentials and transition elements

**Electric vehicles** An electrically powered bus in Sweden

Electric cars and vans would dramatically reduce atmospheric pollution. So, why haven't they yet supplanted petrol and diesel-driven vehicles?

Their widespread adoption has been stunted by the absence of a battery able to provide sufficient motive power for a sufficient length of time.

Much is being done to rectify this, with the development of lightweight batteries that can run a vehicle for at least 500 km between charges. It is hoped that these batteries can be recharged many times using cheap-rate electricity.

There has also been progress with the development of fuel cells that use hydrogen or petrol in a reaction with oxygen. No burning occurs and electric current is generated directly. Appropriate technology does exist, but it is largely exotic and costly.

## 1 REDOX REACTIONS

Redox is a type of chemical reaction that involves electron transfer. At its simplest, one atom loses electrons and another atom gains electrons. Atoms of metals lose electrons, usually to attain a stable electron configuration, and atoms of non-metals gain electrons, usually to attain a stable electron configuration.

### HALF-EQUATIONS

A typical redox reaction is the reaction of the metal magnesium with dilute hydrochloric acid to form magnesium chloride (Fig 25.1):

$$Mg(s) + 2HCl(aq) \rightarrow MgCl_2(aq) + H_2(g)$$

dilute hydrochloric acid

magnesium ribbon

**Fig 25.1** The redox reaction between magnesium and hydrochloric acid forms hydrogen and aqueous magnesium chloride

This reaction is a redox reaction because an electron transfer occurs. An atom of magnesium loses two electrons during the reaction, but it is not immediately obvious where these electrons go. Hydrochloric acid is a strong acid and it is completely dissociated into aqueous chloride ions and aqueous hydrogen ions. It is the hydrogen ions that gain electrons: two hydrogen ions gain one electron each, eventually to form a hydrogen molecule. The hydrogen ions are said to be **reduced** to hydrogen. The chloride ions do not take part in the reaction at all. They are sometimes called **spectator ions**, to indicate that they are present in the solution during the reaction, but take no part in it.

The overall redox reaction can be represented by two half-equations, one to

**STRONG ACIDS AND DISSOCIATION ARE COVERED ON PAGES 365 TO 369.**

show the reduction and the other to show the oxidation:

Oxidation half-equation:  $Mg(s) \rightarrow Mg^{2+}(aq) + 2e^-$
Reduction half-equation:  $2H^+(aq) + 2e^- \rightarrow H_2(g)$

By combining these two half-equations, an ionic equation can be written, which does not show the spectator ions:

$$Mg(s) + 2H^+(aq) \rightarrow Mg^{2+}(aq) + H_2(g)$$

However complicated a redox reaction may be, it is always possible to write two half-equations, one for the oxidation and the other for the reduction.

## WORKING OUT THE HALF-EQUATIONS

Before writing down the two half-equations, the species being oxidized and the species being reduced must be established. This is best achieved using oxidation numbers (see pages 106 to 110). During oxidation, the oxidation number of an atom increases. During reduction, it decreases. The change in oxidation number also gives the number of electrons lost (oxidation) or gained (reduction).

Consider the reaction of aqueous iron(II) ions with acidified manganate(VII) ions. The overall ionic equation is:

$$5Fe^{2+}(aq) + MnO_4^-(aq) + 8H^+(aq) \rightarrow 5Fe^{3+}(aq) + Mn^{2+}(aq) + 4H_2O(l)$$

The oxidation state of Fe changes from +2 to +3 (Fig 25.2). This is oxidation and involves the loss of one electron. The oxidation state of Mn changes from +7 to +2. This is reduction and involves the gain of five electrons. We are now in a position to write the half-equations.

**Fig 25.2** The reaction of acidified manganate(VII) ions, $MnO_4^-(aq)$, with aqueous iron(II) ions, $Fe^{2-}(aq)$. The oxidation numbers that change during the reaction are shown in red

**SEE QUESTION 1.**

The oxidation involves conversion of $Fe^{2+}$ into $Fe^{3+}$, and is written as:

$$Fe^{2+}(aq) \rightarrow Fe^{3+}(aq) + e^-$$

The reduction involves conversion of the manganate(VII) ion into the manganese(II) ion. This involves the gain of five electrons per manganate(VII) ion:

$$5e^- + MnO_4^-(aq) + 8H^+(aq) \rightarrow Mn^{2+}(aq) + 4H_2O(l)$$

Note that the half-equations are balanced, both in terms of their symbols and in terms of the charge. Note also that the aqueous hydrogen ions are not spectator ions, but are needed for the reduction to occur, and that the hydrogen ion is not itself reduced.

## ELECTRON TRANSFER

The two half-equations clearly demonstrate the electron-transfer processes that occur during a redox reaction. The processes shown by the half-equations cannot take place independently, because reduction cannot occur without the simultaneous occurrence of oxidation. It is best to think of the half-equations as a 'book-keeping' exercise to show where electrons are lost and where they are gained.

**? TEST YOURSELF**

**A** Write down the half-equations that correspond to the following overall reactions:
**a)** $Cu(s) + 2Ag^+(aq) \rightarrow 2Ag(s) + Cu^{2+}(aq)$;
**b)** $6Fe^{2+}(aq) + 14H^+(aq) + Cr_2O_7^{2-}(aq)$ $\rightarrow 6Fe^{3+}(aq) + 2Cr^{3+}(aq) + 7H_2O(l)$

## ELECTROLYSIS REACTIONS

**Electrolysis is the decomposition of a liquid by the passage of an electric current**, which enters and leaves the compound through electrodes. Electrolysis is necessarily a redox reaction, since it involves electron transfer. The reactions represented by the two half-equations occur in different parts of an electrolytic cell. Reduction always occurs at the cathode (negative electrode) and oxidation always occurs at the anode (positive electrode).

Consider the electrolysis of molten lead(II) iodide. The two half-equations are:

Oxidation half-equation: $2I^- \rightarrow I_2 + 2e^-$ occurs at the anode
Reduction half-equation: $Pb^{2+} + 2e^- \rightarrow Pb$ occurs at the cathode

In electrolysis, these half-equations are referred to as **electrode reactions**.

## OXIDIZING AGENTS

An **oxidizing agent**, or **oxidant**, is the name given to the substance or species that gains electrons during a redox reaction. In effect, it oxidizes another substance by gaining electrons from it. So an oxidizing agent is reduced during a redox reaction.

Typically, oxidizing agents are non-metal elements or compounds that have an element in a high positive oxidation state. For example, fluorine is an oxidizing agent, as is potassium dichromate(VI) in which chromium has an oxidation state of +6. In every reaction, the oxidizing agent gains electrons. Fluorine molecules gain electrons to form fluoride ions, and dichromate(VI) ions gains electrons to form chromium(III) ions:

$$F_2 + 2e^- \rightarrow 2F^-$$

$$Cr_2O_7^{2-} + 14H^+ + 6e^- \rightarrow 2Cr^{3+} + 7H_2O$$

Note that these two half-equations are written to show that the oxidizing agent gains electrons.

Sometimes, we want to compare the oxidizing power of substances. That is, we want to compare the ability of oxidizing agents to gain electrons. For example, fluorine is the better oxidizing agent of the two shown above, but it is impossible to tell just from the half-equation. This is dealt with on page 548.

## REDUCING AGENTS

A **reducing agent**, or **reductant**, is the name given to the substance or species that loses electrons during a redox reaction. In effect, it reduces another substance by losing electrons to it. So a reducing agent is oxidized during a redox reaction.

Typically, reducing agents are metal elements or compounds that have an element in a low, positive oxidation state or a negative oxidation state. For example, potassium is a reducing agent, as is the sulphide ion in which sulphur has an oxidation state of −2. In every reaction, the reducing agent loses electrons. A potassium atom loses an electron to form a potassium ion, and a sulphide ion loses two electrons to form a sulphur atom.

$$K \rightarrow K^+ + e^-$$
$$S^{2-} \rightarrow S + 2e^-$$

Note that the two half-equations are written to show that the reducing agent loses electrons.

Sometimes, we want to compare the reducing power of substances. That is, we want to compare the ability of reducing agents to lose electrons. For example, potassium is the better reducing agent of the two shown above, but it is impossible to tell this just from the half-equations. This also is discussed on page 549.

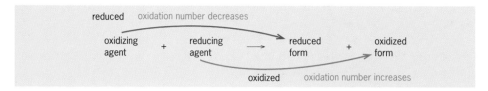

This also is discussed on page 549.

Fig 25.3 A summary of the action of oxidizing agents and reducing agents

# 2 SIMPLE CELLS

It has been known for two centuries that a potential difference is generated when two dissimilar metal electrodes are put into an electrolyte. What takes place in this so-called electrochemical cell is a redox reaction, in which oxidation occurs at one electrode and reduction at the other.

This poses a problem, since in a redox reaction an electron transfer must occur. Given that the two half-reactions (oxidation and reduction) are at different places, the transfer of electrons cannot take place during collision between the oxidizing and the reducing agents. Instead, the electrons are transferred via an external circuit (a metal wire). This may be regarded as electrolysis in reverse. Instead of a reaction occurring because of the passage of an electric current, an electric current (electron transfer) is produced because a redox reaction is occurring.

## DANIELL CELL

In 1836, the English chemist John Fredrick Daniell constructed a simple battery that employed this idea. He used zinc and copper plates as the electrodes and two different electrolytes – aqueous copper(II) sulphate and aqueous zinc sulphate (Fig 25.4). (It is now known that in the Daniell cell zinc is in equilibrium with aqueous zinc ions and copper is in equilibrium with copper(II) ions.) When the two metal plates are connected by a metal wire, a current flows through the wire. The current is a result of the gain and loss of electrons that occur in the cell.

During the operation of the cell, the zinc plate forms zinc ions and loses electrons. This corresponds to the oxidation half-equation:

$$Zn(s) \rightarrow Zn^{2+}(aq) + 2e^-$$

Meanwhile, at the copper plate, copper(II) ions gain electrons to form copper. This corresponds to the reduction half-equation:

$$Cu^{2+}(aq) + 2e^- \rightarrow Cu(s)$$

Now, a redox reaction can occur only if electrons can be transferred. Only when the two plates are connected by a metal wire can this happen. The zinc plate, at which oxidation takes place, is the anode; the copper, at which reduction takes place, is the cathode.

The overall reaction that occurs in the cell is a combination of the two half-equations:

$$Zn(s) + Cu^{2+}(aq) \rightarrow Zn^{2+}(aq) + Cu(s)$$

When zinc is placed directly into aqueous copper(II) sulphate, the same reaction takes place, but the electron transfer occurs during a collision between a zinc atom and a copper(II) ion (Fig 25.5).

Fig 25.4 A Daniell cell consists of a copper plate surrounded by saturated aqueous copper(II) sulphate solution and a zinc plate surrounded by saturated aqueous zinc sulphate. The aqueous zinc sulphate must be added carefully so that it floats on top of the more dense aqueous copper(II) sulphate solution, but does not mix with it. When the two different metals are connected by a wire, a current flows. Eventually, the cell stops producing a potential difference as the solutions mix through diffusion

insulated wire

zinc electrode

aqueous zinc sulphate

aqueous copper(II) sulphate

copper electrode

Fig 25.5 When zinc is dipped into aqueous copper(II) sulphate, a spontaneous redox reaction occurs to form copper and aqueous zinc sulphate

## ANOTHER SIMPLE CELL

It is possible to arrange the reaction between magnesium and hydrochloric acid as two half-reactions that occur in separate places, with the electrons being transferred via an external metal wire.

The arrangement is shown in Fig 25.6. Note the salt bridge. This allows ions in the two electrolytes to migrate from one half-cell to the other without the electrolytes themselves mixing. Without a salt bridge, there would not be a complete circuit for the charge to flow round, and so no current would be produced. A salt bridge can be as simple as a piece of filter paper dipped in saturated aqueous potassium nitrate. Potassium nitrate is usually used because it does not react with most of the electrolytes used in electrochemical cells.

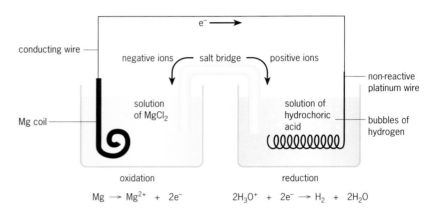

**Fig 25.6** The oxidation of magnesium ions and reduction of hydrogen ions occur in separate parts of the cell. There is no reaction unless both the salt bridge and the external wire are in place

$$Mg \longrightarrow Mg^{2+} + 2e^-$$

$$2H_3O^+ + 2e^- \longrightarrow H_2 + 2H_2O$$

## CELL POTENTIAL

For a current to flow, there must be a potential difference between the two halves of the cell. This is known as the **cell potential** or the electromotive force (e.m.f.). It is given the symbol $E_{cell}^{\ominus}$, and is measured in volts. The cell potential is independent of the amount of each substance present in the cell, but is dependent on each of their concentrations. The effect of concentration on cell potentials and electrode potentials is described on page 551.

### MEASURING THE CELL POTENTIAL WITH A VOLTMETER

The potential of a cell can be measured using a high-resistance voltmeter, as shown in Fig 25.7. This simple set-up does not quite measure the true maximum cell potential. As the current flows from one electrode to the other, the circuit heats up because of the resistance of the wire. So, the cell transfers some of its energy as heat rather than as electric current.

To measure the true maximum cell potential possible, no energy should be transferred as heat. That is, the potential difference should be measured when no current is flowing. This is why a high-resistance voltmeter must be used, so that almost no current is drawn from the cell.

**Fig 25.7** A cell that consists of a zinc electrode dipped into aqueous zinc ions, and a copper electrode dipped into aqueous copper(II) ions, should have a cell potential of 1.10 V (measured on a high-resistance voltmeter) when solutions of 1.0 mol dm$^{-3}$ are used a temperature of 298 K

## ELECTRODE POTENTIALS

Generally, the term 'electrode' is used to describe the conductor that allows the passage of electric current in and out of the cell. In some applications, the electrode is inert, as in the use of graphite or platinum for the electrolysis of acidified water. In other applications, it actually takes part in a reaction. In work on electrochemical cells, however, the term 'electrode' is extended to include what is known as a half-cell. So it refers not only to the conductor, but also to the conducting solution in which it is placed.

As explained earlier, a cell consists of an oxidation and a reduction reaction. Therefore, the overall cell potential should be the sum of the potential of the oxidation process and the potential of the reduction process. The potential of the oxidation process is called the **oxidation potential** of the anode, which, for convenience, is shortened to $E_{oxid}^{\ominus}$, or oxidation potential. Likewise, the potential of the reduction process, which is called the **reduction potential** of the cathode, is shortened to $E_{red}^{\ominus}$, or the reduction potential. So, in an electrochemical cell:

$$E_{cell}^{\ominus} = E_{oxid}^{\ominus} + E_{red}^{\ominus}$$

## STANDARD ELECTRODE POTENTIALS
The word 'standard' in the term **standard electrode potential** signifies that the electrode potential has been measured (or calculated) under standard conditions, which are defined as 298 K, 101 kPa (1 atmosphere) and all aqueous solutions at a concentration of $1.0$ mol dm$^{-3}$. (Standard conditions in thermochemistry are discussed on page 167.)

## STANDARD REDUCTION POTENTIAL
This is the potential difference between a cathode and the solution into which it is dipped, measured under standard conditions. Its symbol is $E_{red}^{\ominus}$. It is impossible to measure an absolute value for $E_{red}^{\ominus}$, so data books give a value of the potential of the cathode with reference to the standard hydrogen electrode, which is assigned a cell potential value of 0.00 V. (The standard hydrogen electrode is discussed in more detail on page 546.) In a Daniell cell, the copper(II) ions are in equilibrium with the copper electrode:

$$Cu^{2+}(aq) + 2e^- \rightleftharpoons Cu(s)$$

The value of $E_{red}^{\ominus}$ gives an indication of the position of this equilibrium. The greater its positive value, the more the position of equilibrium lies to the right (i.e. to the side of the reduced from).

## STANDARD OXIDATION POTENTIAL
This is the potential difference between an anode and the solution into which it is dipped, measured under standard conditions. Again, it is impossible to measure absolute values for $E_{oxid}^{\ominus}$, so a value is used that is the potential of the anode with reference to the standard hydrogen electrode, which is assigned a cell potential value of 0.00 V. In a Daniell cell, the zinc ions are in equilibrium with the zinc anode:

$$Zn(s) \rightleftharpoons Zn^{2+}(aq) + 2e^-$$

The value $E_{oxid}^{\ominus}$ gives an indication of the position of this equilibrium. The greater its positive value, the more the position of equilibrium lies to the right (i.e. to the side of the oxidized form).

## MEASURING ELECTRODE POTENTIALS
As already stated, it is not possible to measure absolute values for the oxidation and reduction potentials. However, if the potential for one of the processes is assigned a value, it is possible to determine a value for any other electrode potential. The chosen process is the redox reaction involving aqueous hydrogen ions and gaseous hydrogen under standard conditions (temperature 298 K, pressure of the hydrogen gas 101 kPa, concentration of the aqueous hydrogen ion 1.00 mol dm$^{-3}$). Both $E_{oxid}^{\ominus}$ and the $E_{red}^{\ominus}$ are arbitrarily given a value of zero. Therefore:

Oxidation: $H_2(g) \rightarrow 2H^+(aq) + 2e^-$ $E_{oxid}^{\ominus} = 0.00$ V
Reduction: $2H^+(aq) + 2e^- \rightarrow H_2(g)$ $E_{red}^{\ominus} = 0.00$ V

**? TEST YOURSELF**

**C** You want to measure the standard reduction potential for the following reaction:

$$Cl_2(g) + 2e^- \rightarrow 2Cl^-(aq)$$

Write down the conditions that you must have to measure the standard reduction potential.

Hence, all other electrode potentials are compared with either the oxidation potential of gaseous hydrogen or the reduction potential of aqueous hydrogen ions. Remember that the oxidation and reduction reactions are part of the reference equilibrium reaction:

$$2H^+(aq) + 2e^- \rightleftharpoons H_2(g)$$

**Fig 25.8** A summary of the values of electrode potentials

| | $E^{\ominus}_{red}$ | | $E^{\ominus}_{red}$ |
|---|---|---|---|
| $F_2 + 2e^- \rightleftharpoons 2F^-$ | $= +2.87$ V | $2F^- \rightleftharpoons F_2 + 2e^-$ | $= -2.87$ V |
| $Cl_2 + 2e^- \rightleftharpoons 2Cl^-$ | $= +1.36$ V | $2Cl^- \rightleftharpoons Cl_2 + 2e^-$ | $= -1.36$ V |
| $Br_2 + 2e^- \rightleftharpoons 2Br^-$ | $= +1.07$ V | $2Br^- \rightleftharpoons Br_2 + 2e^-$ | $= -1.07$ V |
| $I_2 + 2e^- \rightleftharpoons 2I^-$ | $= +0.54$ V | $2I^- \rightleftharpoons I_2 + 2e^-$ | $= -0.54$ V |
| $Sn^{4+} + 2e^- \rightleftharpoons Sn^{2+}$ | $= +0.15$ V | $Sn^{2+} \rightleftharpoons Sn^{4+} + 2e^-$ | $= -0.15$ V |
| $2H^+ + 2e^- \rightleftharpoons H_2$ | $= 0.00$ V | $H_2 \rightleftharpoons 2H^+ + 2e^-$ | $= 0.00$ V |
| $Zn^{2+} + 2e^- \rightleftharpoons Zn$ | $= -0.76$ V | $Zn \rightleftharpoons Zn^{2+} + 2e^-$ | $= +0.76$ V |
| $Mn^{2+} + 2e^- \rightleftharpoons Mn$ | $= -1.18$ V | $Mn \rightleftharpoons Mn^{2+} + 2e^-$ | $= +1.18$ V |
| $Mg^{2+} + 2e^- \rightleftharpoons Mg$ | $= -2.38$ V | $Mg \rightleftharpoons Mg^{2+} + 2e^-$ | $= +2.38$ V |
| $K^+ + e^- \rightleftharpoons K$ | $= -2.92$ V | $K \rightleftharpoons K^+ + e^-$ | $= +2.92$ V |

oxidizing power decreases

reducing power increases

## STANDARD HYDROGEN ELECTRODE

The standard hydrogen electrode is a half-cell that allows $H_2(g)$ to be in equilibrium with $H^+(aq)$. As an electrode the half-cell must also allow electric current to flow in and out. This poses a problem, because gaseous hydrogen is not an electrical conductor. The problem is overcome by using platinum foil (which is inert) as the conducting part of the electrode (Fig 25.9).

hydrogen at 298 K and 101 kPa

salt bridge

1.0 mol dm$^{-3}$ H$^+$(aq) at 298 K

platinum electrode

**Fig 25.9** In the standard hydrogen electrode, hydrogen gas under a pressure of 101 kPa is in contact with aqueous hydrogen ions at a concentration of 1.00 mol dm$^{-3}$. The platinum electrode allows electrical current to enter and leave the hydrogen electrode. The platinum does not react with the hydrogen or the aqueous hydrogen ions. The aqueous hydrogen ions are normally provided by dilute sulphuric acid of the appropriate concentration

## USING THE STANDARD HYDROGEN POTENTIAL TO MEASURE OTHER STANDARD ELECTRODE POTENTIALS

Since the reference equilibrium reaction involves $H_2(g)$ and $H^+(aq)$, it is possible to determine the standard electrode potential of other half-cells just by combining them with a standard hydrogen electrode, as shown in Fig 25.10.

high-resistance voltmeter (reading will be 0.34 V)

e$^-$

negative ions    salt bridge    positive ions

$H_2(g)$ (101 kPa)

anode

cathode

$Cu^{2+}(aq)$ 1.0 mol dm$^{-3}$

$H^+(aq)$ 1.0 mol dm$^{-3}$

hydrogen electrode

$Cu^{2+}$/Cu electrode

**Fig 25.10** A hydrogen electrode is connected to a copper/copper(II) ion electrode using a salt bridge. A high-resistance voltmeter measures the electrode potential of the cell. The arrows show the electron flow from the hydrogen electrode to the copper electrode via the external wire

In Fig 25.10, the two half-equations are:

At the anode:     $H_2(g) \rightarrow 2H^+(aq) + 2e^-$

At the cathode:   $Cu^{2+}(aq) + 2e^- \rightarrow Cu(s)$

Remember: oxidation occurs at the anode and reduction at the cathode. The value of $E_{cell}^{\ominus}$ is +0.34 V, and:

$$E_{cell}^{\ominus} = E_{oxid}^{\ominus} + E_{red}^{\ominus}$$

Therefore:

$$+0.34 = 0.00 + E_{red}^{\ominus}$$

The $E_{red}^{\ominus}$ for the copper electrode is +0.34 V. So $E_{oxid}^{\ominus} = -0.34$ V. It is a general rule that, for the same electrode, the values of $E_{red}^{\ominus}$ and $E_{oxid}^{\ominus}$ have the same magnitude but are of opposite sign.

It is also worth stressing that the electrode potentials do not depend on the amounts of reactants and products involved in redox equations. So, for instance, the reduction potential for:

$$Cu^{2+}(aq) + 2e^- \rightarrow Cu(s)$$

is the same as that for:

$$2Cu^{2+}(aq) + 4e^- \rightarrow 2Cu(s)$$

## CELL CONVENTION

It is cumbersome to keep drawing the electrodes used in electrochemical cells. Therefore, a system of notation has been devised whereby a whole cell can be described on a single line. The anode is written on the left and the cathode on the right. A single vertical bar distinguishes components that are in different phases: for example, a solid electrode and the aqueous ions with which it is in contact. The salt bridge is shown by two dashed vertical bars. Thus, the cell of Fig 25.10 can be represented by:

$$Pt(s) \mid H_2(g) \mid H^+(aq) \parallel Cu^{2+}(aq) \mid Cu(s)$$

Note that the half-cell with the greater positive value of $E_{red}^{\ominus}$ is put on the right.
For the Daniell cell, the notation is:

$$Zn(s) \mid Zn^{2+}(aq) \parallel Cu^{2+}(aq) \mid Cu(s)$$

## OTHER WAYS TO MEASURE STANDARD ELECTRODE POTENTIALS

The standard hydrogen electrode does not always have to be used to measure the standard electrode potential of another half-cell. When it is known for certain that a copper electrode has a standard reduction potential of 0.34 V, then it can be used to determine the standard electrode potential of another half-cell.

**SEE QUESTION 2.**

**✓ REMEMBER THIS**

Sometimes a comma is used to separate components that are in the same phase.

**Fig 25.11** To determine the standard electrode potential for magnesium, two half-cells are required. One cell contains a copper strip and 1.0 mol dm⁻³ copper(II) ions, and the other a magnesium strip and 1.0 mol dm⁻³ magnesium ions. A high-resistance voltmeter records the cell potential, from which the electrode potential of magnesium can be determined:

$$\begin{array}{ccc} E_{cell}^{\ominus} & = & E_{oxid}^{\ominus} + E_{red}^{\ominus} \\ +2.71 & & E_{oxid}^{\ominus} + 0.34 \end{array}$$

So:

$$E_{oxid}^{\ominus} = +2.37 \text{ V}$$

It is possible to experimentally determine any electrode potential, provided a suitable cell is set up

An example of this method is illustrated in Fig 25.11. Magnesium is more reactive than copper. That is, it loses electrons more easily than copper. This makes the magnesium half-cell the anode, since this is where oxidation occurs. So, the notation for the cell in Fig 25.11 is:

$$Mg(s) \mid Mg^{2+}(aq) \parallel Cu^{2+}(aq) \mid Cu(s)$$

## ELECTRODES THAT INVOLVE GASES OR SOLUTIONS

Many redox half-reactions do not involve a metal, but it is still possible to construct a half-cell using the reagents shown in the half-equation. For example, with the aid of gas electrodes, gases can be involved in redox half-equations. Since gases do not conduct electricity, an inert metal (usually platinum) has to be used as the conducting part of the gas electrode. Platinum does not react with dilute acids or most aqueous solutions. For example, the chlorine electrode shown in Fig 25.12 allows $Cl_2(g)$ and $Cl^-(aq)$ to be in equilibrium:

$$Cl_2(g) + 2e^- \rightleftharpoons 2Cl^-(aq)$$

Some half-cells involve only solutions. Take, for example, a half-cell that involves the acidified manganate(VII) ions. It must allow the following equilibrium to be set up:

$$MnO_4^-(aq) + 8H^+(aq) + 5e^- \rightleftharpoons Mn^{2+}(aq) + 4H_2O(l)$$

None of the species given in the equation can act as the electrode. Therefore, a platinum electrode is again needed. Despite the stoichiometry of the equation, all the concentrations of the aqueous species should be 1.0 mol dm$^{-3}$, the temperature 298 K and the pressure 101 kPa, for the half-cell to be at standard conditions.

chlorine at 298 K and 101 kPa

salt bridge

1.0 mol dm$^{-3}$ Cl$^-$(aq) at 298 K

platinum electrode

**Fig 25.12** A chlorine electrode consists of chlorine gas at 101 kPa, aqueous chloride ions at a concentration of 1.0 mol dm$^{-3}$ and an inert platinum electrode to conduct current into or out of the solution

## 3 USING STANDARD ELECTRODE POTENTIALS

direction of electron flow

anode

cathode

**Fig 25.13** A summary of electron flow in an electrochemical cell

Knowledge of electrode potentials allows predictions to be made about oxidizing and reducing agents, and about electrolysis products. It also allows speculation on the feasibility of redox reactions.

When the external circuit in an electrochemical cell is closed, electrons flow from the anode to the cathode. This is always the direction of flow of electrons. The reason is that oxidation, accompanied by the release of electrons, always takes place at the anode. The electrons then travel along the external circuit until, at the cathode, they are used in the reduction process.

### COMPARING OXIDIZING AGENTS

To compare oxidizing agents, it is necessary to look at the $E_{red}^{\ominus}$ of the species involved. This is because an oxidizing agent must accept electrons so as to be reduced. Since the figures are all comparative, the most powerful oxidizing agent is the species with the highest positive (or lowest negative) $E_{red}^{\ominus}$.
(The significance of the sign and size of electrode potentials is summarized in Fig 25.8, page 546.)

## EXAMPLE

**Q** Which is the more powerful oxidizing agent, $F_2$ or $Cl_2$?

**A** The relevant reduction half-equations and electrode potentials (given in Appendix 4) are:

$$F_2 + 2e^- \rightleftharpoons 2F^- \qquad E_{red}^{\ominus} = +2.87 \text{ V}$$
$$Cl_2 + 2e^- \rightleftharpoons 2Cl^- \qquad E_{red}^{\ominus} = +1.36 \text{ V}$$

So fluorine is the more powerful oxidizing agent.

## ? TEST YOURSELF

**F** In each of the following pairs, which is the better oxidizing agent:
**a)** $Cu^{2+}$ or $Cr^{2+}$;
**b)** Acidified $MnO_4^-$ or acidified $Cr_2O_7^{2-}(aq)$?

## COMPARING REDUCING AGENTS

To compare reducing agents, it is necessary to look at the $E_{oxid}^{\ominus}$ of the species involved. This is because a reducing agent must give away electrons so as to be oxidized. Since the figures are all comparative, the most powerful reducing agent is the species with the highest positive (or lowest negative) value of $E_{oxid}^{\ominus}$.

## EXAMPLE

**Q** Which is the better reducing agent, $I^-$ or $Cl^-$?

**A** The relevant oxidation half-equations and electrode potentials (given in Appendix 4) are:

$$I_2 + 2e^- \rightleftharpoons 2I^- \qquad E_{red}^{\ominus} = +0.54 \text{ V} \quad \text{so} \quad E_{oxid}^{\ominus} = -0.54 \text{ V}$$
$$Cl_2 + 2e^- \rightleftharpoons 2Cl^- \qquad E_{red}^{\ominus} = +1.36 \text{ V} \quad \text{so} \quad E_{oxid}^{\ominus} = -1.36 \text{ V}$$

So the iodide ion is the better reducing agent, since its $E_{oxid}^{\ominus}$ has a lower negative value than that of the chloride ion.

## ? TEST YOURSELF

**G** In each of the following pairs, decide which is the better reducing agent:
**a)** Ni or $Sn^{2+}$;
**b)** $F^-$ or Au.

## ELECTROCHEMICAL SERIES AND ELECTRODE POTENTIALS

Metals are reducing agents because they lose electrons. The most reactive metals lose electrons easily, whereas the least reactive metals lose electrons with difficulty. The electrochemical series ranks metals in order of their reactivities, with the most reactive metal at the top. Since the most reactive metals are also the best reducing agents, electrode potentials can be used to deduce the electrochemical series, as Table 25.1 shows.

Note that there is a good correlation between the position of the metal in the electrochemical series and its $E_{oxid}^{\ominus}$.

### Table 25.1 Electrochemical series

| Metal | Half-equation | Reduction potential $E_{red}^{\ominus}$/V | Oxidation potential $E_{oxid}^{\ominus}$/V |
|---|---|---|---|
| Potassium | $K^+ + e^- \rightleftharpoons K$ | −2.92 | +2.92 |
| Calcium | $Ca^{2+} + 2e^- \rightleftharpoons Ca$ | −2.87 | +2.87 |
| Magnesium | $Mg^{2+} + 2e^- \rightleftharpoons Mg$ | −2.38 | +2.38 |
| Aluminium | $Al^{3+} + 3e^- \rightleftharpoons Al$ | −1.66 | +1.66 |
| Zinc | $Zn^{2+} + 2e^- \rightleftharpoons Zn$ | −0.76 | +0.76 |
| Iron | $Fe^{2+} + 2e^- \rightleftharpoons Fe$ | −0.44 | +0.44 |
| Lead | $Pb^{2+} + 2e^- \rightleftharpoons Pb$ | −0.13 | +0.13 |
| Hydrogen | $2H^+ + 2e^- \rightleftharpoons H_2$ | 0.00 | 0.00 |
| Copper | $Cu^{2+} + 2e^- \rightleftharpoons Cu$ | +0.34 | −0.34 |
| Silver | $Ag^+ + e^- \rightleftharpoons Ag$ | +0.80 | −0.80 |

## PREDICTING THE PRODUCTS OF ELECTROLYSIS

As mentioned on page 542, electrolysis involves two half-reactions, which are known as electrode reactions. In solutions that contain several ions, it can be difficult to predict which ion will react at the anode and which at the cathode. Since oxidation always occurs at the anode, it is possible to compare the relevant $E^{\ominus}_{oxid}$ to predict which ion is most likely to react there. In the same way, it is possible to predict which ion will react at the cathode by comparing the relevant $E^{\ominus}_{red}$. There is, however, one major drawback with this approach, which is that the electrode potentials are quoted for standard conditions, which involve concentrations of 1.0 mol dm$^{-3}$ of aqueous ionic species. Therefore, unless these are the conditions of electrolysis, there will always be some doubt about accuracy of the prediction. On page 551, the effect of concentration on electrode potential is considered, and the same approach can be used to make better predictions about the electrolysis products of an aqueous solution.

### ? TEST YOURSELF

**a)** Predict the products of the electrolysis of aqueous zinc nitrate solution using inert electrodes of graphite.

**b)** The electrolysis of concentrated aqueous sodium chloride with inert electrodes gives chlorine and hydrogen. Is this what you would expect from inspection of the relevant electrode potentials? If it is not, suggest possible reasons for the discrepancy.

### EXAMPLE

**Q** Predict the products of the electrolysis of aqueous copper(II) nitrate using carbon electrodes.

**A** Ions present: $Cu^{2+}(aq)$ and $NO_3^-(aq)$, and from water $H^+(aq)$ and $OH^-(aq)$.

Ions attracted to cathode: $Cu^{2+}(aq)$ and $H^+(aq)$
Possible cathode reactions: $Cu^{2+}(aq) + 2e^- \rightarrow Cu(s)$ $E^{\ominus}_{red} = +0.34$ V
$\qquad\qquad\qquad\qquad\qquad 2H^+(aq) + 2e^- \rightarrow H_2(g)$ $E^{\ominus}_{red} = 0.00$ V

By inspection, the more favourable reduction involves copper(II) ions, since its electrode potential has the higher positive value.

Ions attracted to anode: $OH^-(aq)$ and $NO_3^-(aq)$
Anode reaction: $\qquad 4OH^-(aq) \rightarrow O_2(g) + 2H_2O(l) + 4e^-$ $E^{\ominus}_{oxid} = -0.40$ V
The nitrate ion is attracted to the anode, but is not oxidized.

## CALCULATING CELL POTENTIALS

### EXAMPLE

**Q** Work out the cell potential for the following cell:

$$Co(s) \mid Co^{2+}(aq) \parallel Cr^{3+}(aq) \mid Cr(s)$$

**A** By the cell convention, the anode is on the left side of the cell, so this is where the oxidation occurs:

$$Co(s) \rightarrow Co^{2+}(aq) + 2e^-$$

From the table of electrode potentials, $E^{\ominus}_{oxid} = +0.28$ V

(since $E^{\ominus}_{red} = -0.28$ V for $Co^{2+}(aq) + 2e^- \rightarrow Co(s)$).

At the cathode, reduction takes place, so the half-reaction must involve the gain of electrons:

$$Cr^{3+}(aq) + 3e^- \rightarrow Cr(s)$$

From the table of electrode potentials, $E^{\ominus}_{red} = -0.74$ V.

Since $E^{\ominus}_{cell} = E^{\ominus}_{oxid} + E^{\ominus}_{red}$,
$$E^{\ominus}_{cell} = +0.28 + (-0.78) = -0.50 \text{ V}$$

**Q** Calculate the cell potential of the cell that has the following overall reaction:

$$Zn + I_2 \rightarrow Zn^{2+} + 2I^-$$

**A** The oxidation half-equation is:

$$Zn \rightarrow Zn^{2+} + 2e^-$$

for which $E^{\ominus}_{oxid} = +0.76$ V

The reduction half-equation is:

$$I_2 + 2e^- \rightarrow 2I^-$$

for which $E^{\ominus}_{red} = +0.54$ V

Using $E^{\ominus}_{cell} = E^{\ominus}_{oxid} + E^{\ominus}_{red}$ gives:

$$E^{\ominus}_{cell} = +0.76 + 0.54 = +1.30 \text{ V}$$

When any simple cell is set up its cell potential, $E_{cell}^{\ominus}$, can be calculated from the electrode potentials of the two half-reactions taking place. $E_{cell}^{\ominus}$ is given by:

$$E_{cell}^{\ominus} = E_{oxid}^{\ominus} + E_{red}^{\ominus}$$

The values of some reduction potentials are given in Appendix 4. The values and those quoted in most data books are standard electrode potentials. Therefore, any cell potential calculated from them will be a standard cell potential. Most data books list only the reduction half-equations together with their $E_{red}^{\ominus}$ values. The oxidation half-equation is the reverse of the reduction half-equation and the corresponding $E_{oxid}^{\ominus}$ values are obtained by changing the sign of the $E_{red}^{\ominus}$ values.

## ELECTRODE POTENTIALS UNDER NON-STANDARD CONDITIONS

So far, the assumption has been that the reactions occur under standard conditions. That is, all solutions are 1.0 mol dm$^{-3}$ in concentration, the temperature is 298 K and the pressure is 101 kPa. This is not the case for most reactions. Even if standard temperature and pressure could be maintained, as soon as the cell reaction starts, the concentrations of the reactants and products would necessarily change. The effect of changing conditions can be predicted qualitatively by assuming that the reaction taking place is an equilibrium and applying Le Chatelier's principle.

### TEMPERATURE AND CELL POTENTIAL

If the cell reaction as written is exothermic, then as the temperature increases the position of equilibrium shifts to the left. This means that the cell potential assumes a lower positive (or larger negative) value.

### CONCENTRATION AND CELL POTENTIAL

If the concentrations of the reactants are increased, then the position of equilibrium shifts to the right and the cell potential assumes a higher positive (or lower negative) value. If the concentration of one of the products increases, then the opposite effect is observed and the cell potential assumes a lower positive (or larger negative) value.

### CONCENTRATION AND ELECTRODE POTENTIAL

Le Chatelier's principle can also be applied to half-cells, to see how $E_{red}^{\ominus}$ or $E_{oxid}^{\ominus}$ values are affected. Consider the equilibrium between hydrogen ions and hydrogen:

$$2H^+(aq) + 2e^- \rightleftharpoons H_2(g) \quad \text{Under standard conditions } E_{red}^{\ominus} = E_{oxid}^{\ominus} = 0.00 \text{ V}$$

If the concentration of the aqueous hydrogen ion is increased, the position of the equilibrium shifts to the right and the reduction potential becomes positive. At the same time, the oxidation potential for the reaction becomes negative.

### PRESSURE AND CELL POTENTIAL

A change in pressure affects only those cell reactions that involve a gas. Essentially, increasing the pressure of a gas increases its concentration. Therefore, the effect of increasing the pressure of a gaseous reactant or a product on cell potential is the same as that with increasing the appropriate concentration.

Consider the hydrogen half-cell equilibrium again:

$$2H^+(aq) + 2e^- \rightleftharpoons H_2(g) \quad \text{Under standard conditions } E_{oxid}^{\ominus} = E_{red}^{\ominus} = 0.00 \text{ V}$$

If the pressure of the hydrogen is increased, the position of the equilibrium shifts to the left. Therefore, the oxidation potential becomes positive and the reduction potential negative.

**SEE QUESTIONS 1, 2 AND 3.**

**IF YOU WISH TO READ MORE ABOUT THE QUANTITATIVE EFFECTS OF CHANGING TEMPERATURE, PRESSURE AND CONCENTRATION ON ELECTRODE POTENTIALS, YOU SHOULD LOOK UP THE NERNST EQUATION IN A MORE ADVANCED TEXTBOOK.**

## ? TEST YOURSELF

Consider the following cell reaction:

$$Zn(s) + 2H^+(aq) \rightarrow Zn^{2+}(aq) + H_2(g)$$

**a)** Calculate the standard cell potential for this reaction.

**b)** Predict qualitatively the effect on $E_{cell}^{\ominus}$ of a cell with:

(i) $[Zn^{2+}(aq)] = 2.0$ mol dm$^{-3}$ $[H^+(aq)] = 1.0$ mol dm$^{-3}$ and pressure of $H_2 = 101$ kPa

(ii) $[Zn^{2+}(aq)] = 1.0$ mol dm$^{-3}$ $[H^+(aq)] = 2.0$ mol dm$^{-3}$ and pressure of $H_2 = 101$ kPa

(iii) $[Zn^{2+}(aq)] = 1.0$ mol dm$^{-3}$ $[H^+(aq)] = 1.0$ mol dm$^{-3}$ and pressure of $H_2 = 202$ kPa

**THERE IS MORE INFORMATION ON SPONTANEOUS REACTIONS ON PAGES 178 AND 574.**

# 4 FEASIBILITY OF REACTIONS

The absolute value of the cell potential determines the feasibility of the reaction. When the cell potential is positive, the reaction is spontaneous. When it is negative, the reaction is not spontaneous. The description 'spontaneous' refers to the tendency of a reaction to occur. It says nothing about the rate of reaction. So it is possible for a spontaneous reaction to be so slow as not to occur. In an electrochemical cell, if $E^{\ominus}_{cell}$ is positive, the reaction is spontaneous; if $E^{\ominus}_{cell}$ is negative, the reaction is not spontaneous.

## ELECTRON FLOW IN THE EXTERNAL CIRCUIT

As already stated, the external flow of electrons is from the anode to the cathode. The cell potential can be used to identify the anode. Provided the cell potential is positive, the cell works spontaneously and the anode is the electrode at which the oxidation occurs.

One battery being developed uses liquid sulphur and sodium. The overall cell reaction is:

$$2Na + S \rightarrow 2Na^+ + S^{2-}$$

The two half-equations are:

$Na \rightarrow Na^+ + 2e^-$ $\qquad E^{\ominus}_{oxid} = +2.71\ V$
$S + 2e^- \rightarrow S^{2-}$ $\qquad E^{\ominus}_{red} = +0.14\ V$

So, the cell potential is +2.85 V. This suggests that the cell reaction is proceeding in the correct direction. Therefore, electrons flow from the sodium electrode to the sulphur electrode.

Now, consider a cell that has a theoretical overall cell reaction of:

$$2Ag + Cu^{2+} \rightarrow 2Ag^+ + Cu$$

The half-equations are:

$Ag \rightarrow Ag^+ + e^-$ $\qquad E^{\ominus}_{oxid} = -0.80\ V$
$Cu^{2+} + 2e^- \rightarrow Cu$ $\qquad E^{\ominus}_{red} = +0.34\ V$

So $E^{\ominus}_{cell} = -0.46$ V, which indicates that, as written, the reaction will not proceed spontaneously. If the equation is reversed, then the cell potential becomes positive, which indicates that electrons will flow from the copper electrode to the silver electrode:

$Cu(s) + 2Ag^+(aq) \rightarrow Cu^{2+}(aq) + 2Ag(s)$ $\qquad\qquad E^{\ominus}_{cell} = +0.46\ V$

## CELL POTENTIAL AND FREE ENERGY

Cells are a source of electrical energy that can be used to do work. A potential of 1 volt imparts 1 joule of energy to a charge of 1 coulomb. If the energy is used to do work, such as running an electric motor, then:

**work = charge $\times$ potential difference**

In this case, the potential difference is the cell potential. Since the cell potential is necessarily positive in order to produce a current, then the work done is negative. This means that work is done on the surroundings.

**? TEST YOURSELF**

**J** **a)** Explain whether each one of the following reactions will occur spontaneously:
(i) $Zn^{2+} + Mg \rightarrow Mg^{2+} + Zn$
(ii) $2Fe^{3+} + 2F^- \rightarrow 2Fe^{2+} + F_2$
**b)** Work out the cell potential for the following cell:
$Mn(s) \mid Mn^{2+}(aq) \quad Fe^{2+}(aq) \mid Fe(s)$
Hence write down the overall cell reaction for the spontaneous reaction that occurs.

**? TEST YOURSELF**

**K** Predict the direction of the electron flow in each of the following electrochemical cells, given the overall cell reaction:
**a)** $Cu + Br_2 \rightarrow Cu^{2+} + Br^-$
**b)** $Cu^{2+} + Cr \rightarrow Cu + Cr^{2+}$

**✔ REMEMBER THIS**

Remember: Electrons have a negative charge.

Consider now the amount of work done when the molar quantities as shown in the overall cell reaction are used:

**work = (charge on number of electrons transferred during cell reaction)** $\times E_{cell}^{\ominus}$

The charge on the number of electrons transferred is the charge on $n$ moles of electrons, where $n$ is the number of moles transferred in the cell reaction. Therefore:

$$\text{work} = -nFE_{cell}^{\ominus}$$

where $n$ is the number of electrons transferred in the cell reaction, and $F$ is the Faraday constant, which is the charge on 1 mole of electrons in coulombs.

The work done according to this equation is the maximum amount of work that can be done. In reality, some electrical energy is transferred by the heating of the wires. The maximum amount of work available from a process carried out under standard conditions is known as the free-energy change, $\Delta G$ ($G$ was chosen to celebrate the American Willard Gibbs). So, in an electrochemical cell:

$$\Delta G = -nFE_{cell}^{\ominus}$$

The unit of $\Delta G$ is J mol$^{-1}$.

> **REMEMBER THIS**
>
> The Faraday constant F is 96 500 coulombs mol$^{-1}$.

---

### ● EXAMPLE

**Q** Calculate the free-energy change for an electrochemical cell based on the following cell reaction:

$$Mn(s) + Cd^{2+}(aq) \rightarrow Mn^{2+}(aq) + Cd(s)$$

**A** From the overall reaction, the two half-equations are:

Oxidation half-equation: $\quad Mn(s) \rightarrow Mn^{2+}(aq) + 2e^{-} \quad E_{oxid}^{\ominus} = +1.18 \text{ V}$
Reduction half-equation: $\quad Cd^{2+}(aq) + 2e^{-} \rightarrow Cd(s) \quad E_{red}^{\ominus} = -0.40 \text{ V}$

Since $E_{cell}^{\ominus} = E_{oxid}^{\ominus} + E_{red}^{\ominus}$, then:
$E_{cell}^{\ominus} = +1.18 - 0.40 = +0.78 \text{ V}$

From the two half-equations, it can be seen that 2 moles of electrons are transferred during the reaction.

So: $\quad \Delta G = -nFE_{cell}^{\ominus}$
$\quad \Delta G = -2 \times 96\,500 \times 0.78 \text{ J mol}^{-1}$
$\quad\quad\quad = -150.5 \text{ kJ mol}^{-1}$

---

> **? TEST YOURSELF**
>
> **L** Calculate the free-energy changes, $\Delta G$, for each of the following reactions:
>
> **a)** $Cd(s) + Pb^{2+}(aq) \rightarrow Cd^{2+}(aq) + Pb(s)$
> **b)** $2Fe^{3+}(aq) + Cu(s) \rightarrow 2Fe^{2+}(aq) + Cu^{2+}(aq)$
> **c)** $Cl_2(g) + 2Br^{-}(aq) \rightarrow Br_2(l) + 2Cl^{-}(aq)$

## FREE ENERGY AND FEASIBILITY OF REACTIONS

Just as the cell potential must be positive for a reaction to occur spontaneously, so the value of $\Delta G$ must be negative. The free-energy change of a reaction can be used in non-electrochemical cells to predict the feasibility of reactions. (On page 568 is a description of the application of free-energy changes in predicting the feasibility of the reduction of some metal oxides.) Any discussion of the feasibility of reactions must be taken with caution, since the $E_{cell}^{\ominus}$ or $\Delta G$ values do not contain any information about the rate of reaction. Therefore, a reaction may have a large negative $\Delta G$, but still not take place because its activation energy is too large.

**ACTIVATION ENERGY IS COVERED ON PAGES 177–180.**

## EXAMPLE

**Q** Predict whether the following cell reaction should proceed spontaneously:

$$Cr^{3+}(aq) + Fe^{2+}(aq) \rightarrow Cr^{2+}(aq) + Fe^{3+}(aq)$$

**A** From the overall cell reaction, the two half-equations are:

Oxidation half-equation:

$Fe^{2+}(aq) \rightarrow Fe^{3+}(aq) + e^-$ $\qquad E_{oxid}^{\ominus} = -0.77 \text{ V}$

Reduction half-equation:

$Cr^{3+}(aq) + e^- \rightarrow Cr^{2+}(aq)$ $\qquad E_{red}^{\ominus} = -0.41 \text{ V}$

Since the cell potential is given by $E_{cell}^{\ominus} = E_{oxid}^{\ominus} + E_{red}^{\ominus}$, then:

$$E_{cell}^{\ominus} = -0.77 - 0.41 = -1.18 \text{ V}$$

A negative cell potential predicts that the reaction should not take place spontaneously.

The free-energy change for this reaction can also be calculated to show that it is positive:

$$\Delta G = -nFE_{cell}^{\ominus}$$
$$= -1 \times 96\,500 \times (-1.18) \text{ J mol}^{-1}$$
$$= +114 \text{ kJ mol}^{-1}$$

The positive value of the free energy change is another indicator that the reaction will not take place spontaneously.

---

### ? TEST YOURSELF

**M** Predict whether the following reactions will take place spontaneously. In each case, work out the cell potential and the free-energy change for the reaction.

**a)** $Cl_2(g) + 2F^-(aq) \rightarrow F_2(g) + 2Cl^-(aq)$

**b)** $MnO_2^- + 5Cr^{2+}(aq) + 8H^+(aq) \rightarrow$
$Mn^{2+}(aq) + 5Cr^{3+}(aq) + 4H_2O(l)$

---

## 5 BATTERIES AND FUEL CELLS

A battery is an electrochemical cell that is used as an energy source. Some kinds of battery have several electrochemical cells in series, and thereby combine the individual cell potentials to give a large potential at the output terminals. Batteries are normally designed to deliver a reasonably large current and to be able to withstand fairly rough handling.

Batteries are of two types: primary cells and secondary cells.

### PRIMARY CELLS

Primary cells cannot be re-charged. They are the most familiar type of battery on sale today.

### LECLANCHÉ OR DRY CELL

The dry cell (Fig 25.15) consists of a zinc anode (also the cell's casing) and a graphite cathode. A paste of ammonium chloride, manganese(IV) oxide and zinc chloride surrounds the graphite electrode. The porous separator acts as a salt bridge.

**Fig 25.14** This Walkman is powered by batteries of the primary-cell type. Typically, they have to be replaced every two or three months

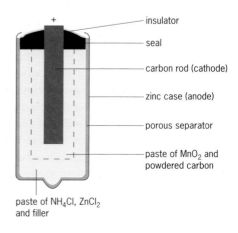

**Fig 25.15** A cross-section through a dry-cell battery

insulator
seal
carbon rod (cathode)
zinc case (anode)
porous separator
paste of $MnO_2$ and powdered carbon

paste of $NH_4Cl$, $ZnCl_2$ and filler

When the cell is connected into an external circuit, electrons flow from the zinc anode to the graphite cathode:

Oxidation half-equation: $Zn \rightarrow Zn^{2+} + 2e^-$

Reduction half-equation: $MnO_2 + NH_4^+ + e^- \rightarrow MnO(OH) + NH_3$

The cell potential is about 1.5 V, which is impossible to verify using standard electrode potentials because the cell reaction occurs under non-standard conditions. The zinc ions present in the paste prevent ammonia being given off by forming an ammine complex, $[Zn(NH_3)_4]^{2+}$. In addition to the nuisance of having to replace this type of battery at frequent intervals, there is the disadvantage that the acidic ammonium chloride corrodes the zinc casing, even when the battery is not being used.

## SECONDARY CELLS

These are batteries that can be recharged and therefore have a much longer life than primary cells. A typical rechargeable battery is the lead–acid battery used in most motor vehicles (Figs 25.16 and 25.17).

**Fig 25.16** Topping up a lead–acid car battery with distilled water

lead grid filled with $PbO_2$, coated with lead(IV) oxide

lead grid filled with spongy lead, coated with lead(II) sulphate

sulphuric acid

**Fig 25.17** A lead–acid storage battery

The lead–acid battery consists of an aqueous electrolyte (30 per cent by volume sulphuric acid) and two sets of plates which form the electrodes. The positive plates are lead grids filled with a paste of spongy lead. The negative plates are lead grids filled with a paste of lead(IV) oxide. At first, the lead reacts with the dilute sulphuric acid to form an insoluble coating of lead(II) sulphate, on both sets of plates:

$$Pb(s) + H_2SO_4(aq) \rightarrow PbSO_4(s) + H_2(g)$$

The battery is initially charged by direct current. During the charging process, the plates become chemically different. A layer of lead(IV) oxide is deposited on each negative plate, but the positive plates retain their layer of lead(II) sulphate. The charging half-equations are:

Oxidation half-equation:

$PbSO_4(s) + 2e^- \rightarrow Pb(s) + SO_4^{2-}(aq)$

Reduction half-equation:

$PbSO_4(s) + 2H_2O(l) \rightarrow PbO_2(s) + 4H^+(aq) + SO_4^{2-}(aq) + 2e^-$

The overall reaction during charging is:

$2PbSO_4(s) + 2H_2O(l) \rightarrow Pb(s) + PbO_2(s) + 4H^+(aq) + 2SO_4^{2-}(aq)$

Electrolysis takes place during charging.

Once the battery is charged, the half-equations can be reversed to provide an external current or flow of electrons. The discharging half-equations are:

Oxidation half-equation:

$Pb(s) + SO_4^{2-}(aq) \rightarrow PbSO_4(s) + 2e^-$

**SEE QUESTION 1.**

? **TEST YOURSELF**

**N** Car batteries are normally described as 12 V batteries. They contain six individual cells, each connected in series.

**a)** Estimate the cell potential of one lead–acid electrochemical cell.

**b)** Use the electrode potentials in Appendix 4 to estimate the cell potential of a lead–acid electrochemical cell.

**c)** Comment on your answers to parts **a)** and **b)**.

Reduction half-equation:

$$PbO_2(s) + 4H^+(aq) + SO_4^{2-}(aq) + 2e^- \rightarrow PbSO_4(s) + 2H_2O(l)$$

The overall discharging reaction is:

$$Pb(s) + PbO_2(s) + 4H^+(aq) + 2SO_4^{2-}(aq) \rightarrow 2PbSO_4(s) + 2H_2O(l)$$

Note that sulphuric acid is used up during the discharging process and so its concentration decreases. This provides a convenient test for a lead–acid battery. Being proportional to concentration, the density of the sulphuric acid in the battery is simply measured with a hydrometer.

## FUEL CELLS

A fuel cell is an electrochemical cell designed so that the reactants, often gases, are replenished continuously. This enables the cell to supply electric current continuously. As its reactants are kept topped up, a fuel cell does not need recharging.

---

**!** **SCIENCE FEATURE** Hydrogen–oxygen fuel cell

A spacecraft needs a continuous supply of electrical energy. This cannot be achieved using conventional batteries, so much of it is provided by a hydrogen–oxygen fuel cell. Essentially, the overall cell reaction is that of oxygen and hydrogen to form water:

$$2H_2(g) + O_2(g) \rightarrow 2H_2O(l)$$

This is a redox reaction, in which electrons are transferred from hydrogen to oxygen. In a fuel cell, this transfer takes place via an external circuit, rather than directly during a collision between particles.

The electrolyte is aqueous sodium hydroxide, which is contained within the cell using porous electrodes. These electrodes allow the passage of water, hydrogen and oxygen.

During their passage through the cathode, oxygen molecules are reduced to hydroxide ions, which enter the electrolyte.

Reduction half-equation:

$$O_2(g) + 2H_2O(l) + 4e^- \rightarrow 4OH^-(aq)$$

During their passage through the anode, hydrogen molecules are oxidized to give water molecules, and thereby hydroxide ions are removed from the electrolyte:

Oxidation half-equation:

$$H_2(g) + 2OH^-(aq) \rightarrow 2H_2O(l) + 2e^-$$

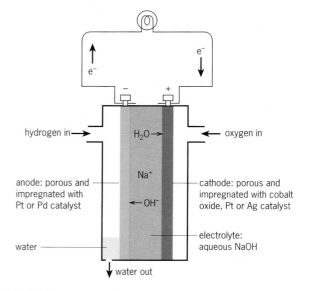

**Fig 25.18** A hydrogen–oxygen fuel cell

---

? **TEST YOURSELF**

**O** Construct the overall cell reaction that takes place within the hydrogen–oxygen fuel cell to convince yourself that all that is happening is the reaction of hydrogen and oxygen to make water.

## 6 ELECTRODE POTENTIALS AND TRANSITION ELEMENTS

The first-row transition elements are a series of metals that share several key characteristics. One of these is that they can have more than one oxidation state in their compounds. This is related to the electron configurations of these elements. The existence of several oxidation states of the transition elements is described fully on pages 518 to 520.

### REDOX REACTIONS AND TRANSITION ELEMENTS

Since transition elements can have several different oxidation states, it is possible for their compounds, and the elements themselves, to take part in redox reactions. That is, the lower oxidation states of a transition element can be converted into one of the higher states by oxidation, and its higher oxidation states can be converted into one of the lower states by reduction. These changes necessarily involve the loss and gain of electrons, so they must be redox reactions.

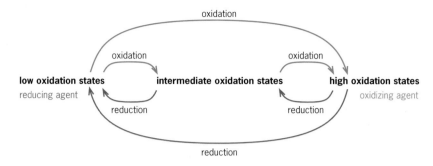

**Fig 25.19** A summary of redox reactions of the transition elements

It is shown on page 545 how $E_{red}^{\ominus}$ and $E_{oxid}^{\ominus}$ can be used to explain redox reactions. So, it follows that the chemistry of transition elements and their compounds involves electrode potentials.

# 7 CHEMISTRY OF SIX FIRST-ROW TRANSITION ELEMENTS

The first-row transition elements start with titanium and finish with copper. (See page 514 for the definition of a transition element.) The important redox reactions of six of these elements are the subject of this section, particular attention being paid to the electrode potentials involved.

## VANADIUM AND ITS COMPOUNDS

**Table 25.2 Electrode potentials of some vanadium species**

| Reduction half-equation | $E_{red}^{\ominus}$/V |
|---|---|
| $V^{2+}(aq) + 2e^- \rightleftharpoons V(s)$ | −1.18 |
| $V^{3+}(aq) + e^- \rightleftharpoons V^{2+}(aq)$ | −0.26 |
| $VO^{2+}(aq) + 2H^+(aq) + e^- \rightleftharpoons V^{3+}(aq) + H_2O(l)$ | +0.34 |
| $VO_2^+(aq) + 2H^+(aq) + e^- \rightleftharpoons VO^{2+}(aq) + H_2O(l)$ | +1.00 |

Vanadium exhibits four common oxidation states, +2, +3, +4 and +5. Under ordinary conditions, vanadium(IV) is considered to be the most stable oxidation state. Vanadium in its lower oxidation states is a reducing agent and in the highest oxidation state it is an oxidizing agent. Therefore the chemistry of vanadium compounds typically involves a conversion from one oxidation state into another.

### VANADIUM IN THE +5 OXIDATION STATE
In the +5 oxidation state, the bonding to the vanadium atom is covalent in character, since it is impossible for a vanadium atom to lose five electrons to form $V^{5+}$. Probably, the most important compound in this oxidation state is vanadium(V) oxide, which is used as a catalyst in the manufacture of sulphuric acid in the contact process.

The aqueous chemistry of vanadium(V) is centred around the vanadyl(V) ion, $VO_2^+(aq)$, and the vanadate(V) ion, $VO_3^-(aq)$.

**Fig 25.20** When the yellow vanadyl(V) ion, $VO_2^+(aq)$, is reduced by zinc, the vanadium is reduced through the +4 and +3 oxidation states, and finally forms the +2 oxidation state. The reduction can be followed by the changes in colour of the solution: +5 yellow, +4 blue, +3 green and +2 purple

## ? TEST YOURSELF

**P** **a)** Construct the overall equation for the reaction between acidified $VO^{2+}(aq)$ and zinc to give $V^{3+}(aq)$.
**b)** By use of the appropriate electrode potentials, show that zinc can reduce acidified $V^{3+}(aq)$ to form $V^{2+}(aq)$.

**SEE QUESTION 4.**

**Fig 25.21** A summary of the reactions of vanadium and its compounds. Reactions that involve oxidation are represented by red arrows, and the reaction that involves reduction by a blue arrow

Since vanadium(V) is the highest oxidation state of vanadium, one would expect it to be highly oxidizing, and to be reduced to one of the lower oxidation states. For example, zinc can reduce $VO_2^+(aq)$ to $V^{2+}(aq)$ in a series of steps that involves the formation of different coloured solutions as the oxidation state of vanadium changes. Each of these steps can be explained by use of the appropriate electrode potential.

The first step is the reduction of vanadyl(V) ion, $VO_2^+(aq)$:

$$Zn(s) \rightarrow Zn^{2+}(aq) + 2e^- \qquad E_{oxid}^{\ominus} = +0.76 \text{ V}$$

$$VO_2^+(aq) + 2H^+(aq) + e^- \rightarrow VO^{2+}(aq) + H_2O(l) \qquad E_{red}^{\ominus} = +1.00 \text{ V}$$

Therefore, $E_{cell}^{\ominus} = +1.76$ V.

The overall equation can be obtained by combining the two half-equations:

$$Zn(s) + 2VO_2^+(aq) + 4H^+(aq) \rightarrow 2VO^{2+}(aq) + 2H_2O(l) + Zn^{2+}(aq)$$

The second step is the reduction of vanadyl(IV) ion, $VO^{2+}(aq)$:

$$VO^{2+}(aq) + 2H^+(aq) + e^- \rightarrow V^{3+}(aq) + H_2O(l) \qquad 0.34E_{red}^{\ominus} = +0.34 \text{ V}$$

Therefore, $E_{cell}^{\ominus} = +1.10$ V.

vanadium(IV) chloride $VCl_4$ $\xleftarrow{Cl_2}$ vanadium V $\xrightarrow{O_2}$ vanadium(V) oxide $V_2O_5$ $\underset{heat}{\overset{H_2O/NH_3(aq)}{\rightleftarrows}}$ ammonium vanadate(V) $NH_4VO_3$ $\xrightarrow[\substack{zinc \\ H_2SO_4}]{excess}$ vanadium(II) sulphate $VSO_4$

## CHROMIUM AND ITS COMPOUNDS

**Table 25.3 Electrode potentials of some chromium species**

| Reduction half-equation | $E_{red}^{\ominus}$/V |
|---|---|
| $Cr^{2+}(aq) + 2e^- \rightleftharpoons Cr(s)$ | −0.91 |
| $Cr^{3+}(aq) + 3e^- \rightleftharpoons Cr(s)$ | −0.74 |
| $Cr^{3+}(aq) + e^- \rightleftharpoons Cr^{2+}(aq)$ | −0.41 |
| $Cr(OH)_3(s) + 3e^- \rightleftharpoons Cr(s) + 3OH^-(aq)$ | −1.30 |
| $[Cr(OH)_4]^-(aq) + 3e^- \rightleftharpoons Cr(s) + 4OH^-(aq)$ | −1.20 |
| $CrO_4^{2-}(aq) + 4H_2O(l) + 3e^- \rightleftharpoons Cr(OH)_3(s) + 5OH^-(aq)$ | −0.13 |

Chromium exhibits three common oxidation states, +2, +3 and +6, of which chromium(III) is the most stable. Chromium is manufactured by the reduction of chromium(III) oxide by a reactive metal (which must be above chromium in the reactivity series). Chromium reacts with dilute acids, such as hydrochloric or sulphuric acid, to give a sky-blue solution that contains the aqueous chromium(II) ion, as indicated by the electrode potential data. The reaction must be carried out in an inert atmosphere to avoid atmospheric oxidation, which would give chromium(III).

## OXIDATION OF AQUEOUS CHROMIUM(III) IONS

Aqueous chromium(III) ions can be oxidized to form chromate(VI). First, they are reacted with aqueous sodium hydroxide to form chromium(III) hydroxide, and then the mixture obtained is boiled with hydrogen peroxide:

$$Cr^{3+}(aq) + 3OH^-(aq) \rightarrow Cr(OH)_3(s)$$

$$2Cr(OH)_3(s) + 3H_2O_2(aq) + 4OH^-(aq) \rightarrow 2CrO_4{}^{2-}(aq) + 8H_2O(l)$$

## CHROMIUM IN THE +6 OXIDATION STATE

The highest oxidation number of a transition element often involves covalent bonding to the transition metal atom. Such is the case with compounds of chromium(VI). These compounds are also very powerful oxidizing agents.

As Table 25.3 shows, $E_{red}^{\ominus}$ for dichromate(VI) has a fairly large positive value, which indicates the ability of dichromate(VI) to oxidize other substances. The equation shows that the product of this reduction is normally the aqueous chromium(III) ion. The use of aqueous dichromate(VI) ion as an oxidizing agent is always associated with the presence of aqueous hydrogen ions (dilute sulphuric acid).

In an alkaline solution, the oxidizing ability is greatly reduced, since dichromate(VI) is converted into chromate(VI). As Table 25.3 shows, $E_{red}^{\ominus}$ for the reduction of the $CrO_4{}^{2-}$ ion has a low negative value:

$$2OH^-(aq) + Cr_2O_7{}^{2-}(aq) \rightleftharpoons 2CrO_4{}^{2-}(aq) + H_2O(l)$$

The conversion of dichromate(VI) into chromate(VI) is pH dependent. In acid, the equilibrium shifts to the left to form orange dichromate(VI). In alkali, the equilibrium shifts to the right to form yellow chromate(VI) ion.

Iron(II) ions are oxidized by acidified $Cr_2O_7{}^{2-}(aq)$ to form $Fe^{3+}(aq)$, and $Cr_2O_7{}^{2-}(aq)$ is reduced to $Cr^{3+}(aq)$. The colour changes in this reaction are complicated, since green $Fe^{2+}(aq)$ reacts with orange $Cr_2O_7{}^{2-}(aq)$ to form orange $Fe^{3+}(aq)$ and blue–green $Cr^{3+}(aq)$:

$$6Fe^{2+}(aq) + Cr_2O_7{}^{2-}(aq) + 14H^+(aq) \rightarrow$$
$$6Fe^{3+}(aq) + 2Cr^{3+}(aq) + 7H_2O(l)$$

Acidified potassium dichromate(VI) can be reduced by zinc in an inert atmosphere to give chromium(II). The reaction must be carried out in an inert atmosphere to prevent oxidation of $Cr^{2+}(aq)$ to $Cr^{3+}(aq)$:

$$4Zn(s) + Cr_2O_7{}^{2-}(aq) + 14H^+(aq) \rightarrow$$
$$4Zn^{2+}(aq) + 2Cr^{2+}(aq) + 7H_2O(l)$$

### ? TEST YOURSELF

**Q** **a)** Write down an equation to show the reaction between chromium metal and dilute sulphuric acid in an inert atmosphere.
**b)** Use the electrode potential to work out $E_{cell}^{\ominus}$ for the following electrochemical cell:
Cr(s) | Cr²⁺(aq) ‖ H⁺(aq) | H₂(g) | Pt
**c)** Suggest why the oxidation of chromium(III) into chromium(VI) is usually carried out in alkaline conditions. Use the electrode potentials in Table 25.3 to help you to explain your answer.

**Fig 25.22** The dichromate(VI) ion has one bridging oxygen atom

**INFORMATION ABOUT ACIDIFIED DICHROMATE(VI) IN THE OXIDATION OF ALCOHOLS AND ALDEHYDES IS GIVEN ON PAGES 295 AND 356.**

**Fig 25.23** A summary of the reactions of chromium and its compounds. Reactions that involve oxidation are represented by red arrows, and those that involve reduction by blue arrows

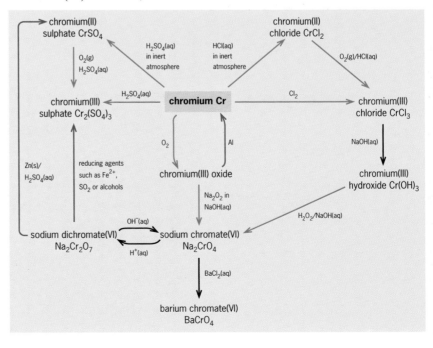

# MANGANESE AND ITS COMPOUNDS

**Table 25.4 Electrode potentials of some manganese species**

| Reduction half-equation | $E_{red}^{\ominus}/V$ |
|---|---|
| $Mn^{2+}(aq) + 2e^- \rightleftharpoons Mn(s)$ | −1.18 |
| $Mn^{3+}(aq) + e^- \rightleftharpoons Mn^{2+}(aq)$ | +1.49 |
| $MnO_2(s) + 2H_2O(l) + 2e^- \rightleftharpoons Mn(OH)_2(s) + 2OH^-(aq)$ | +0.05 |
| $MnO_2(s) + 4H^+(aq) + 2e^- \rightleftharpoons Mn^{2+}(aq) + 2H_2O(l)$ | +1.23 |
| $MnO_4^-(aq) + e^- \rightleftharpoons MnO_4^{2-}(aq)$ | +0.56 |
| $MnO_4^-(aq) + 2H_2O(l) + 3e^- \rightleftharpoons MnO_2(s) + 4OH^-(aq)$ | +0.59 |
| $MnO_4^-(aq) + 4H^+(aq) + 3e^- \rightleftharpoons MnO_2(s) + 2H_2O(l)$ | +1.67 |

**?** **TEST YOURSELF**

**R** **a)** Use the electrode potentials to suggest why manganese(III) is considered to be a very unstable oxidation state with respect to the +2 oxidation state.

**b)** Predict whether you would expect manganese to react with dilute hydrochloric acid. Explain your answer using electrode potentials.

Manganese exhibits several common oxidation states, including +2, +4, +6 and +7. The +2 oxidation state is generally considered to be the most stable under normal conditions.

## MANGANESE(IV) OXIDE AS AN OXIDIZING AND A REDUCING AGENT

Manganese(IV) oxide has an intermediate oxidation number. Therefore, it can act both as a reducing agent and as an oxidizing agent. Manganese(IV) oxide can oxidize hydrochloric acid to form chlorine:

$$MnO_2(s) + 4HCl(aq) \rightarrow MnCl_2(aq) + 2H_2O(l) + Cl_2(g)$$

Manganese(IV) oxide can also be oxidized to give manganate(VI). This is usually achieved by heating the oxide with an oxidizing agent, such as potassium nitrate or potassium chlorate(VI), with potassium hydroxide.

---

**!** **SCIENCE FEATURE** Lithium–manganese(IV) oxide cell

The lithium–manganese(IV) oxide cell – popularly referred to as the lithium cell – cannot have a water-based electrolyte because the lithium would react with the water to form hydrogen. The great advantage of using lithium as one of the electrodes is that its oxidation potential is very large, which gives the battery a high voltage. The battery can therefore deliver a low current for a long time, which makes it an ideal power supply for such items as watches and heart pacemakers.

Oxidation half-equation at the anode:
$$Li \rightarrow Li^+ + e^-$$

Reduction half-equation at the cathode:
$$Li^+ + MnO_2 + e^- \rightarrow LiMnO_2$$

**Fig 25.24** A lithium-manganese(IV) oxide battery

---

## HIGHER OXIDATION STATES OF MANGANESE

Both the +6 and the +7 oxidation states are highly oxidizing, since the manganese can be reduced to manganese(IV) or manganese(II). Aqueous manganate(VI) is a green solution, which is easily oxidized to give the familiar purple colour of the manganate(VII) ion. The manganate(VII) ion can be prepared directly by reacting manganese(II) ions with oxidizing agents such as sodium bismuthate, $NaBiO_3$, or lead(IV) oxide. This reaction is used in the estimation of the percentage of manganese in a sample of steel. The steel is reacted with dilute acid to form a solution that contains aqueous iron(II) ions and aqueous manganese(II) ions. The aqueous manganese(II) ions are oxidized to give purple manganate(VII). The concentration of the manganate(VII) can be determined using ultraviolet–visible spectroscopy (see pages 608 to 609).

## VOLUMETRIC ANALYSIS USING POTASSIUM MANGANATE(VII)

Aqueous potassium manganate(VII) is used in laboratories as an oxidizing agent for organic preparations, volumetric analysis and qualitative analysis.

Acidified potassium manganate(VII) is used to test for reducing agents: it reacts to form manganese(II) ions and the distinct colour of the manganate(VII) ions disappears. For example, iodide ions reduce acidified manganate(VII) ions to form manganese(II) ions and iodine.

In volumetric analysis, acidified potassium manganate(VII) is used to determine the concentration of reducing agents. Since it changes colour during the titration, there is no need to have an indicator. If potassium manganate(VII) is used in the burette, the end-point is the first appearance of a purple-pink colour. A typical example (Fig 25.25) would be the reaction of aqueous iron(II) ions with acidified potassium manganate(VII):

$$5Fe^{2+}(aq) + MnO_4^-(aq) + 8H^+(aq) \rightarrow 5Fe^{3+}(aq) + Mn^{2+}(aq) + 4H_2O(l)$$

### ● EXAMPLE

**Q** A solution of a moss-killer contained aqueous iron(II) ions. A 25.0 cm$^3$ sample of this solution was acidified with dilute sulphuric acid and titrated with 0.0200 mol dm$^{-3}$ potassium manganate(VII). It was found that 21.0 cm$^3$ of the KMnO$_4$(aq) was needed to react fully with the Fe$^{2+}$(aq) in the solution of the moss-killer. Calculate the concentration of Fe$^{2+}$(aq) in the solution of the moss-killer.

**A** $5Fe^{2+}(aq) + MnO_4^-(aq) + 8H^+(aq) \rightarrow 5Fe^{3+}(aq) + Mn^{2+}(aq) + 4H_2O(l)$

moles of MnO$_4^-$ = volume in dm$^3$ × concentration

$$= 0.0210 \times 0.0200 = 4.20 \times 10^{-4}$$

From the equation:

moles of Fe$^{2+}$ = 5 × moles of MnO$_4^-$ = 2.10 × 10$^{-3}$

Therefore:

$$[MnO_4^-(aq)] = \frac{\text{moles of MnO}_4^-}{\text{volume of MnO}_4^- \text{ (in dm}^3)}$$

$$= \frac{2.10 \times 10^{-3}}{0.0250} = 0.0840 \, \text{mol dm}^{-3}$$

In this description of the oxidizing abilities of manganate(VII), references are made to acidified conditions. This is important, because the pH at which the reaction is carried out affects the nature of the reduction product. If the pH is increased so that the reaction is carried out in neutral or alkaline conditions, then manganese(IV) oxide is produced instead of manganese(II) ions. As Table 25.4 shows, in alkaline conditions there is a different half-equation.

**? TEST YOURSELF**

**S** **a)** Write down:
(i) the half-equation to show the reduction of acidified manganate(VII) ions to give manganese(II) ions;
(ii) the oxidation of aqueous iodine ions to form aqueous iodine.
**b)** Hence construct the ionic equation for the reaction between acidified manganate(VII) ions and aqueous iodide ions.

**Fig 25.25** A titration involving potassium manganate(VII) and iron(II) ions

**FOR MORE INFORMATION ABOUT THE CONCENTRATION OF AQUEOUS SOLUTIONS, READ PAGE 139.**

**? TEST YOURSELF**

**T** Potassium manganate(VII) was titrated with acidified iron(II) sulphate solution. 25.0 cm$^3$ of KMnO$_4$(aq) required 35.9 cm$^3$ of 0.100 mol dm$^{-3}$ FeSO$_4$(aq) to react fully. Calculate the concentration of MnO$_4^-$(aq).

**Fig 25.26** A summary of the reactions of manganese and its compounds. Reactions that involve oxidation are represented by red arrows, and those that involve reduction by blue arrows

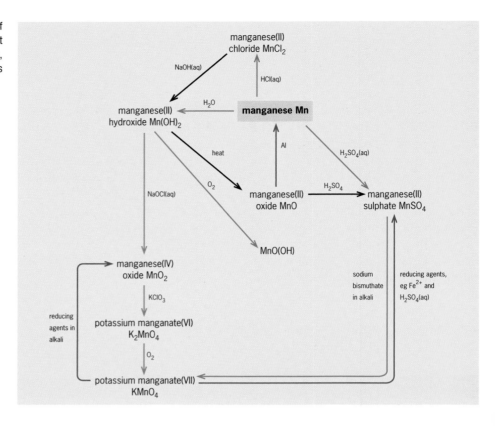

# IRON AND ITS COMPOUNDS

**Table 25.5 Electrode potentials of some iron species**

| Reduction half-equation | $E_{red}^{\ominus}$/V |
|---|---|
| $Fe^{2+}(aq) + 2e^- \rightleftharpoons Fe(s)$ | −0.44 |
| $Fe^{3+}(aq) + 3e^- \rightleftharpoons Fe(s)$ | −0.04 |
| $Fe^{3+}(aq) + e^- \rightleftharpoons Fe^{2+}(aq)$ | + 0.77 |
| $[Fe(CN)_6]^{3-}(aq) + e^- \rightleftharpoons [Fe(CN)_6]^{4-}(aq)$ | + 0.36 |
| $Fe(OH)_3(s) + e^- \rightleftharpoons Fe(OH)_2(s) + OH^-(aq)$ | −0.56 |

Iron exhibits two common oxidation states, +2 and +3, both of which are ionic. Iron does form another oxidation state, +6, but it occurs in only a few compounds.

## RELATIVE STABILITY OF THE +2 AND +3 OXIDATION STATES

As the electrode potentials in Table 25.5 suggest, iron reacts with dilute acids, such as sulphuric acid, to form iron(II) salts:

$$Fe(s) + 2H^+(aq) \rightarrow Fe^{2+}(aq) + H_2(g) \qquad E_{cell}^{\ominus} = +0.44\ V$$

The iron(II) ions are very susceptible to aerial oxidation to form iron(III) ions. The stability of these two oxidation states is very much dependent on the pH. In acidic conditions, aqueous iron(II) ions are oxidized to give iron(III) ions. The appropriate half-equations are:

$$Fe^{2+}(aq) \rightarrow Fe^{3+}(aq) + e^- \qquad E_{oxid}^{\ominus} = -0.77\ V$$
$$O_2(g) + 2H^+(aq) + e^- \rightarrow 2H_2O(l) \qquad E_{red}^{\ominus} = +1.23\ V$$

In alkaline conditions, the oxidation is even more favourable:

$$Fe(OH)_2(s) + OH^-(aq) \rightarrow Fe(OH)_3(s) + e^- \qquad E^{\ominus}_{oxid} = +0.56 \text{ V}$$
$$O_2(g) + 2H_2O(l) + 2e^- \rightarrow 4OH^-(aq) \qquad E^{\ominus}_{red} = +0.40 \text{ V}$$

The relative stability of the +2 and +3 oxidation states can also be affected by the formation of complexes. In Table 25.5, note that the reduction potential for the hexacyanoferrate(III) ion, $[Fe(CN)_6]^{3-}(aq)$, is different from that for the hexaaquairon(III) ion, $[Fe(H_2O)_6]^{3+}(aq)$.

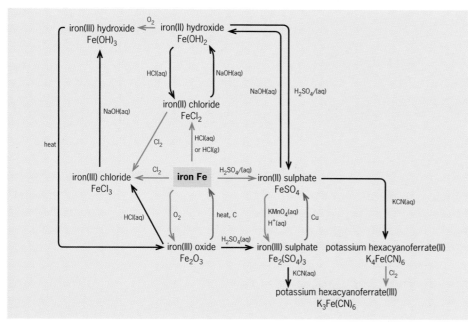

### ❓ TEST YOURSELF

**U** **a)** (i) Write down the overall equation for the aerial oxidation of acidified iron(II) ions.
(ii) What is the $E^{\ominus}_{cell}$ value for this reaction?
**b)** What is the $E^{\ominus}_{cell}$ value for the oxidation of iron(II) hydroxide into iron(III) hydroxide in alkaline conditions?

**V** Compare the $E^{\ominus}_{cell}$ values for the oxidation of the hexaaquairon(II) ion, $[Fe(H_2O)_6]^{2+}(aq)$, by chlorine with those for the oxidation of aqueous hexacyanoferrate(II), $[Fe(CN)_6]^{4-}(aq)$, by chlorine.

**Fig 25.27** A summary of the reactions of iron and its compounds. Reactions that involve oxidation are represented by red arrows, and those that involve reduction by blue arrows

## RUSTING OF IRON

Any unpainted iron object shows evidence of rusting. The surface of iron forms a flaky orange solid, the composition of which is best described as hydrated iron(III) oxide. Rusting is, in fact, a complicated electrochemical process that occurs on the surface of iron.

**Fig 25.28** How rusting occurs

Fig 25.28 summarizes some of the complex electrochemical processes that take place when iron rusts. When iron is in contact with a drop of water, a redox reaction occurs:

Oxidation half-equation: $\quad Fe(s) \rightarrow Fe^{2+}(aq) + 2e^-$
Reduction half-equation: $\quad O_2(g) + 2H_2O(l) + 4e^- \rightarrow 4OH^-(aq)$

These two reactions take place in different areas of the iron, which results in the formation of an anodic region and a cathodic region. The electrons move through the metal from the anodic region to the cathodic region. The circuit is

**Fig 25.29** The railway bridge over the Tyne at Newcastle is an iron structure which has to be painted to prevent rust damage

completed by ions that move through the water. Without the water, the circuit is not complete and rusting cannot take place. Essentially, the water is acting as a salt bridge. If the water contains electrolytes, such as sodium chloride, then the concentration of ions in the droplet is higher and so the rate of rusting increases. The iron(II) ions and the hydroxide in the water droplet are precipitated as iron(II) hydroxide, which is further oxidized to form hydrated iron(III) oxide or rust:

$$Fe^{2+}(aq) + 2OH^-(aq) \rightarrow Fe(OH)_2(s)$$

$$4Fe(OH)_2(s) + O_2(g) + 2H_2O(l) \rightarrow 4Fe(OH)_3(s)$$

## RUST PROTECTION

Any layer that is impervious to water protects iron from rusting. However, the protection provided by paints and other widely used materials is of limited duration, because of such factors as physical damage, chemical deterioration and weathering. So, alternative methods of rust protection use electrochemical principles instead. One of the best known involves giving the iron a protective layer of zinc by a process called galvanization, which is described below.

Iron pipes and tanks in the ground are often protected by a block of magnesium or zinc (Fig 25.30). The block of magnesium (or zinc) is attached to the iron object and, since magnesium (or zinc) has a higher positive oxidation potential than iron, it oxidizes in preference (sacrificially) to the iron. The iron object acts as the cathode in this method.

Tin-plated cans used in the food industry provide another example of electrochemical protection. The oxidation potential for tin has a lower positive value than that for iron. Therefore, tin is less likely to react with moist oxygen and so protects the surface of the iron. There is one drawback with tin plating. It is easily scratched to reveal the iron, in which case the can rusts very rapidly. What happens is that once the iron is in contact with moist air, it gives sacrificial protection to the less reactive tin.

## GALVANIZING IRON

As already stated, a coating of zinc protects iron from rusting. Zinc-coated iron is known as galvanized iron.

Galvanized iron is protected mainly because, in the electrochemical cell formed by galvanization, the zinc is preferentially oxidized. (The oxidation potential for zinc has a larger positive value than that for iron.) However, were it not for another reaction taking place at the same time, all the zinc would eventually be oxidized and rusting could start. The zinc hydroxide produced reacts with carbon dioxide in the air to form a layer of a zinc hydroxide–zinc carbonate compound that adheres firmly to the iron to give further protection.

## COBALT

SEE QUESTION 2.

magnesium anode:

$$Mg \rightarrow Mg^{2+} + 2e^-$$

iron pipe (cathode):

$$O_2 + 2H_2O + 4e^- \rightarrow 4OH^-$$

**Fig 25.30** Sacrificial protection

water drop     $O_2(g)$

$Zn^{2+}$   $Zn(OH)_2$

$$Zn \rightarrow Zn^{2+} + 2e^- \quad O_2 + 2H_2O + 4e^- \rightarrow 4OH^-$$

zinc anode       iron cathode

**Fig 25.31** Galvanized iron. The protective layer of zinc forms an electrochemical cell with the overall cell reaction:

$$2Zn + O_2 + 2H_2O \rightarrow 2Zn(OH)_2$$

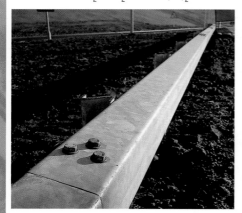

**Fig 25.32** Galvanized iron does not rust as fast as iron on its own. The coating of zinc corrodes in preference to the iron, in the course of which it forms a further protective layer over the iron

**Table 25.6 Electrode potentials of some cobalt species**

| Reduction half-equation | $E_{red}^{\ominus}$/V |
|---|---|
| $Co^{2+}(aq) + 2e^- \rightleftharpoons Co(s)$ | −0.28 |
| $Co^{3+}(aq) + e^- \rightleftharpoons Co^{2+}(aq)$ | +1.82 |
| $[Co(NH_3)_6]^{3+}(aq) + e^- \rightleftharpoons [Co(NH_3)_6]^{2+}(aq)$ | +0.10 |
| $Co(OH)_3(s) + e^- \rightleftharpoons Co(OH)_2(s) + OH^-(aq)$ | +0.17 |

Cobalt exhibits only two common oxidation states, +2 and +3, both of which are ionic in nature. As expected from the reduction potential of $Co^{2+}(aq)$, cobalt reacts slowly with dilute acids to form aqueous cobalt(II) ion:

$$Co(s) + 2H^+(aq) \rightarrow Co^{2+}(aq) + H_2(g) \qquad\qquad E^{\ominus}_{cell} = +0.28\ V$$

## REDOX REACTIONS INVOLVING COBALT(III)

The absence of a higher oxidation state suggests that redox reactions are not particularly significant, but this would be some way from the truth. Just a glance at the electrode potential for $Co^{3+}(aq)$ in Table 25.6 indicates that it should be easy to reduce aqueous cobalt(III) ions to give aqueous cobalt(II) ions. In aqueous solution, water reduces cobalt(III) ions to give cobalt(II) ions and oxygen:

$$Co^{3+}(aq) + e^- \rightarrow Co^{2+}(aq) \qquad\qquad E^{\ominus}_{red} = +1.82\ V$$

$$2H_2O(l) \rightarrow 4H^+(aq) + O_2(g) + 4e^- \qquad\qquad E^{\ominus}_{oxid} = -1.23\ V$$

The overall equation is:

$$4Co^{3+}(aq) + 2H_2O(l) \rightarrow 4Co^{2+}(aq) + 4H^+(aq) + O_2(g) \quad E^{\ominus}_{cell} = +0.59\ V$$

As already mentioned, pH can affect the relative stabilities of oxidation states, an example of which is provided by cobalt. $Co^{3+}(aq)$ is much more stable in basic conditions than in acidic conditions. This means that the oxidation of $Co^{2+}(aq)$ to form $Co^{3+}(aq)$ must be carried out in the presence of aqueous sodium hydroxide. The oxidizing agent of choice is hydrogen peroxide.

The formation of complex ions can also affect the stability of an oxidation state. This is exemplified by the stability of aqueous hexaamminecobalt(III), $[Co(NH_3)_6]^{3+}(aq)$, as compared with the aqueous hexaaquacobalt(III) ion, $[Co(H_2O)_6]^{3+}(aq)$. As Table 25.6 shows, the redox potential for the hexaammine complex is only just above zero, and therefore it is much more difficult to reduce than $Co^{3+}(aq)$.

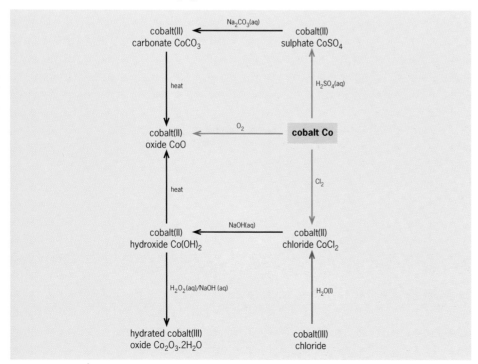

**Fig 25.33** A summary of the reactions of cobalt and its compounds. Reactions involving oxidation are represented by red arrows, and those involving reduction by blue arrows

# COPPER

The main oxidation states of copper are +1 and +2. Both are ionic oxidation states, and involve the loss of either one electron or two electrons per copper atom.

## REACTIONS OF COPPER

| Table 25.7 Reduction potentials involving some copper species | |
|---|---|
| **Reduction half-equation** | $E_{red}^{\ominus}$/V |
| $Cu^{2+}(aq) + 2e^- \rightleftharpoons Cu(s)$ | + 0.34 |
| $Cu^+(aq) + e^- \rightleftharpoons Cu(s)$ | + 0.52 |
| $Cu^{2+}(aq) + e^- \rightleftharpoons Cu^+(aq)$ | + 0.15 |
| $[Cu(CN)_2]^-(aq) + e^- \rightleftharpoons Cu(s) + 2CN^-(aq)$ | + 0.43 |
| $[Cu(NH_3)_2]^+(aq) + e^- \rightleftharpoons Cu(s) + 2NH_3(aq)$ | −0.12 |

The reduction potentials for both $Cu^+(aq)$ and $Cu^{2+}(aq)$ indicate that copper does not react with dilute acids, such as sulphuric or hydrochloric, to form hydrogen. The presence of dissolved oxygen in the acid does, however, allow dilute acids to oxidize copper to give the copper(II) ion:

$$Cu(s) + 2H^+(aq) + \tfrac{1}{2}O_2(g) \rightarrow Cu^{2+}(aq) + H_2O(l)$$

$$E_{cell}^{\ominus} = E_{oxid}^{\ominus} + E_{red}^{\ominus}$$

$$= -0.34 + (+1.23)$$

$$= +0.89 \text{ V}$$

## STABILITY OF THE +1 OXIDATION STATE

Aqueous copper(I) compounds often disproportionate to give copper and copper(II), but insoluble copper(I) compounds, such as the halides, are much more stable:

$$Cu_2SO_4(aq) \rightarrow Cu(s) + CuSO_4(aq)$$

## STABILITY OF COPPER(II) COMPOUNDS

Copper(II) compounds are not really considered to be oxidizing agents, but there a few examples of their reduction to copper(I) compounds. Aldehydes reduce complexed copper(II) ions into copper(I) oxide, which forms a red-brown precipitate. This is the basis of Benedict's or Fehling's test for aldehydes and reducing sugars.

**READ MORE ABOUT THE BENEDICT'S OR FEHLING'S TEST ON PAGE 357.**

Iodide ion reduces aqueous copper(II) ions to give a white precipitate of copper(I) iodide, and so copper(II) iodide does not exist:

$$2Cu^{2+}(aq) + 4I^-(aq) \rightarrow 2CuI(s) + I_2(s)$$

This reaction is the basis of the estimation of the concentration of copper(II) ions in solution, since the iodine produced can be titrated against aqueous sodium thiosulphate.

**Fig 25.34** A summary of the reactions of copper and its compounds. Reactions that involve oxidation are represented by red arrows, and those that involve reduction by blue arrows

# 8 EXTRACTION OF METALS

Several different methods of extracting metals have already been covered. On page 11 is a description of the reduction of iron(III) oxide with carbon in the blast furnace, and Chapter 22 contains a description of the electrolytic extraction of the alkali metals and of aluminium. In this section, some of the free-energy changes associated with the extraction of metals, such as aluminium and iron, are reviewed.

Although this section is not really connected with electrode potentials, it does use the concept of free energy, $\Delta G$, introduced on page 552.

## FREE-ENERGY CHANGE

The second law of thermodynamics states:

> **Any spontaneous change that occurs in the universe must be accompanied by an increase in the entropy of the universe.**

On page 143, entropy is described as a measure of the disorder of a system.

Reactions proceed spontaneously if they are energetically downhill ($\Delta H$ is negative) and if the process leads to more disorder ($\Delta S$ is positive). These two ideas of entropy change and enthalpy change can be combined to give the free-energy change of a reaction, $\Delta G$, expressed as:

$$\Delta G = \Delta H - T\Delta S$$

As already stated, reactions in which $\Delta G$ is negative are spontaneous. Therefore, provided the enthalpy change, the entropy change and the temperature of a reaction are known, it can be predicted whether a reaction will be spontaneous.

### REMEMBER THIS

Free energy is the maximum available work that could be carried out on the surroundings, and a reaction should occur spontaneously when the free-energy change is negative.

### REMEMBER THIS

The description of a reaction as 'spontaneous' does not mean that the rate of reaction has to be fast. In fact, the rate of reaction may be so slow that the reaction does not actually take place. In this case, one of the reactants is said to be kinetically stable, but energetically unstable.

### ● EXAMPLE

**Q** Predict whether the reaction between hydrogen and chlorine is spontaneous at 298 K:

$$H_2(g) + Cl_2(g) \rightarrow 2HCl(g) \quad \begin{aligned} \Delta H^{\ominus} &= -185 \text{ kJ mol}^{-1} \\ \Delta S^{\ominus} &= +141 \text{ J K}^{-1} \text{ mol}^{-1} \end{aligned}$$

**A** For these figures, there is no need to do a calculation.

Since $\Delta H^{\ominus}$ is negative and $-T\Delta S^{\ominus}$ is negative (absolute temperature cannot be negative),

then $\Delta G^{\ominus} = \Delta H^{\ominus} - T\Delta S^{\ominus}$ must also be negative. Therefore, the reaction is spontaneous.

**Q** Predict whether the thermal dissociation of ammonium chloride will proceed spontaneously at 298 K:

$$NH_4Cl(s) \rightarrow NH_3(g) + HCl(g) \quad \begin{aligned} \Delta H^{\ominus} &= +176 \text{ kJ mol}^{-1} \\ \Delta S^{\ominus} &= +284 \text{ J K}^{-1} \text{ mol}^{-1} \end{aligned}$$

**A** $\Delta G^{\ominus} = \Delta H^{\ominus} - T\Delta S^{\ominus}$

$\quad = +176 - 298 \, (+284 \times 10^{-3})$

$\quad = +91 \text{ kJ mol}^{-1}$

[Remember: kJ must be used throughout.]

Since the free-energy change is positive, the reaction cannot be spontaneous at 298 K.

## ENTROPY AND SPONTANEOUS REACTIONS

The third law of thermodynamics states:

> **At absolute zero the entropy of a pure substance is zero.**

As the temperature increases, so does the entropy of the substance. That is, the more the particles move, the greater the entropy becomes. Therefore, gases have a much greater entropy than liquids, and liquids have a much greater entropy than solids. This allows the entropy change of a reaction to be predicted just by looking at the number of moles of gas shown in the equation. No matter what the enthalpy change of a reaction is, if the entropy change is positive, the reaction will be spontaneous for some temperature range.

**READ MORE ABOUT ENTROPY AND SPONTANEOUS REACTIONS ON PAGES 178 AND 179.**

## FREE-ENERGY CHANGE AND TEMPERATURE

Fig 25.35 shows that the free-energy change for a reaction, $\Delta G$, alters with the temperature. Assuming that the enthalpy change is a constant, then the change in the value of $\Delta G$ results from the effect of entropy. This means that a reaction involving a positive entropy change always has a range of temperatures in which the reaction is spontaneous.

**Fig 25.35** The four graphs show how the free-energy change, $\Delta G$, depends on temperature. They refer to every combination of $\Delta H$ and $\Delta S$. The blue sections indicate that the reaction should be spontaneous, since $\Delta G$ is negative. Note that when $\Delta S$ is negative and $\Delta H$ is positive, the reaction cannot be spontaneous

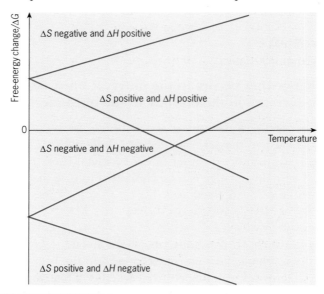

## EXAMPLE

**Q** Predict the temperature range over which the reaction between nitrogen and oxygen becomes spontaneous:

$$N_2(g) + O_2(g) \rightarrow 2NO(g) \quad \begin{array}{l} \Delta H = +180 \text{ kJ mol}^{-1} \\ \Delta S = +25 \text{ J K}^{-1} \text{ mol}^{-1} \end{array}$$

**A** Assuming that both $\Delta H$ and $\Delta S$ are constants, the reaction becomes spontaneous when $\Delta G = 0$.

That is:

$$0 = \Delta H - T\Delta S$$
$$0 = +184 - (T \times 25 \times 10^{-3})$$

[Note: energy is measured in kJ throughout.]

Therefore:

$$T = \frac{184}{25 \times 10^{-3}} = 7360 \text{ K}$$

The temperature range is therefore any temperature above 7360 K.

## ? TEST YOURSELF

**W** Predict whether the following reactions are spontaneous at 298 K:

**a)** $3O_2(g) \rightarrow 2O_3(g)$
$\Delta H = +286$ kJ mol$^{-1}$
$\Delta S = -137$ J K$^{-1}$ mol$^{-1}$

**b)** $2SO_2(g) + O_2(g) \rightarrow 2SO_3(g)$
$\Delta H = -198$ kJ mol$^{-1}$
$\Delta S = -187$ J K$^{-1}$ mol$^{-1}$

In the Example, in reality, both the enthalpy change and the entropy change are not constants. So the actual temperature calculated is not reliable. However, the calculation does demonstrate that this reaction only becomes spontaneous at high temperatures.

## ELLINGHAM DIAGRAMS

Since the free-energy change is a function of temperature, one way to describe the change is to draw an Ellingham diagram. An Ellingham diagram shows how the free-energy change for the oxidation of an element depends on temperature. Note that in each case the free-energy change refers to the equation with 1 mole of oxygen. Consider, for example, the oxidation of carbon to carbon monoxide:

$$2C(s) + O_2(g) \rightarrow 2CO(g)$$

The entropy change for this reaction is positive, since the 2 moles of gas are made from 1 mole of gaseous reactant. It follows that, as the temperature increases, the free-energy change for this reaction has increasing negative values, as shown in Fig 25.36.

The oxidation of carbon to carbon dioxide is much more difficult to assess, since there is 1 mole of gaseous reactant and 1 mole of gaseous product:

$$C(s) + O_2(g) \rightarrow CO_2(g)$$

The entropy change for this reaction has a very small negative value. Therefore, the free-energy change for the oxidation of carbon dioxide remains fairly constant.

For the oxidation of metals by gaseous oxygen to form metal oxides, it is much easier to assess the entropy change because there are no gaseous products. So, the entropy change must be negative. It follows from this that the free-energy change for such reactions assumes smaller and smaller negative values as the temperature increases.

# FREE-ENERGY CHANGES DURING THE EXTRACTION OF METALS

The extraction of a metal by reduction of its oxide with carbon can be considered to be the sum of two reactions. The first reaction is the oxidation of carbon, and the second is the decomposition of the oxide to give the metal and oxygen.

For example, the reduction of iron(II) oxide to give iron has an overall reaction of:

$$2FeO + 2C \rightarrow 2Fe + 2CO$$

This can be written as the combination of two reactions:

$2C + O_2 \rightarrow 2CO$     $\Delta G$ is negative
$2FeO \rightarrow 2Fe + O_2$     $\Delta G$ is positive because it is the opposite of that shown in the Ellingham diagram.

Note that the overall equation is written so that it involves 1 mole of oxygen in each of the reactions.

A reaction is spontaneous when the free-energy change is negative. The minimum temperature at which this happens is given by the point of intersection of the two appropriate lines on the Ellingham diagram. Above that temperature, the magnitude of the negative free energy of oxidation of carbon is greater than the magnitude of the positive free-energy change of decomposition.

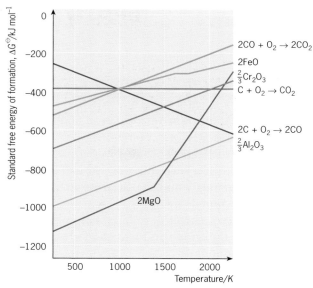

**Fig 25.36** An Ellingham diagram for the formation of some oxides

**SEE QUESTION 5.**

## ❓ TEST YOURSELF

**✗ a)** Which metals in the Ellingham diagram in Fig 25.36 could be obtained by reduction with carbon at 1500 °C?

**b)** The free-energy changes at various temperatures for the following reaction are shown in Table 25.8:

$$Ti(s) + O_2(g) \rightarrow TiO_2(s)$$

**Table 25.8**

| Temperature/K | Free-energy change, $\Delta G$/kJ mol$^{-1}$ |
|---|---|
| 300 | −856 |
| 600 | −801 |
| 1200 | −690 |
| 2000 | −542 |

The free energy changes at various temperatures for the following reaction are shown in Table 25.9:

$$2C(s) + O_2(g) \rightarrow 2CO(g)$$

(i) Plot both sets of data on graph paper.
(ii) Use your results to predict the likelihood of obtaining titanium by the reduction of titanium(IV) oxide with carbon.

**Table 25.9**

| Temperature/K | Free-energy change, $\Delta G$/kJ mol$^{-1}$ |
|---|---|
| 300 | −276 |
| 600 | −330 |
| 1200 | −440 |
| 2000 | −582 |

## ▶ EXTENSION Extraction of titanium

Titanium has a reduction potential of −1.63 V, so it should be a reactive metal. However, it reacts with neither dilute acids nor hot aqueous alkalis. This lack of reactivity combined with its low density favours the use of titanium in the aerospace industry. The former Soviet Union developed many military aircraft using titanium alloys, despite the prohibitive cost of the metal.

Titanium is extracted from its ore, rutile (TiO$_2$), but reduction by carbon does not work because the free-energy change for this reduction is positive, even at high temperatures. Instead, titanium is prepared by the reduction of titanium(IV) chloride with sodium or magnesium. Rutile is first concentrated to produce a mixture that contains in excess of 90% titanium(IV) oxide. The TiO$_2$ is then chlorinated in the presence of coke at 900–1000 °C.

$$TiO_2 + 2Cl_2 + 2C \rightarrow TiCl_4 + 2CO$$

$$TiO_2 + 2Cl_2 + C \rightarrow TiCl_4 + CO_2$$

The titanium(IV) chloride is then reduced by either magnesium or sodium at a temperature usually higher than 1000 °C. This reduction is a batch process and often takes 36–50 hours to complete. The reduction is continued until enough titanium that meets a particular requirement can be isolated. When more titanium is wanted, the reaction is carried out again:

$$2Mg + TiCl_4 \rightarrow 2MgCl_2 + Ti$$

$$4Na + TiCl_4 \rightarrow 4NaCl + Ti$$

## SUMMARY

After studying this chapter, you should know that:

- A redox reaction involves the transfer of electrons from a reducing agent to an oxidizing agent.
- All redox reactions can be written as a sum of two half-equations, one for oxidation and the other for reduction.
- An electrochemical cell is a way to carry out a redox reaction such that the two half reactions take place separately and the electrons are transferred via an external circuit.
- In an electrochemical cell, oxidation always takes place at the anode and reduction always takes place at the cathode.
- An electrode potential is the potential difference between an electrode and the ion-containing solution into which the electrode dips. All electrode potentials are compared to the standard hydrogen electrode, which is assigned a standard electrode potential of 0.00 V.
- The standard hydrogen electrode involves the reversible half-reaction of aqueous hydrogen ions and electrons to form hydrogen gas.
- The chemistry of transition elements can be explained by using electrode potentials.

- The cell potential of an electrochemical cell is the sum of the oxidation potential and the reduction potential. A process that has a positive cell potential takes place spontaneously.
- A primary cell is a battery that cannot be recharged. A secondary cell can be recharged.
- The higher oxidation states of the transition elements are oxidizing agents, and the lower oxidation states are reducing agents.
- Electrode potentials and cell potentials are affected by changes in concentration of the aqueous species, temperature, pressure of gases, pH and the complexing of ions.
- Free-energy change is a measure of the maximum available work that can be done on the surroundings. When a process has a negative free-energy change, it should take place spontaneously.
- The standard free-energy change, $\Delta G$, is related to the standard cell potential, $E^{\ominus}_{cell}$, by the equation $\Delta G = -nFE^{\ominus}_{cell}$.
- The free-energy change, $\Delta G$, is related to the enthalpy change, $\Delta H$, entropy change, $\Delta S$, and the absolute temperature, $T$, by the following equation $\Delta G = \Delta H - T\Delta S$.

## ? EXAM QUESTIONS

1 The reaction taking place when a lead–acid storage battery is recharged is:

$$2PbSO_4(s) + 2H_2O(l) \rightarrow Pb(s) + PbO_2(s) + 2H_2SO_4(aq)$$

a) Describe what happens to $Pb^{2+}$ ions in terms of oxidation and reduction during this process. Use oxidation numbers to support your answer.

b) During discharge of this battery, write a half-equation and specify the states for the reaction taking place at the anode.

IBO Chemistry Standard Level Paper 3 Nov 1998 QF1a,b

2 Use the following redox potential data to answer the questions that follow.

| Electrode reaction | $E^{\ominus}$/V |
|---|---|
| $Fe(OH)_3 + e^- \rightleftharpoons Fe(OH)_2 + OH^-$ | −0.56 |
| $Fe^{2+} + 2e^- \rightleftharpoons Fe$ | −0.44 |
| $Fe^{3+} + 3e^- \rightleftharpoons Fe$ | −0.04 |
| $H^+ + e^- \rightleftharpoons \frac{1}{2}H_2$ | 0.00 |
| $\frac{1}{2}O_2 + H_2O + 2e^- \rightleftharpoons 2OH^-$ | +0.40 |
| $Fe^{3+} + e^- \rightleftharpoons Fe^{2+}$ | +0.77 |
| $\frac{1}{2}Cl_2 + e^- \rightleftharpoons Cl^-$ | +1.36 |
| $Co^{3+} + e^- \rightleftharpoons Co^{2+}$ | +1.82 |

a) (i) Explain why the reaction of iron metal with dilute, aqueous hydrochloric acid gives iron(II) chloride and not iron(III) chloride.

(ii) Suggest how an aqueous solution of iron(III) chloride could be made from iron metal. Justify your answer.

b) Suggest, in outline, the stages in the rusting of iron, using the data above.

c) At 25 °C, a cell is constructed as follows:

$$Pt|Co^{2+}(aq, 1\ mol\ dm^{-3}), Co^{3+}(aq, 1\ mol\ dm^{-3})\|Fe^{3+}(aq, 1\ mol\ dm^{-3}), Fe^{2+}(aq, 1\ mol\ dm^{-3})|Pt$$

Write the equation for the reaction that takes place when a current is allowed to flow. Justify your answer.

Edexcel 6083/P.01 Module Test 3 June 2000 Q1

3 A student set up the electrochemical cell shown in Fig 25.Q3.

**Fig 25.Q3**

copper foil

silver foil

V

salt bridge

1.0 mol dm$^{-3}$ copper(II) sulphate

1.0 mol dm$^{-3}$ silver nitrate

# ? EXAM QUESTIONS

You are provided with the following standard electrode potentials:

$$Cu^{2+}/Cu \quad E^{\ominus} = +0.34 \text{ V}$$
$$Ag^{+}/Ag \quad E^{\ominus} = +0.80 \text{ V}$$

**a)** How could the student have made the salt bridge?

**b)** Write half-equations to show the reactions that occurred in: **(i)** the $Cu/Cu^{2+}$ half cell; **(ii)** the $Ag/Ag^{+}$ half cell.

**c)** Write an equation for the overall cell reaction.

**d)** **(i)** Calculate the standard cell potential for this cell. **(ii)** Identify the electrode at which reduction occurs. Explain your answer.

**e)** The student found that the e.m.f. obtained for this cell was less than the calculated value. Suggest two reasons for this.

OCR Specimen Paper 2815 Transition Metals Q1

**4**

**a)** Vanadium exists in several oxidation states, which are relatively easily interconverted. The compound most often used as a starting reagent in redox reactions of vanadium has the formula $NH_4VO_3$. When this salt is dissolved in acidic solution, the $VO_3^{-}(aq)$ ions are converted into $VO_2^{+}(aq)$ ions. **(i)** Write an equation for the conversion of $VO_3^{-}$ into $VO_2^{+}$ ions in acidic solution. Give the oxidation state of vanadium in these ions and the colour of the solution formed. **(ii)** Give the name of a reagent that will reduce the $VO_2^{+}$ ion in acidic solution to vanadium(IV), but no further. Write a half-equation for the reduction process.

**b)** If zinc is added to an acidic solution of $VO_2^{+}$ ions the final vanadium-containing ion formed is $[V(H_2O)_6]^{2+}$. Give the oxidation state of vanadium in this ion and the colour of the solution formed.

**c)** What important industrial use of transition metal compounds such as vanadium(V) oxide probably depends on the ease of conversion between oxidation states?

Edexcel 6081/P.01 Module Test 1 June 1999 Q1b,c(part),d

**5**

**a)** Explain why the plots (Fig 25.Q5) of change of free energy, $\Delta G$, with temperature, $T$, for the reactions below have such different gradients:

$$2C(s)+O_2(g) \rightarrow 2CO(g) \quad \Delta G = (-224 - 0.18T) \text{ kJ mol}^{-1}$$
$$C(s)+O_2(g) \rightarrow CO_2(g) \quad \Delta G = (-394 - 0.000008T) \text{ kJ mol}^{-1}$$

**Fig 25.Q5**

**b)** For the reaction
$$\tfrac{4}{3}Al(s) + O_2(g) \rightarrow \tfrac{2}{3}Al_2O_3(s)$$
$$\Delta G = (-1115 - 0.21T) \text{ kJ mol}^{-1}$$
predict, graphically, using Fig 25.Q5, the minimum temperature required for the reduction of alumina with graphite.

**c)** Calculate the free-energy change, $\Delta G$, for the overall reaction that occurs at 1300 K for the extraction of aluminium in the Hall–Héroult cell.
$$\tfrac{1}{2}Al_2O_3 + \tfrac{3}{4}C \rightarrow Al + \tfrac{3}{4}CO_2$$

[NEAB Chemistry Specimen Questions CH7 1996 q.]

# KEY SKILLS **ASSIGNMENT**

## MOLYBDENUM AND ITS COMPOUNDS

Molybdenum is a member of the second series of transition elements. It is in the same group of the Periodic Table as chromium, and has the electron configuration $[Kr]4d^55s^1$.

**1** **a)** Suggest why molybdenum has the electron configuration $[Kr]4d^55s^1$ rather than $[Kr]4d^45s^2$.
Hint: Read page 515.

**b)** What oxidation states would you expect to be shown by molybdenum in its compounds?

**c)** For each oxidation state, write down whether it is reducing, oxidizing or both.

Molybdenum can be manufactured from molybdenum(VI) oxide. This oxide is obtained by roasting the sulphide ore, $MoS_2$, in air to give molybdenum(VI) oxide.

**2** Construct an equation to show the reaction of the sulphide ore with oxygen from the air.

Pure molybdenum is obtained by reducing molybdenum(VI)

# KEY SKILLS **ASSIGNMENT**

oxide with hydrogen at high temperature. Reduction using carbon cannot be used, since the molybdenum formed in the reduction then reacts with excess carbon to form an unwanted carbide of molybdenum.

**3** Construct an equation to show the reduction of molybdenum(VI) oxide with hydrogen.

A major use of molybdenum is as an alloy with iron, ferromolybdenum. Extremely hard and strong, even at high temperatures, this alloy is used in filament supports, in heating elements for furnaces, and for drill bits. Ferromolybdenum can be made from molybdenum(VI) oxide without having to make pure molybdenum. A mixture of iron(III) oxide and molybdenum(VI) oxide is reduced using aluminium powder to give a ferromolybdenum that is at least 60 per cent molybdenum by mass. The process is carried out in batches, and at the height of the reduction the temperature inside the reaction vessel reaches 2000 °C.

**4** **a)** Suggest why it is cheaper to make ferromolybdenum directly from molybdenum(VI) oxide and iron(III) oxide rather than from pure iron and pure molybdenum.
**b)** Write balanced equations to show the reduction of:
    (i) iron(III) oxide with aluminium;
    (ii) molybdenum(VI) oxide with aluminium.
**c)** Why is such a high temperature needed for the manufacture of ferromolybdenum? Use ideas about free energy.

Most of the ferromolybdenum made in the United Kingdom used to come from a factory in Stowmarket, Suffolk, which is near to the major seaport at Felixstowe. This plant imported all of its molybdenum(VI) oxide from the United States. Almost half of the ferromolybdenum made in Stowmarket was exported back to the United States.

**5** It is usual to site a chemical factory as near as possible to the sources of its raw materials and to its major customers.
**a)** Suggest possible reasons why a major ferromolybdenum factory was sited in the United Kingdom rather than the United States.
**b)** The plant in the United Kingdom has now been closed and production of molybdenum is carried out mainly in the United States. Suggest possible reasons for the closure of the plant in the United Kingdom.

Molybdenum is a relatively unreactive metal, but it does react with halogens to form halides, with oxygen to form oxides, and with sulphur to form molybdenum(IV) sulphide. Molybdenum(IV) sulphide is used as a solid lubricant and as an additive in greases and oils. It has a layered structure with very weak forces between the layers, so its properties resemble those of graphite.

**6** Write an equation to show the formation of molybdenum(IV) sulphide from sulphur and hence calculate the maximum mass of molybdenum(IV) sulphide that can be made from 1.00 kg of molybdenum.

Table 26.A1 shows six reduction potentials for molybdenum-containing species.

**Table 26.A1 Six standard electrode potentials for molybdenum-containing species**

| Half-reaction | Reduction potential ($E_{red}^{\ominus}$)/V |
| --- | --- |
| $Mo^{3+} + 3e^- \rightarrow Mo$ | −0.20 |
| $MoO_2^+ + 4H^+ + 2e^- \rightarrow Mo^{3+}$ | 0.00 |
| $MoO_4^{2-} + 4H^+ + 2e^- \rightarrow MoO_2 + 2H_2O$ | −1.40 |
| $MoO_3 + 6H^+ + 6e^- \rightarrow Mo + 3H_2O$ | +0.10 |
| $MoO_4^{2-} + 4H_2O + 6e^- \rightarrow Mo + 8OH^-$ | −0.97 |

**7** Consider the following reaction:
$$Mo(s) + 3H^+(aq) \rightarrow Mo^{3+}(aq) + 3\tfrac{1}{2}H_2$$
**a)** Calculate the cell potential for this process and hence deduce whether the reaction is feasible.
**b)** Calculate the free-energy change for this reaction.

**8** Suggest a reducing agent that could reduce aqueous molybdate(VI) ion, $MoO_4^{2-}(aq)$:
**a)** to molybdenum(IV) oxide;
**b)** to molybdenum.
Explain your answers.

**9** **a)** An aqueous solution of molybdenum(III) sulphate is electrolysed using inert electrodes. Predict, with reasons, the products of this electrolysis.
**b)** $H_2MoO_4(aq)$ is reacted with excess zinc and dilute sulphuric acid. Predict the products of this reaction. Explain your answer.

# 26

# Reaction kinetics – a more detailed consideration

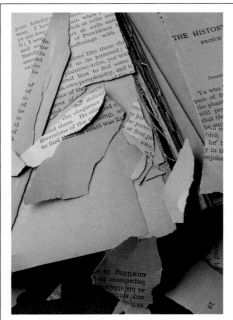

**Paper deteriortion** The long-term effect of hydrogen ions on paper

Unlike industrial chemists, who mostly want to accelerate chemical reactions, curators of the world's great libraries are desperate to stop reactions. If they fail, they will be the keepers of a growing collection of crumbling paper. Millions of books, documents, drawings and photographs are already falling to pieces. Many rare and valuable publications are close to total disintegration as their paper becomes brittle and starts to crumble. This applies not just to publications of many centuries ago, but to those as recent as the mid-1800s. What is the cause, and is it preventable?

Untreated paper is too porous to print on. Ink would seep into the pores to produce fuzzy print. So it has to be sized, a process that fills in the pores with a compound that leaves the paper with a smooth surface. Since 1850, the majority of book paper has been sized with aluminium sulphate – and this is the culprit.

Aluminium ions in the paper react with moisture to produce hydrogen ions. These hydrogen ions break down the cellulose molecules that the paper is composed of, and so it becomes brittle and eventually crumbles.

Neutralizing the acidity caused by hydrogen ions will halt this slow decomposition, and there are ways of doing this that do not damage the books. However, the cost can be prohibitive – sometimes over £1000 per book.

Paper manufacturers are helping to prevent this problem in future by producing more acid-free paper treated with sizing agents that do not produce hydrogen ions. Papers treated with these should have a life expectancy of well over 200 years – nearly ten times that of the crumbling paper.

The rate of the reaction that destroys these millions of publications is certainly slow, but already too many important and irreplaceable books have been damaged. Understanding reaction kinetics has helped to save some of the world's most treasured volumes, but for others it is too late.

## 1 WHY BOTHER WITH REACTION KINETICS?

Reaction kinetics is the study of rates of reaction. In Chapter 1 we first discussed the factors that affect the rate of a chemical reaction (page 8). We studied reaction rates in more detail in Chapter 8 (page 184) and considered how **collision theory** could explain why some factors, such as concentration and temperature, could accelerate or reduce rates of reaction.

We describe a reaction as **spontaneous** if it tends to occur. You met this concept first on pages 178 and 179. A very important spontaneous reaction is that between petrol and air in a car engine (see page 217). Petrol and air do not react in a petrol tank at 25 °C because the reaction rate is so slow that it cannot be measured. But petrol explodes in air when the energy from a spark is added, as in the internal combustion engine. Being spontaneous does not mean that a reaction is necessarily fast.

The reaction between nitrogen and hydrogen is another spontaneous one, yet nothing seems to happen at room temperature. However, under the conditions of the Haber process, ammonia is produced – one of the world's most important industrial chemical reactions. Industrial chemists need to know just how fast they can make a reaction proceed. A reaction that is too slow is unlikely to be a commercial proposition.

With a knowledge of the kinetics of a reaction, you can also understand how the reaction takes place and which species are involved in each step. The series of steps involved in a reaction is called the **reaction mechanism**. When the reaction mechanism is known, then ways to alter the rate of reaction by changing the conditions can be determined. This information is, of course, essential to industrial chemists seeking cheaper ways to bring about higher yields. Also important is that information about the reaction mechanism of a drug in the body can tell doctors how long the drug will remain effective before another dose is required. This often helps pharmaceutical companies to produce drugs with fewer side effects from unwanted reactions.

THE HABER PROCESS IS COVERED ON PAGES 18–22.

## 2 RATES OF REACTION AND THEIR MEASUREMENT

This is a definition for the rate of a reaction:

> **The rate of a reaction is either the change in the concentration of product formed per unit of time, or the change in the concentration of reactant used per unit of time.**

The reaction rate for a reactant that is consumed is negative.

Fig 26.1 shows how the concentration of a reactant changes during the course of a typical chemical reaction. Note that concentration is denoted by the use of square brackets. So, the concentration of a reactant is shown as [reactant]. The similar graph in Fig 26.2 shows the increase in concentration of a product of a chemical reaction. However, in any given reaction, the decrease in the concentration of the reactant is not necessarily the same as the increase in concentration of the product. For example, if 2 moles of reactant produce 1 mole of product, then the initial concentration of the reactant in Fig 26.1 would be twice that of the final concentration of the product in Fig 26.2.

**Fig 26.1** The decrease in the concentration of a reactant, [reactant], during the course of a typical chemical reaction

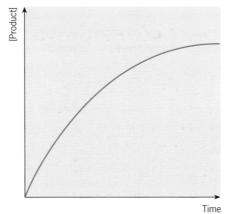

**Fig 26.2** The increase in the concentration of a product, [product], during the course of a typical chemical reaction

## ? TEST YOURSELF

**B** What is the initial rate of the reaction shown in Fig 26.3?
Hint: The tangent at time zero has been drawn for you.

## FINDING THE RATE OF REACTION AT A PARTICULAR TIME

Look again at Fig 26.2. If the rate of reaction were constant, the graph would be a straight line. Clearly, it is changing with time and will eventually fall to zero when the reaction is finished. The graph shows the change of concentration during the course of the reaction and is therefore known as a rate curve. The rate of reaction at any instant in time is given by the gradient of the curve at that instant.

The gradient at any point on a curve is found by drawing the tangent to the curve at that point and taking its gradient. Fig 26.3 shows two tangents. The tangent at time zero is called the initial rate, which occurs when the reactants are first mixed. The gradient of this tangent is the steepest of any taken along the rate curve, which means that the reaction is fastest at the start. The reaction rate at 50 seconds is calculated as follows:

$$\text{rate at 50 s} = \text{gradient of tangent at 50 s} = \frac{3.0 \text{ mol}}{40 \text{ s}} = 0.075 \text{ mol s}^{-1}$$

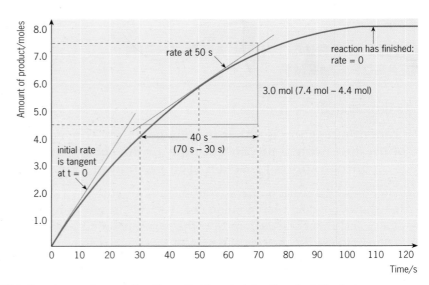

**Fig 26.3** The rate curve for a reaction. The method for calculating the rate at 50 s is shown

## MEASURING REACTION RATES

When the rate of a reaction is being studied, it is crucial to know what the reactants are, what the products are and what state these substances are in. This is conveniently found in the **stoichiometric equation**. This is a balanced chemical equation which states the amount of each reactant that reacts and the amount of each product formed (see page 7). For example, the stoichiometric equation for the formation of nitrogen dioxide, an atmospheric pollutant, from nitrogen monoxide and oxygen in a car exhaust is:

$$2NO(g) + O_2(g) \rightarrow 2NO_2(g)$$

Although this does not detail the steps by which the reaction occurs, it clearly states that only one product is formed and that only two reactants are involved. It also states that three volumes of reactants produce two volumes of product.

Provided we know the stoichiometric equation, we can decide on how to measure the concentration changes that occur as the reaction proceeds. In the above case there is a reduction in volume, which we could use to measure changes in the rate of reaction with time. If the reaction is carried out at a constant volume in a pressure vessel, then change in pressure could be measured.

**Fig 26.4** The reaction between nitrogen monoxide and oxygen could be followed by measuring the variation of pressure with time

pressure gauge

$2NO(g) + O_2(g) \rightarrow 2NO_2(g)$

## ? TEST YOURSELF

**C** What is the name given to the steps by which a chemical reaction takes place?
If you want to check your answer, see page 574.

Anything that changes during a reaction and can be measured may be used to determine a reaction rate, provided that it is proportional in some way to the concentration of a particular reactant or product.

## MEASURING CHANGE IN VOLUME OF GAS PRODUCED

The reaction between dilute hydrochloric acid and magnesium ribbon produces hydrogen according to this stoichiometric equation:

$$Mg(s) + 2HCl(aq) \rightarrow MgCl_2(aq) + H_2(g)$$

The volume of hydrogen produced could be used to follow (monitor) changes in the rate of this reaction. Apparatus for doing this is shown in Fig 26.5.

## CHANGES IN COLOUR

A colorimeter can be used to measure the change in colour of a reaction. This instrument (see pages 608 to 609) measures the amount of electromagnetic radiation absorbed by substances in the visible part of the spectrum. In the reaction between zinc and aqueous copper(II) sulphate, for example, the blue coloration of copper(II) sulphate disappears as the colourless zinc sulphate solution is formed:

$$Zn(s) + CuSO_4(aq) \rightarrow ZnSO_4(aq) + Cu(s)$$
$$\text{blue solution} \qquad \text{colourless solution}$$

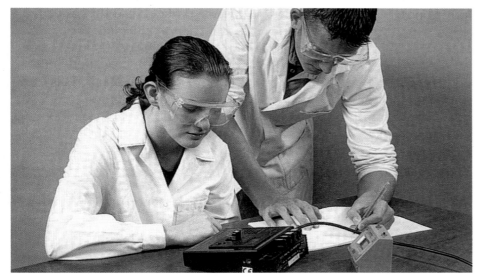

**Fig 26.6** When one of the reactants or products is coloured, the reaction can be followed by measuring the change in absorbance with a colorimeter

## CHANGES IN ELECTRICAL CONDUCTIVITY

Many reactions involve a change of conductivity because the number of ions in the reaction mixture changes during the reaction. In the following reaction, the number of ions decreases as the products are formed and the conductivity of the reaction can be followed using a conductivity cell (Fig 26.7):

$$BrO_3^-(aq) + 6I^-(aq) + 6H^+(aq) \rightarrow 3I_2(s) + 3H_2O(l) + Br^-(aq)$$

## CHEMICAL ANALYSIS

All the techniques so far described follow a reaction continuously and do not interfere with the progress of the reaction. Chemical analysis, however, involves taking samples of the reaction mixture at regular intervals of time. The reaction in the sample is stopped as soon as it is withdrawn. This can be

**REMEMBER THIS**

The rate of a reaction may be defined as the change in concentration of product formed or the change in concentration of reactant used per unit of time.

**Fig 26.5** The change in volume of this reaction due to the evolution of hydrogen gas can be used to measure the reaction rate

**? TEST YOURSELF**

**D** As hydrogen is being lost in the Mg/HCl(aq) reaction, what other properties change during the course of the reaction and how can they be measured?

**Fig 26.7** A conductivity cell

ESTERS ARE COVERED ON PAGE 395.

## ? TEST YOURSELF

**E** A 10.0 cm³ sample of the reaction solution of methyl methanoate and aqueous sodium hydroxide was withdrawn after 2.00 minutes and titrated with $1.0 \times 10^{-3}$ mol dm⁻³ dilute hydrochloric acid. 15.0 cm³ of HCl(aq) was required to neutralize the OH⁻ ions remaining.
**a)** How many moles of H⁺(aq) ions would be present in 15 cm³ HCl(aq)?
**b)** Write down the ionic equation for the neutralization of H⁺(aq) ions by OH⁻(aq).
**c)** How many moles of OH⁻(aq) ions were present in the 10.0 cm³ sample?
**d)** What was the concentration of OH⁻(aq) ions after 2.00 minutes?
Hint: If you are not sure how to do these calculations, look at page 139.

**F** Suggest a method by which you could follow the rates of the following reactions:
**a)** $SO_2Cl_2(g) \rightarrow SO_2(g) + Cl_2(g)$
**b)** $2H_2O_2(aq) \rightarrow 2H_2O(l) + O_2(g)$
**c)** $BrO_3^-(aq) + 5Br^-(aq) + 6H^+(aq) \rightarrow 3Br_2(aq) + 3H_2O(l)$
**d)** $CH_3Br + OH^- \rightarrow CH_3OH + Br^-$
**e)** $CH_3COCH_3 + I_2(aq) \rightarrow CH_3COCH_2I(aq) + H^+(aq) + I^-(aq)$
Use a different method for each reaction, if you can.

accomplished by rapid cooling, by removing one of the reactants or the catalyst, or by diluting the reaction mixture. The process of stopping a reaction (or slowing it down to a rate of almost zero) is called **quenching**.

Methyl ethanoate, $CH_3COOCH_3$, an ester, is an important industrial solvent. It can be hydrolysed using aqueous sodium hydroxide:

$$CH_3COOCH_3(l) + NaOH(aq) \rightarrow CH_3COO^-Na^+(aq) + CH_3OH(aq)$$

The rate of reaction can be followed by monitoring the concentration of sodium hydroxide as it is used up during the course of the reaction. Several identical reaction solutions are set to react. The temperature is kept constant, since changes in temperature affect the reaction rate. They are quenched by dilution with ice-cold water, at different time intervals from the start of the reaction. The concentration of sodium hydroxide that remains in each reaction solution is determined by titrating it with an acid, such as dilute hydrochloric acid.

**Fig 26.8** These students are titrating a reaction mixture. The reaction has been quenched by adding ice while the titration is performed

### ! SCIENCE FEATURE SONOCHEMISTRY

In the 1920s it was first discovered that ultrasound – sound with a frequency above 18 kHz – produces chemical effects. The study of sonochemistry, as it is called, did not really take off until the 1980s, when reliable and inexpensive ultrasound generators became readily available. Now, there is a host of interesting and important applications of ultrasound in chemistry.

rarefaction                    compression

direction of travel

**Fig 26.9** The propagation of a sound wave through a liquid. As the wave travels through the liquid, it creates alternate regions of high pressure (compression) and low pressure (rarefaction). These travelling pressure changes are exceptionally rapid when the liquid is excited by ultrasound

When a liquid is excited by ultrasound, the rapid changes in pressure (see Fig 26.9) produce an effect known as cavitation. As the rarefactions travel through the liquid, they pull its molecules apart to produce tiny cavities or bubbles. The compressions cause the bubbles to collapse, which releases tremendous amounts of energy. It has been estimated that the temperature near the collapse may be about 7000 K, which is the temperature at the surface of the Sun. Even higher temperatures – up to $2 \times 10^6$ K – may be generated as cavitation bubbles implode. The pressure created could make a gas as dense as a metal. However, the rate of cooling is astonishing at $10^{10}$ K per second, so overall the liquid does not become hot.

▶

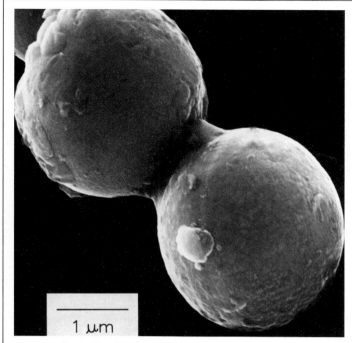

**Fig 26.10** Cavitation bubbles produced by ultrasonic waves

These localized energy hot spots are used to increase the rates of chemical reactions. They also produce highly reactive radicals. For example, the water molecule can be torn apart to produce H• and OH• radicals. Radicals have unpaired

electrons, which makes them highly reactive. So these radicals can combine to produce hydrogen gas and hydrogen peroxide. The OH• radical is also a potent oxidizing agent, which can react with other chemicals placed in the water. (For more about radicals – often called free radicals – see page 266.)

Early on, sonochemistry led to the production of catalysts that have particles so minute that they are called nanostructured catalysts. (A nanometre is $10^{-9}$ metres.) The surface area of these catalyst particles is huge. Nowadays, special combinations of metals in catalysts produced with the aid of ultrasound are making chemical processes more efficient, and it is likely that alternatives to platinum-based catalysts will be found.

Tailoring polymer molecules with ultrasound to enhance particular properties is another exciting prospect. The polymer chains are dissolved in a solvent, in which they are subjected to the awesome energy of cavitation bubbles. The chains break into smaller structures that, under the action of the bubbles, recombine to form different monomers in blocks along the chains.

There is so much unexplored potential in sonochemistry. Another possibility is that some organochlorine pollutants in water supplies could be broken down by ultrasound into harmless products. In a contrasting application, tiny haemoglobin spheres have been synthesized and may make it possible to produce artificial blood.

## 3 RATE EQUATIONS

Figs 26.1 and 26.2 show that the rate of reaction is fastest at the beginning of a reaction. This is because the concentrations of the reactants are at their highest value at the start. As the concentrations of the reactants fall, so does the rate of reaction. The **rate equation** states the relationship between the rate of reaction and the concentration of each reactant. It is also known as the **rate law**. As shown later, the rate equation helps chemists to work out how a reaction takes place. The dependence of the rate of reaction on the concentration of a reactant can be expressed mathematically in the form:

$$\text{rate} \propto [\text{reactant}]^n$$

which gives the rate equation as:

$$\text{rate} = k \times [\text{reactant}]^n \quad \text{or} \quad \text{rate} = k[\text{reactant}]^n$$

where:

- $k$ is a constant called the **rate constant**: the larger the rate constant, the faster the reaction,
- $n$ is called the **order of the reaction** with respect to the given reactant:

**The order of the reaction is the power to which the concentration of the reactant is raised in the experimentally determined rate equation.**

**Table 26.1 The effect of changing the concentration on the rate of reaction between calcium carbonate and hydrochloric acid**

| Mixture | [HCl(aq)]/ mol dm$^{-3}$ | Relative rate |
|---------|------------------------|---------------|
| A | 2.0 | 8 |
| B | 1.0 | 4 |
| C | 0.5 | 2 |
| D | 0.25 | 1 |

The definition of order of reaction states that the rate equation must be determined by experiment. It has nothing to do with the stoichiometric equation. For example, the equation for the reaction of dilute hydrochloric acid with marble chips is:

$$CaCO_3(s) + 2HCl(aq) \rightarrow CaCl_2(aq) + CO_2(g) + H_2O(l)$$

Yet, when experiments are done to find how the rate of reaction varies with the concentration of each reactant, it is discovered that the rate is directly proportional to the concentration of hydrochloric acid (Table 26.1). So, when the concentration doubles, so does the reaction rate. In this case:

$$rate \propto [HCl(aq)]^1$$

Therefore, the rate equation is:

$$rate = k[HCl(aq)]^1 = k[HCl(aq)]$$

(since power 1 is conventionally omitted)

This reaction is **first order** with respect to hydrochloric acid, because the concentration is raised to the power 1 in the experimentally determined rate equation.

**? TEST YOURSELF**

**H** Explain how the calcium carbonate/hydrochloric acid reaction can be followed.

**● EXAMPLE**

**Q** An experiment is carried out to determine the reaction kinetics of this reaction:

$$BrO_3^-(aq) + 6I^-(aq) + 6H^+(aq) \rightarrow 3I_2(s) + 3H_2O(l) + Br^-(aq)$$

The results obtained are given in Table 26.2.

**Table 26.2 Results of a study of the kinetics of the reaction between bromate(V) ions and iodide ions in acid solution**

| Mixture | [BrO$_3^-$(aq)]/mol dm$^{-3}$ | [I$^-$(aq)]/mol dm$^{-3}$ | [H$^+$(aq)]/mol dm$^{-3}$ | Relative initial rate |
|---------|------------------------------|--------------------------|--------------------------|----------------------|
| A | 0.10 | 0.60 | 0.60 | 1 |
| B | 0.20 | 0.60 | 0.60 | 2 |
| C | 0.20 | 1.20 | 0.60 | 4 |
| D | 0.20 | 0.60 | 1.20 | 8 |

Work out the orders of reaction with respect to each reactant, and hence write the overall rate equation.

**A** Comparing the relative rates for A and B shows that the rate in B doubles when [BrO$_3^-$(aq)] doubles. Note that the other concentrations remain the same. Therefore:

$$rate \propto [BrO_3^-(aq)]$$

Comparing the relative rates for B and C shows that [BrO$_3^-$(aq)] and [H$^+$(aq)] are the same in both experiments, while [I$^-$(aq)] doubles in C. This doubles the rate in C. Therefore:

$$rate \propto [I^-(aq)]$$

Comparing the relative rates of B and D shows that, while the other two concentrations stay the same, doubling [H$^+$(aq)] quadruples the rate in D. Therefore:

$$rate \propto [H^+(aq)]^2$$

Combining the three reaction rates gives:

$$rate \propto [BrO_3^-(aq)] \times [I^-(aq)] \times [H^+(aq)]^2$$

So, the rate equation is:

$$rate = k[BrO_3^-(aq)][I^-(aq)][H^+(aq)]^2$$

The reaction is **first order** with respect to [BrO$_3^-$(aq)], because in the rate equation the power to which this concentration is raised is 1.

It is also **first order** with respect to [I$^-$(aq)], because the power to which this concentration is raised is also 1.

The reaction is **second order** with respect to [H$^+$(aq)], because in the rate equation the power to which this concentration is raised is 2.

The overall order of the reaction is the sum of the individual orders, which gives 1+1+2 = 4.

Hence, the reaction is **fourth order**.

# DETERMINING THE RATE EQUATION BY THE INITIAL-RATE METHOD

In the previous worked example, the reaction rates quoted are initial reaction rates. These are the rates at the start of the reaction. They were found by doing four separate experiments and plotting rate curves for each experiment. The tangent at time zero is drawn for each experiment and the gradient of each tangent is determined. This was the procedure followed to find the initial rate in Fig 26.3, page 576.

In the next Example, the initial-rate method is used to find the rate equation and the rate constant of a reaction that is believed to occur in the exhaust gases of car engines to produce the atmospheric pollutant nitrogen dioxide. In each experiment, the concentration of only one reactant is varied.

> **? TEST YOURSELF**
>
> **I** When using the initial-rate method, chemists often stop the experiment long before the reaction is complete. Why is this?

## ● EXAMPLE

**Q** The reaction between nitrogen monoxide and oxygen was investigated using the initial rate method. The stoichiometric equation is:

$$2NO(g) + O_2(g) \rightarrow 2NO_2(g)$$

The results obtained at a particular temperature are given in Table 26.3 below.
**a)** Deduce the orders of the reaction with respect to $NO(g)$ and $O_2(g)$.
**b)** Write the expression for the rate equation.
**c)** Determine the overall order of the reaction.
**d)** Work out the value of the rate constant, showing its units.

**Table 26.3**

| Experiment | [NO(g)]/mol dm$^{-3}$ | [O$_2$(g)]/mol dm$^{-3}$ | Initial rate of production of NO$_2$(g)/mol dm$^{-3}$ s$^{-1}$ |
|---|---|---|---|
| 1 | $1.00 \times 10^{-3}$ | $3.00 \times 10^{-3}$ | $4.00 \times 10^{-4}$ |
| 2 | $1.00 \times 10^{-3}$ | $6.00 \times 10^{-3}$ | $8.00 \times 10^{-4}$ |
| 3 | $2.00 \times 10^{-3}$ | $3.00 \times 10^{-3}$ | $1.60 \times 10^{-3}$ |

**A a)** Comparing experiments 1 and 3, the concentration of NO(g) doubles, which quadruples the rate of reaction.

The concentration of $O_2(g)$ is constant. Therefore:

$$\text{rate} \propto [NO(g)]^2$$

So, the reaction is **second order** with respect to NO(g).

Comparing experiments 1 and 2, the concentration of $O_2(g)$ is doubled, which doubles the rate of reaction.

The concentration of NO(g) is the same in both experiments. Therefore:

$$\text{rate} \propto [O_2(g)]$$

So, the reaction is **first order** with respect to $O_2(g)$.

**b)** Since rate $\propto [NO(g)]^2[O_2(g)]$, the rate equation is:

$$\text{rate} = k[NO(g)]^2[O_2(g)]$$

Note that on this occasion the stoichiometric coefficients (the numbers of moles of each substance in the balanced equation) match those of the orders of the reaction. But do remember that the rate equation can only be determined by experiment.

**c)** The overall order of the reaction is the sum of the powers in the experimentally determined rate equation. This is 2+1 = 3. ([$O_2(g)$] is to the power 1.) So, the overall order is third.

**d)** The rate equation is: rate = $k[NO(g)]^2[O_2(g)]$ which gives:

$k = \dfrac{\text{rate}}{[NO(g)]^2[O_2(g)]}$ and units of $k = \dfrac{\text{mol dm}^{-3}\,\text{s}^{-1}}{(\text{mol dm}^{-3})^2(\text{mol dm}^{-3})}$

Now substitute one of the sets of values into the equation to determine $k$. For example, take those for experiment 1:

$k = \dfrac{4.00 \times 10^{-4}}{(1.00 \times 10^{-3})^2 \times 3.00 \times 10^{-3}}$ and units of $k = \text{dm}^6\,\text{mol}^{-2}\,\text{s}^{-1}$

Therefore:

$$k = 1.33 \times 10^5 \text{ dm}^6\,\text{mol}^{-2}\,\text{s}^{-1}$$

Note that dm$^6$ is placed first when quoting the units. This follows the convention whereby a unit raised to a positive power is placed first.

**J** The antiseptic chlorophenol is made using $SO_2Cl_2(g)$. The latter breaks down according to the equation:

$$SO_2Cl_2(g) \rightarrow SO_2(g) + Cl_2(g)$$

Three experiments were performed to determine the rate equation at a particular temperature. The results obtained are given in Table 26.4.

**Table 26.4 Kinetic data used to determine the rate equation for the decomposition of $SO_2Cl_2(g)$**

| $[SO_2Cl_2(g)]$/mol dm$^{-3}$ | Initial rate of formation of $Cl_2(g)$/mol dm$^{-3}$ s$^{-1}$ |
|---|---|
| 0.02 | $4.4 \times 10^{-7}$ |
| 0.04 | $8.8 \times 10^{-7}$ |
| 0.06 | $1.32 \times 10^{-6}$ |

Deduce the rate equation for this reaction and calculate the rate constant. stating the units of $k$.

## ZERO ORDER AND OTHER ORDERS

In the Examples so far, reactions have been first or second order with respect to each reactant. Another common order of reaction is **zero order** (sometimes called **zeroth order**):

> **A zero-order reaction is one in which the concentration of a reactant has no effect on the rate of reaction.**

For example, the decomposition of ammonia on a tungsten wire takes place according to the equation:

$$2NH_3(g) \xrightarrow{W} N_2(g) + 3H_2(g)$$

The rate equation for this reaction is:

$$\text{rate} = k[NH_3(g)]^0 = k \qquad (\text{since } [\text{reactant}]^0 = 1)$$

It does not matter what the concentration of ammonia is, the rate is always the same at a particular temperature.

Fig 26.11 illustrates the effect of reactant concentration on the rate of reaction for zero, first and second orders. For a zero-order reaction, the concentration has no effect on the rate, and hence the graph is a horizontal line. For a first-order reaction, the rate is directly proportional to the reactant concentration, which gives a straight-line graph. Since rate = $k$[reactant], the gradient is equal to the rate constant, $k$. For a second-order reaction, where:

$$\text{rate} \propto [\text{reactant}]^2$$

the graph is a curve.

The reaction of propanone ($CH_3COCH_3$) with iodine is interesting. Although in the stoichiometric equation hydrogen ions are not involved as reactants, they catalyse the reaction. As the reaction itself produces these ions, the reaction is said to be **auto-catalysed** (self-catalysed):

$$CH_3COCH_3(aq) + I_2(aq) \rightarrow CH_3COCH_2I(aq) + H^+(aq) + I^-(aq)$$

The rate equation for this reaction can be written:

$$\text{rate} = k[CH_3COCH_3(aq)]^1[H^+(aq)]^1[I_2(aq)]^0$$
$$= k[CH_3COCH_3(aq)] [H^+(aq)] \qquad (\text{since } [\ ]^0 = 1)$$

Since the concentration of iodine has no effect on the rate, the reaction is zero order with respect to iodine. But it is first order with respect to both propanone and hydrogen ions.

An order of reaction need not be zero, first or second order. An order of reaction may be a fraction, but this aspect lies outside the scope of this book.

**Fig 26.11** The effect of reactant concentration on the rate of zero, first- and second-order reactions

**K** What is the overall order for the reaction of iodine with propanone?

**L** The rate equations are given for the following reactions. Give the order with respect to each reactant and the overall order for the reaction.
**a)** $BrO_3^-(aq) + 3SO_3^{2-}(aq) \rightarrow Br^-(aq) + 3SO_4^{2-}(aq)$
rate = $k[BrO_3^-(aq)][SO_3^{2-}(aq)]$
**b)** $2H_2(g) + 2NO(g) \rightarrow 2H_2O(g) + N_2(g)$
rate = $k[NO(g)]^2[H_2(g)]$
**c)** $NO_2(g) + CO(g) \rightarrow NO(g) + CO_2(g)$
rate = $k[NO_2(g)]^2$
Work out the units of the rate constant for each reaction. In all cases, concentration is measured in mol dm$^{-3}$, and rate in mol dm$^{-3}$ s$^{-1}$.

# HALF-LIVES AND FIRST-ORDER REACTIONS

The half-life for a chemical reaction is defined as follows:

> **The half-life of a chemical reaction is the time taken for the concentration of a reactant to decrease to half of its initial value.**

In the case of first-order reactions, the half-life is constant. This means that, whatever the starting concentration of a reactant, it will always take the same time for this concentration to be halved.

The decomposition of hydrogen peroxide using manganese(IV) oxide catalyst is a first-order reaction (Fig 26.12):

$$H_2O_2(aq) \rightarrow H_2O(l) + O_2(g)$$

$$rate = k[H_2O_2(aq)]$$

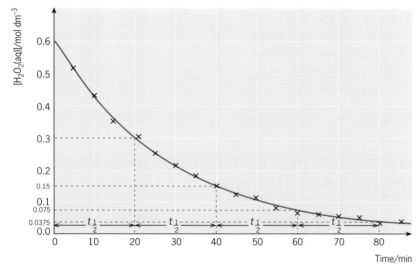

Four successive half-lives from the graph in Fig 26.12 are given in Table 26.5. Note that the four half-lives are all constant. This is a characteristic of first-order reactions.

Using successive half-lives may involve very small concentrations, which can lead to inaccuracies. In Fig 26.12, the time taken for other concentrations of $H_2O_2(aq)$ to decrease by half (e.g. 0.5 mol dm$^{-3}$ falling to 0.25 mol dm$^{-3}$) could have been used.

Therefore, half-lives can be used to identify a first-order reaction from a single kinetics experiment. If the half-life of a reactant is known, the rate constant of a first-order reaction can be calculated using the following relationship between the half-life and the rate constant:

$$k = \frac{0.693}{t_{\frac{1}{2}}}$$

You do not have to know how to derive this equation, but you should note its usefulness in calculating rate constants.

> ### ? TEST YOURSELF
>
> **M** How can the decomposition of hydrogen peroxide reaction be followed?
> Hint: One of the products is a gas.

**Fig 26.12** The change in concentration of hydrogen peroxide with time in the catalytic decomposition of hydrogen peroxide, using $MnO_2$ as the catalyst at 20 °C

**Table 26.5 Four successive half-lives from the decomposition of hydrogen peroxide**

| Initial concentration $H_2O_2(aq)$/mol dm$^{-3}$ | Half-life $t_{\frac{1}{2}}$/min |
| --- | --- |
| 0.60 | 20 |
| 0.30 | 20 |
| 0.15 | 20 |
| 0.075 | 20 |

> ### ? TEST YOURSELF
>
> **N** The determination of the catalytic decomposition of hydrogen peroxide as a first-order reaction can be made with a single experiment (this page). Another method used to study the kinetics of this reaction involves initial rates. Explain how you would study this reaction using the initial-rates method. How would you show that the reaction is first order?

> ### ? TEST YOURSELF
>
> **O** Radioactive decay is another first-order reaction, because successive half-lives are constant. The half-life of sodium-24 is 15 hours.
> **a)** Starting with 1 kg of a radioactive sample of sodium-24, how much of it will be left after 30 hours?
> **b)** What is the rate constant for this decay?

**Fig 26.13** Concentration–time graphs for zero, first- and second-order reactions

## CONCENTRATION–TIME GRAPHS

When a reaction is zero order with respect to a reactant, the reactant concentration has no effect on the rate of reaction:

$$\text{rate} = k[\text{reactant}]^0 = k \qquad (\text{since } [\text{reactant}]^0 \text{ is } 1)$$

In Fig 26.13, the concentration–time graph for a zero-order reaction shows a constant decline in concentration as the reaction proceeds.

For a first-order reaction:

$$\text{rate} = k[\text{reactant}]^1 = k[\text{reactant}] \qquad (\text{since } [\text{reactant}]^1 \equiv [\text{reactant}])$$

As already stated, irrespective of the starting concentration, the half-life has a fixed value, so the concentration–time graph has the characteristic shape shown in Fig 26.13.

For a second-order reaction:

$$\text{rate} = k[\text{reactant}]^2$$

When the starting rate is the same as that of a first-order reaction, the concentration–time graph approaches zero much more slowly, as Fig 26.13 shows. This is because in a second-order reaction the half-life increases as the concentration diminishes, unlike in a first-order reaction for which the half-life is independent of the concentration.

## 4 RATE EQUATIONS IN THE DETERMINATION OF REACTION MECHANISMS

As stated earlier, one of the reasons for studying the kinetics of a reaction is to obtain the information that enables chemists to decide how the reaction occurs – the **reaction mechanism**. Although some reactions do take place in one step, most proceed in a series of steps that lead to the formation of products. In each step, bonds are broken or made.

Rate equations provide evidence of what the reaction mechanism might be. For example, knowledge of the likely reaction mechanism of ozone depletion in the upper atmosphere by chlorofluorocarbons (CFCs) proved to be crucial in helping governments to agree to ban their manufacture.

For a sequence of bond-breaking and bond-making steps to occur, the reactants must collide with sufficient energy. This is the **collision theory** of chemical reactions, which is discussed in more detail in Chapter 8 (pages 179 and 184). Each step in a reaction mechanism usually involves one or two molecules.

$$H_2C \overset{\overset{\textstyle CH_2}{\diagup\diagdown}}{\underline{\qquad}} CH_2 \longrightarrow CH_3CH = CH_2$$

**Fig 26.14** The conversion of cyclopropane into propene

Consider a proposed single-step reaction mechanism: the high-temperature conversion of cyclopropane into propene (Fig 26.14). The experimentally derived rate equation for this reaction is:

$$\text{rate} = k[\text{cyclopropane}]$$

The reaction is first order. So, given the single-step mechanism, only one molecule of cyclopropane can be involved. Hence, the energy given to this molecule must cause it to shake itself apart. This is called a **unimolecular step**. The **molecularity** of a step is the number of species (atoms, molecules or ions) involved.

THERE IS MORE ABOUT THE REACTION MECHANISM OF OZONE DEPLETION BY CFCS IN THE ASSIGNMENT ON PAGE 593.

### ? TEST YOURSELF

**P** The conversion of cyclopropane into propene is sometimes called an isomerization reaction. Can you explain why?
Hint: If you are not sure, refer to page 198.

It is unusual to find reactions that have just one step. Most involve at least two, even when they are first-order reactions. For example, the stoichiometric equation for the hydrolysis of 2-bromo-2-methylpropane is given on the right, and its rate equation is:

$$\text{rate} = k[(CH_3)_3CBr]$$

The stoichiometric equation gives no clue as to the rate equation, which always has to be determined by experiment. A proposed mechanism that fits this rate equation is:

$$(CH_3)_3CBr \xrightarrow{\text{slow}} (CH_3)_3C^+ + Br^-$$
$$(CH_3)_3C^+ + OH^- \xrightarrow{\text{fast}} (CH_3)_3COH$$

It is the slower step that governs the rate of reaction. It is called the **rate-determining step**. Suppose you are throwing a party and send out invitations. You have three people helping you. You write the invitations, which is the slow step. One friend folds them, which does not take long, while another friend puts them in envelopes. The last person addresses the envelopes and sticks on the stamps. It is your step that takes the longest. It does not matter how fast your friends work, the rate is determined by the slowest step. Writing the invitations is the rate-determining step. It is the same with chemical reactions: the slowest step determines the rate of the entire reaction.

The rate-determining step for this hydrolysis reaction is unimolecular because it involves only one molecule. The fast step may be referred to as **bimolecular** because it involves two reacting species. Although the stoichiometry of the overall reaction does not give the rate equation, the stoichiometry of the rate-determining step of the reaction can.

In the rate-determining step:

$$(CH_3)_3CBr \xrightarrow{\text{slow}} (CH_3)_3C^+ + Br^-$$

the rate of reaction is governed by the breaking of a chemical bond in the reactant molecule. As only this molecule is involved, the rate is proportional to the concentration. The greater the concentration, the faster the reaction. So, the rate equation for this step is:

$$\text{rate} = k[(CH_3)_3CBr]$$

which is also the rate of the overall reaction. The overall order is zero for $OH^-$ ions, since these react after the rate-determining step and their concentration therefore has no effect on the reaction rate. Reacting species that occur before the slow step may feature in the overall rate equation.

This reaction is sometimes called an $S_N1$ reaction. S signifies that it is a substitution reaction, N that it is nucleophilic, and 1 that it is first order with one molecule involved in the rate-determining step. For more information on this reaction, turn to page 275.

Whatever reaction mechanism is proposed, the sum of its steps must equal the overall stoichiometric equation. In the case of the hydrolysis of 2-bromo-2-methylpropane, it does (Fig 26.15).

## ? TEST YOURSELF

**Q** $NO_2F$ is an explosive compound. The mechanism for its production is thought to be:

$$NO_2 + F_2 \xrightarrow{\text{slow}} NO_2F + F$$
$$NO_2 + F \xrightarrow{\text{fast}} NO_2F$$

What is the overall stoichiometric equation for this reaction?

$(CH_3)_3CBr \xrightarrow{\text{slow}} (CH_3)_3C^+ + Br^-$

$(CH_3)_3C^+ + OH^- \xrightarrow{\text{fast}} (CH_3)_3COH$

_____

$(CH_3)_3CBr + OH^- \longrightarrow (CH_3)_3COH + Br^-$

**Fig 26.15** Hydrolysis of 2-bromo-2-methylpropane

## ? TEST YOURSELF

**R** 2-bromo-2-methylpropane is called a tertiary halogenoalkane. Primary halogenoalkanes, such as 1-bromobutane, $CH_3CH_2CH_2CH_2Br$, are believed to react by the mechanism displayed in Fig 26.16.

**Fig 26.16** The $S_N2$ reaction mechanism for the hydrolysis of 1-bromobutane

The rate equation is:  $\text{rate} = k[CH_3CH_2CH_2CH_2Br][OH^-]$

Why do you think this is called an $S_N2$ reaction?

## WORKING OUT ANOTHER MECHANISM

The reaction of carbon monoxide with nitrogen dioxide is believed to take place in the exhaust gases of car engines. To determine the mechanism, first write down the stoichiometric equation:

$$NO_2(g) + CO(g) \rightarrow NO(g) + CO_2(g)$$

Note that this equation gives no information about how the reaction occurs, so the next stage is to carry out experiments to determine the overall rate equation. This has been found to be:

$$\text{rate} = k[NO_2]^2$$

The reaction is zero order with respect to carbon monoxide, which means it does not take part in the rate-determining step.

Now a mechanism has to be proposed that is both consistent with the rate equation and, when the steps are added together, gives the overall stoichiometric equation. This is really an educated guess:

$$NO_2 + NO_2 \xrightarrow{\text{slow}} N_2O_4$$
$$\underline{N_2O_4 + CO \xrightarrow{\text{fast}} NO + CO_2 + NO_2}$$
$$NO_2(g) + CO(g) \rightarrow NO(g) + CO_2(g)$$

The slower step is the rate-determining step. This involves a collision between two nitrogen dioxide molecules, so the rate equation for this step is rate = $k[NO_2]^2$, and the overall rate equation is also rate = $k[NO_2]^2$. Carbon monoxide appears in the fast step, so its concentration does not affect the rate. Therefore, the proposed mechanism is consistent with the rate equation and the steps when added together give the overall equation.

However, when this reaction is analysed, a short-lived intermediate, $NO_3$, is discovered. Also, no $N_2O_4$ is found. This means that the educated guess was incorrect and another mechanism must be proposed, such as:

$$NO_2 + NO_2 \xrightarrow{\text{slow}} NO_3 + NO$$
$$\underline{NO_3 + CO \xrightarrow{\text{fast}} NO_2 + CO_2}$$
$$NO_2(g) + CO(g) \rightarrow NO(g) + CO_2(g)$$

This still gives a bimolecular rate-determining step and the rate equation is consistent with this. Also, the overall stoichiometric equation is obtained when the two steps are added together. This is a good example of a mechanism being proposed that is consistent with the rate equation, but is not consistent with other experimental evidence and therefore has to be amended accordingly. No-one can prove that this mechanism *is* correct. What can be said is that, based on current evidence, it is thought to be likely.

Sometimes mechanisms are proposed that have a fast step followed by a slow step. In such cases, a reaction intermediate builds up and could be sufficiently long-lived to be detected. This would give another clue to the mechanism.

## PREDICTING RATE EQUATIONS FROM STEPS IN THE REACTION MECHANISM

Table 26.6 shows three possible steps in reaction mechanisms, together with the rate equations that can be derived from them. The molecularity (number of molecules) of each step and the order are the same. So when the step is unimolecular, it is a first-order reaction. Similarly, a bimolecular reaction gives a second-order reaction. On very rare occasions, a step that contains three molecules is thought to be involved, but this requires three particles to collide simultaneously, which is improbable. When these are the slow steps, they govern the overall rate of the reaction. Therefore, the overall rate equation can be predicted to be the same as this rate-determining step.

 **REMEMBER THIS**

The rate equation can be predicted from the rate-determining step, but not from the overall stoichiometric equation.

**Table 26.6 Possible steps in reaction mechanisms**

| Step | Molecularity | Rate equation |
|---|---|---|
| A → products | Unimolecular | Rate = $k[A]$ |
| A + A → products | Bimolecular | Rate = $k[A]^2$ |
| A + B → products | Bimolecular | Rate = $k[A][B]$ |

---

**! SCIENCE FEATURE Explosive reactions and Nobel Prizes**

For centuries, gunpowder was the only high explosive. Then, in 1846, the compounds nitrocellulose (gun cotton) and nitroglycerine were made. Both compounds have very low activation energies for their decomposition and so are highly unstable. (For information about activation energies see Chapter 8, pages 177 to 180.)

$$4 \begin{matrix} CH_2O-NO_2 \\ | \\ CHO-NO_2 \\ | \\ CH_2O-NO_2 \end{matrix} (l) \longrightarrow \begin{matrix} 12CO_2(g) + 10H_2O(g) \\ + 6N_2(g) + O_2(g) \end{matrix}$$

nitroglycerine

**Fig 26.18(a)** Four moles of liquid give 29 moles of gases. So it is the expansion of the hot gases from a small volume of liquid that produces the explosion

**Fig 26.17** Alfred Nobel (1833–1896) in his laboratory

**Fig 26.18(b)** An engraving from *The Illustrated London News* in 1884 publicizing dynamite made at the Nobel Explosives Company in Ardeer, Ayrshire

**Fig 26.18(c)** Dynamite is sometimes used to demolish unwanted buildings and other structures, such as these cooling towers. The explosive has to be correctly positioned to ensure that the building will collapse in on itself and not be scattered over the surrounding area

The Swedish Nobel family were manufacturers of explosives. In 1863, Alfred Nobel invented a detonator that would set off liquid nitroglycerine. However, in 1864 a nitroglycerine explosion at the Nobel factory killed Alfred's younger brother, Emil, and four other people. Alfred set about making nitroglycerine safe, and within four years he had produced dynamite. (The name is derived from *dynamis*, the Greek for 'power'.)

In dynamite, the nitroglycerine is absorbed by an inert solid called kieselguhr, which renders it stable until it is set off by a detonator. In 1875, Nobel produced gelignite – an even more powerful explosive. This, too, contained nitroglycerine. This time it was mixed with nitrocellulose in a gel. These explosives were soon put to use in civil engineering projects, such as blasting routes for new roads and excavating the ground for canals and the foundations of buildings.

Alfred Nobel was convinced that the only way to stop wars was to produce an explosive so powerful that no-one would dare use it. In this he did not succeed, but in his will he left most of the vast fortune he had amassed from selling explosives to the establishment of five prizes to be awarded annually for outstanding achievements in chemistry, physics, physiology or medicine, literature and peace. In 1969, a sixth prize was added for economics. These Nobel prizes are now acknowledged to be the world's most prestigious accolade. They cannot be awarded posthumously, and no more than three people can share each prize.

# 5 EFFECT OF TEMPERATURE ON THE RATE CONSTANT

As Fig 26.19 shows, increasing the temperature increases the proportion of molecules that have the minimum activation energy needed for them to react when they collide. This means that the rate of reaction increases with temperature. For every 10 °C rise, the rate of most reactions approximately doubles.

TO REMIND YOURSELF ABOUT ENERGY DISTRIBUTIONS IN MOLECULES OF A GAS, SEE PAGE 186 IN CHAPTER 8.

✔ **REMEMBER THIS**

The rate equation can only be determined by doing experiments.

**SEE QUESTIONS 3 AND 4.**

**Fig 26.19** Maxwell–Boltzmann distribution curve at temperatures $T_1$ and $T_2$

At the higher temperature, $T_2$, a greater proportion of molecules has at least the activation energy

Consider the reaction:

$$A + B \rightarrow C + D$$

for which the rate equation is:

$$\text{rate} = k[A]^m[B]^n$$

If the concentrations of A and B are kept constant, then the rate constant must increase with increasing temperature.

This can be clearly seen if we consider how the rate constant varies for the decomposition of hydrogen iodide (Table 26.7). The stoichiometric equation is:

$$2HI(g) \rightarrow H_2(g) + I_2(g)$$

with a rate equation:

$$\text{rate} = k[HI(g)]^2$$

From Table 26.7, the rate constant increases with increasing temperature.

**Table 26.7 Variation of the rate constant over a range of temperatures**

| Temperature/K | Rate constant, k/ dm³ mol⁻¹ s⁻¹ |
|---|---|
| 550 | $2.9 \times 10^{-7}$ |
| 600 | $7.2 \times 10^{-6}$ |
| 650 | $1.1 \times 10^{-4}$ |
| 700 | $1.2 \times 10^{-3}$ |
| 750 | $1.1 \times 10^{-2}$ |
| 800 | $7.1 \times 10^{-2}$ |

? **TEST YOURSELF**

**U** If a 10 °C rise in temperature doubles the rate of a chemical reaction at constant concentrations, what is the effect on the rate constant, $k$?

**SEE QUESTIONS 5, 6 AND 7.**

▶ **EXTENSION The Arrhenius equation**

The Swedish chemist Svante Arrhenius was the first to discover a mathematical relationship between temperature and the rate constant. The **Arrhenius equation** is:

$$k = Ae^{-E_a/RT}$$

where $k$ is the rate constant.

$A$, also a constant, is the collision frequency. However, when molecules collide, they must have the correct orientation to react, a factor that includes the collision frequency. $A$ is sometimes called the Arrhenius constant.

The term $e^{E_a/RT}$ gives the fraction of collisions that have the minimum activation energy at a particular temperature, where $E_a$ is the activation energy in joules, $R$ is the gas constant 8.31 J K⁻¹ mol⁻¹ (see page 125), $T$ is the absolute temperature in K and e means it is an exponential relationship. It expresses the condition that an increase in the temperature increases the reaction rate exponentially.

Another way to express the Arrhenius equation is to take ▶

natural logarithms of both sides:

$$\ln k = \ln A - \frac{E_a}{RT}$$

A straight-line graph of $\ln k$ against $1/T$ can be plotted, as in Fig 26.20. Its gradient is $-E_a/R$, so $E_a$ can be determined.

$$-\frac{E_a}{R} = \text{gradient}$$
$$= \frac{\Delta \ln k}{\Delta 1/T}$$

**Fig 26.20** The activation energy can be determined from the gradient of this graph

**Fig 26.21** Svante Arrhenius (1859–1927). As well as for his equation, he is famous for formulating the idea that electrolytes dissociate in solution to give charged particles

**Fig 26.22** Sir George Porter (born 1920)

Sir George Porter was born in 1920 in Stainforth, Yorkshire. He was studying for his chemistry degree at Leeds University when World War II broke out. Along with other undergraduates, he was asked to take up radio physics during the last year or so of his course. At the time, he was told only that it had 'very high priority'. On graduating in 1941, when radar was in its infancy, he became a radar officer in the Royal Navy.

Sir George has said that his time in the Navy convinced him that he wanted to do scientific research, and in 1945 he began a research career at Cambridge University. He was research student to Professor Ronald Norrish, who did much to establish the study of reaction kinetics.

Sir George was given the task of investigating short-lived free radicals (highly reactive species with at least one unpaired electron). In this research, Norrish had been using an old army searchlight. During a visit to a factory to collect a new bulb for the searchlight, Sir George saw flashbulbs being made. This was the turning point in his research, when he realized that the pulse techniques he had used in radar could be applied to the problem he had been set.

Together with Norrish, he used rapid flashes of light to investigate species that were very short-lived. Another pulse of light fired shortly after the first was used to identify the reactive intermediates by absorption spectroscopy.

By 1950, Sir George could identify reactive intermediates that lasted for a microsecond. In 1967 he observed intermediates that lasted for only nanoseconds ($10^{-9}$ s) through using laser beam flashes. By 1975, he had reduced this to picoseconds ($10^{-12}$ s). In 1966 he received the Nobel Prize together with Norrish and another chemist, Max Eigen, for his work on extremely fast chemical reactions.

Sir George became Director of the Royal Institution in 1967, where he did much to continue the work of Michael Faraday in making science accessible to young people through the Christmas Lectures and other promotional work. He was knighted in 1972.

## SUMMARY

After studying this chapter, you should know that:
- Reaction kinetics is the study of rates of reactions.
- The rate of a reaction is the change in concentration of a product, or the change in concentration of a reactant, per unit of time.
- Concentration is indicated by square brackets, as in [reactant].
- The gradient of a graph of reactant (or product) concentration against time gives the rate of the reaction at any particular instant in time. It is usually worked out using tangents. The initial rate is the gradient at time zero.
- Anything that changes during a reaction, which can be measured, may be used to determine a reaction rate provided that it is proportional in some way to the concentration of a particular reactant or product.
- Rate of reaction = $k[A]^m[B]^n$

  is the rate equation for the reaction:

  $$A + B \rightarrow C + D$$

  where $m$ and $n$ are the orders of the reaction with respect to A and B, and $k$ is the rate constant.

- The overall rate equation must be experimentally determined. It cannot be worked out from the stoichiometric equation.
- When a reaction is zero order with respect to a reactant, the reactant's concentration has no effect on the rate of the reaction.
- The half-life of a chemical reaction is the time taken for the concentration of a reactant to fall to half of its initial value. First-order reactions have constant half-lives.
- The series of steps involved in a reaction is called the reaction mechanism.
- Rate equations provide evidence for proposed reaction mechanisms.
- The rate-determining step is the slowest step in a reaction.
- A unimolecular rate-determining step produces a first-order reaction, while a bimolecular rate-determining step produces a second-order reaction.
- Increasing the temperature increases the rate constant. The relationship between the rate constant and temperature is given by the Arrhenius equation, $k = Ae^{-E_a/RT}$.

## ? EXAM QUESTIONS

**1** [For **a(i)** refer to Chapter 8]
High-altitude aircraft produce nitrogen monoxide molecules, which react with ozone in the upper atmosphere. The nitrogen dioxide produced in this reaction can then react with oxygen atoms, which are also present in the atmosphere at this level.

1.  $NO + O_3 \rightarrow NO_2 + O_2$      Fast
2.  $NO_2 + O \rightarrow NO + O_2$      Slow

**a)** **(i)** State two conditions that determine whether a reaction occurs when two molecules collide.
**(ii)** Write an equation for the overall reaction occurring here.
**(iii)** Explain the role of the nitrogen monoxide in the reaction.
**b)** Table 26.Q1 shows how the initial rate of reaction 2 above varies with the concentrations of $NO_2$ and O.

**Table 26.Q1**

| $[NO_2]$/mol dm$^{-3}$ | [O]/mol dm$^{-3}$ | Initial rate/mol dm$^{-3}$ s$^{-1}$ |
|---|---|---|
| $4.33 \times 10^{-12}$ | $1.07 \times 10^{-14}$ | $2.59 \times 10^{-16}$ |
| $2.17 \times 10^{-12}$ | $5.35 \times 10^{-15}$ | $6.48 \times 10^{-17}$ |
| $2.17 \times 10^{-12}$ | $1.07 \times 10^{-14}$ | $1.30 \times 10^{-16}$ |

**(i)** Deduce the order of reaction with respect to $NO_2$ and O, showing your reasoning clearly.

**(ii)** Write a rate equation for the reaction.
**(iii)** Use the equation in **(ii)** and the first line of data in Table 26.Q1 to find the value of the rate constant, $k$, for the reaction. Include the units of $k$.

OCR 6820 General and Physical Chemistry March 2000 Q4

**2** [Refer to Chapter 8]
Hydrogen reacts with iodine at 450 °C to give hydrogen iodide. The results from several experiments designed to find the rate equation for the reaction are given In Table 26.Q2.

**Table 26.Q2**

| Initial $[I_2]$/mol dm$^{-3}$ | Initial $[H_2]$/mol dm$^{-3}$ | Relative initial rate |
|---|---|---|
| 0.001 | 0.001 | 1 |
| 0.003 | 0.001 | 3 |
| 0.001 | 0.004 | 4 |

**(a)** **(i)** Find the order of reaction with respect to each of the reactants.
**(ii)** Write the rate equation for the reaction.
**(iii)** Explain why rate equations cannot be written from the stoichiometric (chemical) equation for the reaction, but must be found experimentally.
**(iv)** What do you understand by the **rate-determining step** in a chemical reaction?

**(v)** Illustrate your answer to **(iv)** by writing out any reaction mechanism of your choice.

**(b)** Reaction rates generally increase with an increase in temperature for two reasons.

**(i)** State what these reasons are.

**(ii)** Sketch a graph [Number of molecules with a given energy on the vertical axis and Energy on the horizontal axis] of the distribution of molecular energies at a given temperature $T_1$ and at a higher temperature $T_2$ and hence explain the increase in the rate.

**(iii)** By use of the graph, or otherwise, explain the effect of a catalyst on the rate of a chemical reaction.

Edexcel Chemistry Specimen Unit Test 5 Jan 2000 Q2

**3** [For **(b)** refer to Chapter 8]

**a)** Benzene diazonium chloride, $C_6H_5N_2Cl$, in acidic solution is stable below 5 °C, but decomposes rapidly in a first-order reaction when heated to 50 °C.
$$C_6H_5N_2Cl(aq) + H_2O(l) \rightarrow C_6H_5OH(aq) + HCl(aq) + N_2(g)$$
Describe an experiment that could be used to follow this reaction. Your answer should include a diagram of the apparatus that you would use and a statement of the measurements that you would make.

**b)** The decomposition of hydrogen peroxide is also a first-order reaction:
$$2H_2O_2(aq) \rightarrow 2H_2O(l) + O_2(g)$$
At 50 °C it is very much slower than the decomposition of benzene diazonium chloride of the same concentration.

**(i)** Write the rate equation for the decomposition of hydrogen peroxide.

**(ii)** Comment on the **relative** values of the activation energies and the rate constants for the decomposition reactions above.

**(iii)** Explain, in terms of collision theory, the effect on reaction rates of increasing the temperature to 60 °C.

**(iv)** The decomposition of hydrogen peroxide is catalysed by iodide ions. Explain why reactions are speeded up by a catalyst.

**c)** Nuclear decay is also a first-order process. $^{131}I$ is a radioactive isotope and it can be used for thyroid and brain-cell imaging. It has a half-life of 8.0 days. If a sample that contains $1.2 \times 10^{-5}$ g of $^{131}I$ is used, what mass of $^{131}I$ will be left after 32 days?

**d)** The iodination of propanone:
$$CH_3COCH_3 + I_2 \xrightarrow{\text{acid catalyst}} CH_2ICOCH_3 + HI$$
was studied with an excess of propanone and an acid catalyst. The variation of concentration of iodine as a function of time is shown in Fig 26.Q3. What is the order of this reaction with respect to iodine? Give your reasoning.

**Fig 26.Q3**

Edexcel 6082/P.01 Module Test 2 Jan 2000 Q2

**4** Table 26.Q4 shows the results of three experiments to investigate the rate of the reaction between compounds **A** and **B**. All three experiments were carried out at the same temperature.

**Table 26.Q4**

|  | Experiment 1 | Experiment 2 | Experiment 3 |
|---|---|---|---|
| Initial concentration of **A**/mol dm$^{-3}$ | $0.50 \times 10^{-3}$ | $1.00 \times 10^{-3}$ | $1.50 \times 10^{-3}$ |
| Initial concentration of **B**/mol dm$^{-3}$ | $1.00 \times 10^{-3}$ | $1.00 \times 10^{-3}$ | $1.50 \times 10^{-3}$ |
| Initial rate/mol dm$^{-3}$ s$^{-1}$ | $0.15 \times 10^{-3}$ | $0.60 \times 10^{-3}$ | $1.35 \times 10^{-3}$ |

Use the data in Table 16.Q4 to deduce the order of reaction with respect to **B** and the order of reaction with respect to **A**. Hence write an overall rate equation for the reaction.

**b)** In a reaction between compounds **C** and **D**, the order of reaction with respect to **C** is two and the order of reaction with respect to **D** is one. Given that the initial rate of reaction is $4.00 \times 10^{-4}$ mol dm$^{-3}$ s$^{-1}$ when the initial concentration of **C** is $2.00 \times 10^{-2}$ mol dm$^{-3}$ and the initial concentration of **D** is $5.00 \times 10^{-2}$ mol dm$^{-3}$, calculate the value of the rate constant for this reaction at this temperature and deduce its units.

**c)** The rate equation for the decomposition of a compound **E** has a rate constant with the unit s$^{-1}$. The rate constant is $4.31 \times 10^3$ s$^{-1}$ at 700 K and $1.78 \times 10^4$ s$^{-1}$ at a temperature $T$. Use this information to deduce the overall order of reaction and whether temperature $T$ is greater or smaller than 700 K.

AQA CH03 Kinetics and Organic Chemistry June 2000 Q2

**5** Bromate(V) ions react with bromide ions in acid solution according to the following equation:
$$BrO_3^-(aq) + 5Br^-(aq) + 6H^+(aq) \rightarrow 3Br_2(aq) + 3H_2O$$
A series of experiments was carried out to find the order of reaction by measuring the initial rates of formation of bromine for different concentrations of

## ? EXAM QUESTIONS

reactants. The results of the experiments are given in Table 26.Q5.

**Table 26.Q5**

| pH | $[Br^-]$/mol dm$^{-3}$ | $[BrO_3^-]$/mol dm$^{-3}$ | Initial rate/mol dm$^{-3}$ s$^{-1}$ |
|---|---|---|---|
| 1.00 | $2.50 \times 10^{-3}$ | $1.25 \times 10^{-3}$ | $2.00 \times 10^{-6}$ |
| 1.00 | $1.25 \times 10^{-3}$ | $2.50 \times 10^{-3}$ | $2.50 \times 10^{-6}$ |
| 1.00 | $2.50 \times 10^{-3}$ | $6.25 \times 10^{-3}$ | $1.00 \times 10^{-6}$ |
| 2.00 | $2.50 \times 10^{-3}$ | $1.25 \times 10^{-3}$ | $2.00 \times 10^{-6}$ |

**a)** Define the term *order of reaction*.
**b)** Deduce the order of reaction with respect to Br$^-$, BrO$_3^-$ and H$^+$, explaining your reasoning clearly.
**c)** Write a rate equation for the reaction, and use it to calculate the value of the rate constant, $k$, including the units.
**d)** **(i)** Draw a graph [with $k$ on vertical axis, and $T$ on horizontal axis] to show how the value of the rate constant, $k$, varies with temperature, $T$.
**(ii)** Explain why the value of the rate constant varies with temperature.
**e)** The reaction of bromate(V) ions with bromide ions above is thought to proceed via the following reaction mechanism.

Step 1   H$^+$ + Br$^-$ → HBr        Fast
Step 2   H$^+$ + BrO$_3^-$ → HBrO$_3$     Fast
Step 3   HBr + HBrO$_3$ → HBrO + HBrO$_2$   Slow
Step 4   HBrO$_2$ + HBr → 2HBrO      Fast
Step 5   HBrO + HBr → H$_2$O + Br$_2$    Fast

Show how this reaction mechanism is consistent with the experimentally determined rate equation.
*OCR 6824/8 Further Physical March 2000 Q5*

**6** Dinitrogen oxide, N$_2$O, is a moderately stable compound that is used as an anaesthetic. However, at high temperatures it decomposes according to the following equation:

$$N_2O(g) \rightarrow N_2(g) + \tfrac{1}{2}O_2(g) \quad \Delta H = -76 \text{ kJ mol}^{-1}$$

**a)** The uncatalysed decomposition reaction is second order.
**(i)** Give the rate expression for this reaction.
**(ii)** At 1073 K the rate constant (velocity constant) for the reaction is 0.30 dm$^3$ mol$^{-1}$ s$^{-1}$. Calculate the rate of the reaction for 0.10 mol dm$^{-3}$ N$_2$O at 1073 K, specifying the units.
**(iii)** At 1148 K the rate of the same reaction increases by a factor of 5.9 (590%). Calculate the value for the rate constant at 1148 K.
**(iv)** Explain qualitatively why the value for the rate constant increases as the temperature increases.
**b)** The decomposition of dinitrogen oxide using gold as a catalyst is a first-order reaction with an activation energy of 121 kJ mol$^{-1}$. The activation

energy of uncatalysed reaction is 240 kJ mol$^{-1}$.
**(i)** Draw a diagram showing the potential energy changes during the uncatalysed and catalysed reaction using the same axes.
**(ii)** Why is the half-life of the catalysed reaction constant?
*IBO Chemistry Higher Level Paper 2 Nov 1998 Q4*

**7** The decomposition of dinitrogen pentoxide, N$_2$O$_5$, has been studied in solution. When an organic solvent is used, one product (nitrogen dioxide) is retained in solution, while oxygen gas is given off as the other product.
The equation for the overall process is:
$$N_2O_5 \rightarrow 2NO_2 + \tfrac{1}{2}O_2$$
**a)** Draw a diagram of the apparatus that could be used to follow the rate of this reaction.
**b)** The rate equation for the reaction is:
$$\text{rate} = k[N_2O_5]$$
A suggested mechanism for the reaction is:
1st step       N$_2$O$_5$ → NO$_2$ + NO$_3$
2nd step   NO$_2$ + NO$_3$ → NO + O$_2$ + NO$_2$
3rd step    NO + NO$_3$ → 2NO$_2$
**(i)** What is the order of reaction with respect to N$_2$O$_5$?
**(ii)** What are the units of the rate constant $k$?
**(iii)** Suggest which is the rate-determining step. Explain your reasoning.
**c)** The rate constant $k$ was measured at different temperatures.
**(i)** Table 26.Q7 shows data for the temperature and rate constant. Most of the data have been converted into values of 1/temperature and ln $k$. Complete the table by calculating the missing values.

**Table 26.Q7**

| Temperature/K | 1/temperature/K$^{-1}$ | $k$ | ln $k$ |
|---|---|---|---|
| 283 | $3.53 \times 10^{-3}$ | $3.83 \times 10^{-6}$ | −12.47 |
| 293 | $3.41 \times 10^{-3}$ | $1.71 \times 10^{-5}$ | −10.97 |
| 303 | $3.30 \times 10^{-3}$ | $6.94 \times 10^{-5}$ | −9.58 |
| 313 | $3.19 \times 10^{-3}$ | $2.57 \times 10^{-4}$ | −8.27 |
| 323 | | $8.78 \times 10^{-4}$ | |

**(ii)** Plot a graph of ln $k$ on the vertical axis against 1/temperature on the horizontal axis.
**(iii)** The relationship between rate constant, $k$, and temperature is given by the equation:

$$\ln k = C - \frac{E_A}{R}\,(1/T)$$

where $T$ is the temperature in Kelvin, $E_A$ is the activation energy in J mol$^{-1}$, and $C$ and $R$ are constants, $R = 8.31$ J K$^{-1}$ mol$^{-1}$ (you do not need to know or find the value of $C$). Use your graph from **(ii)** to calculate the activation energy, $E_A$, for the reaction.
*Edexcel Chemistry (Nuffield) Specimen Unit Test 4 Jan 2000 Q3*

# KEY SKILLS **ASSIGNMENT**

## REACTION KINETICS AND THE BANNING OF CFCS

In the early 1970s there was concern that high-flying jet aircraft, such as Concorde, would damage the stratospheric ozone layer. The ozone layer shields the lower atmosphere and all terrestrial life from harmful ultraviolet radiation. The source of this concern was an understanding of the reaction kinetics of ozone and nitrogen monoxide. One of the mechanisms proposed was:

**Step 1**  $NO(g) + O_3(g) \xrightarrow{\text{slow}} NO_2(g) + O_2(g)$

**Step 2**  $NO_2(g) + O(g) \xrightarrow{\text{fast}} NO(g) + O_2(g)$

The rate equation for this reaction was found to be:

$$\text{Rate of reaction} = k[NO(g)][O_3(g)]$$

Oxygen atoms can form in the stratosphere from molecular oxygen. The energy to break the bonds in the oxygen molecule is provided by ultraviolet radiation:

$$O_2(g) \xrightarrow{hf} O(g) + O(g)$$

**1** **a)** What is meant by the terms mechanism and reaction kinetics?
**b)** Explain which of the above steps is the rate-determining step in the reaction between NO and $O_3$.
**c)** Explain what is meant by the term *homogeneous catalyst* and decide which species in the mechanism is a homogenous catalyst.

**2** **a)** Work out the stoichiometric equation for the reaction involving NO and $O_3$.
**b)** Why is the proposed mechanism consistent with the rate equation?
**c)** What is the overall order for the reaction and what is the order with respect to each of the following: ozone, atomic oxygen and nitrogen monoxide?

The effect of jet aircraft on the ozone layer was found to be insignificant – there were not enough aircraft to inflict lasting harm. However, in 1973 Professor Sherry Rowland and his colleague Dr Mario Molina began to investigate chlorofluorocarbons, or CFCs as they are now known.

As they are such stable compounds, CFCs were used extensively as aerosol propellants, refrigerants and in insulation foam. So Rowland and Molina decided to find out how long CFC molecules, such as CFC-11 and CFC-12, the two most common gases (Fig 26.A1), could remain in the lower atmosphere without reacting, and what might happen when they are subjected to the high-energy ultraviolet radiation of the stratosphere. The answer to the first part of their investigation was that, because of their stability, CFCs could remain without reacting for many centuries. It was the answer to the second part that began to excite their interest.

**Fig 26.A1**

The breakdown of CFCs by ultraviolet radiation produces chlorine atoms:

$$CCl_2F_2 \xrightarrow{hf} CF_2Cl + Cl$$

Rowland and Molina then began to wonder what could happen to the chlorine atoms and concluded that they would react with the ozone, according to the following mechanism:

**Step 1**  $Cl(g) + O_3(g) \xrightarrow{\text{slow}} ClO(g) + O_2(g)$

**Step 2**  $ClO(g) + O(g) \xrightarrow{\text{fast}} Cl(g) + O_2(g)$

**3** **a)** Predict a rate equation from the mechanism involving chlorine atoms.
**b)** Work out the stoichiometric equation.

**4** Ozone can react with NO or with Cl. The activation energy for Step 1 for the reaction with chlorine atoms is 2 kJ mol$^{-1}$, and the activation energy for Step 2 for the reaction with nitrogen monoxide is 12 kJ mol$^{-1}$.
**a)** What is meant by the term activation energy? (Hint: see Chapter 8.)
**b)** Which is the more effective catalyst, NO or Cl?
**c)** Which reaction has the greater rate constant – the reaction with NO or that with Cl?
**d)** Explain how Cl acts as a catalyst.

**5** Predict a rate equation for the destruction of ozone catalysed by the chlorine atom.

On average, one chlorine atom can destroy 100 000 ozone molecules. Such destructive power is awesome when related to the millions of tonnes of CFCs that had been produced up to the time of Rowland and Molina's research.

When they calculated the likely effect on the ozone layer, they could not believe their figures. They consulted Hal Johnson, who had done many calculations for the effect on the ozone layer of nitrogen monoxide from high-flying aircraft. He, too, knew that chlorine atoms could react with ozone, but did not realize what the source of chlorine atoms might be. This confirmed Molina and Rowland's results and in 1974 they published their predictions in the scientific journal *Nature*.

# KEY SKILLS **ASSIGNMENT**

The United States were quick to ban the production of CFCs, but many countries, among them Britain, France and Japan, did not, and continued with production.

However, in 1985 the British scientist Joe Farman and his team, who were working in the Antarctic, discovered huge losses in the ozone layer. This led to concerted international action to ban the production of CFCs world-wide.

**Fig 26.A2** Sherwood Rowland receives his Nobel Prize from Sweden's King Carl Gustaf in 1995

In 1995, Sherwood Rowland and his colleague Mario Molina were awarded the Nobel Prize for Chemistry. They shared it with Paul Crutzen, a Dutch chemist, who had demonstrated a link between soil bacteria and the production of dinitrogen oxide gas, $N_2O$. This gas also reacts in the stratosphere to form nitrogen monoxide (NO):

$$N_2O(g) + O \rightarrow 2NO(g)$$

**6** Why do scientists publish their findings in journals?

The research of Rowland and Molina can be classed as pure research, as there was no apparent commercial benefit. Much of the funding for science is for targeted research. This type of research is designed to yield commercial benefit and to find useful applications.

**7** Choose another scientific discovery or invention and decide whether it was the product of pure or targeted research. If you were in charge of Britain's scientific budget, what proportion would you devote to pure research? Explain your reasoning.

**Fig 26.A3** The hole in the ozone layer over Antarctica

More information on ozone and CFCs can be found on page 283.

# 27

# Colour in organic and inorganic compounds

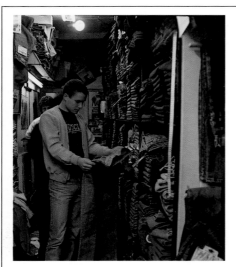

**Denim jeans** These seem to be here to stay

The blue of denim jeans has a long and colourful history. Levi Strauss emigrated to America in 1850 during the gold rush, taking with him some heavy cotton canvas called *serge de Nîmes*. Instead of making tents with the canvas, he used the hard-wearing cloth to make trousers for the gold miners. This was the start of a multi-million pound industry.

The blue dye Levi Strauss used was indigo, which has been used as a dye for at least 3000 years. In 1850, all dyes were extracted from plants or animals, and indigo came almost exclusively from indigo-bearing plants grown in huge plantations in India. The indigo market brought in £4 000 000 of revenue per annum, a fortune at that time.

Six years after Strauss dyed his first pairs of denim jeans, a chance discovery in England by William Perkin led to the world's first synthetic dye. In 1880, Adolph von Baeyer synthesized indigo, and by 1897 synthetic indigo was being manufactured. Within a few years the cultivation of natural indigo had ceased, wiping out one of India's major export industries.

Indigo is a dye that fades, and during the first half of the twentieth century its popularity waned as chemists synthesized new dyes that did not fade. However, in the 1960s blue denim jeans came back into fashion. The fact that synthetic indigo fades with wear became part of its attraction.

The synthesis of indigo is a chemical success story, with chemists first determining indigo's structure and then finding ways to manufacture it much more cheaply than it could be extracted from plants.

## 1 WHY ARE THINGS COLOURED?

Colours play an enormous part in your life. Your eyes detect the colours of objects and send messages to your brain, providing you with a constant stream of information. The colour of traffic lights helps to control traffic flow and the blue flashing light of the emergency services alerts you to their presence. The colour of a food determines how appetizing it appears and may indicate how fresh it is. The colours of your clothes make a statement about your personality.

Advertisers are well aware of how we are influenced by colours. They use warm colours such as orange and red to make us feel at home with a product, whereas blue is a colder colour used to give the hint of sophistication. Whether we realize it or not we are very influenced every waking moment by colour. So just what is colour?

We know from Chapter 3 that visible light is electromagnetic radiation with wavelengths between approximately 400 nm and 700 nm. Visible light forms a very small part of the electromagnetic spectrum. Isaac Newton, in 1666, was the first to realize that what we perceive as white light (such as sunlight) is actually made up of seven colours – red, orange, yellow, green, blue, indigo and violet. These seven colours form the visible spectrum shown in Fig 27.1.

**Fig 27.1** The visible part of the electromagnetic spectrum is formed by passing light through a prism

Newton passed sunlight through a prism to obtain the visible spectrum. He also demonstrated that when this spectrum of colours passed through a second prism it produced white light once more. Thomas Young and George Palmer, 150 years later, independently suggested that receptors in our eyes are sensitive to blue, green or red light and that different stimulations of these three colour receptors (now called cones) enable us to perceive all the different colours. In 1861, James Clerk Maxwell, a Scottish physicist, combined beams of these three coloured lights to produce white light.

The three colours – blue, green and red – are called **additive primary colours** because they cannot be produced by the combination of other coloured lights. However, yellow can be made from the addition of green and red lights, cyan from green and blue lights and magenta from red and blue (Fig 27.2). Yellow, cyan and magenta are called **secondary colours**. Colour televisions make use of the three additive primary colours. The inside of the screen is coated with about a million phosphor dots that glow either blue, red or green when they are struck by an electron beam. These dots combine to form coloured images.

The reason why the world appears so colourful is that the multitude of chemical compounds around us absorb and reflect different wavelengths from the light that falls onto them. For example, a leaf appears green because it absorbs red light and reflects light of other wavelengths. A substance that is white reflects all visible wavelengths of electromagnetic radiation. However, a substance that appears black absorbs all visible wavelengths. A substance appears coloured if it absorbs some of the electromagnetic radiation from white light, but not all of it.

When a compound absorbs wavelengths of one particular colour, a **complementary colour** appears. Pairs of complementary colours are represented on the colour wheel in Fig 27.3. The coloured wavelengths that are absorbed lie opposite their complementary colours. For example, crystals of hydrated copper(II) sulphate appear blue because they absorb light in the orange region of the spectrum.

YOU CAN READ ABOUT RECEPTORS IN THE EYE ON PAGE 308.

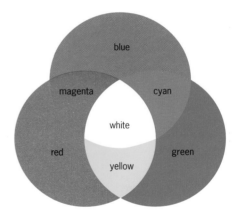

**Fig 27.2** Mixing coloured light produces secondary colours and white light

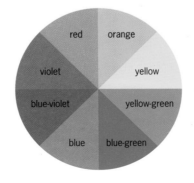

**Fig 27.3** A colour wheel. Complementary colours are opposite one another on this colour wheel

**? TEST YOURSELF**

**A** A solution of potassium dichromate(VI), $K_2Cr_2O_7$, appears orange. What colour does the dichromate(VI) ion absorb?

**Fig 27.4** Potassium dichromate(VI) solution

## ! SCIENCE FEATURE Dyes and pigments

Dyes and pigments differ in one important way: dyes are soluble in the medium in which they are applied and pigments are insoluble. Humans have used dyes and pigments since early times. Some cave paintings discovered in southern France and northern Spain are up to 30 000 years old. The artists of these paintings used mineral pigments to colour them:

● iron(III) oxide provided the red colour;
● iron(II) carbonate provided the yellow colour;
● either soot or manganese(IV) oxide were used for black.

The pigments were applied to surfaces by first mixing them into a paste using mud or oil. Neanderthal tribes painted the bodies of their dead using iron(III) oxide, also known as red ochre. The Ancient Egyptians used some chemical reactions to extend the range of pigments available. For example, red lead, $Pb_3O_4$, was produced by heating together lead and white lead, $PbCO_3.Pb(OH)_2$.

**Fig 27.5** The prehistoric painting of a bison in the caves at Altamira, northern Spain

Pigments are spread as a surface layer in paints and in coloured plastic articles (in these they are dispersed throughout the plastic). Until the nineteenth century, nearly all pigments were inorganic. Artists' paints used to contain lead and chromium compounds, but these are now known to pose a health hazard. They have been replaced by safer red and yellow organic pigments that can been synthesized.

In contrast to pigments, dyes are soluble. Dye molecules attach themselves to the molecules of the substance they are colouring. In some cases, dye molecules use ionic or covalent bonds, but more usually they attach themselves by hydrogen bonds or induced dipole–induced dipole forces (van der Waals forces or non-polar forces). Sometimes a fabric is first treated with a mordant, an intermediate substance that bonds to the fabric and to the dye. The mordant most commonly used in ancient times was potassium aluminium sulphate (also called potash alum), but other metal salts were also used. The metal ions bonded to the fabric and – through the formation of

complexes – to the dye molecules. Potassium aluminium sulphate may have been the first chemical to be purified to avoid contaminants, such as iron(III) salts, which introduced other colours into the dyeing process.

The dyeing of fabrics has a very long history and until the end of the nineteenth century dyes came from animals or plants. This meant that colouring clothes could be an expensive business. One of the most highly prized dyes in the days of the Roman Empire was Tyrian purple. A quarter of a million Mediterranean molluscs were required to produce just 30 g of the dye. Only members of the Emperor's family could wear togas dyed with Tyrian purple, so it became known as royal purple.

**Fig 27.6(a)** The structural formula of Tyrian purple

Early Britons used the woad plant, which contains the chemical indigo, to dye their clothes blue. About 3000 years ago, Mexicans started to use a red dye, cochineal, produced from crushing the Coccus insect. It was not until the sixteenth century that the Spanish brought cochineal to Europe. The production of this dye was a painstaking process that involved collecting about 150 000 insects by hand for every kilogram of dye produced.

**Fig 27.6(b)** The insects that provide cochineal dye are tiny

Another red dye, alizarin, was used by the Ancient Egyptians. It came from the roots of the madder plant, which was very widely cultivated. One of the reasons why many military uniforms were red was the availability of this dye and the fact that it did not fade very quickly in sunlight.

# 2 FLAME TESTS AND ATOMIC EMISSION SPECTROSCOPY

You have probably carried out some flame tests on inorganic compounds yourself. The presence of certain metal ions gives rise to characteristic colours. Fig 27.7 shows the characteristic flame colour caused by the potassium cation, which shows that the substance being tested must contain a potassium compound. Table 27.1 gives the characteristic flame colours of some other cations.

| Table 27.1 Characteristic flame colours of metal ions in Groups 1 and 2 | | | |
|---|---|---|---|
| **Group 1:** | | **Group 2:** | |
| **Metal ion** | **Flame colour** | **Metal ion** | **Flame colour** |
| $Li^+$ | Deep red | $Be^{2+}$ | No colour |
| $Na^+$ | Yellow | $Mg^{2+}$ | No colour |
| $K^+$ | Lilac | $Ca^{2+}$ | Brick red |
| $Rb^+$ | Red | $Sr^{2+}$ | Deep red |
| $Cs^+$ | Blue | $Ba^{2+}$ | Apple green |

**Fig 27.7** The lilac colour indicates that the compound in the flame contains potassium

In Chapter 3, we said that flame colours resulted from electron transitions. The energy from the Bunsen flame causes electrons in the cation to jump up to higher energy levels by absorbing amounts of energy. As they fall back down to their original energy levels they each release a certain amount of energy. If the energy corresponds to a frequency (and wavelength) in the visible spectrum, a colour is observed. You will notice from Table 27.1 that some Group 2 cations do not produce a coloured flame. Many ions do not emit light with frequencies in the visible part of the spectrum, so these do not colour the flame.

When viewed through a spectroscope, the flame colour produced by a particular cation gives rise to a series of coloured lines. This series of lines is called a **line emission spectrum**. Other emission lines are found in other parts of the spectrum, such as the ultraviolet and infrared. A line emission spectrum is characteristic of a particular cation and can be used to identify elements in compounds, rather like fingerprints can be used to identify individuals. It is possible to determine the amount of an element present in a sample by measuring the intensities of the different emission lines in its spectrum.

If bright white light is shone onto a coloured flame and observed through a spectroscope, a series of black lines result, forming a **line absorption spectrum**. The metal ions in the flame remove the frequencies from the white light, which they would normally emit when excited.

> ✔ **REMEMBER THIS**
>
> Remember that $E = h \times f$, where $E$ is energy, $h$ is Planck's constant and $f$ is the frequency. Light is electromagnetic radiation; you can read more about the nature of electromagnetic radiation in Chapter 3

YOU CAN SEE SOME LINE EMISSION SPECTRA ON PAGE 50.

SEE QUESTION 1.

**Fig 27.8** Emission and absorption spectra for sodium. Notice that the frequencies are the same in the emission and absorption spectra because electrons absorb photons of a particular frequency when they become excited and release photons of the same frequency when they return to their ground state

**YOU CAN READ MORE ABOUT STEEL MANUFACTURE ON PAGES 12–15.**

**Fig 27.9** Atomic emission spectrometers are used in the steel industry to determine the presence and percentage content of certain elements in steel

Atomic emission spectroscopy is used in the steel manufacturing process to determine the composition of different steels. Up to 20 elements are used in steels; the amount and presence of these elements has a profound effect on the properties of a particular steel.

Atomic emission spectroscopy is also used to monitor the levels of potassium and sodium in blood. To do this for sodium the wavelength chosen is 589 nm. This is the wavelength of one of the bright yellow emission lines, which is best seen in Fig. 3.4 on page 47. The atomic emission spectrometer is then calibrated using solutions of known sodium ion concentrations. The higher the concentration, the more intense the emission line. The blood serum is then analysed and the intensity of the emission reading at 589 nm is compared with the intensities from known concentrations of sodium ions used to calibrate the instrument. This allows the concentration of sodium ions in the blood serum to be calculated.

## 3 DIAZONIUM COMPOUNDS AND AZO DYES

The first synthetic dye, **mauve**, was accidentally discovered in 1856 by an 18-year-old Englishman, William Perkin. Perkin's discovery led to a search for new dyes and prompted the start of the organic chemical industry. One of the chemicals Perkin used in the production of mauve was phenylamine (Fig 27.10).

Phenylamine is an **aromatic amine** (**arylamine**). It has a benzene ring with an $NH_2$ (amino) functional group directly attached to it.

### DIAZOTIZATION

Six years after the discovery of mauve, Johann Peter Griess used aromatic amines to produce **diazonium compounds** (also known as **diazonium salts**). Diazonium compounds are produced when nitrous acid, $HNO_2$, reacts with an aromatic amine. The reaction mixture must be kept below 5 °C because the diazonium salt is unstable. Nitrous acid is also unstable and it has to be made *in situ* by reacting sodium nitrite with concentrated hydrochloric acid:

$$NaNO_2(aq) + HCl(aq) \rightarrow HNO_2(aq) + NaCl(aq)$$

The nitrous acid reacts with the aryl amine and more of the concentrated hydrochloric acid to give the diazonium salt (Fig 27.11):

phenylamine            benzenediazonium chloride

**Fig 27.11**

This reaction is called a **diazotization**. The usual way to carry this out in the laboratory is to add a cold aqueous solution of sodium nitrite to a solution of phenylamine dissolved in concentrated hydrochloric acid. The reaction mixture is kept below 5 °C by using ice.

---

### ? TEST YOURSELF

**B** Work out the following frequency and the energy of the emission line chosen to analyse sodium in blood. Hint: $c = f\lambda$ and $E = hf$. To remind yourself about the use of these equations, see Chapter 3.

**YOU CAN READ MORE ABOUT PERKIN'S DISCOVERY OF MAUVE IN THE FEATURE BOX ON PAGE 602.**

**Fig 27.10** The structural formula of phenylamine

### ? TEST YOURSELF

**C** What is meant by the term functional group?

**FOR MORE INFORMATION ON FUNCTIONAL GROUPS, SEE PAGE 206.**

### ? TEST YOURSELF

**D** Write the equations for the production of nitrous acid and the diazonium salt formed from phenylamine using sulphuric acid.

The benzene ring stabilizes the diazonium ion by **delocalization** of its electrons. Alkylamines, such as ethylamine ($C_2H_5NH_2$), produce diazonium salts that decompose immediately, even below 5 °C.

Above 5 °C, the solution of benzenediazonium compound rapidly decomposes, giving off nitrogen gas according to the equation:

**Fig 27.12** The benzenediazonium ion

$$C_6H_5N_2^+Cl^- + H_2O \rightarrow C_6H_5OH + N_2 + HCl$$

**YOU CAN FIND OUT MORE ABOUT DELOCALIZATION ON PAGE 330.**

**? TEST YOURSELF**

**E** Why are solid benzenediazonium compounds explosive?
Hint: think about the relative volumes of solid and gas produced on decomposition.

## THE PRODUCTION OF AZO DYES USING COUPLING REACTIONS

A diazonium ion may act as an **electrophile** and attack the benzene ring of another compound. This reaction is called a **coupling reaction**. The first coupling reaction was observed by Griess: an aromatic amine coupled with a diazonium salt to produce a stable **azo compound**.

**✔ REMEMBER THIS**

An electrophile is a species that accepts a lone pair of electrons to form a covalent bond. It is attracted to an electron rich centre. See pages 311 and 334.

phenylamine                    yellow azo dye

**Fig 27.13**

In the 1870s, coupling reactions involving alkaline solutions of phenols were first observed (Fig 27.14a).

phenol                         orange azo dye

**Fig 27.14(a)** The coupling reaction of a diazonium salt with an alkaline solution of phenol

**Fig 27.14(b)** The orange azo dye produced when benzenediazonium chloride couples with phenol

**? TEST YOURSELF**

**F** **a)** Why must a diazo coupling reaction be carried out below 5 °C?
**b)** What happens to phenol molecules when they react with aqueous sodium hydroxide? Draw the structure of the species produced. Hint: Look back to page 343 to help you answer this question.
**c)** Why is the species you have drawn in **b)** more reactive with electrophiles than with benzene?

**SEE QUESTION 2.**

The functional group of an azo compound is the azo group, $-N=N-$. In a compound $R-N=N-R'$, if the R groups are aromatic (aryl) groups, the azo group becomes part of the delocalized systems of the adjoining benzene rings. This gives extra stability to the molecule and is also responsible for the bright colours of azo dyes.

naphthalen-2-ol                    red azo dye

**Fig 27.14(c)** The coupling reaction of a diazonium salt with alkaline naphthalen-2-ol

## SCIENCE FEATURE Perkin and the start of the dyestuffs industry

William Perkin was the son of a carpenter in East London. As William had some artistic talent, his father hoped his son would train as an architect. However, a fascination with chemistry led him to become a student at the Royal College of Chemistry. At the age of 18 years he made an accidental discovery that was to prove to be the start of not only the dyestuffs industry but also the whole organic chemicals industry.

**Fig 27.15** William Henry Perkin (1838–1907) and the original bottle of dye that he produced

At the Royal College of Chemistry, Perkin became an assistant of Professor August von Hofmann, a German chemist. Hofmann suggested that he might like to try synthesize the important, naturally occurring drug quinine, which was used to treat malaria. The starting material was coal tar, because it was known to contain arylamines.

At this time, the structures of organic chemicals had not been deduced. However, the molecular formula of quinine was known, as was the empirical formula of the arylamine he intended to use. Perkin thought that if he oxidized the arylamine he would produce the reaction:

$$2C_{10}H_{13}N + 3[O] \rightarrow \underset{\text{quinine}}{C_{20}H_{24}N_2O_2} + H_2O$$

All Perkin managed to produce was a dirty brown precipitate. Undeterred, he set about oxidizing a simpler compound, phenylamine. This time he produced a black precipitate that, when dried and dissolved in ethanol, produced a brilliant purple solution. The product's structure was nothing like that of quinine, but Perkin was quick to see its potential as a dye. He dyed some silk with the purple dye and sent it to a firm of dyers.

The reply came back, 'If your discovery does not make the goods too expensive, it is decidedly one of the most valuable that has come out for a very long time.'

One of the most important properties of the dye was that it was 'fast' – it didn't fade or change colour, even when

exposed to light and air. Perkin named the dye mauve after a French flower.

He left the Royal College of Chemistry and, with the help of his father and brother, set about building a factory for the large-scale production of mauve. Perkin used coal tar as the starting material because it was a cheap, plentiful by-product of the coal gas industry. Fig 27.16 shows the steps involved in his synthesis of mauve.

**Fig 27.16** A flow chart showing Perkin's synthesis of mauve

He produced benzene by fractionally distilling coal tar. He then nitrated the benzene to produce nitrobenzene. The nitrobenzene was reduced using a mixture of iron filings and ethanoic acid to form phenylamine. This phenylamine he then oxidized using acidified potassium dichromate(VI).

Queen Victoria wore a dress dyed with mauve to the International Exhibition of 1862. *Punch,* the satirical magazine, proclaimed that policeman could be heard telling people to 'get a mauve on!'

It was not long before other dyes of different colours were made using phenylamine. Although the synthesis of dyes began in England, Germany soon became the centre of the dyestuffs

industry. Perkin's mauve was superseded because it cost too much to produce. One of its last applications was in the production of the stamp shown in Fig 27.17(a).

**Fig 27.17(a)** The 1d stamp, printed in 1881, was dyed with Perkin's mauve

At 36 years of age, Perkin sold his factory and returned to his first love, chemistry research. He made many important contributions to organic chemistry, including the production of a perfume from coal tar. However, the synthesis of the quinine molecule was not finally achieved until 1944, almost 90 years after Perkin's attempt.

**Fig 27.17(b)** A sketch by William Perkin of his first synthetic dye factory which made Perkin's mauve

**G**
**a)** What reagents are used to nitrate benzene?

**b)** Perkin used ethanoic acid and iron filings to produce phenylamine from nitrobenzene. What reagents are used to produce phenylamine from nitrobenzene in the laboratory?
Hint: If you have trouble answering these questions, see Fig 15.47 on page 338.

**H**
**a)** If a substance reflects all light with frequencies in the visible range of the spectrum, what colour will it be?

**b)** If a substance absorbs all the radiation in the visible spectrum, what colour will it be?

SEE QUESTIONS 3 AND 4.

## 4 COLOUR IN ORGANIC MOLECULES

What gives rise to the bright colours of azo dyes? We have already seen on page 597 that colour results from electron transitions. If a substance absorbs electromagnetic radiation with frequencies in the visible spectrum, then it will be coloured. This is because the substance reflects or transmits only part of the visible spectrum.

The absorption of a photon (a quantum of electromagnetic radiation) causes an electron to jump from its **ground state** (the lowest possible energy level) to an **excited state** (a higher energy level). If the difference in energy between the ground state and the excited state is equivalent to a photon with a frequency (or wavelength) in the visible part of the spectrum, the substance will absorb at that frequency when white light falls on it. Only outer-shell electrons are excited by visible and ultraviolet radiation. Electrons from inner shells are held much more firmly by the nucleus, so they require much more energy to become excited.

> **REMEMBER THIS**
>
> $E = h \times f$, where $E$ is the energy of the photon, $h$ is Planck's constant and $f$ is the frequency of radiation. A photon with a particular energy has a corresponding frequency. You can read more about photons and the excitation of electrons in Chapter 3.

▶ **EXTENSION** Electron transitions in organic molecules and ultraviolet–visible absorption

On page 81, we discussed how atoms bond together by the overlap of atomic orbitals to form covalent bonds. A σ **bond** (sigma bond) forms when two atomic orbitals overlap at one point. There is always a σ bond between two covalently bonded atoms.

Double and triple bonds consist of not only a σ bond, but also π **bonds** (pi bonds). π bonds form by the sideways overlap of p orbitals. A π bond is usually at a higher energy level than a σ bond. Electron transitions between these bonded orbitals in organic molecules gives rise to absorptions, either in the visible part of the spectrum, which leads to colour, or in the ultraviolet part of the spectrum, which causes the compound to appear colourless. **Lone pairs (non-bonded pairs)** of electrons are also found in orbitals and these, too, contribute electron transitions that occur in the ultraviolet–visible part of the spectrum.

Ethane (Fig 27.18) is composed of σ bonds and electron transitions in these orbitals occur in the ultraviolet part of the spectrum, because more energy is required to excite the electrons in these bonding orbitals.

Fig 27.18 Bonding in ethane comprises only σ bonds

### CONJUGATED SYSTEMS
In ethene, we have a π bond as well as a σ bond. Ethene still appears colourless, because the double bond is restricted to the two carbon atoms (Fig 27.19).

Fig 27.19 Bonding in ethene. A π bond is formed by the sideways overlap of the p orbitals

(Continued on page 604)

> **EXTENSION** Electron transitions in organic molecules and ultraviolet–visible absorption (continued from page 603)

However, in buta-1,3-diene, $CH_2=CH-CH=CH_2$, there are two double bonds. The p orbitals can overlap so that the two $\pi$ bonds interact with one another to form a **delocalized system**. The system is **conjugated**. Such systems have alternating double and single bonds and the delocalization that occurs lowers the energy of some of the electron transitions.

**Fig 27.20** Conjugation in buta-1,3-diene

Although buta-1,3-diene absorbs in the ultraviolet region, its maximum absorption wavelength is 220 nm (compared with 185 nm in ethene) and the **intensity** of absorption also increases. The intensity depends on how many photons are absorbed from the radiation that falls on the substance.

> ✔ **REMEMBER THIS**
>
> $E = h \times f$ and $c = f \times \lambda$. So the higher the wavelength, $\lambda$, *the lower the* frequency, $f$, and the lower the energy, $E$ ($c$ is the speed of light and $h$ is Planck's constant).

> ❓ **TEST YOURSELF**
>
> **I** What wavelength range is light in the visible spectrum? Hint: If you are not sure look back to page 48.

Carotene (Fig 27.21) is a vitamin found in carrots that is used as a food colorant. Its conjugated system involves 11 double bonds, which shifts its maximum absorption into the blue part of the visible spectrum. As blue light is removed from the white light that falls on carotene it appears orange. The part of a molecule responsible for absorbing coloured radiation is called the **chromophore**. It is usually an extended delocalized electron system.

**Fig 27.21** The skeletal formula of carotene showing its extended conjugated system, which forms the chromophore

Orange is the complementary colour to blue. Refer back to the colour wheel on page 597 to remind yourself about complementary colours.

The more conjugated a system the smaller the gap between energy levels, which means the absorptions that excite electrons here occur at longer wavelengths.

### COLOUR IN AZO COMPOUNDS

The first commercially successful azo dye was Chrysoidine (Fig 27.22). The azo group, $-N=N-$, acts as a 'delocalization bridge' between the two benzene rings to form an extended delocalized system, which is the chromophore.

**Fig 27.22** Chrysoidine, an orange azo dye

> ❓ **TEST YOURSELF**
>
> **J** In which region of the electromagnetic spectrum does Chrysoidine absorb?

The two amine functional groups of Chrysoidine have lone pairs of electrons on the nitrogen atoms. These lone pairs interact with the delocalized system. The nature of the functional groups that interact with the chromophore can dramatically alter the colour of the azo dye molecule by causing a shift in the electrons of the chromophore. This electron shift alters the energy required to promote them into an excited state, and so shifts the wavelength of light absorbed.

**Fig 27.23** Blue azo dye. The substituent groups on the left of the molecule tend to accept electrons, while those on the right are electron-donating groups. This shifts the electrons of the chromophore considerably, making the molecule appear blue.

**SEE QUESTION 5.**

# 5 pH INDICATORS

Methyl orange is an azo compound that has different colours depending on the pH of the solution it is in. This property means that it is used as an indicator in acid–base reactions. Adding or removing an $H^+$ ion causes an electron shift in the molecule, and so alters the wavelength at which methyl orange absorbs.

On page 377, we saw that indicators change colour depending on the concentration of $H^+(aq)$ ions in the solution. Each indicator changes colour at a specific pH; using the correct indicator, any neutralization reaction can be monitored to its end point. Methyl orange changes colour between pH 3.2 and pH 4.4. It can be used to determine the neutralization point of a strong acid and a weak base.

yellow above pH 4.4          red below pH 3.2

**Fig 27.24** Methyl orange is used as an acid–base indicator. Its colour depends on the pH of the solution it is in.

Phenolphthalein is colourless below pH 8.2, but above this value it changes to purple. This indicator can be used in titrations with strong alkalis and weak acids.

**? TEST YOURSELF**

**K** What feature of the structure of phenolphthalein in alkaline solution makes it coloured?

**Fig 27.25** The ultraviolet–visible spectra of phenolphthalein at pH 1 and pH 13.

(left graph)
pH = 1
phenolphthalein in acid solution is:

It appears colourless

Wavelength/nm

(right graph)
pH = 13
phenolphthalein in alkaline solution is:

It appears magenta

Wavelength/nm

---

## ▶ EXTENSION Ultraviolet and visible spectroscopy

In Chapter 11, we discussed two types of spectroscopy that involve absorption of electromagnetic radiation by a substance under investigation. In infrared spectroscopy, light is absorbed in the infrared part of the spectrum because of the increased vibration of different bonds within a molecule. In nuclear magnetic resonance spectroscopy, radio waves are absorbed through the excitation of nuclei within molecules.

Ultraviolet and visible spectroscopy is possible because outer electrons of atoms or ions in compounds absorb in the ultraviolet or visible part of the spectrum when they are excited. Compounds that absorb *only* in the ultraviolet part of the spectrum are colourless. In the spectrometer, a beam of electromagnetic radiation passes through a monochromator, which selects varying wavelengths. The beam is then split and one beam passes through a solution of the substance under investigation, while the

other passes through the pure solvent.

The spectra produced usually have broad absorption bands, in contrast to atomic absorption spectra in which gaseous atoms and ions have definite sharp lines. The broad absorption bands occur because in solution a number of vibrational and rotational energy levels are possible for each energy level of the electrons. You can read more about vibrational energy levels on page 242.

**Fig 27.26** A simplified schematic diagram of how an ultraviolet–visible spectrometer works

contains the substance under test, dissolved in a solvent

chart recorder showing absorption spectrum of test substance

sample cell

monochromator (selects wavelength)

source of electromagnetic radiation

reference cell

detector

compares the intensity of light transmitted by the sample cell and reference cell

contains pure solvent

(Continued on page 606)

▶ **EXTENSION** Ultraviolet and visible spectroscopy [continued from page 605]

### INTERPRETING ULTRAVIOLET–VISIBLE ABSORPTION SPECTRA

The horizontal axis of an ultraviolet–visible absorption spectrum gives the wavelength in nanometres (nm). The shape of the absorption peak is usually characteristic of a particular compound and so the spectrum can be used to help identify the compound.

**Fig 27.27** The ultraviolet–visible absorption spectrum of benzene. Notice that all the absorption appears in the ultraviolet part of the spectrum, so the compound appears colourless

**Fig 27.28** The absorption spectrum of a blue dye. Red, orange and yellow light is absorbed by the dye, so blue light is transmitted. (See the colour wheel on page **597**)

However, a more common use of this type of spectroscopy is to measure concentrations accurately from the intensity of absorption on the $y$-axis. For example, the uptake of a drug at different sites around the body can be monitored using ultraviolet–visible spectroscopy by taking samples from these sites and analyzing their solutions.

In the steel industry, the amount of trace metal, such as manganese, can be analysed by reacting the metal to form an identifiable coloured ion and measuring the absorption of its solution. In this case, the manganese may be oxidized to form the purple $MnO_4^-$ ion.

The food industry also uses ultraviolet–visible spectroscopy to determine how much nitrite has been added to meat.

**Fig 27.29** The ultraviolet–visible spectrum of $[Cu(NH_3)_4(H_2O)_2]^{2+}$. The solution of this complex appears blue–violet

## 6 TRANSITION METAL IONS AND COLOUR

A characteristic of transition metals is that many of their compounds are coloured (see Chapter 24). Transition metal compounds are responsible for the colours in gemstones, stained glass windows and pottery glazes. From reading this chapter, you will realize that transition metal ions appear coloured because they absorb some wavelengths in the visible spectrum, but transmit or reflect the rest. The electrons absorb photons of a specific wavelength, become excited and jump to higher energy levels. In compounds of transition elements, colour results from a difference in energies between d orbitals.

## d–d TRANSITIONS

You are probably wondering how d orbitals can be at different energy levels. When we discussed d orbitals in Chapter 24, they were drawn at the same energy level (**degenerate**) on energy level diagrams. (See Fig 24.2, page 515.)

However, this is only true for gaseous transition metal ions. When ligands bond to transition metal ions they cause a splitting of the energy level of the d orbitals. The energy difference between the two sets of d orbitals is often such that the wavelength of photons absorbed is in the coloured part of the spectrum.

Fig 27.30(a) d orbitals are all at the same energy level in a gaseous $Cu^{2+}$ ion. We say they are degenerate

Fig 27.30(b) In $[Cu(H_2O)_6]^{2+}$, the water ligands cause the d orbitals to split, so they are now non-degenerate

Fig 27.30(c) An electron is excited from a lower energy level d orbital to a higher energy level d orbital

Copper(I) compounds are white because the $Cu^+$ ion has a $3d^{10}$ outer electron configuration. Since the 3d subshell is full, no transition of electrons between d orbitals can occur. The same is true of scandium(III) compounds, which have no electrons in the 3d subshell ($3d^0$). But not all colour in transition metals results from d–d transitions; sometimes electrons can jump from the ligand to the metal. This is called **charge transfer** or **electron transfer**, and it is responsible for the bright colours of Prussian blue and chrome yellow.

## FACTORS THAT AFFECT d–d SPLITTING AND COLOUR

The colour of a transition metal complex depends chiefly on the central metal cation. Each transition metal is different and has a different nuclear charge. The larger the nuclear charge, the more firmly electrons are held in their d orbitals. This affects the energy levels of the split d orbitals and thus the amount of energy required to excite an electron from a lower energy d orbital to a higher energy d orbital. This in turn affects the colour.

The oxidation state of the metal also affects the splitting of the d orbitals and, as a result, the colour. This is well illustrated by vanadium complexes in aqueous solution.

| Table 27.2 The relationship between the oxidation state of vanadium and the colour of its complexes | | |
|---|---|---|
| **Ion** | **Oxidation state** | **Colour of aqueous solution** |
| $VO_2^+$ | +5 | Yellow |
| $VO^{2+}$ | +4 | Blue |
| $V^{3+}$ | +3 | Green |
| $V^{2+}$ | +2 | Violet |

The nature of the ligand also has an effect on d–d splitting. Different ligands cause different separations of energy between d orbitals. The

**? TEST YOURSELF**

**P** The difference in energy of the two sets of d orbitals in $[Cu(H_2O)_6]^{2+}$ causes absorption at the red end of the spectrum. What colour is the $[Cu(H_2O)_6]^{2+}$ ion? Hint: Look at the colour wheel on page 597.

**✔ REMEMBER THIS**

A **complex** is formed when a central metal atom or ion is surrounded by species that donate lone pairs of electrons. A species that donates a lone pair of electrons is called a **ligand**. You can read more about this on page 522.

**? TEST YOURSELF**

**Q** **a)** What is the full electron configuration of $Zn^{2+}$?
**b)** Explain why zinc compounds are white.

**? TEST YOURSELF**

**R** Using the colour wheel on page 597, work out which colour is being absorbed by each of the vanadium ions in Table 27.2.

**YOU CAN SEE THESE COLOURS ON PAGE 558.**

**Fig 27.31** Absorption of wavelengths of visible light by $[Cu(H_2O)_6]^{2+}$

**Fig 27.33** Left: An aqueous solution of copper(II) sulphate contains the $[Cu(H_2O)_6]^{2+}$ ion. Right: In this solution, concentrated ammonia has been added, which replaces some water ligands to form the ammine complex $[Cu(NH_3)_4(H_2O)_2]^{2+}$. The ammonia ligand increases the energy gap between the split d orbitals. This affects the energy of the electron transition and therefore the colour of the complex solution

### ? TEST YOURSELF

**S** $[CuCl_4]^{2-}$(aq) is a yellow colour. Sketch its ultraviolet–visible spectrum.

FIG 24.23 ON PAGE 526 IS ANOTHER EXAMPLE OF CIS–TRANS ISOMERISM THAT AFFECTS THE COLOUR OF A COMPLEX.

SEE QUESTIONS 6, 7 AND 8.

**spectrochemical series** is a list of ligands arranged in order of their ability to cause d–d splitting:

$$I^- < Br^- < Cl^- < F^- < OH^- < H_2O < (CO_2)^{2-} < NH_3 < en < CN^-$$

smallest splitting ⟶ greatest splitting
smallest energy gap ⟶ greatest energy gap
longest wavelength ⟶ shortest wavelength

$[Cu(H_2O)_6]^{2+}$ absorbs at the red end of the spectrum, which makes an aqueous solution of copper(II) ions appear pale blue (Fig 27.31).

However, by substituting $H_2O$ with $NH_3$ ligands to form the ion $[Cu(NH_3)_4(H_2O)_2]^{2+}$, the difference in energy of the d orbital split increases. This shifts the absorption into the yellow part of the spectrum, which makes the solution deep blue.

**Fig 27.32** (ABOVE) The ultraviolet–visible spectrum of $[Cu(H_2O)_6]^{2+}$. Compare this with the ultraviolet–visible spectrum of $[Cu(NH_3)_4(H_2O)_2]^{2+}$ in Fig 27.29

Another factor that has an effect on d–d splitting is the number of each type of ligand in the complex:

$[Cr(H_2O)_6]^{3+}$  $[Cr(H_2O)_5Cl]^{2+}$  $[Cr(H_2O)_4Cl_2]^+$
violet  green  dark green

The arrangement of ligands around the central metal cation can also have an effect the colour of a complex (Fig 27.34).

*cis* isomer is violet     *trans* isomer is green

**Fig 27.34** The *cis* and *trans* isomers of $[Co(NH_3)_4Cl_2]^+$

## USING A COLORIMETER TO DETERMINE THE FORMULA OF A COMPLEX ION

A colorimeter is a simple form of visible spectrometer. A fixed band of wavelengths of visible light is selected using a filter. This is then passed through the solution under test. The light that has not been absorbed by the solution is transmitted to a photocell. The light generates an electric current, which is measured by a meter. The more light that is transmitted by the solution, the greater the electric current produced. However, most colorimeter meters are calibrated so that they record the light absorbed by the solution rather than the light transmitted. The absorbance of a solution is proportional to the concentration of the coloured compound in the solution.

**Fig 27.35** The workings of a colorimeter

white light source

colour filter     solution under test     photocell     meter

In Chapter 24 (page 531) we noted that the addition of aqueous potassium thiocyanate to a solution of aqueous iron(III) ions $[Fe(H_2O)_6]^{3+}$ causes a distinctive blood-red complex to form. We can use the colorimeter to determine the formula of this complex.

The first task is to select a filter for the colorimeter that will allow the solution to absorb most strongly. Since the iron thiocyanate complex is red it will absorb most in the blue wavelengths of the spectrum, so a blue filter is chosen. Solutions to be analysed are next made up to give different molar proportions of $Fe^{3+}(aq)$ and $SCN^-(aq)$: 10:0, 9:1, 8:2, 7:3, 6:4, 5:5, etc.

If the formula of the complex is $[Fe(H_2O)_5(SCN)]^{2+}$, then 1 mole of $[Fe(H_2O)_6]^{3+}$ would need to be added to 1 mole of $SCN^-$ ions, that is the 5:5 mixture. This is, in fact, the correct formula. If these different molar proportions were placed in turn into the colorimeter, a graph similar in shape to that in Fig 27.36 would be obtained. You will notice that the maximum absorbance corresponds to the molar proportions:

$$5Fe^{3+}(aq):5SCN^-(aq)$$

which is the 1 mole:1 mole ratio. Table 27.3 shows how the different solutions may be prepared.

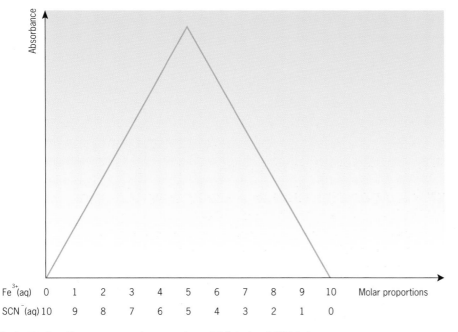

**Fig 27.36** Absorbance versus molar proportions of $Fe^{3+}(aq)$ and $SCN^-(aq)$

---

**? TEST YOURSELF**

**T**   **a)** What is the colour of the $[Fe(H_2O)_6]^{3+}$ ion?
To remind yourself turn to page 527
**b)** Which colour is it absorbing?
**c)** What is its shape?

---

**Table 27.3 Mixtures of 0.1 mol dm⁻³ of solutions of $Fe^{3+}(aq)$ and $SCN^-(aq)$ that could be used to determine the formula of the iron thiocyanate complex using a colorimeter**

| Volume of 0.1 mol dm⁻³ $Fe^{3+}(aq)/cm^3$ | Volume of 0.1 mol dm³ $SCN^-(aq)/cm^3$ |
|---|---|
| 10 | 0 |
| 9 | 1 |
| 8 | 2 |
| 7 | 3 |
| 6 | 4 |
| 5 | 5 |
| 4 | 6 |
| 3 | 7 |
| 2 | 8 |
| 1 | 9 |
| 0 | 10 |

---

**? TEST YOURSELF**

**U**   **a)** Write the method you would use to determine the formula of the nickel(II)–EDTA (**e**thylene**di**amine**te**tra**a**cetic ion) complex. You are provided with 0.05 mol dm⁻³ solutions of $Ni^{2+}(aq)$ and EDTA.
**b)** The formula of the complex under investigation is $[Ni(EDTA)]^{2-}$. Sketch the graph of absorbance against molar proportions you would expect to obtain from the colorimeter readings.

## ⚠ SCIENCE FEATURE The colours of gemstones

Some of the most beautiful and highly prized examples of the colours that transition metal ions can impart are found in gemstones. In many cases, the transition metal ions are actually impurities. For example, the deep red of rubies comes from traces of $Cr^{3+}$ ions embedded in a lattice of aluminium oxide. Since $Cr^{3+}$ ions have the same charge as $Al^{3+}$ ions and are about the same size, they occupy about 5% of the $Al^{3+}$ positions.

Emeralds are coloured green because $Cr^{3+}$ ions are present. This time the $Cr^{3+}$ ions are embedded in a lattice that contains large silicate anions $Si_6O_{18}^{12-}$ together with the cations $Be^{2+}$ and $Al^{3+}$.

**Fig 27.37** The deep red of these rubies results from the presence of $Cr^{3+}$ ions

**Fig 27.38** The green colour of the emeralds in this specimen is caused by $Cr^{3+}$ ions present as an impurity in $Be_3Al_2Si_6O_{18}$

So how can $Cr^{3+}$ impart different colours? The answer lies in the splitting of the d orbitals. There are three 3d electrons in $Cr^{3+}$ (Fig 27.39). One of these d electrons becomes excited by absorbing a photon of light with a wavelength in the visible spectrum. The wavelengths that are transmitted through the gemstone are predominantly red, which is why a ruby is red. In the case of emerald, the different environment of the $Cr^{3+}$ ion leads to a different energy gap between the two sets of d orbitals, one that allows the transmission of blue–green light while absorbing violet, yellow and red wavelengths.

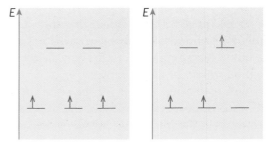

**Fig 27.39** Excitation of electrons in $Cr^{3+}$. $Cr^{3+}$ has three d electrons. The energy difference between the two sets of d orbitals in emerald or ruby causes the absorption of some of the visible spectrum wavelengths as photons excite the electrons to the higher energy d orbitals.

Other precious gemstones are coloured by different transition metal ions. The blue of sapphire results from $Ti^{4+}$, $V^{3+}$ or $Co^{3+}$ ions. Aquamarine, named after the colour of a tropical sea, owes its pale blue colour to $Fe^{3+}$ ions. The $Fe^{3+}$ ion also gives a yellow colour to topaz and a red colour to garnet. The purple of amethyst is caused by $Mn^{3+}$ ions.

## ⚠ SCIENCE FEATURE The amazing Monastral Blue

Through your studies in science you may have realized that many important discoveries occur by accident. An example in this chapter is the discovery of the first synthetic dye by William Perkin, which he called mauve (see the Science Feature on page 602). The English writer Horace Walpole invented a term for this in 1754; he called accidental discoveries that lead to happy outcomes *serendipity*. The word has its origin in a fairy tale about the three princes of Serendip who were constantly making chance but fortunate discoveries. The story of the blue pigment Monastral Blue is an example of serendipity.

The story begins in 1928 when a researcher at Scottish Dyes Ltd, A. G. Dandridge, noticed blue crystals on the lid of a glass-lined iron vat designed to make the compound phthalimide from molten phthalic anhydride and ammonia. Phthalimide is a white compound and was a necessary compound in the manufacture of some dyestuffs. Sometimes, when the glass lining became damaged, similar blue crystals would form on the sides of the

vat and contaminate the white molten phthalimide. It is fortunate that Dandridge was curious enough to examine and analyze these crystals. He realized that they were made of a compound of iron and he named it iron phthalocyanine (Fig 27.40). He also understood that it had potential as a pigment.

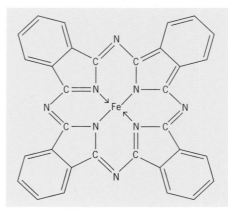

**Fig 27.40** Iron phthalocyanine

▶

**Fig 27.41** Steam locomotives were often painted with Monastral Blue.

The ring structure that surrounds the iron atom is very similar to those found in nature, called **porphyrins**. Iron porphyrin occurs in haemoglobin (Feature box on page 525) and the plant pigment chlorophyll also contains a porphyrin ring. The ring structures

contribute to the colour of the molecules. If you examine the iron phthalocyanine's structure closely, you will see it has a conjugated structure that forms a large delocalized system.

Once the structure of iron phthalocyanine was known researchers at Imperial College, London, set about making other phthalocyanines, the best of which was copper phthalocyanine, first made in 1934. This intense blue pigment goes under the trade name Monastral Blue. It is used extensively by printers, paint manufacturers and to colour plastics. It is non-toxic and is even used to stain cells in the sclera membrane of the eye to facilitate laser surgery. The reason it is so useful in laser treatments is that Monastral Blue strongly absorbs in the infrared region of the spectrum.

### ? TEST YOURSELF

**a)** Porphyrins and phthalocyanines are examples of *chromophores*, which have extensive *conjugated systems*. Explain the meanings of the terms in italics.
**b)** Iron phthalocyanine is a blue colour. What wavelengths of light is it absorbing?
**c)** Draw a possible visible spectrum for iron phthalocyanine.

## SUMMARY

- Electron energy levels exist in atoms and molecules. Electron transitions between these levels emit or absorb radiation.
- Colour arises when a substance absorbs or emits electromagnetic radiation partly in the visible spectrum.
- Absorption spectra record the wavelengths of photons that excite electrons and so cause them to jump to higher energy levels. Emission spectra record the wavelengths of photons emitted when excited electrons return to lower energy levels.
- Atomic emission spectroscopy can be used to determine the concentrations of metal ions in biological fluids, such as blood serum.
- Dyes are soluble in the medium in which they are applied; pigments are insoluble.
- Azo dyes are produced by coupling reactions, such as that between benzenediazonium chloride and phenol.
- The functional group of an azo compound is the azo group, –N=N–.
- Electron transitions in organic molecules give rise to ultraviolet or visible absorptions, because the molecules contain double or triple bonds, delocalized systems or lone pairs of electrons
- Ultraviolet–visible spectroscopy is possible because

when the outer electrons of atoms or ions in compounds are excited they absorb in the ultraviolet or visible part of the spectrum.
- Conjugated systems have alternating double and single bonds; the delocalization that occurs lowers the energy of electron transitions, so conjugated compounds may appear coloured.
- The part of a molecule responsible for absorbing coloured radiation is called the chromophore. It is usually an extended delocalized electron system.
- The colour changes of acid–base indicators results from a change in the delocalization of the chromophore when $H^+$ ions are added or removed.
- The colour of a transition element's ions results from a difference in energies between d orbitals caused by the ligands around the metal cation. This splitting allows d–d electron transitions.
- Factors that affect d–d splitting include the oxidation state of the metal, the nature and number of the ligands, and the arrangement of the ligands around the metal cation.
- Complexes of $Zn^{2+}$ and $Cu^+$ are white because the 3d subshell is full.
- The formula of a complex can be determined using colorimetry.

## ? EXAM QUESTIONS

**1**

**a)** What colour, if any, is produced in flame tests on the following compounds:
(i) sodium nitrate; (ii) barium carbonate;
(iii) magnesium sulphate?

**b)** Explain how flame colours are produced.

**2**

**a)** Write an equation for the reaction of phenol with benzenediazonium chloride.

**b)** Give the conditions for this reaction.

**c)** What can the product be used for?

**3**

**a)** Phenylamine is formed by the reduction of nitrobenzene. Mixtures of phenylamine and nitrobenzene could be separated by fractional distillation, but another way is to shake the mixture with dilute hydrochloric acid, separate the aqueous and organic layers, and make the aqueous layer strongly alkaline with sodium hydroxide solution.
(i) Give the reagents and conditions that are used to reduce nitrobenzene to phenylamine.
(ii) Give the structural formula of the organic substance that is found in the aqueous phase after shaking with acid.
(iii) What is the effect on the organic species of making this aqueous layer strongly alkaline?

**b)** Phenylamine is used to make azo dyes. It is first converted into benzenediazonium chloride.
(i) State the reagents and the conditions needed to convert phenylamine into benzenediazonium chloride.
(ii) Give reasons for the temperature conditions employed.
(iii) Write an equation, using structural formulas, to represent the reaction.
(iv) Addition of the benzenediazonium chloride solution to an alkaline solution of 2-naphthol gives the scarlet dye phenylazo-2-naphthol (naphthalen-2-ol). Give the structural formula of phenylazo-2-naphthol.

Edexcel 6084/P.01 Module Test 4 Jan 1999 Q4a,b

**4** 4-Methylphenylamine is an aromatic amine.

**a)** Draw the structure for 4-methylphenylamine.

**b)** 4-Methylphenylamine reacts with aqueous sodium nitrite and dilute hydrochloric acid to form a diazonium salt.
(i) State the conditions for this reaction.
(ii) Draw the structure of the diazonium salt formed. Give a name of a substance that will react with the diazonium salt to form a di-azo dye.

**5** 4-Aminobenzenesulphonic acid is used in the manufacture of a range of sulphanilamide drugs, which are used to treat pneumonia and related diseases. The structure of the compound is shown in Fig 27.Q5.

**Fig 27.Q5** $H_2N$—⬡—$SO_3H$

The sulphonic acid group has the same structure as part of the sulphuric acid molecule.

**a)** 4-Aminobenzenesulphonic acid reacts with nitrous acid to form a diazonium salt.
(i) Give the names of the reagents needed for this reaction.
(ii) Give the conditions for this reaction.
(iii) Draw the structure of the diazonium salt formed.
(iv) Give the name of a substance that will react with the diazonium salt to form an azo-dye.

**b)** Explain why azo-dyes are coloured.

**6**

**a)** State four factors that determine the colour of a particular transition-metal complex.

**b)** Give the shape of the complex ion $[Cu(H_2O)_6]^{2+}$ and state the type of electronic transition responsible for the blue colour of this ion.

**c)** Give the formula and shape of the ion responsible for the greenish-yellow colour of the solution obtained when $[Cu(H_2O)_6]^{2+}$ is treated with an excess of concentrated hydrochloric acid.

**d)** The manganate(VII) ion, $[MnO_4]^-$, is purple. Give the full electronic configuration of manganese in this oxidation state, and explain how the colour arises.

AQA CH08 Chemistry of Colour Module Test June 2000 Q2(not d)

**7** The visible absorption spectrum of an aqueous solution that contains the tetrachlorocobaltate(II) ion is shown in fig 27.Q7.

**Fig 27.Q7**

**a)** What is the formula of the tetrachlorocobaltate(II) ion?

**b)** Suggest a possible shape for the tetrachlorocobaltate(II) ion.

**c)** The tetrachlorocobaltate(II) ion absorbs visible light.
**(i)** Explain why the solution appears blue when viewed in daylight.
**(ii)** Explain how the tetrachlorocobaltate(II) ion absorbs visible light.

**8** The colour circle in Fig 27.Q8 is useful for relating the absorption spectrum of a dye to its observed colour.

**a)** What is the term used to describe colours in the opposite sectors of the circle?

**b)** Deduce the observed colour of a dye solution that has an absorption maximum at 570 nm and state the colour of the light absorbed by the dye.

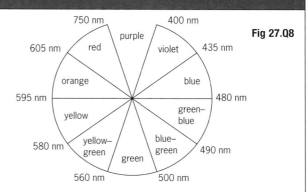

**Fig 27.Q8**

**c)** Why in the colour circle in Fig 27.Q8 are there no wavelengths associated with the colour purple?

AQA CH08 Chemistry of Colour Module Test June 2000 Q3a(not iii)

# KEY SKILLS **ASSIGNMENT**

## HOW ARE DYES STUCK ONTO TEXTILES?

**The dye of blue denim jeans is indigo and it is called a vat dye. A vat dye is usually soluble in its reduced form, but when oxidized becomes insoluble and precipitates in the pores of the denim cotton fibres. This property means that vat dyes do not wash out of clothes.**

(To learn a little more about the history of blue denim and indigo, read the Opener of this chapter.)

Vat dyes are particularly effective for cotton fibres and other fabrics that contain cellulose. The large number of hydroxyl groups on the cellulose molecules mean that the fabric readily absorbs water and hence the water-soluble dye.

soluble reduced form of indigo which is a pale yellow

↓ O₂ from air

insoluble indigo which precipitates in the cotton fibre

**Fig 27.A1** The soluble and insoluble forms of indigo

**1** **a)** The reduced form of indigo is pale yellow, while indigo is blue. Both molecules have conjugated systems. What is meant by the term conjugated system?

**b)** Which form of indigo absorbs the longest wavelength of light?

**c)** Which form of indigo absorbs the highest energy photons in the visible spectrum?

**2** **a)** Cellulose is a polymer made up of glucose units. Draw the repeat unit of cellulose.

**b)** What type of bonding occurs between water molecules and the hydroxyl group? (See page 290.)

Another type of cotton dye is called a **direct dye**. Direct dyes are long planar molecules that can lie alongside the cellulose polymer chains, and form intermolecular hydrogen bonds and induced dipole–induced dipole forces. Such large dye molecules are made water-soluble using $SO_3^-Na^+$ groups. Direct dyes are usually diazo or triazo dyes (Fig 27.A2).

**Fig 27.A2** C.I. Direct Brown 138, a direct dye. C.I. are the initials for the Colour Index, an internationally recognized publication that lists and categorizes almost 40 000 commercial pigments and dyes

**3** Why is Direct Brown called a triazo dye?

**4** What is meant by the term chromophore?

A much simpler azo compound is produced by starting with phenylamine to produce the benzenediazonium ion. This undergoes a coupling reaction with phenol to produce a bright orange azo compound.

# KEY SKILLS **ASSIGNMENT**

**5** Explain how you would synthesize this orange azo compound in the laboratory, stating the conditions required. Write balanced equations for the reactions you describe.

The intermolecular forces between direct dyes and cotton are relatively weak, so the water-fastness of the dye is poor and some of the dye comes out when the material is washed. However, in 1956 chemists at ICI produced a dye that would form covalent bonds with the cellulose fibres of cotton. The covalent bonds are formed by reactive groups that first bond to the dye molecules (Fig 27.A3). These groups then react with the hydroxyl groups on the cellulose fibres (Fig 27.A4). The dye is water-fast and the colour does not run when the textile is washed because covalent bonds have formed with the cellulose fibres. Such a dye is called **fibre-reactive**.

Fig 27.A3 Trichlorotriazine reacts with a dye molecule

**Fig 27.A4** The reactive chlorotriazine group forms a covalent bond with cotton

**Fig 27.A5** Reactive Red 2

**6** **a)** Draw the dye molecule shown in Fig 27.A5 before the reactive group has bonded to it.
**b)** Which functional groups make the dye water-soluble?

**7** Why does the covalent bond between the dye and cotton make it water-fast?

Wool and silk contain many amino functional groups ($NH_2$). These groups are basic and react with acid dyes that contain sulphonic acid groups ($SO_3H$) to form ionic bonds (Fig 27.A6).

**Fig 27.A6** Acidic dyes bond to fibres that contain basic $NH_2$ groups, so this type of dye is used for wool and silk.

**Fig 27.A7** C.I. Acid Orange 7 is a salt of the sulphonic acid, which makes it more soluble.

Poly(propenenitrile) fibres (acrylics) contain $CO_2H$ and $SO_3H$ groups, which are acidic. These can form ionic bonds with basic dyes.

**Fig 27.A8** Basic dyes form ionic bonds with acid functional groups on poly(propenenitrile) fibres

The last group of dyes we consider are **disperse dyes**. All the other dye groups are water-soluble when applied, but polyester fibres do not form hydrogen bonds with water and are known as **hydrophobic** (water-hating). They do not allow water molecules to penetrate them. Disperse dyes are a fine suspension of dye particles that are absorbed by the fibres and held there by induced dipole–induced dipole forces and some hydrogen bonding. As the dye is only sparingly soluble it tends to stay in the fibres, and so it is water-fast.

**8** Draw the structures of the monomers that make up the polymer fibre chains of nylon, wool, polyester and poly(propenenitrile). Chapter 14 and Chapter 20 will help you.

# 28

# Ammonia, bases and food production

# Ammonia, bases and food production

**28**

**Population explosion** The graph shows the population explosion that has taken place since the early 1800s

With the world's population increasing at more than 250 000 people per day, the question arises: How will enough food for these extra mouths be produced? The overall output of staple foods, such as grain, has risen significantly since 1950 through better farming methods and the use of fertilizers. But in many African and Latin American countries food production is actually falling.

As crops grow, they take essential elements such as nitrogen and phosphorus from the soil. Farmers in poorer countries cannot afford fertilizers to restore these elements to the soil. As a result, crop yields fall and in time the soil becomes unable to sustain useful growth.

In richer countries, the pressure to build homes, factories and roads on good agricultural land has meant farmers have sought to obtain higher crop yields from smaller amounts of land by using pesticides and fertilizers. But the use of these agrochemicals has its problems: the over-use of fertilizers has polluted groundwater, and hence rivers which feed into reservoirs.

There are moves to develop fertilizers that could be used in smaller amounts to maintain maximum crop yields. These would also reduce the harmful effects on the environment. If the manufacturers do manage to find the right formula, and can convince farmers everywhere that these new fertilizers are effective, then there could be a better chance of meeting the needs of the world's ever-swelling population.

**Food source** Grains are the world's leading food crops, but the increase in production has not kept up with the increase in demand

## 1 THE NITROGEN CYCLE

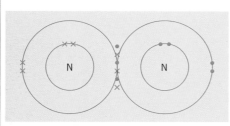

**Fig 28.1** Dot-and-cross diagram for a nitrogen molecule

For nearly all manufactured fertilizers, the starting substance is ammonia – a compound of hydrogen and nitrogen. Nitrogen is readily available since it forms about 80 per cent by volume of the lower atmosphere. However, it is a very unreactive gas because of the strong triple covalent bond in each nitrogen molecule (Fig 28.1). The bond energy for this triple bond is 994 kJ mol$^{-1}$.

All plants and animals contain nitrogen in amino acids, polypeptides, proteins and nucleic acids. Plants cannot obtain nitrogen directly from the air. Instead, they take up nitrogen-containing compounds from the soil. The **nitrogen cycle** (Fig 28.2) describes the way that nitrogen and its compounds circulate within the environment and living organisms.

Fig 28.2 shows the processes that remove nitrogen-containing compounds from the soil. There must be a constant supply of nitrates to the soil for plants to grow healthily. Much of the nitrogen cycle has taken place naturally for many millions of years. In recent times, however, human activity has added some extra stages to the cycle. The advent of intensive farming has affected the

YOU CAN READ ABOUT AMINO ACIDS, POLYPEPTIDES AND PROTEINS IN CHAPTER 19.

delicate balance of the nitrogen cycle. Intensive farming removes more nitrogen-containing compounds than are naturally put back into the soil, so it has become necessary to supplement the supply of nitrogen by adding synthetic fertilizers such as ammonium nitrate.

# NITROGEN FIXATION

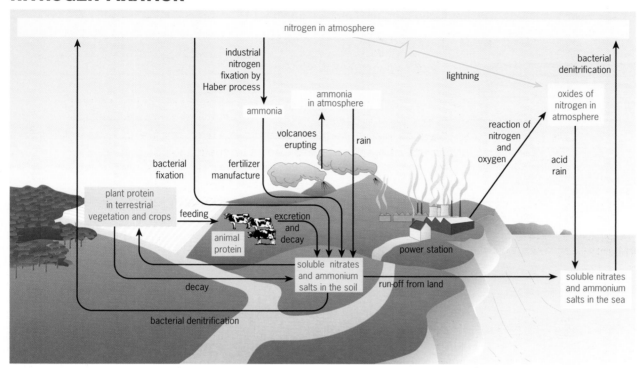

Fig 28.2 The nitrogen cycle

Processes that convert atmospheric nitrogen into nitrogen-containing substances are a very important part of the nitrogen cycle. They are known as **nitrogen-fixation** processes. Nitrogen fixation occurs by natural processes and as a result of human activity. The most important synthetic nitrogen-fixation process is the manufacture of ammonia. This chapter shows how the synthesis of such a simple molecule has had a profound effect on agriculture and the chemical industry.

## LEGUMINOUS PLANTS AND NITROGEN FIXATION

Some plants, such as peas and clover, are able to convert atmospheric nitrogen into ammonium ions, $NH_4^+$, by the action of bacteria found in root nodules. The ammonium ions may be directly absorbed into the roots and metabolized by plants to form plant protein. Alternatively, the ammonium ions may be oxidized by nitrifying bacteria to give nitrate ions:

$$NH_4^+(aq) + 1\tfrac{1}{2}O_2(g) \rightarrow NO_2^-(aq) + 2H^+(aq) + H_2O(l)$$

$$NO_2^-(aq) + \tfrac{1}{2}O_2(g) \rightarrow NO_3^-(aq)$$

The aqueous nitrate ions can then be absorbed through the roots by osmosis and later assimilated into plant protein.

## LIGHTNING AND NITROGEN FIXATION

Molecular nitrogen only reacts if enough energy is available to break the very strong triple covalent bond. Lightning provides sufficient energy to break the bond, and allows the direct combination of nitrogen and oxygen to form nitrogen monoxide:

$$N_2(g) + O_2(g) \rightarrow 2NO(g)$$

Fig 28.3 The bacteria in root nodules of peas, clover and beans convert atmospheric nitrogen into ammonium ions and nitrates

**Table 28.1** $K_c$ at different temperatures for the equilibrium $N_2(g) + 3H_2(g) \rightleftharpoons 2NH_3(g)$

| Temperature/K | $K_c$/dm⁶ mol⁻² |
|---|---|
| 298 | $4.05 \times 10^8$ |
| 400 | $4.39 \times 10^3$ |
| 500 | $5.98 \times 10$ |
| 600 | $4.03$ |
| 700 | $2.56 \times 10^{-1}$ |
| 800 | $2.98 \times 10^{-2}$ |
| 900 | $5.45 \times 10^{-3}$ |

Nitrogen monoxide oxidizes into nitrogen dioxide even at low temperatures. Nitrogen dioxide may be further oxidized in the presence of water and oxygen to give dilute nitric acid, a solution that contains nitrate ions. These processes are represented by the equations below.

$$NO(g) + \tfrac{1}{2}O_2(g) \rightarrow NO_2(g)$$

$$2NO_2(g) + H_2O(l) + \tfrac{1}{2}O_2(g) \rightarrow 2HNO_3(aq)$$

**Fig 28.4** The enormous quantity of electrical energy available within a lightning bolt is sufficient to allow nitrogen to react with oxygen. This is one of the most important natural nitrogen-fixation processes.

## 2 THE HABER PROCESS AND NITROGEN FIXATION

In Chapter 1, we describe the manufacture of ammonia by the Haber process.

$$N_2(g) + 3H_2(g) \rightleftharpoons 2NH_3(g)$$

Ammonia is a starting material in the production of many fertilizers. The presence of the triple bond in each nitrogen molecule means that nitrogen and hydrogen react very slowly under normal conditions. You will also notice that the reaction is reversible. The choice of conditions used in the manufacture of ammonia determines the rate of reaction and the equilibrium position of the reaction mixture.

The position of equilibrium for the reaction between nitrogen and hydrogen can be estimated by looking at the numerical value of the equilibrium constant. The equilibrium constant, $K_c$, for this reaction is represented by the equation:

$$K_c = \frac{[NH_3(g)]^2}{[H_2(g)]^3[N_2(g)]}$$

The concentrations refer to the molar concentrations at equilibrium.

Table 28.1 illustrates that at low temperatures the numerical value of $K_c$ is higher – the equilibrium position lies well to the right. At higher temperatures the equilibrium position lies to the left, the side of the reactants nitrogen and hydrogen. In contrast, the rate of reaction is very slow at low temperatures, but accelerates at higher temperatures. This conflict between rate of reaction and position of equilibrium is explored in some detail in Chapter 1.

In a gas-phase reaction it is not always appropriate to measure the concentrations of the reactants and the products. It is often easier to measure the total pressure of the equilibrium mixture and the partial pressures of the reactants and of the products.

### PARTIAL PRESSURE

For a mixture of gases it is convenient to express the composition in terms of **partial pressures** rather than concentrations.

**The partial pressure of a gas in a container of several gases is the pressure that particular gas would have if it were the only gas present in the container.**

Consider a mixture of neon and helium gases. The partial pressure of neon, $p$Ne, is proportional to the **mole fraction** of neon, $x$Ne, in the mixture, and:

$$pNe = xNe \times P$$

where $P$ is the total pressure of the mixture of gases.

The mole fraction of neon is the number of moles of neon gas in the mixture divided by the sum of the number of moles of each gas present:

$$xNe = \frac{nNe}{nHe + nNe}$$

where $n$Ne = number of moles of neon and $n$He = number of moles of helium.

WE INTRODUCED THE CONCEPT OF THE
MOLE FRACTION ON PAGE 134.

✔ **REMEMBER THIS**

In a gaseous mixture, the mole ratio of each gas equals the ratio of the partial pressures of the gases.

● **EXAMPLE**

**Q** A mixture of gases at a total pressure of $1.0 \times 10^5$ Pa contains 3.4 mol of carbon dioxide, 4.5 mol of oxygen and 1.1 mol of nitrogen. What is the partial pressure of each gas in the mixture?

**A** The mole fraction of $CO_2 = \dfrac{3.4}{3.4 + 4.5 + 1.1} = 0.38$

So: $pCO_2 = 0.38 \times 1.0 \times 10^5$ Pa
$\quad\quad\quad = 3.8 \times 10^4$ Pa

The mole fraction of $O_2 = \dfrac{4.5}{3.4 + 4.5 + 1.1} = 0.50$

So: $pO_2 = 0.50 \times 1.0 \times 10^5$ Pa
$\quad\quad\quad = 5.0 \times 10^4$ Pa

The mole fraction of $N_2 = \dfrac{1.1}{3.5 + 4.5 + 1.1} = 0.12$

So: $pN_2 = 0.12 \times 1.0 \times 10^5$ Pa
$\quad\quad\quad = 1.2 \times 10^4$ Pa

? **TEST YOURSELF**

**C** A cylinder contains 2.0 mol of oxygen gas, and 5.6 mol of gaseous dinitrogen oxide. The pressure inside the cylinder is 450 kPa. What is the partial pressure of oxygen in the cylinder?

LATER IN THIS CHAPTER, WE DISCUSS
THE FACTORS THAT AFFECT THE
POSITION THE EQUILIBRIUM REACHES
AND THE NUMERICAL VALUE OF THE
EQUILIBRIUM CONSTANT.

## THE EQUILIBRIUM CONSTANT, $K_p$

In reactions that involve gaseous reactants or products, the equilibrium constant is often stated in terms of the partial pressures of the gases present. The expression for this equilibrium constant resembles that for $K_c$, except that the partial pressures of the gases in the equilibrium mixture are used instead of the molar concentrations of the components in the mixture. The expression for the equilibrium constant, $K_p$, for the Haber process is:

$$K_p = \frac{(pNH_3)^2}{(pH_2)^3(pN_2)}$$

where $p$NH$_3$ is the partial pressure of ammonia gas.

Notice that the partial pressures are raised to the same powers as the concentrations in the expression for $K_c$. The equilibrium constant $K_p$ has both a numerical value and a unit. The higher the numerical value, the greater the proportion of ammonia in the equilibrium mixture and the more the position of equilibrium lies on the right-hand side. If the partial pressures are measured in Pascals, the units of $K_p$ for this equilibrium will be Pa$^{-2}$. The numerical value of this equilibrium constant will not be affected by changes in concentration and pressure.

If an equilibrium involves solids, liquids and solutions, then only the gases in the process are used in the $K_p$ expression. The unit for the equilibrium constant in the reaction:

$$NH_4Cl(s) \rightleftharpoons NH_3(g) + HCl(g) \quad K_p = (pNH_3)(pHCl) \quad \text{is} \quad Pa \times Pa = Pa^2.$$

? **TEST YOURSELF**

**D** Write down expressions for the equilibrium constant, $K_p$, for each of the following equilibria:
**a)** $H_2(g) + I_2(g) \rightleftharpoons 2HI(g)$;
**b)** $2NO(g) + O_2(g) \rightleftharpoons 2NO_2(g)$;
**c)** $CaCO_3(s) \rightleftharpoons CaO(s) + CO_2(g)$.

**E** For each of the equilibria in Test Yourself **D** state the units of the equilibrium constant $K_p$. Assume that the partial pressures are all measured in Pa.

**SEE QUESTIONS 2, 3 AND 4.**

## ? TEST YOURSELF

**F** A mixture of nitrogen and hydrogen were heated at a constant temperature in a reaction vessel until equilibrium was reached. The partial pressures of each gaseous component in the reaction mixture at equilibrium were:

$pNH_3 = 8.3 \times 10^5$ Pa
$pN_2 = 3.5 \times 10^4$ Pa
$pH_2 = 1.1 \times 10^5$ Pa

Calculate the numerical value for the equilibrium constant, $K_p$.

## ? TEST YOURSELF

**G** Sulphur dioxide and oxygen were mixed in a mole ratio of 2:1 at a total initial pressure of 300 kPa. The mixture was held at a temperature of 430 °C in the presence of a suitable catalyst until it reached equilibrium:

$$2SO_2(g) + O_2(g) \rightleftharpoons 2SO_3(g)$$

At equilibrium, the partial pressure of sulphur trioxide was 190 kPa.

**a)** Calculate the initial partial pressures of $SO_2$ and $O_2$.

**b)** Calculate the partial pressures of $SO_2$ and $O_2$ at equilibrium.

**c)** Calculate the value of $K_p$ for this equilibrium.

## ◯ EXAMPLE

**Q** Fritz Haber studied the reaction between hydrogen and ammonia. He mixed nitrogen and hydrogen in a mole ratio of 1:3, at 300 °C and in the presence of a catalyst. The reaction was allowed to reach equilibrium. At equilibrium, the total pressure of the equilibrium mixture was 800 kPa and the partial pressure of the ammonia was 75 kPa. Calculate the numerical value for $K_p$ under these conditions.

**A** As in other equilibrium questions, we need to start with a balanced equation. Under each term we write the partial pressure at equilibrium.

$$N_2(g) + 3H_2(g) \rightleftharpoons 2NH_3(g)$$

Partial pressures at equilibrium:   $x$   $3x$   75 kPa

If the partial pressure of nitrogen is $x$ kPa, then the partial pressure of hydrogen must be $3x$. This follows because the initial mole ratio was 1:3, which remains unchanged throughout the reaction, since the stoichiometry of the reaction indicates that for every mole of nitrogen that reacts, three moles of hydrogen react.

The total pressure is the sum of the partial pressures:

$$x + 3x + 75 = 800 \text{ kPa}$$

So:                    $x = 181$ kPa

Therefore:              $pN_2 = 181$ kPa

and                    $pH_2 = 543$ kPa.

Substituting these values into the expression for the equilibrium constant gives:

$$K_p = 1.93 \times 10^{-7} \text{ kPa}^{-2}.$$

## MANUFACTURING AMMONIA

Most manufacturers of ammonia use:

● a temperature of about 450 °C;
● a pressure of between 200 and 400 atmospheres ($2 \times 10^4$ kPa and $4 \times 10^4$ kPa);
● an iron catalyst with potassium hydroxide promoter.

The conditions are chosen for economic reasons. A typical factory produces up to 1500 tonnes of ammonia per day. Some manufacturers use pressures as high as 1000 atmospheres, but this presents technological difficulties in the construction of the plant. The use of better catalysts has allowed some modern plants to reduce the operating pressures to as low as 80 atmospheres.

## LE CHATELIER'S PRINCIPLE AND THE HABER PROCESS

Le Chatelier's principle predicts how changes in temperature, pressure and concentration will affect a system at equilibrium. Fig 28.5 summarizes the effects on the equilibrium position when the conditions are changed in the Haber process.

### THE POSITION OF EQUILIBRIUM

Le Chatelier's principle suggests that low-temperature and high-pressure conditions would maximize ammonia production. In terms of rate of reaction, if the temperature is too low the reaction proceeds too slowly. Therefore, manufacturers use a combination of a moderate temperature and a catalyst to ensure that the rate of ammonia production is high.

High pressure gives a fast rate of reaction for a gas-phase reaction, but using higher pressures drastically increases the cost of building and running the plant.

## EQUILIBRIUM CONSTANTS

The equilibrium constants $K_c$ and $K_p$ do not change when the concentration of any species in the equilibrium changes. However, the numerical values will change with temperature. For an endothermic reaction, as the temperature increases the numerical value of the equilibrium constant increases. For an exothermic reaction, as the temperature rises the numerical value of the equilibrium constant decreases. Catalysts have no effect whatsoever on the numerical value of the equilibrium constant.

## 3 MANUFACTURING NITRIC ACID

Some of the ammonia that is produced in the Haber process is used to manufacture nitric acid. The **Ostwald** process uses cheap materials – ammonia, air and water – to manufacture nitric acid (Fig 28.6).

**Fig 28.6** A flow diagram showing the Ostwald process for manufacturing nitric acid

The flow diagram illustrates the three key steps in the manufacture of nitric acid:

- oxidation of ammonia to give nitrogen monoxide, NO;
- oxidation of nitrogen monoxide to give nitrogen dioxide;
- disproportionation of dinitrogen tetroxide to form nitric acid.

Each of these reactions is an equilibrium process, so the compromise between rate and yield is an important economic consideration.

During the manufacturing process, the oxidation state of nitrogen changes from –3 in ammonia to +5 in nitric acid. This change also occurs in the bacterial oxidation of ammonium ions to nitrate ions in the soil.

## OXIDATION OF AMMONIA

Ammonia can be oxidized to give nitrogen monoxide by reacting excess air and ammonia at 900 °C and about 10 atmospheres, using a platinum–rhodium catalyst (page 535).

$$4NH_3(g) + 5O_2(g) \rightleftharpoons 4NO(g) + 6H_2O(g) \quad \Delta H = -909 \text{ kJ mol}^{-1}$$

The ammonia must be purified to remove any catalytic poisons because, although the starting materials are cheap, the catalyst itself is very expensive. Although the conditions are a compromise between rate and yield, a 96 per cent conversion can be achieved with them.

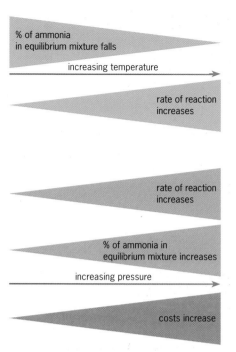

**Fig 28.5** The effects of changing conditions on the Haber process

**SEE QUESTIONS 1, 2 AND 3.**

✔ **REMEMBER THIS**

In economic terms, the best conditions for the Haber process are a compromise between achieving a high rate of reaction and a high percentage of ammonia in the equilibrium mixture.

Not all ammonia produced by the Haber process is used to make fertilizers; nitric acid and polyamides, such as nylon, are other important derivatives.

**?** **TEST YOURSELF**

**H** a) Using Le Chatelier's principle, predict the conditions that would give the maximum conversion of ammonia to nitrogen monoxide.
**b)** Predict conditions that would give the maximum rate of reaction.
**c)** Comment on the conditions actually used for the catalytic oxidation.
**d)** Write down an expression for the equilibrium constant, $K_p$, for the oxidation of ammonia.

**I** At 25 °C and 100 kPa, the partial pressures in an equilibrium mixture of $N_2O_4$ and $NO_2$ are $pN_2O_4 = 70$ kPa and $pNO_2 = 30$ kPa.
**a)** Calculate the numerical value for the equilibrium constant $K_p$ and state the units for the constant.
**b)** Describe the effect on the numerical value of $K_p$ if the temperature of the equilibrium mixture increases, assuming the total pressure remains at 100 kPa.
**c)** Describe the effect on the numerical value of $K_p$ if the total pressure of the equilibrium mixture increases, assuming the temperature remains at 25 °C.

**J** Le Chatelier's principle suggests that the conversion of $NO_2$ into $N_2O_4$ should be carried out at high pressure. Suggest advantages of carrying out this reaction at low pressures.

**K** Write down the oxidation state of nitrogen in: **a)** $N_2O_4$; **b)** $HNO_3$; **c)** NO.

**L** a) By using the change in oxidation state, confirm that the reaction of dinitrogen tetroxide, $N_2O_4$, with water is a disproportionation.
**b)** Explain why the reaction of $N_2O_4$ with water is carried out at low temperature.

## OXIDATION OF NITROGEN MONOXIDE

The gases from the oxidation of ammonia are extremely hot and they must be cooled down to about 25 °C. At this temperature, extra air is added to the nitrogen monoxide. Two exothermic reactions take place to finally form dinitrogen tetroxide, $N_2O_4$.

$$2NO(g) + O_2(g) \rightleftharpoons 2NO_2(g) \qquad \Delta H = -115 \text{ kJ mol}^{-1}$$

$$2NO_2(g) \rightleftharpoons N_2O_4(g) \qquad \Delta H = -58 \text{ kJ mol}^{-1}$$

Low temperatures are used for both of these reactions. According to Le Chatelier's principle, lower temperatures favour a shift in the equilibrium position to the right because both of the reactions are exothermic. Le Chatelier's principle also predicts that high pressures promote product formation because there are fewer moles of gas on the right-hand side of the equation. In practice, a pressure of only 4 to 10 atmospheres is sufficient to obtain a reasonable yield and to move the gases around the pipes within the plant. Higher pressures only increase the operating costs within the chemical plant and have little effect on shifting the position of equilibrium, which already lies well to the right-hand side.

## FORMATION OF NITRIC ACID

In the final stage of the manufacture of nitric acid, dinitrogen tetroxide ($N_2O_4$) reacts with water to form nitric acid ($HNO_3$) and nitrogen monoxide (NO).

$$3N_2O_4(g) + 2H_2O(l) \rightleftharpoons 4HNO_3(aq) + 2NO(g) \qquad \Delta H = -103 \text{ kJ mol}^{-1}$$

This is an example of disproportionation, since nitrogen in $N_2O_4$ is both oxidized and reduced during the reaction. The nitrogen monoxide produced can be recycled so that it is not wasted.

The nitric acid formed by the reaction still contains some water. This water can be removed by treatment with concentrated sulphuric acid.

**Fig 28.7** Nitric acid is used to make explosives such as TNT and nitroglycerine

## 4 FERTILIZERS AND THE NITROGEN CYCLE

All plants require essential elements for healthy growth. The three most important elements required are nitrogen, phosphorus and potassium (NPK). When fertilizers are sold they often quote an NPK value, which gives the percentage by mass of each of these three elements in the fertilizer.

As you can see from Table 28.2, no single compound gives all three of the essential elements. To provide all three essential elements, commercial fertilizers are normally a mixture of at least two chemicals.

**Table 28.2 Some chemical fertilizers and their NPK ratios. All the figures in this table can be worked out from the percentage composition by mass of the compounds**

| Fertilizer | Formula | %N | %P | %K |
|---|---|---|---|---|
| Ammonium nitrate | $NH_4NO_3$ | 35 | 0 | 0 |
| Ammonium phosphate | $(NH_4)_3PO_4$ | 28 | 21 | 0 |
| Potassium nitrate | $KNO_3$ | 39 | 0 | 14 |
| Ammonium sulphate | $(NH_4)_2SO_4$ | 21 | 0 | 0 |
| Urea | $(NH_2)_2CO$ | 47 | 0 | 0 |

Fertilizers need to be soluble in water so that they can be absorbed through the roots of plants. It is also preferable if they can be manufactured in pellet form to enable easy application to farm land.

## MANUFACTURE OF FERTILIZERS

Most fertilizers are manufactured by reacting either ammonia or nitric acid with another compound (Fig 28.8). Ammonia is a base. It reacts with acids to give ammonium salts, some of which are listed in Table 28.2. For example, ammonia and phosphoric acid react to form ammonium phosphate, which contains two of the essential elements needed by plants.

Problems arise when fertilizers such as ammonium nitrate are applied to alkaline soils. Under these conditions, ammonia can be formed either as a gas or as aqueous ammonia:

$$(NH_4)_2SO_4(s) + 2OH^-(aq) \rightarrow 2NH_3(g) + SO_4^{2-}(aq) + 2H_2O(l)$$

Addition of an excess of calcium oxide or calcium hydroxide to neutralize acidic soils can have the same effect on fertilizers that contain the ammonium ion; again, ammonia gas is liberated and the soil loses available nitrogen.

**? TEST YOURSELF**

**M** A 1 kg bag of fertilizer is made up of 500 g of potassium nitrate and 500 g of ammonium phosphate. Calculate the NPK ratio for this bag of fertilizer.

**? TEST YOURSELF**

**N** Write down the equation to show the formation of the fertilizer ammonium nitrate from nitric acid and aqueous ammonia.

**SEE QUESTION 4.**

**Fig 28.8** Some reactions of ammonia that give rise to many fertilizers

**Fig 28.9** Ammonium nitrate fertilizer being sprayed onto a field. In some parts of the world, liquid ammonia is pumped directly into the ground as a fertilizer. This can be quite dangerous – liquid ammonia can cause chemical burns and irritate the respiratory tract

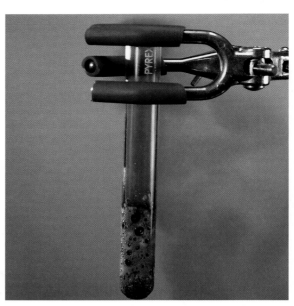

**Fig 28.10** Some reactions of nitric acid. Concentrated nitric acid acts as an oxidizing agent, whereas dilute nitric acid acts as a strong acid. Notice that dilute nitric acid reacts with copper to form nitrogen monoxide rather than hydrogen

**Fig 28.11** Concentrated nitric acid oxidizes the non-reactive metal copper to form a blue solution of aqueous copper(II) nitrate. The brown gas is nitrogen dioxide

## OVER-USE OF FERTILIZERS

Although fertilizers can increase crop yields, it is important to use the correct amount of fertilizer to ensure that most of it is absorbed by the plants. If too much fertilizer is applied, it is leached from the ground by water and eventually it can reach rivers and even domestic water supplies. This is a particular problem with nitrate fertilizers, which are more likely to dissolve into the soil solution. In East Anglia there is some concern because the nitrate level in domestic water supplies exceeds the acceptable concentration of $50\,\text{mg}\,\text{dm}^{-3}$ recommended by the European Union.

Much research into the level of fertilizer run-off and leaching is still required. It is not only the amount and type of fertilizer used that affect nitrate levels – the time of year, type of crop, soil type and underlying bedrock are all factors that must be considered.

## EUTROPHICATION

Leaching of fertilizers from the ground by water causes aqueous nitrate and phosphate ions to enter rivers. Nitrate ions and, to a greater extent, phosphate ions help the growth of all plants in a river. Green algae reproduce in huge numbers using the nitrate and phosphate nutrients and form an algal bloom that covers the surface and clogs up the river.

The algal bloom prevents sunlight from reaching plants that grow beneath the surface of the water. Consequently, these underwater plants cannot photosynthesize and they die. Colonies of bacteria feed on the decaying plant material. The bacteria's respiration uses up most of the dissolved oxygen. As a result, other aerobic forms of life, such as fish, die. The whole process – which results in the death of a river – is known as **eutrophication**.

**Fig 28.12** Excess phosphate and nitrate nutrients in river water cause the accelerated growth of green algae. The algae form a thick surface layer, which causes underwater plants and fish to die. Increased algal growth in the sea near river estuaries is also of concern

## ORGANIC FERTILIZERS

The process of bacterial decay provides a route by which animal and plant material may be recycled into ammonium ions and nitrate ions. Many farmers now use organic fertilizers, such as manure, fish blood and bone. These fertilizers derive from living organisms rather than the synthetic fixation of nitrogen via the Haber process. Organic farming uses neither pesticides nor herbicides, so it produces crops that contain fewer toxins. However, the use of organic fertilizers, such as manure or silage, can still cause water pollution and eutrophication if they are allowed to enter a river.

Excessive levels of nitrates in tap water can cause methaemoglobinaemia (blue-baby syndrome) in bottle-fed babies. In this syndrome, fetal haemoglobin stays in the baby's blood after birth instead of being replaced at the usual rate by normal haemoglobin.

Fetal haemoglobin has a greater affinity for nitrogen monoxide (NO) than normal haemoglobin; as a result the oxygen-carrying efficiency of the blood is reduced. Breast-fed babies are at a much lower risk of this happening. The NO that enters a bottle-fed baby's blood is thought to come from nitrite and nitrate ions in the water that the milk powder is mixed with.

It is believed that the nitrate and nitrite ions in tap water are probably from fertilizer run-off that enters the water supply, but this is by no means the whole story. There are nitrates in many vegetables, particularly leaf and root crops, and other products such as beer. Nitrite ions are also in the body as a result of the biochemical reduction of nitrate ions.

Nitrates are added to cured meats to prevent bacterial spoilage and food poisoning, as the labels on the packaging indicate. An average-sized ham may contain several grams of potassium nitrate or sodium nitrate. Most cured meats also contain sodium nitrite or potassium nitrite to inhibit the growth of the toxin-producing bacterium *Clostridium botulinum*.

Some scientists also think that nitrates in our diet are responsible for some stomach cancers. When nitrite ions reach the stomach, the acidic conditions are sufficient to produce nitrous acid, which in turn can undergo protonation to form $[H_2ONO]^+$.

$$NO_2^-(aq) + H^+(aq) \rightleftharpoons HNO_2(aq)$$

$$HNO_2(aq) + H^+(aq) \rightleftharpoons [H_2ONO]^+(aq)$$

The protonated form of nitrous acid reacts with nitrogen-containing compounds in food to produce suspected carcinogens (chemicals that cause cancer).

Scrutiny of nitrate and nitrite ions in food preservation and tap water will continue until we fully understand the implications for our health.

**Fig 28.13** Sodium nitrite is responsible for the pink colour of ham. It inhibits the growth of dangerous bacteria

## 5 AMMONIA AS A BASE

Ammonia is one of the most important bases. In aqueous conditions, an ammonia molecule can accept a proton from a water molecule to form an ammonium ion and a hydroxide ion:

$$NH_3(aq) + H_2O(l) \rightleftharpoons NH_4^+(aq) + OH^-(aq)$$

The presence of OH$^-$ ions means that the solution is alkaline and that ammonia shows basic properties.

The basic properties of ammonia allow it to be identified by:
- its reaction with hydrogen chloride, with which it forms a white smoke of ammonium chloride;
- its effect on moist red litmus, which turns blue.

We discuss the basic characteristics of ammonia in more detail later in the chapter.

YOU CAN READ MORE ABOUT
CONJUGATE ACIDS ON PAGE 367.

**Fig 28.14** A dot-and-cross diagram for ammonia, water, the ammonium ion and the hydroxide ion

### BASE DISSOCIATION CONSTANT

In Chapter 17, we defined Brønsted–Lowry acids as proton donors and Brønsted–Lowry bases as proton acceptors. Fig 28.14 shows that when ammonia dissolves in water, ammonia behaves as a Brønsted–Lowry base. Look at the equation:

$$NH_3(aq) + H_2O(l) \rightleftharpoons NH_4^+(aq) + OH^-(aq)$$

$NH_3$ accepts a proton from $H_2O$ to form $NH_4^+$, which explains why $NH_3$ is considered a base. In this case, $H_2O$ is the conjugate acid and donates a proton to the $NH_3$ base.

Although one of the four bonds in $NH_4^+$ is a dative covalent bond (see Fig 28.14), the four bonds are indistinguishable.

The reaction of ammonia with water is an equilibrium process, so it is possible to write an expression for the equilibrium constant:

$$K_c = \frac{[NH_4^+(aq)][OH^-(aq)]}{[NH_3(aq)][H_2O(l)]}$$

We have included water in the expression because it is a reactant, not just a solvent. However, since water is a solvent to all components of the equilibrium, its concentration is almost constant and can be ignored. By removing the water term from the expression, we can write down a new equilibrium constant called the **base dissociation constant, $K_b$**.

$$K_b = \frac{[NH_4^+(aq)][OH^-(aq)]}{[NH_3(aq)]}$$

The numerical value for $K_b$ is often very small, so $pK_b$, the negative logarithm to base 10 of the base dissociation constant, is sometimes used instead:

$$\mathbf{pK_b = -log_{10}K_b}$$

## STRONG AND WEAK BASES

We have already described the reaction of ammonia with water to form aqueous hydroxide ions. In aqueous solutions, a **base** is a substance that increases the concentration of $OH^-(aq)$ either by reacting with water or by adding extra $OH^-$ ions. Sodium hydroxide is a base, since it fully dissociates in water to form aqueous sodium ions and aqueous hydroxide ions:

$$NaOH(s) + \xrightarrow{H_2O(l)} Na^+(aq) + OH^-(aq)$$

Bases, such as sodium hydroxide, that dissolve in water are called **alkalis**.

Sodium hydroxide and ammonia show an important difference in their behaviour with water. Sodium hydroxide fully dissociates in water, so it is known as a **strong base**; ammonia only partially dissociates and forms a dynamic equilibrium, so it is known as a **weak base**:

$$NH_3(aq) + H_2O(l) \rightleftharpoons NH_4^+(aq) + OH^-(aq)$$

The position of equilibrium lies almost completely on the side of the undissociated $NH_3$, so a solution of a weak base contains only a small proportion of hydroxide ions.

## BASE STRENGTH OF AMINES

Amines are a class of organic molecules closely related to ammonia. Amines are derived from ammonia by the replacement of one, two or three hydrogen atoms by alkyl and or aryl groups (Fig 28.15).

SEE QUESTIONS 4 AND 8.

✔ **REMEMBER THIS**

Do not confuse the idea of strong and weak bases with concentrated and dilute bases. Strong and weak in this context refer to the extent of dissociation of the base in water.

✔ **REMEMBER THIS**

Amines are called primary, secondary or tertiary, depending on how many alkyl or aryl groups are bonded to the nitrogen atom; for example, diethylamine is a secondary amine.

Fig 28.15 The structure of some amines

Amines are bases because they have a lone pair of electrons on the nitrogen atom that can be donated to $H^+(aq)$. This means they are proton acceptors:

$$R^1R^2R^3N(l) + H_2O(l) \rightleftharpoons R^1R^2R^3NH^+(aq) + OH^-(aq)$$

$R^1$, $R^2$ and $R^3$ can be H, an aryl or an alkyl group.

The strength of amines can be compared through their base dissociation constants. More dissociation occurs in a strong base, so it has a greater base dissociation constant.

**Table 28.3 The base dissociation constant of some amines**

| Amine | Formula | Base dissociation constant, $K_b$/mol dm$^{-3}$ | p$K_b$ |
|---|---|---|---|
| Ammonia | $NH_3$ | $1.8 \times 10^{-5}$ | 4.7 |
| Methylamine | $CH_3NH_2$ | $4.38 \times 10^{-4}$ | 3.4 |
| Ethylamine | $C_2H_5NH_2$ | $5.6 \times 10^{-4}$ | 3.3 |
| Diethylamine | $(C_2H_5)_2NH$ | $1.3 \times 10^{-3}$ | 2.9 |
| Phenylamine | $C_6H_5NH_2$ | $3.8 \times 10^{-10}$ | 9.4 |

The base strengths of the amines in Table 28.3 can be explained in terms of the availability of the lone pair on nitrogen; the more available the lone pair, the stronger the base. In methylamine and ethylamine, the alkyl groups are electron-releasing groups, which make the lone pair on the nitrogen atom more available for donation, and therefore for accepting protons. Substituting another hydrogen atom with an alkyl group further increases the availability of the lone pair. Therefore, diethylamine is a stronger base than ethylamine.

Phenylamine is a much weaker base than ammonia because the lone pair on the nitrogen atom is delocalized into the $\pi$ system of the benzene ring. This makes the lone pair less available for donation to a $H^+(aq)$.

All amines react with dilute acids, such as hydrochloric acid, to give a salt that is soluble in water. Phenylamine reacts with dilute hydrochloric acid to give a solution of phenylammonium chloride:

$$C_6H_5NH_2(l) + HCl(aq) \rightleftharpoons C_6H_5NH_3^+(aq) + Cl^-(aq)$$

**TEST YOURSELF**

O **a)** Write an equation to show the dissociation of methylamine in water.
**b)** Write an expression for the base dissociation constant for methylamine.

**TEST YOURSELF**

P **a)** Write an equation for the reaction of aqueous ammonia with sulphuric acid.
**b)** Write an equation to show the reaction between ethylamine and dilute hydrochloric acid.

**SEE QUESTIONS 5, 6 AND 7.**

## pH OF AN ALKALINE SOLUTION

In Chapter 17, we defined the pH of an aqueous solution as the negative logarithm to base 10 of the hydrogen-ion concentration:

$$pH = -\log_{10}[H^+(aq)]$$

We can determine the pH of a solution by knowing the hydrogen-ion concentration. When bases dissolve in water they increase the aqueous hydroxide-ion concentration. To find out what effect this has on the aqueous hydrogen-ion concentration, we must study the acid–base behaviour of water.

## SELF-IONIZATION OF WATER

Water is considered to be a covalent substance, but even in the most pure sample of water there is a very small concentration of $H^+(aq)$ and $OH^-(aq)$. These ions are a result of **self-ionization**:

$$\underset{\text{acid}}{H_2O(l)} + \underset{\text{base}}{H_2O(l)} \rightleftharpoons \underset{\text{conjugate acid}}{H_3O^+(aq)} + \underset{\text{conjugate base}}{OH^-(aq)}$$

The equation shows that two molecules of water react with one another; one behaves as a base and the other as an acid. In this way, a proton is transferred

✔ **REMEMBER THIS**

Remember: $H^+(aq)$ is used to represent $H_3O^+(aq)$.

**? TEST YOURSELF**

**Q** The ionization of water is an endothermic process.
**a)** Using Le Chatelier's principle, predict what will happen to the position of equilibrium as the temperature of the water increases.
**b)** What will happen to the numerical value of the ionic product of water as the temperature of the water increases?
**c)** The electrical conductivity of water increases with increasing temperature. Suggest a reason why.

**? TEST YOURSELF**

**R** A sample of stomach acid has an aqueous hydrogen concentration of $5.78 \times 10^{-3}$ mol dm$^{-3}$. Calculate the OH$^-$(aq) concentration in the sample.

**S** The pH of an aqueous solution at 298 K is 7.6.
**a)** Calculate the value of $[H^+(aq)]$.
**b)** Hence calculate the value of $[OH^-(aq)]$.

**T** The pH of pure water at 298 K is 7.0. Calculate the concentrations of the H$^+$(aq) and of the OH$^-$(aq) in pure water.

from one water molecule to the other one. The self-ionization is an equilibrium process and is often written in a simpler form:

$$H_2O(l) \rightleftharpoons H^+(aq) + OH^-(aq)$$

The position of this equilibrium lies very much to the left. The concentration of H$^+$(aq) and of OH$^-$(aq) is extremely small.

## IONIC PRODUCT OF WATER

The equilibrium constant for the ionization of water is:

$$K_c = \frac{[H^+(aq)][OH^-(aq)]}{[H_2O(l)]}$$

Since the concentration of water is so large, it is considered to be constant. The expression simplifies to:

$$K_w = [H^+(aq)][OH^-(aq)]$$

where $K_w$ **is the ionic product of water**. The ionic product is defined as the product of the concentrations, in mol dm$^{-3}$, of aqueous hydrogen ions and of aqueous hydroxide ions in water. Its unit is mol$^2$ dm$^{-6}$.

$K_w$ is a constant at a particular temperature, so it will not change even if the concentration of H$^+$(aq) changes. At 298 K, the ionic product of water is $1.0 \times 10^{-14}$ mol$^2$ dm$^{-6}$. This means that the concentrations of aqueous hydroxide ions and aqueous hydrogen ions are mathematically linked. In any aqueous solution, if one of the two concentrations is known then the other one can be calculated.

**◉ EXAMPLE**

**Q** The concentration of hydrogen ions in a sample of tap water at a temperature of 298 K is $2.5 \times 10^{-8}$ mol dm$^{-3}$. Calculate the concentration of aqueous hydroxide ions in the water sample.

**A** Use the ionic product of water:

$$K_w = [H^+(aq)][OH^-(aq)]$$

Substitute into this equation the hydrogen ion concentration and the numerical value for the ionic product:

$$1.0 \times 10^{-14} = 2.5 \times 10^{-8} \times [OH^-(aq)]$$

Rearranging this equation gives:

$$[OH^-(aq)] = \frac{1.0 \times 10^{-14}}{2.5 \times 10^{-8}}$$
$$= 4.0 \times 10^{-7} \text{ mol dm}^{-3}$$

### pOH

The concentration of OH$^-$(aq) in pure water is $1.0 \times 10^{-7}$ mol dm$^{-3}$. In most aqueous solutions, $[OH^-(aq)]$ has a small value, so it is convenient to use a logarithmic scale like the one used to measure the aqueous hydrogen ion concentration. The scale that is used is called the **pOH scale**. pOH is the negative logarithm to base 10 of the hydroxide ion concentration in mol dm$^{-3}$:

$$\mathbf{pOH = -log_{10}[OH^-(aq)]}$$

There is a simple connection between the pH and pOH of an aqueous solution. The sum of the pH and the pOH values is 14:

$$pOH + pH = 14$$

This equation is very useful in calculating the pH of alkaline solutions.

## DETERMINING THE pH OF AQUEOUS SOLUTIONS OF STRONG BASES

The pH of a strong base, such as aqueous sodium hydroxide, can be determined simply if the overall concentration of the base is known. We have described two ways that link the $H^+(aq)$ and the $OH^-(aq)$ concentrations, and either can be used to calculate the pH of a basic solution.

### ● EXAMPLE

**Q** What is the pH of 0.150 mol dm$^{-3}$ aqueous potassium hydroxide?

**A** Potassium hydroxide is a strong base that fully dissociates into its constituent aqueous ions:

$$KOH(aq) \rightarrow K^+(aq) + OH^-(aq)$$

The stoichiometry of the equation shows that in 0.150 mol dm$^{-3}$ aqueous potassium hydroxide, $[OH^-(aq)] = 0.150$ mol dm$^{-3}$. Using the ionic product of water gives:

$$[H^+(aq)] = \frac{K_w}{[OH^-(aq)]} = \frac{1.00 \times 10^{-14}}{0.150} = 6.67 \times 10^{-14}\,mol\,dm^{-3}$$

So:

$$pH = -\log_{10}(6.67 \times 10^{-14})$$
$$= 13.2$$

The next Example shows another way to work out the pH of an alkaline solution. This time it uses the idea of the pOH value of a solution. You must decide for yourself which of the two methods for calculating the pH of an alkaline solution you find easier to use.

### ● EXAMPLE

**Q** What is the pH of 0.200 mol dm$^{-3}$ aqueous barium hydroxide?

**A** Aqueous barium hydroxide is a strong base. It fully dissociates in water to give $Ba^{2+}(aq)$ and $OH^-(aq)$:

$$Ba(OH)_2(aq) \rightarrow Ba^{2+}(aq) + 2OH^-(aq)$$

1 mol of barium hydroxide contains 2 mol of hydroxide ions, so:

$$[OH^-(aq)] = 2 \times [Ba(OH)_2(aq)] = 0.400\ mol\,dm^{-3}$$

$$pOH = -\log_{10}(0.400) = 0.40$$

Using the relationship between pOH and pH gives:

$$pH = 14 - pOH = 13.60$$

**Table 28.4 pH and pOH values for aqueous solutions**

| [H$^+$(aq)] | [OH$^-$(aq)] | pH | pOH |
|---|---|---|---|
| 10 | $1 \times 10^{-15}$ | −1 | 15 |
| 1 | $1 \times 10^{-14}$ | 0 | 14 |
| $1 \times 10^{-1}$ | $1 \times 10^{-13}$ | 1 | 13 |
| $1 \times 10^{-2}$ | $1 \times 10^{-12}$ | 2 | 12 |
| $1 \times 10^{-3}$ | $1 \times 10^{-11}$ | 3 | 11 |
| $1 \times 10^{-4}$ | $1 \times 10^{-10}$ | 4 | 10 |
| $1 \times 10^{-5}$ | $1 \times 10^{-9}$ | 5 | 9 |
| $1 \times 10^{-6}$ | $1 \times 10^{-8}$ | 6 | 8 |
| $1 \times 10^{-7}$ | $1 \times 10^{-7}$ | 7 | 7 |
| $1 \times 10^{-8}$ | $1 \times 10^{-6}$ | 8 | 6 |
| $1 \times 10^{-9}$ | $1 \times 10^{-5}$ | 9 | 5 |
| $1 \times 10^{-10}$ | $1 \times 10^{-4}$ | 10 | 4 |
| $1 \times 10^{-11}$ | $1 \times 10^{-3}$ | 11 | 3 |
| $1 \times 10^{-12}$ | $1 \times 10^{-2}$ | 12 | 2 |
| $1 \times 10^{-13}$ | $1 \times 10^{-1}$ | 13 | 1 |
| $1 \times 10^{-14}$ | 1 | 14 | 0 |

### ? TEST YOURSELF

**U** Calculate the pH of each of the following aqueous solutions at 298 K:
**a)** a solution with a pOH of −0.34,
**b)** a solution with [OH$^-$(aq)] = 3.67 × 10$^{-4}$ mol dm$^{-3}$;
**c)** 0.100 mol dm$^{-3}$ aqueous caesium hydroxide, CsOH;
**d)** 2.00 × 10$^{-3}$ mol dm$^{-3}$ aqueous calcium hydroxide, Ca(OH)$_2$.

**V** The pH of an aqueous solution of calcium hydroxide is 12.1. Calculate the concentrations in mol dm$^{-3}$ of each of the following ions in the solution:
**a)** H$^+$(aq);
**b)** OH$^-$(aq);
**c)** Ca$^{2+}$(aq).

## ● EXAMPLE

**Q** Calculate the pH of 0.10 mol dm$^{-3}$ aqueous ammonia ($K_b = 1.8 \times 10^{-5}$ mol dm$^{-3}$).

**A**

$$NH_3(aq) + H_2O(l) \rightleftharpoons NH_4^+(aq) + OH^-(aq)$$

| | | | |
|---|---|---|---|
| At start: | 0.10 mol dm$^{-3}$ | 0 mol dm$^{-3}$ | 0 mol dm$^{-3}$ |
| At equilibrium: | 0.10 – $x$ mol dm$^{-3}$ | $x$ mol dm$^{-3}$ | $x$ mol dm$^{-3}$ |

Using the expression for $K_b$: $1.8 \times 10^{-5} = \dfrac{x \times x}{0.10 - x}$

Ammonia is a weak base, so the position of equilibrium lies to the left. Therefore, $x$ is small compared to 0.10, so:

$$1.8 \times 10^{-5} = \frac{x \times x}{0.10}$$

Solving this equation gives:

$$x = 1.34 \times 10^{-3} \text{ mol dm}^{-3}$$

Since $x$ is [OH$^-$(aq)], then pOH = 2.88 and using the expression pOH + pH = 14 gives the pH = 11.12.

## DETERMINING THE pH OF WEAK BASES

Provided that you know the p$K_b$ or $K_b$ value of a weak base, it is easy to calculate the pH of an aqueous solution.

## ? TEST YOURSELF

**W** Use the $K_b$ value in Table 28.3 to calculate the pH of 0.020 mol dm$^{-3}$ methylamine.

SEE QUESTIONS 5 AND 6.

## ▶ EXTENSION The titration of carbonate ion with dilute hydrochloric acid

In Chapter 17, various titration curves that involve strong and weak acids are described. Titrating sodium carbonate with aqueous hydrochloric acid shows two rapid changes in pH in the titration curve (Fig 28.16).

When sodium carbonate reacts with hydrochloric acid, first hydrogencarbonate ion is formed and then carbonic acid, which decomposes to give carbon dioxide.

$$Na_2CO_3(aq) + HCl(aq) \rightarrow NaCl(aq) + NaHCO_3(aq)$$
$$NaHCO_3(aq) + HCl(aq) \rightarrow NaCl(aq) + H_2CO_3(aq)$$
$$H_2CO_3(aq) \rightarrow CO_2(g) + H_2O(l)$$

In the first reaction, $CO_3^{2-}$(aq) acts as a base and accepts an aqueous proton, H$^+$(aq). In the second reaction, $HCO_3^-$(aq) acts a base. The two-step reaction means that the carbonate ion has two $K_b$ values. With the use of appropriate indicators, it is possible to titrate sodium carbonate to form sodium hydrogencarbonate (using phenolphthalein) or sodium carbonate (methyl orange).

**Fig 28.16** Titration curve for the titration of 25 cm$^3$ of 0.100 mol dm$^{-3}$ Na$_2$CO$_3$ with 0.100 mol dm$^{-3}$ HCl

## BUFFERS

On page 378 is given a description of how buffer solutions can resist a change of pH when small volumes of acid or alkali are added to them. Many buffers are made from a mixture of a weak acid and its conjugate base. Another form of buffer can be made from a mixture of a weak base and its conjugate acid. Ammonia is a weak base and its conjugate acid is the ammonium ion, so a mixture of NH$_3$(aq) and NH$_4$Cl(aq) is a buffer solution.

When H$^+$(aq) is added to this buffer solution it reacts with NH$_3$(aq), so the pH of the solution hardly changes:

$$NH_3(aq) + H^+(aq) \rightarrow NH_4^+(aq)$$

If hydroxide ions are added, the equilibrium below shifts to the right so as to minimize the increase:

$$OH^-(aq) + NH_4^+(aq) \rightleftharpoons NH_3(aq) + H_2O(l)$$

There is a vast excess of NH$_4^+$(aq) to react with any extra OH$^-$(aq) added to the buffer. The NH$_4^+$(aq) is provided by the complete dissociation of the ionic salt NH$_4$Cl.

## CALCULATING THE pH OF A BUFFER SOLUTION

### ⬤ EXAMPLE

**Q** Calculate the pH of a buffer solution that contains 0.10 mol dm$^{-3}$ ammonia and 0.20 mol dm$^{-3}$ ammonium chloride.

**A** The $NH_3(aq)$ will be in equilibrium with its conjugate acid:

$$NH_3(aq) \; + \; H_2O(l) \; \rightleftharpoons \; NH_4^+(aq) \qquad + \qquad OH^-(aq)$$

At start:      0.10 mol dm$^{-3}$          0.20 mol dm$^{-3}$      0 mol dm$^{-3}$
                                          (from the dissociation
                                          of $NH_4Cl$)

At equilibrium:   0.10 − $x$ mol dm$^{-3}$      0.20 + $x$ mol dm$^{-3}$     $x$ mol dm$^{-3}$

Since $NH_3(aq)$ is a weak base, the equilibrium lies almost completely on the left.

Therefore, $x$ is very small and can be ignored compared to 0.10 and 0.20.
At equilibrium:   $[NH_3(aq)] = 0.10$ mol dm$^{-3}$
                   $[NH_4^+(aq)] = 0.20$ mol dm$^{-3}$
                   $[OH^-(aq)] = x$ mol dm$^{-3}$

Substituting these values into the expression for the base dissociation constant gives:

$$K_b = 1.8 \times 10^{-5} = \frac{0.20 \times x}{0.10}$$

Solving for $x$ gives:      $x = 9 \times 10^{-6}$ mol dm$^{-3}$
So                         pOH = 5.0 and pH = 9.0.

### ⁇ TEST YOURSELF

✗ Using the value for $K_b$ in Table 28.3 calculate the pH of a buffer solution that is 0.200 mol dm$^{-3}$ methylamine and 0.100 mol dm$^{-3}$ methylammonium chloride.

### ▶ EXTENSION The use of aqueous ammonia to identify metal ions

Aqueous ammonia is in equilibrium with aqueous ammonium ions and aqueous hydroxide ions. The aqueous hydroxide ions react with any aqueous metal ions present to form insoluble precipitates of the metal hydroxides.

These hydroxides have characteristic colours, as shown in Table 28.5, and they help chemists to identify metal ions in solution. The hydroxides formed sometimes redissolve in excess ammonia through the formation of ammine complexes with the central metal ions. You can read more about complex ions on page 520.

**Table 28.5**

| Metal ion | Effect of adding aqueous ammonia | Effect of adding excess ammonia | Chemical equations |
|---|---|---|---|
| Aluminium | White ppt | White ppt | $Al^{3+}(aq) + 3OH^-(aq) \rightarrow Al(OH)_3(s)$ |
| Calcium | None | None | |
| Chromium(III) | Green ppt | Green ppt | $Cr^{3+}(aq) + 3OH^-(aq) \rightarrow Cr(OH)_3(s)$ |
| Cobalt(II) | Blue ppt | Redissolves giving a brownish–yellow solution | $Co^{2+}(aq) + 2OH^-(aq) \rightarrow Co(OH)_2(s)$ |
| | | | $Co(OH)_2(s) + 6NH_3(aq) \rightarrow [Co(NH_3)_6]^{2+}(aq) + 2OH^-(aq)$ |
| Copper(II) | Light-blue ppt | Redissolves giving a dark blue solution | $Cu^{2+}(aq) + 2OH^-(aq) \rightarrow Cu(OH)_2(s)$ |
| | | | $Cu(OH)_2(s) + 4NH_3(aq) \rightarrow [Cu(NH_3)_4]^{2+}(aq) + 2OH^-(aq)$ |
| Iron(II) | Green ppt | Green ppt | $Fe^{2+}(aq) + 2OH^-(aq) \rightarrow Fe(OH)_2(s)$ |
| Iron(III) | Red–brown ppt | Red–brown ppt | $Fe^{3+}(aq) + 3OH^-(aq) \rightarrow Fe(OH)_3(s)$ |
| Lead(II) | White ppt | White ppt | $Pb^{2+}(aq) + 2OH^-(aq) \rightarrow Pb(OH)_2(s)$ |
| Lithium | None | None | |
| Magnesium | White ppt | White ppt | $Mg^{2+}(aq) + 2OH^-(aq) \rightarrow Mg(OH)_2(s)$ |
| Manganese(II) | White ppt | White ppt | $Mn^{2+}(aq) + 2OH^-(aq) \rightarrow Mn(OH)_2(s)$ |
| Nickel(II) | Green ppt | Redissolves giving a blue solution | $Ni^{2+}(aq) + 2OH^-(aq) \rightarrow Ni(OH)_2(s)$ |
| | | | $Ni(OH)_2(s) + 6NH_3(aq) \rightarrow [Ni(NH_3)_6]^{2+}(aq) + 2OH^-(aq)$ |
| Potassium | None | None | |
| Silver | Brown ppt | Redissolves to give a colourless solution | $2Ag^+(aq) + 2OH^-(aq) \rightarrow Ag_2O(s) + H_2O(l)$ |
| | | | $Ag_2O(s) + H_2O(l) + 4NH_3(aq) \rightarrow 2[Ag(NH_3)_2]^+(aq) + 2OH^-(aq)$ |
| Sodium | None | None | |
| Zinc | White ppt | Redissolves giving a colourless solution | $Zn^{2+}(aq) + 2OH^-(aq) \rightarrow Zn(OH)_2(s)$ |
| | | | $Zn(OH)_2(s) + 6NH_3(aq) \rightarrow [Zn(NH_3)_6]^{2+}(aq) + 2OH^-(aq)$ |

## 6 AMMONIA AND AMINES AS NUCLEOPHILES

We have already described ammonia and amines as bases, since they can donate the lone pair of electrons on nitrogen to accept a proton. The same lone pair on nitrogen also allows ammonia and amines to behave as nucleophiles. The lone pair is donated to make a covalent bond with an electron-deficient centre, normally a carbon atom.

**YOU CAN READ MORE ABOUT NUCLEOPHILIC SUBSTITUTION IN CHAPTER 12.**

### NUCLEOPHILIC SUBSTITUTION

Ammonia reacts with halogenoalkanes to form amines. An excess of ammonia is used to ensure that a primary amine is formed. Normally a halogenoalkane is heated with excess ammonia in ethanol solution in a sealed tube. Ammonia will react with iodoethane to form ethylamine:

$$CH_3CH_2I + NH_3 \rightarrow CH_3CH_2NH_2 + HI$$

The excess ammonia reacts with the hydrogen iodide produced to form ammonium iodide, which helps to drive the reaction to completion.

Ethylamine will react with iodoethane to form diethylamine, which is another reason why an excess of ammonia is used to try and ensure that the product of the first nucleophilic substitution does not react with the iodoethane reactant:

$$CH_3CH_2NH_2 + CH_3CH_2I \rightarrow (CH_3CH_2)_2NH + HI$$

If an excess of the halogenoalkane is used, then a tertiary amine and finally a quaternary ammonium salt will be formed:

$$3CH_3CH_2I + NH_3 \rightarrow (CH_3CH_2)_3N + 3HI$$
$$4CH_3CH_2I + NH_3 \rightarrow (CH_3CH_2)_4N^+I^- + 3HI$$

### TEST YOURSELF

**Y** Predict the products of the following reactions:
a) iodomethane and excess ammonia;
b) 1-bromobutane and ethylamine;
c) 2-chloropropane and methylamine.

**SEE QUESTIONS 6 AND 7.**

### TEST YOURSELF

**Z** Predict the products of the reactions between:
a) $CH_3COCl$ and $CH_3CH_2NH_2$;
b) $C_6H_5COCl$ and $CH_3NH_2$;
c) $(CH_3CO)_2O$ and $NH_3$.

### NUCLEOPHILIC ADDITION AND ELIMINATION WITH AMMONIA AND AMINES

We described the chemistry of acyl chlorides and acid anhydrides in Chapter 18. Both have very polar carbonyl groups and react by condensation with ammonia or amines to form amides. The reaction is often carried out in the presence of excess ammonia or amine to react with the hydrogen chloride or hydrochloric acid that is formed as a co-product. The mechanism of the reaction is known as addition–elimination, as shown in Fig 28.17.

**Fig 28.17** The addition–elimination mechanism for the condensation reaction between propanoyl chloride and ammonia

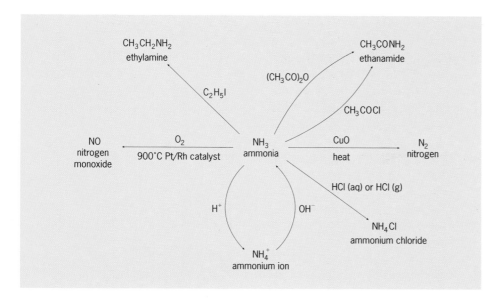

**Fig 28.18** Some important reactions of ammonia

# 7 USE OF PHOSPHORUS COMPOUNDS IN AGRICULTURE

In this chapter, we have concentrated on the chemistry of nitrogen, but phosphorus compounds also have an important role to play in increasing food production. Unlike nitrogen, which is available from the unreactive reservoir of atmospheric nitrogen, phosphorus occurs naturally in a combined form as phosphate rocks. Elemental phosphorus does not appear in the phosphorus cycle. Nitrogen exists in protein, DNA and RNA, whereas phosphorus is principally found in cell membranes as phospholipids.

## PHOSPHATE FERTILIZERS

Some fertilizers contain phosphorus(V) oxide as a source of phosphorus. When it is applied to soil, phosphorus(V) oxide reacts with water to form phosphoric acid. This, in turn, reacts with basic components of the soil, such as metals, to form phosphates.

Rocks that contain phosphates are plentiful. However, many are insoluble or only sparingly soluble, so they release aqueous phosphate ions to the soil very slowly. Therefore, phosphate rocks are treated with sulphuric acid to make soluble calcium hydrogenphosphate. When controlled amounts of sulphuric acid are added, a solid called superphosphate forms. Superphosphate is a mixture of calcium sulphate and calcium dihydrogenphosphate. This fertilizer is rich in soluble phosphate:

$$Ca_3(PO_4)_2(s) + 2H_2SO_4(aq) + 5H_2O(l) \rightarrow Ca(H_2PO_4)_2.H_2O(s) + 2CaSO_4.2H_2O(s)$$

## HERBICIDES

Agrochemicals not only increase crop yields by supplying the essential elements, but they can also kill off unwanted plants or weeds, and so reduce the competition for the essential nutrients. Such agrochemicals are known as herbicides. Herbicides are often polychlorinated compounds, but there are examples of herbicides that contain phosphorus.

> **? TEST YOURSELF**
>
> **A** Calculate the percentage by mass of phosphorus in superphosphate, $Ca(H_2PO_4)_2.2CaSO_4.5H_2O(s)$.

$$HO-\underset{\underset{OH}{|}}{\overset{\overset{O}{\|}}{P}}-CH_2-NHCH_2COOH$$
glyphosate

$$NH_2CH_2COOH$$
glycine

**Fig 28.19** The herbicide glyphosate is a derivative of the amino acid glycine

Fig 28.20 Insecticides help farmers to reduce the number of insects that feed on crops

Fig 28.21 Two organophosphorus insecticides

**YOU CAN FIND OUT MORE ABOUT PHEROMONES IN THE ASSIGNMENT ON PAGE 88.**

## INSECTICIDES

Insecticides are used to kill insects that reduce crop yields by feeding on the crops before they are harvested. It is estimated that over £2500 million is spent world-wide each year on insecticides; more than 15 per cent of crops planted are lost to feeding insects.

Reducing the amount of crop lost to insect feeding causes a dramatic increase in crop yield without the need for extra fertilizers or extra agricultural land. Insecticides are often toxic – not only to insects, but also to mammals and birds – so there are concerns about their continued use. A good insecticide must have the following characteristics:
- it must act against all the insects that infest the particular crop;
- it must not damage the crop;
- it must persist for a sufficient time to avoid the need for repeat spraying or application;
- it must not react with herbicides that may be used on the crop;
- it must be safe to use and must leave no harmful residues.

These characteristics are difficult to achieve, so a compromise is often reached. Future development of new insecticides will try to achieve all five characteristics.

Organophosphorus compounds are often used as insecticides (Fig 28.21). These organophosphorus insecticides are less persistent in nature than polychlorinated compounds, since they can be broken down by bacterial enzymes in the soil.

Future developments in insecticides include the development of insect antifeedants, which do not kill the insect directly, but inhibit feeding so that the insect dies through starvation. Another possibility is to develop an insecticide that uses an aggregation pheromone to bring all the insects to a particular location at which they are treated with another insecticide. Using this method, most of the crop remains unaffected by the application of the insecticide.

## SUMMARY

After studying this chapter, you should know the following:
- Nitrogen is a very unreactive element because of the presence of a triple covalent bond in its molecule.
- Nitrogen, phosphorus and potassium are essential elements for plant growth.
- The nitrogen cycle describes the processes involved in the passage of nitrogen and its compounds through both the living and the non-living environment.
- The Haber process is the only viable synthetic way of converting atmospheric nitrogen to nitrogen-containing compounds. In the Haber process, nitrogen and hydrogen react to form ammonia. The synthesis typically takes place at 450 °C and 200 atmospheres, in the presence of an iron catalyst.
- In an industrial process, a compromise is always made between the position of equilibrium and the rate of reaction to ensure that the chemical is produced using the most economic conditions.
- An increase of temperature shifts the position of equilibrium to the left-hand side and decreases the numerical value of the equilibrium constant in an exothermic reaction.
- An increase in pressure shifts the position of equilibrium to the side of the stoichiometric equation that has the smaller number of moles of gaseous substances, but has no effect on the numerical value of the equilibrium constant.
- For an equilibrium that involves at least one gaseous component, you can write an equilibrium constant, $K_p$, based upon partial pressures.
- Changes in pressure and concentration have no effect on the numerical value of the equilibrium constant.
- The partial pressure of a gas in a mixture of gases is equal to the product of its mole fraction and the total pressure.
- Ammonia is a weak base and reacts with acids, such as sulphuric acid, phosphoric acid and nitric acid, to form salts that are used as fertilizers.

# SUMMARY

- Ammonia is converted into nitric acid by its catalytic oxidation to nitrogen monoxide and the subsequent reaction of nitrogen monoxide with oxygen and water.
- The ionic product of water, $K_w$, is the product of the molar aqueous hydrogen-ion concentration and the molar aqueous hydroxide-ion concentration in water.
- Strong bases fully dissociate when dissolved in water, whereas weak bases form an equilibrium mixture. The equilibrium constant for a reaction between a weak base and water is called the base dissociation constant, $K_b$.

- The larger the value of $K_b$, the stronger the base. Phenylamine is a weaker base than ammonia, which is a weaker base than ethylamine.
- Herbicides and insecticides are used to increase crop yields; they are often organophosphorus compounds.
- Amines react as nucleophiles with halogenoalkanes, acyl chlorides and acid anhydrides.

# ? EXAM QUESTIONS

**1** Fig 28.Q1 shows the nitrogen cycle.

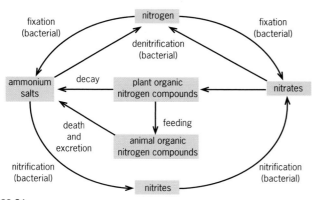

**Fig 28.Q1**

**a)** Which processes lead to an increase in the concentration of nitrogen compounds that can be utilized by plants?

**b)** The process of bacterial nitrification may be represented by the following equations:
$$NH_4^+(aq) + 1\tfrac{1}{2}O_2(g) \rightarrow NO_2^-(aq) + 2H^+(aq) + H_2O(l)$$
$$NO_2^-(aq) + \tfrac{1}{2}O_2(g) \rightarrow NO_3^-(aq)$$
Give the oxidation state of nitrogen in each of the three nitrogen-containing species involved.

**c)** With reference to the diagram of the nitrogen cycle, explain why it has become necessary to supplement naturally formed nutrients with manufactured fertilizers.

**d)** The basis of fertilizer production is the Haber process, which is shown in the following equation:
$$N_2(g) + 3H_2(g) \rightleftharpoons 2NH_3(g) \quad \Delta H = -92 \text{ kJ mol}^{-1}$$
**(i)** Which part of the nitrogen cycle does it mimic?
**(ii)** The percentage of ammonia in the equilibrium mixture under different conditions of temperature and pressure is given in Table 28Q1.

**Table 28.Q1**

| Pressure/$10^5$ Pa | Temperature/K | | | |
|---|---|---|---|---|
| | 473 | 573 | 673 | 773 |
| 10 | 51 | 15 | 4 | 1 |
| 100 | 32 | 53 | 25 | 11 |
| 300 | 93 | 71 | 47 | 24 |
| 500 | 95 | 84 | 65 | 42 |

Describe and explain the effects of both temperature and pressure on the equilibrium mixture.

**e)** Briefly state the main causes of eutrophication.

OCSEB Environmental Chemistry June 1996 127/27 q. 4

**2** Ethanol, $C_2H_5OH$, is an important industrial chemical. Around 200 000 tonnes are manufactured in Britain each year. The usual method of manufacture is by the hydration of ethene with steam in the presence of a phosphoric acid catalyst:
$$C_2H_4(g) + H_2O(g) \rightleftharpoons C_2H_5OH(g)$$
Table 28.Q2 gives details of various experiments carried out under different conditions in order to measure the percentage of ethene converted into ethanol at equilibrium.

**a) (i)** State Le Chatelier's principle.

**Table 28.Q2**

| Experiment | Mole ratio of reactants ethene:steam | Temperature /°C | Pressure /atm | % ethene converted |
|---|---|---|---|---|
| A | 1:1 | 300 | 50 | 32 |
| B | 1:2 | 300 | 50 | 40 |
| C | 1:3 | 300 | 50 | 50 |
| D | 1:2 | 300 | 60 | 46 |
| E | 1:2 | 300 | 70 | 55 |
| F | 1:2 | 250 | 50 | 42 |
| G | 1:2 | 350 | 50 | 38 |

## ? EXAM QUESTIONS

**(ii)** Use the data in the Table 28.Q2 to explain how the position of the equilibrium varies with changes in concentration, temperature and pressure. Show clearly your reasoning and identify which experiments you would use to assist in each of your deductions.

**(iii)** Predict, with justification, whether the forward reaction is exothermic or endothermic.

**b)** Write an expression for the equilibrium constant, $K_p$, for this reaction. Calculate a value for $K_p$ under the conditions used in experiment B in the Table 28.Q2.

**c)** Phosphoric acid, suspended on silica, is used as a catalyst in this reaction. State the effect of the catalyst on the percentage of ethene converted into ethanol at equilibrium.

UCLES Modular: How Far How Fast March 1996 4826 q. 6

**3**

**a)** Water gas is produced by the action of steam on coke.

$C(s) + H_2O(g) \rightleftharpoons CO(g) + H_2O(g) \quad \Delta H^\ominus = +142 \text{ kJ mol}^{-1}$

The equilibrium constant in terms of partial pressure, $K_p$, is given by the expression:

$$K_p = \frac{(pH_2) \times (pCO)}{(pH_2O)}$$

**(i)** What is meant by the term *partial pressure*?

**(ii)** Explain why the equilibrium constant $K_p$ is independent of the amount of carbon present.

**(iii)** Steam is passed over coke at $1000\,°C$ and $1.01 \times 10^5$ Pa. The percentage of steam that has decomposed at equilibrium is found to be 89.0%. Calculate the value of $K_p$ at this temperature.

**(iv)** State and explain how the value of $K_p$ would change if the temperature was decreased.

**b)** Water gas is used in the industrial manufacture of methanol. A mixture of carbon monoxide and hydrogen is passed over a copper-based catalyst.

$CO(g) + 2H_2(g) \rightleftharpoons CH_3OH(g) \quad \Delta H^\ominus = -91 \text{ kJ mol}^{-1}$

Explain, giving your reasoning, the factors that affect the choice of temperature and pressure to be used in this manufacture.

OCR 6824/8 Further Physical Chemistry June 2000 Q4

**4**

**a)** The covalent compound urea, $(NH_2)_2C=O$, is commonly used as a fertilizer in most of the European Union, whereas in the UK the most popular fertilizer is ionic ammonium nitrate, $NH_4NO_3$.

**(i)** Calculate the percentage of available nitrogen in urea.

**(ii)** Apart from the nitrogen content, suggest **two** advantages of using urea as a fertilizer compared with using ammonium nitrate.

**b)** The ammonium ion in water has an acid dissociation constant, $K_a = 5.62 \times 10^{-10}$ mol dm$^{-3}$. The conjugate acid of urea has $K_a = 0.66$ mol dm$^{-3}$. Use these data to explain which of ammonia or urea is the stronger base.

**c)** With moderate heating the ammonium nitrate volatilizes reversibly:

$NH_4NO_3(s) \rightleftharpoons NH_3(g) + HNO_3(g) \quad \Delta H = +171 \text{ kJ mol}^{-1}$

**(i)** State why the expression for $K_p$ for the reversible change does not include ammonium nitrate.

**(ii)** 6.00 g of ammonium nitrate was gently heated in a sealed vessel until equilibrium was reached. The equilibrium constant was found to be 15.7 atm$^2$ under these conditions. Calculate the partial pressure of ammonia present at equilibrium and, hence, the percentage of the ammonium nitrate which has dissociated.

(One mole of gas under these conditions exerts a pressure of 50 atm.)

Edexcel Module Test (Synoptic) June 2001 Q4a,c,e

**5** The expression for $K_b$ for a base **A** in water is:

$$K_b = \frac{[XH^+(aq)][OH^-(aq)]}{[X(aq)]}$$

**a)** A weak base has $pK_b = 4.0$. Calculate the pH of a 0.01 mol dm$^{-3}$ aqueous solution of this base.

**b)** Compound **B** has three nitrogen-containing functional groups. Explain which functional group you expect to be the most basic and which the least basic.

**B**

**Fig 28.Q5**
OCR 6825 Further Organic Chemistry March 2000 Q4

**6** Hydrazine, $N_2H_4$, is an example of a Brønsted–Lowry base.

$N_2H_4(l) + H_2O(l) \rightarrow N_2H_5^+(aq) + OH^-(aq)$

**a)** **(i)** What is meant by the term Brønsted–Lowry base?

**(ii)** What is the conjugate acid of hydrazine?

**(iii)** State and explain the effect of dilution on the position of equilibrium.

**b)** An aqueous solution of hydrazine has a pH of 10.82. Calculate the hydroxide ion concentration in the solution. [$K_w = 1.00 \times 10^{-14}$ mol$^2$ dm$^{-6}$.]

**c)** Suggest the formula for the product of the reaction of hydrazine with: **(i)** hydrochloric acid; **(ii)** iodomethane; **(iii)** ethanoyl chloride.

**7** Ammonia, $NH_3$, ethylamine, $CH_3CH_2NH_2$, methylamine, $CH_3NH_2$, and phenylamine, $C_6H_5NH_2$, are all amines.

**a)** Explain why amines are bases.

**b)** Arrange, with reasons for your choice, the amines in order of base strength.

**c)** Explain why methylamine forms an alkaline solution in water.

**d)** Phenylamine reacts with dilute hydrochloric acid. Write an equation for this reaction.

**e)** Ethylamine reacts with bromoethane to give a number of products.
**(i)** Write down the structure of three of the possible products that can be formed.
**(ii)** What type of reaction is taking place?

**8**

**a)** In cold aqueous sodium hydroxide, ammonium salts produce the following equilibrium:
$$NH_4^+(aq) + OH^-(aq) \rightleftharpoons NH_3(aq) + H_2O(l)$$
**(i)** Identify the two acid–base conjugate pairs in equilibrium.
**(ii)** Explain the effect of raising the temperature on this equilibrium.

**c)** Pure **liquid** ammonia ionizes as follows:
$$NH_3(l) + NH_3(l) \rightleftharpoons NH_4^+(am) + NH_2^-(am)$$
where (am) represents solutions in liquid ammonia. The Brønsted–Lowry theory of acid–base behaviour applies to solutions in liquid ammonia.
**(i)** Suggest why ammonium salts behave as acids in liquid **ammonia**.
**(ii)** The salt sodium amide, $Na^+NH_2^-$, reacts with ammonium chloride in liquid ammonia in an acid–base reaction. Write an equation to represent the reaction.

Edexcel CH6 Synoptic Paper June 1999 Q2c,d

# KEY SKILLS **ASSIGNMENT**

## THE MANUFACTURE AND PURIFICATION OF SYNTHESIS GAS

Ammonia is made from nitrogen and hydrogen:

$$N_2(g) + 3H_2(g) \rightleftharpoons 2NH_3(g)$$

Typical conditions for the reaction are a temperature of 450 °C and a pressure of 200 to 400 atmospheres, although some factories use pressures as high as 1000 atmospheres. An iron catalyst is also used to accelerate the reaction. To ensure that there is little wastage of nitrogen and hydrogen it is important that the mole ratio (hence volume ratio) of hydrogen to nitrogen is 3:1.

**1** Suggest why it is important to have a 3:1 mole ratio of hydrogen to nitrogen in the Haber process.

Synthesis gas is a mixture of hydrogen and nitrogen used in the Haber process. In the United Kingdom most synthesis gas is obtained from methane, water and air. A complex series of reactions converts these materials into carbon monoxide, carbon dioxide, hydrogen and nitrogen. Synthesis gas is made once the carbon dioxide and carbon monoxide are removed. Fig 28.A1 shows the processes involved.

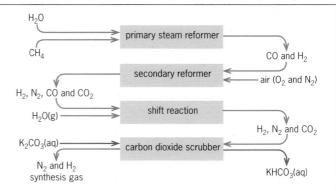

**Fig 28.A1** The production of synthesis gas for the Haber process

**2** Give advantages in using raw materials, such as air and water, in an industrial process.

The first reaction in the process is called primary steam reforming. In this reaction, methane and steam are heated in the presence of a nickel catalyst at a temperature of about 1100 K and 30 atmospheres pressure:

$$CH_4(g) + H_2O(g) \rightleftharpoons CO(g) + 3H_2(g)$$

# KEY SKILLS **ASSIGNMENT**

**3**
**a)** Write down an expression for the equilibrium constant, $K_p$, for the primary steam reforming.

**b)** The partial pressures at equilibrium are as follows: $pCH_4 = 80$ kPa, $pH_2O = 80$ kPa, $pCO = 710$ kPa and $pH_2 = 2130$ kPa. Calculate a numerical value, and state the units, for $K_p$.

**c)** Primary steam reforming is endothermic. Use Le Chatelier's principle to predict the conditions that make the position of equilibrium move to the right-hand side.

**d)** Comment on the actual conditions used for the primary steam-reforming step.

The next stage is the secondary steam reforming. At the end of this stage, a mixture of nitrogen, hydrogen, carbon dioxide and carbon monoxide is produced. Most of the unconverted methane from the primary reforming stage is converted into hydrogen and carbon dioxide, while at the same time oxygen is removed as oxides:

$$2H_2(g) + O_2(g) \rightarrow 2H_2O(g)$$
$$CH_4(g) + H_2O(g) \rightleftharpoons CO(g) + 3H_2(g) \quad \Delta H = +210 \text{ kJ mol}^{-1}$$

**4**
Predict the reaction conditions needed to obtain a high yield from the reaction of methane and steam.

Carbon dioxide and carbon monoxide must then be removed so that only hydrogen and nitrogen remain. The shift reaction converts carbon monoxide into carbon dioxide by reaction with steam at 670 K using an iron-oxide catalyst.

**5**
**a)** Construct the equation for the reaction between carbon monoxide and steam.

**b)** Suggest why it is important to remove all the carbon monoxide before the gases enter the Haber process.

Finally, carbon dioxide is removed in a scrubber, which involves reaction with aqueous potassium carbonate:

$$K_2CO_3(aq) + H_2O(l) + CO_2(g) \rightarrow 2KHCO_3(aq)$$

It is possible to regenerate the aqueous potassium carbonate by heating the aqueous potassium hydrogencarbonate with steam.

**6**
Suggest why it is useful to be able to regenerate the aqueous potassium carbonate.

After scrubbing, the gases present are nitrogen and hydrogen. Any small traces of carbon monoxide or carbon dioxide are removed by reaction with hydrogen using a nickel catalyst. This results in the formation of a trace of methane, $CH_4$.

**7**
Construct equations to show the reactions of:
**a)** carbon monoxide with hydrogen;
**b)** carbon dioxide with hydrogen.

**8**
Nitrogen is also obtained by the fractional distillation of liquid air; hydrogen is obtained from the cracking of hydrocarbons. Suggest advantages of making nitrogen and hydrogen in the process described earlier in the passage, rather than from the fractional distillation and cracking routes.

**9**
A company wants to build a plant to make ammonia starting from methane and air. Discuss the geographical location that the company should choose to build the plant.

# 29

# Group 4 and Group 6 elements

# Group 4 and Group 6 elements

**Silicon** Fundamental to computers

Silicon compounds are at the centre of materials science and vital to the glass, cement and quartz industries. The element silicon is now being used on a smaller scale in the manufacture of semiconductors, solar cells and microchips. The impact of silicon in computing has been enormous. The introduction of micromachines that involve silicon technology means that the applications of silicon-based materials will continue to expand.

Specialized silicon compounds are being developed to withstand harsh conditions. For example, materials in spacecraft have to perform over a wide temperature range and while bombarded by high levels of cosmic radiation.

Among the materials that involve silicon that chemists have been developing are:
● ceramics, such as silicon carbide, that can withstand extremely high temperatures;
● glasses, some of which are used to store radioactive waste;
● silicones that can be used as oils or water-resistant sealants;
● aerogels that are solids with a density only four times that of air.

Almost all of these materials contain silicon–oxygen single bonds in their structure. This bond is fundamental to large areas of materials science.

## 1 GROUP 4 AND GROUP 6 ELEMENTS

THE CHEMISTRY OF OXYGEN AND OXIDES IS BEEN COVERED IN DETAIL IN EARLIER CHAPTERS, SEE PAGES 450–456.

This chapter describes the chemistry of some non-metals, a semi-metal and some metals. We consider the elements in Group 4, one of the most diverse groups of elements in the Periodic Table. At the top of Group 4 is carbon, a non-metal; at the bottom of the group is lead, a metal. We identify some of the trends and patterns within the group, and concentrate on the chemistry of silicon and its manufacturing applications.

In addition, we describe some aspects of the chemistry of sulphur, one of the non-metals in Group 6.

## 2 GROUP 4 ELEMENTS

**Table 29.1 Group 4 Elements**

| Element | Symbol |
|---|---|
| Carbon | C |
| Silicon | Si |
| Germanium | Ge |
| Tin | Sn |
| Lead | Pb |

When we described the properties of the halogens in Chapter 23 and the elements of Groups 1 and 2 in Chapter 22, we were able to emphasize the similarity in chemical properties and the predictable variation of physical properties with increasing atomic number. This is not really possible with the elements of Group 4, because the elements vary from a typical non-metal at the top of the group to typical metals at the bottom of the group.

### OCCURRENCE OF THE ELEMENTS

Carbon occurs naturally in the Earth's crust as carbonate minerals, crude oil, natural gases and coals. Furthermore, all living organisms contain compounds of carbon. Elemental carbon occurs in small quantities as graphite or diamond deposits.

Silicon comprises almost a quarter of the Earth's crust and is almost always found combined with oxygen. Clays, sands and most of the components of soils contain compounds of silicon and oxygen. Despite the abundance of silicon in the Earth's crust it is still a difficult element to isolate. The extraction process for most of the Group 4 elements is quite similar. The favoured method is to reduce the oxide of the Group 4 element using carbon or a reducing agent of similar reactivity:

$$2PbO(s) + C(s) \rightarrow 2Pb(s) + CO_2(g)$$

**SOME OF THE MINERALS THAT CONTAIN SILICON AND OXYGEN ARE DESCRIBED LATER IN THE CHAPTER.**

**Fig 29.1** Clays are complex silicates and aluminosilicates. Silicates contain strong silicon–oxygen bonds

## ▶ EXTENSION The manufacture of silicon

Silicon is manufactured by the high temperature reduction of silica (silicon dioxide) with either magnesium or carbon:

$$SiO_2 + 2C \rightarrow 2CO + Si$$

The silicon obtained by the reduction of silica is not pure enough to use in the manufacture of semiconductors, microchips and solar cells. It is purified by converting silicon into silicon(IV) chloride, and then reducing it back to silicon:

$$Si + 2Cl_2 \rightarrow SiCl_4$$
$$SiCl_4 + 2Mg \rightarrow Si + 2MgCl_2$$

It is easy to remove the soluble magnesium chloride from the silicon by washing with hot water. A further process, called zone refining, yields very pure silicon (Fig 29.2).

inert atmosphere

silicon rod

solidifying silicon

molten silicon

melting silicon

moving heat source

region where impurities eventually concentrate

**Fig 29.2** The zone-refining process. Zone refining is used to make ultra-pure silicon. A rod of silicon is heated at one end by a moving heat source. The heat source produces a thin cross section of molten silicon. As the heat source moves downwards so does the cross section of molten silicon, taking dissolved impurities with it. Eventually, the impurities are concentrated at the bottom of the rod while the rest of the silicon is ultra-pure

## ? TEST YOURSELF

**A** Write down the equation for the reaction of magnesium with silicon dioxide.

**B** Suggest reasons why silicon is very expensive to isolate, even though the raw material silicon dioxide is both cheap and plentiful.

**C a)** Why must zone refining take place in an inert atmosphere rather than in air?
**b)** Suggest a gas that could be used to provide the inert atmosphere.

**READ ABOUT THE BLAST FURNACE ON PAGE 11.**

## EXTRACTION OF TIN AND LEAD

Both tin and lead are extracted by reduction of an oxide (lead(II) oxide or tin(IV) oxide) with carbon. In both cases, the chemistry of the reduction is similar to that in the blast furnace during the extraction of iron.

## PHYSICAL PROPERTIES OF GROUP 4 ELEMENTS

**Table 29.2 Physical properties of Group 4 elements**

|  | Carbon | Silicon | Germanium | Tin | Lead |
|---|---|---|---|---|---|
| Atomic number | 6 | 14 | 32 | 50 | 82 |
| Electron configuration | $1s^2\,2s^2\,2p^2$ | $1s^2\,2s^2\,2p^6\,3s^2\,3p^2$ | $[Ar]3d^{10}\,4s^2\,4p^2$ | $[Kr]4d^{10}\,5s^2\,5p^2$ | $[Xe]4f^{14}\,5d^{10}\,6s^2\,6p^2$ |
| Covalent atomic radius/pm | 77 | 117 | 122 | 140 | 146 |
| Electronegativity | 2.5 | 1.7 | 2.0 | 1.7 | 1.6 |
| First ionization energy/kJ $mol^{-1}$ | 1090 | 786 | 762 | 707 | 716 |
| Melting point/ °C | 3570 (diamond) | 1414 | 959 | 232 | 328 |
| Boiling point/ °C | Sublimes above 3700 | 3309 | 2837 | 2687 | 1755 |
| Density/kg $m^{-3}$ | 3510 (diamond); 2220 (graphite) | 2330 | 5360 | 7310 (grey tin); 5750 (white tin) | 11350 |
| Electrical conductivity | Good (graphite); Poor (diamond) | Semiconductor | Semiconductor | Good | Good |
| Structure of solid | Giant molecular | Giant molecular | Giant molecular | Giant metallic | Giant metallic |

**READ ABOUT THE STRUCTURE AND PROPERTIES OF DIAMOND, GRAPHITE AND SILICON ON PAGES 154–157.**

**? TEST YOURSELF**

**D** Write balanced equations for the reduction by carbon of:
**a)** lead(II) oxide, PbO;
**b)** tin(IV) oxide, $SnO_2$.

**E** Lead(II) oxide is obtained from galena, PbS. Galena is roasted in air and forms lead(II) oxide and sulphur dioxide.
**a)** Suggest one environmental problem with this roasting process.
**b)** Write down the balanced chemical equation for the roasting process.

**SEE QUESTIONS 1 AND 2.**

**READ MORE ABOUT METALLIC BONDING AND THE PROPERTIES OF METALS ON PAGES 152–154.**

**? TEST YOURSELF**

**F** What properties of tin and lead make them suitable for producing an alloy to be used in soldering?

We described in Chapter 7 how the physical properties of an element are related to its structure. Giant molecular structures have high melting points and simple molecular structures have low melting points. Giant metallic structures conduct heat and electricity very well, but molecular structures in general do not.

The metallic character of the elements increases as the atomic (proton) number increases. With a larger atomic radius, the outer electrons are farther from the nucleus and are better shielded by inner shells of electrons. As a result, tin and lead have structures that involve positive lead and tin ions in a sea of delocalized electrons.

Tin and lead have the typical physical properties of metals. They are:
● dense;
● lustrous;
● good conductors of heat and electricity;
● malleable;
● ductile.

Tin and lead have relatively low melting points, which indicates that the metallic bonding is not as strong as in the transition elements. Transition elements are able to use d electrons in the sea of delocalized electrons, but tin and lead use only p electrons from the 5p and 6p subshells, respectively.

**Fig 29.3** An alloy of tin and lead can be used to mend broken electrical connections

Silicon carbide is a very useful ceramic material. It has an extremely high melting point and it is extremely hard. These two properties explain its use as an abrasive.

The structure of silicon carbide is similar to that of diamond (Fig 29.4), and as a result the two substances share many physical properties.

Research is under way to produce silicon carbide fibres that could be incorporated into a composite material. Such a fibre would have an extremely high tensile-fracture strength.

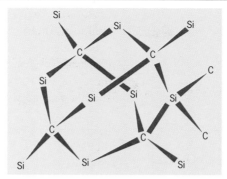

**Fig 29.4** The structure of silicon carbide. Silicon carbide has a diamond-like structure; each silicon atom is surrounded by four carbon atoms and each carbon atom is surrounded by four silicon atoms

# 3 CHEMICAL PROPERTIES OF THE ELEMENTS OF GROUP 4

You can see from Table 29.2 that all the elements in Group 4 have four electrons in the outer shell. So one atom of the element could either gain or lose four electrons to attain a stable electron configuration. It is virtually impossible for any atom to lose four electrons to form an $X^{4+}$ ion. Firstly, the ion's high charge density would polarize any anion in its vicinity and so would form a covalent bond. Secondly, the energy required to remove four electrons is extremely large.

This means that an atom of a Group 4 element must gain four electrons to give the $X^{4-}$ ion. But the ion is not very stable. Four electrons are more likely to be gained by covalent bonding, giving the elements an oxidation state of +4 in some of their compounds. In fact, almost all the compounds of silicon and carbon are covalent with an oxidation state of +4.

## THE INERT PAIR EFFECT

Germanium, tin and lead form both ionic compounds and covalent compounds. In ionic compounds, an atom of these elements loses two electrons to form $Ge^{2+}$, $Sn^{2+}$ or $Pb^{2+}$ ions. For example, if a lead atom loses two electrons it attains an electron configuration that involves a filled outer sub-shell, $6s^2$. The two electrons in the 6s orbital are known as the **inert pair**, because they are in the outer shell of electrons but are not used in bonding. As a result lead, germanium and tin can show two oxidation states:
● the ionic, +2 oxidation state;
● the covalent, +4 oxidation state.
It is impossible for carbon and silicon atoms to behave in the same way, because the loss of two electrons does not give a sufficiently stable electron configuration.

**READ MORE ABOUT CHARGE DENSITY AND POLARIZATION OF ANIONS ON PAGES 84 AND 480.**

**Table 29.3 The ionization energies of Group 4 elements**

| Element | First ionization energy/kJ $mol^{-1}$, $X(g) \rightarrow X^+(g) + e^-$ | Second ionization energy/kJ $mol^{-1}$, $X^+(g) \rightarrow X^{2+}(g) + e^-$ | Energy needed to make 1 mole of $X^{2+}(g)$ ions from X atoms/kJ $mol^{-1}$ $X(g) \rightarrow X^{2+}(g) + 2e^-$ |
|---|---|---|---|
| Carbon | 799 | 2420 | |
| Silicon | 786 | 1580 | |
| Germanium | 762 | 1540 | |
| Tin | 707 | 1410 | |
| Lead | 716 | 1450 | |

## METALLIC CHARACTER IN GROUP 4

We describe in Chapter 7 how the loss of electrons from an atom to form a cation is characteristic of a metal atom and the gain of electrons to form either an anion or a covalent bond is characteristic of a non-metal atom. The loss of electrons results in a +2 oxidation state and shows the metallic character of germanium, tin and lead. The metallic character of the elements increases with increasing atomic number, so that lead has the most metallic character of all the Group 4 elements.

Elements with higher atomic (proton) numbers have more shielding electron shells between the outer electrons and the nucleus. This means that the outer electrons experience a reduced effective nuclear charge. In addition, the outer electrons are further away from the nucleus, so an atom of lead can easily lose the two 6p electrons.

Carbon and silicon show no evidence whatsoever of metallic character; in fact, the +2 oxidation state is virtually absent from the chemistry of both elements.

## NON-METALLIC CHARACTER IN GROUP 4

The non-metallic character of the elements decreases as atomic (proton) number increases. However, there is evidence of some non-metallic character as far down the group as lead. Lead(IV) compounds and, in particular, the lead(IV) halides are covalent in nature; they involve a lead atom that gains four electrons through covalent bonding to achieve a stable electron configuration.

## REACTIONS OF GROUP 4 ELEMENTS

Whether metallic or non-metallic character is considered, the elements in Group 4 are members of a relatively unreactive group. The elements will react with non-metals, such as oxygen and the halogens, but such reactions normally require heating.

### REACTIONS WITH OXYGEN

Most Group 4 elements react with air or oxygen when heated to form a dioxide in which the element has a +4 oxidation state. Carbon forms a gaseous dioxide, whereas the other elements form solid dioxides:

$$C(s) + O_2(g) \rightarrow CO_2(g)$$
$$X(s) + O_2(g) \rightarrow XO_2(s)$$

where X is Si, Ge and Sn.

Lead forms lead(II) oxide when it is heated in air:

$$2Pb(s) + O_2(g) \rightarrow 2PbO(s)$$

Even if lead(IV) oxide were to form, it would thermally decompose at Bunsen flame temperatures to give lead(II) oxide.

### REACTIONS WITH HALOGENS

All the elements except carbon react when heated with halogens to give a halide:

$$X(s) + 2Cl_2(g) \rightarrow XCl_4(l)$$

where X = Si, Ge and Sn. Lead forms the chloride $PbCl_2$, rather than the tetrachloride, because of the thermal instability of lead(IV) chloride:

$$Pb(s) + Cl_2(g) \rightarrow PbCl_2(s)$$

**SEE QUESTION 2.**

**? TEST YOURSELF**

**J** Draw a dot-and-cross diagram to show the covalent bonding in lead(IV) chloride. You need only draw the electrons in the outermost shell.

**Fig 29.5** Charcoal is mainly carbon. It burns to give carbon dioxide in a highly exothermic reaction

**? TEST YOURSELF**

**K** Write down equations for the reactions between each pair of elements:
**a)** lead and iodine; **b)** tin and fluorine;
**c)** germanium and bromine.

# 4 OXIDES OF GROUP 4 ELEMENTS

Most of the Group 4 elements form two oxides. These are an oxide of formula XO with the element in the +2 oxidation state and an oxide of formula $XO_2$ with the element in the +4 oxidation state.

## $XO_2$ OXIDES (THE DIOXIDES)

All Group 4 elements form a dioxide with the empirical formula $XO_2$, where X is C, Si, Ge, Sn or Pb. These dioxides are often referred to as the element(IV) oxide, such as silicon(IV) oxide. There is a distinct change in the chemical properties of these oxides as the atomic (proton) number of the Group 4 element increases. This reflects a change in the stability of the +4 oxidation state.

READ ABOUT WHY CARBON DIOXIDE FORMS A SIMPLE MOLECULE AND SILICON DIOXIDE FORMS A GIANT MOLECULE ON PAGES 459–460.

READ MORE ABOUT QUARTZ AND SILICON DIOXIDE ON PAGES 459–460.

**Table 29.4 The dioxides of Group 4 elements**

| Oxide | Formula | Structure and bonding | Acid–base behaviour | Thermal stability | Oxidizing power |
|---|---|---|---|---|---|
| Carbon dioxide | $CO_2$ | Simple molecular | Acidic | Does not decompose on heating | Very weak oxidizing agent |
| Silicon(IV) oxide | $SiO_2$ | Giant molecular | Acidic | Does not decompose on heating | increasing oxidizing power |
| Germanium(IV) oxide | $GeO_2$ | Giant molecular | Amphoteric | Does not decompose on heating | |
| Tin(IV) oxide | $SnO_2$ | Giant molecular with ionic character | Amphoteric | Does not decompose at Bunsen burner temperatures | |
| Lead(IV) oxide | $PbO_2$ | Giant molecular with ionic character | Amphoteric | Decomposes on heating | Powerful oxidizing agent |

## THERMAL STABILITY OF THE DIOXIDES

The thermal stability of the dioxides decreases as the atomic (proton) number of the Group 4 element increases. Strong heating of lead(IV) oxide gives oxygen and lead(II) oxide:

$$2PbO_2(s) \rightarrow 2PbO(s) + O_2(g)$$

Silicon(IV) oxide is very thermally stable; it is often a component of ceramics. One crystalline form of silicon dioxide is quartz. Quartz is a hard, brittle, clear and colourless solid. It is used for architectural decorations, semi-precious jewels, optical instruments and as a frequency controller in radio transmitters.

Quartz melts at 1600 °C to form a viscous liquid. When this liquid cools, it does not crystallize and the internal structure remains random. A substance with this type of structure is known as a glass.

## OXIDIZING POWER OF THE DIOXIDES

Since Group 4 elements form two sets of oxides, $XO_2$ (oxidation state of X +4) and XO (oxidation state of X +2); it is possible to convert $XO_2$ into XO, which is a reduction. This means that the dioxides $XO_2$ are oxidizing agents. Lead(IV) oxide and tin(IV) oxide are the most powerful of these oxidizing agents and both can oxidize hydrochloric acid to form chlorine:

$$PbO_2(s) + 4HCl(aq) \rightarrow PbCl_2(aq) + 2H_2O(l) + Cl_2(g)$$

SEE QUESTION 5.

**Fig 29.6** Radioactive material is easily stored in glass enclosed in steel. As long as it remains dry, any radiation is confined by this process (called vitrification)

## ? TEST YOURSELF

**M** Explain why lead(IV) oxide is a better oxidizing agent than silicon dioxide. Hint: think about the relative stabilities of the +2 oxidation states.

## ? TEST YOURSELF

**L** Write down the equation for the reaction of aqueous sodium silicate, $Na_2SiO_3$, with dilute hydrochloric acid.
Hint: in this reaction the sodium silicate behaves in a similar way to sodium carbonate.

Aerogels are an exciting new class of synthetic materials that have an extremely low density. They consist of 10 per cent silica or silicon(IV) oxide, the same material found in sand, and 90 per cent gas. They have such a low density that they can float on whisked egg white. In addition, they have unusual optical, thermal, electrical and acoustic properties.

It has been known for some time that an aqueous solution of sodium silicate and hydrochloric acid react to form a gelatinous solid. Heating this mixture eventually gives a solid called silica gel, which has a large surface area and readily adsorbs water. Aerogels are similar in nature, but are solids with gas, rather than liquid, trapped within a cage of atoms.

Fig 29.7 Silica gel absorbs water from the air to make a dry atmosphere

When aerogels were first discovered, to prepare them was a long and difficult process that involved using potentially dangerous methanol to make the gel. Later on, a method was developed to make an aerogel filled with carbon dioxide. Such aerogels could be made into dry pellets that contain strings of silica tetrahedra encircling the gas spaces.

Aerogels are excellent insulators because there are fewer solid atoms to conduct heat than in conventional insulators and the gas pores are too small for the gas inside to conduct heat effectively. Some aerogels have one-fifth of the thermal conductivity of poly(phenylethene) and others have a density that is only four times that of air.

One application of aerogels is in windows. A layer of aerogel can be sandwiched between two layers of glass to improve the insulation properties of the glass. The aerogel also scatters light and produces a frosted glass effect.

Research is underway to improve the thermal insulation properties of aerogels still further by removing all the gas from the pores, to leave a vacuum. There is also scope to make aerogel catalysts – the reaction between gases takes place within the gel itself.

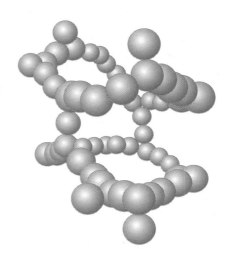

Fig 29.8 An aerogel is made of silica chains that encircle air-filled pores. The circles in the diagram represent $SiO_4$ tetrahedra

## ACID–BASE REACTIONS OF THE DIOXIDES

In Chapter 21 we describe the acid–base reactions of oxides of the elements in Period 3. Covalent oxides are often acidic and ionic oxides are basic. Most metal oxides are ionic and most non-metal oxides are covalent, so we can say that metal oxides are normally basic and non-metal oxides are often acidic.

On this basis, we would expect most Group 4 dioxides to show some acidic behaviour, because bonding in carbon dioxide and silicon dioxide is covalent, and even in lead(IV) oxide there is a high degree of covalent bonding. This acidic behaviour is shown by the reactions of the dioxides. The test for carbon dioxide, in which lime water (aqueous calcium hydroxide) turns milky, is an example of an acidic reaction of carbon dioxide. The precipitate formed is calcium carbonate. In the same way, sodium hydroxide pellets react with carbon dioxide to form sodium carbonate:

$$2NaOH(s) + CO_2(g) \rightarrow Na_2CO_3(s) + H_2O(l)$$

Limestone is added to blast furnaces. In these, it is decomposed into calcium oxide, which reacts with $SiO_2$ (sand) impurity to remove it as calcium silicate, $CaSiO_3$. This is an example of silicon dioxide behaving as an acidic oxide.

 **REMEMBER THIS**

If carbon dioxide is bubbled through lime water for several minutes, a white precipitate, $CaCO_3(s)$, is formed first. This reacts with further carbon dioxide to form a colourless solution of calcium hydrogencarbonate.

**SEE QUESTIONS 3 AND 5.**

Even lead(IV) oxide shows some acidic character when it reacts with concentrated aqueous sodium hydroxide to form the plumbate(VI) ion.

$$PbO_2(s) + 2NaOH(aq) \rightarrow Na_2PbO_3(aq) + H_2O(l)$$

As the dioxides of lead and tin also have a high degree of ionic character, they also show basic properties, but their reactions with acids are often complicated by subsequent redox reactions. Lead(IV) oxide and tin(IV) oxide are known as **amphoteric** oxides, since they have both acidic and basic properties.

## XO OXIDES (THE MONOXIDES)

Table 29.5 The oxides of Group 4 elements (SiO is omitted from the table because it does not exist under standard conditions)

| Oxide | Formula | Structure and bonding | | Acid–base behaviour | |
|---|---|---|---|---|---|
| Carbon monoxide | CO | Simple molecular | | Neutral | |
| Germanium(II) oxide | GeO | Giant ionic | Increasing ionic character | Amphoteric | Increasing basic character |
| tin(II) oxide | SnO | Giant ionic | | Amphoteric | |
| lead(II) oxide | PbO | Giant ionic | | Amphoteric | |

? TEST YOURSELF

N Write down the equation for the reaction of carbon dioxide with lime water to form insoluble calcium carbonate.

**SEE QUESTIONS 3 AND 5.**

**FOR MORE INFORMATION ABOUT BLAST FURNACES, SEE PAGE 11.**

Carbon monoxide has completely different properties to those of the other monoxides formed by Group 4 elements. It is a covalently bonded gaseous compound that is a neutral oxide. Germanium(II) oxide, tin(II) oxide and lead(II) oxide are ionically bonded solid oxides that show increasing basic properties as the ionic character of the oxide increases.

**Fig 29.9** Carbon monoxide is a simple molecule and is iso-electronic with a nitrogen molecule. It has two covalent bonds and one dative covalent bond

## ! SCIENCE FEATURE Metal carbonyls

Carbon monoxide forms a range of compounds with transition metals. They are known as carbonyls. Carbon monoxide molecules act as ligands and donate pairs of electrons to an empty orbital of an atom of the transition element. An interesting feature of many of these carbonyls is that the oxidation state of the transition element is often 0.

**Fig 29.10** The structure of tetracarbonylnickel(0). Tetracarbonylnickel(0) is a tetrahedral neutral complex. Carbon monoxide ligands form dative bonds with the central nickel atom

The first metal carbonyl was discovered by Ludwig Mond in 1889. It was tetracarbonylnickel(0), $Ni(CO)_4$, a colourless, volatile liquid. The discovery of $Ni(CO)_4$ opened up the way for large-scale production of pure nickel.

Carbon monoxide reacts with nickel at 50 °C to give volatile $Ni(CO)_4$, which evaporates to leave behind impurities:

$$Ni(s) + 4CO(g) \rightarrow Ni(CO)_4(l)$$

The $Ni(CO)_4$ decomposes at 200 °C to yield pure nickel and CO. Fortunately for Mond, the discovery of stainless steel at that time created a demand for large quantities of nickel, so he was able to develop a large manufacturing plant that produced pure nickel for a ready-made market.

**Fig 29.11** Some other metal carbonyls. Some metal carbonyls contain bridging carbon monoxide ligands

**Fig 29.12** Nickel prepared via $Ni(CO)_4$ is used as an alloy with tungsten and other metals to make the engines for this airliner

**Fig 29.13** A summary of the properties of Group 4 oxides

## OXIDATION OF THE MONOXIDES

Since lead(IV) oxide is the least thermally stable of the dioxides, lead forms the most stable monoxide.

The monoxides CO, GeO and SnO all readily react with oxygen to give the respective dioxides. For example, carbon monoxide combusts to give carbon dioxide, and anhydrous tin(II) oxide reacts with oxygen in the air to give $SnO_2$:

$$2SnO(s) + O_2(g) \rightarrow 2SnO_2(s)$$

Lead(II) oxide will not form lead(IV) oxide when heated in air, but at $400\,°C$ with prolonged heating it will form red lead (see Test Yourself **P**).

## ACID–BASE REACTIONS OF THE MONOXIDES

The oxides become more basic as the atomic (proton) number of the Group 4 element increases. This is because bonding in the oxide becomes more ionic in character, and ionic oxides show basic properties. Even though the basic character increases, lead(II) oxide is not a basic oxide; it is amphoteric because it reacts with acids and bases. In acids, lead(II) oxide forms lead(II) salts and with aqueous alkalis it forms plumbate(II), $PbO_2^{2-}$:

$$PbO(s) + 2HNO_3(aq) \rightarrow Pb(NO_3)_2(aq) + H_2O(l)$$
$$PbO(s) + 2NaOH(aq) \rightarrow Na_2PbO_2(aq) + H_2O(l)$$

| Monoxides | | | Dioxides | | | | |
|---|---|---|---|---|---|---|---|
| CO | simple molecular | neutral oxide | $CO_2$ | simple molecular | | acidic | |
| | | | $SiO_2$ | giant molecular | thermal | | oxidizing |
| GeO | | | $GeO_2$ | | stability | | power |
| SnO | | | $SnO_2$ | | decreases | | increases |
| PbO | giant ionic | amphoteric | $PbO_2$ | giant molecular/ionic | | amphoteric | |

# 5 CHLORIDES OF GROUP 4 ELEMENTS

Earlier in the chapter we describe how the stability of Group 4 dioxides (with the element in the +4 oxidation state) decreases as the atomic (proton) number of the element increases, whereas the stability of the monoxide (element in the +2 oxidation state) increases. Exactly the same trend in stability is observed with the two sets of chlorides, $XCl_4$ (with the element in the +4 oxidation state) and $XCl_2$ (with the element in the +2 oxidation state). The +4 oxidation state involves covalent bonding and the +2 oxidation state involves ionic bonding.

## THE TETRACHLORIDES

**Table 29.6 Group 4 tetrachlorides**

| Formula | Structure and bonding | Bond length of X–Cl/nm | Thermal stability | Action of cold water |
|---|---|---|---|---|
| $CCl_4$ | Simple molecular | 0.177 | Thermally stable | None |
| $SiCl_4$ | Simple molecular | | thermal stability decreases | Rapidly hydrolysed to form $SiO_2$ and HCl |
| $GeCl_4$ | Simple molecular | increases | | Rapidly hydrolysed to form $GeO_2$ and HCl |
| $SnCl_4$ | Simple molecular | | | Rapidly hydrolysed to form $SnO_2$ and HCl |
| $PbCl_4$ | Simple molecular | | Thermally decomposes at room temperature to form $PbCl_2$ and $Cl_2$ | Rapidly hydrolysed to form $PbO_2$ and HCl |

With the exception of carbon, the tetrachlorides ($XCl_4$) are named element(IV) chloride. In the case of carbon, the name tetrachloromethane is preferred to show its relationship to the hydrocarbon methane. All of the tetrachlorides have a simple molecular structure and so have low boiling points and melting points.

## THERMAL STABILITY OF THE TETRACHLORIDES

The thermal stability of the tetrachlorides decreases with increasing atomic (proton) number of the Group 4 element. This trend in thermal stability is explained in terms of the bond lengths and bond strengths of the X–Cl bond. We have described that a covalent bond is formed by the overlap of atomic orbitals; in the case of the Pb–Cl bond, there is very little overlap because of the difference in energy between the orbitals involved. With very little overlap, the covalent bond is extremely weak and is therefore easy to break. The Pb–Cl bond is much longer than the other X–Cl bonds, which contributes to the weakness of the Pb–Cl bond.

## HYDROLYSIS OF THE TETRACHLORIDES

In Chapters 21 and 23 we describe the hydrolysis of covalent chlorides. With the exception of tetrachloromethane, all the tetrachlorides are readily hydrolysed to form the dioxide as a precipitate and either hydrogen chloride or hydrochloric acid, depending on the amount of water available during the hydrolysis:

$$XCl_4(l) + 2H_2O(l) \rightarrow XO_2(s) + 4HCl(aq)$$

where X is Si, Ge, Sn and Pb.

## CHLORIDES WITH THE ELEMENT IN THE +2 OXIDATION STATE

We have already described the inert-pair effect to explain why germanium, tin and lead can have an oxidation state of +2. The three elements form chlorides of the general formula $XCl_2$ in this oxidation state. Germanium(II) chloride is not really ionic and anhydrous tin(II) chloride has a great deal of covalent character. The only solid chloride for which the term ionic is really appropriate is lead(II) chloride.

## REDUCING PROPERTIES

Germanium(II) chloride and tin(II) chloride are both reducing agents. Aqueous solutions of tin(II) chloride are best stored with a little tin metal to prevent aerial oxidation. Aqueous acidified tin(II) chloride reduces iron(III) ions and dichromate(VI) ions:

$$3Sn^{2+}(aq) + 14H^+(aq) + Cr_2O_7{}^{2-}(aq) \rightarrow 2Cr^{3+}(aq) + 3Sn^{4+}(aq) + 7H_2O(l)$$

The aqueous $Sn^{4+}$ ions have too high a charge density to exist on their own and so they form complex hydrated ions.

Lead(II) chloride is not a reducing agent, as shown by the following electrode potentials:

$$Pb^{4+}(aq) + 2e^- \rightleftharpoons Pb^{2+}(aq) \qquad E_{red} = +1.69 \text{ V}$$

$$Sn^{4+}(aq) + 2e^- \rightleftharpoons Sn^{2+}(aq) \qquad E_{red} = +0.15 \text{ V}$$

The highly positive $E_{red}$ for the reduction of $Pb^{4+}(aq)$ indicates that lead(IV) is very easily reduced to $Pb^{2+}$ and so the reverse reaction, $Pb^{2+}$ to $Pb^{4+}$, will be much more difficult than the corresponding reaction of $Sn^{2+}$ to $Sn^{4+}$.

**Fig 29.14** The shape of the covalent tetrachlorides. The tetrachlorides of Group 4 elements are all simple, tetrahedral molecules

**? TEST YOURSELF**

**Q** Write an equation to show the thermal decomposition of lead(IV) chloride.

**SEE QUESTIONS 1 AND 3.**

**READ WHY TETRACHLOROMETHANE CANNOT BE HYDROLYSED ON PAGE 454.**

**✔ REMEMBER THIS**

Solutions that contain aqueous iodide ions and lead(II) ions react to form yellow lead(II) iodide precipitate:

$$Pb^{2+}(aq) + 2I^-(aq) \rightarrow PbI_2(s)$$

This could be used as a test for $Pb^{2+}(aq)$.

**? TEST YOURSELF**

**R** Construct an ionic equation to show the reaction of $Fe^{3+}(aq)$ with $Sn^{2+}(aq)$ to form $Fe^{2+}(aq)$.

**S** Write an ionic equation for the reaction that occurs when chlorine bubbles through a solution of acidified aqueous tin(II) chloride. Suggest the name(s) of the product(s).
Hint: you can read about the reactions of chlorine in Chapter 23.

## ▶ EXTENSION The silanes

Carbon forms an almost infinite number of compounds that contain carbon–hydrogen single bonds. Among these compounds are the alkanes, which have the general formula $C_nH_{2n+2}$. Silicon forms a range of compounds similar in terms of general formula to the alkanes, which are collectively called the silanes. The chain length of carbon–carbon bonds is not limited. However, the longest chain silane that has been isolated at present has six silicon atoms.

All silanes are simple molecular compounds and the single bonds are arranged tetrahedrally about the central silicon atoms.

The vacant 3d orbitals in silicon atoms make the properties of the silanes different from those of the corresponding alkanes. The enthalpy changes of formation of silane ($SiH_4$) and methane indicate that silane is considerably less stable with respect to the elements silicon and hydrogen than methane is to carbon and hydrogen. In fact, the enthalpy change of formation of silane is positive.

$$Si(s) + 2H_2(g) \rightarrow SiH_4(g) \qquad \Delta H_f = +34 \text{ kJ mol}^{-1}$$
$$C(s) + 2H_2(g) \rightarrow CH_4(g) \qquad \Delta H_f = -75 \text{ kJ mol}^{-1}$$

Silane spontaneously ignites in air to form silicon dioxide and water:

$$SiH_4(g) + 2O_2(g) \rightarrow SiO_2(s) + 2H_2O(g) \qquad \Delta H_f = -1429 \text{ kJ mol}^{-1}$$

The enthalpy change of combustion of silane is much higher than that of methane ($-802$ kJ mol$^{-1}$), because of the weakness of the Si–H single bonds broken and the strength of the Si–O bonds formed in silicon dioxide. Silane reacts with aqueous sodium hydroxide to form hydrogen and the silicate ion.

$$SiH_4(g) + 2OH^-(aq) + H_2O(l) \rightarrow SiO_3^{2-}(aq) + 4H_2(g)$$

Again, the formation of the very strong silicon–oxygen single bond acts as a driving force for this reaction. In contrast, methane is inert to aqueous sodium hydroxide.

**Fig 29.15** The structures of some silanes. Notice the similarity to the homologous series called the alkanes

**Fig 29.16** The dot-and-cross diagram for silane

**Fig 29.17** The tetrahedral arrangement of the silane molecule

---

## ? TEST YOURSELF

**T** **a)** What is the molecular formula of the silane that contains six silicon atoms per molecule?
**b)** Draw a possible structural formula for this silane.

## ? TEST YOURSELF

**U** The relative molecular mass of a silicone fluid with the structure shown in Fig 29.20 and an R group of $CH_3$ is approximately 1500. Estimate the number of silicon atoms there are in the polymer chain.

## ! SCIENCE FEATURE Silicones

Silicones are polymers that have a backbone of alternating silicon and oxygen atoms with carbon-based side chains. They are characteristically non-toxic and are stable over a wide range of temperatures.

**Fig 29.18** Silicones have a polymer chain of alternate oxygen and silicon atoms joined by a single bond. The side chains R and R$^1$ are alkyl groups or aryl groups

**Fig 29.19** Silicones are manufactured by the hydrolysis of dichlorosilanes followed by condensation of the resulting dihydroxysilanes. A molecule of water is eliminated during each condensation step

## ! SCIENCE FEATURE Silicones

The side chain shown in Fig 29.18 can be varied to develop a wide range of silicones with different properties. If the side chain of the silicone is a methyl group, then the resulting polymer is a liquid. These polymers are linear with no cross-linking between polymer chains. This means that the polymer chains only have weak intermolecular forces between them, so the silicone has a relatively low melting point.

**Fig 29.20** The structure of silicone fluid. A silicone fluid has an R group such as a methyl or ethyl group. There are no cross-links with other polymer chains and only weak intermolecular forces exist between the polymer chains

Fig 29.20 shows that there are no double bonds in the polymer chain, so all the bonds in the polymer molecule can freely rotate. Such a feature gives this type of silicone the physical properties required to make the non-stick coating on pans.

Another use of silicone fluids is to protect masonry from rain damage. Silicone bonds to masonry and the hydrophobic (water-hating) methyl groups repel water molecules from the surface of the building.

An advantage of silicones is the way that the side chain can be manipulated to change the property of the material. If the side chains contain more reactive groups – such as alkenes or hydroxyl groups – then some degree of cross-linking between the polymer chains can occur. Providing the cross-linking is not extensive, the silicone will be solid at room temperature, but still retain a great deal of flexibility and resistance to water.

Silicones are electrical insulators and they can be used to insulate electric cables, as long as an inert filler is added to increase the rigidity of the material.

**Fig 29.21** Many bathroom sealants used to seal the sides of baths and showers are made from silicone rubber

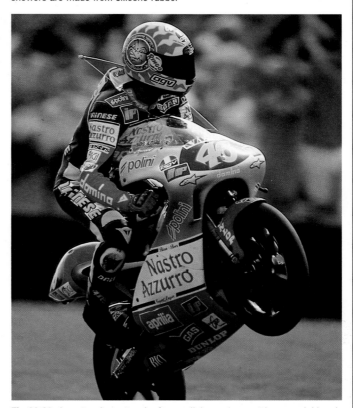

**Fig 29.22** An extensive network of cross-links are present in more rigid and less flexible silicone resins. Some racing helmets are made of silicone resin mixed with glass fibre

## 6 OXY-SALTS

The metal elements of Group 4 form oxy-salts such as lead(II) nitrate and lead(II) carbonate, which contain the cation $Pb^{2+}$. The non-metal elements also form oxy-salts, but they are present as the anion rather than as the cation. So carbon forms carbonates and silicon forms silicates. There are examples of the other elements in Group 4 that form anions similar to carbonates, such as plumbate(II), which illustrates the amphoteric nature of lead and its compounds.

**Fig 29.23** Lead(II) chromate(VI) is used to make the yellow lines on roads

## OXY SALTS OF LEAD

Most lead(II) salts are insoluble, but two of the oxy-salts – lead(II) ethanoate, $Pb(CH_3COO)_2$, and lead(II) nitrate, $Pb(NO_3)_2$ – are soluble in water. They are often used in qualitative analysis to precipitate anions as insoluble lead(II) compounds.

When aqueous lead(II) nitrate is added to any soluble sulphate, white lead(II) sulphate forms as a precipitate:

$$Pb^{2+}(aq) + SO_4^{2-}(aq) \rightarrow PbSO_4(s)$$

Lead(II) chromate(VI) forms as a yellow precipitate when aqueous lead(II) nitrate is added to aqueous sodium chromate(VI):

$$Pb^{2+}(aq) + CrO_4^{2-}(aq) \rightarrow PbCrO_4(s)$$

## CARBONATES

Carbonates are derived from the weak acid carbonic acid, formed when carbon dioxide dissolves in water.

$$CO_2(aq) + H_2O(l) \rightleftharpoons H_2CO_3(aq)$$

Carbonates are salts that contain the $CO_3^{2-}$ ion and hydrogencarbonates contain the $HCO_3^-$ ion. Only carbonates of the Group 1 elements and ammonium carbonate are soluble in water. Therefore, most carbonates are prepared by precipitation:

$$M^{2+}(aq) + CO_3^{2-}(aq) \rightarrow MCO_3(s)$$

where M is a metal in the +2 oxidation state.

Metals that form a positive ion with a high charge density, such as $Fe^{3+}$, $Al^{3+}$ and $Cr^{3+}$, do not form carbonates. In solution, these ions polarize water molecules to form sufficient aqueous hydrogen ions to react with the carbonate ion to form carbon dioxide:

$$CO_3^{2-}(s) + 2H^+(aq) \rightarrow CO_2(g) + H_2O(l)$$

## THERMAL DECOMPOSITION

With the exception of the Group 1 carbonates, all carbonates thermally decompose to give carbon dioxide. The energy changes that accompany this decomposition are discussed in detail in Chapter 22.

## REACTION WITH ACID

Treating any carbonate with acid evolves carbon dioxide and produces a salt. Lead(II) carbonate reacts with dilute nitric acid to form lead(II) nitrate, water and carbon dioxide:

$$PbCO_3(s) + 2HNO_3(aq) \rightarrow Pb(NO_3)_2(aq) + CO_2(g) + H_2O(l)$$

## SILICATES

Carbonates contain the anion $CO_3^{2-}$; the corresponding silicon compounds, called the silicates, have a very different structure and contain more complex negative ions. The simplest silicate structure contains the tetrahedral $SiO_4^{4-}$ ion.

**READ ABOUT HYDROGENCARBONATES ON PAGE 478.**

## ? TEST YOURSELF

**V** Write an equation to show the decomposition of lead(II) carbonate.

**SEE QUESTION 4.**

The structure of the anion in silicates is based on an $SiO_4$ tetrahedron that contains covalent silicon–oxygen single bonds. If two $SiO_4$ tetrahedral units share an oxygen atom, the silicate contains the $Si_2O_7^{6-}$ ion.

**Fig 29.24** Minerals such as olivine, $Mg_2SiO_4$, and zircon, $ZrSiO_4$, contain the tetrahedral silicate ion, $SiO_4^{4-}$

---

**! SCIENCE FEATURE The asbestos problem**

Asbestos is a silicate mineral that is fibrous. It has long been used for fireproofing and insulating buildings. Since it became known that exposure to asbestos fibres can cause serious respiratory disorders, such as asbestosis, asbestos has been removed from many buildings.

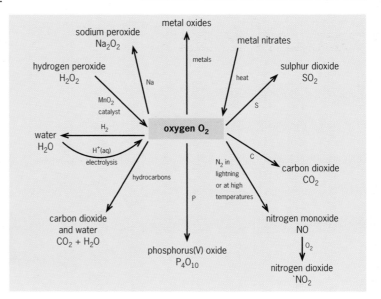

**Fig 29.25** A scanning electron micrograph of the fibres of blue asbestos. The lung disease asbestosis is caused by breathing in these fibres

**Fig 29.26** Removing asbestos has become a priority to avoid exposure to asbestos fibres. Many buildings, including schools, have been temporarily closed during the controlled removal of asbestos. This sack of asbestos will be buried in a landfill site

---

# 7 GROUP 6 – OXYGEN AND SULPHUR

Group 6 contains two well-known elements – oxygen and sulphur – together with three less well-known elements – selenium, tellurium and polonium. (We do not discuss selenium, tellurium and polonium, because their chemistry is well beyond the scope of this book.)

The chemistry of oxygen and oxides is described in earlier chapters, but Figs 29.27 and 29.28 give some general information about oxygen and oxides.

The discussion in this chapter concentrates on sulphur.

**Table 29.7 The elements of Group 6**

| Element | Symbol | Electron configuration |
|---------|--------|------------------------|
| Oxygen | O | $1s^2\ 2s^2\ 2p^4$ |
| Sulphur | S | $1s^2\ 2s^2\ 2p^6\ 3s^2\ 3p^4$ |
| Selenium | Se | $[Ar]3d^{10}\ 4s^2\ 4p^4$ |
| Tellurium | Te | $[Kr]4d^{10}\ 5s^2\ 5p^4$ |
| Polonium | Po | $[Xe]4f^{14}\ 5d^{10}\ 6s^2\ 6p^4$ |

| | |
|---|---|
| **Basic oxides** | Ionic oxides, contain $O^{2-}$. React with acids to give salt and water e.g. $Na_2O$, $CuO$ |
| **Amphoteric oxides** | Contain $O^{2-}$. React with both acids and alkalis to give salts, e.g. $Al_2O_3$, $PbO$, $ZnO$ |
| **Acidic oxides** | Covalent oxides. Often react with bases to give salts e.g. $CO_2$, $SO_2$, $SO_3$ |
| **Neutral oxides** | Covalent oxides, often having a non-metal in a low oxidation state e.g. $CO$, $NO$ |

**Fig 29.27** A summary of the oxides

**DETAILS ON THE CHEMISTRY OF OXYGEN CAN BE FOUND ON THE FOLLOWING PAGES:**
**REACTIONS OF OXYGEN WITH ELEMENTS ON PAGE 456.**
**BASIC OXIDES ON PAGE 459.**
**ACIDIC OXIDES ON PAGE 460.**

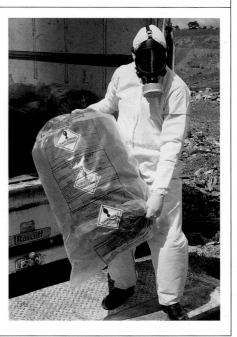

**Fig 29.28** Some of the reactions involving oxygen

**Fig 29.29** About two-thirds of the sulphur found in the atmosphere comes from sulphurous gases emitted by volcanoes

## OCCURRENCE OF SULPHUR

Sulphur is an element that has been known to humans since the earliest times. It was once known as brimstone. Sulphur deposits are widespread, as are sulphur's compounds with metals and non-metals. Many of the ores of metals are sulphides, a combination of a metal and sulphur.

## CHEMICAL PROPERTIES OF SULPHUR

Sulphur has the electron configuration $1s^2\ 2s^2\ 2p^6\ 3s^2\ 3p^4$; it has six electrons in its outermost shell. To obtain a stable octet of electrons, the sulphur atom gains two electrons either by sharing electrons with the atoms of a non-metal to form covalent bonds or by gaining electrons lost by metal atoms to form ionic bonds. The ionic compounds contain $S^{2-}$ ions and are sulphur's equivalent to oxides. The oxidation number of sulphur in a sulphide is $-2$.

A sulphur atom can also expand its octet by accepting electrons into its vacant 3d orbitals to form compounds in which sulphur has an oxidation state of $+4$ or $+6$.

**Fig 29.30** Some dot-and-cross diagrams of sulphur compounds. In $H_2S$ and $Na_2S$, sulphur has a stable octet and an oxidation number of $-2$. In $SO_2$ and $SO_3$, the sulphur atom has expanded its octet and as a result the oxidation numbers of sulphur are $+4$ in $SO_2$ and $+6$ in $SO_3$

## SULPHUR DIOXIDE AND SULPHUROUS ACID

Sulphur dioxide is produced when almost any compound that contains sulphur burns in air. In Chapter 10 we discussed the formation of sulphur dioxide in power stations and indicated how it is implicated in the formation of acid rain.

**Fig 29.31** Sulphur dioxide is a bent or V-shaped molecule

### ACIDIC PROPERTIES OF SULPHUR DIOXIDE

Sulphur dioxide dissolves in water to form sulphurous acid, $H_2SO_3$(aq), a moderately strong acid. An aqueous solution of sulphurous acid normally has about 25 per cent of the acid molecules dissociated to form $HSO_3^-$ ions. Only a small proportion of these ions dissociate further to form $SO_3^{2-}$ ions:

$$SO_2(aq) + H_2O(l) \rightleftharpoons H_2SO_3(aq)$$

$$H_2SO_3(aq) \rightleftharpoons HSO_3^-(aq) + H^+(aq)$$

$$HSO_3^-(aq) \rightleftharpoons SO_3^{2-}(aq) + H^+(aq)$$

Alkalis, such as aqueous potassium hydroxide, neutralize sulphurous acid to give the metal sulphite:

$$2KOH(aq) + H_2SO_3(aq) \rightarrow K_2SO_3(aq) + 2H_2O(l)$$

It is also possible to obtain potassium hydrogensulphite by controlling the mole ratios of acid and alkali that react.

### REDUCING PROPERTIES OF SULPHUR DIOXIDE

Sulphurous acid and sulphur dioxide are both reducing agents. Sulphur dioxide is used as an antioxidant to preserve food: it is oxidized easily,

### ? TEST YOURSELF

**W** What is the mole ratio of potassium hydroxide to sulphurous acid needed to make:
**a)** potassium sulphite;
**b)** potassium hydrogensulphite.

**SEE QUESTION 7.**

removing oxygen in the air that could have oxidized the food. Aqueous chlorine, aqueous dichromate(VI) and aqueous iron(III) are all reduced by sulphurous acid. The reaction of moist sulphur dioxide and acidified potassium dichromate(VI) is a **test for sulphur dioxide**:

$$Cr_2O_7^{2-}(aq) + 3SO_2(aq) + 2H^+(aq) \rightarrow 2Cr^{3+}(aq) + 3SO_4^{2-}(aq) + H_2O(l)$$

Filter paper previously dipped into acidified potassium dichromate(VI) changes colour from orange to a blue–green in the presence of sulphur dioxide.

## SULPHITES

Salts of sulphurous acid are known as sulphites. These salts react with dilute acid to release sulphur dioxide:

$$K_2SO_3(s) + 2HNO_3(aq) \rightarrow 2KNO_3(aq) + SO_2(g) + H_2O(l)$$

Many solid sulphites thermally decompose when heated:

$$ZnSO_3(s) \rightarrow ZnO(s) + SO_2(g)$$

In this way, sulphite salts behave in a similar way to carbonates.

## SULPHUR TRIOXIDE

Sulphur trioxide is a typical acidic oxide and it dissolves in water to form the strong acid sulphuric acid:

$$H_2O(l) + SO_3(g) \rightarrow H_2SO_4(aq)$$

It also reacts directly with solid bases to form sulphates; barium oxide reacts to form barium sulphate:

$$BaO(s) + SO_3(g) \rightarrow BaSO_4(s)$$

# 9 SULPHURIC ACID

Sulphuric acid is one of the world's key industrial chemicals. Justus von Liebig wrote in 1843 that the consumption of sulphuric acid is a barometer of a nation's commercial prosperity. Sulphuric acid's wide range of uses means that Liebig's statement still holds today. Sulphuric acid is widely used in the oil and petroleum industry. It is also involved in making:

● fertilizers, such as ammonium phosphate and superphosphate;
● paints, pigments and dyes;
● soaps and detergents;
● plastics and fibres;
● general chemicals, such as salts and battery acid.

## THE CONTACT PROCESS

Sulphuric acid is manufactured using the contact process. It was patented in 1831 by Peregrine Phillips and later improved by Rudolf Messel. The contact process is an efficient and economical way of converting relatively cheap raw materials into sulphuric acid.

## RAW MATERIALS

The raw materials chosen in the United Kingdom are sulphur, air and water. Water and air are very cheap raw materials, but most sulphur needs to be imported.

About 10 per cent of the sulphur comes from the petroleum industry;

**?** **TEST YOURSELF**

**X** Write down the equations for the reduction by sulphur dioxide of:
**a)** aqueous chlorine;
**b)** aqueous iron(III) ions.
Hint: Use the half equations in Appendix 4 to help.

**Fig 29.32** Gaseous sulphur trioxide is a trigonal planar molecule

THE CONTACT PROCESS IS REALLY JUST THE CATALYTIC OXIDATION OF SULPHUR DIOXIDE TO FORM SULPHUR TRIOXIDE, RATHER THAN ALL THE PROCESSES THAT TAKE PLACE TO MAKE SULPHURIC ACID sulphur is an impurity that must be removed from crude oil before the oil can be processed. Selling this impurity adds to the economic efficiency of an oil refinery.

The three stages of the manufacture of sulphuric acid:

● stage one, the oxidation of molten sulphur to give sulphur dioxide:

$$S + O_2 \rightarrow SO_2 \quad \Delta H = -297 \text{ kJ mol}^{-1}$$

● stage two, the actual contact process, the reaction of sulphur dioxide with oxygen to give sulphur trioxide:

$$SO_2(g) + \tfrac{1}{2}O_2(g) \rightleftharpoons SO_3(g) \quad \Delta H = -98 \text{ kJ mol}^{-1}$$

● stage three, the reaction of sulphur trioxide with water:

$$SO_3(g) + H_2O(l) \rightarrow H_2SO_4(l) \quad \Delta H = -130 \text{ kJ mol}^{-1}$$

In theory, it is a simple process. However, there are a number of problems involved with these three stages that must be overcome before the yield of sulphuric acid is economically viable.

## OXIDATION OF SULPHUR

In the combustion of molten sulphur the enthalpy change of reaction is negative. The temperature used ensures that the rate of reaction is sufficient to provide an economic rate of production of sulphur dioxide. The rate of reaction must be controlled to ensure that the reaction does not go too fast because of the release of heat energy during the reaction.

## THE CONTACT PROCESS – CATALYTIC CONVERSION OF SULPHUR DIOXIDE INTO SULPHUR TRIOXIDE

The second stage is the oxidation of sulphur dioxide to give sulphur trioxide. The reaction is very slow and needs a catalyst.

Originally, the precious metal platinum was used, but now manufacturers use vanadium(V) oxide. The reaction is exothermic, so Le Chatelier's principle suggests that the forward reaction is favoured by low temperatures. The volume of products is smaller than that of the reactants, so high pressure also favours the forward reaction. To maintain an excess of reactants and continually remove the product from the reaction mixture is another way to shift the equilibrium position to the right.

A low temperature slows down the rate of reaction and high pressures are expensive to maintain, so a compromise set of conditions is chosen that gives an economic yield of sulphur trioxide. Manufacturers use a temperature of between 450 °C and 600 °C. A pressure of around 10 atmospheres is sufficient to push the gases around the plant. There is no need to use a high pressure in this reaction, despite the Le Chatelier prediction, because even at low pressure the position of equilibrium lies well to the right.

At least three catalyst chambers or beds are used to ensure maximum conversion of sulphur dioxide (Fig 29.33). The conversion yield can exceed 98 per cent. As well as being economically viable, this high yield has environmental benefits, because very little sulphur dioxide waste enters the atmosphere.

## THE HYDRATION OF SULPHUR TRIOXIDE

The reaction of sulphur trioxide with water does not work with large quantities of sulphur trioxide. The normal way to dissolve a gas industrially is to use a counter flow system in which water is sprayed downwards and gas flows upwards. In the case of sulphur trioxide, this technique causes clouds of

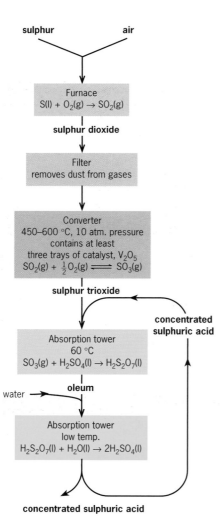

**Fig 29.33** Flow chart to show the manufacture of sulphuric acid

sulphuric acid droplets to form. These droplets are difficult to condense, so a different method is used.

The sulphur trioxide is dissolved in concentrated sulphuric acid to form oleum, $H_2S_2O_7$. The oleum is diluted with measured amounts of water to give concentrated sulphuric acid:

$$SO_3(g) + H_2SO_4(l) \rightarrow H_2S_2O_7(l)$$

$$H_2S_2O_7(l) + H_2O(l) \rightarrow 2H_2SO_4(l)$$

Both of the processes are exothermic and are carried out at low temperature.

Throughout the plant, the flow of energy is vital. The reactions are exothermic and the conditions needed vary from high to low temperatures. Complex heat exchangers ensure that little heat is wasted.

SEE QUESTIONS 7, 8 AND 9.

## REACTIONS OF SULPHURIC ACID

Sulphuric acid has a wide range of reactions in addition to those expected of a strong acid. They include:
- redox reactions with both dilute and concentrated sulphuric acid;
- dehydration reactions;
- sulphonation reactions;
- reactions as a non-volatile acid.

YOU CAN READ ABOUT THE USE OF SULPHURIC ACID IN THE DEHYDRATION OF ALCOHOLS ON PAGE 310.

YOU CAN READ ABOUT SULPHONATION ON PAGE 338.

Fig 29.34 Some of the reactions of sulphuric acid

**TEST YOURSELF**

Y Look at Fig 29.34. In which reactions does sulphuric acid act as:
a) an acid;
b) an oxidizing agent?

## REACTIONS AS A NON-VOLATILE ACID

Sulphuric acid reacts with the salts of many other acids to liberate the acid. We have already described in Chapter 23 the reaction of concentrated sulphuric acid with sodium chloride, in which it forms hydrogen chloride. In the same way, concentrated sulphuric acid reacts when heated with sodium nitrate to form nitric acid:

$$NaNO_3(s) + H_2SO_4(l) \rightarrow NaHSO_4(s) + HNO_3(l)$$

**TEST YOURSELF**

Z What is the name of the acid formed when concentrated sulphuric acid reacts with:
a) potassium benzoate;
b) magnesium ethanoate?

## SULPHATES

Sulphuric acid forms a class of oxy-salts called sulphates. Sulphates are generally soluble in water and are relatively resistant to thermal decomposition. Many sulphates do not change when heated at Bunsen burner temperatures, but at higher temperatures they decompose to form the corresponding oxide.

We describe the trend of solubility of the sulphates of the Group 2 elements in Chapter 22. The lack of solubility of barium sulphate in water is the basis of the **test for aqueous sulphate** ions.

SEE QUESTIONS 6 AND 9.

**Table 29.8 The tests for sulphate, carbonate, sulphite and thiosulphate ions**

| | Add aqueous barium nitrate followed by excess dilute nitric acid |
|---|---|
| Sulphate | White precipitate of barium sulphate that does not redissolve in the acid |
| Carbonate | White precipitate of barium carbonate that redissolves in acid with the formation of carbon dioxide |
| Sulphite | White precipitate of barium sulphite that redissolves slowly and forms sulphur dioxide |
| Thiosulphate | White precipitate of barium thiosulphate that changes to a yellow precipitate on addition of acid, with the formation of sulphur dioxide |

**FOR MORE INFORMATION ON THIOSULPHATES, SEE PAGE 110.**

## THIOSULPHATES

Thiosulphates contain the $S_2O_3^{2-}$ ion. This ion is really a sulphate ion in which one oxygen atom has been replaced by a sulphur atom. On reaction with dilute acids, thiosulphates form yellow sulphur as a precipitate:

$$S_2O_3^{2-}(aq) + 2H^+(aq) \rightarrow S(s) + H_2O(l) + SO_2(g)$$

## SUMMARY

After studying this chapter, you should know that:

- The metallic character of the elements in Group 4 increases with atomic number, and the non-metallic character decreases with atomic number.
- The stability of the +4 oxidation state in Group 4 decreases with increasing atomic number, and the stability of the +2 oxidation state increases with increasing atomic number.
- All the tetrachlorides of Group 4 elements except carbon are readily hydrolysed by cold water to form the corresponding dioxide and HCl.
- The thermal stability of the Group 4 tetrachlorides decreases (as the stability of the +4 oxidation state decreases and the strength of the covalent bond with chlorine weakens) with increasing atomic number of the Group 4 element.
- The tetrachlorides of Group 4 are all simple molecular covalent molecules.
- The thermal stability of the Group 4 dioxides decreases (as the stability of the +4 oxidation state decreases) with increasing atomic number of the Group 4 element.

- The acidic character of Group 4 oxides decreases (as the ionic character increases) with increasing atomic number of the Group 4 element.
- The oxidizing power of the dioxides of Group 4 elements increases (as the stability of the +4 oxidation state decreases) with increasing atomic number of the element.
- The ionic character and the basic character of the monoxides of Group 4 elements increase (as the stability of the +2 oxidation state increases) with increasing atomic number of the element.
- Sulphuric acid is manufactured by the catalysed oxidation of sulphur dioxide by oxygen, followed by a controlled two-step hydration with water.
- The presence of sulphates, sulphites, thiosulphates and carbonates in solution can be tested by using aqueous barium chloride followed by dilute hydrochloric acid or by using aqueous barium nitrate solution followed by dilute nitric acid.

## ? EXAM QUESTIONS

1 This question concerns Group IV of the Periodic Table.
a) Name the type of bonding and structure found in diamond and in lead.
b) Use your answers in a) to explain the differences in the melting points and electrical conductivities in diamond and in lead.

c) (i) Explain why $CCl_4$ is unaffected by water, but $SiCl_4$ reacts with it.
   (ii) Suggest why Si can form the complex ion $[SiF_6]^{2-}$, whereas C cannot form $[CF_6]^{2-}$.
d) Describe and explain the trend in the most stable oxidation state in Group IV oxides as the group is descended.

OCR 4822 Trends and Patterns June 2000 Q3

## ? EXAM QUESTIONS

**2 a)** Describe and explain the following variation of physical properties within Group 4:
**(i)** electrical conductivity of the elements;
**(ii)** melting points of the elements.

**b)** Explain why the metallic character of the elements in Group 4 increases with increasing atomic (proton) number.

**c)** Explain why the principal oxidation state of lead in its compounds is +2, but that of silicon in its compounds is +4.

**3** Explain each of the following observations.

**a)** Both lead(II) oxide and lead(IV) oxide react with warm concentrated hydrochloric acid to form lead(II) chloride.

**b)** Lead(II) oxide reacts with both aqueous sodium hydroxide and dilute nitric acid, whereas carbon monoxide does not react with aqueous sodium hydroxide or dilute nitric acid.

**c)** Tetrachloromethane does not react with water, but silicon(IV) chloride immediately forms a white precipitate and an acidic solution.
In each case, write balanced equations to illustrate your answers.

**4** Domestic water sometimes contains aqueous lead(II) ions:

**a)** Explain how the aqueous lead(II) ions could have entered the water supply.

**b)** Describe how you could confirm the presence of aqueous lead(II) ions in tap water.

**c)** Samples of lead(II) salts were heated strongly in a Bunsen flame. State the name of the solid product formed and the name of the gas formed in each case when each of the following salts were heated:
**(i)** lead(II) nitrate;
**(ii)** lead(II) carbonate;
**(iii)** lead(II) sulphate.

**5** Oxides may be classified as acidic, amphoteric, basic or neutral.

**a)** Give an example of each type of oxide and describe the reaction, if any, of your chosen oxides with dilute sodium hydroxide.

**b)** Group 4 elements form two series of oxides, one with formula $XO$ and the other with formula $XO_2$.
**(i)** Explain why the basic character of the oxides $XO$ increases with increasing atomic number of X.
**(ii)** Suggest why tin(II) oxide is more basic in character than tin(IV) oxide.

**c)** Describe and explain the variation in thermal stability of the Group 4 oxide, $XO_2$.

**6 a)** Describe the reaction of sulphuric acid with each of the following (in each case write an equation for the reaction):
**(i)** the reaction of dilute sulphuric acid with magnesium;
**(ii)** the reaction of concentrated sulphuric acid with sucrose, $C_{12}H_{22}O_{11}$;
**(iii)** the reaction of concentrated sulphuric acid with sodium bromide.

**b)** Concentrated sulphuric acid reacts when heated with carbon to form carbon dioxide and sulphur dioxide:
**(i)** Write an equation for the reaction.
**(ii)** The reaction is an example of a redox reaction. By using oxidation numbers identify the reducing agent and the oxidizing agent in the reaction.

**7** Sulphur forms two oxides, $SO_2$ and $SO_3$.

**a)** By considering both bonding and non-bonding electrons, suggest shapes for these two simple molecules.

**b)** Sulphur trioxide is a key intermediate in the production of sulphuric acid from sulphur in the Contact process. Describe the reactions by which it is formed and explain how it is converted into sulphuric acid in the later stages of the process. Write equations where appropriate and explain why the particular conditions are used.

**c)** Gardeners sometimes fumigate their greenhouses to rid them of pests and moulds by burning a sulphur 'candle'. A gaseous concentration of sulphur dioxide of 50 p.p.m. (parts per million) by volume is effective. Calculate how many grams of sulphur a gardener needs to burn in order to produce a concentration of 50 p.p.m. of $SO_2$ in a greenhouse that measures 2 m × 3 m × 4 m. [Volume of 1 mol of gas = 24 dm$^3$ (0.0024 m$^3$) at r.t.p.]

OCR 9254 Linear Syllabus Paper 1 June 2000 Q5

**8** The reaction between sulphur dioxide and oxygen is a dynamic equilibrium:
$$2SO_2 + O_2 \rightleftharpoons 2SO_3 \quad \Delta H = -196 \text{ kJ mol}^{-1}$$

**a)** Explain what is meant by **dynamic equilibrium**.

**b)** Give the effect on the rate of this reaction, and the effect on the position of equilibrium, of:
**(i)** increasing the temperature;
**(ii)** increasing the pressure.

**c)** This reaction is one of the steps in the industrial production of sulphuric acid. The normal operating conditions are:
**(i)** a temperature of 450 °C,

**(ii)** a pressure of 2 atmospheres,

**(iii)** the use of a catalyst.

Justify the use of each of these conditions.

**d)** Give the name of the catalyst used.

**e)** Give one large-scale use of sulphuric acid.

Edexcel 6242/P.01 Unit Test 2 June 2001 Q4

**9** **a)** Describe the large scale production of sulphuric acid from sulphur, giving essential conditions and equations where relevant.

**b)** State two large-scale uses of sulphuric acid.

**c)** Sulphuric acid reacts with hydrogen bromide and hydrogen iodide as shown below.

$$H_2SO_4 + 2HBr \rightarrow 2H_2O + SO_2 + Br_2$$
$$H_2SO_4 + 8HI \rightarrow 4H_2O + H_2S + 4I_2$$

What is the oxidation state of sulphur in each of $H_2SO_4$, $SO_2$ and $H_2S$?

UCLES Modular: Trends and Patterns Spring 1996 q. 6

# KEY SKILLS **ASSIGNMENT**

## SILICATES AND ALUMINOSILICATES

**Soil is a mixture of water, organic materials, air and inorganic materials. The inorganic materials can be classified as sand, silt and clay.**

Sand is composed almost entirely of silica or silicon(IV) oxide and consists of particles with a diameter of between 0.05 mm and 2 mm. Silt has much the same chemical composition as sand, but contains some silicate and aluminosilicates. The particle size of silt is between 0.002 mm and 0.05 mm in diameter. Clay is almost entirely silicates and aluminosilicates and has the smallest particle size, being less than 0.002 mm diameter.

**1** Draw the structure of silicon(IV) oxide.

The silicates are based on a tetrahedral $SiO_4$ unit. Fig 29.A1 shows part of the structure of a silicate. Cations bind together the silicate chains.

**Fig 29.A1** The structure of a silicate chain. Silicate chains are bound together by cations that occupy the space between the chains. The empirical formula of this silicate is $SiO_3{}^{2-}$, so a five-silicon unit will carry a net charge of 10–

**2** If the cation present in the silicate drawn in Fig 29.A1 is a magnesium ion, what is the empirical formula of the silicate?

Aluminosilicates are also based on a $SiO_4$ tetrahedral unit, but sometimes the silicon atom is replaced by an aluminium atom.

**Fig 29.A2** The chain structure of an aluminosilicate. In this chain some silicon atoms are replaced by aluminium atoms. The net charge on this five silicon/aluminium unit is 12–

**3** List the similarities and the differences between the silicate drawn in Fig 29.A1 and the aluminosilicate in Fig 29.A2.

Aluminosilicates can facilitate ion exchange in soils. This is a surface action and involves cations binding at the surface of the clay. Other cations, such as protons, can replace the cations at the surface of the clay, and often release cations that are toxic to plants.

**4** Give two reasons why sand cannot facilitate ion exchange in soils.

**5** The oxidation number of aluminium in an aluminosilicate is +3 and that of silicon is +4. Explain why the number of cations that can bind with the silicate increases when some aluminium atoms replace silicon atoms.

**6** Suggest why acid rain can increase the amounts of toxic aqueous metal ions within the soil.

**7** Suggest why acid rain will have little effect on soils with a high sand content.

Clays should be a potential source of both elemental aluminium and elemental silicon. However, aluminium is manufactured by the electrolytic decomposition of aluminium oxide derived from bauxite (impure aluminium oxide) and silicon is made from the reduction of silica or silicon(IV) oxide by carbon or magnesium.

**8** Clays are far more widespread than silica or bauxite. Suggest why clay is not used as a raw material in the manufacture of either silicon or aluminium.

**9** Give the names of two materials that are manufactured from clay.

# APPENDIX 1: SIGNIFICANT FIGURES AND DECIMAL PLACES

Students and teachers alike have difficulty with the use of significant figures and decimal places in calculations. This appendix contains simple rules that will enable you to give answers to calculations to the appropriate number of decimal places or significant figures.

## Significant figures

Whenever you make a measurement of a physical quantity in the laboratory, such as the temperature of a liquid or the mass of a test tube, there will be an uncertainty in the measurement. When you use an electronic balance to measure the mass of a test tube, the last figure on the balance reading almost always fluctuates. One moment it may be 16.41 g and the next it is 16.40 g. This shows that the last figure in the mass must be uncertain. We say that 16.40 g has four significant figures and the last figure, in this case the zero, is uncertain.

Whenever you measure a physical quantity, always quote the figure to include the first one which is uncertain. Sometimes this will be a zero, but it must be quoted.

## Determining the number of significant figures in a measurement

In this book you must assume that the measurements have been determined correctly and that they have the correct number of significant figures. To determine the number of significant figures, use the following rules.

● Find the first non-zero digit on the left and count the total number of digits,
  e.g. 0.00234 has three significant figures and 234.12 has five significant figures.
● If the number has a decimal point, count all the digits to the right, even if they are zero,
  e.g. 0.120 has three significant figures.
● If the number is written in standard form, do not count the exponential part of the number,
  e.g. $1.23 \times 10^{-3}$ has three significant figures and $9.560 \times 10^6$ has four significant figures.

These rules do not apply to quantities such as 100 m. Does this have one, two or three significant figures? Also, if you cannot tell, for instance, if you did not make the measurement yourself, then the number of significant figures is uncertain.

Other numbers are exact, for example there are 1000 cm$^3$ in 1 dm$^3$. There is no uncertainty with the 1000; it cannot be anything else.

**Table A1.1**

| Measurement | Number of significant figures |
|---|---|
| 24.0 °C | 3 |
| $5.67 \times 10^{-5}$ m$^3$ | 3 |
| 0.00560 moles | 3 |
| 2500 cm$^3$ | uncertain |

## Decimal places

To determine the number of decimal places in a number, simply count the number of digits to the right of the decimal place. Even zeros will be counted if they are significant. So if the mass of a sample of copper(II) oxide is measured as 0.300 g (we assume that the mass was determined to the nearest milligram), then this number has three decimal places. Remember, the number of significant figures is normally different from the number of decimal places.

## Calculations

You must be very careful during a calculation not to quote too many or too few decimal places or significant figures. Do not just copy the answer from your calculator. Think about how many significant figures or decimal places should be used. These are some simple rules.

● When adding or subtracting numbers, it is the number of decimal places that is important. Decide which number in the calculation has the least number of decimal places. This tells you the number of decimal places in the answer.
  For example, the $M_r$ of NO is 14.0 + 16.0 = 30.0.
● When multiplying or dividing numbers, it is the number of significant figures that is important. Decide which number in the calculation has the least number of significant figures. This tells you how many significant figures in the answer.
  For example, the number of moles of carbon in
  $$14.20\,\text{g} = \frac{14.2}{12.0} = 1.183333$$
  but the answer should be quoted only to three significant figures: 1.18.
● Remember that exact numbers are just that, and so do not affect the number of significant figures.

# APPENDIX 2: THE PERIODIC TABLE

Relative atomic masses given in brackets refer to the isotopic mass of the most abundant isotope of the elements concerned.

Key:
atomic no
**symbol**
name
relative atomic mass

metal | non-metal

| Period | Group 1 | 2 | | 3 | 4 | 5 | 6 | 7 | 0 |
|---|---|---|---|---|---|---|---|---|---|
| 1 | 1 **H** hydrogen 1.0 | | | | | | | | 2 **He** helium 4.0 |
| 2 | 3 **Li** lithium 6.9 | 4 **Be** beryllium 9.0 | | 5 **B** boron 10.8 | 6 **C** carbon 12.0 | 7 **N** nitrogen 14.0 | 8 **O** oxygen 16.0 | 9 **F** fluorine 19.0 | 10 **Ne** neon 20.2 |
| 3 | 11 **Na** sodium 23.0 | 12 **Mg** magnesium 24.3 | | 13 **Al** aluminium 27.0 | 14 **Si** silicon 28.1 | 15 **P** phosphorus 31.0 | 16 **S** sulphur 32.1 | 17 **Cl** chlorine 35.5 | 18 **Ar** argon 39.9 |
| 4 | 19 **K** potassium 39.1 | 20 **Ca** calcium 40.1 | (transition metals) | 31 **Ga** gallium 69.7 | 32 **Ge** germanium 72.6 | 33 **As** arsenic 74.9 | 34 **Se** selenium 79.0 | 35 **Br** bromine 79.9 | 36 **Kr** krypton 83.8 |
| 5 | 37 **Rb** rubidium 85.5 | 38 **Sr** strontium 87.6 | (transition metals) | 49 **In** indium 114.8 | 50 **Sn** tin 118.7 | 51 **Sb** antimony 121.8 | 52 **Te** tellurium 127.6 | 53 **I** iodine 126.9 | 54 **Xe** xenon 131.3 |
| 6 | 55 **Cs** caesium 132.9 | 56 **Ba** barium 137.3 | (transition metals) | 81 **Tl** thallium 204.4 | 82 **Pb** lead 207.2 | 83 **Bi** bismuth 209.0 | 84 **Po** polonium (209) | 85 **At** astatine (210) | 86 **Rn** radon (222) |
| 7 | 87 **Fr** francium (223) | 88 **Ra** radium (226) | (transition metals) | | | | | | |

Transition metals (Periods 4–7, Groups 3–12):

| 21 **Sc** scandium 45.0 | 22 **Ti** titanium 47.8 | 23 **V** vanadium 50.9 | 24 **Cr** chromium 52.0 | 25 **Mn** manganese 54.9 | 26 **Fe** iron 55.9 | 27 **Co** cobalt 58.9 | 28 **Ni** nickel 58.7 | 29 **Cu** copper 63.5 | 30 **Zn** zinc 65.4 |
|---|---|---|---|---|---|---|---|---|---|
| 39 **Y** yttrium 88.9 | 40 **Zr** zirconium 91.2 | 41 **Nb** niobium 92.9 | 42 **Mo** molybdenum 95.9 | 43 **Tc** technetium (98) | 44 **Ru** ruthenium 101.1 | 45 **Rh** rhodium 102.9 | 46 **Pd** palladium 106.4 | 47 **Ag** silver 107.9 | 48 **Cd** cadmium 112.4 |
| 57 **La** lanthanum 138.9 | 72 **Hf** hafnium 178.5 | 73 **Ta** tantalum 181.0 | 74 **W** tungsten 183.9 | 75 **Re** rhenium 186.2 | 76 **Os** osmium 190.2 | 77 **Ir** iridium 192.2 | 78 **Pt** platinum 195.1 | 79 **Au** gold 197.0 | 80 **Hg** mercury 200.6 |
| 89 **Ac** actinium (227) | 104 **Unq** unniliquadium (261) | 105 **Unp** unnilpentium (262) | 106 **Unh** unnilhexium (263) | 107 **Uns** unnilseptium (262) | 108 **Uno** unniloctium (265) | 109 **Une** unnilennium (266) | | | |

Lanthanides:

| 58 **Ce** cerium 140.1 | 59 **Pr** praseodymium 140.9 | 60 **Nd** neodymium 144.2 | 61 **Pm** promethium (145) | 62 **Sm** samarium 150.4 | 63 **Eu** europium 152.0 | 64 **Gd** gadolinium 157.3 | 65 **Tb** terbium 158.9 | 66 **Dy** dysprosium 162.5 | 67 **Ho** holmium 164.9 | 68 **Er** erbium 167.3 | 69 **Tm** thulium 168.9 | 70 **Yb** ytterbium 173.0 | 71 **Lu** lutetium 175.5 |
|---|---|---|---|---|---|---|---|---|---|---|---|---|---|

Actinides:

| 90 **Th** thorium 232.0 | 91 **Pa** protactinium (231) | 92 **U** uranium 238.1 | 93 **Np** neptunium (237) | 94 **Pu** plutonium (244) | 95 **Am** americium (243) | 96 **Cm** curium (247) | 97 **Bk** berkelium (247) | 98 **Cf** californium (251) | 99 **Es** einsteinium (254) | 100 **Fm** fermium (253) | 101 **Md** mendelevium (256) | 102 **No** nobelium (254) | 103 **Lr** lawrencium (257) |
|---|---|---|---|---|---|---|---|---|---|---|---|---|---|

# APPENDIX 3: IONIZATION ENERGIES

## First ionization energies of the elements

These values are in kJ mol$^{-1}$.

| 1 | 2 | 3 | 4 | 5 | 6 | 7 | 8 | 9 | 10 | 11 | 12 | 13 | 14 | 15 | 16 | 17 | 18 |
|---|---|---|---|---|---|---|---|---|----|----|----|----|----|----|----|----|----|
| **H** 1310 | | | | | | | | | | | | | | | | | **He** 2370 |
| **Li** 519 | **Be** 900 | | | | | | | | | | | **B** 799 | **C** 1090 | **N** 1400 | **O** 1310 | **F** 1680 | **Ne** 2080 |
| **Na** 494 | **Mg** 736 | | | | | | | | | | | **Al** 577 | **Si** 786 | **P** 1060 | **S** 1000 | **Cl** 1260 | **Ar** 1520 |
| **K** 418 | **Ca** 590 | **Sc** 632 | **Ti** 661 | **V** 648 | **Cr** 653 | **Mn** 716 | **Fe** 762 | **Co** 757 | **Ni** 736 | **Cu** 745 | **Zn** 908 | **Ga** 577 | **Ge** 762 | **As** 966 | **Se** 941 | **Br** 1140 | **Kr** 1350 |
| **Rb** 402 | **Sr** 548 | **Y** 636 | **Zr** 669 | **Nb** 653 | **Mo** 694 | **Tc** 699 | **Ru** 724 | **Rh** 745 | **Pd** 803 | **Ag** 732 | **Cd** 866 | **In** 556 | **Sn** 707 | **Sb** 833 | **Te** 870 | **I** 1010 | **Xe** 1170 |
| **Cs** 376 | **Ba** 502 | **La** 540 | **Hf** 531 | **Ta** 577 | **W** 770 | **Re** 762 | **Os** 841 | **Ir** 887 | **Pt** 866 | **Au** 891 | **Hg** 1010 | **Tl** 590 | **Pb** 716 | **Bi** 774 | **Po** 812 | **At** | **Rn** 1040 |
| **Fr** 381 | **Ra** 510 | **Ac** 669 | | | | | | | | | | | | | | | |

| **Ce** 665 | **Pr** 556 | **Nd** 607 | **Pm** 556 | **Sm** 540 | **Eu** 548 | **Gd** 594 | **Tb** 648 | **Dy** 657 | **Ho** | **Er** | **Tm** | **Yb** 598 | **Lu** 481 |
|---|---|---|---|---|---|---|---|---|---|---|---|---|---|
| **Th** 674 | **Pa** | **U** 385 | **Np** | **Pu** | **Am** | **Cm** | **Bk** | **Cf** | **Es** | **Fm** | **Md** | **No** | **Lr** |

## Ionization energies of selected elements

**Table A1.2**

| Element | Atomic (proton) number | First ionization energy/kJ mol$^{-1}$ | Second ionization energy/kJ mol$^{-1}$ | Third ionization energy/kJ mol$^{-1}$ | Fourth ionization energy/kJ mol$^{-1}$ |
|---|---|---|---|---|---|
| K | 19 | 420 | 3070 | 4600 | 5860 |
| Ca | 20 | 590 | 1150 | 4940 | 6480 |
| Sc | 21 | 630 | 1240 | 2390 | 7110 |
| Ti | 22 | 660 | 1310 | 2720 | 4170 |
| V | 23 | 650 | 1370 | 2870 | 4600 |
| Cr | 24 | 650 | 1590 | 2990 | 4770 |
| Mn | 25 | 720 | 1510 | 3250 | 5190 |
| Fe | 26 | 760 | 1560 | 2960 | 540 |
| Co | 27 | 760 | 1640 | 3230 | 5100 |
| Ni | 28 | 740 | 1750 | 3390 | 5400 |
| Cu | 29 | 750 | 1960 | 3350 | 5690 |
| Zn | 30 | 910 | 1730 | 3830 | 6190 |

# APPENDIX 4: STANDARD ELECTRODE POTENTIALS

| Electrode reaction | $E^{\ominus}/V$ | Electrode reaction | $E^{\ominus}/V$ |
|---|---|---|---|
| $Ag^+ + e^- \rightleftharpoons Ag$ | +0.80 | $Hg^{2+} + 2e^- \rightleftharpoons Hg$ | +0.85 |
| $Ag^{2+} + e^- \rightleftharpoons Ag^+$ | +1.98 | $2Hg^{2+} + 2e^- \rightleftharpoons Hg_2^{2+}$ | +0.91 |
| $AgBr + e^- \rightleftharpoons Ag + Br^-$ | +0.07 | $Hg_2Cl_2 + 2e^- \rightleftharpoons 2Hg + 2Cl^-$ | +0.27 |
| $AgCN + e^- \rightleftharpoons Ag + CN^-$ | -0.04 | $I_2 + 2e^- \rightleftharpoons 2I^-$ | +0.54 |
| $Ag(CN)_2^- + e^- \rightleftharpoons Ag + 2CN^-$ | -0.38 | $2HOI + 2H^+ + 2e^- \rightleftharpoons I_2 + 2H_2O$ | +1.45 |
| $AgCl + e^- \rightleftharpoons Ag + Cl^-$ | +0.22 | $2IO_3^- + 12H^+ + 10e^- \rightleftharpoons I_2 + 6H_2O$ | +1.19 |
| $AgI + e^- \rightleftharpoons Ag + I^-$ | -0.15 | $K^+ + e^- \rightleftharpoons K$ | -2.92 |
| $Ag(NH_3)_2^+ + e^- \rightleftharpoons Ag + 2NH_3$ | +0.37 | $Li^+ + e^- \rightleftharpoons Li$ | -3.04 |
| $Ag_2O + H_2O + 2e^- \rightleftharpoons 2Ag + 2OH^-$ | +0.34 | $Mg^{2+} + 2e^- \rightleftharpoons Mg$ | -2.38 |
| $Al^{3+} + 3e^- \rightleftharpoons Al$ | -1.66 | $Mn^{2+} + 2e^- \rightleftharpoons Mn$ | -1.18 |
| $Al(OH)_4^- + 3e^- \rightleftharpoons Al + 4OH^-$ | -2.35 | $MnO_2 + 4H^+ + 2e^- \rightleftharpoons Mn^{2+} + 2H_2O$ | +1.23 |
| $As + 3H^+ + 3e^- \rightleftharpoons AsH_3$ | -0.38 | $MnO_4^- + e^- \rightleftharpoons MnO_4^{2-}$ | +0.56 |
| $H_3AsO_4 + 2H^+ + 2e^- \rightleftharpoons H_3AsO_3 + H_2O$ | +0.56 | $MnO_4^- + 4H^+ + 3e^- \rightleftharpoons MnO_2 + 2H_2O$ | +1.67 |
| $Au^+ + e^- \rightleftharpoons Au$ | +1.68 | $MnO_4^- + 8H^+ + 5e^- \rightleftharpoons Mn^{2+} + 4H_2O$ | +1.52 |
| $Au^{3+} + 3e^- \rightleftharpoons Au$ | +1.50 | $N_2 + 8H^+ + 6e^- \rightleftharpoons 2NH_4^+$ | +0.27 |
| $H_3BO_3 + 3H^+ + 3e^- \rightleftharpoons B + 3H_2O$ | -0.73 | $HNO_2 + H^+ + e^- \rightleftharpoons NO + H_2O$ | +0.99 |
| $Ba^{2+} + 2e^- \rightleftharpoons Ba$ | -2.90 | $NO_3^- + 2H^+ + e^- \rightleftharpoons NO_2 + H_2O$ | +0.81 |
| $Ba^{2+} + 2e^- \rightleftharpoons Be$ | -1.85 | $NO_3^- + 3H^+ + 2e^- \rightleftharpoons HNO_2 + H_2O$ | +0.94 |
| $BiO^+ + 2H^+ + 3e^- \rightleftharpoons Bi + H_2O$ | +0.28 | $NO_3^- + 4H^+ + 3e^- \rightleftharpoons NO + 2H_2O$ | +0.96 |
| $Br_2 + 2e^- \rightleftharpoons 2Br^-$ | +1.07 | $2NO_3^- + 10H^+ + 8e^- \rightleftharpoons N_2O + 5H_2O$ | +1.11 |
| $2HOBr + 2H^+ + 2e^- \rightleftharpoons Br_2 + 2H_2O$ | +1.59 | $2NO_3^- + 12H^+ + 10e^- \rightleftharpoons N_2 + 6H_2O$ | +1.24 |
| $2BrO_3^- + 12H^+ + 10e^- \rightleftharpoons Br_2 + 6H_2O$ | +1.52 | $NO_3^- + 10H^+ + 8e^- \rightleftharpoons NH_4^+ + 3H_2O$ | +0.87 |
| $CO_2 + H^+ + e^- \rightleftharpoons \frac{1}{2}H_2C_2O_4$ | -0.49 | $Na^+ + e^- \rightleftharpoons Na$ | -2.71 |
| $Ca^{2+} + 2e^- \rightleftharpoons Ca$ | -2.87 | $Ni^{2+} + 2e^- \rightleftharpoons Ni$ | -0.25 |
| $Cd^{2+} + 2e^- \rightleftharpoons Cd$ | -0.40 | $Ni(NH_3)_6^{2+} + 2e^- \rightleftharpoons Ni + 6NH_3$ | -0.51 |
| $Ce^{4+} + e^- \rightleftharpoons Ce^{3+}$ | +1.45 | $H_2O_2 + 2H^+ + 2e^- \rightleftharpoons 2H_2O$ | +1.77 |
| $Cl_2 + 2e^- \rightleftharpoons 2Cl^-$ | +1.36 | $O_2 + 4H^+ + 4e^- \rightleftharpoons 2H_2O$ | +1.23 |
| $2HOCl + 2H^+ + 2e^- \rightleftharpoons Cl_2 + 2H_2O$ | +1.64 | $O_2 + 2H^+ + 2e^- \rightleftharpoons H_2O_2$ | +0.68 |
| $2ClO_3^- + 12H + 10e^- \rightleftharpoons Cl_2 + 6H_2O$ | +1.47 | $O_2 + 2H_2O + 4e^- \rightleftharpoons 4OH^-$ | +0.40 |
| $CO^{2+} + 2e^- \rightleftharpoons CO$ | -0.28 | $O_3 + 2H^+ + 2e^- \rightleftharpoons O2 + H_2O$ | +2.07 |
| $CO^{3+} + e^- \rightleftharpoons Co^{2+}$ | +1.82 | $P + 3H^+ + 3e^- \rightleftharpoons PH_3$ | -0.04 |
| $Co(NH_3)_6^{2+} + 2e^- \rightleftharpoons Co + 6NH_3$ | -0.43 | $H_3PO_4 + 2H^+ + 2e^- \rightleftharpoons H_3PO_3 + H_2O$ | -0.28 |
| $Cr^{2+} + 2e^- \rightleftharpoons Cr$ | -0.91 | $Pb^{2+} + 2e^- \rightleftharpoons Pb$ | -0.13 |
| $Cr^{3+} + 3e^- \rightleftharpoons Cr$ | -0.74 | $PB^{4+} + 2e^- \rightleftharpoons PB^{2+}$ | +1.69 |
| $Cr^{3+} + e^- \rightleftharpoons Cr^{2+}$ | -0.41 | $PbO_2 + 4H^+ + 2e^- \rightleftharpoons Pb^{2+} + 2H_2O$ | +1.47 |
| $Cr_2O_7^{2-} + 14H^+ + 6e^- \rightleftharpoons 2Cr^{3+} + 7H_2O$ | +1.33 | $PbO_2 + H_2O + 2e^- \rightleftharpoons PBO + 2OH^-$ | +0.28 |
| $Cs^+ + e^- \rightleftharpoons Cs$ | -2.92 | $Ra^{2+} + 2e^- \rightleftharpoons Ra$ | -2.92 |
| $Cu^+ + e^- \rightleftharpoons Cu$ | +0.52 | $Rb^+ + e^- \rightleftharpoons Rb$ | -2.92 |
| $Cu^{2+} + 2e^- \rightleftharpoons Cu$ | +0.34 | $S + 2e^- \rightleftharpoons S^{2-}$ | -0.51 |
| $Cu^{2+} + e^- \rightleftharpoons Cu^+$ | +0.15 | $S + 2H^+ + 2e^- \rightleftharpoons H_2S$ | +0.14 |
| $Cu^{2+} + I^- + e^- \rightleftharpoons CuI$ | +0.86 | $SO_4^{2-} + 4H^+ + 2e^- \rightleftharpoons H_2SO_3 + H_2O$ | +0.17 |
| $Cu(NH_3)_4^{2+} + 2e^- \rightleftharpoons Cu + 4NH_3$ | -0.05 | $S_2O_8^{2-} + 2e^- \rightleftharpoons 2SO_4^{2-}$ | +2.01 |
| $2D^+ + 2e^- \rightleftharpoons D_2$ | -0.003 | $S_4O_6^{2-} + 2e^- \rightleftharpoons 2S_2O_3^{2-}$ | +0.09 |
| $F_2 + 2e^- \rightleftharpoons 2F^-$ | +2.87 | $Sb + 3H^+ + 3e^- \rightleftharpoons SbH_3$ | -0.51 |
| $Fe^{2+} + 2e^- \rightleftharpoons Fe$ | -0.44 | $SbO^+ + 2H^+ + 3e^- \rightleftharpoons Sb + H_2O$ | +0.21 |
| $Fe^{3+} + 3e^- \rightleftharpoons Fe$ | -0.04 | $Sn^{2+} + 2e^- \rightleftharpoons Sn$ | -0.14 |
| $Fe^{3+} + e^- \rightleftharpoons Fe^{2+}$ | +0.77 | $Sn^{4+} + 2e^- \rightleftharpoons Sn^{2+}$ | +0.15 |
| $Fe(CN)_6^{3-} + e^- \rightleftharpoons Fe(CN)_6^{4-}$ | +0.36 | $V^{2+} + 2e^- \rightleftharpoons V$ | -1.2 |
| $FeO_4^{2-} + 8H^+ + 3e^- \rightleftharpoons Fe^{3+} + 4H_2O$ | +2.20 | $V^{3+} + e^- \rightleftharpoons V^{2+}$ | -0.26 |
| $2H^+ + 2e^- \rightleftharpoons H_2$ | 0.00 | $VO^{2+} + 2H^+ + e^- \rightleftharpoons V^{3+} + H_2O$ | +0.34 |
| $H_2 + 2e^- \rightleftharpoons 2H^-$ | -2.25 | $VO^{2+} + 2H^+ + e^- \rightleftharpoons VO^{2+} + H_2O$ | +1.00 |
| $Hg_2^{2+} + 2e^- \rightleftharpoons 2Hg$ | +0.79 | $VO^{3-} + 4H^+ + e^- \rightleftharpoons VO^{2+} + 2H_2O$ | +1.00 |
| | | $Zn^{2+} + 2e^- \rightleftharpoons Zn$ | -0.76 |

# PHOTOGRAPH CREDITS

## Photo acknowledgements

Every effort has been made to contact the holders of copyright material, but if any have been inadvertently overlooked the publishers will be pleased to make the necessary arrangements at the first opportunity. The publishers would like to thank the following for permission to reproduce photographs (T = Top, B = Bottom, C = Centre, L = Left, R = Right):

Ace Photoagency, 3.A1, P Freytag, 3.1, D Kerwin, 3.12, Barry, 14.51; Allsport/P Rondeau, 14.47, M Hewitt, 17.31, M Thompson, 20.A3, M Cooper, 29.22; Ancient Art and Architecture Collection Ltd, 11.7; Associated Press AP, 7.24, p.212, 11.2, NOAA, 10.19; Argonne National Laboratory, managed and operated by the University of Chicago for the US Department of Energy, 4.19a; BASF, 1.21; BBC, 26.22; Barnaby's Picture Library, 16.19; Doris J Beck, Dept of Biological Sciences, Bowling Green State University, 24.A1, 24.A2; Biopol: Trademark and Property of Monsanto plc, p.430T; www.JohnBirdsall.co.uk, 5.19, 17.1, p.388, p.498, 25.14; Blackpool Pleasure Beach Ltd, 1.14D; Chris Bonington Picture Library, 18.38; British Aerospace Airbus, 22.34; British Steel, p.2CR, 1.3, 1.4, 1.16, 27.9; Broadleaf Design and Marketing, 10.32; Bruce Coleman Ltd/K Rushby, p.234T, N Myers, 11.1, H Lange, 13.3, K Taylor, 17.14; Camera Press/R Crane, 4.25; Chubb Fire Ltd, 12.8; Collections/B Boswell, 29.1; Martyn Chillmaid, 3.8, 6.40a, 10.22; Dr A Clifford and Professor K D Bartle, University of Leeds, 6.A1, 6.A2; Complete Weed Control, Ltd, 28.9; Daimler-Benz, 166CR; P Duckers, Assoc. Member of the Guild of Railway Artists, 27.41; Dupont, 20.21, 20.23, 20.A6, 20.A7; Thomas Eisner and Daniel Aneshansley, Cornell University, 8.Q4; AIP Emilio Segre Visual Archives/ William G Myers Collection, 2.7; J Henriksson/FLT-PICA bild AB, 26.A2; Vivien Fifield, 2.3a, 2.11a, 3.7, 3.11a; Ford Motor Company Ltd, 10.21b; Fundamental Photos, NYC/K Brochmann, p.574; GSF Photo Library, 4.2b; Leslie Garland Photo Library, 22.19, 22.29, 24.1, p.540, 25.29, 25.32, 29.3; Tony Waltham/Geophotos, p.2TL, 1.8, 1.9, 1.12a,b,c, e, 10.26, 14.41, 15.55, p.410, 21.38, p.466C, 22.1; Gettyimages, 1.12d, 7.15, p.166T, 8.21, 8.Q2, 10.9, 14.2, 14.37R, 15.26, 15.40, 18.24, 21.34, p.466T, 22.36, 23.6, 23.16, 29.5; Peter Gould, 6.20a, Sally and Richard Greenhill, 12.17, 20.17, p.596, Greenpeace UK/Hodson, p.430C; Michael Holford, 27.5; Holt Studios International/N Cattlin, 1.22, 17.12, 17.15, 23.1, 28.3; IBM UK, 3.16b; IBM Almaden Research Center, 15.8; ICCE/Boulton, 11.25L; ICI Chemicals and Polymers, 9.A1; Kos Picture Source,

14.54; Andrew Lambert Photographs, 4.26c, 4.27c, 4.28a, 4.29c, 4.31c, 4.32b, 4.33b, 4.34b, 4.37b, 4.38b, 5.16, 5.17, 5.18, 5.19, 5.21, 5.22, 6.20b, d, 6.23, p.148CR, 7.8, 7.9, 7.20R, p.192TR, 11.4, 11.19, 11.20, 11.48, 12.1, 12.7, 12.28, 13.30, 13.33, 14.16, 14.37L, 15.13, 15.41, 16.6, 16.16, 16.30, 16.32, 16.34, 17.3, 17.10, 17.11, 17.21, 18.3, 18.12, 18.18, 18.19, 19.A2, 20.12b, 20.19, 20.26, 21.1, 21.2, 21.16, 21.19, 21.22, 21.23, 21.24, 21.25, 21.26, 21.27, 21.35, 21.36, 22.3, 22.4, 22.5, 22.6, 22.9, 22.10, 22.12, 22.13, 22.14, 22.15, 22.20, 22.35, 23.2, 23.3, 23.10, 23.11, 23.13, 23.14, 23.15, 24.9, 24.25, 24.27, 24.28, 24.29, 24.30, 24.31, 24.32, 24.33, 24.34, 25.5, 25.7, 25.16, 25.20, 25.25, 26.6, 26.8, 27.4, 27.7, 27.14bR, 27.33, 28.11, 28.13, 29.7, 29.21, 29.23; Kristen Lehman/yachtShots, 20.A8; B Barbey/Magnum Photos, 9.28; NASA, 7.25, p.166CL, 8.18, 21.33, 22.11; NHPA/S Dalton, 17.13, A Ackerley, 17.26; Sassoferratto – *The Virgin and Child Embracing*, © National Gallery, London, 11.46; Natural History Museum, London, 7.3, 7.20L; Natural Visions/Heather Angel, 27.6b; The Nobel Foundation, 26.17; Oxford Scientific Films Ltd/H Rogers, 4.A2; 'PA' News Photo Library, 8.20, 18.17, 24.11, 28.7; Panos Pictures/P Fryer, 18.2; B Masini/Phototake, NYC, 20.9; Vinegar and Oil set by Simone ten Homple, 1991, reproduced with permission of the P and O Makower Trust, 24.8; Popperfoto, 2.8, Popperfoto/Reuter, 12.40, 26.18c; Premaphotos/K G Preston-Mafham, 16.14; Princeton University, 2.16; Quadrant/Auto Express, 11.25(inset); Reading Buses, 18.25; Rex Features Ltd, 7.17, p.192BR, 27.14bL; Charlotte Roberts, Calvin Wells Laboratory, Department of Archaeological Sciences, University of Bradford, 19.19; Rolls-Royce plc, 29.12; Ann Ronan at Image Select, 2.1, 2.13, 3.2b, 3.3, 5.3, 5.5, 26.18b; Gavin Rowe, 20.33; By permission of the President and Council of the Royal Society, 11.A1; Science Photo Library, 4.46, 5.1, 6.2, 6.3, 6.9, 11.26, 14.13, p.448, 24.20, 26.21, 27.15L, J King-Holmes, 1.1, 18.4L, Department of Clinical Radiology, Salisbury District Hospital, 1.14A, A Bartel, 1.14B, T Kinsbergen, 1.14C, NASA, p.28, 2.15, 10.14, NOAA, 2.17, 3.14, 12.39, 26.A3, P Plailly, 2.20, p.90C, ROE, p.46, R Megna, Fundamental Photos, 3.9a, S Camazine, 3.16a, G Tompkinson, p.68, p.148B, 11.40, P Plailly, Eureilos, 4.24c, C Priest, 4.36, D Parker, p.90T, 7.14, 19.27, J-L Charmer, 5.2, V Habbick, Visions, 5.20, Kairos/Latin Stock, p.118, C D Winters, 6.21, 6.40b, BSIP LECA, 6.22, G Muller, Struers, GmbH, 7.7, Rosenfeld Images Ltd, p.192C, 21.32, 22.26, ESA, 10.15, US Geological Survey, 10.16, A and H-Frieder Michler, 10.21a, S Fraser, 11.39, p.514, Ed Young, 11.42, CNRI, p.286, J Durham,

# ANSWERS TO TEST YOURSELF QUESTIONS

## CHAPTER 1

**A** (a) $CaCO_3 \rightarrow CaO + CO_2$;
(b) $2Mn + O_2 \rightarrow 2MnO$;
(c) $4P + 5O_2 \rightarrow P_4O_{10}$

**B** (a) 56; (b) 2; (c) 3.5

**C** (a) 1.0 mol of iron; 0.50 mol of calcium; 2.0 mol of copper; 0.125 (0.1 to 1 sig. fig.) mol of sulphur; (b) 120 g of calcium; b 6 g of neon

**D** (a) $P_4O_{10}$, 284; (b) $O_2$, 32; (c) $CaSiO_3$, 116

**E** (a) 90; b) 342; (c) 102

**F** (a) $Fe_2O_3 + 3C \rightarrow 3CO + 2Fe$; (b) 0.525 tonnes

**G** 96%

**H** (a) $Fe_3O_4 + 4CO \rightarrow 3Fe + 4CO_2$; (b) 7.24 tonnes; (c) 89.8%

**I**
(a) Exothermic
$S(s) + O_2(g)$ ———
$\Delta H = -297$ kJ mol$^{-1}$
——— $SO_2(g)$

(b) Endothermic
——— $CaO(s) + CO_2(g)$
$\Delta H = +178$ kJ mol$^{-1}$
$CaCO_3(s)$ ———

(c) Exothermic
$Fe_2O_3(s) + 3CO(g)$ ———
$\Delta H = -27$ kJ mol$^{-1}$
——— $2Fe(s) + 3CO_2(g)$

**J** Does not decompose. Melting point in excess of 2000°C; does not react with substances present in the blast furnace

**K** (a) To ensure thorough mixing of reactants; larger surface area; faster rate of reaction; (b) Allows easy passage of gases through the furnace from bottom to top

**L** Continuous processes need less labour, no need to close and reset reactor vessel, the product formed has a more consistent quality

**M** (a) $Mg + S \rightarrow MgS$; (b) 0.24 kg; (c) Incomplete reaction; may react with any oxygen or oxide still present in molten iron; (d) Larger surface area; can be dispersed throughout the molten iron; so faster rate of reaction; (e) Iron can be tapped off from beneath the molten magnesium sulphide; molten magnesium sulphide can be skimmed off the top

**N** (a) Rate of reaction becomes so low that the reaction appears to stop; (b) 2.00 mol; (c) 1.56 mol

**O** (a) Decreases, since concentration of HI decreases; (b) Shifts to the right to minimize the decrease in concentration of HI

**P** (a) and (c)

**Q** Rate of reaction too slow at low temperatures

**R** (a) Position of equilibrium shifts to the left (the endothermic direction) to minimize the temperature increase; (b) Position of equilibrium shifts to the left (the endothermic direction) to minimize the temperature increase; (c) Position of equilibrium shifts to the right (the endothermic direction) to minimize the temperature increase

**S** (a) Position of equilibrium shifts to the left since there are fewer moles of gas on the left and so this will minimize the increase in pressure; (b) No effect on position of equilibrium since the number of moles of gas on both sides of the equation is the same; (c) Position of equilibrium shifts to the right since there are fewer moles of gas on the right and so this will minimize the increase in pressure

**T** (a) Reaction is endothermic so that an increase in temperature will shift the position of equilibrium to the right and absorb heat and so minimize the increase in temperature; (b) Position of equilibrium shifts to the left because there are fewer moles of gas on the left and so this will minimize the increase in pressure; (c) The higher the pressure the faster the rate of reaction because there are more collisions per second; easier to move gases around a plant under pressure

**U** (a) Increases surface area and so a faster rate of reaction; heterogeneous catalyst with surface action so needs the greatest amount of surface to be available; (b) Shifts the equilibrium to the right the next time the gas goes through the synthesis reaction vessel

## CHAPTER 2

**A** 2000

**B** (a) $\alpha$, positive; $\beta$, negative; $\gamma$, neutral; (b) $\beta$ has a smaller mass than $\alpha$ particles

**C** 3 km

**D** Number of protons = number of electrons = $Z$ = 11

**E** (a) Iron, $A$ = 56, $Z$ = 26, 26 protons, 26 electrons and 30 neutrons; (b) Mercury, $A$ = 200, $Z$ = 80, 80 protons, 80 electrons and 120 neutrons

**F** $^{12}_{6}C$, $A$ = 12, $Z$ = 6, 6 protons, 6 electrons and 6 neutrons; $^{13}_{6}C$, $A$ = 13, $Z$ = 6, 6 protons, 6 electrons and 7 neutrons; $^{14}_{6}C$, $A$ = 14, $Z$ = 6, 6 protons, 6 electrons and 8 neutrons

**G** $A_r$, Br = 80.0; $A_r$, Mg = 24.3

**H** $^{4}_{2}He + ^{4}_{2}He \rightarrow ^{8}_{4}Be$ + energy; the element is beryllium

**I** $^{226}_{88}Ra \rightarrow ^{4}_{2}He + ^{222}_{86}Rn$; the element is radon

**J** (a) Since $\alpha$ particles can only penetrate a few centimetres in air they will not escape from the detector; (b) The americium oxide in the detector will still be radioactive

**K** $^{90}_{38}Sr \rightarrow ^{0}_{-1}e + ^{90}_{39}Y$; the element is yttrium

**L** (a) $^{235}_{92}U \rightarrow ^{232}_{90}Th + ^{4}_{2}He$;
(b) $^{131}_{53}I \rightarrow ^{131}_{54}Xe + ^{0}_{-1}e$;
(c) $^{32}_{14}Si \rightarrow ^{32}_{15}P + ^{0}_{-1}e$

**M** $9.63 \times 10^{-16}$ $(1 \times 10^{-15})$ g

## CHAPTER 3

**A** $5.50 \times 10^{-7}$ m; green

**B** (a) $3.0 \times 10^{-2}$ m;
(b) $2.0 \times 10^{-35}$ J;
(c) $1.2 \times 10^{-11}$ J

**C** Since frequency of gamma radiation is greater than that of ultraviolet, a photon of gamma radiation will have more energy than that of ultraviolet

**D** Higher energy end of spectrum

**E** (a)

(b) $n = 4 \rightarrow n = 3$

**F** Yellow

**G** Formula, $2n^2$; number of electrons, 50

**H**

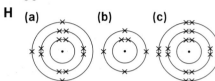

**I** (a) $Mg^{10+}(g) \rightarrow Mg^{11+}(g) + e^-$;
(b) $Mg^{10+}$ has a noble gas configuration $1s^2$, which does not have any inner-shell shielding electrons so the two electrons are very strongly attracted to the nucleus; in $Mg^{9+}$ the electron configuration is $1s^22s^1$, which has an inner shell of shielding electrons, so the electron in the outer shell is much less firmly attracted to the nucleus than the outer electrons in $Mg^{10+}$

**J**

First electron lost is from the outer shell and has the lowest ionization energy, because it is shielded by two sets of inner shells. It is also the furthest electron away from the nucleus. The next eight electrons lost all have the same number of inner shielding electrons, but the

ionic radius decreases as the charge on the ion increases, so there is more attraction towards the nucleus. The number of electrons is decreasing, but the nuclear charge is the same, so there is a greater attraction for the remaining electrons. The last two electrons lost are the inner-shell electrons that are not shielded and are very close to the nucleus. The large increases in ionization energy occur when an electron is removed from a new shell, so there are fewer shielding shells of electrons

**K** (a) The enthalpy change when one electron is lost from each atom in a mole of gaseous atoms, M, to form gaseous positive ions, $M^+(g)$;
(b) Lithium has a core charge of +1 and an inner shielding shell of two electrons, whereas neon has a core charge of +8 and an inner shielding shell of two electrons; this means that the outer electrons in a neon atom are attracted much more strongly to the nucleus than those in lithium. The atomic radius of lithium is also greater than that of neon

**L** (a) Same number of inner-shell shielding electrons, but the core charge of magnesium is +2 whereas in sodium it is +1, so outer electrons in magnesium are much more firmly attracted and closer to the nucleus; (b) (i) In magnesium the electron is removed from the 3s orbital whereas in aluminium the electron is removed from the 3p energy level, which is higher in energy (and further away from the nucleus) than the 3s energy level and so requires less energy; (ii) In phosphorus each 3p orbital contains just one electron so that there is no electron–electron repulsion in these orbitals, but in sulphur one 3p orbital contains two electrons so that there is some electron–electron repulsion and therefore it is easier to remove

one electron from a sulphur atom than from a phosphorus atom

**M** Be $(1s^22s^2)$ has a full 2s subshell, whereas B $(1s^22s^22p^1)$ has an extra electron in the 2s subshell. The 2p subshell has a higher energy level and the electron in this subshell is further away from the nucleus so that it requires less energy to be removed from the atom

**N** Na, $1s^22s^22p^63s^1$;
Cl, $1s^22s^22p^63s^23p^5$;
Al, $1s^22s^22p^63s^23p^1$

**O** As, $1s^22s^22p^63s^23p^63d^{10}4s^24p^3$;
Se, $1s^22s^22p^63s^23p^63d^{10}4s^24p^4$;
Br, $1s^22s^22p^63s^23p^63d^{10}4s^24p^5$

## CHAPTER 4

**A** $ns^2np^6$

**B** $Na^+$ has an electron configuration of Ne; $Cl^-$ has an electron configuration of Ar

**C**
(a) KF $\quad K^+ \quad \begin{bmatrix} \times\times \\ \times \ F \ \times \\ \bullet \ \times\times \end{bmatrix}^-$

(b) MgO $\quad Mg^{2+} \quad \begin{bmatrix} \times\times \\ \bullet \ O \ \times \\ \bullet \ \times\times \end{bmatrix}^{2-}$

(c) $CaCl_2 \quad Ca^{2+} \quad \begin{bmatrix} \times\times \\ \times \ Cl \ \times \\ \bullet \ \times\times \end{bmatrix}^- \quad \begin{bmatrix} \times\times \\ \times \ Cl \ \times \\ \times\times \end{bmatrix}^-$

(b) $Na_2O \quad \begin{matrix} Na^+ \\ Na^+ \end{matrix} \quad \begin{bmatrix} \times\times \\ \bullet \ O \ \times \\ \bullet \ \times\times \end{bmatrix}^{2-}$

**D** (a) 7; (b) 1; (c) 2

**E** $Na^+$ is isoelectronic with Ne; $Cl^-$ is isoelectronic with Ar

**F** (a) K, $1s^22s^22p^63s^23p^64s^1 \rightarrow$
$K^+, 1s^22s^22p^63s^23p^6$ or
K, $1s^22s^22p^63s^23p^64s^1 \rightarrow$
K, $[Ar]4s^1 \rightarrow K^+, [Ar]$
F, $1s^22s^22p^5 \rightarrow F^-, 1s^22s^22p^6$ or
F, $[He]2s^22p^5 \rightarrow F^-, [Ne]$
(b) Mg, $1s^22s^22p^63s^2 \rightarrow$
$Mg^{2+} \ 1s^22s^22p^6$ or
Mg, $[Ne]3s^2 \rightarrow Mg^{2+}, [Ne]$
$2 \times Cl, 1s^22s^22p^63s^2 \ 3p^5 \rightarrow$
$2 \times Cl^-, 1s^22s^22p^63s^23p^6$ or
$2 \times Cl, [Ne]3s^23p^5 \rightarrow 2 \times Cl^-, [Ar]$
(c) $2 \times Al, 1s^22s^22p^63s^23p^1 \rightarrow$
$2 \times Al^{3+}, 1s^22s^22p^6$ or
$2 \times Al, [Ne]3s^23p^1 \rightarrow 2 \times Al^{3+}, [Ne]$
$3 \times O, 1s^22s^22p^4 \rightarrow 3 \times O^{2-},$
$1s^22s^22p^6$ or

3 × O, [He]$2s^22p^4$ → 3 × $O^{2-}$, [Ne]

**G** It is exothermic, energy is transferred into the surroundings when a gaseous fluorine atom gains an electron

**H** (a) (b) (c)

**I** (a)

O=O

(b)

H
C=C
H    H
H    H

(c)

H—C≡N

**J**

**K** (a) (b)

O=S=O

**L**

F—B←N—H

**M** Outer shell of the carbon atom is surrounded by one single bond (a bonding pair of electrons) and a triple bond (which acts as one bonding pair of electrons) and no lone pairs. To minimize repulsion of the bonding pairs of electrons the electron pairs are arranged as far away from each other as possible, i.e. at an angle of 180°

**N** Trigonal planar because the central carbon atom is surrounded by two bonding pairs of electrons from single bonds and a double bond pair which is counted as one region of electron density. Three 'bonding pairs' repel each other to minimize repulsion to give a trigonal planar shape

**O**

**P** Tetrahedral (outer shell of aluminium surrounded by four bonding pairs which repel each other equally to minimize repulsion)

**Q** Beryllium ion surrounded by four bond pairs that equally repel each other to minimize repulsion and so the shape is tetrahedral

**R** (a) Four bond pairs and no lone pairs

(b) Four bond pairs and no lone pairs

(c) Three bond pairs and one lone pair

(d) V-shaped with a bond angle of about 120°. Two double bond pairs and one lone pair.

**S** Because the hybrid orbitals are a result of overlap between one s orbital and two p orbitals

**T** F–H, O–H, N–H, Cl–C, H–P, and C–I, F–F; the first atom is slightly negative and the second atom is slightly positive – the last three bonds are not polar

**U** +6; core charge remains constant but as the atomic number decreases so does the atomic radius and the number of inner shielding electron shells, so the nucleus has a greater attraction for the electron pair within a covalent bond

**V** Aluminium chloride

## CHAPTER 5

**A** Unreactive so found as elements in the Earth's crust rather than compounds or ores easy to reduce using carbon

**B** Discovery of radioactivity

**C** Difference in $A_r$ between the first two elements is the same or nearly the same as the difference between $A_r$ of the second two elements; difference in atomic number between first two elements is the same as that between the second two elements

**D** (a) They are in atomic number order; (b) Argon ($A_r$ = 40.0) and potassium ($A_r$ = 39.1); tellurium ($A_r$ = 127.6) and iodine ($A_r$ = 126.9)

**E** Selenium

**F** Rf after the scientist named Rutherford; Db after the place in Russia, Dubna, where research was carried out; Sg after the scientist named Seaborg; Bh after the scientist named Bohr; Hs after the scientist named Hass; Mt after the scientist named Meitner

**G** N, $1s^22s^22p^3$; Cl, $1s^22s^22p^63s^23p^5$; Ca, $1s^22s^22p^63s^23p^64s^2$; Ti, $1s^22s^22p^63s^23p^63d^24s^2$

**H** (a) 7; (b) 3; (c) 4

**I** (a) p; (b) p; (c) d; (d) d

**J** (a) Increase; (b) Decrease

**K** (a) (i) OH = 96 pm; (ii) NH = 100 pm; (b) Atomic radii are determined from single bonds; (c) Does not form $He_2$ so cannot measure a He–He bond

**L** As the atomic number increases, even though the nuclear charge increases there are more shielding shells of electrons, so that the electrons in the outer shell are less firmly attracted towards the nucleus and are further away from the nucleus

**M** **(a)** Greater nuclear charge; **(b)** (i) S has equal numbers of electrons and protons, $S^{2-}$ has two more electron than protons so the outer electrons are less firmly attracted to nucleus; (ii) As the atomic number increases, atoms have same number of inner shell shielding electrons, but increased nuclear charge

**N** Both have the same number of shielding shells of electrons, but sodium has a larger atomic radius and less protons in the nucleus, i.e. a smaller nuclear charge

**O** No shielding shells of electrons, the smallest atomic radius and one more proton than the only other atom with no shielding shells

**P** **(a)** The np sub-shell is higher in energy than the ns sub-shell, so that less energy is needed to remove an electron from an element in Group 3, which has one electron in an np orbital whereas a Group 2 element does not; from Group 3 to Group 4 there is no increase in shielding, but an increase in the nuclear charge so the outer electron is more firmly attracted to the nucleus with a Group 4 element; **(b)** Elements in Group 5 have one electron in each of the three outer p orbitals, so minimizing electron–electron repulsion within an orbital, whereas in an element of Group 6 one outer p orbital has to have two electrons and so there is electron–electron repulsion in that orbital. A Group 7 element has an atom with the same number of shielding shells of electrons as the element in Group 6, but a greater nuclear charge, so the outer electrons are held more strongly

**Q** Has a full 3d sub-shell

**R** Francium has the largest atomic radius and the most number of shielding shells of all the Group 1 elements, so it has the lowest first-ionization energy and thus a francium atom can lose electrons very easily

**S** **(a)** Oxidation; **(b)** Oxidation; **(c)** Reduction; **(d)** Neither; **(e)** Oxidation; **(f)** Oxidation; **(g)** Reduction

**T** **(a)** $Ca \rightarrow Ca^{2+} + 2e^-$, $Cl_2 + 2e^- \rightarrow 2Cl^-$; **(b)** $Mg \rightarrow Mg^{2+} + 2e^-$, $F_2 + 2e^- \rightarrow 2F^-$; **(c)** $Mg \rightarrow Mg^{2+} + 2e^-$, $O_2 + 4e^- \rightarrow 2O^{2-}$; **(d)** $Na \rightarrow Na^+ + e^-$, $N_2 + 6e^- \rightarrow 2N^{3-}$

**U** **(a)** $Al \rightarrow Al^{3+} + 3e^-$, $O_2 + 4e^- \rightarrow 2O^{2-}$, $4Al + 3O_2 \rightarrow 4Al^{3+} + 6O^{2-}$; **(b)** (i) $Zn + 2Fe^{3+} \rightarrow Zn^{2+} + 2Fe^{2+}$; (ii) $MnO_4^- + 8H^+ + 5Fe^{2+} \rightarrow Mn^{2+} + 4H_2O + 5Fe^{3+}$; (iii) $Cr_2O_7^{2-} + 14H^+ + 6Fe^{2+} \rightarrow 2Cr^{3+} + 6Fe^{3+} + 7H_2O$

**V** **(a)** Al is +3 and O is –2; **(b)** Ca is +2 and Cl is –1; **(c)** Mg is +2 and N is –3; **(d)** Cu is +2 and Cl is –1; **(e)** K is +1 and S is –2; **(f)** Ba is +2 and F is –1

**W** **(a)**

**(b)** Chlorine; **(c)** C is +4 and Cl is –1

**X** **(a)** H is +1, O is –2 and N is +5; **(b)** (i) H is +1, O is –2 and Cl is +5; (ii) It is the oxidation number of the chlorine

**Y** **(a)** K is +1, S is +6 and O is –2; **(b)** K is +1, S is +4 and O is –2; **(c)** +2; **(d)** –1; **(e)** +2; **(f)** +2.5

**Z** **(a)** +6; **(b)** Ferrate(VI); **(c)** Mn is +7 in $MnO_4^-$ and Mn is +6 in $MnO_4^{2-}$; **(d)** (i) +2; (ii) –3; (iii) +2

**A′** **(a)** Oxygen; **(b)** Carbon

**B′** Sodium atoms lose electrons (oxidation) to form sodium ions; the electrons lost are gained by hydrogen molecules (reduction) to form hydride ions

**C′** **(a)** (i) Cl in $Cl_2$ changes from 0 to –1, Cl in $PCl_3$ has no change, P changes from +3 to +5; (ii) $Cl_2$;

**(b)** (i) Cu from +2 to +1, I from –1 to 0, oxidizing agent is $Cu^{2+}$, reducing agent is $I^-$; (ii) Mn from +4 to +2, Cl from –1 to 0, oxidizing agent is $MnO_2$, reducing agent is HCl (iii) Cu from 0 to +2, O from 0 to –2, oxidizing agent is $O_2$, reducing agent is Cu (iv) Cl from 0 to +1 and to –1, oxidizing agent is $Cl_2$, reducing agent is $Cl_2$

**D′** Oxidation state of Cl in $Cl_2$ is 0, in NaCl is –1 and in $NaClO_3$ is +5; chlorine, $Cl_2$, is oxidized to give $NaClO_3$ and reduced to give NaCl; simultaneous oxidation and reduction of the same element is disproportionation

**E′** Selenium is a non-metal, forms $Se^{2-}$, forms an acidic oxide, will be a solid at room temperature and pressure, reacts with metals to form ionic compounds, is less reactive than sulphur, is a poor oxidizing agent, is a poor conductor of heat and electricity, and has a hydride with the formula $H_2Se$

## CHAPTER 6

**A** 2:1:2 ($H_2$:$O_2$:$H_2O(g)$)

**B** Using Avogadro's hypothesis, 1 molecule of hydrogen reacts with $\frac{1}{2}$ molecule of oxygen to make one molecule of water; assuming both water and hydrogen are diatomic, the formula of water must be $H_2O$

**C** 25 °C

**D** 0.25 mole

**E** $1.7 \times 10^3$ dm$^3$ (to 2 sig. fig.)

**F** 29 dm$^3$

**G** 25 °C (298 K) and 101 kPa

**H** 2 dm$^3$

**I** 100 cm$^3$

**J** **(a)** 298 K; **(b)** 373 K; **(c)** 223 K

**K** 63.7 cm$^3$

**L** **(a)** $4.9 \times 10^4$ cm$^3$; **(b)** $3.1 \times 10^4$ dm$^3$

**M** 146

**N** So that it can be subtracted from the volume of the gaseous compound under test, otherwise this volume will include the small volume of air

**O** 46

**P** **(a)** There is no change in the mean kinetic energy of the particles. Particles will be more spread out, so that fewer particles collide per unit area of container wall per second, i.e. pressure decreases; **(b)** $2.41 \times 10^{22}$; **(c)** If the amount of gas increases, the number of particles increases, and so to keep the number of particles that collide with the container wall per second constant, the volume of gas will have to increase to lengthen the distance between particles. If the amount doubles then the volume will need to double to keep the particles the same distance apart and thus keep the number of collisions per second the same

**Q** **(a)** Steam is a gas, so particles are very spread out with lots of space between them, and when it condenses the particles are close together with only very little space between them; **(b)** Lots of space between particles in a gas, but little space between particles in a liquid; **(c)** More particles per unit volume in a liquid, since the particles are much closer together than in a gas; **(d)** Particles in a liquid are free to move and are not in fixed positions; **(e)** Particles in a solid are not free to move. but particles in gases and liquids are free to move

**R** $\Delta H_m$ is much smaller than $\Delta H_b$

**S** **(a)** 2; **(b)** 1

**T** **(a)** 190 to 210 K; **(b)** −83 to −63 °C

**U**

$$\overset{\delta+}{H}\,\underset{\underset{H^{\delta+}}{|}}{\overset{N^{\delta-}}{|}}\,H^{\delta-}$$

$$\overset{\delta+}{H}\,\underset{\underset{H^{\delta+}}{|}}{\overset{N^{\delta-}}{|}}\,H^{\delta+}$$

**V** Suggestions could include: much less liquid water present in the oceans, so fewer ocean currents such as the Gulf Stream; weather may be drier since less water is available to be evaporated; there may not be a polar ice-cap at the North Pole, since it will sink under the surface of the water once formed; possibly warmer weather as a result; sea-water may become saltier and so it would be less likely that marine life could survive

**W** **(a)** −240 °C; **(b)** Helium

**X** Ethanol = 0.8 and water = 0.2

**Y** 0.2

**Z** Increases

**A′** **(a)** (i) 0.2 mole; (ii) 0.125 mole; (iii) 1 mole; (iv) $5 \times 10^{-3}$ mole; (v) $2.5 \times 10^{-3}$ mole; **(b)** (i) 1.6 mol dm$^{-3}$; (ii) 1.0 mol dm$^{-3}$

**B′** **(a)** 159.6 g; **(b)** 26.5 g; **(c)** 3.5 g

**C′** 0.951 mol dm$^{-3}$ and 57.06 g/cm$^3$

**D′** **(a)** Increase; **(b)** Decrease; **(c)** Decrease; **(d)** Increase; **(e)** Increase; **(f)** Decrease

## CHAPTER 7

**A** Particles of a liquid are in random motion, collide with one another and with the container wall, are not in an ordered pattern, are attracted to one another, but less so than in a solid, are moving faster than in a solid and slower than in a gas and the distance between particles is very small; particles of a gas are in random motion, collide with one another and with the container wall, are not in an ordered pattern, are only weakly attracted towards each other, are moving faster than in a liquid and the distance between particles is large

**B** Particles of a solid gain energy and vibrate more energetically until eventually the inter-particle force is broken and the particles become free to move around; as soon as the substance has melted the particles will move in any direction (random motion), but they will still be close together

**C** **(a)** $Zn(s) + 2Ag^+(aq) \rightarrow Zn^{2+}(aq) + 2Ag(s)$; **(b)** $Zn(s) + Pb^{2+}(aq) \rightarrow Zn^{2+}(aq) + Pb(s)$

**D** **(a)** 12; **(b)** 12; **(c)** 8

**E** Stronger metallic bonding because of electrostatic attraction between two moles of delocalized electrons for each mole of $Mg^{2+}$, i.e. $Mg^{2+}$ higher charge of cation, smaller ionic radius and more delocalized electrons

**F** Aluminium has 3 moles of electrons per mole of metal compared to 2 moles for Mg and 1 mole for Na

**G** Carbon dioxide

**H** **(a)** The layers of carbon atoms can easily slide over each other because of the weak van der Waals forces between each layer; **(b)** Has a very high melting point; **(c)** Diamond, since each carbon atom is bonded covalently to four other carbon atoms, but in graphite a carbon atom is only covalently bonded to three others

**I** **(a)** Silicon is a giant molecule with each silicon atom bonded covalently to four other silicon atoms; covalent bonds are much stronger than intermolecular attractions so that a large amount of energy is needed to break the many covalent bonds; **(b)** (i) Giant molecular structure

(ii)

$$\overset{\displaystyle C}{\underset{\displaystyle C}{\overset{|}{\underset{|}{C-Si-C}}}}$$

**J** As the temperature is increased, more thermal energy is available so that a greater fraction of the electrons can be promoted to an excited energy level

**K** **(a)** Si has four electrons in its outer shell and As has five electrons; **(b)** Each As atom will provide one extra electron, since only four electrons are needed for bonding; these extra electrons are easier to excite

**L** **(a)** $1s^2 2s^2 2p^1$; **(b)** Each B atom will provide one electron hole since it has one less electron in its outer shell than Si

**M** Astatine, since it has a larger molecule with more electrons than the other halogens, so it is more likely to have an asymmetric distribution of electrons and hence there are stronger induced dipole–induced dipole attractions between molecules

**N** All have weak intermolecular forces that are induced dipole–induced dipole attractions, but $P_4$ and $S_8$ have larger molecules and more electrons than $N_2$ and $O_2$, so an asymmetric distribution of electrons is easier in $P_4$ and $S_8$ and thus the intermolecular forces in $P_4$ and $S_8$ are stronger and they have much higher melting points and boiling points than $N_2$ and $O_2$

**O** Radon

## CHAPTER 8

**A** The energy released cannot be used to do useful work

**B** The enthalpy change of combustion will be different if gaseous water is produced, and also under standard conditions water must be a liquid

**C**

Energy/kJ mol$^{-1}$

$H_2(g) + \frac{1}{2}O_2(g)$

$\Delta H = -286$ kJ mol$^{-1}$

$H_2O(l)$

Extent of reaction

Energy/kJ mol$^{-1}$

$C_3H_8(g) + 5O_2(g)$

$\Delta H = -2219$ kJ mol$^{-1}$

$4H_2O(l) + 3CO_2(g)$

Extent of reaction

**D (a)** 4.2 kJ/4200 J;
**(b)** 84 kJ/84 000 J

**E (a)** Metal is a much better conductor of heat than glass;
**(b)** Temperature rise is a difference in temperature and, since absolute temperature is temperature in °C + 273, then the difference in temperature on both scales is identical; **(c)** Use a lid, calculate the energy absorbed by the metal calorimeter using the specific heat

capacity of the metal, have the calorimeter at the optimum height above the burner

**F** Incomplete combustion; loss of energy to the surroundings; no account taken of the energy absorbed by the calorimeter itself; experiment not carried out under standard conditions; possible loss of liquid fuel through evaporation

**G (a)** (i) Lower value; (ii) Weigh inside the bomb calorimeter so that no vapour can escape; **(b)** (i) Insulated container which prevents energy loss by conduction from the water into the surroundings, water in the calorimeter completely surrounds the chamber where the fuel is burnt so almost all the energy will be transferred into the water; (ii) To ensure that the temperature of the water is constant all through the calorimeter

**H** 51.7 kJ mol$^{-1}$

**I** $N_2$ has an extremely high bond energy since it has a triple covalent bond

**J (a)** H–H and O=O since they are the bonds in the molecules $H_2$ and $O_2$, which are necessarily identical, C=O figure refers to the C=O bond in carbon dioxide and these must also be identical; **(b)** +2061 kJ mol$^{-1}$; **(c)** Bond energies are quoted for one mole of gaseous bonds so that all the reactants and products must be gases; if methanol is not a gas in the calculation then the enthalpy change for $CH_3OH(l)$ to $CH_3OH(g)$ must be accounted for

**K (a)** By experiment $\Delta H = -286$ kJ mol$^{-1}$ and by calculation using bond energies it is $\Delta H = -243$ kJ mol$^{-1}$ (the calculation uses average O–H bond energy, not the bond energy for O–H in $H_2O$); **(b)** The energy or enthalpy of the products is higher than the energy or enthalpy of the reactants

**L** −658 kJ mol$^{-1}$

**M (a)** The fire will not be put out, and often it will become worse because more gaseous petrol is produced;

**(b)** Methanol has a low energy-density value

**N (a)** $C_8H_{18}(l) + 12\frac{1}{2}O_2(g) \rightarrow 8CO_2(g) + 9H_2O(l)$; **(b)** Per mole of $C_8H_{18}$ 16 moles of C=O and 18 moles of O–H

**O (a)** 50 432 kJ kg$^{-1}$;
**(b)** (i) 12 000 dm$^3$; (ii) 1.35 litres

**P (a)**

**(b)**

**c)** It is not exothermic and it does not involve the reaction of one mole of $N_2$

**Q (a)** $C(s) + O_2(g) \rightarrow CO_2(g)$; **(b)** $C(s) + 2H_2(g) \rightarrow CH_4(g)$; **(c)** Both transfer energy to the surroundings and so are more stable than the elements that make them up

**R** The standard state of carbon is graphite since it is energetically more stable than diamond

**S (a)** −104 kJ mol$^{-1}$;
**(b)** −110 kJ mol$^{-1}$;
**(c)** It is impossible to react carbon with oxygen without at least making some carbon dioxide

**T (a)** Enthalpy change of formation of $N_2$ is 0 since it is an element;
**(b)** −129 kJ mol$^{-1}$

## CHAPTER 9

**A** Methane, $CH_4$

**B** Loss of four electrons from a carbon atom to form $C^{4+}$ requires too much energy, and a gain of four electrons to form $C^{4-}$ also requires too much energy

**C** Methane

Water

**D** Dative covalent bond/coordinate bond

**E** $C_8H_{18}$

**F** Ethane

Pentane

**G** (a) $C_7H_{14}$ and $C_9H_{18}$ are unsaturated

(b)

Both have the same molecular formula

**H** (a)

(b) Same molecular formula

**I** (a)

(b)

(c)

or

**J** (i)

(ii)

(iii)

$CH_3$—$CH_2$—$CH$=$CH_2$

(iv)

**K** (a) The free rotation about C–C bonds means this is an alternative way to draw butane; (b) This has a molecular formula of $C_5H_{12}$ rather than $C_4H_{10}$

**L** $CH_3CH_2CH_2CH_2CH_2CH_3$; $(CH_3)_2CHCH_2CH_2CH_3$; $CH_3CH_2CH(CH_3)CH_2CH_3$; $CH_3CH(CH_3)CH(CH_3)CH_3$; $CH_3C(CH_3)_2CH_2CH_3$

**M** (a) $C_4H_{10}$; (b) It has molecular formula $C_4H_6$ not $C_4H_{10}$

**N** (a)

(b) Alcohol

**O** Isomers

**P** $C_9H_{20}$

**Q** (a) Sulphur dioxide; (b) Acid rain

**R** (a) Kerosene and/or diesel oil;
(b) (i) $C_{12}H_{26} \rightarrow CH_3(CH_2)_8CH_3 + C_2H_4$; (ii) $C_{12}H_{26} \rightarrow CH_3CH(CH_3)(CH_2)_5CH_3 + C_3H_6$

**S** On breaking a C–C bond both carbon atoms need to form an extra covalent bond, but there are no extra hydrogen atoms available, so a carbon=carbon double bond is formed to provide the hydrogen atoms

**T** (a) Hexane, cyclohexane, benzene. $CH_3(CH_2)_4CH_3 \rightarrow C_6H_{12} + H_2$; $C_6H_{12} \rightarrow C_6H_6 + 3H_2$; (b) Hydrogen

**U** (a) Because the molecular formula of the product is the same as the reactant, but they have different structures; (b) Unbranched hydrocarbons fit into a zeolite pore, but branched hydrocarbons do not

## CHAPTER 10

**A** $C_7H_{16}$, $CH_3CH_2CH_2CH_2CH_2CH_2CH_3$; $C_8H_{18}$, $CH_3CH_2CH_2CH_2CH_2CH_2CH_2CH_3$

**B** (a) $CH_3CH_2CH_2CH_2CH_2CH_2CH_2 CH_2CH_3$

(b) Decane

**C** (a) Pentyl; (b) $C_nH_{2n+1}$

**D** $CH_3CH_2CH_2CH_3$

**E** (a) It is another way of drawing methylbutane; (b) The methyl group can only be on carbon 2, if it was on carbon 1 then the molecule is pentane

**F** $CH_3C(CH_3)_2CH(C_2H_5)CH_2CH_2CH_3$; $CH_3CH_2CH_2CH(C_3H_7)CH_2CH_2CH_3$

**G** Two

**H** (a) Methane, ethane, propane and butane; (b) (i) Boiling point is about 216 °C and density is about 0.748 g cm$^{-3}$; (ii) 18

**I** The $\delta+$ end of the molecule and the $\delta-$ end are quite large, so there is a fairly strong electrostatic intermolecular attraction

**J** As the atomic number increases a noble gas atom has more electrons and a greater atomic radius, so that the formation of an instantaneous dipole is easier and thus the strength of the induced dipole–induced dipole intermolecular attractive force increases

**K** (a) It is hotter in Spain so the petrol is more volatile, and as temperature increases vapour pressure increases; (b) Use branched-chain alkanes

**L** (a) Exothermic; (b) Spontaneous reactions are ones that have a tendency to occur (and the entropy change is positive), activation energy is the minimum energy per mole of particles needed for a collision between particles to lead to a reaction; (c) The reaction of alkanes and oxygen is exothermic so alkanes and oxygen have more enthalpy than carbon dioxide and water (energetically unstable), but the activation energy is sufficiently high that at room temperature the rate of reaction between petrol and oxygen is very slow (kinetically stable); **(d)** Reacts with oxygen to release lots of energy (large value of energy released per gram of fuel), easily stored, easily transported, non-toxic; readily available

**M (a)** $C_nH_{2n+1}OH$; **(b)** Same chemical properties, and physical properties vary with increasing $M_r$ (boiling point, melting point and density all increase with increasing $M_r$)

**N** $C_{10}H_{22} + 15\frac{1}{2}O_2 \rightarrow 10CO_2 + 11H_2O$

**O (a)** Visible light; **(b)** Absorbed by coloured substances and by gases in the atmosphere

**P** Plants that can photosynthesize using up carbon dioxide are not present on Mars and Venus

**Q (a)** In summer months, during which green plants grow very fast, there is a high rate of photosynthesis that uses up carbon dioxide; **(b)** In winter months there is an increase in the mass of fuel burnt and the rate of growth by green plants is low, so not much photosynthesis occurs

**R (a)** EU nations agreed in Kyoto in December 1997 to reduce emissions of greenhouse gases to that of 1990 by the year 2012. This is about an 8% reduction per annum; **(b)** (i) 700 g; (ii) $C_8H_{18} + 12\frac{1}{2}O_2 \rightarrow 8CO_2 + 9H_2O$; (iii) 2160 g; (iv) $8.6 \times 10^{11}$ g

**S (a)** The catalyst is a solid but the reactants are gases, so a large surface area is required to maximize the number of collisions between reactant and catalyst, and also the catalyst works by adsorption (surface action) so it needs a large surface area; **(b)** This gives an even more larger surface area

**T** At the start of a journey when the engine is cold or when the car is stationary.

**U (a)** $C_6H_6 + 7\frac{1}{2}O_2 \rightarrow 6CO_2 + 3H_2O$; **(b)** Greenhouse effect

**V** $N_2 + O_2 \rightarrow 2NO$

**W** Greenhouse effect

**X** $NO_2$ is regenerated at the end of the reaction steps, which speeds up the rate of reaction and the mechanism of the oxidation of $SO_2$ is changed

**Y** $1s^22s^22p^4$, when occupying the three 2p atomic orbitals the electrons are arranged so as to minimize the electron–electron repulsion that occurs if two electrons are in the same orbital, so only one orbital, the $2p_x$, is occupied by two electrons while the other two, $2p_y$ and $2p_z$, have one electron in each, and so it has two unpaired electrons

## CHAPTER 11

**A** Organic compounds that contain mostly carbon and hydrogen are more likely to be completely combusted, and the amounts of products of the combustion, $H_2O$ and $CO_2$, can be easily determined

**B** For ethane, $CH_3$, and for ethanoic acid, $CH_2O$

**C (a)** $C_5H_{11}$; **(b)** $C_2H_5O_2N$

**D** $CH_2O$

**E (a)** $C_6H_6$; **(b)** $CH_2O_2$; **(c)** $C_8H_{10}N_4O_2$

**F** Prevents collisions between positive ions and other particles, and avoids undue contamination of the sample with air, since the molecules in air will also be ionized by electron bombardment

**G** 72.8

**H** $CH_2O_2$

**I (a)** Since all the particles have a charge of +1 then $m/e = m$ since $e = 1$; **(b)** $C_2H_4^+$; **(c)** $m/e = 43$, $C_3H_7^+$

**J (a)** If a molecule with all paired electrons loses one electron then it must end up with one unpaired electron, indicated by a dot for the

a free radical and a '+' for the positive ion, but simplified it is important to show the '+' and so $C_4H_{10}^+$ is also used; **(b)** Only positive particles are detected by the mass spectrometer and $CH_3\bullet$ is neutral

**K (a)** CO or $C_2H_4$; **(b)** $CH_3CH_2O\bullet$; **(c)** $C_6H_5\bullet$

**L (a)** $OH\bullet$; **(b)** $m/e = 45$ is $CH_2O+$, $m/e = 29$ is $CHO^+$, $m/e = 28$ is $CO^+$
**(c)**

**M (a)** 10; **(b)** (i) $m/e = 49$ is $CH_2{}^{35}Cl^+$ and $m/e = 51$ is $CH_2{}^{37}Cl^+$; (ii) $m/e = 64$ is $CH_3CH_2{}^{35}Cl^+$ and $m/e = 66$ is $CH_3CH_2{}^{37}Cl^+$; (iii) $CH_3\bullet$; **(c)** $m/e$ values are shown next to the peaks:

**(d)** (i) $m/e = 74$ is $({}^{37}Cl_2)^+$, $m/e = 72$ is $({}^{35}Cl^{37}Cl)^+$, $m/e = 70$ is $({}^{35}Cl_2)^+$; (ii) ratio of ${}^{35}Cl:{}^{37}Cl$ is 3:1 so the ratio of $({}^{35}Cl_2)^+:({}^{35}Cl^{37}Cl)^+:({}^{37}Cl_2)^+$ is 9:6:1 so the peak at $m/e = 74$ is the smallest

**N (a)** $H_2$ does not have a dipole moment even during vibrations; **(b)** Only one vibration is possible with a diatomic molecule; **(c)** Bond energy of HCl is greater than that of HI

**O** A is the hydrogen bond in O–H stretch, and B is the C–O stretch

**P (a)** O–H; **(b)** Confusion between the stretching frequency for O–H in water and in ethanol; **(c)** Propanone also contains C–H bonds so that the C–H stretching frequency at 3000 cm$^{-1}$ can result from ethanol or propanone; **(d)** 0.02%

**Q** Mass numbers are all odd

**R** Proton magnetic resonance

**S (a)** 1; **(b)** 1; **(c)** 2

**T** Signal at $\delta = 2.1$ (will be a doublet) integrated trace three protons, and

signal at δ = 9.7 (will be a quartet) integrated trace one proton
**U** CHO protons have four arrangements and $CH_3$ protons have two peaks
**V** Compound Y
**W** (a) Pencil will not dissolve in any solvent used; (b) So that as the solvent moves up the chromatography paper it will not evaporate; (c) 0.15 and 0.50; (d) It is not soluble in the mobile phase
**X** Adsorption
**Y** (a) The liquid must not boil when the GLC column is in a hot oven; (b) GLC works by the difference in solubility between the liquid and the carrier gas.
**Z** Since retention time is dependent on the flow rate, temperature, packing of the column, and length of the column, any data-base used must have the same GLC characteristics as used in the chromatographic separation
**A'** Adsorption

## CHAPTER 12
**A** Five atoms (four Cl and one I)
**B** (a) C is 1,2-dibromo-1,2-dichloroethane and D is 1,1-dibromo-1,2-dichloroethane;
(b) (i)

(ii)

(iii)

(iv)

(v)

**C** (a) Perfluorobutane or decafluorobutane;

(b) 2-Chlorobutane;
(c) Perchlorohexane
**D** (a) (i) $CF_3Cl$; (ii) $C_2F_2Cl_4$;
(b) Tetrachlorodifluoroethane
**E** This property means that they can smother and prevent oxygen from reaching a flame in a fire
**F** (a) About 52 °C; (b) Stronger intermolecular attraction in 1-chlorobutane; (c) Isomers that have branched chains are less able to pack efficiently with one another and so, overall, the intermolecular attraction is weaker than in an unbranched isomer
**G**

**H** $CH_3\bullet$ is the only free radical

**I** (a) 79 kJ mol$^{-1}$; (b) Fluorine
**J** There is no net change when a methyl free radical collides with a methane molecule,
$CH_3\bullet + CH_4 \rightarrow CH_4 + CH_3\bullet$
**K** (a) Have a small mole ratio of $Cl_2$:$CH_4$ and carry out chlorination over a short time period; (b) Have a high mole ratio $Cl_2$:$CH_4$ and carry out reaction over a long time period
**L** Any two reactions that involve two free radicals present in the reaction mixture, for example,
$Cl\bullet + CH_3\bullet \rightarrow CH_3Cl$ and
$Cl\bullet + CHCl_2\bullet \rightarrow CHCl_3$
**M** Initiation step, $Cl_2 \rightarrow 2Cl\bullet$
Propagation steps:
$Cl\bullet + C_2H_6 \rightarrow C_2H_5\bullet + HCl$
$C_2H_5\bullet + Cl_2 \rightarrow C_2H_5Cl + Cl\bullet$
$Cl\bullet + C_2H_5Cl \rightarrow C_2H_4Cl\bullet + HCl$
$C_2H_4Cl\bullet + Cl_2 \rightarrow C_2H_4Cl_2 + Cl\bullet$
$Cl\bullet + C_2H_4Cl_2 \rightarrow C_2H_3Cl_2\bullet + HCl$
$C_2H_3Cl_2\bullet + Cl_2 \rightarrow C_2H_3Cl_3 + Cl\bullet$
$Cl\bullet + C_2H_3Cl_3 \rightarrow C_2H_2Cl_3\bullet + HCl$

$C_2H_2Cl_3\bullet + Cl_2 \rightarrow C_2H_2Cl_4 + Cl\bullet$
$Cl\bullet + C_2H_2Cl_4 \rightarrow C_2HCl_4\bullet + HCl$
$C_2HCl_4\bullet + Cl_2 \rightarrow C_2HCl_5 + Cl\bullet$
$Cl\bullet + C_2HCl_5 \rightarrow C_2Cl_5\bullet + HCl$
$C_2Cl_5\bullet + Cl_2 \rightarrow C_2Cl_6 + Cl\bullet$
Termination steps:
$Cl\bullet + C_2H_5\bullet \rightarrow C_2H_5Cl$
$C_2H_5\bullet + C_2H_5\bullet \rightarrow C_4H_{10}$
**N** Initiation step: $F_2 \rightarrow 2F\bullet$
Propagation steps:
$F\bullet + CH_4 \rightarrow CH_3\bullet + HF$
$CH_3\bullet + F_2 \rightarrow CH_3F + F\bullet$
$F\bullet + CH_3F \rightarrow CH_2F\bullet + HF$
$CH_2F\bullet + F_2 \rightarrow CH_2F_2 + F\bullet$
$F\bullet + CH_2F_2 \rightarrow CHF_2\bullet + HF$
$CHF_2\bullet + F_2 \rightarrow CHF_3 + F\bullet$
$F\bullet + CHF_3 \rightarrow CF_3\bullet + HF$
$CF_3\bullet + F_2 \rightarrow CF_4 + F\bullet$
Termination steps:
$F\bullet + CH_3\bullet \rightarrow CH_3F$
$CH_3\bullet + CH_3\bullet \rightarrow C_2H_6$
**O** 1-bromohexane

2-bromohexane

3-bromohexane

**P** Bubble chlorine into refluxing methylbenzene in the presence of ultraviolet light over a long period of time or excess chlorine over a long time, $C_6H_5CH_3 + 3Cl_2 \rightarrow C_6H_5CCl_3 + 3HCl$
**Q** (a) 2-Iodo-3-methylhexane;
(b) (i) Ethanol, red phosphorus and iodine; (ii) Hexan-3-ol, sodium bromide and concentrated sulphuric acid, or hexan-3-ol, concentrated hydrobromic acid and concentrated sulphuric acid, or hexan-3-ol and phosphorus(III) bromide
**R** Fluoromethane

**T** **(a)** NH$_2^-$ nucleophile

$$\left[ H \overset{\times\times}{\underset{\bullet\bullet}{N}} H \right]^-$$

**(b)** CN$^-$ nucleophile

$$\left[ \overset{\bullet\bullet}{C} \overset{\times}{\underset{\bullet\bullet}{N}} \overset{\times}{\times} \right]^-$$

**(c)** BCl$_3$ not a nucleophile

**T** **(a)** +90 kJ mol$^{-1}$; **(b)** −80 kJ mol$^{-1}$
**U** Pentan-2-ol
**V** **(a)**

**(b)**

**(c)**

**W**

**X** **(a)** Substitution:
CH$_3$CH$_2$CH$_2$I + NaOH → CH$_3$CH$_2$CH$_2$OH + NaI
Elimination:
CH$_3$CH$_2$CH$_2$I + NaOH → CH$_3$CH=CH$_2$ + NaI + H$_2$O

**(b)** Pent-1-ene
**Y**

(outer electrons shown only)

**Z** A homogeneous catalyst

**CHAPTER 13**

**A**

**(a)** **(b)** **(c)**

**B** **(a)**

primary (i)

secondary (ii)

primary and secondary (iii)

secondary (iv)

**(b)** Geraniol is primary, nerol is primary, linalool is tertiary

**C** **(a)** Dihydric
**(b)**

and

and

**(c)** Geraniol is monohydric, nerol is monohydric, linalool is monohydric, maltose is polyhydric

**D** 97 °C
**E** **(a)** Both have strong intermolecular hydrogen bonding, but butan-1-ol is a larger molecule with more electrons so it has stronger van der Waals forces as well; **(b)** Decane and decan-1-ol are structurally similar since they both have a long-chain alkyl group that is non-polar. As a result both have van der Waals forces of approximately the same strength. The extra hydrogen bonding in decan-1-ol is not significant, since the molecules are large. In methane the intermolecular force is a weak van der Waals

force, since it is a small molecule, but in methanol it forms strong hydrogen bonding

**F** Nonan-1-ol is more soluble in an alkane solvent than water because it has a very long non-polar alkyl group that can form van der Waals intermolecular bonds with the solvent molecules; methanol is a polar molecule and cannot easily form van der Waals forces – it forms hydrogen bonds with water molecules when dissolved in water

**G** The chain form has a free aldehyde group but the ring form does not; the ring form has a C–O–C bond, which is not present in the chain form

**H** **(a)** The glucose molecule has an aldehyde at carbon number 1, but sorbitol has a hydroxyl group; **(b)** sorbitol can form many intermolecular hydrogen bonds between the hydroxyl groups and water molecules

**I** **(a)** C$_2$H$_4$ + H$_2$SO$_4$ → CH$_3$CH$_2$OSO$_3$H;
**(b)** CH$_3$CH$_2$OSO$_3$H + H$_2$O → CH$_3$CH$_2$OH + H$_2$O;
**(c)** Cyclopentanol
**J** CO$_2$ + 3H$_2$ → CH$_3$OH + H$_2$O
**K** **(a)** 2-Bromopentane;
**(b)** Iodocyclohexane;
**(c)** 1-Chloropropane
**L** **(a)** Reflux, under heating, ethanol with PI$_3$ (or phosphorus and iodine);
**(b)** Reflux, under heating, pentan-2-ol with PCl$_5$ or SOCl$_2$, or reflux, under heating, pentan-2-ol, NaCl(s) and H$_2$SO$_4$(l); **(c)** Reflux under heating 3-methylhexan-2-ol with PI$_3$ (or phosphorus and iodine)
**M** CH$_3$OCH$_3$ (methoxymethane)
**N** **(a)**

**(b)**

**(c)** But-2-ene

**O**

CH$_3$CH$_2$CH$_2$CH$_2$—C—O—H + [O] → CH$_3$CH$_2$CH$_2$CH$_2$—C + H$_2$O

**P** Geraniol is oxidized to compound **A** and nerol is oxidized to compound **B**

**A**

CH₃ ...CHO ... H₃C CH₃

**B**

CH₃ ...CHO ... H₃C CH₃

**Q**

H–C–C–C–C–C–C ... O ... H–C–H

**R** (a) $CH_3CH_2CH_2CH_2OH + 2[O] \rightarrow$ $CH_3CH_2CH_2COOH + H_2O$
(b)

**S** (a) Cyclohexanone;
(b) $CH_3CHOHCH_2CH_3 + [O] \rightarrow$ $CH_3COCH_2CH_3 + H_2O$
**T** Possible oxidation products include

(structures) CH₂OH ... CHO ... and

**U** (a) Butanal; (b) Butanone;
(c) No reaction
**V** $CH_3CHOHCH_2CH_2CH_3$
**W** The highly electronegative chlorine atom withdraws electrons away from the O–H bond, which makes it weaker than the O–H bond in ethanol; the weaker the O–H bond the stronger the acid
**X** $2Na + 2CH_3OH \rightarrow 2CH_3O^-Na^+ + H_2$; Sodium methoxide

## CHAPTER 14
**A** (a) $C_7H_{16}$;
(b) $C_{10}H_{22} \rightarrow C_2H_4 + C_8H_{18}$
**B** $C_2H_4 + 3O_2 \rightarrow 2CO_2 + 2H_2O$;
$C_3H_6 + 4\frac{1}{2}O_2 \rightarrow 3CO_2 + 3H_2O$
**C** (a) $CH_4 + 2O_2 \rightarrow CO_2 + 2H_2O$;
(b) Keep the mole ratio of oxygen to methane in the ratio of 1:2, since combustion requires a ratio of 2:1
**D** Each carbon atom is surrounded by one double bond (which counts as

one bond pair) and two single bonds (two bond pairs) and no lone pairs. The three 'bond pairs' repel each other to minimize repulsion and so take up a trigonal planar arrangement, which has an angle of 120° between the bond pairs
**E** (a)

(i) ... (ii) ... (iii)

(b) (i) Oct-1-ene;
(ii) 3-Ethyl-2-methylpent-2-ene;
(iii) Cyclohexene
**F** (a)

(i) ... (ii)

(b) (i) *trans*-Hept-3-ene;
(ii) *cis*-2,3-Dichloropropent-2-ene
(c) Hex-2-ene and hex-3-ene
**G** 265 kJ mol⁻¹
**H** (a) Molecules are larger and have more electrons so that the strength of the induced dipole–induced dipole (van der Waals) intermolecular force increases;
(b) about 60 °C
**I** (a) (i) Ethanol; (ii) Propan-1-ol or propan-2-ol; (iii) Cycloheptanol;
(b) (i) Pent-1-ene, *cis*-pent-2-ene and *trans*-pent-2-ene; (ii) *Pent*-2-ene
**J** Hex-1-ene, *cis*-hex-2-ene and *trans*-hex-2-ene

(structures)

**K** Lower bond energy than σ bond
**L** $Br^+$ and $CH_3^+$

**M**

**N** (structure with CH₃ and +)

**O** (a) 2-Bromobutane;
(b) Bromocyclohexane;
(c) 2-Bromopentane
**P** To make a primary alcohol the carbocation formed during the addition must be a primary carbocation, but these are less stable than secondary carbocations and so the addition of a proton to an alkene normally occurs via a secondary or tertiary carbocation rather than a primary one
**Q** (a) (b) (c)

(structures) Br Br ... Cl ... Cl Cl

**R**

(structures) Cl ... Cl ... Cl Cl

**S** The carbocation formed can react with either a bromide ion or water to give either $CH_3CHBrCHBrCH_3$ or $CH_3CHOHCHBrCH_3$
**T** 5
**U** (a) Propane-1,2-diol; (b) Butane-2,3-diol; Cyclopentane-1,2-diol
**V** (a) Ethanoic acid; (b) Ethanoic acid and propanoic acid;
(c) Pentanedioic acid
**W** (a) 60°; (b) Bond angle should be about 109° so that the electron pairs in the C–C bond are very close to each other and will repel strongly
**X** Carbon dioxide and water, $C_2H_4O + 3O_2 \rightarrow 2CO_2 + 2H_2O$
**Y** (a) $CH_2OHCH_2OCH_3$;
(b) $CH_2OHCH_2OCH(CH_3)_2$;
(c) $CH_2OHCH_2NH_2$
**Z** (a) Propene with $Cl_2$, $H_2O$, $C_3H_6$ and $C_6H_6$; (b) Propene → poly(propene)

## CHAPTER 15

**A** (a) [structure: methylbenzene] (b) [structure: chlorobenzene]

(c) [structures: ethylbenzene] and [1,2-dimethylbenzene] and [1,3-dimethylbenzene] or [1,4-dimethylbenzene]

**B** (a) (i) $C_6H_{10} + H_2 \rightarrow C_6H_{12}$;
(ii) $C_6H_8 + 2H_2 \rightarrow C_6H_{12}$;
(iii) $C_6H_6 + 3H_2 \rightarrow C_6H_{12}$;
(b) (i) $-120$ kJ mol$^{-1}$;
(ii) $-240$ kJ mol$^{-1}$; (iii) $-360$ kJ mol$^{-1}$

**C** (a) $C_6H_6 + 7\frac{1}{2}O_2 \rightarrow 6CO_2 + 3H_2O$;
(b) It would be less than that calculated for Kekule's structure since it requires more energy to break the bonds in a benzene ring than three C=C and three C–C bonds

**D** (a) $C_{10}H_8$; (b) $C_{12}H_9$

**E** (a) [1,4-dimethylbenzene] (b) [1,2-dimethylbenzene] (c) [1,3-dimethylbenzene]

(b) Because it will be 1,3-dimethylbenzene since the position numbers are as low as possible

**F** (a) [ethylbenzene] (b) [1,2-diethylbenzene] (c) [1,3-diethylbenzene]

**G** (a) (i) [1,3-chloro-iodobenzene] (ii) [1,4-chloro-iodobenzene]

(b) Position numbers must be as low as possible so it must be 1,2-dichlorobenzene

**H** (a) (i) [structure: C6H5-CH2-CH2-OH] (ii) [structure: C6H5-CH=CH2]

(b) 1-Phenylpropan-1-one, 1-phenylpropan-2-one, 3-phenylpropan-1-ol, 3-phenylpropanal

**I** (a) [2,4,6-trichlorophenol] (b) [2,4,6-trimethylphenol]

**J** $C_6H_6$, $C_{10}H_8$, $C_{12}H_9$ and $C_{14}H_{10}$

**K** (a) Methylbenzene
(b)

**L** (a) $C_6H_6 + Cl^+ \rightarrow C_6H_5Cl + H^+$
(b) [mechanism: benzene + NO2+ → intermediate → nitrobenzene + H+]

**M** $-1$

**N** $H^+ + AlCl_4^- \rightarrow HCl + AlCl_3$

**O** [dot-and-cross diagrams of Br, Cl, Al reaction]

**P** [structure: methylbenzene]

**Q** (a) $C_6H_5COCH_2CH_3$;
(b) $CH_3CH_2CH_2COCl$

**R** $C_6H_5SO_3H + 3NaOH \rightarrow$
$C_6H_5O^-Na^+ + Na_2SO_3 + 2H_2O$

**S** $C_{10}H_{18}$

**T** $C_6H_6 + 3Cl_2 \rightarrow C_6H_6Cl_6$

**U** (a)
(i) [aniline + 3Cl2 → 2,4,6-trichloroaniline + 3HCl]
(ii) [aniline + 3Br2 → 2,4,6-tribromoaniline + 3HBr]

(b)
(i) [2-chloronitrobenzene and 4-chloronitrobenzene]
(ii) [1-chloro-2,6-dinitrobenzene and 1-chloro-2,4-dinitrobenzene]
(iii) [1,3-dinitrobenzene]

**V** (a) [2-nitromethylbenzene] and [4-nitromethylbenzene]
(b) [2,4-dinitromethylbenzene] and [2,6-dinitromethylbenzene]

**W** (a) $CH_3CH_2COOC_6H_5$;
(b) $C_6H_5COOC_6H_5$

**X** (a) 1,4-Dimethylbenzene;
(b) (i) $C_6H_5COOH$; (ii) $C_6H_5COOH$;
(iii) [benzene-1,2,4-tricarboxylic acid]

**Y** (a) [2-chloromethylbenzene] and [4-chloromethylbenzene]

(b) $C_6H_5CH_2Cl$, $C_6H_5CHCl_2$ and $C_6H_5CCl_3$

## CHAPTER 16

**A** (a) Aldehydes are citral and $CH_3(CH_2)_9CHO$, and ketones are menthone and oestrone; (b) Phenol

**B** (a) (i) Both have the same molecular formula, $C_3H_6O$, and the name for such compounds is structural isomers;
(b) $CH_3CH_2CH_2CHO$;
(c) $CH_3CH_2COCH_2CH_3$

**C** (a) Hydrogen bonding
(b) [diagram: hydrogen bonding between ethanal and water]

(c) The benzene ring is non-polar and cannot form hydrogen bonds with water molecules

**D** $C_6H_5CHO + HCN \rightarrow C_6H_5CH(CN)OH$

[structure: benzaldehyde + HCN → cyanohydrin]

**E** (a) Ethanal gives ethanol, $CH_3CH_2OH$, and propanone gives propan-2-ol, $CH_3CHOHCH_3$;
(b) Ethanol is a primary alcohol and propan-2-ol is a secondary alcohol;
(c) $CH_3COCH_3 + 2[H] \rightarrow$
$CH_3CHOHCH_3$

**F**

**G** (a) Propan-1-ol; (b) 2-Methylhexan-2-ol; (c) Butan-2-ol

**H** Alkyl, cycloalkyl or aryl

**I** (a) $H^-$;
(b) $CH_3CHO + 2[H] \rightarrow CH_3CH_2OH$, $CH_3COCH_3 + 2[H] \rightarrow CH_3CHOHCH_3$

**J** Methanal (and ethanal)

**K** (a) $CH_3CHO$; (b) $C_6H_5COCH_3$; (c) $C_2H_5COCH_3$; (d) $C_2H_5OH$

## CHAPTER 17

**A** (a) $HCl(g) + H_2O(l) \rightarrow H_3O^+(aq) + Cl^-(aq)$; (b) $HI(g) + H_2O(l) \rightarrow H_3O^+(aq) + I^-(aq)$

**B** $HBr(g) + aq \rightarrow H^+(aq) + Br^-(aq)$, $HI(g) + aq \rightarrow H^+(aq) + I^-(aq)$

**C** (a) $F^-$; (b) $OH^-$; (c) $SO_4^{2-}$

**D** (a) $H_3O^+$; (b) $OH^-$; (c) $H_2SO_4$

**E** (a) $HNO_2(aq) = H^+(aq) + NO_2^-(aq)$;
(b) $K_a = \dfrac{[H^+(aq)][NO_2^-(aq)]}{[HNO_2(aq)]}$;
(c) Ethanoic acid

**F** (a) (i) 9.9; (ii) 4.8; (b) Acid is diprotic and the protons are lost sequentially, the first proton being transferred to a water molecule before the second proton

**G** (a) The stronger the acid the weaker the conjugate base;
(b) Acid strength HI > HCl > HF, since the H–I bond is the weakest bond and easiest to break and the H–F bond is the strongest

**H** (a) (i) 0.0; (ii) 4.0; (iii) 0.98; (b) 2.2

**I** Both are $6.6 \times 10^{-7}$ mol dm$^{-3}$

**J** 2.37

**K** (a) (i) Pentanoic acid; (ii) Butanoic acid; (iii) Pentanedioic acid

(b) (i) $ClCH_2CH_2COOH$
(ii)      (iii)      (iv)

**L** (a) $C_nH_{2n+1}COOH$; (b) (i) 185 °C; (ii) No observable trend

**M** (a) Calcium ethanoate and water; (b) Calcium propanoate and hydrogen; (c) Magnesium ethanoate, carbon dioxide and water; (d) Calcium ethanaote, carbon dioxide and water; (e) Sodium benzoate, carbon dioxide and water

**N** (a) $Ca(s) + 2CH_3COO(aq) \rightarrow Ca^{2+}(CH_3COO^-)_2(aq) + H_2(g)$;
(b) Ethanoic acid is a weak acid and hydrochloric acid is a strong acid, so there is a greater concentration of hydrogen ions with dilute hydrochloric acid than with ethanoic acid and, since the rate of reaction is proportional to the concentration of hydrogen ions, reaction with hydrochloric acid is faster

**O** $K_a = \dfrac{[H^+(aq)][In^-(aq)]}{[HIn(aq)]}$

**P** (a) Methyl orange, bromocresol green or methyl red;
(b) Phenolphthalein = 9.15 and thymolphthalein = 9.9

**Q** $K_{a1} = \dfrac{[H^+(aq)][COOHCOO^-(aq)]}{[(COOH)_2]}$
$K_{a2} = \dfrac{[H^+(aq)][(COO^-)_2(aq)]}{[COOHCOO^-(aq)]}$

**R** $[C_2H_5COOH] = 0.750$ mol dm$^{-3}$ and $[C_2H_5COO^-] = 0.350$ mol dm$^{-3}$

**S** The concentration decreases

**T** (a) Sodium benzoate is ionic and the negative carboxylate groups can form strong electrostatic attractions with polar water molecules, whereas benzoic acid is not ionic and as a result forms much weaker electrostatic interactions with water molecules;
(b) Many foods contain water or are stored in water, so to control acidity the additive must dissolve in water

**U** (a) $CH_3CH_2CHO + [O] \rightarrow CH_3CH_2COOH$;
(b) $CH_3CH_2CH_2CH_2OH + 2[O] \rightarrow CH_3CH_2CH_2COOH + H_2O$

**V** (a) Propan-1-ol and sodium ethanoate; (b) Ethanol and benzoic acid; (c) Ammonia and sodium propanoate; (d) Benzoic acid and ammonium chloride

**W** (a) $CH_3COOH + PCl_5 \rightarrow CH_3COCl + HCl + POCl_3$; (b) $C_6H_5COOH + SOCl_2 \rightarrow C_6H_5COCl + SO_2 + HCl$

## CHAPTER 18

**A** Electronegativity is the ability of an atom in a covalent bond to attract the bonding pair of electrons towards itself

**B** Methyl ethanoate

Propyl propanoate

Pentyl butanoate

Methyl benzoate

**C** (a)
(i) $K_c = \dfrac{[ester][water]^2}{[acid][alcohol]^2} = \dfrac{[conc][conc]^2}{[conc][conc]^2}$
so all the units cancel out, as $(conc)^0$ means no units;

(ii) $K_c = \dfrac{[N_2O_4]}{[NO_2]^2} = \dfrac{(conc)}{(conc)^2} = (conc)^{-1}$
$= (mol\ dm^{-3})^{-1}$, so the units are dm$^3$ mol$^{-1}$; (b) (i) $CH_3COOH + C_5H_{11}OH = CH_3COOC_5H_{11} + H_2O$;
(ii) $K_c = \dfrac{[CH_3COOC_5H_{11}][H_2O]}{[CH_3COOH][C_5H_{11}OH]}$; (iii) 4.15

**D** (a) 0.67; (b) 0.67; (c) 0.33; (d) 0.33

**E** Reflux under heating excess ethanol with carboxylic acid in the presence of a little concentrated sulphuric acid as a catalyst and remove the water as it is formed

**F** Increasing the temperature shifts the position of equilibrium towards the left; increasing catalyst concentration has no effect on the position of equilibrium; increasing alcohol concentration shifts the position of equilibrium towards the right

**G** (a) Ethanol and ethanoic acid; (b) Ethanol and butanoic acid; (c) Prop-2-en-1-ol and ethanoic acid

**H** To minimize the increase of ethanol the position of the equilibrium shifts towards the right to try to use up ethanol

**I** (a) Ethanol and sodium ethanoate; (b) Propan-1-ol and sodium propanoate; (c) Ethanol and sodium benzoate

**J** (a) $CH_3CH_2CH_2CH_2OH$ and $CH_3COOH$; (b) $C_6H_5OH$ and $C_6H_5COOH$; (c) $CH_3CH_2OH$ and $(COOH)_2$

**K** To shift the position of equilibrium onto the product side to try and minimize the excess methanol

**L**

**M** (a) $C_6H_5COCl + H_2O \rightarrow C_6H_5COOH + HCl$; (b) $CH_3CH_2COCl + H_2O \rightarrow CH_3CH_2COOH + HCl$

**N** (a) $C_6H_5COCl + 2NaOH \rightarrow C_6H_5COO^-Na^+ + NaCl + H_2O$; (b) C–Cl in chlorobenzene is much stronger than the C–Cl in benzoyl chloride because the delocalized $\pi$ system of the benzene ring is extended over the C–Cl bond and so is much more difficult to break; (c) Carbonyl carbon is much more electron deficient than the carbon atom in the C–Cl bond in chlorobenzene

**O** (a)

(b) (i) Sodium ethanoate, water and sodium chloride; (ii) Ethanamide and ammonium chloride; (iii) benzamide and hydrogen chloride (or ammonium chloride); (iv) Methyl benzoate and hydrogen chloride; (v) Benzoic anhydride and sodium chloride

**P**

**Q** (a) $(CH_3CO)_2O + 2NH_3 \rightarrow CH_3CONH_2 + CH_3COO^-NH_4^+$; (b) $(CH_3CO)_2O + CH_3CH_2OH \rightarrow CH_3COOCH_2CH_3 + CH_3COOH$; (c) $(CH_3CO)_2O + C_6H_5NH_2 \rightarrow CH_3CONHC_6H_5 + CH_3COOH$; (d) $(CH_3CO)_2O + H_2O \rightarrow 2CH_3COOH$

**R** $CH_3CH_2CH_2CH_2CONH_2$ or $C_5H_{11}ON$

**S** $CH_3CONHCH_2CH_3$

**T** (a)

(b)

# CHAPTER 19

**A** (a)

(b)

**B** (a) Lysine, serine, aspartic acid and glutamic acid; (b) Leucine and isoleucine are structural isomers;

c) (i)

(ii) secondary alcohol; (iii) The R group in serine has one less carbon atom and it is a primary alcohol

**C** 2-Hydroxypropanoic acid

**D** (a) An enantiomer is one of a pair of optical isomers, its mirror image is non-superimposable;

(b)

**E**

**F**

**G** (a) $NH_2CH_2COOH + NaOH \rightarrow NH_2CH_2COO^-Na^+ + H_2O$, or $^+NH_3CH_2COO^- + NaOH \rightarrow NH_2CH_2COO^-Na^+ + H_2O$; (b) $HOOCCHRNH_2 + HCl \rightarrow HOOCCHRNH_3^+Cl^-$, or $^-OOCCHRNH_3^+ + HCl \rightarrow HOOCCHRNH_3^+Cl^-$ where $R = (CH_3)_2CHCH_2-$

**H** Aspartic acid and glutamic acid

**I**

and

**J** 4527

**K**

**L** Tyr–Gly–Gly–Phe–Met

**M** **(a)** Lysine, serine, aspartic acid and glutamic acid; **(b)** Alanine, valine, leucine, isoleucine, proline and phenylalanine

# CHAPTER 20

**A** Two substances, one with a double bond (or triple bond), react together to form one product in which the double bond has been changed into a single bond (or the triple bond is changed into a double bond or a single bond)

**B** **(a)**

**(b)** Polymer is a hydrocarbon that is fully saturated with no carbon–carbon double bonds

**C** Breaking a covalent bond so that one electron from the bonding shared pair goes to each atom

**D** Linear polymers can lie closer to one another, which allows more contact between molecules so that the induced dipole–induced dipole forces of attraction can operate over a larger surface, and hence intermolecular force is stronger than in a branched chain, in which the molecules cannot lie as close to one another

**E** **(a)**

**(b)** Syndiotactic poly(propene) will have a higher melting point, will be more dense and will be harder than atactic poly(propene)

**F** $C_6H_5CH=CH_2$

**G** $nCF_2=CF_2 \rightarrow -(CF_2CF_2)_n-$

**H** $CH_3COOH + CH_3CH_2OH = CH_3COOCH_2CH_3 + H_2O$

**I** **(a)** (i)

(ii) $CH_3OH$
(iii) Base hydrolysis

Acid hydrolysis

**(b)**

**J** When blended with cotton fibres perspiration will be absorbed, which will make the fabric more comfortable to wear

**K** $nCH_3CHOHCOOH \rightarrow -(OCH(CH_3)CO-)_n- + nH_2O$

**L** $H_2NCH_2CH_2CH_2CH_2CH_2NH_2$ and $HOOCCH_2CH_2CH_2CH_2CH_2CH_2CH_2COOH$

**M** NaOH(aq)/reflux under heating, $H_2NCH_2CH_2CH_2CH_2CH_2NH_2$ and $Na^+{}^-OOCCH_2CH_2CH_2CH_2COO^-Na^+$, 6 mol dm$^{-3}$ HCl(aq)/reflux under heating; $Cl^-H_3N^+CH_2CH_2CH_2CH_2CH_2CH_2NH_3^+Cl^-$ and $HOOCCH_2CH_2CH_2CH_2COOH$

**N** **(a)** (i) Condensation; (ii) Thermosetting; **(b)** High melting point, strong and rigid, relatively low density and a good thermal conductor

**O**

**P** Intermolecular hydrogen bonds

**Q** Thermoplastics have weak intermolecular forces, such as induced dipole–induced dipole interactions, which are easily broken and overcome so that the melting point is low; once melted the polymer chains can take up new positions, and on cooling the intermolecular forces are easily reformed

**R** It cannot be remoulded because it decomposes when heated strongly, so the plastic cannot be recycled

**S** Collection of the bottles, e.g. fuel; cleaning of bottles; relabelling of bottles, melting and remoulding

**T** Issues include: toxic gases, such as hydrogen chloride and hydrogen cyanide, will be produced during incineration; high temperatures are often required for the incineration process to prevent the formation of toxic materials; incineration means that fewer landfill sites are needed for plastic waste; need to sort out household waste; less crude oil needed for fuels; energy locked in the bonds of plastics is not wasted

# CHAPTER 21

**A** **(a)** Na is +1 and Cl is –1; **(b)** Mg is +2 and Cl is –1; **(c)** Al is +3 and Cl is –1; **(d)** Si is +4 and Cl is –1; **(e)** P is +3 and Cl is –1; **(f)** P is +5 and Cl is –1; **(g)** S is +1 and Cl is –1

**B** **(a)** Oxygen; **(b)** Carbon

**C** **(a)** 0 to +4; **(b)** 0 to –1; **(c)** Chlorine, as it gains electrons and its oxidation number decreases; **(d)** Silicon, since it loses electrons and its oxidation number increases

**D** **(a)** P from +3 to +5 and Cl from 0 to –1; **(b)** Chlorine

**E** $Mg(s) + Cl_2(g) \rightarrow MgCl_2(s)$; $2Al(s) + 3Cl_2(g) \rightarrow Al_2Cl_6(s)$; $Si(s) + 2Cl_2(g) \rightarrow SiCl_4(l)$; $P_4(s) + 6Cl_2(g) \rightarrow 4PCl_3(l)$; $P_4(s) + 10Cl_2(g) \rightarrow 4PCl_5(s)$; $2S(l) + Cl_2(g) \rightarrow S_2Cl_2(l)$

**F** LiCl and NaCl both have a face-centred cubic structure, but CsCl has a body-centred cubic structure;

coordination number of $Li^+$ and $Na^+$ = 6, but that of $Cs^+$ = 8; change in structure because of the increase in the ratio of the cation radius to the anion radius from LiCl to CsCl

**G**

**H** $BCl_3$ has a trigonal planar shape so that all the individual bond dipoles cancel out

**I** Ionic attraction between $PCl_4^+$ and $PCl_6^-$ in $PCl_5$ is stronger than the permanent dipole–permanent dipole intermolecular attraction in $PCl_3$

**J** **(a)** $2Mg(s) + O_2(g) \rightarrow 2MgO(s)$; $4Al(s) + 3O_2(g) \rightarrow 2Al_2O_3(s)$; $Si(s) + O_2(g) \rightarrow SiO_2(s)$; $P_4(s) + 5O_2(g) \rightarrow P_4O_{10}(s)$; $S(s) + O_2(g) \rightarrow SO_2(g)$; **(b)** $P_4O_{10}$, $SO_2$, $SO_3$ and $Cl_2O_7$; **(c)** $Na_2O$ is sodium oxide, MgO is magnesium oxide, $Al_2O_3$ is aluminium oxide, $SiO_2$ is silicon(IV) oxide or silicon dioxide; $SO_2$ is sulphur dioxide or sulphur(IV) oxide; $SO_3$ is sulphur trioxide or sulphur(VI) oxide; $P_4O_6$ is phosphorus(III) oxide; $P_4O_{10}$ is phosphorus(V) oxide; $Cl_2O_7$ is chlorine(VII) oxide

**K** **(a)** Acids; **(b)** Sodium nitrate; **(c)** Sodium chloride

**L** Diamond contains all carbon atoms and $SiO_2$ contains silicon and oxygen; coordination number of all atoms is four in diamond, but in $SiO_2$ it is four around Si and two around O; the C–C bond length is shorter than the Si–O bond length in $SiO_2$

**M** **(a)** $SO_2(g) + H_2O(l) = H_2SO_3(aq)$, sulphurous acid; **(b)** $P_4O_6(s) + 6H_2O(l) \rightarrow 4H_3PO_3(aq)$, phosphoric(III) acid; **(c)** $P_4O_{10}(s) + 6H_2O(l) \rightarrow 4H_3PO_4(aq)$, phosphoric(V) acid

## CHAPTER 22

**A** **(a)** (i) ~23 °C; (ii) ~680 °C; (iii) ~275 nm; (iv) ~187 nm;

**(b)** Strength of metallic bonding decreases as atomic (proton) number increases; **(c)** Rubidium, caesium and francium

**B** **(a)** Reacts (recombines) to give sodium chloride, $2Na + Cl_2 \rightarrow 2NaCl$; **(b)** Chlorine will react with iron to give iron(III) chloride

**C** Electrolysis of molten lithium chloride

**D** **(a)** $Li^+$ is $1s^2$, $K^+$ is $1s^22s^22p^63s^23p^6$, $Rb^+$ is $1s^22s^22p^63s^23p^63d^{10}4s^24p^6$; **(b)** Na atoms require less energy to lose an electron than Li atoms, as the outer electrons are less strongly attracted to the nucleus because of the increased atomic radius and the increased number of shielding inner shells of electrons

**E** ~360 kJ mol$^{-1}$

**F** **(a)** $2Na(s) + 2H_2O(l) \rightarrow 2NaOH(aq) + H_2(g)$, $2K(s) + 2H_2O(l) \rightarrow 2KOH(aq) + H_2(g)$; **(b)** Francium sinks in water, and there is an extremely rapid (possibly explosively) exothermic reaction to form hydrogen and an alkaline solution of francium hydroxide, $2Fr(s) + 2H_2O(l) \rightarrow 2FrOH(aq) + H_2(g)$

**G** **(a)** $2Na(s) + O_2(g) \rightarrow Na_2O_2(s)$; **(b)** $2K(s) + O_2(g) \rightarrow K_2O_2(s)$; **(c)** $Rb(s) + O_2(g) \rightarrow RbO_2(s)$

**H** $Rb_2O(s) + H_2O(l) \rightarrow 2RbOH(aq)$

**I** Density is ~4500 kg m$^{-3}$, melting point is ~640 °C, boiling point is ~1500 °C (difficult to estimate since there is no obvious trend)

**J** $Ba(s) + 2H_2O(l) \rightarrow Ba(OH)_2(aq) + H_2(g)$; $Sr(s) + 2H_2O(l) \rightarrow Sr(OH)_2(aq) + H_2(g)$

**K** As you go down the group, the atomic radius increases and the number of shielding shells increases, so the attraction between the positive nucleus and the outer shell electrons decreases; the numerical addition of the first- and second-ionization energies is smaller down the group, so electrons are lost more easily from the atoms

**L** **(a)** Magnesium atoms lose electrons so are oxidized, hydrogen ions gain electrons so are reduced; **(b)** Magnesium changes from 0 to +2, which is oxidation, hydrogen changes from +1 to 0, which is reduction

**M** **(a)** $Ca(s) + 2HCl(aq) \rightarrow CaCl_2(aq) + H_2(g)$; **(b)** $Ba(s) + 2HCl(aq) \rightarrow BaCl_2(aq) + H_2(g)$; **(c)** $Mg(s) + 2CH_3COOH(aq) \rightarrow Mg(CH_3COO)_2(aq) + H_2(g)$

**N** **(a)** $Mg(OH)_2(s) + 2HCl(aq) \rightarrow MgCl_2(aq) + 2H_2O(l)$; **(b)** $CaCO_3 + H_2SO_4 \rightarrow CaSO_4 + H_2O$, $CaCO_3 + 2HNO_3 \rightarrow Ca(NO_3)_2 + CO_2 + H_2O$

**O** **(a)** $Mg^{2+}(g) + O^{2-}(g) \rightarrow MgO(s)$; **(b)** $Mg^{2+}(g) + SO_4^{2-}(g) \rightarrow MgSO_4(s)$; **(c)** $2Fe^{3+}(g) + 3SO_4^{2-}(g) \rightarrow Fe_2(SO_4)_3(s)$; **(d)** $3Mg^{2+}(g) + 2N^{3-}(g) \rightarrow Mg_3N_2(s)$

**P** $-3454$ kJ mol$^{-1}$

**Q** **(a)** $Li_2CO_3(s) \rightarrow Li_2O(s) + CO_2(g)$; **(b)** $2LiNO_3(s) \rightarrow Li_2O(s) + 2NO_2(g) + O_2(g)$; **(c)** $2FrNO_3(s) \rightarrow 2FrNO_2(s) + O_2(g)$

**R** **(a)** Solvent is water and solute is alcohol; **(b)** Solvent is alcohol and solute is water

**S** **(a)** $Mg^{2+}(g) + aq \rightarrow Mg^{2+}(aq)$; **(b)** $SO_4^{2-}(g) + aq \rightarrow SO_4^{2-}(aq)$

**T** NaCl is $+5$ kJ mol$^{-1}$, LiCl is $-32$ kJ mol$^{-1}$

**U** $MgSO_4$ is 337 g dm$^{-3}$, $CaSO_4$ is $1.9 \times 10^{-1}$ g dm$^{-3}$, $SrSO_4$ is $1.4 \times 10^{-3}$ g dm$^{-3}$, $BaSO_4$ is $2.6 \times 10^{-4}$ g dm$^{-3}$

**V** **(a)** $K_{sp}(CaSO_4) = [Ca^{2+}(aq)] [SO_4^{2-}(aq)]$; **(b)** $K_{sp}(BaSO_4) = [Ba^{2+}(aq)][SO_4^{2-}(aq)]$; **(c)** $K_{sp}(Ag_2CrO_4) = [Ag^+(aq)]^2[CrO_4^{2-}(aq)]$; **(d)** $K_{sp}(PbCl_2) = [Pb^{2+}(aq)][Cl^-(aq)]^2$

**W** **(a)** $2.0 \times 10^{-10}$ mol$^2$ dm$^{-6}$; **(b)** $[Ca^{2+}(aq)] = 1.25 \times 10^{-2}$ mol dm$^{-3}$, $[OH^-(aq)] = 2 \times 1.25 \times 10^{-2} = 2.50$ mol dm$^{-3}$, $K_{sp}(Ca(OH)_2) = [Ca^{2+}(aq)] [OH^-(aq)]^2 = (1.25 \times 10^{-2})(2.50 \times 10^{-2})^2 = 7.81 \times 10^{-6}$ mol$^3$ dm$^{-9}$

**X** (a) $7.5 \times 10^{-3}$ mol dm$^{-3}$;
  (b) $2.1 \times 10^{-4}$ mol dm$^{-3}$;
  (c) $5.3 \times 10^{-4}$ mol dm$^{-3}$

**Y** $2 \times 10^{-9}$ mol dm$^{-3}$

**Z** No the ionic product $1.0 \times 10^{-9}$ is less than the $K_{sp}$

**A¢** (a) $Al^{3+} + 3e^- \rightarrow Al$; (b) Some $F^-$ will be oxidized at the anode to form $F_2$ and some of the fluorine will react with moisture in the atmosphere to give HF

**B¢** No, since aluminium carbonate would have to be formed in aqueous conditions and in aqueous solution $Al^{3+}$(aq) is highly acidic because $Al^{3+}$ polarizes water molecules to form $H^+$(aq) and so the solution will react with carbonate ions to give carbon dioxide and water; also $Al^{3+}$ has a very high charge density so it will polarize the carbonate ion, weakening the covalent bonds in the anion so it will decompose

## CHAPTER 23

**A** Fluorine will be a pale yellow gas, and astatine will be a black solid

**B** (a) (i) 85; (ii) 7; (iii) Melting point ~210 °C, boiling point ~280 °C, covalent radius ~150 pm, electronegativity ~2.2;
  (b) Electronegativity is the ability of an atom in a covalent bond to attract the bonding pair of electrons; in the halogens, the increased number of inner shielding shells and same core charge of +7 means the effective nuclear charge decreases as atomic number increases, so the nucleus has a less powerful attraction for electrons in the shared bonding pair

**C** (a) $1s^22s^22p^63s^23p^6$;
  (b) $1s^22s^22p^63s^23p^63d^{10}4s^24p^6$

**D** (a) No reaction; (b) $Cl_2$(aq) + $2I^-$(aq) $\rightarrow I_2$(aq) + $2Cl^-$(aq); (c) No reaction; (d) $I_2$(aq) + $2At^-$(aq) $\rightarrow At_2$(s) + $2I^-$(aq)

**E** (a) (i) $2Fe$(s) + $3Cl_2$(g) $\rightarrow 2FeCl_3$(s);
  (ii) $Cu$(s) + $Cl_2$(g) $\rightarrow CuCl_2$(s);
  (iii) $2Fe$(s) + $3F_2$(g) $\rightarrow 2FeF_3$(s);

(b) (i) Chromium(III) chloride;
  (ii) Zinc chloride; (iii) Barium iodide

**F** (a) Oxidation states of $Cl_2$, HCl and HOCl are 0, −1 and +1, respectively; (b) Oxidation number of chlorine has increased and decreased, showing that one atom in $Cl_2$ has been oxidized and the other reduced

**G** (a) $Cl_2$(aq) + $SO_2$(aq) + $2H_2O$(l) $\rightarrow 2Cl^-$(aq) + $SO_4{}^{2-}$(aq) + $4H^+$(aq);
  (b) Astatine is the least powerful oxidizing agent of the halogens, and iodine also cannot oxidize $Fe^{2+}$(aq)

**H** $1.05 \times 10^{-3}$ mol dm$^{-3}$

**I** Iodine changes oxidation state from 0 to −1, whereas the oxidation state of sulphur changes from +2 to +2.5. So $I_2$ is reduced and $S_2O_3{}^{2-}$ is oxidized

**J** (a) +4 in $XeF_4$ and +6 in $XeF_6$;
  (b) $Xe + 2F_2 \rightarrow XeF_4$, $Xe + 3F_2 \rightarrow XeF_6$

**K** $PBr_3$(l) + $3H_2O$(l) $\rightarrow H_3PO_3$(aq) + $3HBr$(aq), n.b. HBr(g) also acceptable

**L** Halogens have high electronegativities so that the halogen end of the H–X (where X is F, Cl, Br and I) will be δ− and the hydrogen δ+

**M** (a) $K_c = \dfrac{[H_2][I_2]}{[HI]^2}$;
  (b) Mole ratio HI:$H_2$:$I_2$ is 8.47:1:1

**N** (a) Sodium chloride, water and carbon dioxide; (b) Magnesium iodide and water; (c) Sodium bromide and water

**O** $KMnO_4$(s) + 8HCl(aq) $\rightarrow$ KCl(aq) + $MnCl_2$(aq) + $4H_2O$(l) + $2\frac{1}{2}Cl_2$(g)

**P** (a) NaI(s) + $H_2SO_4$(l) $\rightarrow NaHSO_4$(s) + HI(g); (b) 2HI(g) + $H_2SO_4$(l) $\rightarrow SO_2$(g) + $I_2$(s) + $2H_2O$(l); (c) Initially, hydrogen astatide and sodium hydrogensulphate, then hydrogen astatide reduces sulphuric acid to form astatine, hydrogen sulphide, sulphur dioxide, sulphur and water; (d) Hydrogen fluoride and sodium hydrogensulphate

## CHAPTER 24

**A** As the atomic number of each element in the period increases, electrons fill the 3d sub-shell rather than the outer sub-shells

**B** (a) Mn is $1s^22s^22p^63s^23p^63d^54s^2$ and Ni is $1s^22s^22p^63s^23p^63d^84s^2$;
  (b) It minimizes electron–electron repulsion within the 3d orbitals

**C** (a) $Co^{2+}$ is $[Ar]3d^7$ and $Co^{3+}$ is $[Ar]3d^6$; (b) $Cu^+$ is $[Ar]3d^{10}$ and $Cu^{2+}$ is $[Ar]3d^9$; (c) $Zn^{2+}$ is $[Ar]3d^{10}$

**D** $Sc^{3+}$ is $1s^22s^22p^63s^23p^6$; $Sc^{3+}$ does not have any 3d electrons, but a transition element needs a partially filled d sub-shell in the electron configuration of one of its oxidation states

**E** Metals form cations; metals normally form basic oxides; metals are reducing agents; metals normally react with non-metals to form ionic compounds; metals are malleable and ductile

**F** (a) $Cr^+$, $[Ar]3d^5$, has a stable half-filled sub-shell with no electron–electron repulsion within a 3d orbital; (b) $Mn^{2+}$, $[Ar]3d^5$, also has a stable half-filled sub-shell with no electron–electron repulsion within a 3d orbital; (c) The third ionization energy refers to an electron lost from the third shell rather than the fourth shell, so there are fewer shielding shells and so a much greater attraction between the outer electrons and the nucleus

**G** (a) $Fe^{3+}$ is $[Ar]3d^5$, which has a stable half-filled sub-shell, whereas $Fe^{2+}$ is $[Ar]3d^6$ and has one 3d orbital with electron–electron repulsion, and so $Fe^{2+}$ will lose electrons to become $Fe^{3+}$; (b) $Mn^{2+}$ is $[Ar]3d^5$ which has a stable half-filled sub-shell, whereas $Mn^{3+}$ is $[Ar]3d^4$ and so the loss of electrons is not compensated by the formation of a much more stable electron configuration;

**c)**

Cu$^+$ [Ar] ↑↓ ↑↓ ↑↓ ↑↓ ↑↓ | 3d | □ 4s

Cu$^{2+}$ [Ar] ↑↓ ↑↓ ↑↓ ↑↓ ↑ | □

**H** Transition-metal atoms always lose electrons in reaction since they often have only two electrons in their outer shell, and metals have a low electronegativity

**I** **(a)** The ability of an atom in a covalent bond to attract the bonding pair of electrons; **(b)** Fluorine

**J** **(a)** (i) Titanium(IV) oxide; (ii) Iron(III) oxide; (iii) Manganese(III) hydroxide; (iv) Chromium(VI) oxide; (v) Vanadium(V) oxide; **(b)** (i) $Cu(OH)_2$; (ii) $MnCO_3$; (iii) $TiCl_4$; (iv) $Cu(NO_3)_2$; (v) $FeBr_2$

**K** **(a)** $Na_2FeO_4$; **(b)** Chromate(VI)

**L** **(a)** +1; **(b)** +2; **(c)** +3; **(d)** +2; **(e)** +3

**M** **(a)** $[Ag(CN)_2]^-$ contains a central $Ag^+$, $[Ag(NH_3)_2]^+$ contains a central $Ag^+$ and $[Ag(S_2O_3)_2]^{3-}$ contains a central $Ag^+$; **(b)** $[Ag(CN)_2]^-$ is linear, $[Ag(NH_3)_2]^+$ is linear and $[Ag(S_2O_3)_2]^{3-}$ is linear

**N** **(a)** (i) Hexaamminecobalt(III); (ii) Hexacyanoferrate(III); (iii) Tetrachlorocuprate(II); (iv) Diamminesilver(I); **(b)** (i) $[Fe(H_2O)_6]^{3+}$; (ii) $[Co(NH_3)_4Cl_2]^+$; (iii) $[CoCl_4]^{2-}$

**O** **(a)** 6; **(b)** Tetradentate; **(c)** Most arteries contain a high concentration of oxyhaemoglobin, in which an oxygen molecule is complexed with the $Fe^{2+}$, which is red in colour

**P** **(a)** *cis*- and *trans*-Tetraamminedichlorochromium(III), respectively;

**(b)**

H$_3$N   Cl     H$_3$N   Cl
    Ni        Ni
H$_3$N   Cl     Cl   NH$_3$

*cis*        *trans*

**Q**

**R** $[Fe(H_2O)_6]^{3+}$ is hexaaquairon(III), $[Fe(OH)(H_2O)_5]^{2+}$ is pentaaquamonohydroxoiron(III) and $[Fe(OH)_2(H_2O)_4]^+$ is tetraaquadihydroxoiron(III)

**S** $[Al(H_2O)_6]^{3+}(aq) = [Al(H_2O)_5(OH)]^{2+}(aq) + H^+(aq)$

**T** **(a)** $[CoCl_4]^{2-}(aq) + 6H_2O(l) = [Co(H_2O)_6]^{2+}(aq) + 4Cl^-(aq)$; **(b)** On heating, water evaporates and so the position of equilibrium shifts to the left side to try to minimize this loss, which results in an increase in $[CoCl_4]^{2-}(aq)$ and so the solution takes on the colour of this complex, which is blue

**U** **(a)** Increases the surface area, which increases rate of reaction; **(b)** Increases surface area and minimizes the mass of platinum, which is very expensive, that needs to be used; **(c)** Although a powder has a large surface area, unless gas can permeate all the way through the powder most of the powder will not be in contact with the gas. In a fluidized bed all the surface area will be available to the gas

**V** In $SO_2$ sulphur is +4 and in $SO_3$ it is +6, as sulphur loses electrons the oxidation state of sulphur increases

**W** The graph is a mirror image of Fig 24.40, starting at the origin with a very small gradient, then the gradient increases and finally, at the end of the reaction, the gradient decreases to zero

## CHAPTER 25

**A** **(a)** $Cu(s) \rightarrow Cu(aq) + 2e^-$, and $Ag^+(aq) + e^- \rightarrow Ag(s)$; **(b)** $Fe^{2+}(aq) \rightarrow Fe^{3+}(aq) + e^-$, and $Cr_2O_7^{2-}(aq) + 14H^+(aq) + 6e^- \rightarrow 2Cr^{3+}(aq) + 7H_2O(l)$

**B** **(a)** $Na^+ + e^- \rightarrow Na$, and $2Cl^- \rightarrow Cl_2 + 2e^-$; **(b)** $Pb^{2+} + 2e^- \rightarrow Pb$, and $2Br^- \rightarrow Br_2 + 2e^-$

**C** 298 K, $[Cl^-(aq)] = 1.0$ mol dm$^{-3}$ and pressure of $Cl_2(g) = 101$ kPa

**D** 0.5 mol dm$^{-3}$

**E** **(a)** $Cu(s) | Cu^{2+}(aq) || Ag^+(aq) | Ag(s)$, electrons flow from copper to silver; **(b)** $Zn(s) | Zn^{2+}(aq) || H^+(aq) | H_2(g) | Pt(s)$, electrons flow from zinc to platinum (hydrogen)

**F** **(a)** $Cu^{2+}$; **(b)** Acidified $MnO_4^-$

**G** **(a)** Ni; **(b)** Au

**H** **(a)** Hydrogen at cathode and oxygen at anode; **(b)** Expect hydrogen at cathode and oxygen at anode, a prediction based on $[OH^-(aq)] = [Cl^-(aq)] = 1.0$ mol dm$^{-3}$, but in concentrated NaCl(aq) the $[Cl^-(aq)]$ and $[OH^-(aq)] << 1.0$ mol dm$^{-3}$

**I** **(a)** +0.76 V; **(b)** (i) Less positive; (ii) More positive; (iii) Less positive

**J** **(a)** (i) $E = +1.62$ V, so spontaneous; (ii) $E = -2.10$ V, so not spontaneous; **(b)** +0.74 V, $Mn(s) + Fe^{2+}(aq) \rightarrow Mn^{2+}(aq) + Fe(s)$

**K** **(a)** From Cu to $Br_2$; **(b)** From Cr to $Cu^{2+}$

**L** **(a)** −52 kJ mol$^{-1}$; **(b)** −83 kJ mol$^{-1}$; **(c)** 56 kJ mol$^{-1}$

**M** **(a)** $E = -1.51$ V, $\Delta G = +291$ kJ mol$^{-1}$, not spontaneous; **(b)** $E = +1.93$ V, $\Delta G = -931$ kJ mol$^{-1}$, spontaneous

**N** **(a)** 2 V; **(b)** +1.60 V; **(c)** The lead–acid battery does not operate under standard conditions, e.g. $[H_2SO_4(aq)] > 1.0$ mol dm$^{-3}$ and the temperature will not be 298 K

**O** $O_2(g) + 2H_2O(l) + 4e^- \rightarrow 4OH^-(aq)$ and $2H_2(g) + 4OH^-(aq) \rightarrow 4H_2O(l) + 4e^-$ gives $O_2(g) + 2H_2(g) + 2H_2O(l) + 4e^- + 4OH^-(aq) \rightarrow 4OH^-(aq) + 4e^- + 4H_2O(l)$, cancelling gives $O_2(g) + 2H_2(g) \rightarrow 2H_2O(l)$

**P (a)** $2VO^{2+}(aq) + 4H^+(aq) + Zn(s) \rightarrow$ $Zn^{2+}(aq) + 2V^{3+}(aq) + 2H_2O(l)$; **(b)** $2V^{3+}(aq) + Zn(s) \rightarrow Zn^{2+}(aq) + 2V^{2+}(aq)$, and $E = +0.50$ V, i.e. the process should be spontaneous

**Q (a)** $Cr(s) + H_2SO_4(aq) \rightarrow CrSO_4(aq) + H_2(g)$; **(b)** $+0.91$ V; **(c)** in acidic conditions $2Cr^{3+}(aq) + 7H_2O(l) \rightarrow Cr_2O_7^{2-}(aq) + 14H^+(aq) + 6e^-$, so $E = -1.33$ V, but in alkaline conditions $Cr(OH)_3(s) + 5OH^-(aq) \rightarrow CrO_4^{2-}(aq) + 4H_2O(l) + 3e^-$, so $E = +0.13$ V, and thus, using the same oxidizing agent, $E$ for the alkaline conditions will be more positive (so it is more feasible) than that for the acidic conditions

**R (a)** The reduction potential for $Mn^{3+}(aq) + e^- = Mn^{2+}(aq)$ is $+1.49$ V, and this highly positive nature of the reduction potential indicates that manganese(III) is a powerful oxidizing agent and so is easily reduced to manganese(II), i.e. manganese(II) is more stable than manganese(III);
**(b)** Manganese should react with $HCl(aq)$ to give hydrogen since the standard electrode potential for $Mn(s) + 2H^+(aq) \rightarrow Mn^{2+}(aq) + H_2(g)$ is positive ($+1.18$ V)

**S (a)** (i) $MnO_4^-(aq) + 8H^+(aq) + 5e^- \rightarrow Mn^{2+}(aq) + 4H_2O(l)$;
(ii) $2I^-(aq) \rightarrow I_2(aq) + 2e^-$;
**(b)** $2MnO_4^-(aq) + 16H^+(aq) + 10I^-(aq) \rightarrow 5I_2(aq) + 2Mn^{2+}(aq) + 8H_2O(l)$

**T** $0.0287$ mol dm$^{-3}$

**U (a)** (i) $Fe^{2+}(aq) + O_2(g) + 4H^+(aq) \rightarrow 2Fe^{3+}(aq) + 2H_2O(l)$; (ii) $+0.46$ V;
**(b)** $+0.96$ V

**V** $E = +0.59$ V for the oxidation of $[Fe(H_2O)_6]^{2+}$, and $E = +1.00$ V for the oxidation of $[Fe(CN)_6]^{4-}$

**W (a)** Not spontaneous, since $\Delta H$ is positive and $\Delta S$ is negative;
**(b)** Spontaneous, since $\Delta G = -142$ kJ mol$^{-1}$

**X (a)** Iron and possibly chromium;
**(b)** (ii) Should be feasible above approximately 1770 K

## CHAPTER 26

**A (a)** Rate of reaction decreases during the course of a reaction;
**(b)** Rate of reaction changes as temperature changes

**B** $0.17$ mol s$^{-1}$

**C** Reaction mechanism

**D** Mass of reaction mixture, measured by placing the reaction container on a balance, and concentration of $H^+(aq)$, measured with a pH meter, since pH will change during reaction

**E (a)** $1.50 \times 10^{-5}$; **(b)** $H^+(aq) + OH^-(aq) \rightarrow H_2O(l)$; **(c)** $1.50 \times 10^{-5}$;
**(d)** $1.50 \times 10^{-3}$ mol dm$^{-3}$

**F (a)** Monitor the increase in pressure if the reaction is carried out in a sealed container; **(b)** Measure the volume of oxygen produced using a gas syringe; **(c)** Monitor the change in conductivity of the reaction mixture and use colorimetry to monitor the appearance of bromine; **(d)** Take aliquots out of the reaction mixture and titrate against an acid of known concentration, and also take aliquots out of the reaction mixture and acidify with $HNO_3(aq)$, then add $AgNO_3(aq)$ to measure the mass of AgBr precipitate formed; **(e)** Monitor the change in conductivity of the solution, and monitor the change in $I_2(aq)$ concentration using colorimetry or by taking out aliquots and titrating with $Na_2S_2O_3(aq)$

**G** $m$ with respect to A, and $n$ with respect to B

**H** By the change in total mass of the reaction mixture and by the change in total volume of carbon dioxide collected

**I** Since the rate at time zero is all that is required; once sufficient results are obtained to be able to construct the gradient at time zero then no more are needed

**J** Rate = $k[SO_2Cl_2]$, and $k = 4.5 \times 10^4$ s$^{-1}$

**K** Second

**L (a)** First order with respect to $BrO_3^-$ and first order with respect to $SO_3^{2-}$, to give a total order of 2, with rate constant units of dm$^3$ mol$^{-1}$ s$^{-1}$; **(b)** Second order with respect to NO and first order with respect to $H_2$, to give a total order of 3, with rate constant units dm$^6$ mol$^{-2}$ s$^{-1}$; **(c)** Second order with respect to $NO_2$ and zero order with respect to CO, giving a total order of 2, with rate constant units dm$^3$ mol$^{-1}$ s$^{-1}$

**M** Loss of mass from the reaction mixture, and volume of oxygen formed

**N** Record the change in either the mass lost by the reaction mixture or volume of oxygen formed at regular time intervals over a set time period, using the same mass of catalyst, same volume of $H_2O_2(aq)$ and same temperature; do experiment five times using different concentrations of $H_2O_2(aq)$; plot graph of time against either the mass loss or volume of oxygen formed for each concentration; determine the initial rate of reaction by drawing a tangent to the curve at time zero; plot the initial rates obtained against the concentration of $H_2O_2(aq)$, and first-order kinetics will give a straight line through the origin

**O (a)** $0.25$ kg; **(b)** $k = 4.62 \times 10^{-2}$ h$^{-1}$

**P** Cyclopropane and propene are structural isomers, since they have the same molecular formula but different structural formulas

**Q** $2NO_2 + F_2 \rightarrow 2NO_2F$

**R** It is a substitution reaction that involves a nucleophile, and it has second-order kinetics, i.e. two particles are involved in the rate-determining step

**S** Mechanism 2, since the slowest step involves a collision between $H_2O_2$ and $Br^-$, i.e. it will be first order with respect to both $H_2O_2$ and $Br^-$

**T (a)** Slow step rate = $k[N_2O_5]$, fast

step rate = $k[NO][NO_3]$; **(b)** $N_2O_5 \rightarrow NO_2 + NO_3$; **(c)** Rate = $k[N_2O_5]$; **(d)** Because NO is not involved in the slowest step of the reaction

**U** It doubles

## CHAPTER 27

**A** Blue

**B** $f = 5.09 \times 10^{14}$, and $E = 3.37 \times 10^{-19}$ J

**C** A group of atoms in a molecule that gives the molecule a characteristic set of reactions

**D** $2NaNO_2(aq) + H_2SO_4(aq) \rightarrow Na_2SO_4(aq) + 2HNO_2(aq)$; $C_6H_5NH_2(l) + H_2SO_4(aq) + HNO_2(aq) \rightarrow C_6H_5N_2^+(aq) + HSO_4^-(aq) + 2H_2O(l)$; or $2C_6H_5NH_2(l) + H_2SO_4(aq) + 2HNO_2(aq) \rightarrow 2C_6H_5N_2^+(aq) + SO_4^{2-}(aq) + 4H_2O(l)$

**E** They decompose very easily to produce nitrogen gas, so that a small volume of solid will produce a large volume of gas

**F** **(a)** At a higher temperature the diazonium salt will react with water to give phenol and nitrogen; **(b)** Phenol reacts in an acid–base reaction to give sodium phenoxide, $C_6H_5OH + OH^- \rightarrow C_6H_5O^- + H_2O$

Sodium phenoxide:

**(c)** A lone pair on the oxygen atom is donated into the system of the benzene ring to extend the system over seven atoms, which increases the electron density in the benzene ring so that the species is more reactive towards electrophiles than benzene is

**G** **(a)** Concentrated sulphuric acid and concentrated nitric acid; **(b)** Tin and concentrated hydrochloric acid

**H** **(a)** White; **(b)** Black

**I** 400–700 nm

**J** Blue

**K** Extended π system over the whole of the molecule

**L** **(a)** So that the only absorbance is due to the solution under test;

**(b)** Many solvent molecules will contain chromophores that absorb ultraviolet and/or visible light, for example carbonyl groups, carbon–carbon double bonds or benzene rings

**M** Yellow

**N**

**O** Red light

**P** Blue

**Q** **(a)** $1s^2 2s^2 2p^6 3s^2 3p^6 3d^{10}$; **(b)** Even though the energy of the 3d orbitals can be split in $Zn^{2+}$ there is no possibility of a d → d transition since all the 3d orbitals are full

**R** $VO_2^+$ absorbs blue–violet. $VO^{2+}$ absorbs orange, $V^{3+}$ absorbs red–violet, $V^{2+}$ absorbs yellow–green

**S**

**T** **(a)** Orange; **(b)** Blue; **(c)** Octahedral

**U** **(a)** Use colorimetry by making up solutions that contain different molar ratios of nickel(II) ion and EDTA, analyse each solution with a colorimeter, and find the solution with the maximum absorbance, which should have the stoichiometric molar ratio of the complex; **(b)** Either an inverted V-shape graph as in Fig 27.36, or the absorbance increases until it reaches a maximum and then remains at the maximum absorbance (this is obtained if the number of moles of one ion is kept constant and the number of moles of the other ion is changed)

**V** **(a)** A chromophore is a group of atoms responsible for the absorption of ultraviolet and/or

visible light, and a conjugated system is an arrangement of atoms with an extended π system over many atoms in which there are alternating single and double bonds; **(b)** $\lambda_{max} = \sim620$ nm; **(c)** Drawing of a spectrum with a $\lambda_{max} = \sim620$ nm

## CHAPTER 28

**A** Reduced, since the oxidation number of nitrogen changes from 0 to –3, i.e. nitrogen gains electrons

**B** **(a)** +3; **(b)** +5; **(c)** +5; **(d)** +2; **(e)** +4

**C** 118 kPa

**D** **(a)** $K_p = \dfrac{(pHI)^2}{(pH_2)(pI_2)}$;

**(b)** $K_p = \dfrac{(pNO_2)^2}{(pNO)^2(pO_2)}$;

**(c)** $K_p = pCO_2$

**E** **(a)** None; **(b)** $Pa^{-1}$; **(c)** Pa

**F** $1.5 \times 10^{-8}$

**G** **(a)** $pSO_2 = 200$ kPa and $pO_2 = 100$ kPa; **(b)** $pSO_2 = 10$ kPa and $pO_2 = 5$ kPa; **(c)** $72.2$ kPa$^{-1}$

**H** **(a)** Low temperature since it is an exothermic reaction, low pressure since the volume of the products is more than the volume of the reactants (more moles of gas on product side), and increase the concentration of oxygen; **(b)** High pressure, high temperature and the addition of a catalyst; **(c)** Very high temperature is used to ensure a rapid rate of reaction, a low pressure is used as this also reduces the operating costs (high-pressure apparatus is expensive), a catalyst is used to increase the rate of reaction, and excess air is used to ensure that the concentration of oxygen is higher than the mole ratio predicted from the equation;

**(d)** $K_p = \dfrac{(pNO)^4(pH_2O)^6}{(pNH_3)^4(pO_2)^5}$

**I** **(a)** For $2NO_2(g) = N_2O_4(g)$, $K_p = 7.7 \times 10^{-2}$ kPa$^{-1}$; **(b)** $K_p$ decreases; **(c)** $K_p$ stays the same

**J** Lower maintenance costs as no need to have the expensive apparatus required to withstand high pressure

**K** (a) +4; (b) +5; (c) +2

**L** (a) $N_2O_4$ (N has +4 oxidation state) is oxidized to $HNO_3$ (N has +5 oxidation state) and reduced to NO (N has +2 oxidation state);
(b) Shifts the position of equilibrium to the right, since the reaction is exothermic, and to ensure that the water remains liquid

**M** N:P:K ratio is 33.5:10.5:7

**N** $HNO_3(aq) + NH_3(aq) \rightarrow NH_4NO_3(aq)$

**O** (a) $CH_3NH_2(aq) + H_2O(l) = CH_3NH_3^+(aq) + OH^-(aq)$;
(b) $K_b = \dfrac{[CH_3NH_3^+(aq)][OH^-(aq)]}{[CH_3NH_2(aq)]}$

**P** (a) $NH_3(aq) + H_2SO_4(aq) \rightarrow NH_4HSO_4(aq)$, or $2NH_3(aq) + H_2SO_4(aq) \rightarrow (NH_4)_2SO_4(aq)$;
(b) $C_2H_5NH_2(aq) + HCl(aq) \rightarrow C_2H_5NH_3^+(aq) + Cl^-(aq)$

**Q** (a) Position of equilibrium shifts towards the right; (b) Increases;
(c) The position of equilibrium moves to the right so $[H^+(aq)]$ and $[OH^-(aq)]$ increase, which are the mobile charge carriers and so the electrical conductivity increases

**R** $1.73 \times 10^{-12}$ mol dm$^{-3}$

**S** (a) $2.51 \times 10^{-8}$ mol dm$^{-3}$;
(b) $3.98 \times 10^{-7}$ mol dm$^{-3}$

**T** $[H^+(aq)] = [OH^-(aq)] = 1.0 \times 10^{-7}$ mol dm$^{-3}$

**U** (a) 14.34; (b) 10.56; (c) 13.00;
(d) 11.60

**V** (a) $7.94 \times 10^{-13}$ mol dm$^{-3}$;
(b) $1.26 \times 10^{-2}$ mol dm$^{-3}$;
(c) $6.3 \times 10^{-3}$ mol dm$^{-3}$

**W** 11.5

**X** 10.9

**Y** (a) Methylamine, $CH_3NH_2$;
(b) N-Ethylbutylamine, $CH_3CH_2CH_2CH_2NHCH_2CH_3$;
(c) N-Methyl-2-propylamine, $(CH_3)_2CHNHCH_3$

**Z** (a) $CH_3CONHCH_2CH_3$;
(b) $C_6H_5CONHCH_3$; (c) $CH_3CONH_2$

**A¢** 10.4%

## CHAPTER 29

**A** $SiO_2 + 2Mg \rightarrow Si + 2MgO$

**B** High temperatures are needed (high energy costs), magnesium is a very expensive metal to buy, and the silicon produced must be ultra-pure, which requires further expensive processing.

**C** (a) To stop silicon oxidizing to form $SiO_2$; (b) Neon, argon or nitrogen

**D** (a) $PbO + C \rightarrow Pb + CO$, or $2PbO + C \rightarrow 2Pb + CO_2$;
(b) $SnO_2 + 2C \rightarrow Sn + 2CO$, or $SnO_2 + C \rightarrow Sn + CO_2$

**E** (a) Acid rain;
(b) $2PbS + 3O_2 \rightarrow 2PbO + 2SO_2$

**F** Low melting points and high electrical conductivity.

**G** Both have a giant molecular structure held together by strong covalent bonds, atoms with a coordination number of 4, a tetrahedral arrangement about each atom in the lattice, and the bond angles in both are 109°; differences are that the C–C bond length is shorter than Si–C, and in diamond each carbon atom is surrounded by four carbon atoms, whereas in SiC a carbon atom is surrounded by four silicon atoms

**H** (a) Sn is $[Kr]4d^{10}5s^25p^2$ and $Sn^{2+}$ is $[Kr]4d^{10}5s^2$;
(b) Ge is $[Ar]3d^{10}4s^24p^2$ and $Ge^{2+}$ is $[Kr]3d^{10}4s^2$;
(c) Si is $1s^22s^22p^63s^23p^2$ and $Si^{2+}$ is $1s^22s^22p^63s^2$;
(d) The inert pairs of electrons are the $5s^2$ and $4s^2$ electrons in $Sn^{2+}$ and $Ge^{2+}$, respectively, which are in the outer shell of electrons but are not used in bonding, whereas $Si^{2+}$ is not formed because silicon uses all four outer electrons in bonding

**I** (a)

| Element | Energy for X(g) $\rightarrow X^{2+}(g)$/kJ mol$^{-1}$ |
|---|---|
| Carbon | 3219 |
| Silicon | 2366 |
| Germanium | 2302 |
| Tin | 2117 |
| Lead | 2166 |

(b) $X^+(g)$ has more protons than electrons so the electrons are held more strongly to the nucleus and so are more difficult to remove than in $X(g)$, in which the number of electrons and protons are equal;
(c) Both have the same core charge of +4 but in Ge there are more inner shielding shells of electrons than in C and since the atomic radius of Ge is greater than that of C the outer electrons are at a greater distance from the nucleus; (d) 1st and 2nd ionization energies of tin are lower than expected, or those for Pb are greater than expected, which is accounted for by presence of inner shell 4f electrons in Pb.

**J**

**K** (a) $Pb + I_2 \rightarrow PbI_2$;
(b) $Sn + 2F_2 \rightarrow SnF_4$;
(c) $Ge + 2Br_2 \rightarrow GeBr_4$

**L** $Na_2SiO_3(aq) + 2HCl(aq) \rightarrow 2NaCl(aq) + SiO_2(s) + H_2O(l)$

**M** $PbO_2$ is less stable than $SiO_2$ (stability of +4 oxidation state decreases down the group), but PbO is much more stable than SiO (stability of +2 oxidation state increases down the group), so that $PbO_2$ can accept electrons (act as an oxidizing agent) to form $Pb^{2+}$, whereas $SiO_2$ will not accept electrons to form $Si^{2+}$

**N** $CO_2(g) + Ca(OH)_2(aq) \rightarrow CaCO_3(s) + H_2O(l)$

**O** (a) $Pb(CH_3COO)_2(aq)$ and $H_2O(l)$;
(b) $Na_2SnO_2(aq)$ and $H_2O(l)$

**P** (a) $Pb_3O_4(s) \rightarrow 3PbO(s)$ (lead(II) oxide) $+ \frac{1}{2}O_2(g)$ (oxygen);
(b) $Pb_3O_4(s) + 8HCl(aq) \rightarrow 3PbCl_2(aq)$ (lead(II) chloride) $+ Cl_2(g)$ (chlorine) $+ 4H_2O(l)$ (water);
(c) $Pb_3O_4(s) + 6NaOH(aq) \rightarrow 2Na_2PbO_2(aq)$ (sodium plumbate(IV)) $+ Na_2PbO_3(aq)$ (sodium

plumbate(VI)) + $3H_2O(l)$ (water)

**Q** $PbCl_4 \rightarrow PbCl_2 + Cl_2$

**R** $2Fe^{3+}(aq) + Sn^{2+}(aq) \rightarrow$
$2Fe^{2+}(aq) + Sn^{4+}(aq)$

**S** $Cl_2(g) + Sn^{2+}(aq) \rightarrow$
$Sn^{4+}(aq) + 2Cl^-(aq)$, tin(IV) chloride

**T** **(a)** $Si_6H_{14}$;
**(b)** $SiH_3SiH_2SiH_2SiH_2SiH_2SiH_3$ or
any other structural isomer

**U** About 20

**V** $PbCO_3(s) \rightarrow PbO(s) + CO_2(g)$

**W** **(a)** 2 mole KOH to 1 mole $H_2SO_3$;
**(b)** 1 mole KOH to 1 mole $H_2SO_3$

**X** **(a)** $Cl_2(aq) + SO_2(g) + 2H_2O(l) \rightarrow$
$2HCl(aq) + H_2SO_4(aq)$;
**(b)** $2Fe^{3+}(aq) + SO_2(g) + 2H_2O(l) \rightarrow$
$2Fe^{2+}(aq) + SO_4^{2-}(aq) + 4H^+(aq)$

**Y** **(a)** With magnesium carbonate,
magnesium, and copper(II) oxide;
**(b)** With copper, hydrogen iodide,
magnesium and benzene

**Z** **(a)** Benzoic acid; **(b)** Ethanoic acid

# INDEX